Lecture Notes in Computer Science 9937

Commenced Publication in 1973
Founding and Former Series Editors:
Gerhard Goos, Juris Hartmanis, and Jan van Leeuwen

Editorial Board

More information about this series at http://www.springer.com/series/7409

Hujun Yin · Yang Gao · Bin Li
Daoqiang Zhang · Ming Yang
Yun Li · Frank Klawonn
Antonio J. Tallón-Ballesteros (Eds.)

Intelligent Data Engineering and Automated Learning – IDEAL 2016

17th International Conference
Yangzhou, China, October 12–14, 2016
Proceedings

 Springer

Editors
Hujun Yin
University of Manchester
Manchester
UK

Yang Gao
Nanjing University
Nanjing
China

Bin Li
Yangzhou University
Yangzhou, Jiangsu
China

Daoqiang Zhang
Nanjing University of Aeronautics
  and Astronautics
Nanjing
China

Ming Yang
Nanjing Normal University
Nanjing
China

Yun Li
Yangzhou University
Yangzhou, Jiangsu
China

Frank Klawonn
Ostfalia University of Applied Sciences
Wolfenbüttel
Germany

Antonio J. Tallón-Ballesteros
University of Seville
Seville
Spain

ISSN 0302-9743   ISSN 1611-3349 (electronic)
Lecture Notes in Computer Science
ISBN 978-3-319-46256-1   ISBN 978-3-319-46257-8   (eBook)
DOI 10.1007/978-3-319-46257-8

Library of Congress Control Number: 2016950754

LNCS Sublibrary: SL3 – Information Systems and Applications, incl. Internet/Web, and HCI

This Springer imprint is published by Springer Nature
The registered company is Springer International Publishing AG
The registered company address is: Gewerbestrasse 11, 6330 Cham, Switzerland

Preface

The core of the IDEAL conference has been evolving from mining, analyzing, and exploring to interpreting and making sense of seas of data; this is particularly meaningful in this big-data and deep-learning era. The IDEAL conference has served its purposes well over the last 18 years, which have witnessed a fast changing world of data science and an increase in the development and deployment of learning and autonomous systems and intelligent bots. It has become one of the leading forums for learning the complexity and hidden dynamics of data-driven real-world problems and turning data into information, knowledge, and solutions. The IDEAL conference attracts international experts, researchers, leading academics, practitioners, and industrialists from the communities of machine learning, computational intelligence, novel computing paradigms, data mining, knowledge management, biology, neuroscience, bio-inspired systems and agents, distributed systems, and robotics. It continues to evolve to embrace emerging topics and exciting trends. This year IDEAL was held in one of the most beautiful historical cities in mainland China, Yangzhou. The conference received 115 submissions, which were rigorously peer-reviewed by the Program Committee members and experts. Only the papers judged to be of highest quality were accepted and included in these proceedings.

This volume contains 68 papers accepted and presented at the 17th International Conference on Intelligent Data Engineering and Automated Learning (IDEAL 2016), held on 12–14 October 2016 in Yangzhou, China. These papers provided a valuable and timely sample of the latest research outcomes in data engineering and automated learning, from methodologies, frameworks, and techniques to applications. The topics presented included evolutionary algorithms, deep learning neural networks, probabilistic modelling, particle swarm intelligence, big data analytics, and applications in regression, classification, clustering, medical and biological modelling and prediction, text processing, and image analysis. IDEAL 2016 also enjoyed outstanding keynote talks from leaders in the field, Xin Yao, Zhihua Zhou, Longbing Cao, and Bo An.

We would like to thank all the people who devoted so much time and effort to the successful running of the conference, in particular the members of the Program Committee and reviewers, as well as the authors who contributed to the conference. We are also very grateful for the hard work of the local organizing team at Yangzhou University, especially Prof. Yun Li, in local arrangements, as well as for the help from

Miss Yao Peng at the University of Manchester in checking through all the camera-ready files. Continued support and collaboration from Springer is also greatly appreciated.

July 2016

Hujun Yin
Yang Gao
Bin Li
Daoqiang Zhang
Ming Yang
Yun Li
Frank Klawonn
Antonio J. Tallón-Ballesteros

Organization

Honorary Chair

Xin Yao   University of Birmingham, UK

General Chairs

Hujun Yin   University of Manchester, UK
Yang Gao   Nanjing University, China
Bin Li   Yangzhou University, China

Program Co-chairs

Daoqiang Zhang   Nanjing University of Aeronautics and Astronautics, China
Ming Yang   Nanjing Normal University, China
Yun Li   Yangzhou University, China
Frank Klawonn   Ostfalia University of Applied Sciences, Germany
Antonio   University of Seville, Spain
     J. Tallón-Ballesteros

International Advisory Committee

Lei Xu (Chair)   Chinese University of Hong Kong, Hong Kong, China
Yaser Abu-Mostafa   CALTECH, USA
Shun-ichi Amari   RIKEN, Japan
Michael Dempster   University of Cambridge, UK
José R. Dorronsoro   Autonomous University of Madrid, Spain
Nick Jennings   University of Southampton, UK
Soo-Young Lee   KAIST, South Korea
Erkki Oja   Helsinki University of Technology, Finland
Latit M. Patnaik   Indian Institute of Science, India
Burkhard Rost   Columbia University, USA
Xin Yao   University of Birmingham, UK

Steering Committee

Hujun Yin (Chair)   University of Manchester, UK
Laiwan Chan (Chair)   Chinese University of Hong Kong, Hong Kong, China
Guilherme Barreto   Federal University of Ceará, Brazil
Yiu-ming Cheung   Hong Kong Baptist University, Hong Kong, China
Emilio Corchado   University of Burgos, Spain

Songcan Chen
Sung-Bae Cho
Andrzej Cichocki
Jacek Cichosz
Stelvio Cimato
André Coelho
Leandro Coelho
Rafael Corchuelo
Francesco Corona
Paulo Cortez
Jose Alfredo F. Costa
Marcelo A. Costa
Raúl Cruz-Barbosa
Ernesto Cuadros-Vargas
Alfredo Cuzzocrea
Bogusław Cyganek
Ernesto Damiani
Nicoletta Dessì
Fernando Díaz
Weishan Dong
Jose Dorronsoro
Gérard Dreyfus
Adrião Duarte
Jochen Einbeck
Florentino Fdez-Riverola
Francisco Ferrer
Carmelo J.A. Bastos Filho
Juan J. Flores
Gary Fogel
Pawel Forczmanski
Felipe M.G. França
Dariusz Frejlichowski
Bogdan Gabrys
Marcus Gallagher
Matiaz Gams
Salvador Garcia
Ana Belén Gil
Maríajosé Ginzo-Villamayor
Fernando Gomide
Petro Gopych
Marcin Gorawski
Juan Manuel Górriz
Lars Graening
Manuel Graña
Maciej Grzenda
Jerzy Grzymala-Busse

Alberto Guillen
Anne Håkansson
Barbara Hammer
Ioannis Hatzilygeroudis
Francisco Herrera
Álvaro Herrero
J. Michael Herrmann
James Hogan
Jaakko Hollmén
Vasant Honavar
Wei-Chiang Samuelson Hong
Iñaki Inza
Konrad Jackowski
Vahid Jalali
Dariusz Jankowski
Joao E. Kogler Jr.
Vicente Julian
Ata Kaban
Juha Karhunen
Miroslav Karny
Rheeman Kil
Sung-Ho Kim
Mario Koeppen
Andreas König
Rudolf Kruse
Lenka Lhotska
Bin Li
Clodoaldo A.M. Lima
Paulo Lisboa
Honghai Liu
Wenjian Luo
José Everardo B. Maia
Urszula Markowska Kaczmar
José F. Martínez
Roque Marín
Giancarlo Mauri
José M. Molina
Tim Nattkemper
Antonio Neme
Ajalmar Rêgo Darocha Neto
Yusuke Nojima
Chung-Ming Ou
Vasile Palade
Stephan Pareigis
Juan Pavón
Carlos Pedreira

Sarajane M. Peres
Javier Bajo Pérez
Jorge Posada
Izabela Rejer
Bernardete Ribeiro
José Riquelme
Ignacio Rojas
José Luis Calvo Rolle
Fabrice Rossi
Regivan Santiago
Jose Santos
Javier Sedano
Ivan Silva
Leandro Augusto Da Silva
Dragan Simic
Michael Small
Ying Tan
Ke Tang
Ricardo Tanscheit

Dante Tapia
Peter Tino
Renato Tinós
Pawel Trajdos
Alicia Troncoso
Eiji Uchino
Marley Vellasco
Alfredo Vellido
José R. Villar
Tzai-Der Wang
Lipo Wang
Wenjia Wang
Dongqing Wei
Michal Wozniak
Wu Ying
Du Zhang
Huiyu Zhou
Andrzej Zolnierek
Rodolfo Zunino

Special Session on Intelligent Analysis for Crowdsourced Software Engineering Problems

Organizers

Bin Li   Yangzhou University, China
Xiaobing Sun   Yangzhou University, China
Shaowei Wang   Queens University, Canada

Special Session on Knowledge Representation, Reasoning, and Online Learning in Big Data Analysis

Organizers

Ming Yang   Nanjing Normal University, China
Yang Gao   Nanjing University, China
Wensheng Zhang   Institute of Automation of Chinese Academy of Sciences, China
Wanqi Yang   Nanjing Normal University, China

Contents

A Simplified Algorithm Based on IF Model

Xianfeng Liu[✉] and Yamin Zuo

Changzhou Vocational Institute of Mechatronic Technology,
Changzhou 213164, Jiangsu, People's Republic of China
{9755724,842054162}@qq.com

Abstract. The ability of the connection between two neurons (synapse) to change in strength in response to neural activity is known as synaptic plasticity. Experimental data has shown that synaptic plasticity can depend upon the relative timings of pre- and postsynaptic neuron spikes – this is known as spike-timing-dependent plasticity (STDP). It is proposed that the traditional IF model with STDP can simulate the natural properties of biological neuron much better. This paper proposes a new algorithm based on the IF model with STDP, and simulated in MATLAB. Experimental results have indicated that algorithm can reflect the performance of nerve cells and has a better bionic function.

Keywords: Neural network · IF model · STDP

1 Introduction

In biology, neural network is a huge and complicated network consisting of biological neurons and used to produce consciousness and behavior. In recent years, because of the rapid development in artificial intelligence, the research of artificial neural network is gotten more and more attentions. As the name implies, artificial neural network can imitate the biological mode of thinking and implement the application of artificial intelligence by building a model through the biological prototype. Integrate-and-fire model (termed as IF model) is a popular and simplified biological neuron model which can describe the relationship between input and output of neurons. However, biological response to the external stimulation is not a simple relationship between input and output, "learning" is also required. The term "learning" refers to an algorithm of connected strength between neurons [1].

Neuron is the basic element of the neural network, which constituted by the dendrites, the cell body and the axon. Figure 1 describes the architecture of a neuron. Cell body, also referred as soma, is the central processing component of the cell. Nucleus is the control unit. The dendrites and the axon are used to deliver information between neurons. The junction between the axon of a presynaptic cell and the dendrite of a postsynaptic neuron is the synapse. It is an essential part in delivering information from one neuron to another. Synaptic plasticity is the ability of the connection between two neurons to change in strength which depend on the relative timings of presynaptic and

Fund Project: Sponsored by Qing Lan Project of Jiangsu Province of China.

H. Yin et al. (Eds.): IDEAL 2016, LNCS 9937, pp. 1–9, 2016.
DOI: 10.1007/978-3-319-46257-8_1

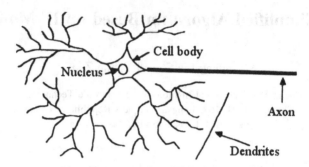

Fig. 1. Components of a neuron [4]

postsynaptic neuron spikes. This phenomenon is known as spike-timing-dependent plasticity (termed as STDP) [2, 3]. STDP means the value of synaptic weight should depend on the relative timings of presynaptic and postsynaptic neuron spikes. Synaptic weight will be potentiated when a presynaptic pulse arrives before the postsynaptic pulse. By contrast, if the synapse receives the postsynaptic pulse first, the weight will be depressed. In general, STDP plays a role of a learning mechanism for generating neural response according to input timing, order and sequence. The IF model with STDP makes a more realistic simulation for biological neural network system.

2 The Mathematical Description of IF Model and Its Simulation in MATLAB

Biological neuron model is often used to describe the properties of neurons in mathematical way. It can describe and predict biological processes accurately. One of the neuron models is the IF model. In this model, the soma is described as a membrane with a capacitor C (as shown in the right graph of Fig. 2).

Figure 2 shows the schematic diagram of IF model, t_{pre}^{f} represents the arriving time of presynaptic pulse, while t_{post}^{f} designates the fire time of postsynaptic pulse. The module inside the dashed circle on the right-hand side is the basic integrate-and-fire circuit. A current $I(t)$ charges the RC circuit. The voltage which crosses the capacitor C is called membrane voltage $v(t)$. The next step is to compare membrane voltage $v(t)$ to a threshold voltage termed as θ. At time t_{post}^{f}, when the membrane voltage $v(t)$ is equal to the threshold voltage θ, an output pulse $\delta\left(t - t_{post}^{f}\right)$ will be generated and the membrane potential will be set to its reset value. Otherwise, if the membrane voltage does not reach the threshold, there will be no output pulse created, just as in the case of biological neuron. The left part of Fig. 2 indicates a low-pass filter. spike $\delta\left(t - t_{pre}^{f}\right)$ could be processed to an input current pulse $\alpha\left(t - t_{pre}^{f}\right)$ [5].

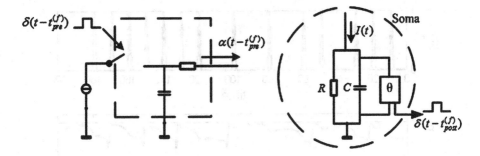

Fig. 2. Schematic diagram of IF model [2]

The membrane potential is given by the following Eq. (1), where $v = v(t)$, $I = I(t)$, R is the resistance, C is the capacitance in the model, v_0 is the resting value of the potential, and $\tau_m = RC$ is the time constant. By solving this differential equation, the form of the membrane potential could be achieved as shown in Eq. (2). And $\hat{t}$ is the last fire of the neuron [6]. In this simulation, an exponential current with a given decay would be used. Therefore, the current $I = I(t)$ could have the form as shown in Eq. (3), where q is the total charge delivered by the spike, τ_s is the time constant of the current's decay and t_j^f contains the fire times of the presynaptic neuron.

$$\dot{v} = -\frac{v - v_0}{\tau_m} + \frac{I}{C} \tag{1}$$

$$v(t) = v_0 + (v_r - v_0) \exp(-\frac{t - \hat{t}}{\tau_m}) + \frac{1}{C} \int_0^{t-\hat{t}} \exp(-\frac{s}{\tau_m}) I(t - s) ds \tag{2}$$

$$I(t - s - t_j^{(f)}) = \begin{cases} \dfrac{q}{\tau_s} \exp\left(-\dfrac{t - s - t_j^{(f)}}{\tau_s}\right), & s \leq t - t_j^{(f)} \\ 0, & \textit{otherwise} \end{cases} \tag{3}$$

Building and simulating an IF model in MATLAB according to the above equations. The parameters can be set as following: $C = 2$, $R = 10$, $v_r = -0.5$, $v_0 = 0$, $q = 2$, $\tau_s = 5$, threshold = 1. Figure 3 shows the simulated result. The input is assumed as random value. The top graph of Fig. 3 shows the spikes of input, also called presynaptic spikes. The middle graph shows the membrane potential, the dash line represents the threshold voltage. And the bottom graph is the spikes of output, also called postsynaptic spikes. According to Fig. 3, when the membrane potential reaches the threshold, an output spike will be generated.

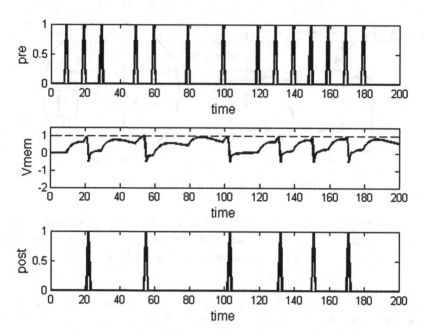

Fig. 3. Simulation of IF model

3   The Mathematical Description of STDP

In the neural network, the function of the synapse is to provide "synaptic weight" to presynaptic spikes. Figure 4 illustrates a simple structure of integrate-and-fire model with synapse. pre_{A1} and pre_{A2} are used to express two presynaptic inputs. W_{A1} is synaptic weight for synapse A1, while W_{A2} represents synapse A2. After crossing these two synapses, the spikes would become $S_1 = \sum pre_{A1} \times W_{A1}$ and $S_2 = \sum pre_{A2} \times W_{A2}$. Neuron A acts as the integrate-and-fire module. Before the firing process, spikes s_1 and

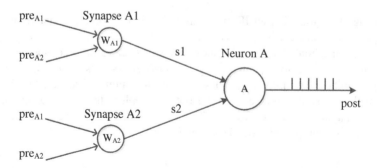

Fig. 4. Structure of IF model with synapse

s_2 would be integrated to $x = \sum_{i=1}^{2} S_i$. And when x is equal to the threshold, a spike *post* will be fired at the output [7–9].

The membrane potential as shown in Eq. (4), the input is f_i, the synaptic weight is w_i and t_c is the time of the last output.

$$u(i) = \frac{1}{C}\left(\sum w_i \times f_i\right) - \frac{1}{C}(I \times (i - t_c)) \tag{4}$$

The basic weight independent STDP learning window is defined in the curve of Fig. 5. Weight independent means the learning window is same for all weight strength. The horizontal axis of the curve represents the time interval between presynaptic and postsynaptic spikes (termed as $\Delta t = t_{pre} - t_{post}$). The vertical axis represents the synaptic weight modification (termed as Δw_{ij}). When a presynaptic pulse arrives before the post synaptic pulse ($\Delta t < 0$), the synaptic weight will be potentiated. By contrast, if the synapse receives the postsynaptic pulse first ($\Delta t > 0$), the weight will be depressed.

Assuming that the presynaptic spike arrival times are named as t_{pre}^{f}, where $f = 1, 2,$ 3,… counts the presynaptic spikes. Similarly, t_{post}^{n} represents the firing times of the postsynaptic neuron, where n = 1, 2, 3,… labels the postsynaptic spikes. Synaptic weight w_{ij} is the weight strength between presynaptic and postsynaptic spikes. Weight modification Δw_{ij} is the total change of weight. Equation (5) is the function of weight change. The shape of STDP learning curve is determined by the exponential decay

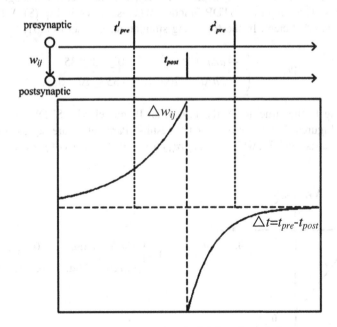

Fig. 5. Weight independent STDP learning window

function $W(x)$. A common choice for the STDP function $W(x)$ is shown in Eqs. (6) and (7). The parameters A_+ and A_- may depend on the current value of the synaptic weight w_{ij}. τ_+ and τ_- are time constant.

$$\Delta w_{ij} = \sum_{f=1}^{N} \sum_{n=1}^{N} W(t_{post}^n - t_{pre}^f) \tag{5}$$

$$W(x) = A_+ \exp(-x/\tau_+) \tag{6}$$

$$W(x) = -A_- \exp(x/\tau_-) \tag{7}$$

4   A Simplified STDP Algorithm Based on IF Model

It is complicated to explain STDP learning curve in a mathematical equation. In order to achieve a much simply MATLAB model of STDP learning, a new simplified STDP algorithm is proposed in this paper. The new model can realize the basic function represented by STDP and simplify the MATLAB code. Additionally, the simplified IF model can be used to design an analogue circuit for STDP. The simplified STDP curve is shown in Fig. 6. The horizontal axis describes the time interval between presynaptic and postsynaptic pulses. The vertical axis expresses the change value of synaptic weight. Notice that the slope of the first quadrant (the potential side) is smaller than which in the third quadrant (the depressed side), it is decided by biological natural characteristics. The simplified STDP function $W(x)$ is shown in Eq. (8). Where a is a weight change coefficient. In the following simulation, a is set to 0.01.

$$W(x) = \begin{cases} -a\Delta t + 35a, & 0 \leq \Delta t \leq 35 \\ -2a\Delta t - 70a, & -35 \leq \Delta t < 0 \end{cases} \tag{8}$$

According to the structure in the Fig. 4, an IF model with STDP can be built in MATLAB. Figures 7 and 8 represent the results. Parameters are set as: input pulse signal pre_{A1} starts from 20, while signal pre_{A2} from 30. The steps of pre_{A1} and pre_{A2} are

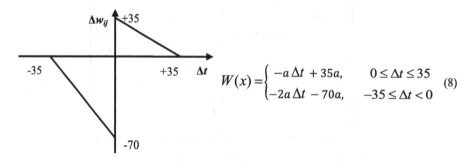

Fig. 6. Simplified STDP learning curve

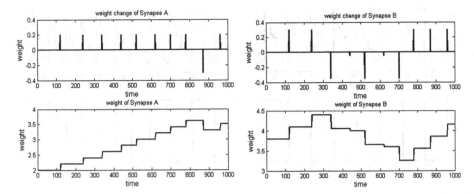

Fig. 7. Simulation result. The top graph describes the weight change of synapse A1. The bottom plot shows the weight of synapse A1.

Fig. 8. Simulation result. The top graph describes the weight change of synapse A2. The bottom plot shows the weight of synapse A2.

20 and 30 respectively. W_{A1} is set to 2 and W_{A2} is 3.8. Notice that the synaptic weight can be only modified after a pair of presynaptic and postsynaptic pulse appearing. Figure 7 shows the details of weight in synapse A1. The initial weight is 2. The top graph shows the change of weight in synapse A1 in every spiking time. The bottom graph shows the weight value in synapse A1. Similarly, Fig. 8 illustrates the details of weight in synapse A2. The initial weight is 3.8.

Due to the STDP learning curve in Fig. 6, the synaptic weight would be potentiated if the presynaptic signal arrives after the postsynaptic signal. By contrast, if the presynaptic signal comes first, the weight change could be a negative value. The weight could only be changed when a pair of presynaptic and post synaptic pulses happens. In the simplified STDP leaning model, the rate is different from every integrating process. This phenomenon indicates that the weight has important effect in the membrane voltage integrating speed.

5 Compare the Simulation Result of Original and New iF Model

The new IF model introduced with STDP uses a simplifier and easier to implement expression to describe. Figures 9 and 10 show the simulation results. The top graph describes the membrane voltage. The bottom plot shows the postsynaptic spikes. The structure of neural network is shown in Fig. 4. In the original IF model, the weight of synapse A1 is 2 while synapse A2 is 3.8. Compare the postsynaptic pulse in these two graphs. In the original model, the rate of integrating membrane voltage could not be changed. While in the model with STDP, the rate is different from every integrating process. When the integrating speed changed, the time of output pulse and the time interval will be changed. The new IF model can be more realistic simulation of biological neural network and the reaction to external stimulate (i.e. output of neuron) will

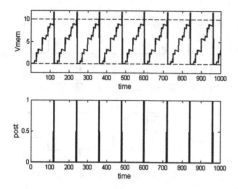

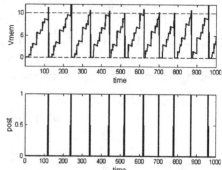

Fig. 9. Simulation of the original IF model. The top graph describes the membrane voltage. The bottom plot shows the postsynaptic spikes.

Fig. 10. Simulation of the New IF model. The top graph describes the membrane voltage. The bottom plot shows the postsynaptic spikes.

be changed in response to synaptic weight. Additionally, the modified IF model is linear, which is more convenient and reasonable for designing a hardware circuit.

6 Conclusion and Future Work

This paper designs a simplifier algorithm of IF model with STDP, and provides an ideal linear STDP learning curve. A MATLAB model was built by this algorithm to simulate the IF mechanism, synapse and STDP learning. In the future work, this new IF model will be used as the neuron in the further simulation to design a STDP learning analogue circuit. This linear curve will be substituted by a real curve which is derived from the experimental data after the electronic circuit is designed.

References

1. Yongpu, L.: Research new IF model and learning rules. Syst. Eng. Electron. **28**(4), 583–586 (2006)
2. Abbott, L.F., Nelson, S.B.: Synaptic plasticity: taming the beast. Nat. Neurosci. Supplement. **3**, 1178–1183 (2000)
3. McCulloch, W.S., Pitts, W.: A logical calculus of the ideas immanent in nervous activity. Bull. Math. Biophys. **5**, 127–147 (1943)
4. Stergiou, C., Siganos, D.: Neural Networks. http://www.doc.ic.ac.uk/~nd/surprise_96/journal/vol4/cs11/report.html#Contents
5. Wulfram, G., Werner, M.K.: Spiking Neuron Models-Single Neurons, Populations, Plasticity, pp. 277–280. Cambridge University Press, Cambridge (2002)
6. Mitte, M.: A simple neuron model-the integrate and fire neuron. http://www.dreamincode.net/forums/topic/72868-a-simple-neuron-model-the-integrate-and-fire-neuron/

7. Indiveri, G.: Modeling selective attention using a neuromorphic analog VLSI device. Neural Comput. **12**, 2857–2880 (2000)
8. Bofill-i-petit, A., Murray, A.F.: Synchrony detection and amplification by silicon neurons with STDP synapses. IEEE Trans. Neural Netw. **15**(5), 1296–1304 (2004)
9. Lin, L.: Research of kinetic model of double inputs with spike time dependent plasticity synapse. Appl. Res. Comput. **4**, 1311–1313 (2012)

Predict Two-Dimensional Protein Folding Based on Hydrophobic-Polar Lattice Model and Chaotic Clonal Genetic Algorithm

Shuihua Wang[1], Lenan Wu[2], Yuankai Huo[1], Xueyan Wu[1],
Hainan Wang[1], and Yudong Zhang[1(✉)]

[1] School of Computer Science and Technology, Nanjing Normal University,
Nanjing 210023, Jiangsu, China
zhangyudong@njnu.edu.cn
[2] School of Information Science and Engineering, Southeast University, Nanjing, China

Abstract. In order to improve the performance of prediction of protein folding problem, we introduce a relatively new chaotic clonal genetic algorithm (abbreviated as CCGA) to solve the 2D hydrophobic-polar lattice model. Our algorithm combines three successful components—(i) standard genetic algorithm (SGA), (ii) clonal selection algorithm (CSA), and (iii) chaotic operator. We compared this proposed CCGA with SGA, artificial immune system (AIS), and immune genetic algorithm (IGA) for various chain lengths. It demonstrated that CCGA had better performance than other methods over large-sized protein chains.

Keywords: Protein folding · Chaotic clonal genetic algorithm · Clonal selection algorithm · Hydrophobic-polar model · Artificial immune system

1 Introduction

Protein folding (PF) is a physical process for a protein chain acquires its 3-dimensional structure [1, 2]. It imposes a challenge to biologists since the problem has an extremely large search space [3]. A successful solution to PF requires to solve two important problems [4]: (i) a series of free residues for the protein chain and (ii) an efficient optimization procedure. Now since the PF data are easily available, the latter is the most difficult thing.

Traditional optimizers will not work for the PF problem, because the model (See Sect. 2) is multimodal and non-differential. Besides, the problem is NP-hard [5]. Hence, advanced global optimizers are introduced to solve it.

The genetic algorithm (GA) and particle swarm optimization [6] are one of the most popular global optimizations. Moreover, swarm intelligence approaches [7]: biogeography-based optimization [8], firefly algorithm [9], artificial bee colony [10], and bacterial chemotaxis optimization [11], have attracted interest from scholars in many fields. Nevertheless, massive researchers had investigated the PF problems merely with GA, since the encoding in GA is more suitable for PF problems.

© Springer International Publishing AG 2016
H. Yin et al. (Eds.): IDEAL 2016, LNCS 9937, pp. 10–17, 2016.
DOI: 10.1007/978-3-319-46257-8_2

Lin and Hsieh [12] presented a Taguchi-genetic algorithm (TGA). Huang, Yang and He [13] employed GA and optimal second structure. Narayanan, Krishnakumar and Judy [14] proposed an enhanced MapReduce framework using parallel genetic algorithm (PGA). However, GA may still encounter problems in converging into global optimal. To improve its performance, we introduced a novel chaotic clonal genetic algorithm (abbreviated as CCGA) in this study. Below we will show how the mechanism of CCGA and how it can be applied to solve PF problem.

2 Two Dimensional Hydrophobic-Polar Model

The hydrophobic-polar (HP) protein folding model [15] is a simplified version for exampling structures of protein folds in space. It stems from the fact that hydrophobic interactions between amino acids residues drive proteins to fold into native structure [16]. In this model, protein chains are composed of two types of residues, viz., polar (P) and hydrophobic (H) [17]. Figure 1 offers an example of a 10-residue chain with energy of −4.

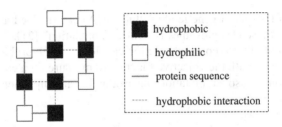

Fig. 1. A 10-residue chain of HPHPPHPHPH

Figure 2 indicates that protein chains can turn at each residue position 90° to either left (L) or right (R) or continue (C) ahead. The first interaction is always set as 'C'. Hence, the protein chain can be presented as 'CLCLLRCCL'. Note that clashes (i.e., residues overlap at the same position) is not allowed. In all, our object is to minimize the following expression

$$E = n(-1) \tag{1}$$

where n represents the number of hydrophobic interactions, E the energy function.

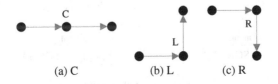

(a) C   (b) L   (c) R

Fig. 2. Turning directions of a chain (C = Continue, L = Left, R = Right)

3 Materials

The materials consisted of five chains with residues sizes of 20, 36, 48, 64, and 85, respectively. Their information are listed in Table 1. The minimum energy was obtained by exhaustion method. The protein sequences can be found in Table 1 in reference [18].

Table 1. Five protein chains

Index	1	2	3	4	5
Size	20	36	48	64	85
Energy	−9	−14	−23	−40	−52

4 Our Optimizor

4.1 Standard Genetic Algorithm

The individuals of standard genetic algorithm (SGA) are encoded as chromosomes [19, 20]. A set of those chromosomes is termed "population" [21]. A random population is created initially to represent solution candidates to PF problem. The energy function is associated with the objective function to measure each candidate [22]. At each step, selection, crossover, mutation, and evaluation are implemented in sequence as in Fig. 3a.

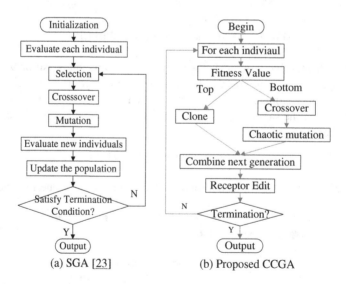

(a) SGA [23]   (b) Proposed CCGA

Fig. 3. Diagram of algorithms

4.2 Two Improvements

The first improvement is we introduce the clonal selection mechanism from clonal selection algorithm (CSA), which is inspired by clonal selection theory of acquired immunity, which offers explanation how B and T lymphocytes enhances their response to antigens. CSAs are commonly applied to optimization fields, such as node detection [24], scheduling [25], feature selection [26], weight training [27], advanced intelligence turning [28], etc.

The next improvement is we introduce the chaos theory, because reproduction operator in clonal selection algorithm (CSA) and crossover operator in standard genetic algorithm (SGA) cannot generate any new variants to the current chromosome. To introduce mutations, we employed a chaotic number generator n_t on current chromosome:

$$n_{t+1} \leftarrow 4n_t\left(1 - n_t\right) \tag{2}$$

where $n_0 \in (0, 1)$ and $n_0 \notin \{0.25, 0.5, 0.75\}$. Figure 4 shows why the initial value of n_t cannot be the value of either 0.25, or 0.5, or 0.75.

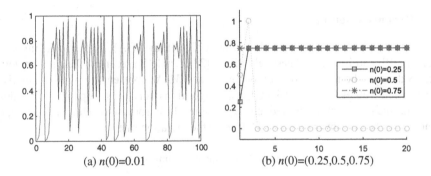

(a) $n(0)=0.01$ (b) $n(0)=(0.25, 0.5, 0.75)$

Fig. 4. Time series of chaotic number

4.3 Our Chaotic Clonal Genetic Algorithm

The chaotic clonal genetic algorithm (CCGA) was proposed with above two improvements. On one hand, the chaotic operator of Eq. (2) was employed to add mutation to current chromosomes, with the aim of guaranteeing the dynamic ergodicity within solution space. On the other hand, the clonal selection and receptor edit mechanisms were included, so that a chromosome with larger affinity will have more chances to be reproduced meanwhile the population size keep unchanged.

The diagram of CCGA was pictured in Fig. 3b. At each step, all the chromosomes are sorted with regards to the fitness values. Afterwards, the whole set was segmented into two parts: top part and bottom part as shown in Fig. 5.

Clone operations will perform over the top part, and crossover and chaotic mutation operations will perform over the bottom part.

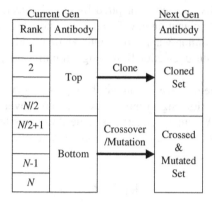

Fig. 5. Diagram of two parts

5   Experiments and Results

5.1   Parameter Setting

We compared the CCGA with standard genetic algorithm (SGA) [29], artificial immune system [30], and immune genetic algorithm [31]. Table 2 presented their parameters. Here S denotes the number of chromosomes, and MAX represents maximum iterative steps. Their values are equivalent for all algorithms for fair comparison. Besides, r denotes the receptor editing frequency, e denotes the chromosome elimination rate, C denotes for the crossover rate, and M denotes the mutation rate.

Table 2. Parameters setting

Approach	C	M	r	e	S	MAX
SGA [29]	0.3	0.08	–	–	1000	50000
AIS [30]	–	1	23	25 %	1000	50000
IGA [31]	0.3	0.15	–	10 %	1000	50000
CCGA (Proposed)	0.3	0.15	23	25 %	1000	50000

5.2   Algorithm Comparison

Table 3 lists the results of all four algorithms. We used "*success rate (SR)*" to measure all methods, which is defined as the ratio of success runs among all 100 runs.

Table 3. SR of different algorithms

Index	1	2	3	4	5
SGA [29]	94 %	2 %	0 %	0 %	0 %
AIS [30]	**100 %**	**99 %**	**81 %**	12 %	0 %
IGA [31]	**100 %**	78 %	22 %	6 %	1 %
CCGA (Proposed)	98 %	95 %	77 %	**14 %**	3 %

(Bold means the best)

Table 3 showed that CCGA achieves SR of 98 %, 98 %, 77 %, 14 %, and 3 % for all five protein-chains. It indicates that the CCGA has similar performance on small-size chains, but as the chain become longer (Index = 4 & 5), the proposed CCGA has better performance than SGA [29], IGA [31], and AIS [30]. The best structures found by CCGA are shown in Fig. 6.

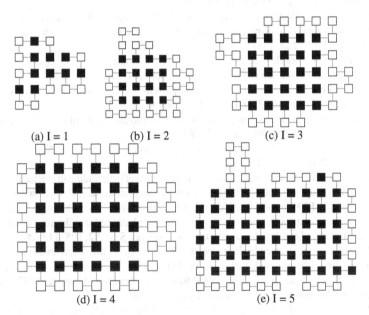

(a) I = 1   (b) I = 2   (c) I = 3

(d) I = 4   (e) I = 5

Fig. 6. The ideal solutions (I = Index)

6   Conclusions and Future Directions

We proposed a new global optimization method—chaotic clonal genetic algorithm (abbreviated as CCGA) based on GA, CSA, and chaotic operator. Experiment results show the superiority of CCGA to recent methods. Future work is composed of two folds. We shall try to increase the prediction performance for large-size chain, and we shall test other advanced global optimization algorithms, such as hybridization of swarm intelligence methods [32, 33].

Acknowledgment. This paper was supported by Natural Science Foundation of Jiangsu Province (BK20150983), Program of Natural Science Research of Jiangsu Higher Education Institutions (15KJB470010), Nanjing Normal University Research Foundation for Talented Scholars (2013119XGQ0061, 2014119XGQ0080), Open Project Program of the State Key Lab of CAD&CG, Zhejiang University (A1616).

References

1. Cumberworth, A., Bui, J.M., Gsponer, J.: Free energies of solvation in the context of protein folding: implications for implicit and explicit solvent models. J. Comput. Chem. **37**, 629–640 (2016)
2. Dumitriu, A., Golji, J., Labadorf, A.T., Gao, B.B., Beach, T.G., Myers, R.H., Longo, K.A., Latourelle, J.C.: Integrative analyses of proteomics and RNA transcriptomics implicate mitochondrial processes, protein folding pathways and GWAS loci in Parkinson disease. BMC Med. Genomics **9**, Article ID: 5 (2016)
3. An, P., Winters, D., Walker, K.W.: Automated high-throughput dense matrix protein folding screen using a liquid handling robot combined with microfluidic capillary electrophoresis. Protein Expr. Purif. **120**, 138–147 (2016)
4. Rivas-Pardo, J.A., Eckels, E.C., Popa, I., Kosuri, P., Linke, W.A., Fernandez, J.M.: Work done by titin protein folding assists muscle contraction. Cell Rep. **14**, 1339–1347 (2016)
5. Santos, J., Villot, P., Dieguez, M.: Emergent protein folding modeled with evolved neural cellular automata using the 3D HP model. J. Comput. Biol. **21**, 823–845 (2014)
6. Ji, G.: A comprehensive survey on particle swarm optimization algorithm and its applications. Math. Probl. Eng. **2015**, Article ID: 931256 (2015)
7. Agarwal, P., Bhatnagar, V., Balochian, S.: Swarm Intelligence and Its Applications 2014. Sci. World J. **2014**, Article ID: 204294 (2014)
8. Wang, S., Yang, J., Liu, G., Du, S., Yan, J.: Multi-objective path finding in stochastic networks using a biogeography-based optimization method. Simulation **92**, 637–647 (2016)
9. Othman, M.M., El-Khattam, W., Hegazy, Y.G., Abdelaziz, A.Y.: Optimal placement and sizing of voltage controlled distributed generators in unbalanced distribution networks using supervised firefly algorithm. Int. J. Electr. Power Energy Syst. **82**, 105–113 (2016)
10. Wu, L.: Magnetic resonance brain image classification by an improved artificial bee colony algorithm. Prog. Electromagn. Res. **116**, 65–79 (2011)
11. Zhang, Y.: Stock market prediction of S&P 500 via combination of improved BCO approach and BP neural network. Expert Syst. Appl. **36**, 8849–8854 (2009)
12. Lin, C.J., Hsieh, M.H.: An efficient hybrid Taguchi-genetic algorithm for protein folding simulation. Expert Syst. Appl. **36**, 12446–12453 (2009)
13. Huang, C.H., Yang, X.B., He, Z.H.: Protein folding simulations of 2D HP model by the genetic algorithm based on optimal secondary structures. Comput. Biol. Chem. **34**, 137–142 (2010)
14. Narayanan, A.G.H., Krishnakumar, U., Judy, M.V.: An enhanced mapreduce framework for solving protein folding problem using a parallel genetic algorithm. In: Satapathy, S.C., Avadhani, P.S., Udgata, S.K., Lakshminarayana, S. (eds.) ICT and Critical Infrastructure: Proceedings of the 48th Annual Convention of Computer Society of India - Vol I. AISC, vol. 248, pp. 241–250. Springer, Heidelberg (2014)
15. Garza-Fabre, M., Rodriguez-Tello, E., Toscano-Pulido, G.: Constraint-handling through multi-objective optimization: the hydrophobic-polar model for protein structure prediction. Comput. Oper. Res. **53**, 128–153 (2015)

16. Giaquinta, E., Pozzi, L.: An effective exact algorithm and a new upper bound for the number of contacts in the hydrophobic-polar two-dimensional lattice model. J. Comput. Biol. **20**, 593–609 (2013)
17. Zhang, Y., Wu, L., Wang, S.: Solving two-dimensional hp model by firefly algorithm and simplified energy function. Math. Probl. Eng. **2013**, Article ID: 398141 (2013)
18. Huo, Y., Zhu, Q., Wang, S., Wu, L.: Polymorphic BCO for protein folding model. J. Comput. Inf. Syst. **6**, 1787–1794 (2010)
19. Zhang, Y., Wang, S., Ji, G., Dong, Z.: Genetic pattern search and its application to brain image classification. Math. Probl. Eng. **2913**, Article ID: 580876 (2013)
20. Al Adwan, F., Al Shraideh, M., Al Saidat, M.R.S.: A genetic algorithm approach for breaking of simplified data encryption standard. Int. J. Secur. Appl. **9**, 295–303 (2015)
21. Gibbs, M.S., Maier, H.R., Dandy, G.C.: Using characteristics of the optimisation problem to determine the genetic algorithm population size when the number of evaluations is limited. Environ. Modell. Softw. **69**, 226–239 (2015)
22. Javaid, W., Tariq, A., Hussain, I.: A comparison of a standard genetic algorithm with a hybrid genetic algorithm applied to cell formation problem. Adv. Mech. Eng. **11**, Article ID: 301751 (2014)
23. Wang, S., Lu, Z., Wei, L., Ji, G., Yang, J.: Fitness-scaling adaptive genetic algorithm with local search for solving the multiple depot vehicle routing problem. Simulation **92**, 601–616 (2016)
24. Sindhuja, L.S., Padmavathi, G.: Replica node detection using enhanced single hop detection with clonal selection algorithm in mobile wireless sensor networks. J. Comput. Netw. Commun. **13**, Article ID: 1620343 (2016)
25. Shui, X.G., Zuo, X.Q., Chen, C., Smith, A.E.: A clonal selection algorithm for urban bus vehicle scheduling. Appl. Soft Comput. **36**, 36–44 (2015)
26. Marinaki, M., Marinakis, Y.: A hybridization of clonal selection algorithm with iterated local search and variable neighborhood search for the feature selection problem. Memetic Comput. **7**, 181–201 (2015)
27. Cozma, P., Dragoi, E.N., Mamaliga, I., Curteanu, S., Wukovits, W., Friedl, A., Gavrilescu, M.: Modelling and optimization of CO2 absorption in pneumatic contactors using artificial neural networks developed with clonal selection-based algorithm. Int. J. Nonlinear Sci. Numer. Simul. **16**, 97–110 (2015)
28. Kim, D.H., Cho, J.H.: Advanced intelligence tuning using hybrid of clonal selection and genetic algorithm, GM and PM. Int. J. Comput. Intell. Appl. **14**, 19 (2015)
29. Lin, C.D., Anderson-Cook, C.M., Hamada, M.S., Moore, L.M., Sitter, R.R.: Using genetic algorithms to design experiments: a review. Qual. Reliab. Eng. Int. **31**, 155–167 (2015)
30. Zhang, Y., Huo, Y.: Artificial immune system for protein folding model. J. Converg. Inf. Technol. **6**, 55–61 (2011)
31. Zhang, T., Ding, Y., Shao, S.: Protein subcellular location prediction based on pseudo amino acid composition and immune genetic algorithm. In: Huang, D.-S., Li, K., Irwin, G.W. (eds.) ICIC 2006. LNCS (LNBI), vol. 4115, pp. 534–542. Springer, Heidelberg (2006)
32. Feng, C., Du, S., Yan, J., Wang, Q., Phillips, P.: Feed-forward neural network optimized by hybridization of PSO and ABC for abnormal brain detection. Int. J. Imaging Syst. Technol. **25**, 153–164 (2015)
33. Phillips, P., Dong, Z., Yang, J.: Pathological brain detection in magnetic resonance imaging scanning by wavelet entropy and hybridization of biogeography-based optimization and particle swarm optimization. Progr. Electromagn. Res. **152**, 41–58 (2015)

From Parzen Window Estimation to Feature Extraction: A New Perspective

Zhong-bao Liu$^{(\boxtimes)}$, Jing Zhang, and Wen-ai Song

School of Software, North University of China, Taiyuan 030051, China
liu_zhongbao@hotmail.com

Abstract. Researches on current feature extraction methods are mainly based on two ways. One originates from geometric properties of high-dimensional datasets and attempt to extract fewer features from the original data space according to a certain criterion. The other originates from dimension reduction deviation and tries to make the deviation between data before and after dimension reduction be as small as possible. However, there exists almost no any study about them from the perspective of the scatter change of a dataset. Based on Parzen window density estimator, the relevant feature extraction methods are thoroughly revisited from a new perspective and the relations between Parzen window and LPP and LDA are built in this paper.

Keywords: Feature extraction · Parzen window · Data distribution characteristics · New perspective

1 Introduction

Pattern analysis often suffers a high-dimensional space, which leads to low recognition accuracy and expensive computational costs. Dimensionality reduction techniques provide a mean to solve above problems. In pattern recognition and image processing, feature extraction is a special form of dimensionality reduction (DR). When the input data is too large to be processed and it is suspected to be notoriously redundant, the input data will be transformed into a reduced representation set of features. Transforming the input data into the set of features is called feature extraction. In recent years, many feature extraction methods (FEM) have been proposed. Principal component analysis (PCA) [1] uses an orthogonal transformation to convert a set of observations of possibly correlated variables into a set of values of uncorrelated variables called principal components. Singular value decomposition (SVD) [2] realizes feature extraction according to the contribution of singular value. Independent component analysis (ICA) [3] finds the independent components by maximizing the statistical independence of the estimated components. Linear discriminant analysis (LDA) [4] finds a linear combination of features which characterizes or separates two or more classes of objects or events. Multi-dimensional scaling (MDS) [5] explores similarities or dissimilarities in data. Locally linear embedding (LLE) [6] attempts to discover nonlinear structure in high dimensional data by exploiting the local symmetries of linear reconstructions. Laplacian eigenmap (LE) [7] is sensitive to the size of neighbors and preserves local manifold. Canonical correlation analysis (CCA) [8] seeks

© Springer International Publishing AG 2016
H. Yin et al. (Eds.): IDEAL 2016, LNCS 9937, pp. 18–27, 2016.
DOI: 10.1007/978-3-319-46257-8_3

to identify and quantify the associations between two sets of variables. Locally preserving projections (LPP) [9] finds a linear projection that optimally preserves the neighborhood structure of the data set.

The above feature extraction methods mainly originate from space geometry and errors between before and after dimension reduction. In the view of space geometry, feature extraction is the process of reducing the feature space according to a certain criterion. In the view of errors between before and after DR, feature extraction aims to keep the deviation of the data before and after DR be as small as possible. While the transformation of the data distribution characteristics during DR is always ignored in many feature extraction methods. Therefore, probability density is introduced in this paper to characterize data distribution, and the process of feature extraction is viewed as the process of preserving feature characteristics. In this paper, the relation between kernel density estimation and current feature extraction methods is discussed and feature extraction and related methods are reviewed from a new perspective.

2 Parzen Window

Current kernel density estimation methods include [10]: Rosenblent, Parzen, Prakasarao, Silverman and so on. Parzen window is one of well-performed estimation methods [11] and used to characterize data distribution.

Parzen window can be defined as follows.

$$p(x) = \sum_{i=1}^{N} \alpha_i K_\delta(x, x_i) \tag{1}$$

$$s.t \sum_{i=1}^{N} \alpha_i = 1, \quad \alpha_i \geq 0 (i = 1, 2, \ldots, N) \tag{2}$$

where $K_\delta(x, x_i)$ is the kernel function with the width parameter δ. $K_\delta(x, x_i)$ should satisfy $K(t) \geq 0$ and $\int K(t)dt = 1$.

Gaussian kernel and Epanechnikov kernel are paid more attention in this paper, and they are the optimal kernel in the sense of minimum variance.

3 Parzen Window and LPP

Given $X = [x_1, x_2, \ldots, x_N] \in R^{n \times N}$ denote a n-dimensional dataset, where $x_i(i = 1, \ldots, N) \in R^n$ and N denote the ith data points and the total number of the dataset.

LPP is a linear projection map that arises by solving a variational problem that optimally preserves the neighborhood structure of the dataset. In the view of density estimation, the process of feature extraction is viewed as the process of preserving feature characteristics based on the data scatter. Feature extraction aims to keep the data distribution before and after DR be as small as possible.

Parzen window can be redefined as follows.

$$p_\delta(x) = \frac{1}{N} \sum_{k=1}^{c} \sum_{j=1}^{N_k} \frac{1}{\sqrt{2\pi}\delta_k} \exp(-\|x - x_{kj}\|^2 / 2\delta_k^2) \tag{3}$$

where the variance δ_k demonstrates the extend of all kinds of the dataset deviating from their center, therefore, δ_k is also called divergence. The transformation of the divergence δ_k can be represented as $\partial p_\delta / \partial \delta_k$. The derivation of (3) can be calculated:

$$\frac{\partial p_\delta}{\partial \delta_k} = \frac{1}{N} \sum_{j=1}^{N_k} \frac{1}{\sqrt{2\pi}\delta_k^2} \exp(-\|x - x_{kj}\|^2 / 2\delta_k^2)(\|x - x_{kj}\|^2 / \delta_k^2 - 1) \tag{4}$$

Given a definite class, the divergence δ_k is a constant. Let N_k and δ_k be represented by N and δ respectively. Ignoring the constants in (4), it can be obtained:

$$\frac{\partial p_\delta}{\partial \delta} = \sum_{j=1}^{N} \|x - x_j\|^2 \exp(-\|x - x_j\|^2 / 2\delta^2) \tag{5}$$

If x belongs to the training dataset, (5) can be transformed to:

$$\frac{\partial p_\delta}{\partial \delta} = \sum_{i,j=1}^{N} \|x_i - x_j\|^2 \exp(-\|x_i - x_j\|^2 / 2\delta^2) \tag{6}$$

In order to preserve the data distribution characteristics before and after DR, $\partial p_\delta / \partial \delta_k$ should be minimized. The objective function is expressed as follows.

$$\min_{W} \mathbf{1}^T W^T \sum_{i,j=1}^{N} \|x_i - x_j\|^2 \exp(W^T(-\|x_i - x_j\|^2 / 2\delta^2)W)W\mathbf{1} \tag{7}$$

Where $\mathbf{1} = [1,1,\ldots,1]^T$ which ensure (6) after projection is still a scalar.

The relation between data points x_i and x_j can be described by $\exp(-\|x_i - x_j\|^2 / 2\delta^2)$, therefore, (7) can be simplified as:

$$\min_{W} \mathbf{1}^T W^T \sum_{i,j=1}^{N} \|x_i - x_j\|^2 \exp(-\|x_i - x_j\|^2 / 2\delta^2)W\mathbf{1} \tag{8}$$

As $\mathbf{1}$ is irrelevant to W, (8) can be rewritten as:

$$\min_{W} W^T X^T L X W \tag{9}$$

where $S_{ij} = \exp(-\|x_i - x_j\|^2/2\delta^2)$, $D_{ii} = \sum_{j=1}^{N} S_{ij}$, $L = D - S$. A constraint is put to D:

$$W^T X^T D X W = 1 \tag{10}$$

Based on above observation, the following optimization problem can be obtained:

$$\min_{W} W^T X^T L X W$$

$$s.t \quad W^T X^T D X W = 1$$

To our convenience, the above optimization problem is called feature extraction method based on Parzen window (FEMPW).

It can be seen from LPP that when x_i and x_j satisfy $\|x_i - x_j\|^2 < \varepsilon$, they will be connected by an edge with weight. The weight function can be expressed as follows.

$$S = \frac{1}{\sqrt{2\pi}\delta} \exp(-\|x - x_j\|^2/2\delta^2) \tag{11}$$

which satisfies

$$\int_{-\varepsilon}^{\varepsilon} \sum_{k=1}^{c} \sum_{j=1}^{N_i} \frac{C}{\sqrt{2\pi}\delta_k} \exp(-\|x - x_{kj}\|^2/2\delta_k^2) dx = 1 \tag{12}$$

where C is a constant.

Solving (12), $C = 1/N(2\Phi(\varepsilon) - 1)$ is obtained, where $\Phi(\varepsilon) = \int_{-\infty}^{\varepsilon} \frac{1}{\sqrt{2\pi}} \exp(-x^2/2) dx$. For a given ε, the value of $\Phi(\varepsilon)$ is obtained by referring to the integral table and C can be calculated. When $\varepsilon \to \infty$, $C = 1/N$. In this case, FEMPW is consistent with LPP.

4   Parzen Window and LDA

4.1   LDA

Let $x_i = [x_{i1}, x_{i2}, \ldots, x_{iN_i}]$ be a $n \times N_i$ matrix and each column is a n-dimensional data points, where $x_{ij} \in R^n, (i = 1, \ldots, c; j = 1, \ldots, N_i)$ and N_i denote the jth data point of the ith class and the size of the ith class, respectively. The mean of all the dataset is $\bar{x} = \frac{1}{N} \sum_{i=1}^{N} x_i$. Let the mean of the ith class be $\bar{x}_i (i = 1, \ldots, c)$, then $\bar{x} = \sum_{i=1}^{c} \frac{N_i}{N} \bar{x}_i$.

4.1.1   In the View of Geometric Properties

In the view of geometric properties, LDA aims to extract the most important low-dimensional features from high-dimensional space in order to separate different classes in low-dimensional space and to keep the relation between data points in each class be as close as possible.

Two scatter matrices, called within-class (S_W) and between-class (S_B), are defined:

$$S_B = \sum_{i=1}^{c} N_i(\bar{x}_i - \bar{x})(\bar{x}_i - \bar{x})^{\mathrm{T}} \tag{13}$$

$$S_W = \sum_{i=1}^{c} \sum_{j=1}^{N_i} (x_{ij} - \bar{x}_i)(x_{ij} - \bar{x}_i)^{\mathrm{T}} \tag{14}$$

LDA is defined as follows.

$$J(W) = \max_{W} \frac{W^T S_B W}{W^T S_W W} \tag{15}$$

4.1.2   In the View of Dimension Reduction Error

(1) Compute the upper bound of two-class classification error rate

In order to make the upper bound of two-class classification error rate is as small as possible, the Bhattacharyya coefficient should be as large as possible. Let J_B denote Bhattacharyya coefficient, it can be obtained:

$$J_B = -\ln \int \sqrt{p(x|\omega_1)p(x|\omega_2)}dx \tag{16}$$

Let two classes obey Gaussian distribution: $x|y = 0 \sim N(\mu_1, \sum_1), x|y = 1 \sim N(\mu_2, \sum_2)$, J_B can be represented as:

$$J_B = \frac{1}{8}(\mu_2 - \mu_1)(\frac{\sum_1 + \sum_2}{2})^{-1}(\mu_2 - \mu_1)^T + \frac{1}{2}\ln\frac{|(\sum_1 + \sum_2)/2|}{|\sum_1|^{\frac{1}{2}} + |\sum_2|^{\frac{1}{2}}} \tag{17}$$

When $\sum_1 = \sum_2 = \Sigma$, (17) can be rewritten as:

$$J_B = \frac{1}{8}(\mu_2 - \mu_1)\Sigma^{-1}(\mu_2 - \mu_1)^T \tag{18}$$

Especially, when $\Sigma = E$, (18) can be rewritten as:

$$J_B = \frac{1}{8}(\mu_2 - \mu_1)(\mu_2 - \mu_1)^T \tag{19}$$

(2) Maximum likelihood estimation

The maximum likelihood estimation is defined as follows.

$$L(\phi, \mu_1, \mu_2, \Sigma) = \log \prod_{i=1}^{N} p(x_i, y_i; \phi, \mu_1, \mu_2, \Sigma) = \log \prod_{i=1}^{N} p(x_i|y_i; \mu_1, \mu_2, \Sigma)p(y_i; \phi) \quad (20)$$

Calculate the derivation of (20) and let the derivatives equal zero, it can be obtained:

$$\Sigma = \sum_{i=1}^{N_1} (x_i - \mu_1)(x_i - \mu_1)^T + \sum_{i=1}^{N_2} (x_i - \mu_2)(x_i - \mu_2)^T \quad (21)$$

Where Σ is the mean of sample feature variance.

In order to make the upper bound of two-class classification error rate and data scatter be as small as possible, the projection W should satisfy:

$$J(W) = \max_{W} \frac{W^T S_B W}{W^T S_W W}$$

where $S_B = (\mu_2 - \mu_1)(\mu_2 - \mu_1)^T$, $S_W = \sum_{i=1}^{N_1} (x_i - \mu_1)(x_i - \mu_1)^T + \sum_{i=1}^{N_2} (x_i - \mu_2)(x_i - \mu_2)^T$.

4.2   Relation Between Parzen Window and LDA

Inspired by Epanechnikov kernel, the following kernel function is given as follow.

$$p_\delta(x) = 1 - \frac{\sum_{i=1}^{c} \|x - \bar{x}_i\|^2}{\delta^2} \left(\frac{\sum_{i=1}^{c} \|x - \bar{x}_i\|^2}{\delta^2} \leq 1\right) \quad (22)$$

Derivation of (22) is as follows.

$$\frac{\partial p_\delta}{\partial \delta} = \frac{2}{\delta^3} \|x - \bar{x}_i\|^2 (i = 1, 2, \ldots, c) \quad (23)$$

(23) demonstrates the transformation of data scatter. In order to preserve the data local characteristics before and after DR be as close as possible, the projection W should satisfy $\partial p_\delta / \partial \delta$ is minimum. (23) can be represented as:

$$\min_{W} \mathbf{1}^T W^T \sum_{j=1}^{N_i} \|x_{ij} - \bar{x}_i\|^2 W\mathbf{1} \quad (24)$$

where the vector $\mathbf{1}$ ensures (24) is still a scalar. The vector $\mathbf{1}$ is irrelevant to $\mathbf{W}$ and all the classes of data are taken into consideration, it can be obtained:

$$\min_{\mathbf{W}} \mathbf{W}^T S_W \mathbf{W} \tag{25}$$

Where $S_W = \sum_{i=1}^{c} \sum_{j=1}^{N_i} \|x_{ij} - \bar{x}_i\|^2$.

The between-class scatter S_B is discussed as follows.
(22) can be rewritten as:

$$p_\delta(x) = 1 - \frac{\sum_{i=1}^{c} \sum_{j=1}^{N_i} \|x_{ij} - \bar{x}_i\|^2}{\delta^2}$$

$$= 1 - \frac{\sum_{i=1}^{c} \|N\bar{x} - \bar{x}_i\|^2}{\delta^2} \leq 1 - \frac{\sum_{i=1}^{c} N^2 \|\bar{x} - \bar{x}_i\|^2}{\delta^2} \leq 1 - \frac{\sum_{i=1}^{c} N_i \|\bar{x} - \bar{x}_i\|^2}{\delta^2}$$

Therefore, it can be obtained:

$$p_\delta(x) \leq 1 - \frac{\sum_{i=1}^{c} N_i \|\bar{x} - \bar{x}_i\|^2}{\delta^2} \tag{26}$$

The derivation of the upper bound of (26) is

$$\frac{\partial p_\delta}{\partial \delta} = \frac{2}{\delta^3} \|\bar{x} - \bar{x}_i\|^2 \quad (i = 1, 2, \ldots, c) \tag{27}$$

In order effectively separate different classes, the projection $\mathbf{W}$ should satisfy $\partial p_\delta / \partial \delta$ is maximized for a given class. It can be obtained:

$$\max_{\mathbf{W}} \mathbf{1}^T \mathbf{W}^T S_B \mathbf{W} \mathbf{1} \tag{28}$$

Where $S_B = \sum_{i=1}^{c} N_i \|\bar{x} - \bar{x}_i\|^2$.

As the vector $\mathbf{1}$ is irrelevant to $\mathbf{W}$, (28) can be rewritten as:

$$\max_{\mathbf{W}} \mathbf{W}^T S_B \mathbf{W} \tag{29}$$

Based on above observation, it can be obtained FLDA:

$$J(\mathbf{W}) = \max_{\mathbf{W}} \frac{\mathbf{W}^T S_B \mathbf{W}}{\mathbf{W}^T S_W \mathbf{W}}$$

5 Experimental Analysis

In this section, the relation between Parzen window and current feature extraction methods is evaluated using pattern classification tasks on the artificial and standard datasets. The recognition process is composed of the following steps. Firstly, the experimental dataset is divided into two parts, one for training and the other for test. Secondly, the optimal projection W is calculated from the training dataset and the test dataset after projection is easily obtained. Finally, the nearest neighbor (NN) algorithm is applied in low-dimensional subspace for classification.

The ORL dataset includes ten different images of each of 40 distinct subjects. For some subjects, the images were taken at different times, varying the lighting, face expressions (open /closed eyes, smiling /not smiling) and face details (glasses /no glasses). First k images of each subject are selected for training and the remaining images are used for test. The value of k is selected from 3, 4, 5, 6, 7, 8 and the reduced dimension is selected from 10, 12, 14, 16, 18, 20, 22, 24, 26, 28, 30. The relations between reduced dimensions and recognition accuracy of PCA, LPP, LDA and FEMPW are shown in Fig. 1 and Table 1 lists the best recognition results and the corresponding optimal reduced dimensions of all algorithms. In Table 1, the unit of recognition accuracy is % and the dimension in the bracket is the corresponding optimal reduced dimensions. "Average" demonstrates the average performance of each algorithm. In the following experiments, PCA + LDA is represented by LDA.

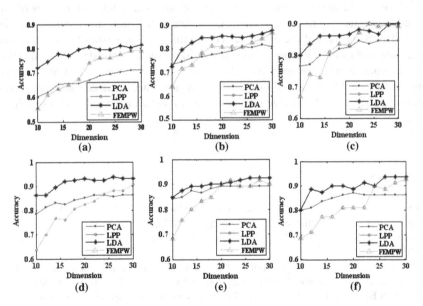

Fig. 1. The relation between reduced dimensions and recognition accuracies of PCA, LPP, LDA and FEMPW on ORL dataset with various values of k: (a) $k = 3$. (b) $k = 4$. (c) $k = 5$. (d) $k = 6$. (e) $k = 7$. (f) $k = 8$.

Table 1. Performance of all algorithms on ORL dataset

Size of train	PCA	LPP	LDA	FEMPW
3	71.1(28)	78.9(28)	81.4(30)	78.9(28)
4	80.8(28)	86.7(30)	87.5(30)	86.7(30)
5	84.5(22)	89.0(24)	90.5(30)	89.0(24)
6	86.3(22)	90.6(30)	95.0(30)	90.6(30)
7	89.2(20)	91.7(22)	92.5(26)	90.0(22)
8	87.3(20)	92.5(30)	93.8(26)	92.5(30)
Average	83.2	88.2	90.1	88.2

From Fig. 1 and Table 1, it can be seen the recognition accuracies of each algorithm grow with the size of training dataset. In the average performance, the recognition accuracy of FEMPW is obviously higher than PCA while a little lower than LDA. Moreover, the recognition accuracy of FEMPW is equivalent to LPP, which is coincidence with the conclusion "when $\varepsilon \rightarrow \infty$, FEMPW is equivalent to LPP".

6 Conclusions

Feature extraction is one of most important tasks in pattern recognition. Most of current feature extraction methods focus on space geometry and errors between before and after DR, while the data distribution characteristics are always ignored. In view of this, probability density is introduced to characterize data distribution, and feature extraction is viewed as the process of preserving data distribution characteristics based on the data scatter. Researches in this paper provide a way to revisit feature extraction and related methods from a new perspective, while only three feature extraction methods, including LPP, LDA, are discussed in this paper and whether other feature extraction methods can be deduced by Parzen window is an important issue to be discussed in the future.

References

1. Camacho, J., Pic, J., Ferrer, A.: Data understanding with PCA: structural and variance information plots. Chemometr. Intell. Lab. Syst. **100**(1), 48–56 (2010)
2. Lipovetsky, S.: PCA and SVD with nonnegative loadings. Pattern Recogn. **42**(1), 68–76 (2009)
3. Radulovic, J., Rankovic, V.: Feedforward neural network and adaptive network-based fuzzy inference system in study of power lines. Expert Syst. Appl. **37**(1), 165–170 (2010)
4. Peter, N.B., Joao, P.H., David, J.K.: Eigenfaces vs. Fisherfaces: recognition using class specific linear projection. IEEE Trans. Pattern Anal. Mach. Intell. **19**(7), 711–720 (1997)
5. Hu, Y.H., He, S.H.: Integrated Evaluation Method. Scientific Press, Beijing (2000)
6. Roweis, S.T., Saul, L.K.: Nonlinear dimensionality reduction by locally linear embedding. Science **290**, 2323–2326 (2000)

7. Laplacian eigenmaps and spectral techniques for embedding and clustering. In: Proceedings of Advances in Neural Information Processing Systems (NIPS), pp. 585–591. MIT Press, Cambridge (2001)
8. Borga, M., Knutsson, H.: Canonical correlation analysis in early vision. In: Proceedings of the 9th European Symposium on Artificial Neural Networks, Bruges, Belgium, pp. 309–314 (2001)
9. He, X.F., Niyogi, P.: Locality preserving projections. In: Advances in Neural Information Processing Systems (NIPS), Vancouver, Canada, pp. 153–160 (2003)
10. Odiowei, P.P., Cao, Y.: Nonlinear dynamic process monitoring using canonical variate analysis and kernel density estimations. IEEE Trans. Ind. Inf. 6(1), 36–45 (2010)
11. Wu, G.M., Huo, J.Q., Wang, X.P.: A new clustering algorithm based on weighted Parzen window. J. Univ. Sci. Technol. Chin. 32(5), 546–551 (2002). (in Chinese)

An Efficient Sparse Optimization Method
for Unfinished Magic Squares

Yali Liang, Xiaohua Xu$^{(\boxtimes)}$, Zheng Liao, and Ping He$^{(\boxtimes)}$

Department of Computer Science, Yangzhou University, Yangzhou, China
arterx@gmail.com, angeletx@gmail.com

Abstract. Magic square is an old and interesting mathematical problem, which has the same value of all the sums of the elements in each row, column and diagonal. An unfinished magic square denotes that it gives us some clues and we need to fill the empty cells. In order to solve the unfinished magic squares more efficiently, we propose a solution based on sparse optimization. Using the properties of magic square, we establish a model of constraint programming. Then we transform the constraints into sparse linear constraints, meanwhile use l_0 norm minimization as the objective function to ensure the sparsity of the solution. Moreover, we use l_1 norm to approximate l_0 norm on the basis of RIP and KGG condition. This paper uses the primal-dual interior point method of linear programming, the branch and bound algorithm of binary programming and dual simplex method of integer linear programming to solve the magic square problems. The experimental results show that dual simplex method of integer linear programming can reach almost 100 % success rate. In addition, we propose a kind of special magic square problem and we apply this idea to construct and solve this problem, and obtain the good results.

Keywords: Magic square · Sparse optimization · Special magic square

1 Introduction

Magic square problem is a complex problem of permutation and combination with a long history [1]. The magic square of order n is a matrix of $n \times n$, which is filled with unrepeated numbers $1, 2, \ldots, n^2$, and the sum of the elements in each row, column, diagonal is a definite number called magic number with $n(n^2 + 1)/2$. An unfinished magic square may look as in Fig. 1. The initial set of occupied cells is called the "clues" of the magic square. The objective is to fill the empty cells.

Magic square was first used in astrology, prophecy and interpretation of philosophy, natural phenomena and human behavior. Only after the 17th century, it was carried out more serious mathematical thinking. In the late 19th century, mathematicians applied magic square to probability theory and mathematical analysis. With the rapid development of computer technology, the magic square has been applied to program design, artificial intelligence, graph theory, game theory, experimental design, and other aspects [2, 3].

The most primitive method of solving the unfinished magic squares is exhaustive method based on combination algorithm [4]. The advantage is that it can find out the

© Springer International Publishing AG 2016
H. Yin et al. (Eds.): IDEAL 2016, LNCS 9937, pp. 28–36, 2016.
DOI: 10.1007/978-3-319-46257-8_4

10			11
4			1
13		3	16
	9	12	6

Fig. 1. An easy unfinished magic square of order 4

optimal solution set within the time limit in the tolerant to recuperate. But, it has the defect of high time complexity as well as the explosion problem. Another method to solve the unfinished magic square is an intelligent optimization algorithm hybrid Immune genetic algorithm and tabu search algorithm [5], which overcome the limitation of exhaustive method with reducing the invalid combination merger and retain the advantages, but it also rely on combination algorithm whose complexity is very high.

In recent years, some scholars have explored the special magic squares. Loly et al. [6] use the Jordan form and singular value decomposition to construct magic square spectra. Nordgren [7] introduces a new class of quasi-regular magic squares which includes regular and most-perfect magic squares. Chen et al. [8] propose the regular sparse anti-magic squares with small odd densities.

This paper proposes a new thought that unfinished magic square problem can be converted to an l_1 norm optimization problem. First, we propose a constraint programming due to the properties of magic square. Then we make the constraints sparse. Moreover, we transform primal problem into a linear programming problem. As the linear programming problem is a classical optimization problem and there are many excellent algorithms, it can be solved quickly [9]. In addition, we propose a new special magic square problem and we apply this idea to construct and solve it.

2 Model to Solve Unfinished Magic Square

For the unfinished magic square of order n, assuming that the solution is a matrix x, and x_{ij} represents the elements in the i-th row of j-th column of x. A simplified form of constraint programming is as follows:

$$\min 1 \text{ s.t.} \begin{cases} x_{ij} \in \{1, 2, \cdots, n^2\}, \\ \text{box}_i(x) = \{1, 2, \cdots, n^2\}, \\ \text{clue}_i(x) = clue_i, \\ \text{rowsum}_i = \sum_j x_{ij} = n(n^2 + 1)/2, \\ \text{colsum}_j = \sum_i x_{ij} = n(n^2 + 1)/2, \\ \text{mdiagsum} = \sum_i x_{ii} = n(n^2 + 1)/2, \\ \text{sdiagsum} = \sum_i x_{i(n+1-i)} = n(n^2 + 1)/2. \end{cases} \tag{1}$$

In formula (1), the value of the objective function is always 1. Constraint functions respectively express that the elements which fill the magic square of order n are unrepeated $1,2,...,n^2$, x needs to satisfy the clues, and all the sums of the elements in each row, column, diagonal is $n(n^2 + 1)/2$.

2.1   Sparse Constraints

Let S denote the solution of an unfinished magic square of order n, the content of cell m is denoted by $S_m \in \{1,2,...,n^2\}$ for $m = 1,2,...,n^2$ with the cells enumerated in row order. Let $i_m = [I(S_m = 1), I(S_m = 2), ..., I(S_m = n^2)]^T$ denote the indicator vector associated with cell m, where $I(S_m = k)$ is the indicator function that is equal to one when $S_m = k$ and zero otherwise. Let x denote a vector of size n^4 constructed as

$$x = [i_1 i_2 \cdots i_{n^2}]^T \tag{2}$$

The box constraint that S should comprise all numbers $1,2,...,n^2$ can be expressed as

$$\left[\underbrace{I_{n^2 \times n^2} I_{n^2 \times n^2} \cdots I_{n^2 \times n^2} I_{n^2 \times n^2}}_{n^2} \right] x = 1_{n^2} \tag{3}$$

where $I_{n^2 \times n^2}$ denotes the $n^2 \times n^2$ identity matrix.

The cell constraints denote that each cell of S should be filled. For example, the first cell S_1 which should be filled with a number of $1,2,...,n^2$ can be expressed as

$$\left[\underbrace{1 1 \cdots 1}_{n^2} \underbrace{0 \cdots 0}_{n^4 - n^2} \right] x = 1 \tag{4}$$

The clues can also be expressed as linear equality constraints on x. For example, the clue that cell S_1 takes the value 5 can be expressed as

$$\left[\underbrace{0 \cdots 0}_{4} 1 \underbrace{0 \cdots 0}_{n^2 - 5} \underbrace{0 \cdots 0}_{n^4 - n^2} \right] x = 1 \tag{5}$$

The sum of the elements in each row must be a magic number with the value n $(n^2 + 1)/2$. Take the first row as an example, it can be expressed as

$$\left[\underbrace{1 2 \cdots n^2 \cdots 1 2 \cdots n^2}_{n^2 \times n} \underbrace{0 \cdots 0}_{n^4 - n^3} \right] x = n(n^2 + 1)/2 \tag{6}$$

The sum of the elements in each column must be a magic number with the value n $(n^2 + 1)/2$. Take the first column as an example, it can be expressed as

$$\left[\underbrace{\underbrace{12 \cdots n^2 0 \cdots 0}_{n^3} \cdots \underbrace{12 \cdots n^2 0 \cdots 0}_{n^3}}_{n}\right] x = n(n^2+1)/2 \tag{7}$$

The sum of the elements in main diagonal must be a magic number with the value n $(n^2 + 1)/2$, it can be expressed as

$$\left[\underbrace{12 \cdots n^2 0 \cdots 0}_{n^3} \underbrace{0 \cdots 0}_{n^2} \underbrace{12 \cdots n^2 0 \cdots 0}_{n^3 - n^2} \cdots \underbrace{0 \cdots 0 12 \cdots n^2}_{n^3}\right] x = n(n^2+1)/2 \tag{8}$$

The sum of the elements in secondary diagonal must be a magic number with the value $n(n^2 + 1)/2$, it can be expressed as

$$\left[\underbrace{0 \cdots 0 12 \cdots n^2}_{n^3} \underbrace{0 \cdots 0 12 \cdots n^2 0 \cdots 0}_{n^3 - 2n^2} \underbrace{12 \cdots n^2 0 \cdots 0}_{2n^2} \cdots \underbrace{12 \cdots n^2 0 \cdots 0}_{n^3}\right] x = n(n^2+1)/2 \tag{9}$$

By the formula (1), all the constraints are transformed into a linear equation set as follows

$$\min 1 \text{ s.t. } A_{eq}x = \begin{bmatrix} A_{box} \\ A_{cell} \\ A_{clue} \\ A_{rowsum} \\ A_{colsum} \\ A_{diagsum} \end{bmatrix} x = b_{eq} = \begin{bmatrix} 1 \\ \vdots \\ 1 \\ n(n^2+1)/2 \\ \vdots \\ n(n^2+1)/2 \end{bmatrix} \tag{10}$$

It should be noted, for magic square of order n, the constraints make x be in the range of real values, and do not provide binary constraint. The advantage is to ensure the constraints are linear. The disadvantage is the number of elements is n^4, the number of equations is $n^2 + n^2 + N_C + n+n + 2 = 2(n^2 + n+1) + N_C$, where N_C denotes the number of clues and $0 \le N_C \le n^2$. Obviously, it is an underdetermined equation. According to (10), the obtained solution x is not necessarily satisfied with the binary format.

2.2 l_0 and l_1 Norm Minimization

Using the concept of sparse optimization, we can transform binary constraint that x is binary into l_0 norm minimization as follows [10]:

$$\min \|x\|_0 \text{ s.t.} \mathbf{A}_{eq}x = b_{eq} \tag{11}$$

In (11), l_0 norm minimization is a NP-hard problem and computational complexity. In some cases, l_0 and l_1 norm are equivalent. According to RIP (Restricted Isometry Property) condition [11], KGG(Kashin Garnaev Gluskin) inequality [12], and a large number of instances of magic squares, (11) can be approximately written as

$$\min \|x\|_1 \text{ s.t. } \mathbf{A}_{eq}x = b_{eq} \tag{12}$$

2.3 Magic Square Nesting Problem

This model can be used to construct and solve some special magic squares. Moreover, we propose a kind of special magic square named magic square nesting problem.

Definition: The center of a magic square of order n nests a small magic square of order $n-2$. For the small magic square, all the sums of the elements in each row, column, and diagonal are equal.

For the small magic square, we add 3 new sum's constraints about the sums of the elements in each row, column and diagonal represented by matrix **B**. Because these sums are equal and unknown, we cannot denote these sums with a fixed value. Then move the elements of the matrix B up and get a new matrix $\mathbf{B}_+$ and that meets the equations $(\mathbf{B} - \mathbf{B}_+)x = 0$.

Due to the properties of magic square nesting problem, the optimization formula is:

$$\min \|x\|_1 \text{ s.t.} \begin{cases} \mathbf{A}_{eq}x = b_{eq} \\ (\mathbf{B} - \mathbf{B}_+)x = 0 \end{cases} \tag{13}$$

2.4 Solving Methods

The formula (12) is a linear programming problem. At present, there are 4 methods for solving linear programming problems: interior point method, simplex method, active set method and trust region method. Because of the existence of equality constraint and boundary constraint, the trust region method is not available. Moreover, the interior point method performed well in many tests [13], so we use interior point method. In order to improve performance, the primal-dual interior point method [14] is adopted. Besides, we use branch and bound algorithm to solve binary programming problem and dual simplex method to solve the integer linear programming problem.

The whole algorithm based on sparse optimization to solve unfinished magic squares is as follows:

Algorithm. sparse optimization for unfinished magic squares

Data: an unfinished magic square matrix S and the order of n

begin

    Step1. initialize x as a zero vector of size n^4;

    Step2. construct a sparse matrix $\mathbf{A}_{eq}$ by the formulas (3)-(9);

    Step3. construct b_{eq};

    Step4. compute x by formula (12);

    Step5. transform x into S;

    return S;

end

The time complexity for step 1 to initialize x as a zero vector of size n^4 is $O(n^4)$. Since it requires Since it requires $O(n^4)$ time to complete the sparse constraints and construct the sparse matrix $\mathbf{A}_{eq}$, time complexity of step 2 is $O(n^4)$. The time complexity for step 3 to construct b_{eq} is $O(n^2)$. The computation of l_1 norm minimization in step 4 requires $O(n^6)$ time. The last step to transform x into S requires $O(n^4)$ time. Our algorithm is efficient for unfinished magic squares.

3 Experiments

In this paper, we use the computer with 2 GHz CPU and 2 G RAM, operating system is Windows 7, and software is Matlab2014a with Optimization Toolbox. The experimental data is obtained from the digital magic square game which includes unfinished magic squares of order 4–7 with 3 levels namely easy, normal and hard. For each order, we solve 200 easy, 100 normal and 50 hard unfinished magic squares. The methods compared with are the mentioned dual simplex method, branch and bound algorithm and primal-dual interior point method. The average success rate and of these algorithms are listed in the following 4 tables, where N_C denotes the number of clues.

Table 1. Results of solving unfinished magic square of order 4

Levels	Dual simplex method		Branch and bound algorithm		Primal-dual interior point method	
	Success rate	Time(s)	Success rate	Time(s)	Success rate	Time(s)
Easy($N_C = 10$)	1.00	0.059	1.00	0.135	1.00	0.028
Normal($N_C = 8$)	1.00	0.056	1.00	0.266	0.90	0.029
Hard($N_C = 6$)	1.00	0.064	1.00	0.315	0.56	0.033

Table 2. Results of solving unfinished magic square of order 5

Levels	Dual simplex method		Branch and bound algorithm		Primal-dual interior point method	
	Success rate	Time(s)	Success rate	Time(s)	Success rate	Time(s)
Easy($N_C = 17$)	1.00	0.055	1.00	0.273	1.00	0.041
Normal($N_C = 15$)	1.00	0.056	1.00	0.386	0.90	0.041
Hard($N_C = 13$)	1.00	0.066	1.00	0.579	0.50	0.051

Table 3. Results of solving unfinished magic square of order 6

Levels	Dual simplex method		Branch and bound algorithm	
	Success rate	Time(s)	Success rate	Time(s)
Easy($N_C = 22$)	1.00	0.111	1.00	0.892
Normal($N_C = 20$)	1.00	0.156	1.00	4.952
Hard($N_C = 18$)	1.00	0.811	1.00	87.861

Table 4. Results of solving unfinished magic square of order 7

Levels	Dual simplex method		Branch and bound algorithm	
	Success rate	Time(s)	Success rate	Time(s)
Easy($N_C = 31$)	1.00	0.265	1.00	75.523
Normal($N_C = 28$)	1.00	1.554	1.00	632.517
Hard($N_C = 25$)	1.00	30.172	1.00	2786.421

As can be seen from the above Tables 1, 2, 3 and 4, the dual simplex method solving integer linear programming obtains a success rate up to 100 % for all levels of unfinished magic squares of each order as same as the branch and bound algorithm. With the rise of the difficulty and the order of magic square, the success rate remains. While the success rate of primal-dual interior point method solving linear programming gradually decreases, that is far below other two methods.

From the time point of view, the computation time of dual simplex method and branch and bound algorithm is related to the order and the difficulty of unfinished magic square, while the operation time of primal-dual interior point method almost has nothing to do with the difficulty of magic square. Moreover, the dual simplex method needs relatively less time consumption than the branch and bound algorithm.

In general, the dual simplex method of integer linear programming is superior to other two methods. This method not only ensures the success rate, but also reduces the time consumption.

In addition, we construct some special magic squares that meet the magic square nesting problem. We select two typical examples those are shown in Fig. 2, where bold blue numbers indicate the elements in the small magic squares.

a. 5 order nests 3 order   b. 6 order nests 4 order

Fig. 2. Two example of magic square nesting problem (Color figure online)

4 Conclusion

In this paper, a new strategy to solve magic square problems based on sparse optimization is proposed. It transforms a constraint programming problem into a sparse linear programming problem. For 3 levels of magic square problems, the success rate of the dual simplex method solving integer linear programming has reached almost 100 %. In addition, we has extended this method to the special magic squares, established a linear programming model and obtained good results.

Many puzzle problems or engineering problems can be represented by the following models: constrained programming, integer programming, or sparse optimization. The mathematical significance of this paper is to propose a transformation strategy based on sparse optimization, which can be transformed into a sparse linear programming problem which is easy to solve.

Acknowledgments. This research was supported in part by the Chinese National Natural Science Foundation under Grant nos. 61402395, 61472343 and 61379066, Natural Science Foundation of Jiangsu Province under contracts BK20151314,BK20140492 and BK20130452, Natural Science Foundation of Education Department of Jiangsu Province under contract 13KJB520026, and the New Century Talent Project of Yangzhou University.

References

1. Gu, Z.: Magic Algorithm and Program Design. Beijing science and technology literature press, Beijing (1993)
2. Wu, H.: Magic Square and Prime Numbers: Entertainment Mathematical Two Classical Propositions. Beijing Science Press, Beijing (2008)
3. Kitajima, A., Kikuchi, M.: Numerous but rare: an exploration of magic squares. PLoS ONE **10**, 49–55 (2015)
4. Lv, Z.: A rapid condition combinatorial algorithm. J. Zhejiang Norm. Univ. (Nat.Sci.) **29**, 52–55 (2006)

5. Lv, Z.: Research On Intelligent Computation Of Magic Square Problem. National University of Defense Technology Engineering Master Thesis (2005)

6. Loly, P., Cameron, I., Trump, W., Schindel, D.: Magic square spectra. Linear Algebra Appl. **430**, 2659–2680 (2009)

7. Nordgren, R.P.: On properties of special magic square matrices. Linear Algebra Appl. **437**, 2009–2025 (2012)

8. Chen, G., Chen, H., Chen, K., Li, W.: Regular sparse anti-magic squares with small odd densities. Discrete Math. **339**, 138–156 (2016)

9. Babu, P., Pelckmans, K., Stoica, P., Li, J.: Linear systems, sparse solutions, and sudoku. Sig. Process. Lett. IEEE **17**, 40–42 (2010)

10. Saab, R., Yilmaz, Ö.: Sparse recovery by non-convex optimization-instance optimality. Appl. Comput. Harmonic Anal. **29**, 30–48 (2008)

11. Needell, D., Tropp, J.A.: CoSaMP: Iterative signal recovery from incomplete and inaccurate samples. Appl. Comput. Harmonic Anal. **26**, 301–321 (2008)

12. Ganguli, S., Sompolinsky, H.: Statistical mechanics of compressed sensing. Phys. Rev. Lett. **104**, 188701 (2010)

13. Kwon, O.M., Park, J.H.: Delay-dependent stability for uncertain cellular neural networks with discrete and distribute time-varying delays. J. Franklin Inst. **345**, 766–778 (2008)

14. Edlund, K., Sokoler, L.E., Jrgensen, J.B.: A primal-dual interior-point linear programming algorithm for MPC. In: Proceedings of Joint 48th IEEE Conference on Decision and Control and 28th Chinese Control Conference, pp. 351–356 (2009)

Enhancing State Space Search for Planning by Monte-Carlo Random Walk Exploration

Qiang Lu[1]([⊠]), You Xu[2], Yixin Chen[2], Ruoyun Huang[2], and Ling Chen[1]

[1] Yangzhou University, Yangzhou 225127, China
{lvqiang,lchen}@yzu.edu.cn
[2] Wahington University in St. Louis, St. Louis 63130, USA
{yx2,chen,ruoyun.huang}@cse.wustl.edu

Abstract. State space search is one of the most important and prover-bial techniques for planning. At the core of state space search, heuristic function largely determines the search efficiency. In state space search for planning, a well-observed phenomenon is that for most of the time during search, it explores a large number of states while the minimal heuristic value has not been reduced. This so called "plateau escape" phenomenon has attracted many interests in heuristic search areas, especially in satis-fiability (SAT) and constraint satisfaction problems (CSP). In planning, the efficiency of many state space search based planners largely depend on how fast they can escape from these plateaus. Therefore, their search per-formance can be improved if we could reduce the plateau escaping time.

In this paper, we propose a Monte-Carlo Random Walk (MRW) assisted plateau escaping algorithm for planning. Specifically, it invokes a Monte-Carlo random search procedure to find an exit when a plateau is detected during the search. We establish a theoretical model to analyze when a Monte-Carlo random search is helpful to state space search in finding plateau exits. We subsequently implement a sequential and a par-allel version of the proposed scheme. Our experimental results not only show the advantages of using random-walk to assist state space search for planning problems, but also validates the performance analysis in the theoretical model.

Keywords: State space search · Plateau exploration · Random walk exploration

1 Introduction

During the last decades, state space search has been extensively studied in plan-ning. As far as we know, state space search is one of the most proverbial and suc-cessful approaches to planning. Typically, State space search employs a heuristic function that maps any state to a real number that estimates the distance to a goal. At the beginning of search, the initial state s_0 is inserted into an *open* list which stores all states to be explored. At each iteration of the search, it fetches a state s with the smallest heuristic value from *open* list and checkes whether it

© Springer International Publishing AG 2016
H. Yin et al. (Eds.): IDEAL 2016, LNCS 9937, pp. 37–45, 2016.
DOI: 10.1007/978-3-319-46257-8_5

is a goal state. If not, it generates all its successors and insertes them into the *open* list. After than, it inserts s into a *closed* list which stores all the searched states and avoids duplicated search. The search stops when a goal state is found.

During the state space search, for any state s, we define a function $h^*(s)$ which represents the smallest heuristic value of all explored states so far before s. Obviously, h^* monotonically decreases with the number of searched states and finally reduces to 0 when a goal is found. In state space search, the quality of the heuristic function largely determines the total number of explored states [11,14, 17]. Unfortunately, even with an almost perfect heuristic function, it still needs to explore an exponential number of states [9]. Given a planning problem, let $\mathcal{S}$ be the set of explored states. Since $|\mathcal{S}|$ is usually an exponential number while $h^*(s_0)$ is a constant number, we have $|\mathcal{S}| \gg h^*(s_0)$. Since $h^*(s)$ is a monotonic function mapping from $\mathcal{S}$ to $[0, h^*(s_0)]$, for most of the state s and its successor s', we have $h^*(s) = h^*(s')$. It means that the minimal heuristic value h^* is not reduced during the most of the search time even it explores a large number of states. The phenomenon is well-known as *plateau exploration*. In planning, the plateau exploration consists of most of the search time in state-of-the-art state space search planners. Therefore, to improve the performance of state space search, it is very important to find a way which can quickly escape from a plateau.

In this paper, we propose a best-first state space search algorithm with a Monte-Carlo Random Walk (MRW) procedure which can assist the search to escape from plateaus efficiently. Specifically, when the state space search falles into a plateau, it calles a MRW procedure to explore the state space and help find an exit state of the plateau.

We make three contributions in this paper. First, we propose an algorithmic framework where random walks are incorporated to help state space search find the exit states of a plateau. Second, we conduct theoretical analysis to study the impact of using random walks on different types of search space models, including tree search, graph search, and search with many dead end states. Third, we implement two variants of the proposed algorithm and compare their performances to a pure state space search planner on various testing domains.

2   Related Work

Plateau escaping techniques have attracted many interests in satisfiability (SAT) and constraint satisfaction problems (CSP) [8,19]. In these areas, a plateau is defined as a set of neighboring variable assignments which have the same number of unsatisfied clauses [6]. Glover and Laguna proposed a plateau search algorithm to avoid falling back to the same states on a plateau in solving CSP and SAT problems [7]. Kautz and Selman developed a random-walk based algorithm, WalkSAT, which can find an exit to escape from plateau [1,3,12].

In planning, the plateau is also known as local minima. Hoffmann presented a detailed analysis of how long the maximum exit distance of a local search is in hill climbing [10]. Benton etc. indicated that g-value plateaus is a great challenge in planning [2]. Space reduction techniques, such preferred operators [16] and partial order reduction [4,5], are also important and effective methods to reduce the time of escaping from a plateau. Besides the traditional techniques of local minima escaping, a recently proposed Monte-Carlo random walk (MRW) algorithm has shown good performance of escaping from plateaus and solving planning problems [13].

Our proposed MRW assisted best-first search (RW-BFS) for planning is inspired by both the MRW approach and the preferred operators techniques. Specifically, since state space search shows good performance in many planning problems, we adopt it as the basic search procedure. When a plateau is detected, we invoke a MRW procedure to assist the search to find an exit efficiently.

3 Best-First State Space Search Assisted with Monte-Carlo Random Walk

In this paper, we tackle classical planning problems that can be formulated using STRIPS or ADL. Given a classical planning task $\mathcal{T}$ with an initial state s_0, $\mathcal{A}_h$ is a state space search algorithm guided by heuristic function h. $\mathcal{S}$ is the set of states explored by $\mathcal{A}_h$.

Definition 1 (Order of states). *Given a state space search procedure $\mathcal{A}_h$ on $\mathcal{S}$ guided by h, the relation $\mathcal{R} \subseteq \mathcal{S} \times \mathcal{S}$ is defined as follows: for any states a and b, $\mathcal{R}(a,b)$ holds if and only if a is explored before b according to $\mathcal{A}_h$. We use $a < b$ to denote $\mathcal{R}(a,b)$.*

It is easy to see that states in $\mathcal{S}$ can be sorted according to $\mathcal{R}$ to a total order. Let $\mathcal{L}$ be an ordered list of all states in $\mathcal{S}$ where the initial state s_0 is the first state and the first goal state found by $\mathcal{A}_h$ is the last state in $\mathcal{L}$, we can define a plateau as follows.

Definition 2 (Plateau and Exit). *Given a heuristic function h, let P be a sub-sequence of $\mathcal{L}$, P is a **plateau** if $|P| > 1$, and for every state $s \in P$, $h(s) \geq l$. A state e_P is an **exit** of P if $h(e_P) < l$ and e_P is an immediate successor of some state in P.*

Now we introduce our MRW assisted best-first state space search (RW-BFS) framework shown in Algorithm 1. Based on a standard best-first state space search procedure, it adds a *plateau detection* after exploring a new state (Line 9). If it confirms that the search was fallen into a plateau, a MRW exploration will be invoked to help find an exit of the current plateau, a state with a smaller heuristic value than h^* (Line 10). Then, it inserts the exit state to the *open* list (Line 11).

The idea of *MRW exploration* is inspired from the Monte-Carlo exploration algorithm introduced in [13]. Given a start state s and the current smallest heuristic value h^*, it uses a t-times **random walks** to search the neighbors of s. The MRW exploration returns a state s' with a smaller h value than h^* if finds one. For each iteration of the **random walk**, it searches n paths where each

---

Algorithm 1. The RW-BFS Algorithm

Input: Initial state s_0

1 $open \leftarrow s_0$;
2 **while** $open$ is not $empty$ **do**
3      $s \leftarrow open$.pop() // s is the state with the minimal heuristic value;
4   **if** s is $goal$ **then return** solution(s_0, s);
5   **if** $h(s) \leq h^*$ **then** $h^* \leftarrow h(s)$;
6   **if** s is not a $dead$ end **then**
7          $closed \leftarrow s$;
8   **foreach** $s_i \in successor(s)$ **do**   evaluate $h(s_i)$; $open \leftarrow (s_i, h(s_i))$;
9   **if** $plateau$ is $detected$ **then**
10          $s' \leftarrow MRW$ $exploration$ (s, h^*);
11         $open$.push(s');
12 **return** no $solution$

---

path contains l actions. Note that the random walk only evaluates the end state of each path. The total number of random walks are restricted by t which makes sure the MRW procedure always returns in a finite time, whether an exit state is found or not.

Plateau Detection. The accuracy of the *plateau detection* is an important factor of the performance of the MAS Algorithm. It can neither be too unresponsive nor too sensitive. The MRW exploration will nearly not be invoked if the detection is too unresponsive which leads to normal state space search. On the other hand, it will be called frequently and hinder the progress of state space search by constant interruptions. Thus, we adopt a balanced detection strategy described as follows. A plateau is detected if either of the two conditions is satisfied:

- The value of h^* has not been reduced for m consecutive states.
- The moving average of the heuristic values of w recent states is higher than $h^* + \delta$ where δ is a threshold that reduces gradually during search.

4   Performance Analysis

Analyzing the performance of a heuristic search algorithm applied to a general planning problem is complex and difficult. In this section, we analyze the performance of the proposed RW-BFS algorithm on a simplified model to draw some insights on when the stochastic search can improve the overall efficiency.

To establish our model, we introduce the following two definitions.

Definition 3 (N-Neighbor). *For any state* $x \in S$, *the N-Neighbor of* x, *denoted by* $N(x, d)$, *is a set of states that are reachable from* x *using* d *or fewer of actions.*

Definition 4 (Plateau Graph). *Given a planning task* $\mathcal{T}$, *a plateau graph* $G_d = (V, E)$ *of state* x *is a simple digraph with* $V = N(x, d)$ *satisfying: (1)*

there is an edge $(s_i, s_j) \in E$ if and only if there is an action that leads s_i to s_j;
(2) for all states s in V, $h(s) \geq h(x)$, and there are no dead end states in V.
An exit state of G_d is a state $s_e \notin G_d$ such that $h(s_e) < h(x)$.

For a given plateau graph G_d, in the following analysis, we compare the number of heuristic evaluations required to escape from G_d for both state space search and random walk algorithms, as heuristic evaluation takes up most of the time for search algorithms. To simplify our analysis, we assume that the random walk is unbiased, meaning that instead of picking the best state among all n cases in **walk**, a random s' is chosen with probability $1/n$ to be s_{min} in **walk**. We further simplify the structure of G_d as a graph where nodes have the same in- and out-degrees. Without loss of generality, we assume that every node in G_d has p successors and q parents, where $p \geq q \in Z^+$.

Lemma 1. Given a plateau graph G_d of x where every node in G_d has an in-degree of $q \geq 1$, if there exists a state $s_e \in N(x, d+1)$ such that $h(s_e) < h(x)$, a state space search procedure, in the worst case, will have to evaluate the heuristic value for $\frac{(p/q)^{d+1}-1}{(p/q-1)}$ states before finding s_e.

Proof. We prove this by using mathematical induction. It is easy to see that this proposition is true when $d = 0, 1$, where we have 1 and $1 + \frac{p}{q}$ states, respectively.

Assuming this proposition is true for all $d < k$, we have $|G_{k-1}| = \frac{(p/q)^k - 1}{(p/q) - 1}$ and $|G_{k-2}| = \frac{(p/q)^{k-1} - 1}{(p/q) - 1}$. Thus, there are $(p/q)^{k-1}$ states that are exactly $k-1$ steps away from x. Since G is a simple graph, according to our assumption, there are $p(p/q)^{k-1}/q = (p/q)^k$ states that are k steps away from a. Thus, we have

$$|G_k| = \frac{(p/q)^k - 1}{(p/q) - 1} + (p/q)^k = \frac{(p/q)^{k+1} - 1}{(p/q) - 1}.$$

Thus this proposition is also true for $d = k$. □

We can see that R can help state space search in finding an exit of G_d if

$$\frac{(p/q)^{d+1} - 1}{(p/q) - 1} \geq \frac{tp^{tl}}{E}.$$

We can derive a necessary condition for the above inequality by replacing the left side as $(p/q)^{d+1}$. In this case, any tl must satisfy $d \leq tl \leq d + \log_p E - (d+1)(1 - \log_p q) - \log_p t$. It is easy to see that as q increases from 1 to p, random walk exploration becomes less effective. The insight is that RW-BFS is more effective when there are not many paths (q is smaller compared to p) that can lead to the same state.

5   Experimental Results

We report results on two variants of RW-BFS, namely, the sequential version RW-BFS$_s$ and the parallel version RW-BFS$_p$. The baseline in our comparison is

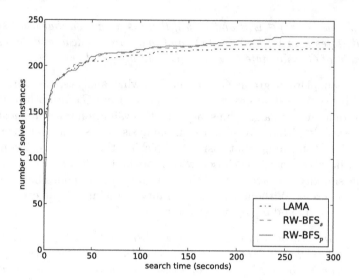

Fig. 1. Number of solved instances in the given time limit.

LAMA [15], a state space search planner. In our experiment, all three planners use the same settings and same heuristic functions for the state space search part. We set $t = 4$ and let $l = 2 * j - 1, n = 200 + 250j$ for $j = 1 \cdots t$ for both RW-BFS$_s$ and RW-BFS$_p$. In RW-BFS$_p$, line 2–9 in Algorithm 1 are running in parallel using $t = 4$ separate threads. The *open* list is shared by these threads so that possible exit states discovered by these threads can be inserted into *open* directly. We also set $m = 3000$, $\delta = 0.2h^*$, and $w = 10$.

We test all domains in IPC-6 [18], including Elevators (Elevator), Openstacks (Open), PARC printer (Parc), Peg solitaire (Peg), Scanalyzer-3D (Scan), Sokoban, Transport (Trans) and Woodworking (Wood). All experiments are conducted on a quad core workstation with a 2.4 GHz CPU and 2 GB memory. The running time limit for each problem is set to 300 s.

Fig. 1 shows the number of problem instances solved by the three planners. We have also run Monte-Carlo random walk (pure stochastic search) on these problems but its performance was poor and did not reported here. Clearly, both RW-BFS$_s$ and RW-BFS$_p$ solve more problem instances than the baseline planner. They solve 233, 210, and 204 instances, respectively. We also give detailed comparisons of three planners on all IPC-6 problems for which *random walk exploration* is invoked. Problems in which *random walk exploration* is not invoked are omitted from our comparison because in this case three algorithms are essentially identical.

To show the contribution and overhead brought by random walk in RW-BFS$_s$, we also report the time spent in random walk (T'), the length of the sub-path found by random exploration in the final solution path (L'), and the number of heuristic evaluations in random exploration procedure. For RW-BFS$_p$, we report the length of the sub-path (L') in the final solution path. We omit the

Table 1. Comparison of the search time ("T"), solution length ("L"), number of heuristic evaluations ("#HE") of LAMA, RW-BFS$_s$ and RW-BFS$_p$. For RW-BFS$_p$, "#HE" is the total number of heuristic evaluations of all threads.

P	LAMA			RW-BFS$_s$						RW-BFS$_p$			
				Total			RW			Total			RW
	T	L	E	T	L	#HE	T'	L'	#HE'	T	L	#HE	L'
elevator-25	-	-	-	44.8	268	41555	11.4	18	11596	57.63	249	51071	13
elevator-26	-	-	-	59.1	261	51712	21.5	36	20086	30.81	265	17723	25
elevator-27	44.5	310	31249	42.4	353	31434	12.4	24	10220	18.49	187	1280	15
elevator-28	-	-	-	25.0	276	16895	6.2	9	4407	14.23	202	12068	10
elevator-29	21.4	311	12797	11.3	308	7007	0.9	1	666	13.82	213	1641	24
elevator-30	-	-	-	104.3	306	63671	32.3	27	21124	53.75	227	12651	21
wood-20	-	-	-	50.3	219	50335	47.7	102	48587	22.42	242	21465	110
wood-23	2.3	25	6013	1.9	28	15076	1.8	7	14740	41.11	25	95400	1
wood-26	22.3	93	19221	5.6	99	16312	4.9	13	15269	12.12	97	12305	15
wood-28	-	-	-	28.6	121	32746	27.9	24	32224	11.16	137	14005	91
wood-29	63.7	103	46700	54.9	124	67946	52.0	25	65589	39.60	144	32093	25
parc-17	154.6	81	5205	53.5	75	1765	6.4	1	154	126.14	93	1842	12
parc-24	114.8	29	6657	156.9	28	8546	12.9	1	301	76.40	29	4533	15
sokoban-23	4.2	249	30485	9.5	230	63234	1.7	2	4040	6.14	258	65120	1
sokoban-24	108.4	279	367882	57.6	312	173380	14.3	4	8915	161.41	313	466868	3
sokoban-25	7.5	177	35097	31.1	391	122051	5.7	5	5159	17.36	245	105773	12
sokoban-26	-	-	-	109.4	417	450086	47.5	10	84417	53.85	583	128279	22
sokoban-27	112.5	81	482420	5.5	158	21916	1.4	4	2441	14.27	157	10719	1
sokoban-28	29.8	450	82203	54.1	436	139494	8.6	0	17230	5.24	465	6197	11
peg-29	8.7	45	37097	20.7	45	161290	11.8	0	123121	9.97	45	13611	0
peg-30	90.9	54	395509	104.3	55	499095	31.2	0	188839	97.71	55	707	0
scan-29	269.6	81	3715	278.8	81	3816	10.0	0	98	246.62	81	3403	0
trans-19	83.1	332	26068	214.6	331	67386	127.8	0	40901	89.35	331	26337	0

time spent in random walk and the number of heuristic evaluations in random exploration procedure because these two metrics are in parallel to the runtime overhead of state space search in RW-BFS$_p$.

We summarize three findings from Table 1. First, these results show that for problems which LAMA cannot solve within 300 s, e.g., elevator 25,26,28. RW-BFS can successfully find a solution in which a substantial portion of the path is generated by random walks. These results clearly show that random walks can assist state space search to escape from plateaus. Second, comparing the performance of two sequential planners LAMA and RW-BFS$_s$ to RW-BFS$_p$, we see that the overhead brought in by alternating between random walk and state space search can be mitigated by using a parallel implementation, at the cost of using more computing cores. Third, according to our performance analysis, if the state space has $q > 1$, the random walk procedure may not be helpful. A closer look at problems in the Pegsol domain reveals that they indeed have $q > 1$, as in this game, there can be multiple moves at each state and there can be multiple action paths arriving at the same state. Results on peg-29 and peg-30 confirmed our analysis that a random walk exploration would not assist the state space search much when q is close to p. In contrast, problems in the Sokoban domain are well-known to have many loops and dead ends. In this case, except for Sokoban-14, all sub-paths generated by the random exploration procedures are all relatively short compared to the solution length, as a longer path would encounter more dead ends or loops.

6   Conclusion

In this paper, we propose a best-first search algorithm assisted with a Monte-Carlo random walk exploration which can help the search escape from plateaus more quickly. Our algorithm has three advantages. First, MRW exploration has a certain probability to find an "exit" state and quickly jump out of a plateau while a best-first search may have to search all states in the plateau. Second, since each iteration of a random walk only evaluates the end state of a path, it saves a plenty of heuristic evaluation time compared with best-first search. Third, MRW exploration requires only a little extra memory. Actually, consider the reduced number of explored states, it can reduce the memory usage of the original best-first search.

Acknowledgments. This work has been supported in part by National Natural Science Foundation of China (Nos. 61502412, 61033009, and 61175057), Natural Science Foundation of the Jiangsu Province (No. BK20150459), Natural Science Foundation of the Jiangsu Higher Education Institutions (No. 15KJB520036), United States NSF grants IIS-0534699, IIS-0713109, CNS-1017701, and a Microsoft Research New Faculty Fellowship.

References

1. Adsit, C., Bradley, K., Heinrich, C.: Walksat: solving Boolean satisfiability via stochastic search (2014)
2. Benton, J., Talamadupula, K., Eyerich, P., Mattmüller, R., Kambhampati, S.: G-value plateaus: a challenge for planning. In: Proceedings of International Conference on Automated Planning and Scheduling, pp. 259–262 (2010)
3. Cai, S., Luo, C., Su, K.: Improving walkSAT by effective tie-breaking and efficient implementation. Comput. J. **58**(11), 2864–2875 (2015)
4. Chen, Y., Xu, Y., Yao, G.: Stratified planning. In: Proceedings of International Joint Conference on Artificial Intelligence (2009)
5. Chen, Y., Yao, G.: Completeness and optimality preserving reduction for planning. In: Proceedings of International Joint Conference on Artificial Intelligence (2009)
6. Frank, J.D., Cheeseman, P., Stutz, J.: When gravity fails: local search topology. J. Artif. Intell. Res. **7**, 249–281 (1997)
7. Glover, F., Laguna, M.: Tabu Search. Kluwer Academic Publishers, Norwell (1997)
8. Hampson, S., Kibler, D.: Plateaus and plateau search in Boolean satisfiability problems: when to give up searching and start again. In: The 2nd DIMACS Implementation Challenge, pp. 437–456 (1993)
9. Helmert, M., Röger, G.: How good is almost perfect. In: Proceedings of AAAI Conference on Artificial Intelligence, pp. 944–949 (2008)

10. Hoffmann, J.: Local search topology in planning benchmarks: a theoretical analysis. In: Proceedings of International Conference on AI Planning and Scheduling, pp. 92–100 (2002)
11. Jones, M.T.: Artificial Intelligence: A Systems Approach. Jones & Bartlett Learning, Massachusetts (2015)
12. Kautz, H., Selman, B.: Pushing the envelope: planning, propositional logic, and stochastic search. In: Proceedings of AAAI Conference on Artificial Intelligence (1996)
13. Nakhost, H., Müller, M.: Towards a second generation random walk planner: an experimental exploration. In: Proceedings of International Joint Conference on Artificial Intelligence (2013)
14. Phillips, M., Narayanan, V., Aine, S., Likhachev, M.: Efficient search with an ensemble of heuristics. In: Proceedings of International Joint Conference on Artificial Intelligence (2015)
15. Richter, K.F., Winter, S.: Computational aspects: how landmarks can be observed, stored, and analysed. Landmarks, pp. 137–173. Springer, Heidelberg (2014)
16. Richter, S., Helmert, M.: Preferred operators and deferred evaluation in satisficing planning. In: Proc. International Conference on Automated Planning and Scheduling (2009)
17. Shleyfman, A., Katz, M., Helmert, M., Sievers, S., Wehrle, M.: Heuristics and symmetries in classical planning. In: Proceedings of AAAI Conference on Artificial Intelligence, pp. 3371–3377 (2015)
18. The Sixth International Planning Competition (2008). http://ipc.informatik. uni-freiburg.de/
19. Xie, F., Müller, M., Holte, R.: Adding local exploration to greedy best-first search in satisficing planning. In: Proceedings of AAAI Conference on Artificial Intelligence, pp. 2388–2394 (2014)

Mining Frequent Trajectory Patterns in Road Network Based on Similar Trajectory

Ming Qiu and Dechang Pi[(✉)]

College of Computer Science and Technology,
Nanjing University of Aeronautics and Astronautics,
29 Jiangjun Road, Nanjing 211106, Jiangsu, People's Republic of China
Qiuml993@163.com, dc.pi@nuaa.edu.cn

Abstract. Mining Trajectory Patterns plays an important role in moving objects. In many practical applications, the movement of objectives is always constrained by spatial space (e.g. road network). Therefore, it has more realistic significance to work on frequent trajectory patterns. This paper proposes a frequent trajectory patterns mining algorithm based on similar trajectory (named SimTraj-PrefixSpan). Since the trajectory data with the same frequent trajectory pattern may be not exactly the same, a trajectory similarity is utilized to measure whether considered trajectories have the same pattern. Computational results on simulated data and real data verify that the proposed algorithm can mine more complete and continuous frequent trajectory patterns, and shows the superiority of the proposed algorithm over traditional trajectory patterns mining algorithms in terms of mining efficiency.

Keywords: Trajectory data mining · Frequent pattern · Pattern mining · Similar trajectory

1 Introduction

With the development of positioning technology and the popularity of mobile devices (e.g. wearable devices), Location Based Services (LBS) [1] become more extensive, and a great deal of trajectory data has been accumulated. In many practical applications, the movement trajectories of spatial objectives are always constrained by spatial space (e.g. road network [2]). Thereby, mining and analyzing the trajectory data [3] in constrained network can provide more and better service to users. Up to now, it is known that mining trajectory data mainly focused on two directions: trajectory clustering and trajectory pattern mining. This paper considers mining frequent trajectory patterns in road network. The traditional frequent pattern mining algorithm can be divided into two categories. One is Aprior [4] based algorithms, which is based on candidate set generation and test. The other is PrefixSpan [5] based algorithms, which is based on divide and conquer algorithm and thought of database projection pattern and correlation algorithms. Recently, Lin [6] proposed a hybrid multilevel search algorithm to mine long frequent pattern, Prabamanieswari [7] proposed the Fuzzy based frequent itemset mining, which reduces the number of scanning database and has better performance than Fuzzy Apriori [8]; Prabha and Lawrance [9] proposed Fuzzy

© Springer International Publishing AG 2016
H. Yin et al. (Eds.): IDEAL 2016, LNCS 9937, pp. 46–57, 2016.
DOI: 10.1007/978-3-319-46257-8_6

based CFP, which mines fuzzy frequent itemsets with less execution time and memory usage, has less execution time and memory usage.

Because of the continuity, periodicity and other special nature of the trajectory data, the traditional frequent pattern mining algorithms are not suitable. Therefore, Miko-lajMorzy [10] proposed the AprioriTraj algorithm based on the Aprior algorithm and Traj-PrefixSpan [11] algorithm based on PrefixSpan algorithm successively. Recently, Wang [12] proposed a method to resolve frequent pattern mining on personal trajectory data by accurately check the road corner; Chen [13] put forward the temporal semantic trajectory pattern named STS-TPs, and proposed a corresponding mining algorithm based on PrefixSpan algorithm.

The research background of frequent pattern mining is mainly divided into two kinds: one is road network; the other is unrestricted space. But the researches about trajectory data mining on road network are mainly based on the clustering of the trajectory data. For example, Han, et al. [14] proposed NEATthat integrated the information of locality, flow and density. This method is faster than existing density-based trajectory clustering approaches.

As the above mentioned, it can be found that previous researches on frequent trajectory patterns mostly focused on mining efficiency. However, due to limitations of positioning technology, the accuracy of location data may be not particularly high, and then users may be localized in the wrong way. In addition, traffic conditions or road repairing can also result in that the same frequent trajectory pattern may be not exactly the same. Therefore, this paper presents a frequent trajectory patterns mining algorithm based on similar trajectory (SimTraj-PrefixSpan), which uses the trajectory similarity to measure whether the trajectories have the same pattern. For the trajectories with the same pattern, the SimTraj-PrefixSpan allows their data is not exactly the same while satisfying the requirement of similarity. The experimental results prove that the SimTraj-PrefixSpan can mine more complete and continuous frequent trajectory patterns.

The remainder of this paper is organized as follows. In the next section, basic definitions about trajectory are presented. In Sect. 3, the detail of SimTraj-PrefixSpan is explained. Section 4 presents the experimental results. The conclusions are made in Sect. 5.

2 Basic Definitions

Definition 1 (Road Network Node). A road network node v expressed by a two-tuple (x, y), where x and y are the latitude and longitude of the road network node respectively.

Definition 2 (Road Network Edge). A road network edge e expressed by a two-tuple $<u, v>$, where u and v are two different road network nodes respectively, and e is the same with $<v, u>$.

Definition 3 (Adjacency). If there is a road network edge is $<u, v>$, then u is adjacent to v.

Definition 4 (Trajectory). Atrajectory T is a sequence composed by a series of adjacent road network nodes, expressed by $T = (v_1, v_2, \ldots, v_n)$, where v_i is adjacent to v_{i+1}.

Definition 5 (Sub Trajectory). There are two trajectories $T = (v_1, v_2, \ldots, v_n)$ and $T' = (v'_1, v'_2, \ldots, v'_m)(m \leq n)$. If there exists a natural number k makes $v'_1 = v_k$, $v'_2 = v_{k+1}, \ldots, v'_m = v_{k+m-1}$, then trajectory T' is the sub trajectory of the trajectory T.

Definition 6 (Prefix). For the trajectory $T = (v_1, v_2, \ldots, v_n)$ and the trajectory $T' = (v'_1, v'_2, \ldots, v'_m)(m \leq n)$, if there are $v'_i = v_i$ for $i \leq m$, then T' is the prefix of T.

Definition 7 (Projection). For the trajectory $T = (v_1, v_2, \ldots, v_n)$ and the trajectory $S = (u_1, u_2, \ldots, u_m)(m \leq n)$, if S is the prefix of the T', T' is the sub trajectory of T, and there doesn't exist longer sub trajectory T'' makes that S is the prefix of the T'', then T' is the projection of T corresponding the trajectory S.

Definition 8 (Postfix). For the trajectory $T = (v_1, v_2, \ldots, v_n)$ and the trajectory $T' = (v'_1, v'_2, \ldots, v'_m)(m \leq n)$, if there are $v'_i = v_{n-m+i}$ for $i \leq m$, then T' is the postfix of T.

Definition 9 (Support). Support is the percentage of the trajectories that contains the trajectory T in the trajectory database D_T at first, but in this paper, support is the percentage of the trajectories that contains the similar trajectory of the trajectory T in the trajectory database D_T. A predetermined threshold, called minimum support, is denoted as *min_support*.

Definition 10 (Frequent Trajectory Pattern). If the support of a trajectory T is not less than the minimum support *min_support*, then T is a frequent trajectory pattern.

3   Frequent Trajectory Pattern Mining Algorithm Based on Similar Trajectory

3.1   Similar Trajectory

In actual applications, the factors such as low precision positioning technology make that the trajectories with the same frequent trajectory pattern may be not identical. Therefore the similar trajectory is used to replace the identical trajectory to mine more frequent trajectory patterns. The common measures of trajectory similarity include Euclidean distance, Hausdroff distance, Frechet distance, the longest common sub sequence, trajectory matching, etc. [15]. On the basis of Euclidean distance, this paper defines a new simpler trajectory similarity measure method, and combines it with the mining algorithm organically to improve the mining efficiency.

Suppose there are two trajectories $T = (v_1, v_2, \ldots, v_n)$ and $S = (u_1, u_2, \ldots, u_n)$, where $v_i = (x_i, y_i)$ and $u_i = (x'_i, y'_i)$. Given a maximum distance of corresponding nodes *max_dist* and a maximum angle of the corresponding sub trajectory *max_angle*, the formula is as follows:

$$dist(v_i, u_i) = \sqrt{(x_i - x_i')^2 + (y_i - y_i')^2} \tag{1}$$

$$angle(v_i, v_{i+1}, u_i, u_{i+1}) = arccos(\frac{(x_{i+1} - x_i) * (x_{i+1}' - x_i') + (y_{i+1} - y_i) * (y_{i+1}' - y_i')}{dist(v_i, v_{i+1}) * dist(u_i, u_{i+1})}) \tag{2}$$

If $dist(v_i, u_i)$ and $dist(v_{i+1}, u_{i+1})$ are not more than *max_dist* and $angle(v_i, v_{i+1}, u_i, u_{i+1})$ is not more than *max_angle*, then $<v_i, v_{i+1}>$ and $<u_i, u_{i+1}>$ are similar trajectories. If for any $i \prec n$, $<v_i, v_{i+1}>$ and $<u_i, u_{i+1}>$ are similar trajectories, then T and S are similar trajectories.

Example: Table 1 is the trajectory data in the database, assuming that *max_dist* = 1 and *max_angle* = 50°. According to the above definition of similar trajectory, only the trajectory of O_1 and the trajectory of O_2 are similar trajectories, because the distance of the trajectory of O_1 and the trajectory of O_5 at time t_5 is 2 more than *max_dist*; the angle of the sub trajectories of the trajectory of O_1 and the trajectory of O_4 from time t_4 to t_5 is 60° more than *max_angle*. Therefore, only the trajectory of O_1 and the trajectory of O_2 are similar trajectories.

Table 1. Some trajectory data in the database

Object	t_1	t_2	t_3	t_4	t_5
O_1	(1,2)	(2,2)	(3,2)	(4,2)	(5,2)
O_2	(1,2)	(2,2)	(3,1)	(4,1)	(5,2)
O_3	(1,3)	(2,2)	(3,1)	(4,1)	(5,0)
O_4	(1,2)	(2,1)	(3,1)	(4,1)	(5,3)

If the trajectory T is a frequent trajectory pattern, then all the sub trajectory of T are frequent trajectory patterns, so we must ensure that all the corresponding sub trajectory of the similar trajectory we defined are frequent trajectory patterns too. According to the definition, if the trajectory T and S are similar trajectories, then for any $i \prec n, <v_i, v_{i+1}>$ and $<u_i, u_{i+1}>$ are similar trajectories, so all the corresponding sub trajectory of T and S are similar trajectories too.

3.2   SimTraj-PrefixSpan Algorithm

In this section we describe the design and implement of SimTraj-PrefixSpan. In order to allow trajectory data is not exactly the same, the similar trajectory based on the distance of corresponding nodes and the angle between the corresponding sub-trajectories is defined. There are three main steps in the algorithm.

1. Compute surrounding nodes. Traverse the road network, and then add the nodes whose distances with node v are not more than *max_dist* into its corresponding surrounding nodes set $N(v)$.

2. Construct the candidate set of the node. Traverse the trajectory database D_T and count the number of each node, add the node that support more than the minimum support *min_support* into the candidate set L_0.
3. Construct projection database. Create a projection database for each candidate node v in L_0, each trajectory in the projection database must contain the prefix $<u>$ formed by one node u in the set $N(v)$.

The last step will be run recursively in the projection database until the current results is empty. Each mining must ensure that the current candidate edge must adjacent to the last edge of the mining result before the recursion. The pseudo-code of SimTraj-PrefixSpan algorithm can be described as Algorithm 1.

```
Algorithm1: SimTraj-PrefixSpan

Input: D_T,The trajectory database;min_support;max_dist;

        max_angle;G=(V,E),The information of road network.

Output: FreTraj,The set of frequent trajectories.

1: for each node v in V do

2:    for each node u in V do

3:       if dist(v,u)<=max_dist then

4:          add u into set N(v);

5:       end if

6:    end for

7:end for

8:for each trajectory T in D_T do

9:    for each node v in T do

10:      for each node u in N(v) do

11:         u.count++;

12:      end for

13:   end for

14:end for

15:L_0={v in V:support(v)>=min_support}

16:call GetFreTrajectories(s,L_0,D_T,min_support)
```

Description:

(1) Steps 1–7: The nodes whose distance with node v are not more than *max_dist* were added into its corresponding surrounding nodes set $N(v)$; Steps 8–15: Calculate the support of each node and add the nodes that support more than the minimum support into the candidate node set L_0.

(2) The *GetFreTrajectories*($L_0, D_T, min\_support$) is an recursive process, in reference to the idea of divide and conquer to mine frequent trajectory patterns. The identical trajectories is replaced by the similar trajectory to measure whether the trajectory has the same pattern. Algorithm is described as follows:

```
Algorithm2: GetFreTrajectories(s,L0,DT,min_support)

Input:s=<u1,u2,...,un>,The prefix trajectory;L0,The set of candidate
node;DT,The trajectory database;min_support

Output:FreTraj,The set of frequent trajectories

1:for each node v in L0 do
2:  if s != <> and v is not adjacent to un then
3:    continue;
4:  end if
5:  inital D'T to empty set;
6:  for each trajectory T=<t0,t1,...,tm> in DT do
7:    for each node v' in N(v) do
8:      if <v'> is the prefix trajectory of T then
9:        if s!= <> and angle(un,v,t0,t1) > max_angle then
10:          continue;
11:        end if
12:        v.count++;
13:        if the postfix trajectory of T over<v'> is T' and T' != <> then
14:          extend T' by v' form T'=<v',t'1,...t'k> and add T' into D'T;
15:        end if
16:      end if
17:    end for
18:  end for
19:  if support(v)>= min_support then
20:    extend s by v to form a frequent trajectory s'= <u1,u2,...,un,v>
          and add s into FreTraj;
21:  end if
22:  if size(D'T)/num_of_trajectories >= min_support then
23:    call GetFreTrajectories(s',L0,D'T,min_support)
24:  end if
25:end for
```

4 Experimental Results and Analysis

The SimTraj-PrefixSpan algorithm and the Traj-PreixSpan algorithm are corded by Java language, and the experimental environment is Microsoft Windows 7 professional operating system with 3.20 GHZ Intel Core i5-3479 CPU processor and 4 GB memory. Then simulated data and real data are used to evaluate the performance of the above considered algorithms.

4.1 Experiment on Simulated Data

Simulated data is generated by Thomas Brinkhoff Road Network Moving Object Data Generator [16] according to the map of Oldenburg, which is a set of trajectory point. The data must be preprocessed before the experiment. The points are transformed to the corresponding roads in the road network, and then expanded to the trajectory data format defined as Definition 4.

Firstly, experiment on the simulated data to study the relationship between the performance of the SimTraj-PrefixSpan algorithm and the parameter max_dist, in which $min\_support = 0.1$ and $max\_angle = 40°$. Figure 1 shows that the number of the frequent trajectory patterns mined by the SimTraj-PrefixSpan algorithm has no increase when max_dist less than 40, and increases faster when max_dist greater than 40, but increases slowly form 70 to 100.

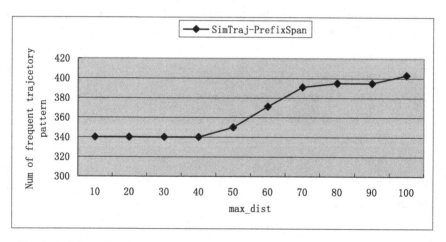

Fig. 1. Relation of the number of the frequent trajectory pattern and maximal distance

Then according to the previous experiment, we set the max_dist to 100 and $min\_support$ to 0.1, and study the relationship between the performance of SimTraj-PrefixSpan and max_angle. Figure 2 shows that the number of frequent trajectory patterns mined by the SimTraj-PrefixSpan algorithm increases steadily with the increase of max_angle. With the increase of max_angle, the similarity of the trajectory is reduced, so that more similar trajectories are believed to have the same frequent trajectory pattern.

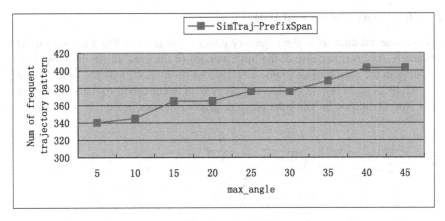

Fig. 2. Relation of the number of the frequent trajectory pattern and maximal angle

Finally, the performance of SimTraj-PrefixSpan and Sim-PrefixSpan with different *min_support* is analyzed. According to the previous experiment, we set *max_dist* = 100 and *max_angle* = 40°. The experimental results with different *min_support* are shown in Fig. 3. From Fig. 3, we can see that the SImTraj-PrefixSpan can mines more frequent trajectory patterns than the Traj-PrefixSpan with different support.

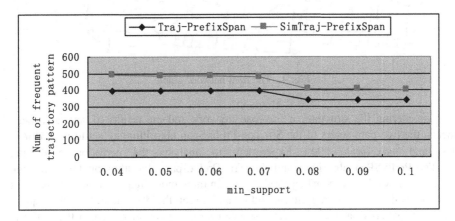

Fig. 3. Relation of the number of the frequent trajectory pattern and minimum support 1

The experiments on the simulated data study the effect of the parameters *max_dist* and *max_angle* on the performance of the SimTraj-PrefixSpan. It can be conclude that when the parameters are small, the similarity of similar trajectories is higher, and mining results of the SimTraj-PrefixSpan algorithm is more close to the Traj-PrefixSpan algorithm. In order to ensure that SimTraj-PrefixSpan algorithm can be used in actual applications, the experiments on the real data are performed in the following section.

4.2   The Experiment on Real Data

The experimental data set is the trajectory data of 536 taxis in San Francisco in May 2008 nearly 30 days. Because of the huge amount of the data, so we only use data of seven days in the region of longitude 122°19'2.64"W to 122°25'31.08"W and latitude 37°45'51.48"N to 37°47'42"N. The trajectory data is converted to the trajectory data format with Definition 3. The road network graph is shown in Fig. 4.

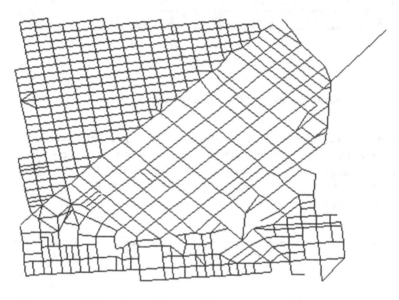

Fig. 4. The road network graph

Considering the structure of the road network and the results of several experiments, the two parameters of the SimTraj-PrefixSpan algorithm are set to $max\_dist = 100$ and $max\_angle = 90°$. Experiment to analysis the performance of the SimTraj-PrefixSPan algorithm and the Traj-PrefixSpan algorithm on the real data. Figure 5 shows that the SimTraj-PrefixSpan can mine more frequent trajectory patterns than the Traj-PrefixSpan algorithm, and it can be seen that the number of the frequent trajectory patterns particularly decrease when $min\_support$ increased to the 0.05. Its reason is that the correlation between each trajectory is not strong.

Then the running time of SimTraj-PrefixSpan and Traj-PrefixSpan are compared, in which $min\_support = 0.02$. Figure 6 shows that the running time increase of the SimTraj-PrefixSpan algorithm is roughly the same with the Traj-PrefixSpan algorithm. So the efficiency of two algorithms is similarity. And the reason why SimTraj-PrefixSpan algorithm need more time than Traj-PrefixSpan algorithm is it spends much time to generator the surrounding point set and spend more time to mine the frequent trajectory patterns.

Experiment on the real data to compare the results of the SimTraj-PrefixSpan algorithm and the Traj-PrefixSpan algorithm, which $max\_dist = 100$, $max\_angle =$

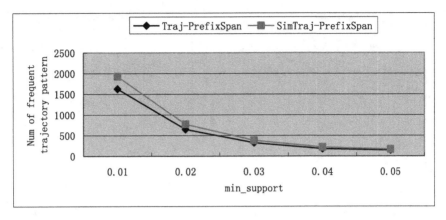

Fig. 5. Relation of the number of the frequent trajectory pattern and minimum support 2

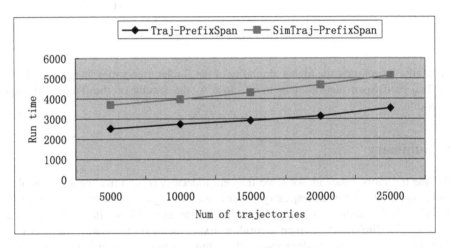

Fig. 6. Relation of the running time and data size.

$90°$ and *min_support* $= 0.02$. The experimental results on real road networks are shown in Fig. 7, where Fig. 7(a) and (b) are the results of Traj-PrefixSpan and SimTraj-PrefixSpan, respectively. From Fig. 7, it can be seen that the frequent trajectory patterns mined by the SimTraj-PrefixSpan algorithm is more complete and continuous than those mined by Traj-PrefixSpan. Therefore, it can be concluded that the performance of SimTraj-PrefixSpan is better than the Traj-PrefixSpan algorithm in term of mining effectiveness.

Through the above experiments on simulated data and real data, it can be observed that the SimTraj-PrefixSpan can mine more complete frequent trajectory patterns than the Traj-PrefixSpan algorithm by replacing the same trajectories to similar trajectories. And we can adjust the parameter *max_dist* and *max_angle* to adjust the similarity of

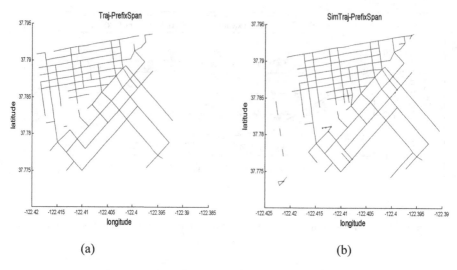

Fig. 7. The results of two algorithms

the similar trajectories, the higher similarity of the similar trajectories, the results of the SimTraj-PrefixSpan algorithm is more close to the results of the Traj-PrefixSpan algorithm.

5 Summary

Due the trajectory data with the same frequent trajectory pattern may be identical, this paper proposed a frequent trajectory patterns mining algorithm based on similar trajectory, which combines the Traj-PrefixSpan algorithm with similar trajectory and proposes the SimTraj-PrefixSpan algorithm. Experiments on simulated data and real data show that the SimTraj-PrefixSpan algorithm can mine more and more complete and more continuous frequent trajectory patterns and its ensure the mining efficiency is similar higher than that of to the mining efficiency of the traditional Traj-PrefixSpan algorithm at the same time.

In the current big data environment, because of the large amount of data, the SimTraj-PrefixSpan algorithm proposed in this paper can not be directly applied to the actual problems. In further study, we intend to optimize the efficiency of the algorithm and implement it with the distributed programming.

Acknowledgements. The research work was supported by National Natural Science Foundation of China (U1433116).

References

1. Chow, C.Y., Mokbel, F.: Query-aware location annoymization for road networks. Geoinformation **15**(3), 571–607 (2011)
2. Nowosielski, A., et al.: Automatic analysis of vehicle trajectory applied to visual surveillance. In: Choraś, R.S. (ed.) Image Processing and Communications Challenges 7. Advances in Intelligent Systems and Computing, vol. 389, pp. 89–96. Springer International Publishing, Switzerland (2016)
3. Zheng, Y.: Trajectory data mining: an overview. ACM Trans. Intell. Syst. Technol. (TIST) **6**(3), 1–41 (2015)
4. Agrawal, R., Srikant, R.: Fast algorithms for mining association rules. In: Proceedings of Very large data bases (VLDB), pp. 487–499. Morgan Kaufmann Publishers, San Francisco (1994)
5. Pei, J., Han, J.-w., Mortazavi-Asl, B., et al.: PrefixSpan: mining sequential patterns efficiently by prefix-projected pattern growth. In: ICDE 2001 Proceedings of the 17th International Conference on Data Engineering. Washington D C, USA, pp. 215–224. IEEE Computer Society (2001)
6. Lin, S.-J., Chen, Y.-C., Yang, D.-L., et al.: Discovering long maximal frequent pattern. In: Proceedings of 2016 Eighth International Conference on Advanced Computational Intelligence (ICACI). IEEE (2016)
7. Prabamanieswari, R.: A combined approach for mining fuzzy frequent itemset. In: Proceedings of International Journal of Computer Applications (IJCA) International Seminar on Computer Vision, pp. 1–5 (2013)
8. Kuok, C.M., Fu, A.W.-C., Wong, M.H.: Mining fuzzy association rules in databases. In: ACM SIGMOD, pp. 41–46 (1998)
9. Suriya Prabha, K., Lawrance, R.: Mining fuzzy frequent itemset using Compact Frequent Pattern (CFP) tree algorithm. In: International Conference on Computing and Control Engineering (ICCCE) (2012)
10. Morzy, M.: Mining frequent trajectories of moving objects for location prediction. In: Perner, P. (ed.) MLDM 2007. LNCS (LNAI), vol. 4571, pp. 667–680. Springer, Heidelberg (2007)
11. Morzy, M.: Prediction of moving object location based on frequent trajectories. In: Levi, A., Savaş, E., Yenigün, H., Balcısoy, S., Saygın, Y. (eds.) ISCIS 2006. LNCS, vol. 4263, pp. 583–592. Springer, Heidelberg (2006)
12. Wang, T., Zhang, D., Zhou, X., et al.: Mining personal frequent routes via road corner detection. IEEE Trans. Syst. Man Cybern. Syst. **46**(4), 445–458 (2015)
13. Chen, C.C., Kuo, C.H., Peng, W.C.: Mining spatial-temporal semantic trajectory patterns from raw trajectories. In: IEEE International Conference on Data Mining Workshop, pp. 1019–1024 (2015)
14. Han, B., Liu, L., Omiecinski, E.R.: Road-network aware trajectory clustering: integrating locality, flow and density. IEEE Trans. Mob. Comput. **14**(2), 416–429 (2015)
15. Magdy, N., Sakr, M.A., Abdelkader, T.M., et al.: Review on trajectory similarity measures. In: IEEE Seventh International Conference on Intelligent Computing and Information Systems (2015)
16. Brinkhoff, T.: Generating network-based moving objects. In: The 12th International Conference on Scientific and Statistical DataBase Management, pp. 253–255. IEEE Computer Society, Washington (2000)

An Effective Genetic Algorithm with Uniform Crossover for Bi-objective Unconstrained Binary Quadratic Programming Problem

Chao Huo[1], Rong-Qiang Zeng[1,2(⊠)], Yang Wang[3], and Ming-Sheng Shang[4]

[1] School of Computer Science and Engineering, University of Electronic Science and Technology of China, Chengdu 610054, Sichuan, People's Republic of China
chaohuo0811@gmail.com
[2] School of Mathematics, Southwest Jiaotong University, Chengdu 610031, Sichuan, People's Republic of China
zrq@home.swjtu.edu.cn
[3] School of Management, Northwestern Polytechnical University, Xi'an 710072, Shanxi, People's Republic of China
sparkle.wy@gmail.com
[4] Chongqing Institute of Green and Intelligent Technology, Chinese Academy of Sciences, Chongqing 400714, People's Republic of China
msshang@cigit.ac.cn

Abstract. The unconstrained binary quadratic programming problem is one of the most studied NP-hard problem with its various practical applications. In this paper, we propose an effective multi-objective genetic algorithm with uniform crossover for solving bi-objective unconstrained binary quadratic programming problem. In this algorithm, we integrate the uniform crossover within the hypervolume-based multi-objective optimization framework for further improvements. The computational studies on 10 benchmark instances reveal that the proposed algorithm is very effective in comparison with the original multi-objective optimization algorithms.

Keywords: Multi-objective optimization · Hypervolume contribution · Genetic Algorithm · Uniform crossover · Unconstrained Binary Quadratic Programming problem

1 Introduction

The Unconstrained Binary Quadratic Programming (UBQP) problem is a classic NP-hard combinatorial problem with a number of applications [15]. The UBQP problem is extended into multi-objective case in [13], where the multiple objectives are to be maximized simultaneously, then the multi-objective UBQP problem can mathematically be formulated as follows [14]:

$$f_k(x) = x'Q^k x = \sum_{i=1}^{n}\sum_{j=1}^{n} q_{ij}^k x_i x_j \tag{1}$$

© Springer International Publishing AG 2016
H. Yin et al. (Eds.): IDEAL 2016, LNCS 9937, pp. 58–67, 2016.
DOI: 10.1007/978-3-319-46257-8_7

where $Q^k = (q_{ij}^k)$ is an $n \times n$ matrix of constants and x is an n-vector of binary (zero-one) variables, i.e., $x_i \in \{0, 1\}$ ($i = 1, \ldots, n$), $k \in \{1, \ldots, m\}$.

UBQP is notable for its ability to formulate a wide range of other practical problems in different fields, such as machine scheduling [1], traffic management [8], computer aided design [12], financial analysis [16], etc. Many heuristic and metaheuristic algorithms have been proposed to tackle UBQP in the literature [11], such as simulated annealing [2], scatter search [3], tabu search [9], memetic algorithms [17], etc.

Moreover, A. Liefooghe et al. [13] proposed a hybrid metaheuristic algorithm to solve the multi-objective UBQP problem, which combines an elitist evolutionary multi-objective optimization algorithm with an effective single-objective tabu search procedure based on the scalarizing function. In [14], they proposed the multi-objective local search algorithms with three different strategies to solve the bi-objective UBQP problem more efficiently.

In this paper, we propose a multi-objective genetic algorithm to solve the bi-objective UBQP problem. This algorithm integrates the uniform crossover within the hypervolume-based multi-objective optimization framework for further improvements. Actually, there are two main components: hypervolume contribution selection and genetic algorithm with uniform crossover. The hypervolume contribution selection procedure iteratively improve the Pareto approximation set until it can not be improved any more. Then, the uniform crossover is used to further improve the whole quality of the Pareto approximation set.

This paper is organized as follows. In the next section, we introduce the basic notations and definitions of multi-objective optimization. In Sect. 3, we briefly review the previous work related to the uniform crossover and the multiparent crossover. Afterwards, we present our proposed multi-objective genetic algorithm in Sect. 4. Section 5 provides the computational results and analyze the behavior of the proposed algorithm. Finally, the conclusions are provided in the last section.

2 Multi-objective Optimization

In this section, we present the basic notations and definitions of multi-objective optimization. Let X denote the search space of the optimization problem under consideration and Z the corresponding objective space. Without loss of generality, we assume that $Z = \Re^n$ and that all n objectives are to be maximized. Each $x \in X$ is assigned exactly one objective vector $z \in Z$ on the basis of a vector function $f : X \to Z$ with $z = f(x)$, and the mapping f defines the evaluation of a solution $x \in X$ [7]. Actually, we are often interested in those solutions that are Pareto optimal with respect to f. The relation $x_1 \succ x_2$ means that the solution x_1 is *preferable* to x_2. The dominance relation between two solutions x_1 and x_2 is often defined as follows [7]:

Definition 1 (Pareto Dominance). A decision vector x_1 is said to dominate another decision vector x_2 (written as $x_1 \succ x_2$), if $f_i(x_1) \geq f_i(x_2)$ for all $i \in \{1, \ldots, n\}$ and $f_j(x_1) > f_j(x_2)$ for at least one $j \in \{1, \ldots, n\}$.

Definition 2 (Pareto Optimal Solution). $x \in X$ is said to be Pareto optimal if and only if there does not exist another solution $x' \in X$ such that $x' \succ x$.

Definition 3 (Pareto Optimal Set). S is said to be a Pareto optimal set if and only if S is composed of all the Pareto optimal solutions.

Definition 4 (Non-Dominated Solution). $x \in S$ $(S \subset X)$ is said to be non-dominated if and only if there does not exist another solution $x' \in S$ such that $x' \succ x$.

Definition 5 (Non-Dominated Set). S is said to be a non-dominated set if and only if any two solutions $x_1 \in S$ and $x_2 \in S$ such that $x_1 \nsucc x_2$ and $x_2 \nsucc x_1$.

Since there does not exist the total order relation among all the solutions in multi-objective optimization, the aim is to generate the Pareto optimal set, which keeps the best compromise among all the objectives. Nevertheless, in most cases, it is impossible to generate the Pareto optimal set in a reasonable time. Therefore, we are interested in finding a non-dominated set which is as close to the Pareto optimal set as possible, and the whole goal is often to identify a good Pareto approximation set.

3   Previous Work

The Genetic Algorithms (GA) have a good potential of solving many different combinatorial optimization problems, which are often integrated into the hybrid metaheuristics as an important part for further improvements [7]. In this section, we provide a literature review of the studies related to the uniform crossover and the multi-parent crossover.

U. Benlic et al. [6] proposed a powerful population-based memetic algorithm for solving quadratic assignment problem, which integrates an effective local optimization algorithm within the evolutionary computing framework based on a uniform crossover. In this algorithm, the uniform crossover is used to further enforce the search capacity of the proposed algorithm. The extensive computational studies reveal that their proposed algorithm is very competitive.

E. D. Paolo et al. [10] proposed an efficient genetic algorithm with uniform crossover for solving the multi-objective airport gate assignment problem. In this algorithm, a new defined uniform crossover operator is used to generate high-quality offsprings, which is crucial to the successful implementations. The extensive simulation studies illustrate the advantages of the proposed GA scheme with the uniform crossover operator.

Zhipeng Lü et al. [15] presented a multi-parent hybrid genetic-tabu algorithm for dealing with unconstrained binary quadratic programming problem, which incorporates the tabu search procedure into the genetic algorithm framework. In this algorithm, a multi-parent crossover operator called MSX is jointly employed with the conventional uniform crossover to generate diversified new solutions. The computational results on 10 large benchmark instances indicate that the proposed algorithm is very effective.

Yang Wang et al. [18] integrated four multi-parent crossover operators (called MSX, Diagonal, U-Scan and OB-Scan) within the memetic algorithm framework for dealing with unconstrained binary quadratic programming problem. Their proposed algorithms apply these crossover operators to further improve the results generated by the tabu search procedure. The experimental results and the analysis on the behavior of the algorithm provide the evidences and the insights as to key role of the crossover operators.

4 Multi-Objective Genetic Algorithm

The Multi-Objective Genetic Algorithm (MOGA) is proposed to solve the bi-objective UBQP problem, which is composed of two main components: hypervolume contribution selection and genetic algorithm with the uniform crossover. The general scheme of this algorithm is described in Algorithm 1.

Algorithm 1. Multi-Objective Genetic Algorithm

    Input: N (Population size)
    Output: A: (Pareto approximation set)
    Step 1 - Initialization: $P \leftarrow N$ randomly generated solutions
    Step 2: $A \leftarrow \Phi$
    Step 3 - Fitness Assignment: Compute a fitness value for each solution $x \in P$
    Step 4:
    while Running time is not reached **do**
      repeat
        1) Hypervolume Contribution Selection: $x \in P$
      until all neighbors of $x \in P$ are explored
        2) $A \leftarrow$ Non-dominated solutions of $A \bigcup P$
        3) Genetic Algorithm: $y \in A$
    end while
    Step 5: Return A

In MOGA, all the solutions in an initial population are randomly generated, i.e., each variable of one solution is randomly assigned a value 0 and 1 (Step 1). Then, each solution is computed a fitness value by the Hypervolume Contribution (HC) indicator defined in [5] (Step 3).

After realizing the fitness assignment, we optimize the initial population with the hypervolume contribution selection procedure in order to obtain a Pareto approximation set. Afterwards, we apply the uniform crossover operator to generate the offsprings, in order to update the Pareto approximation set A for further improvements.

4.1 Hypervolume Contribution Selection

In order to generate a set of efficient individuals used for generating high-quality offsprings with the uniform crossover operator, we apply the Hypervolume

---

Algorithm 2. Hypervolume Contribution Selection (HCS)

Steps:
  1) $x^* \leftarrow$ one randomly chosen unexplored neighbors of x
  2) $P \leftarrow P \bigcup x^*$
  3) compute x^* fitness: $HC(x^*, P)$
  4) update all $z \in P$ fitness values
  5) $\omega \leftarrow$ worst solution in P
  6) $P \leftarrow P \backslash \{\omega\}$
  7) update all $z \in P$ fitness values
  8) **if** $\omega \neq x^*$, Progress $\leftarrow$ True

Contribution Selection (HCS) procedure presented in Algorithm 2 to the initial population.

For the UBQP problem, we flip the value 0 (or 1) of the kth variable of each solution $x \in P$ to 1 (or 0) to obtain a new solution x^* as the neighbor of x. Then, we compute the objective function values of this new solution with the fast incremental neighborhood evaluation formula [15] below:

$$\Delta_i = (1 - 2x_i)(q_{ii} + \sum_{j \in N, j \neq i, x_j = 1} q_{ij}) \qquad (2)$$

Actually, the move value Δ_i can be computed in linear time, which makes the local search procedure more efficiently.

In the HCS procedure, one solution x^*, which is one of the unexplored neighbors of x in the population P, is assigned to a fitness value by the HC indicator. If x^* is dominated, the fitness values of all the solutions in P keep unchanged. If x^* is non-dominated, we need to update the fitness values of non-dominated neighbors of x^* in the objective space.

Afterwards, the solution ω with the worst fitness value is removed from the population P. If ω is dominated, the fitness values of the other solutions keep unchanged. If ω is non-dominated, the fitness values of the non-dominated neighbors of ω need to be updated. The whole procedure will repeat until the termination criterion is satisfied.

4.2 Genetic Algorithm

The uniform crossover operator is widely used to solve many combinatorial optimization problems, such as quadratic assignment problem [6], gate assignment problem [10], single-objective UBQP problem [15], and so on. In this work, we employ the uniform crossover operator to improve the Pareto approximation set A generated by the HCS procedure. The exact steps are given in Algorithm 3.

In the genetic algorithm, we randomly select two non-dominated solutions as two parents from the Pareto approximation set A. For the crossover process, we employ the standard uniform crossover operator to recombine the selected parents.

Algorithm 3. Genetic Algorithm (GA)

Steps:
1) randomly select two parents: $y_i \in A$ and $y_j \in A$
2) generate an offspring $z \leftarrow \text{Crossover}(y_i, y_j)$
3) $z' \leftarrow \text{Mutation}(z)$
4) $A \leftarrow \text{HCS}(z')$

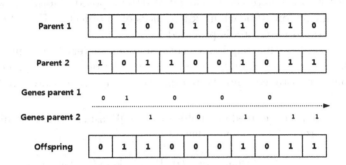

Fig. 1. An example of the uniform crossover for UBQP.

Furthermore, the elements of the parents are scanned from left to right, and each element in the offspring keeps the value x_i (0 or 1) in either of these two parents with the equal probability. The unassigned variables in the offspring is given 0 or 1 randomly. An example of this crossover process is illustrated in Fig. 1.

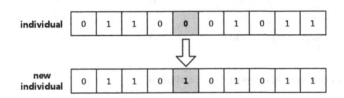

Fig. 2. An example of the mutation for UBQP.

Afterwards, we apply the mutation procedure to the generated offspring by randomly flipping the value of one variable to the opposite value (0 or 1), in order to generate a new offspring. An example of the mutation procedure is illustrated in Fig. 2. Then, we insert this new offspring into the Pareto approximation set A with the HC indicator for further improvements.

5   Experiments

In this section, we first present the parameter settings for the MOGA algorithm. Then, we introduce a performance assessment protocol used to evaluate the effectiveness of multi-objective optimization algorithm. Finally, we provide the computational results and the performance analysis.

5.1 Parameters Settings

The MOGA algorithm is programmed in C++ and compiled using Dev-C++ 5.0 compiler on a PC running Windows 7 with Core 2.50 GHz CPU and 4 GB RAM. In order to evaluate the efficiency of our proposed algorithm, we carry out the experiments on 10 benchmark instances of bi-objective UBQP problem, which are generated by the tools provided in [13].

Besides, this algorithm requires to set a few parameters, we mainly discuss two important ones: the running time and the population size. The precise information about the instances and the parameter settings is presented in Table 1.

Table 1. Parameter settings used for bi-objective UBQP instances: instance dimension (D), population size (P) and running time (T).

	Dimension (D)	Population (P)	Time (T)
bubqp_1000_01	1000	10	100''
bubqp_1000_02	1000	10	100''
bubqp_2000_01	2000	20	200''
bubqp_2000_02	2000	20	200''
bubqp_3000_01	3000	30	300''
bubqp_3000_02	3000	30	300''
bubqp_4000_01	4000	40	400''
bubqp_4000_02	4000	40	400''
bubqp_5000_01	5000	50	500''
bubqp_5000_02	5000	50	500''

5.2 Performance Assessment Protocol

In this paper, we evaluate the effectiveness of multi-objective optimization algorithms using a test procedure that has been undertaken with the performance assessment package provided by Zitzler et al.[1].

The quality assessment protocol works as follows: we first create a set of 20 runs with different initial populations for each algorithm and each benchmark instance. Afterwards, we calculate the reference set PO^* in order to determine the quality of k different sets $A_0 \ldots A_{k-1}$ of non-dominated solutions. Furthermore, we define a reference point $z = [w_1, w_2]$, where w_1 and w_2 represent the worst values for each objective function in $A_0 \cup \cdots \cup A_{k-1}$. Then, the evaluation of a set A_i of solutions can be determined by finding the hypervolume difference between A_i and PO^* [19], and this hypervolume difference has to be as close as possible to zero.

---

[1] http://www.tik.ee.ethz.ch/pisa/assessment.html.

5.3 Computational Results

In this subsection, we provide the computational results obtained by our proposed MOGA algorithm, the indicator-based multi-objective local search algorithm (IBMOLS) proposed in [4] and the hypervolume-based multi-objective local search algorithm (HBMOLS) proposed in [5].

The computational results are summarized in Table 2. Each line in this table contains a value both **in bold** and **in grey box**, which is the best result obtained on the considered instance. The another two values refer to that the corresponding algorithms are statistically outperformed by the algorithm which obtains the best result (with a confidence level greater than 95 %).

Table 2. The computational results on bi-objective UBQP problem obtained by the algorithms: IBMOLS, HBMOLS and MOGA

Instance	Algorithm		
	IBMOLS	HBMOLS	MOGA
bubqp_1000_01	0.527656	0.510992	**0.179459**
bubqp_1000_02	0.184199	0.116860	**0.116035**
bubqp_2000_01	0.560875	0.516558	**0.220095**
bubqp_2000_02	0.618150	0.634531	**0.112132**
bubqp_3000_01	0.570740	0.557831	**0.213661**
bubqp_3000_02	0.639882	0.653772	**0.134274**
bubqp_4000_01	0.533230	0.568231	**0.103679**
bubqp_4000_02	0.629849	0.627149	**0.069898**
bubqp_5000_01	0.506434	0.541808	**0.136980**
bubqp_5000_02	0.630504	0.636295	**0.044327**

According to Table 2, we can see that all the best results are obtained by MOGA, which statistically outperforms the algorithms IBMOLS and HBMOLS on all the instances. Moreover, the most significant result is achieved on the instance bubqp_5000_02, where the average hypervolume difference value obtained by MOGA is much smaller (about 15 times) than the values obtained by IBMOLS and HBMOLS.

Compared with IBMOLS and HBMOLS, we can see the evident contribution of the uniform crossover in MOGA. We assume that IBMOLS and HBMOLS could be often trapped in the local optima after a certain amount of running time, since they only focus on the local search procedure without any strategy used for jumping out of the local optima.

For MOGA, the uniform crossover is used to generate two new solutions, which are inserted into the Pareto approximation set for further improvements. In fact, these new generated solutions make MOGA have a chance to jump out of the local optima. The computational results indicate the uniform crossover evidently improve the quality of the Pareto approximation set. Therefore, MOGA has a better performance on all the instances.

6   Conclusion

In this paper, we have presented a simple and effective multi-objective genetic algorithm for the well-known bi-objective unconstrained binary quadratic programming problem. The proposed algorithm has combined the hypervolume-based optimization with the standard uniform crossover for the further improvements on the Pareto approximation set. We have evaluated this algorithm on the set of 10 benchmark instances, and the computational results indicate that the MOGA algorithm performs very well on these instances.

Acknowledgments. The work in this paper was supported by the Fundamental Research Funds for the Central Universities (Grant No. A0920502051408-25), supported by the Research Foundation for International Young Scientists of China (Grant No. 61450110443), supported by the Scientific Research Foundation for the Returned Overseas Chinese Scholars (Grant No. 2015S03007) and supported by National Natural Science Foundation of China (Grant No. 61370150 and 71501157).

References

1. Alidaee, B., Kochenberger, G.A., Ahmadian, A.: 0–1 quadratic programming approach for the optimal solution of two scheduling problems. Int. J. Syst. Sci. **25**, 401–408 (1994)
2. Alkhamis, T.M., Hasan, M., Ahmed, M.A.: Simulated annealing for the unconstrained binary quadratic pseudo-Boolean function. Eur. J. Oper. Res. **108**, 641–652 (1998)
3. Amini, M., Alidaee, B., Kochenberger, G.: A scatter search approach to unconstrained quadratic binary programs. New Meth. Optim. **108**, 317–330 (1999)
4. Basseur, M., Liefooghe, A., Le, K., Burke, E.: The efficiency of indicator-based local search for multi-objective combinatorial optimisation problems. J. Heuristics **18**(2), 263–296 (2012)
5. Basseur, M., Zeng, R.-Q., Hao, J.-K.: Hypervolume-based multi-objective local search. Neural Comput. Appl. **21**(8), 1917–1929 (2012)
6. Benlic, U., Hao, J.-K.: Memetic search for the quadratic assignment problem. Expert Syst. Appl. **42**, 584–595 (2015)
7. Coello, C.A., Lamont, G.B., Van Veldhuizen, D.A.: Evolutionary Algorithms for Solving Multi-Objective Problems (Genetic and Evolutionary Computation). Springer, New York (2007)
8. Gallo, G., Hammer, P.: Quadratic knapsack problems. Math. Program. **12**, 132–149 (1980)

9. Glover, F., Kochenberger, G., Alidaee, B.: Adaptive memory tabu search for binary quadratic programs. Manag. Sci. **44**, 336–345 (1998)
10. Hu, X.-B., Paolo, E.D.: An efficient genetic algorithm with uniform crossover for the multi-objective airport gate assignment problem. Multi Objective Memetic Algorithm **171**, 71–89 (2009)
11. Kochenberger, G., Hao, J.-K., Glover, F., Lewis, M., Lü, Z., Wang, H., Wang, Y.: The unconstrained binary quadratic programming problem: a survey. J. Comb. Optim. **28**, 58–81 (2014)
12. Krarup, J.: Computer aided layout design. Math. Program. Study **9**, 75–94 (1978)
13. Liefooghe, A., Verel, S., Hao, J.-K.: A hybrid metaheuristic for multiobjective unconstrained binary quadratic programming. Appl. Soft Comput. **16**, 10–19 (2014)
14. Liefooghe, A., Verel, S., Paquete, L., Hao, J.-K.: Experiments on local search for bi-objective unconstrained binary quadratic programming. In: Proceedings of the 8th International Conference on Evolutionary Multi-criterion Optimization (EMO 2015), Guimarães, Portugal, pp. 171–186 (2015)
15. Lü, Z., Hao, J.-K., Glover, F.: A study of memetic search with multi-parent combination for UBQP. In: Prodeedings of the 10th International Conference on Evolutionary Computation in Combinatorial Optimization (EvoCOP 2010), Istanbul, Turkey, pp. 154–165 (2010)
16. McBride, R.D., Yormark, J.S.: An implicit enumeration algorithm for quadratic integer programming. Manag. Sci. **26**, 282–296 (1980)
17. Merz, P., Katayama, K.: Memetic algorithms for the unconstrained binary quadratic programming problem. BioSystems **78**, 99–118 (2004)
18. Wang, Y., Lü, Z., Hao, J.-K.: A study of multi-parent crossover operators in a memetic algorithm. In: Prodeedings of the 11th International Conference on Parallel Problem Solving from Nature (PPSN XI), Krakow, Poland, pp. 556–565 (2010)
19. Zitzler, E., Thiele, L.: Multiobjective evolutionary algorithms: a comparative case study and the strength pareto approach. Evol. Comput. **3**, 257–271 (1999)

Face Recognition Based on Structural Incoherence and Low Rank Projection

Hefeng Yin and Xiaojun Wu[✉]

School of IoT Engineering, Jiangnan University,
Wuxi, People's Republic of China
yinhefeng@126.com, wu_xiaojun@jiangnan.edu.cn

Abstract. To solve the problem that both training and test samples are corrupted due to occlusion and disguise during face recognition, a new method which is based on low rank matrix recovery with structural incoherence and low rank projection (LRSI_LRP) is presented. First, the training images are decomposed into a set of clean images and sparse errors via LRSI, and the derived clean images from distinct classes are forced to be as independent as possible by introducing structural incoherence regularization term into robust PCA. Then a low-rank projection matrix is learned based on the original training images and the recovered clean ones, and this low-rank projection matrix can correct corrupted test samples by projecting them onto their corresponding underlying subspaces. Finally, the corrected test samples are classified based on sparse representation-based classification (SRC). Experimental results on AR and Extended Yale B databases verify the efficacy and robustness of the proposed method.

Keywords: Face Recognition · Low-rank matrix recovery · Structural incoherence · Low-rank projection matrix · Sparse Representation-based Classification

1 Introduction

Face recognition (FR) has been an active topic for researchers in the areas of computer vision and image processing due to its wide applications in practical applications, such as video surveillance, access control and human-computer interaction. For FR, conventional approaches such as Eigenfaces [1], Fisherfaces [2], or Laplacianfaces [3] can be applied to reduce the dimension of the face images, then the derived low-dimensional features are fed into classifiers to obtain the final recognition result. Nevertheless, these techniques are not robust to outliers and their performance will be degraded dramatically when test samples are corrupted due to occlusion or disguise.

Recently, sparse representation based classification (SRC) [4] has caught lots of attention due to its impressive results in robust FR. SRC expresses each test image as a sparse linear combination of training image data by solving an l_1-minimization problem, then the classification is performed by checking which class yields the least reconstruction error. Unlike traditional approaches such as Eigenface and Fisherface, SRC does not need an explicit feature extraction phase. The superior performance

© Springer International Publishing AG 2016
H. Yin et al. (Eds.): IDEAL 2016, LNCS 9937, pp. 68–78, 2016.
DOI: 10.1007/978-3-319-46257-8_8

reported in [4] suggests that it is a promising direction for FR. However, SRC forms the dictionary by directly using all the training images, thus the generated dictionary may have a huge size, which is detrimental to the following sparse solver. In addition, using original training samples as a dictionary could not fully take advantage of the discriminative information exist in the training data. To learn a more compact dictionary, Aharon et al. [5] proposed the K-SVD algorithm by generalizing the K-means clustering process. However, K-SVD focuses on only the representational power of the dictionary without considering its capability for discrimination. By incorporating the classification error into the objective function, Xu et al. [6] proposed a supervised within-class-similar discriminative dictionary learning (SCDDL) algorithm for FR. Majumdar [7] put forward discriminative label consistent dictionary learning (DLCDL) algorithm by adding a new label consistence term to K-SVD. Zheng et al. [8] presented a discriminative dictionary learning method via Fisher discrimination K-SVD algorithm.

Though SRC and its variants are robust to test sample with occlusion and corruption, the performance of them might be degenerated when some training and test samples are both corrupted. Using low-rank matrix recovery (LRMR) for denoising has attracted much attention recently. In particular, Wright et al. [9] presented the robust PCA (RPCA) method, which aims to recover a low-rank matrix from corrupted observations. Furthermore, Liu et al. [10] proposed low-rank representation (LRR) method. Based on LRMR, many approaches are presented for robust FR. Chen et al. [11] proposed a low-rank matrix approximation algorithm with structural incoherence (LRSI) for robust FR. Zheng et al. [12] proposed a LRMR algorithm with Fisher discriminant regularization. Du et al. [13] presented a graph regularized low-rank sparse representation recovery (GLRSRR) model to recover the clean training samples from the corrupted ones.

Approaches based on LRMR can effectively deal with the situation that both the original training and test samples might be corrupted. However, these methods cannot correct the corrupted test images for image classification if the test images are corrupted. Recently, deep learning achieves very promising results for FR [14]. Despite its impressive performance, it remains unclear how to design a good architecture to adapt to a specific classification task due to the lack of theoretical guidance [15]. To alleviate the above-mentioned limitations, we propose a robust FR method based on LRSI and low rank projection matrix (LRSI_LRP). The main difference between LRSI_LRP and the method in [16] is that a low-rank projection matrix between the original training samples and the recovery results is learned to correct the corrupted test samples in LRSI_LRP, and the low rank projection matrix is also critical to improve the performance of FR besides LRSI [17]. Experimental results obtained on the AR database and Extended Yale B database validate the efficacy and robustness of the proposed method.

The remainder of this paper is structured as follows. Section 2 reviews related work on SRC and LRMR. In Sect. 3, we present our proposed method. Experimental results on two databases are presented in Sect. 4. Finally, Sect. 5 concludes this paper.

2   Related Work

2.1   Sparse Representation Based Classification

Suppose that we are given n training images from C distinct classes, and each class i has n_i images. Let $X = [X_1, X_2, \cdots, X_C] \in \mathbb{R}^{m \times n}$ be the training set, where $X_i \in \mathbb{R}^{m \times n_i}$ contains training images of the ith class as its columns, and m is the dimension of each image. Given a test image $y \in \mathbb{R}^m$, SRC calculates the sparse representation α of y, which is computed via the l_1 minimization process over X. More specifically, SRC solves the following optimization problem for deriving the sparse representation α

$$\min_{\alpha} \|y - X\alpha\|_2^2 + \lambda \|\alpha\|_1 \tag{1}$$

where $\lambda > 0$ is a parameter that balances reconstruction error and sparsity.

Let α_i be a vector in $\mathbb{R}^n$ with nonzero entries as those in α that are associated with class i. Once (1) is solved, the test input y is classified according to the following rule:

$$identity(y) = arg \min_{i} \|y - X_i \alpha_i\|_2^2 \tag{2}$$

2.2   Low Rank Matrix Recovery

Although SRC achieves impressive results, its performance may be degenerated in practical applications due to occlusions and corruptions exist in the training samples. Recent researches indicate that the above problems can be alleviated to some extent via LRMR. Robust PCA is a representative work of LRMR, and it seeks to decompose corrupted observations into two matrices, one is a low-rank matrix and the other is the associated sparse error matrix. More specifically, to derive the low-rank approximation of the input data matrix X, RPCA minimizes the rank of matrix A while reducing the l_0-norm of E. As the optimization of rank function and l_0-norm is highly nonconvex, we can get the following tractable convex optimization surrogate by replacing the rank function with nuclear norm $\|A\|_*$ (i.e., the sum of the singular values of A), and the l_0-norm with the l_1-norm.

$$\min_{A,E} \|A\|_* + \beta \|E\|_1, \ s.t. X = A + E \tag{3}$$

3   FR Based on Structural Incoherence and Low Rank Projection

Our proposed method, LRSI_LRP, consists of two critical components: LRMR with structural incoherence and low rank projection. In the following part, the objective function and optimization procedure of LRSI are presented. Then the low rank projection matrix is described. Finally the detailed procedure of LRSI_LRP is outlined.

3.1   Low Rank Matrix Recovery with Structural Incoherence

Although RPCA can process the original data matrix X and obtain a low-rank matrix A for better representation ability, the derived matrix A does not contain sufficient discriminating information. Inspired by [18], we can promote the incoherence between the derived low-rank matrices of different classes for classification tasks. The introduction of such incoherence would force the resulting low-rank matrices to be as independent as possible. Therefore, the derived low rank matrices can form a better dictionary for classification purposes. Based on the formulation of RPCA in (3), a regularization term can be added to the objective function to enforce the incoherence between the low-rank matrices. The objective function of LRSI is:

$$\min_{A_i,E_i} \sum_{i=1}^{C} \left\{ \|A_i\|_* + \beta\|E_i\|_1 \right\} + \eta \sum_{j\neq i} \left\|A_j^T A_i\right\|_F^2, s.t. X_i = A_i + E_i \qquad (4)$$

In (4), the first term performs the standard low-rank decomposition of the data matrix X. The second term sums up the Frobenius norms between each pair of the low-rank matrices A_i and A_j, which is penalized by the parameter η balancing the low-rank matrix recovery and matrix incoherence.

Problem (4) can be solved class by class, thus we get the following optimization problem for different classes:

$$\min_{A_i,E_i} \|A_i\|_* + \beta\|E_i\|_1 + \eta \sum_{j\neq i} \left\|A_j^T A_i\right\|_F^2, s.t. X_i = A_i + E_i \qquad (5)$$

According to Cauchy-Schwarz inequality,

$$\left\|A_j^T A_i\right\|_F^2 \leq \|A_j\|_F^2 \|A_i\|_F^2 \qquad (6)$$

We can replace the term $\eta \sum_{j\neq i} \left\|A_j^T A_i\right\|_F^2$ in (5) with $\eta' \|A_i\|_F^2$, where $\eta' = \eta \sum_{j\neq i} \|A_j\|_F^2$. Thus Problem (5) can be reformulated as

$$\min_{A_i,E_i} \|A_i\|_* + \beta\|E_i\|_1 + \eta' \|A_i\|_F^2, s.t. X_i = A_i + E_i \qquad (7)$$

Augmented Lagrange multipliers (ALM) is applied to solve the above problem, and the augmented Lagrangian function for (7) is

$$L(A_i, E_i, Y_i, \mu, \eta') = \|A_i\|_* + \beta\|E_i\|_1 + \eta' \|A_i\|_F^2 + <Y_i, X_i - A_i - E_i > + \frac{\mu}{2}\|X_i - A_i - E_i\|_F^2$$

where Y_i is Lagrange multipliers, and $\mu > 0$ is a penalty parameter.

1) Updating A_i with fixed E_i, Y_i

$$
\begin{aligned}
A_i^{k+1} &= arg\,\min_{A_i} \|A_i\|_* + \eta'\|A_i\|_F^2 + <Y_i^k, X_i - A_i - E_i^k> +\frac{\mu^k}{2}\|X_i - A_i - E_i^k\|_F^2 \\
&= arg\,\min_{A_i} \|A_i\|_* + \left(\eta' + \frac{\mu^k}{2}\right) <A_i, A_i> - \mu^k <X_i - E_i^k + \frac{Y_i^k}{\mu^k}, A_i> \quad (8) \\
&= arg\,\min_{A_i} \varepsilon\|A_i\|_* + \frac{1}{2}\|X_a - A_i\|_F^2
\end{aligned}
$$

where $\varepsilon = (2\eta' + \mu^k)^{-1}$ and $X_a = \mu^k \varepsilon \left(X_i - E_i^k + \frac{Y_i^k}{\mu^k}\right)$.

2) Updating E_i with fixed A_i, Y_i

$$
\begin{aligned}
E_i^{k+1} &= \beta\|E_i\|_1 + <Y_i^k, X_i - A_i^{k+1} - E_i> +\frac{\mu^k}{2}\|X_i - A_i^{k+1} - E_i\|_F^2 \\
&= arg\,\min_{E_i} \varepsilon'\|E_i\|_1 + \frac{1}{2}\|X_e - E_i\|_F^2
\end{aligned} \quad (9)
$$

where $\varepsilon' = \frac{\beta}{\mu^k}$ and $X_e = X_i - A_i^{k+1} + \frac{Y_i^k}{\mu^k}$.

3.2   Low Rank Projection Matrix

Consider a set of the principal components $Y = [y_1, y_2, \cdots, y_n] \in \mathbb{R}^{m \times n}$ obtained from the original training samples $X = [x_1, x_2, \cdots, x_n] \in \mathbb{R}^{m \times n}$, where the sparse corruptions have been efficiently removed. Each column of these matrices is of length m. A key property when studying a linear projection P is the low rank projection between X and Y, which can project any data point x onto its underlying subspace, and thus produce accurate recovery results P_x. The recovery result is considered to have been drawn from a union of multiple low-rank subspaces. Consequently, it is reasonable to assume that P is a low-rank matrix. We can learn the low-rank projection P by solving the following optimization problem:

$$
\min_P \|P\|_*, \quad s.t. Y = PX \quad (10)
$$

Suppose P^* is the optimal solution of the above problem, then $P^* = YX^+$ is the unique minimizer to (10), where X^+ is the pseudo-inverse of X. More concretely, the pseudo-inverse is defined by $X^+ = V\Sigma^{-1}U^T$, where $U\Sigma V^T$ is the skinny SVD of X. By using this low rank projection matrix, the principal components and error of any new sample x can be expressed by P^*x and $x - P^*x$, respectively.

3.3 Detailed Procedure of LRSI_LRP

Based on LRSI and low rank projection matrix, we present our proposed method, LRSI_LRP. It mainly consists of three steps: first, recovery results from original training samples are obtained by LRSI. Second, low rank projection matrix can be learned to correct the test samples. After applying PCA to the recovery results and the corrected test samples, SRC is exploited to classify the corrected test samples. For completeness of presentation, we outline the detailed procedure of LRSI_LRP in Algorithm 1, which summarizes the procedure of integrating low-rank matrix recovery and SRC.

Algorithm 1 FR via Structural Incoherence and Low Rank Projection

Input: Training samples $X = [x_1, x_2, \cdots, x_n] \in \mathbb{R}^{m \times n}$ for C classes, test sample $y \in \mathbb{R}^m$ and balance parameters β, η.

1 **for** $i = 1; i \leq C; i++$

Solve the following optimization problem:

$$\min_{A_i, E_i} \|A_i\|_* + \beta\|E_i\|_1 + \eta \sum_{j \neq i} \|A_j^T A_i\|_F^2, \ s.t. X_i = A_i + E_i$$

end for

2 Obtain the recovery results from C classes, $Y = [A_1, A_2, \cdots, A_C]$;

3 Calculate the low rank projection matrix: $P^* = YX^+$;

4 Correct the test sample y: $y_r = P^*y$;

5 Calculate principal components W of Y: $W = PCA(Y)$;

6 Project Y and y_r onto W: $D = W(Y), y_p = W(y_r)$;

7 Obtain the sparse coefficients of y_p on D: $\min_{\alpha} \|y_p - D\alpha\|_2^2 + \lambda\|\alpha\|_1$

8 **for** $i = 1; i \leq C; i++$

$\quad e_i = \|y_p - D_i\alpha_i\|_2^2;$

end for

Output: $identity(y) = \arg\min_i(e_i)$.

---

4 Experiments

To verify the efficacy of the proposed algorithm, experiments are conducted on the AR database and Extended Yale B database. Our approach is compared with several other algorithms including nearest neighbor (NN), LRC [19], SRC [4], CRC [20], LRR_SRC [16] and LRSI [11]. In our experiments, we apply the *SolveDALM* function to solve the l_1 minimization problem with $\lambda = 0.001$.

4.1 FR with Real Face Occlusion

The AR database consists of over 4000 frontal images of 126 subjects. For each individual, 26 pictures are taken in two separated sessions. In each session, 13 images are available for each individual. These images contain three images with sunglasses, another three with scarves, and the remaining seven with illumination and expression changes. In the experiment, we choose a subset of the AR database that contains 50 male subjects and 50 female subjects. Following the protocol in [11], experiments are conducted under three different scenarios:

Sunglasses: We first consider occluded training samples due to the presence of sunglasses, which occluded about 20 % of the face image. We use seven unobscured images and one image with sunglasses (randomly chosen) from session 1 for training (eight training images per class), and the remaining unobscured images (all from session 2) and the rest of the images with sunglasses (two taken at session 1 and three at session 2) for testing (twelve test images per class).

Scarf: In this scenario, we consider occluded training samples due to the presence of scarves, which occluded about 40 % of the face image. The choice of training and test set is similar to that for **Sunglasses** case.

Sunglasses+Scarf: In the last scenario, the training samples are occluded by sunglasses and scarves, which is more challenging than the above two scenarios. Seven unobscured images, two corrupted images (one with sunglasses and one with scarf) from session 1 are used for training (nine training images per class), and the rest are used for testing (seventeen test images per class).

In this experiment, we compare our LRSI_LRP with other approaches for feature dimensions of 25, 50, 100, 200, and 300. Parameters for LRSI_LRP are $\beta = 0.02$, $\eta = 0.001$ in this experiment. Since there are at most three obscured images for each type of occlusion available in session 1, we repeat our experiment for each scenario three times and report the averaged accuracy and standard deviation. Tables 1, 2 and 3 summarize the results. Table 1 shows that LRSI_LRP outperforms the other methods for all dimensions. From Table 2, we can see that LRSI_LRP achieves the best result for images occluded by scarves. Moreover, Table 3 shows that LRSI_LRP achieves the best performance as the percentage of occlusion increases. LRSI_LRP consistently outperforms the compared approaches under different scenarios, which validates the robustness of our proposed method.

In order to vividly illustrate the effectiveness of the proposed method, Figs. 1(a) and (b) give part of the original training images and their recovered ones via LRSI. As we can see from Figs. 1(a) and (b), variations such as expression, illumination and occlusions are eliminated to some extent after LRSI, and the corrected training samples

Table 1. Recognition accuracy (%) of different methods on the AR database with the occlusion of sunglasses.

Dim.	25	50	100	200	300
NN	53.25 ± 0.38	61.22 ± 1.51	64.86 ± 0.71	66.06 ± 1.19	66.03 ± 1.19
LRC	59.64 ± 1.91	69.28 ± 1.04	72.69 ± 0.82	74.56 ± 0.88	74.64 ± 0.86
SRC	55.94 ± 0.98	69.92 ± 1.69	77.47 ± 0.38	81.17 ± 0.98	83.39 ± 0.54
CRC	50.86 ± 1.84	72.67 ± 1.59	83.44 ± 1.06	88.14 ± 0.99	89.47 ± 0.94
LRR_SRC	52.31 ± 2.51	74.72 ± 3.08	85.03 ± 1.73	87.64 ± 0.63	87.11 ± 1.58
LRSI	51.44 ± 0.67	73.14 ± 1.01	84.11 ± 0.54	88.33 ± 0.36	89.53 ± 0.61
LRSI_LRP	72.58 ± 1.52	84.19 ± 0.63	90.17 ± 0.22	90.81 ± 0.29	90.86 ± 0.32

Table 2. Recognition accuracy (%) of different methods on the AR database with the occlusion of scarves.

Dim.	25	50	100	200	300
NN	43.25 ± 3.84	51.00 ± 4.34	54.14 ± 4.76	55.67 ± 4.77	56.14 ± 4.73
LRC	53.67 ± 2.54	63.47 ± 2.17	67.58 ± 2.30	68.64 ± 1.84	68.58 ± 1.80
SRC	47.69 ± 3.88	61.08 ± 4.65	70.19 ± 4.32	76.50 ± 2.55	78.36 ± 2.13
CRC	44.33 ± 1.75	69.42 ± 2.75	82.89 ± 1.80	88.78 ± 1.14	89.72 ± 0.63
LRR_SRC	47.86 ± 0.29	74.08 ± 0.94	85.11 ± 0.85	87.94 ± 0.59	86.78 ± 1.05
LRSI	43.06 ± 0.51	70.67 ± 1.17	82.67 ± 0.60	87.47 ± 0.92	88.72 ± 0.25
LRSI_LRP	70.11 ± 2.16	85.22 ± 0.21	89.86 ± 0.19	90.11 ± 0.32	90.06 ± 0.17

Table 3. Recognition accuracy (%) of different methods on the AR database with the occlusion of sunglasses and scarves.

Dim.	25	50	100	200	300
NN	43.10 ± 2.46	51.31 ± 3.36	55.27 ± 3.42	57.04 ± 3.29	57.08 ± 3.34
LRC	53.71 ± 0.89	64.06 ± 0.79	68.20 ± 1.45	69.94 ± 1.35	70.10 ± 1.10
SRC	47.90 ± 2.77	61.37 ± 4.31	70.84 ± 2.95	75.73 ± 2.24	78.04 ± 2.31
CRC	42.31 ± 1.35	66.29 ± 3.06	79.59 ± 2.21	87.08 ± 0.97	87.84 ± 1.49
LRR_SRC	45.31 ± 1.24	69.82 ± 1.93	83.06 ± 0.80	86.14 ± 1.29	86.18 ± 0.93
LRSI	41.49 ± 1.41	66.20 ± 0.89	79.53 ± 0.16	86.69 ± 0.68	87.78 ± 0.44
LRSI_LRP	70.06 ± 1.08	83.26 ± 0.87	89.21 ± 0.37	89.62 ± 0.12	89.59 ± 0.17

have a better representation ability than the original images used as a dictionary for SRC. Figures 1(c) and (d) show part of the original test images and the recovered ones by low rank projection matrix. Similarly, we can see that variations in the original test images are removed. Thus, improved recognition accuracy can be expected by using these recovered images.

(a) Some original training samples

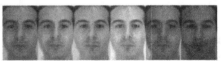

(b) Recovery results by LRSI

(c) Some test samples

(d) Recovered test samples by LRPM

Fig. 1. Recovery results by LRSI and low rank projection matrix.

4.2   FR with Random Pixel Corruption

Extended Yale B database is used for training and testing in this subsection, and this database consists of 2414 frontal-face images of 38 individuals (about 64 images per subject) captured under various laboratory-controlled lighting conditions. These images have size of 192 × 168 pixels. We use two settings for the percentage of corrupted samples, 10 % and 20 %. Under both settings, 10 % of the pixels in each selected image are corrupted. To corrupt any chosen pixel, its observed value is replaced by a uniformly distributed random value in the range [0, 1] and pixels are randomly chosen for each sample. Figure 2 shows some representative examples of the training and test images with random pixel corruption.

Half of the images are used for training, and the remaining for testing. The feature dimensions of the face images are selected to be 50, 100, and 300 for the purpose of fair comparison with the other approaches. Parameters for LRSI_LRP are $\beta = 0.1$, $\eta = 0.001$ for this experiment. All experiments run 10 times and the averaged accuracy and standard deviation are reported. Comparison of different approaches is listed in Table 4 for the different percentages of pixel corruptions. At lower dimensions, our approach outperforms other methods, as the percentage of corrupted samples increases, LRSI_LRP still has superior performance over other competing approaches. For example, LRSI_LRP achieves recognition accuracy of 93.35 % and 92.57 % when the data dimension is 50. This represents an improvement of 15.92 % and 16.64 %, respectively, over LRSI. Even when the dimension is 300, the proposed method outperforms LRSI by 1.40 % and 2.72 %, respectively. This shows that the proposed method is much more robust to random pixel corruption than other compared methods.

Fig. 2. Face images with random pixel corruption.

Table 4. Recognition accuracy (%) of different methods on the Extended Yale B database with random pixel corruption.

Dim.	10 % samples are corrupted			20 % samples are corrupted		
	50	100	300	50	100	300
NN	59.32 ± 1.19	70.00 ± 0.48	76.57 ± 0.75	57.85 ± 0.98	68.73 ± 0.47	75.99 ± 0.77
LRC	92.15 ± 0.12	93.99 ± 0.11	95.20 ± 0.30	91.11 ± 0.18	92.86 ± 0.06	94.20 ± 0.18
SRC	87.78 ± 0.29	92.49 ± 0.80	95.44 ± 0.31	86.73 ± 0.30	91.40 ± 0.38	94.41 ± 0.29
CRC	86.69 ± 0.42	94.24 ± 0.24	98.00 ± 0.12	85.01 ± 0.88	92.86 ± 0.77	96.95 ± 0.30
LRR_SRC	83.78 ± 0.70	92.35 ± 0.42	97.41 ± 0.52	81.94 ± 1.10	91.18 ± 0.81	96.02 ± 0.48
LRSI	77.43 ± 0.84	88.93 ± 1.47	96.60 ± 0.39	75.93 ± 0.80	85.50 ± 0.92	94.27 ± 0.05
LRSI_LRP	93.35 ± 0.38	96.60 ± 0.53	98.00 ± 0.55	92.57 ± 0.59	95.35 ± 0.38	96.99 ± 0.17

5 Conclusions

A robust FR method based on LRSI and low rank projection matrix is proposed to tackle the situation that both training and test samples contain corrupted samples. By using low rank matrix recovery with structural incoherence, which enforces the atoms learned for distinct classes to be as independent as possible, a more suitable dictionary for SRC is obtained. Furthermore, a low-rank projection matrix is learned to correct the corrupted test samples. Experimental results obtained on two publicly available databases validate the efficacy and robustness of our proposed method. In future, we will investigate FR under small sample size (SSS) condition, even with single training sample per person (STSPP), which is of great importance in practical applications.

Acknowledgments. This work was supported by the National Natural Science Foundation of China under Grant No. 61373055 and the Innovation Project of Graduate Education of Jiangsu Province under Grant No. KYLX_1123.

References

1. Turk, M., Pentland, A.: Eigenfaces for recognition. J. Cogn. Neurosci. **3**(1), 71–86 (1991)
2. Belhumeur, P.N., Hespanha, J.P., Kriegman, D.J.: Eigenfaces vs. Fisherfaces: recognition using class specific linear projection. IEEE Trans. Pattern Anal. Mach. Intell. **19**(7), 711–720 (1997)
3. He, X., Yan, S., Hu, Y., Niyogi, P., Zhang, H.: Face recognition using Laplacianfaces. IEEE Trans. Pattern Anal. Mach. Intell. **27**(3), 328–340 (2005)
4. Wright, J., Yang, A.Y., Ganesh, A., Sastry, S.S., Ma, Y.: Robust face recognition via sparse representation. IEEE Trans. Pattern Anal. Mach. Intell. **31**(2), 210–227 (2009)
5. Aharon, M., Elad, M., Bruckstein, A.: K-SVD: an algorithm for designing overcomplete dictionaries for sparse representation. IEEE Trans. Sig. Process. **54**(11), 4311–4322 (2006)
6. Xu, L., Wu, X., Chen, K., Yao, L.: Supervised within-class-similar discriminative dictionary learning for face recognition. J. Vis. Commun. Image Represent. **38**, 561–572 (2016)
7. Majumdar, A.: Discriminative label consistent dictionary learning. In: IEEE International Conference on Image Processing, pp. 1016–1020 (2015)
8. Zheng, H., Tao, D.: Discriminative dictionary learning via Fisher discrimination K-SVD algorithm. Neurocomputing **162**, 9–15 (2015)
9. Wright, J., Ganesh, A., Rao, S., Peng, Y., Ma, Y.: Robust principal component analysis: exact recovery of corrupted low-rank matrices via convex optimization. In: Proceedings of Advances in Neural Information Processing Systems, pp. 2080–2088 (2009)
10. Liu, G., Lin, Z., Yan, S., Sun, J., Yu, Y., Ma, Y.: Robust recovery of subspace structures by low-rank representation. IEEE Trans. Pattern Anal. Mach. Intell. **35**(1), 171–184 (2013)
11. Chen, C.F., Wei, C.P., Wang, Y.C.F.: Low-rank matrix recovery with structural incoherence for robust face recognition. In: IEEE Conference on Computer Vision and Pattern Recognition, pp. 2618–2625 (2012)
12. Zheng, Z., Yu, M., Jia, J., Liu, H., Xiang, D., Huang, X., Yang, J.: Fisher discrimination based low rank matrix recovery for face recognition. Pattern Recogn. **47**(11), 3502–3511 (2014)

13. Du, H., Zhang, X., Hu, Q., Hou, Y.: Sparse representation-based robust face recognition by graph regularized low-rank sparse representation recovery. Neurocomputing **164**, 220–229 (2015)

14. Sun, Y., Wang, X., Tang, X.: Sparsifying neural network connections for face recognition. In: Conference on Computer Vision and Pattern Recognition, pp. 4856–4864 (2016)

15. Hu, G., Yang, Y., Yi, D., Kittler, J., Christmas, W., Li, S.Z., Hospedales, T.: When face recognition meets with deep learning: an evaluation of convolutional neural networks for face recognition. In: Proceedings of the IEEE International Conference on Computer Vision Workshops, pp. 142–150 (2015)

16. Wei, C.P., Chen, C.F., Wang, Y.C.F.: Robust face recognition with structurally incoherent low-rank matrix decomposition. IEEE Trans. Image Process. **23**(8), 3294–3307 (2014)

17. Zhang, X., Hao, S., Xu, C., Qian, X., Wang, M., Jiang, J.: Image classification based on low-rank matrix recovery and naive Bayes collaborative representation. Neurocomputing **169**, 110–118 (2015)

18. Ramirez, I., Sprechmann, P., Sapiro, G.: Classification and clustering via dictionary learning with structured incoherence and shared features. In: IEEE Conference on Computer Vision and Pattern Recognition, pp. 3501–3508 (2010)

19. Naseem, I., Togneri, R., Bennamoun, M.: Linear regression for face recognition. IEEE Trans. Pattern Anal. Mach. Intell. **32**(11), 2106–2112 (2010)

20. Zhang, L., Yang, M., Feng, X.: Sparse representation or collaborative representation: which helps face recognition? In: IEEE International Conference on Computer Vision, pp. 471–478 (2011)

A Note on the k-NN Density Estimate

Jie Ding[1,2,3(✉)] and Xinshan Zhu[4]

[1] School of Information Engineering,
Yangzhou University, Yangzhou 225127, China
jieding@yzu.edu.cn
[2] State Key Laboratory of Software Development Environment,
Beihang University, Beijing, China
[3] State Key Laboratory for Novel Software Technology,
Nanjing University, Nanjing, China
[4] School of Electrical Engineering and Automation,
Tianjin University, Tianjin 300072, China
xszhu@tju.edu.cn

Abstract. k-NN (k Nearest Neighbour) density estimate as a nonparametric estimation method is widely used in machine learning or data analysis. The convergence problem of k-NN approach has been intensively investigated. In particular, the equivalence of convergence in weak or strong sense (i.e. in probability sense or in almost surely sense) has been respectively developed. In this note, we will show that the k-NN estimator converges in probability is equivalent to converge in the L^2 sense. Moreover, some relevant asymptotic results about the expectations of k-NN estimator will be established.

Keywords: k-NN density estimate · Equivalence · L^2 convergence

1 Introduction

The k-nearest neighbor (k-NN) method, which has been widely used in machine learning or statistical learning, has attracted much attention since its birth. k-NN density estimation, which is first introduced in [5], is a well known and simple density estimation procedures [4]. k-NN density estimate is one kind of kernel estimate. A kernel estimate on R^d is

$$f_n(x) = (nh_n)^{-1} \sum_{i=1}^{n} K((x - X_i)/h_n), \tag{1}$$

where $X_1, X_2, \cdots, X_n$ are sampled independently from a density f on R^d, K is a kernel density function, and h_n is a window width.

Let K be bounded with compact support, then [3]:

(1) [Weak version.] the following statements are equivalent:
  (A) $f_n(x) \to f(x)$ in probability, as $n \to \infty$, for almost all x;
  (B) $\lim_{n\to\infty} h_n = 0, \lim_{n\to\infty} nh_n = \infty$;
  (C) $\int |f_n(x) - f(x)| dx \to 0$ in probability, as $n \to \infty$;

H. Yin et al. (Eds.): IDEAL 2016, LNCS 9937, pp. 79–88, 2016.
DOI: 10.1007/978-3-319-46257-8_9

(2) [Strong version.] the following statements are equivalent if $nh_n^d/\log\log n$ is semimonotone:

(A) $f_n(x) \to f(x)$ almost surely, as $n \to \infty$, for almost all x;

(B) $\lim_{n\to\infty} h_n = 0, \lim_{n\to\infty} nh_n^d/\log\log n = \infty$.

More work on the equivalence of weak, strong and complete convergence are presented in [1,2]. A survey can be found in [6,7].

In the context of k-NN, the statement (1)(B), i.e. (B) in weak version, becomes $\lim_{n\to\infty} k_n/n = 0, \lim_{n\to\infty} k_n = \infty$, where k_n is the number of neighbours that a point x has. It is well known that L^2 convergence implies probability convergence, while probability convergence cannot imply L^2 convergence in general. In this note, we will show that they are equivalent, for k-NN density estimates. That is, we will prove that (1)(A) in weak version (i.e. $f_n(x) \to f(x)$ in probability, as $n \to \infty$, for almost all x) is equivalent to the following

(D) $E[f_n(x) - f(x)]^2 \to 0$ as $n \to \infty$, for almost all x.

This is the main contribution of this paper. Meanwhile, some relevant properties of k-NN density estimates will be established in this paper, and used to prove this equivalence.

The remainder of this note is structured as follows. Section 2 presents some preliminary results, based on which the main conclusions, particularly the equivalence convergence result, are proved in Sect. 3. We conclude this note in Sect. 4.

2   Preliminary Results

As we have mentioned, there are N independent nodes $\{X_i\}_{i=1}^n$ distributed on R^d with an identical density function $f(x)$, where $d > 0$ is an integer. We suppose that $f(x)$ is continuous on R^d. According to classical measure theory, for any $\epsilon > 0$, there exists a compact region $\Omega \subset R^d$ such that $\int_\Omega f > 1 - \epsilon$, and $f(x)$ is positive on Ω. In addition, there exists $M > 0$, such that $\frac{1}{M} \leq f(x) < M$ for any $x \in \Omega$. Furthermore, because Ω is bounded, so the diameter of Ω, defined by $d(\Omega) = \sup_{x,y\in\Omega} d(x,y)$, is finite, i.e. $d(\Omega) < \infty$, where $d(x,y)$ denotes the distance between points x and y. Moreover, there exists $\beta > 0$ such that $\Omega_\beta = \{x \in \Omega : d(x, \partial\Omega) \geq \beta\} \subset \Omega$ and $\int_{\Omega_\beta} > 1 - 2\epsilon$.

In the following, we denote

$$u(x,t) = \int_{B(x,t)\cap\Omega} f(y)\mathrm{d}y. \tag{2}$$

$$F_x^N(t) = 1 - (1 - u(x,t))^{N-1}. \tag{3}$$

$$F_{x,(k)}^N(t) = 1 - \sum_{i=0}^{k-1} C_{N-1}^i [u(t)]^i [1 - u(t)]^{N-1-i}. \tag{4}$$

Clearly, $u(x, t)$ is the probability of a given node falling into region $B(x, t) \cap \Omega$. In the following, if there is no confusion, $u(x, t)$ is usually denoted by $u(t)$.

The probability of at least one node (k nodes, respectively) among $N - 1$ nodes falling into $B(x, t) \cap \Omega$ is $F_x^N(t)$ ($F_{x,(k)}^N(t)$, respectively).

Given a node, namely X, let T ($T_{(k)}$, respectively) be the distance between node X and its nearest neighbour (kth nearest neighbour, respectively) among other $N - 1$ nodes. Clearly, T and $T_{(k)}$ are random variables, satisfying

$$\Pr\left(T \leq t | X = x\right) = F_x^N(t), \quad \Pr\left(T_{(k)} \leq t | X = x\right) = F_{x,(k)}^N(t).$$

Some preliminary results are presented below.

Lemma 1. $\forall \eta > 0, \exists \alpha < \beta, \forall t < \alpha, \forall x \in \Omega_\beta$, then

$$u(t)\left(1 - \frac{\eta}{M}\right) \leq f(x)\mathrm{Vol}(B(x, t)) \leq u(t)(1 + \eta M). \tag{5}$$

Proof. Because $f(x)$ is continuous on compact set Ω_β, for any $\eta > 0$, there exists a $\alpha > 0$ and $\alpha < \beta$, such that for any $t < \alpha$,

$$\frac{\int_{B(x,t)} f(y)\mathrm{d}y}{\mathrm{Vol}(B(x,t))}(1 - \eta/M) \leq f(x) \leq \frac{\int_{B(x,t)} f(y)\mathrm{d}y}{\mathrm{Vol}(B(x,t))}(1 + \eta M).$$

This leads to the conclusion (5).

Lemma 2. *Suppose* $m \in Z^+$, *then for almost all* $x \in \Omega$,

$$\lim_{N \to \infty} \int_0^\infty [N\mathrm{Vol}(B(x,t))]^m \, \mathrm{d}F_x^N(t) = \frac{m!}{[f(x)]^m}. \tag{6}$$

Proof. Let $x \in \Omega_\beta$, and $\alpha > 0$ be sufficiently small.

(1) Case $m = 1$.

$$\begin{aligned}
\limsup_{N \to \infty} N \int_\alpha^\infty \mathrm{Vol}(B(x,t))\mathrm{d}F_x^N(t) &\leq \limsup_{N \to \infty} N\mathrm{Vol}(B(x, d(\Omega))) \int_\alpha^\infty \mathrm{d}F_x^N(t) \\
&= \limsup_{N \to \infty} N\mathrm{Vol}(B(x, d(\Omega)))\mathrm{d}F_x^N(t)|_\alpha^\infty \\
&= \limsup_{N \to \infty} N\mathrm{Vol}(B(x, d(\Omega)))\left[1 - u(\alpha)\right]^{N-1} \\
&= 0.
\end{aligned}$$

Therefore,

$$\lim_{N \to \infty} \int_\alpha^\infty N\mathrm{Vol}(B(x,t))\mathrm{d}F_x^N(t) = 0.$$

According to Lemma 1,

$$\int_0^\alpha \text{Vol}(B(x,t))dF_x^N(t)$$

$$= -\frac{1}{f(x)} \int_0^\alpha f(x)\text{Vol}(B(x,t))\frac{\partial[1-u(t)]^{N-1}}{\partial t}dt$$

$$\leq -\frac{1+\eta M}{f(x)} \int_0^\alpha u(t)\frac{\partial[1-u(t)]^{N-1}}{\partial t}dt$$

$$= -\frac{1+\eta M}{f(x)} \left\{ u(t)[1-u(t)]^{N-1}\big|_0^\alpha - \int_0^\alpha u(t)[1-u(t)]^{N-1}du(t) \right\}$$

$$= -\frac{1+\eta M}{f(x)} \left\{ u(\alpha)[1-u(\alpha)]^{N-1} + \frac{1}{N}[1-u(\alpha)]^N - \frac{1}{N} \right\}.$$

Similarly, we have

$$\int_0^\alpha \text{Vol}(B(x,t))dF_x^N(t) \geq -\frac{1-\eta/M}{f(x)} \left\{ u(\alpha)[1-u(\alpha)]^{N-1} + \frac{1}{N}[1-u(\alpha)]^N - \frac{1}{N} \right\},$$

leading to

$$\frac{1-\eta/M}{f(x)} \leq \liminf_{N\to\infty} \int_0^\alpha \text{Vol}(B(x,t))dF_x^N(t)$$

$$\leq \limsup_{N\to\infty} \int_0^\alpha \text{Vol}(B(x,t))dF_x^N(t) \leq \frac{1+\eta M}{f(x)}.$$

Let α tend to zero, we have

$$\lim_{\alpha\to 0}\limsup_{N\to\infty} N\int_0^\infty \text{Vol}(B(x,t))dF_x^N(t)$$

$$= \lim_{\alpha\to 0}\limsup_{N\to\infty} N\left\{ \left(\int_0^\alpha + \int_\alpha^\infty \right) \text{Vol}(B(x,t))dF_x^N(t) \right\}$$

$$\leq \lim_{\alpha\to 0} \frac{1+\eta M}{f(x)}$$

$$= \frac{1}{f(x)}.$$

Similarly, we can obtain

$$\lim_{\alpha\to 0}\liminf_{N\to\infty} N\int_0^\infty \text{Vol}(B(x,t))dF_x^N(t) \geq \lim_{\alpha\to 0} \frac{1-\eta/M}{f(x)} = \frac{1}{f(x)}.$$

Therefore,

$$\lim_{N\to\infty}\int_0^\infty N\mathrm{Vol}(B(x,t))\mathrm{d}F_x^N(t)=\frac{1}{f(x)},\forall x\in\Omega_\beta \tag{7}$$

(2) Case $m=2$. Noting that

$$\int_0^\alpha (\mathrm{Vol}B(x,t))^2\mathrm{d}F_x^N(t)$$

$$=-\frac{1}{(f(x))^2}\int_0^\alpha [f(x)B(x,t)]^2\mathrm{d}[1-u(t)]^{N-1}$$

$$\leq -\left[\frac{1+\eta M}{f(x)}\right]^2\int_0^{u(\alpha)}u^2\mathrm{d}(1-u)^{N-1}$$

$$=-\left[\frac{1+\eta M}{f(x)}\right]^2\left\{u^2(1-u)^{N-1}\big|_0^{u(\alpha)}-2\int_0^{u(\alpha)}u(1-u)^{N-1}\mathrm{d}u\right\}$$

$$=-\left[\frac{1+\eta M}{f(x)}\right]^2\left\{u(\alpha)^2(1-u(\alpha))^{N-1}+\frac{2}{N}\int_0^{u(\alpha)}ud(1-u)^N\right\}$$

$$=-\left[\frac{1+\eta M}{f(x)}\right]^2\left\{u(\alpha)^2(1-u(\alpha))^{N-1}+\frac{2}{N}\left[u(\alpha)(1-u(\alpha))^N+\frac{1}{N+1}\left[(1-u(\alpha))^{N+1}-1\right]\right]\right\}.$$

It is clearly to see that

$$\limsup_{N\to\infty}N^2\int_0^\alpha [\mathrm{Vol}(B(x,t))]^2\mathrm{d}F_x^N(t)\leq 2\left[\frac{1+\eta M}{f(x)}\right]^2,$$

then by a similar calculation to the case $m=1$, we have

$$\lim_{N\to\infty}N^2\int_0^\infty [\mathrm{Vol}(B(x,t))]^2\mathrm{d}F_x^N(t)=\frac{2}{(f(x))^2},a.s.$$

For the case of $m\geq 3$, the proof is similar and thus omitted.

This convergence is uniform with respect to $x\in\Omega_\beta$. Noting that $\int_{R^d\setminus\Omega_\beta}f<2\epsilon$. So $N\int_0^\infty \mathrm{Vol}(B(x,t))\mathrm{d}F_x^N(t)$ is in fact almost uniformly and thus almost surely, convergent to $\frac{1}{f(x)}$ on R^d, as $N\to\infty$. The proof is complete.

By Eq. (4), we have that

$$\mathrm{d}F_{x,(k)}^N=-\mathrm{d}[1-u(t)]^{N-1}$$

$$-\sum_{i=1}^{k-1}C_{N-1}^i\left\{i[u(t)]^{i-1}[1-u(t)]^{N-1-i}\mathrm{d}u(t)+[u(t)]^i\mathrm{d}[1-u(t)]^{N-1-i}\right\}. \tag{8}$$

Then, we similarly have the following

Lemma 3. *For any $k\geq 1$ and almost all $x\in\Omega$, we have*

$$\lim_{N\to\infty}N\int_0^\infty \mathrm{Vol}(B(x,t))\mathrm{d}F_{x,(k)}^N(t)=\frac{k}{f(x)}. \tag{9}$$

Proof. Suppose $i \geq 1$, and $\alpha > 0$ be sufficiently small, then

$$\int_0^\alpha \text{Vol}(B(x,t))[u(t)]^i d[1 - u(t)]^{N-1-i}$$

$$= \frac{1}{f(x)} \int_0^\alpha f(x) \text{Vol}(B(x,t))[u(t)]^i d[1 - u(t)]^{N-1-i}$$

$$\leq \frac{1 + \eta M}{f(x)} \int_0^\alpha [u(t)]^{i+1} d[1 - u(t)]^{N-1-i}$$

$$= \frac{1 + \eta M}{f(x)} \int_0^{u(\alpha)} u^{i+1} d[1 - u]^{N-1-i}$$

$$= \frac{1 + \eta M}{f(x)} \left\{ P(N)[1 - u(\alpha)]^{N-1-i} + \frac{(N-i-1)!(i+1)!}{N!}[(1 - u(\alpha)^N) - 1] \right\},$$

where the $P(N)$ is polynomial function of N. Therefore,

$$\limsup_{N \to \infty} NC_{N-1}^i \int_0^\alpha \text{Vol}(B(x,t))[u(t)]^i d[1 - u(t)]^{N-1-i} = \frac{1 + \eta M}{f(x)}(i+1).$$

By a similar calculation, we have

$$\limsup_{N \to \infty} NC_{N-1}^i \int_0^\alpha \text{Vol}(B(x,t))[u(t)]^{i-1}[1 - u(t)]^{N-1-i} du(t) = \frac{1 + \eta M}{f(x)}.$$

By the formaulae (8),

$$\frac{1 - \eta/M}{f(x)} \left(1 + \sum_{i=1}^{k-1}(i+1) \right) - \frac{1 + \eta M}{f(x)} \sum_{i=1}^{k-1} i$$

$$\leq \liminf_{N \to \infty} -NC_{N-1}^i \int_0^\alpha \text{Vol}(B(x,t)) d\left[(1 - u(t))^{N-1}\right]$$

$$- \limsup_{N \to \infty} \sum_{i=1}^{k-1} C_{N-1}^i \int_0^\alpha \text{Vol}(B(x,t))[u(t)]^{i-1}[1 - u(t)]^{N-1-i} du(t)$$

$$+ \liminf_{N \to \infty} -\sum_{i=1}^{k-1} C_{N-1}^i \int_0^\alpha \text{Vol}(B(x,t))[u(t)]^i d[1 - u(t)]^{N-1-i}$$

$$\leq \liminf_{N \to \infty} N \int_0^\alpha \text{Vol}(B(x,t)) dF_{x,(k)}^N(t)$$

$$\leq \limsup_{N \to \infty} N \int_0^\alpha \text{Vol}(B(x,t)) dF_{x,(k)}^N(t)$$

$$\leq \limsup_{N \to \infty} -NC_{N-1}^i \int_0^\alpha \text{Vol}(B(x,t)) d\left[(1 - u(t))^{N-1}\right]$$

$$- \liminf_{N \to \infty} \sum_{i=1}^{k-1} C_{N-1}^i \int_0^\alpha \text{Vol}(B(x,t))[u(t)]^{i-1}[1 - u(t)]^{N-1-i} du(t)$$

$$+ \limsup_{N \to \infty} - \sum_{i=1}^{k-1} C_{N-1}^i \int_0^\alpha \mathrm{Vol}(B(x,t))[u(t)]^i \mathrm{d}[1 - u(t)]^{N-1-i}$$

$$= \frac{1 + \eta M}{f(x)} \left(1 + \sum_{i=1}^{k-1}(i+1) \right) - \frac{1 + \eta/M}{f(x)} \sum_{i=1}^{k-1} i.$$

Therefore

$$\lim_{\alpha \to 0} \lim_{N \to \infty} N \int_0^\alpha \mathrm{Vol}(B(x,t)) \mathrm{d}F_{x,(k)}^N(t) = \frac{k}{f(x)}.$$

Similarly, we have

$$\limsup_{N \to \infty} N \int_\alpha^\infty \mathrm{Vol}(B(x,t)) \mathrm{d}F_{x,(k)}^N(t) = 0,$$

which leads to

$$\lim_{N \to \infty} N \int_0^\infty \mathrm{Vol}(B(x,t)) \mathrm{d}F_{x,(k)}^N(t) = \frac{k}{f(x)}, a.s.$$

The proof is complete.

3   Main Results

The following two theorems are about the expectations of T and $T_{(k)}$.

Theorem 1. *For almost all $x \in \Omega$, we have*

$$\lim_{T \to \infty} E_T \left[(N\mathrm{Vol}(B(X,T)))^m \,|\, X = x \right] = \frac{m!}{[f(x)]^m}, \tag{10}$$

Proof. Equation (10) can be immediately obtained by Lemma 2.

Theorem 2. *For any fixed $k \geq 2$, and almost all $x \in \Omega$,*

$$\lim_{N \to \infty} E_{T_{(k)}} \left(N\mathrm{Vol}(B(x,T_{(k)})) \right) = \frac{k}{f(x)}, \tag{11}$$

$$\lim_{N \to \infty} E_{T_{(k)}} \left(N\mathrm{Vol}(B(x,T_{(k)}))^2 \right) = \frac{k^2 + k}{f(x)^2}. \tag{12}$$

Proof. The first equation follows from Lemma 3. The second equation holds because

$$\lim_{N \to \infty} E_{T_{(k)}} \left[N\mathrm{Vol}(B(x,T_{(k)})) \right]^2 = \frac{2 + \sum_{i=1}^{k-1}[(i+2)(i+1) - i(i+1)]}{[f(x)]^2} = \frac{k^2 + k}{[f(x)]^2}.$$

The last equation can be directly derived from the first two equations.

Let

$$\hat{g}_N(x) = \frac{N\mathrm{Vol}(B(x, T_{(k)}))}{k}, \tag{13}$$

then according to Theorem 3,

$$\lim_{N \to \infty} \mathrm{E}\hat{g}_N(x) = \frac{1}{[f(x)]}. \tag{14}$$

$$\lim_{N \to \infty} \mathrm{E}[\hat{g}_N(x)]^2 = \frac{k^2 + k}{k^2[f(x)]^2}. \tag{15}$$

Let $\{k_N\}$ be a subsequence of $\{N\}$, satisfying that

$$k_N \longrightarrow \infty, \frac{k_N}{N} \longrightarrow 0, \quad N \longrightarrow \infty, \tag{16}$$

then by a similar calculation, we have

$$\mathrm{E}\left[\hat{g}_N(x) - \frac{1}{f(x)}\right]^2 \to 0, N \to \infty.$$

That is, $\hat{g}_N(x) \to \frac{1}{f(x)}$ in L^2 sense, as $N \to \infty$, for almost all $x \in \Omega$. If we set

$$\hat{f}_N(x) = \frac{1}{\hat{g}_N(x)} = \frac{k_N}{N\mathrm{Vol}(B(x, T_{k_N}))}, \tag{17}$$

then $\hat{f}_N(x)$ is the famous $k-$NN estimator of $f(x)$. So we have obtained

Theorem 3. *If $\{k_N\}$ satisfies the condition (16), then for almost all x in Ω,*

$$\hat{g}_N(x) \xrightarrow{L^2} \frac{1}{f(x)}, \quad N \longrightarrow \infty; \tag{18}$$

or

$$\hat{f}_N(x) \xrightarrow{L^2} f(x), \quad N \longrightarrow \infty. \tag{19}$$

The following lemma is well known.

Lemma 4. *(Theorem 3.4.6 in [8]). Let $p \geq 1$ and $\{h, h_n, n \geq 1\} \subset L^p$, then the following two conditions are equivalent:*

(1) $(\mathrm{E}\|h_n - h\|^p)^{1/p} \to 0, n \to \infty$;
(2) $h_n \to h$ in probability sense, and $(\mathrm{E}|h_n|^p)^{1/p} \to (\mathrm{E}|h|^p)^{1/p}, n \to \infty$.

Using this lemma, we will prove the following equivalence convergence conclusion.

Theorem 4. *For almost all $x \in \Omega$, the following conclusions are equivalent:*

(1) $k_N \to \infty$, *as* $N \to \infty$.
(2) $\hat{g}_N(x)$ *converges to* $\frac{1}{f(x)}$ *in* L^2 *sense, as* $N \to \infty$.
(3) $\hat{g}_N(x)$ *converges to* $\frac{1}{f(x)}$ *in probability sense, as* $N \to \infty$.

Proof. (1) $\implies$ (2) $\implies$ (3) is clear. The left is to prove that (3) $\implies$ (1). If k_N dose not tend to infinity, then there exists a K and a subsequence of $\{k_N\}$, denote by $\{k_{N'}\}$, such that $k_{N'} = K$ for any N'. So according to formula (13) and (15),

$$\lim_{N' \to \infty} \mathrm{E}[\hat{g}_{N'}(x)]^2 = \frac{K^2 + K}{K^2[f(x)]^2} \neq \frac{1}{[f(x)]^2} = \mathrm{E}\left[\frac{1}{f(x)}\right]^2.$$

By Lemma 4, we obtain that $\hat{g}_N(x)$ does not converge to $\frac{1}{f(x)}$ in L^2 sense. This contradicts the hypothesis. Therefore, formulae (3) $\implies$ (1) holds.

Follow the method given above, we can analogously prove that

Theorem 5. *For almost all $x \in \Omega$, the following conclusions are equivalent:*

(1) $k_N \to \infty$, *as* $N \to \infty$.
(2) $\hat{f}_N(x)$ *converges to* $f(x)$ *in* L^2 *sense, as* $N \to \infty$.
(3) $\hat{f}_N(x)$ *converges to* $f(x)$ *in probability sense, as* $N \to \infty$.

In general, L^2 convergence is stronger than probability convergence. However, they are equivalence for k-NN density estimator.

4   Conclusions

In this paper we have established some properties for the k-NN density estimator. More importantly, these properties have been utilised to reveal the equivalence of k-NN density estimate convergence in probability and L^2 sense.

Acknowledgements. The authors acknowledge the financial support by the National NSF of China under Grant No. 61472343, as well as the Open Fund of the State Key Laboratory of Software Development Environment under Grant No. SKLSDE-2015KF-05, Beihang University.

References

1. Devroye, L.: The equivalence of weak, strong and complete convergence in l_1 for kernel density estimates. Ann. Stat. **11**(3), 896–904 (1983)
2. Devroye, L.: A note on the l_1 consistency of variable kernel estimates. Ann. Stat. **13**(3), 1041–1049 (1985)
3. Devroye, L., Penrod, C.S.: The consistency of automatic kernel density estimates. Ann. Stat. **12**(4), 1231–1249 (1984)

4. Don, O., Loftsgaarden, C.P.Q.: A nonparametric estimate of a multivariate density function. Ann. Math. Stat. **36**(3), 1049–1051 (1965)
5. Fix, E., Hodges, J.L.: Discriminatory analysis, nonparametric discrimination, consistency properties. Int. Stat. Rev. **57**(3), 238–247 (1989)
6. Scott, D.W.: Multivariate Density Estimation: Theory, Practice, and Visualization, 2nd edn. Wiley, Toronto (2015)
7. Wied, D., Weißbach, R.: Consistency of the kernel density estimator: a survey. Stat. Pap. **53**(1), 1–21 (2012)
8. Yan, J.: Lectures on Measure Theory, 2nd edn. Science Press, Beijing (2004). (in Chinese)

Visualizing MOOC User Behaviors: A Case Study on XuetangX

Tiantian Zhang and Bo Yuan[✉]

Intelligent Computing Lab, Division of Informatics, Graduate School at Shenzhen,
Tsinghua University, Shenzhen 518055, People's Republic of China
2573546543@qq.com, yuanb@sz.tsinghua.edu.cn

Abstract. The target of KDD CUP 2015 is to use the MOOC (Massive Open Online Course) user dataset provided by XuetangX to predict whether a user will drop a course. However, despite of the encouraging performance achieved, the dataset itself is largely not well investigated. To gain an in-depth understanding of MOOC user behaviors, we conduct two case studies on the dataset containing the information of 79,186 users and 39 courses. In the first case study, we use visualization techniques to show that some courses are more likely to be simultaneously enrolled than others. Furthermore, a set of association rules among courses are discovered using the Apriori algorithm, confirming the practicability of using historical enrollment data to recommend courses for users. Meanwhile, clustering analysis reveals the existence of clear grouping patterns. In the second case study, we examine the influence of two user factors on the dropout rate using visualization, providing valuable guidance for maintaining student learning activities.

Keywords: User behavior · Visualization · Association rule · Clustering · MOOC

1 Introduction

The MOOC (Massive Open Online Course) refers to a new type of online courses aimed at unlimited participation and open access via the web [1]. It is the result of a recent development in distance education [2], which was first introduced in 2008 and emerged as a popular learning mode in 2012 [3]. Most importantly, MOOCs build on the active engagement of millions of students who self-organize their participation according to their individual learning goals, prior knowledge and skills as well as common interests [4]. MOOCs provide not only traditional course materials such as lecture slides, question sets and reading materials but also course videos, online self-testing, and interactive forums to support community interactions among students, instructors, and teaching assistants [1]. Meanwhile, as a brand new education mode based on the concept of "Internet + Education", MOOCs break the limitation of space and time and users can conduct individual learning anytime and anywhere.

User behavior analysis is an important factor for designing and implementing an effective MOOC, which can make learning a convenient, rewarding and personalized experience. Meanwhile, the ability of MOOCs to generate a tremendous amount of data opens up unprecedented opportunities for educational research [5]. Researchers can get

© Springer International Publishing AG 2016
H. Yin et al. (Eds.): IDEAL 2016, LNCS 9937, pp. 89–98, 2016.
DOI: 10.1007/978-3-319-46257-8_10

better acquainted with users' learning behavior and learning outcomes by taking advantage of data analytics methods and propose insightful suggestions on curriculum improvement. Educators can also provide personalized guidance and recommend appropriate materials to learners to improve the quality of learning.

As a new education mode, existing research on MOOCs is relatively limited. In 2012, an article in *Science* made an introduction of MOOCs and predicted that they will change the future of education [6]. Afterwards, an article in *Nature* discussed the development and trend of MOOCs in 2013 [7]. In the last two years, MOOCs have received more and more attentions from the research community. Some examples are: evaluating the geographic data in MOOCs [8]; analyzing student submissions to help instructors understand the problem solving process [9]; using the records of learning activity to develop a conceptual framework for understanding how users engage with MOOCs [10]; investigating and improving the peer assessment mechanism [11, 12]; using timely interventions to improve online learning [13]. Researchers have also proposed a framework to classify posts in discussion forums [14] and established a large scale collaborative data analytics platform named MoocViz to analyze the data from different courses and MOOC platforms [15]. There are also studies on user behavior to speculate which MOOC platforms are easy to use [16] as well as on dropout prediction and user retention [17, 18].

In the above, there are very few studies on courses themselves. However, it is not surprising that students often do not know which courses are the most beneficial ones for them and which other courses they also need to enroll to make the most of their study. Our work focuses on the data of user enrollment on MOOCs to investigate the relationships among courses and make proper course recommendation for students. Another critical issue is the high dropout rate. For all MOOC platforms, there is a phenomenon that a large number of users may enroll in a course but most of them will drop the course somewhere before the end. In this paper, we also study the connection between learning behavior and course completion rate, which will assist in designing and implementing effective MOOCs to maintain and encourage learning activities.

Section 2 provides a detailed description of the data source. Section 3 contains the case study on course analysis and presents the association rules among courses and shows the clustering pattern of courses. Section 4 presents another case study on factors affecting the dropout rate. This paper is concluded in Sect. 5 with some discussions and directions for future work.

2   Preliminaries

The high dropout rate has been a major issue in MOOC platforms and some reports show that the certification rates may be less than 10 %. Due to the inherent characters of online learning, users take far less obligation than in the normal classroom. Also, there are many distracting factors that may prevent users from consistent learning. After all, users may enroll in a course with totally different intentions and motivations (e.g., browsing vs. earning certification). In order to maintain and encourage students' learning activities, it is important to predict their likelihood of dropout so that retention measures can be taken as necessary. In KDD Cup 2015, an anonymous dataset was provided by XuetangX, the largest MOOC platform in China. The target of the competition is to predict whether a user will drop a course based on his or her prior activities within a time period of 30 days. If a user leaves no records for a specific course in the log during the next 10 days, the case is claimed as a dropout. Despite of the promising performance achieved by competition participants, there is still a lack of clear understanding of the dataset itself. In this paper, we conduct in-depth analysis of the dataset using visualization and other data analytics techniques to identify relationships among courses, which can help students make right decisions on enrollment. We also examine some factors associated with the dropout rate, providing valuable information to course instructors and platform operators.

The dataset contains the information of 79,186 users and 39 courses, with 120,542 enrollment records in total. Each record has a binary label indicating its dropout status with "1" indicating that the user will drop the course. The dataset is unbalanced with 79.29 % enrollment records labeled as "1" and 20.71 % labeled as "0". It also provides 8,157,277 user behavior logs within 30 days after the course starts, which contain the detailed user activities such as solving problems, watching videos or engaging in discussion. Figure 1 (left) shows the statistics of all courses where many courses have high enrollment numbers but all courses also feature high dropout rates ranging from 66.09 % to 92.93 %. Figure 1 (right) visualizes the starting time of each course where all courses roughly start from one of two time points, possibly corresponding to the two semesters in practice.

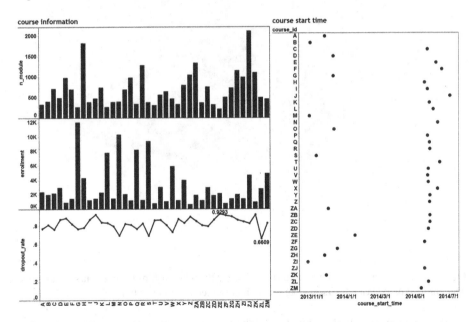

Fig. 1. The statistics of all courses: the number of modules, enrollment numbers and the dropout rate of each course (left) and the starting time of each course (right).

3   Course Analysis

From the enrollment records, we extracted two attributes username and course ID to explore the relationships among courses. There were many users enrolling in only a single course and around 27.81 % users enrolled in at least two courses. So we finally selected the enrollment records of these 22,021 users as the data source.

The relationship between courses and users was visualized using the network diagram. For the sake of clarity and readability, we further selected the records of users who enrolled in at least 6 courses and visualized them using *Gephi*. In Fig. 2, nodes with label represent courses while nodes without label represent users and the enrollment status is indicated by arcs. The size of each course node is measured by its indegree (enrollment numbers). It is clear that: (1) courses had different levels of enrollment (popularity); (2) course nodes were roughly grouped into some clusters, indicating that certain courses were often enrolled simultaneously by the same user. Based on this preliminary observation, we will conduct further investigation using association rules learning and clustering analysis to reveal more insightful information.

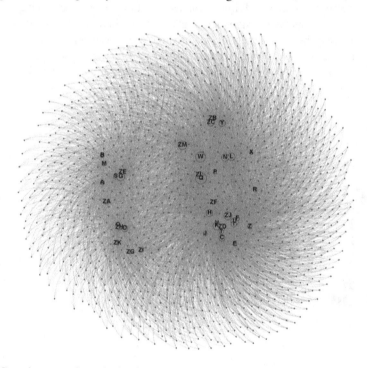

Fig. 2. Complex network graph visualization showing the relationship between courses and users who enrolled in at least 6 courses. Courses are represented by nodes with label.

We used the Apriori algorithm to find the frequent course sets and association rules among courses. The minimum support value and confidence value were set to 1.5 % and 20 %, respectively. Table 1 shows association rules with confidence greater than 30 %.

The maximum confidence value was up to 43.6 %, which means that if a user enrolls in course ZA, there is a possibility of 43.6 % that he/she will also enroll in course G. This finding is likely to be valuable in practice and these association rules can be employed by MOOC platforms to provide course recommendation services to students based on their enrollment records. By doing so, MOOC platforms are expected to become more personalized and user-friendly. Figure 3 shows all association rules among courses and the arrow from X to Y represents a rule X → Y. Furthermore, the thickness of the arrow represents the confidence value of the rule.

Table 1. Association rules among courses(confidence value > 30 %)

Rule	Confidence	Rule	Confidence
('ZC') → ('Y')	0.301	('S') → ('G')	0.370
('L') →('N')	0.308	('O') → ('G')	0.375
('O') → ('D')	0.308	('A') → ('G')	0.382
('ZH') → ('G')	0.309	('D') → ('G')	0.402
('ZC') → ('W')	0.310	('ZE') → ('G')	0.406
('B') → ('S')	0.340	('C') → ('ZJ')	0.431
('ZD') → ('U')	0.357	('ZA') → ('G')	0.436

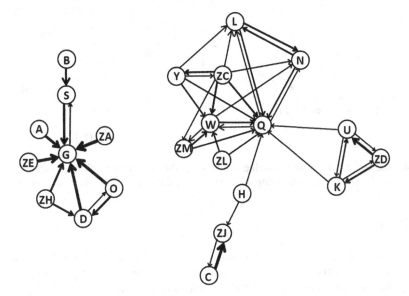

Fig. 3. Association rules among courses. The arrow from X to Y represents the rule X→Y. The thickness of the arrow represents the confidence of the corresponding rule.

We also conducted course clustering using the k-means and hierarchical clustering. The Jaccard distance (Eq. 1) was used to measure the dissimilarity between courses M and N, depending on their enrollment sets.

$$d_J(M, N) = 1 - \frac{|M \cap N|}{|M \cup N|} = 1 - \frac{|M \cap N|}{|M| + |N| - |M \cap N|} \tag{1}$$

Figure 4 shows the graphical results of course clustering using different techniques. For the k-means clustering, we used multi-dimensional scaling to transform the distance matrix of courses into a coordinate matrix. As shown in Fig. 4(a), courses were grouped into three clusters, marked by different colors. Figure 4(b) shows the dendrogram of the hierarchical clustering with the average link criterion. Although the number of clusters can be varied as necessary, it indicates the existence of three major clusters. These two clustering results are largely consistent, apart from a few discrepancies. Furthermore, if the real course names/contents are available, we can expect to reveal more underlying patterns by matching the clustering results with the properties of courses and make more principled course recommendations.

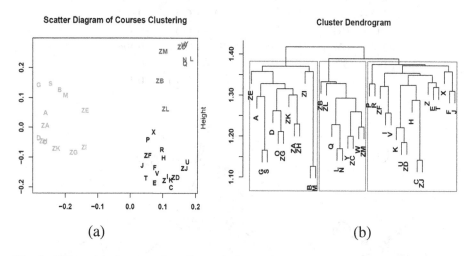

(a)   (b)

Fig. 4. The results of course clustering analysis: (a) the clustering result by k-means; (b) the dendrogram by hierarchical clustering with the average link criterion. (Color figure online)

4   Dropout Analysis

In this section, we present some analysis on the dropout pattern. We extracted the first and the last time records that users logged into the courses and their dropout label as well as the numbers of courses that they enrolled and dropped from user logs.

Based on the date of last login, we calculated the distribution of dropout students. Figure 5(top) shows the dropout rate curves of selected courses within 30 days after the starting of courses. Each curve represents a course and the dropout rate is defined as the number of dropout students who stopped learning the corresponding course at a specific date, divided by the total number of students enrolled in that course. Overall, the dropout rates of most courses were at peaks within 3 days after the course started. After that, the dropout rates declined sharply and finally reached a relatively steady state, which is consistent with the tendency of all courses, as shown in Fig. 5(bottom) where the dropout rate is defined as the number of enrollment cancellations with a specific last login date, divided by the total enrollment numbers. This shows that most users who finally dropped the course actually stopped learning after only a few days. However, there were some courses showing different patterns. For example, the black curve increased abruptly at the 16th day and the red curve maintained a low dropout rate within the first half period but increased to its peak at the 22nd day.

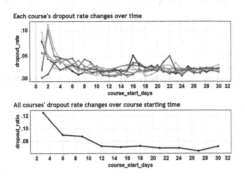

Fig. 5. Distribution of enrollment cancellations based on the last date of login

Similarly, based on the date of first login, the relationship between dropout rate and the starting time of learning is shown in Fig. 6. Within the first week, the dropout rate increased monotonically, indicating that if a user started leaning the course later, it will be more likely that he or she would drop the course. Note that the dropout rate here is defined as the number of enrollment cancellations with a specific first login date, divided by the total enrollment numbers with the same first login date. The relatively low dropout rate at the beginning may be contributed to the common sense that well-organized and self-motivated users were often get themselves ready for studying long before the course started. During the middle period (between the 10th day and the 20th day), the dropout rate was stable and fluctuated slightly but at the end of the curve, the dropout rate dropped rather unexpectedly. A possible explanation is that some late starting users still needed extra times to figure out whether the specific course was suitable for them and continued to engage in learning activities within the next 10 days after the 30th day.

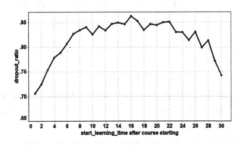

Fig. 6. Relationship between dropout rate and the first date of login

Another factor related to dropout is the number of courses enrolled. In Fig. 7, the horizontal axis represents the total number of courses enrolled and the vertical axis represents the average number of courses dropped. The simple linear regression was used to identify possible relationships. The red line is based on students who only enrolled in courses in the same semester while the black line is based on students who enrolled in courses in both two semesters. Firstly, it is evident that there was a linear relationship between the number of courses dropped and the number of courses enrolled. In other words, the probability of a user dropping a course was not directly influenced by the total number of courses enrolled.

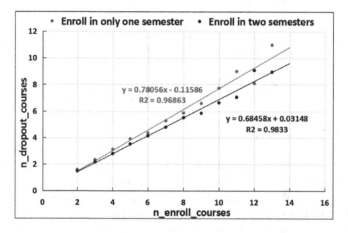

Fig. 7. The results of linear regression on the average number of courses dropped and the number of courses enrolled. It shows the correlation coefficients and the R^2 values.

Furthermore, the correlation coefficients were 78.06 % and 68.46 % for the red line and the black line, respectively, both with large R^2 values. This shows that, given the same number of courses enrolled, students enrolling in two semesters were less likely to drop courses than those enrolling in only a single semester. Consequently, we can conclude that reduced stress in study can help improve the course completion rate.

5   Conclusion

In this work, we conducted a series of MOOC user behavior analysis based on the dataset from KDD CUP 2015. In the course analysis, we found that some courses were likely to be enrolled simultaneously. Based on this finding, we discovered a set of association rules among courses so that it is possible to use the enrollment data to recommend courses for users. Furthermore, we also used the k-means and hierarchical clustering techniques to group all courses into three clusters. In the dropout analysis, we identified some factors related to the dropout rate such as the first and last login dates as well as the distribution of courses over semesters. In general, early starting users and users distributing courses in different semesters are both less likely to drop the course compared to other students.

As more and more universities are contributing high-quality courses to MOOC platforms and even offer online degree programs, people from all over the world will benefit greatly from the popularity of MOOCs. As a new education mode, MOOCs provide unprecedented opportunities for educators and researchers but also face many challenges such as the high dropout rate and low completion rate. In the future, we will conduct further study on personalized recommendation and user behavior prediction to help build more effective MOOC platforms.

Acknowledgement. This work was partially supported by the research foundation (QTone Education) of the Research Center for Online Education, Ministry of Education, P.R. China.

References

1. Massive Open Online Course. https://en.wikipedia.org/wiki/Massive_open_online_course
2. Bozkurt, A., Akgun-Ozbek, E., Yilmazer, S., Erdogdu, E., Ucar, H., Guler, E., Sezgin, S., Karadeniz, A., Sen-Ersoy, N., Goksel-Canbek, N., Dincer, G.D., Ari, S., Aydin, C.H.: Trends in distance education research: a content analysis of journals 2009-2013. Int. Rev. Res. Open Distrib. Learn. **16**(1), 330–363 (2015)
3. Pappano, L.: The year of the MOOC. N. Y. Times **2**(12), 2012 (2012)
4. McAuley, A., Stewart, B., Siemens, G., Cormier, D.: The MOOC model for digital practice (2010). http://www.elearnspace.org/Articles/MOOC_Final.pdf
5. Breslow, L., Pritchard, D.E., DeBoer, J., Stump, G.S., Ho, A.D., Seaton, D.T.: Studying learning in the worldwide classroom: research into edX's first MOOC. Res. Pract. Assess. **8**, 13–25 (2013)
6. Stein, L.A.: Casting a wider net. Science **338**(6113), 1422–1423 (2012)
7. Waldrop, M.M.: Campus 2.0. Nature **495**(7440), 160–163 (2013)
8. Nesterko, S.O., Dotsenko, S., Hu, Q., Seaton, D., Reich, J., Chuang, I., Ho, A.: Evaluating geographic data in MOOCs. In: NIPS Workshop on Data Driven Education (2013)
9. Han, F., Veeramachaneni, K., O'Reilly, U.M.: Analyzing millions of submissions to help MOOC instructors understand problem solving. In: NIPS Workshop on Data Driven Education (2013)
10. Anderson, A., Huttnocher, D., Kleinberg, J., Leskovec, J.: Engaging with massive online courses. In: Proceedings of the 23rd International Conference on World Wide Web, pp. 687–698. ACM (2014)
11. DíezPeláez, J., Rodríguez, Ó.L., Betanzos, A.A., Troncoso, A., Rionda, A.B.: Peer assessment in MOOCs using preference learning via matrix factorization. In: NIPS Workshop on Data Driven Education (2013)
12. Shah, N.B., Bradley, J.K., Parekh, A., Wainwright, M., Ramchandran, K.: A case for ordinal peer-evaluation in MOOCs. In: NIPS Workshop on Data Driven Education (2013)
13. Williams, J.J., Williams, B.: Using interventions to improve online learning. In: NIPS Workshop on Data Driven Education (2013)

14. Stump, G.S., DeBoer, J., Whittinghill, J., Breslow, L.: Development of a framework to classify MOOC discussion forum posts: methodology and challenges. In: NIPS Workshop on Data Driven Education (2013)
15. Dernoncourt, F., Halawa, S., O'Reilly, U.: MoocViz: a large scale, open access, collaborative, data analytics platform for MOOCs. In: NIPS Workshop on Data Driven Education (2013)
16. Pireva, K., Imran, A.S., Dalipi, F.: User behavior analysis on LMS and MOOC. In: IEEE Conference on e-learning, e-Management and e-Services, pp. 21–26 (2015)
17. Yang, D., Sinha, T., Adamson, D., Rose, C.P.: Turn on, Tune in, Drop out: anticipating student dropouts in massive open online courses. In: NIPS Workshop on Data Driven Education (2013)
18. Balakrishnan, G.K.: Predicting student retention in massive open online courses using Hidden Markov Models. Technical report No. UCB/EECS-2013-109, University of California, Berkeley (2013)

Modified-DBSCAN Clustering for Identifying Traffic Accident Prone Locations

Chenlu Qiu[✉], Huiying Xu, and Yongqiang Bao

Traffic Management Research Institute of the Ministry of Public Security,
Wuxi 214151, Jiangsu, China
echotosusan@gmail.edu

Abstract. Road traffic accidents, especially expressway traffic accidents, have become a severe problem in China. Under this condition, identification of road traffic accident prone locations is in urgent need. This work proposes a modification of DBSCAN clustering algorithm with parameters ε and minPts carefully chosen for identifying traffic accident prone locations. Experimental results on traffic accident datasets of three national expressways are given, demonstrating the effectiveness of the proposed algorithm.

Keywords: Traffic accident prone locations · DBSCAN clustering · Kurtosis

1 Introduction

With the rapid economic growth in China, China has been experiencing a high speed urbanization and motorization. Urbanization and motorization result in increasing numbers of motor vehicles and inexperienced vehicle drivers, which lead to increasing number of road traffic accidents and deaths. According to the World Health Organization report, in year 2015, the total number of fatalities in China is among the highest in the world. National official statistics suggest that most fatal car crashes occurs in expressways. Under these conditions, road traffic accidents, especially expressway traffic accidents, have become a severe problem in China. Traffic accidents, deaths and injuries are predictable and preventable, but the identification of accident prone locations is a prerequisite to analyze causes and take effective countermeasures. Accident prone location is a loosely defined term referring to black spots and/or areas susceptible to having a greater than average number of accidents. A lot of research has been done to identify road traffic accident prone locations, which are mainly statistical approaches such as [1–4]. Other approaches includes [5–7], etc.

Accident frequency method [1] is a conventional statistical approach which utilizes the number of accidents as an indicative index. In specified time and

Y. Bao—This research was supported by Central Public-interest Scientific Institution Basal Research Fund 2016SJA07 and National Key Technologies R&D Program 2014BAG01B04.

H. Yin et al. (Eds.): IDEAL 2016, LNCS 9937, pp. 99–105, 2016.
DOI: 10.1007/978-3-319-46257-8_11

scope, if the accident number is greater than a threshold, it is identified as accident prone location. Similarly, accident rate method [2] uses accident rate (the number of accidents per unit road length, per million vehicle kilometers or per unit traffic flow) as an indicative index. Multidisciplinary method [3] computes the distribution of accident frequency and rate, taking the number of accidents as horizontal coordinate axis and the number of accident rate as longitudinal coordinate axis. The whole coordinates can be divided into four areas: high accident rate and high accident frequency area, high accident rate but low accident frequency area, low accident rate but high accident frequency area, accident rate and low accident frequency area. Road sections fall into the first area can be then classified as accident prone locations, and road sections fall into the last area can be classified as comparatively safe areas. Sections fall into the other two areas are subject to further analysis. Accident cumulative frequency curve method [4] calculates accident cumulative frequency curve, taking the number of accidents per unit length as horizontal coordinate axis and the cumulative frequency of occurrence greater than that accident frequency as longitudinal coordinate axis. Usually, small amount of accidents scatter over large proportion road segmentations while large amount of accidents concentrate in small proportion of road segmentation. The higher the accident frequency is, the lower the cumulative frequency of occurrence is. Therefore, accident prone sections are then be distinguished if the cumulative frequency is small and the accident frequency is high. In all above methods [1–4], road unit length is an essential parameter to divide road into sections for calculating accident frequency and/or rate, resulting in the leakage of accident prone locations. To eliminate this drawback, a step size which is much smaller than the unit length should be introduced to search nearby areas exhaustively. However, it cannot fundamentally address this problem and it would dramatically increase the computing burden especially when the step size is small.

Statistical approaches suffer from the drawback of leakage as it divides road into road sections of unit length. Approaches based on clustering [8] do not have this issue as no road segmentation is needed. Density-Based Spatial Clustering of Applications with Noise (DBSCAN) [9] is one of the most popular density based clustering algorithm. It needs two parameters, radius that delimitates the neighbourhood area denoted by ε and minimum number of points required to form a dense region denoted by minPts. Starting with an arbitrary data point, DBSCAN counts the number of points within radius ε. If it has more than minPts points in its ε neighbourhood, it is a core point. If it has no more than minPts points in its ε neighbourhood but is in the neighborhood of a core point, it is a border point. Points which is neither a core point nor a border point are considered as noise points. DBSCAN continues with checking all other points till all points are classified. A cluster is formed as a core point together with all points that are density reachable from it w.r.t. ε and minPts. If two core points are within a radius of ε of each other, then two clusters are joined. The two parameters ε and minPts must be specified by the user. However, there is no clear criterion to determine meaningful or optimal ε and minPts in many applications.

This paper proposes a modification of DBSCAN clustering algorithm with parameters ε and minPts carefully chosen for identification of accident prone locations. The way of choosing ε and minPts is different from [8] where ε is set by a domain expert, and the minimum threshold value of minPts is adaptively chosen according to the accident cumulative frequency curve. The rest of the paper is organized as follows. The proposed algorithm is presented in Sect. 2, where the criterion of determining "optimal" ε and minPts according to the kurtosis of accident rate curve are discussed. Experimental results are given in Sect. 3. Conclusion and future directions are given in Sect. 4.

2 Modified DBSCAN Clustering for Identifying Expressway Traffic Accident Prone Locations

Let $\mathcal{D} := \{d_1, d_2, \cdots, d_m\}$ denote a traffic accident dataset where m is the total number of accidents and d_i denotes the geographic coordinates of the ith accident. Euclidean distance is used to define the distance for a pair of points. Given user specified parameters ε and minPts, DBSCAN clustering algorithm for identifying traffic accident prone locations proceed as follows.
$\mathcal{C} \leftarrow$ **DBSCAN**$(\mathcal{D}, \varepsilon, \text{minPts})$:

1. **Find the density of each point:** estimate the density around each point by counting the number of points within its ε neighborhood.
2. **Identify core points by thresholding:** find all points that have a density greater than minPts.
3. **Form clusters from the core points:** form clusters by joining core points if they are density reachable w.r.t. ε and minPts.
4. **Assign border points to clusters:** assign border points to all reachable clusters.
5. **Output the clusters:** output all clusters $\mathcal{C}$,

$$\mathcal{C} = \{C_1, C_2, \cdots, C_n\}$$

where C_i is the ith clustered points' set and n is the total number of clusters.

In above algorithm, border points are assigned to all reachable clusters and therefore the clustering result is deterministic. We make this necessary modification for identifying accident prone road sections. An alternative way is to skip this step treating all border points as noise points. This is different from the original DBSCAN algorithm [9] in which border points are arbitrarily assigned to clusters.

Define $L_i := \max_{p,q} \text{dist}(p, q), \forall p, q \in C_i$ where $\text{dist}(p, q)$ is the Euclidean distance between p and q. In practice, clustered accident prone road sections of arbitrary large length is undesirable, i.e., we need $\max_i(L_i) \le L_{\max}$. Note that high density cluster should be completely enclosed within low density clusters

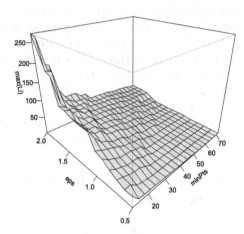

Fig. 1. $\max_i(L_i)$ versus ε and minPts

for DBSCAN clustering. Therefore, $\max_i(L_i)$ is proportional to ε given minPts, and $\max_i(L_i)$ is inverse proportional to minPts given ε. This is also verified by numerical experiment as shown in Fig. 1, plotting $\max_i(L_i)$ versus ε and minPts for an accident dataset collected in one national expressway.

We now discuss the criterion of determining ε and minPts. Let

$$\mathcal{S}_\varepsilon := \{\varepsilon_0, \varepsilon_0 + \Delta_\varepsilon, \varepsilon_0 + 2\Delta_\varepsilon, \cdots\} \tag{1}$$

$$\text{minPts}_0 := \frac{m}{\text{length of entire road}} \times \varepsilon_0 \tag{2}$$

$$\mathcal{S}_{\text{minPts}} := \{\text{minPts}_0, 2\text{minPts}_0, \cdots, 10\text{minPts}_0\} \tag{3}$$

where $\mathcal{S}_\varepsilon$ is an increasing order sequence of ε starting from ε_0 with step size Δ_ε; $\frac{m}{\text{length of entire road}}$ is the average accident rate; and $\mathcal{S}_{\text{minPts}}$ is an increasing order sequence of minPts whose elements is proportional to minPts_0. Let $\mathcal{S}_{\varepsilon,\text{minPts}} = \{(\varepsilon, \text{minPts})\}$ be an empty set. The proposed algorithm proceeds as follows.
$\mathcal{C}^* \leftarrow$ **Modified-DBSCAN**$(\mathcal{D}, \mathcal{S}_\varepsilon, \mathcal{S}_{\text{minPts}})$:

1. **Find all combinations of ε and minPts satisfying $\max_i(L_i) \leq L_{\max}$:**
   for all $\varepsilon \in \mathcal{S}_\varepsilon$ and minPts $\in \mathcal{S}_{\text{minPts}}$, do
   (i) $\mathcal{C} \leftarrow$ DBSCAN$(\mathcal{D}, \varepsilon, \text{minPts})$,
   (ii) If $\max_i L_i \leq L_{\max}$, $\mathcal{S}_{\varepsilon,\text{minPts}} \leftarrow \mathcal{S}_{\varepsilon,\text{minPts}} \cup \{(\varepsilon, \text{minPts})\}$.
2. **Compute kurtosis of accident rate distribution for all clustered road sections:**
   for all $(\varepsilon, \text{minPts}) \in \mathcal{S}_{\varepsilon,\text{minPts}}$, compute

$$N(C_i) \leftarrow \text{ number of points in cluster } C_i$$

$$\rho_i(\varepsilon, \text{minPts}) \leftarrow \frac{N(C_i)}{\max_{d \in C_i} d - \min_{d \in C_i} d}$$

$$\bar{\rho}(\varepsilon, \text{minPts}) \leftarrow \frac{1}{n} \sum_{i=1}^{n} \rho_i(\varepsilon, \text{minPts})$$

$$\text{Kurt}(\varepsilon, \text{minPts}) \leftarrow \frac{\sum_i (\rho_i - \bar{\rho})^4}{(\sum_i (\rho_i - \bar{\rho})^2)^2}$$

3. **Find the optimal parameters ε^* and minPts*:**

$$(\varepsilon^*, \text{minPts}^*) \leftarrow \arg \max_{(\varepsilon, \text{minPts}) \in \mathcal{S}_{\varepsilon, \text{minPts}}} \text{Kurt}(\varepsilon, \text{minPts})$$

4. **Output the clusters:**

$$\mathcal{C}^* \leftarrow \text{DBSCAN}(\mathcal{D}, \varepsilon^*, \text{minPts}^*)$$

In Modified-DBSCAN algorithm, step 1 finds the combinations of ε and minPts such that the constraints $\max_i(L_i) \leq L_{\max}, \forall i$ are satisfied. Step 2 computes kurtoses of the accident rate distribution curve for all valid combinations of ε and minPts. Kurtosis is a useful measure of the "peakedness" and "tailedness" of a distribution curve. In practice, clustering results giving accident rate distribution curve with steep peak and flat tail are more desirable. Therefore, the combinations of ε and minPts giving the largest kurtosis is chosen as the optimal parameters denoted by ε^* and minPts* in step 3. Step 4 outputs the DBSCAN clustering result with $\varepsilon = \varepsilon^*$ and minPts = minPts*.

We should also mention that the length of clustered accident prone road sections varies. It does not necessary equals or be proportional to the radius ε. This is different from statistical approaches in which the identified accident prone road sections are always of length proportional to unit length.

3   Experimental Results

We apply Modified-DBSCAN algorithm to identify accident prone locations for three national expressways using the accident data between March 2015 and February 2016. In all experiments, $L_{\max} = 20$ kilometers(km), $\varepsilon_0 = 0.5$ km, $\Delta_\varepsilon = 0.1$ km, and $\mathcal{S}_\varepsilon = \{0.5, 0.6, \cdots, 2\}$ km. We use R_1 to denote the ratio of road section length to the total length, and R_2 to denote the ratio of accidents occurring in each road section to the total number of accidents. The ratio between R_2 and R_1 are computed for illustration purpose, which is proportional to the accident rate of each road section.

Table 1 lists top five accident prone road sections identified by Modified-DBSCAN for the first national expressway arranging in decreasing order of accident rate. As can be seen, the first ranking road section is of length about 0.5 Km

Table 1. Accident prone road sections identified by Modified-DBSCAN

Rank	Road length	R_1	R_2	R_2/R_1
1	0.501 km	0.04 %	0.78 %	1950
2	6.006 km	0.50 %	8.69 %	1738
3	4.601 km	0.38 %	5.10 %	1342
4	1.721 km	0.14 %	0.82 %	586
5	8.001 km	0.67 %	3.58 %	534
Total	–	1.73 %	18.97 %	–

Table 2. Accident prone road sections identified by Modified-DBSCAN

Rank	Road length	R_1	R_2	R_2/R_1
1	1.197 km	0.12 %	12.42 %	10350
2	2.004 km	0.20 %	3.38 %	1690
3	3.301 km	0.33 %	5.50 %	1666
4	1.005 km	0.10 %	1.51 %	1510
5	1.051 km	0.11 %	1.58 %	1436
Total	–	0.86 %	24.39 %	–

Table 3. Accident prone road sections identified by Modified-DBSCAN

Rank	Road length	R_1	R_2	R_2/R_1
1	0.001 km	< 0.01 %	0.26 %	> 2600
2	4.001 km	0.18 %	3.21 %	1783
3	6.101 km	0.27 %	4.17 %	1544
4	1.001 km	0.05 %	0.60 %	1200
5	5.101 km	0.22 %	2.60 %	1181
Total	–	0.73 %	10.84 %	–

accounting only 0.04 % of the total road length. However, about 0.78 % of total accidents occur in this section and the ratio $R_2/R_1 = 1590$ is the highest among all clustered road sections. In sum, 18.97 % of total accidents occur in the top five road sections accounting only 0.67 % of the total road length, showing the effectiveness of proposed clustering algorithm.

Similarly, Tables 2 and 3 list top five accident prone road sections for the other two national expressways. It is worth mentioning that about 0.26 % of total accidents occur in the same place in the third national expressway, ranking first in all clustered road sections. It is an accident prone black spot for which additional attention should be taken.

4 Conclusion and Future Directions

This work proposes a modification of DBSCAN clustering algorithm for identifying road traffic accident prone locations. In the proposed algorithm, kurtosis is used as a useful measure of "peakedness" and "tailedness" for the accident rate distribution curve, and ε and minPts giving the largest kurtosis are chosen as global density parameters. The experimental results on accident datasets of three major national expressways demonstrate the effectiveness of the proposed algorithm.

A well known disadvantage of DBSCAN is that it cannot cluster data well with varies densities, since the parameters ε and minPts cannot be chosen appropriately for all clusters. To address this issue, one possible way is to apply Shared Nearest Neighbor (SNN) algorithm [10], which not only finds the dense clusters, but also can find clusters of low or medium density. Besides this, we would also explore the spacial and temporal characteristics for traffic accident datasets. In doing this, some distance function computing the temporal similarity between pairs of points need to be developed.

References

1. Oppe, S.: Development of traffic and traffic safety: global trends and incidental fluctuations. Accid. Anal. Prev. **23**(5), 413–422 (1991)
2. Wright, C.C., Abbess, C.R., Jarrett, D.F.: Estimating the regression-to-mean effect associated with road accident black spot treatment: towards a more realistic approach. Accid. Anal. Prev. **20**(3), 199–214 (1988)
3. Larsen, L.: Methods of multidisciplinary in-depth analyses of road traffic accidents. J. Hazard. Materials **111**(1–3), 115–122 (2004)
4. Fang, S., Guo, Z., Yang, Z.: A new identification method for accident prone location on highway. J. Traffic Transp. Eng. **1**, 91–98 (2001)
5. Yannis, G., Gollas, J., Kanellaidis, G.: A comparative analysis of the potential of international road accident data piles. IATSS Res. **22**, 111–120 (2000)
6. Saccomanno, F.F., Grossi, R., Greco, D., Mehmood, A.: A mehmood identifying black spots along highway SS107 in Southern italy using two models. J. Transp. Eng. **127**(6), 515–522 (2001)
7. Qin, X., Ng, M., Reyes, P.E.: Identifying crash-prone locations with quantile regression. Accid. Anal. Prev. **42**(6), 1531–1537 (2010)
8. Xie, L., Chaozhong, W., Lyn, N., Gao, Y.: Accident-prone section identification approach based on improved clustering algorithm. J. Wuhan Univ. Technol. (Transp. Sci. Eng.) (4), 904–908 (2014)
9. Ester, M., Kriegel, H.-P., Sander, J., Xiaowei, X.: A density-based algorithm for discovering clusters in large spatial databases with noise. In: Proceedings of the Second International Conference on Knowledge Discovery and Data Mining, pp. 226–231 (1996)
10. Ertoz, L., Steinbach, M., Kumar, V.: Finding clusters of different sizes, shapes, and densities in noisy, high dimensional data. In: Proceedings of Second SIAM International Conference on Data Mining, pp. 47–58 (2003)

A Novel Link Prediction Algorithm
Based on Spatial Mapping in PPI Network

Qiang-Mei Wu[1], Wei Liu[1,2(✉)], Hai-yan Hong[1], and Ling Chen[1,3]

[1] Institute of Information Engineering, Yangzhou University, Yangzhou, China
yzliuwei@126.com
[2] Jiangsu Co-Innovation Center for Prevention and Control of Important Animal
Infectious Diseases and Zoonoses, Yangzhou, China
[3] National Key Lab of Novel Software Tech, Nanjing University,
Nanjing, China

Abstract. Most life activities are finished by protein-protein interactions(PPI). So far, there have been many methods proposed for link prediction in the PPI network. However, these prediction methods only use single information. This paper proposes a novel algorithm to predict the potential interactions based on the topological and attribute similarity between proteins. This paper also studies the way of balancing attribute similarity and topological similarity. The experimental results on yeast PPI network show that the proposed algorithm has higher accuracy and good biometric characteristic.

Keywords: Protein-protein interactions · Link predict · Spatial mapping

1 Introduction

Protein-Protein Interaction(PPI) is the correlation between protein molecules. The problem of PPI prediction [1] is a problem of link prediction in complex network essentially. At present, there are many effective link prediction methods, but most of them only consider part of the information when constructing weighted network. However, it is far from enough to predict protein interaction information when just one kind of information is taken into account. And some other methods just simply research in the unweighted network: for example, Saito et al. [2, 3] firstly put forward using the topological relations between a pair of interacting proteins and neighboring nodes to assess the reliability of protein interactions. Chen et al. [4] also proposed a method IRAP based on network topology. Goldberg et al. ascertained a threshold according to the characteristics of actual network, and thought that the protein interaction is reliable above this threshold. In addition, there are CD-Dist [5] and Fsweight [6] etc. However, the prediction method of PPI network without making full use of biological attribute similarity is not complete and it is difficult to get a high precision.

Based on the above shortcomings, this paper proposes a new method to predict the potential interactions based on the two information. We respectively map the topology information and attribute information to different spatial with different linear transformation. In order to make the two information organically integration, the mapping spaces are as close as possible. The method mentioned above has changed the way that

© Springer International Publishing AG 2016
H. Yin et al. (Eds.): IDEAL 2016, LNCS 9937, pp. 106–113, 2016.
DOI: 10.1007/978-3-319-46257-8_12

only depends on a certain type of information as the dominant approach. The experimental results on yeast data have proved that the fusion of the two information of PPI network make the interaction prediction results more accurate.

2 Similarity Matrix Reflecting Topology Information

It has been noted that if two nodes have more common neighbors, there may be an edge between them in the network. Based on this, we calculate the topological similarity between two proteins denoted as $s_t \in [0, 1]$ In the graph G, for an edge $e = \langle v_i, v_j \rangle \in E$, the topology similarity $s_t(v_i, v_j)$ is defined as:

$$s_t(v_i, v_j) = 2p/(d(v_i) + d(v_j))\qquad(1)$$

where p represents the number of common neighbors between v_i and v_j; $d(v_i)$ and $d(v_j)$ denote the degrees of v_i and v_j respectively. Formula (1) reflects the cross extent of topological structure between node v_i and v_j in the network.

For example, there is a PPI network as shown in the left of Fig. 1, according to the above definition we can construct the right similarity matrix A reflecting the PPI network topology similarity. The higher the value, the higher the topological similarity, the more likely there is a potential link.

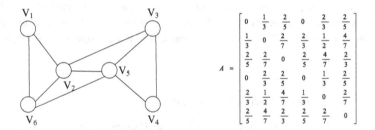

$$A = \begin{bmatrix} 0 & \frac{1}{3} & \frac{2}{5} & 0 & \frac{2}{3} & \frac{2}{5} \\ \frac{1}{3} & 0 & \frac{2}{7} & \frac{2}{3} & \frac{1}{2} & \frac{4}{7} \\ \frac{2}{5} & \frac{2}{7} & 0 & \frac{2}{5} & \frac{4}{7} & \frac{2}{3} \\ 0 & \frac{2}{3} & \frac{2}{5} & 0 & \frac{1}{3} & \frac{2}{5} \\ \frac{2}{3} & \frac{1}{2} & \frac{4}{7} & \frac{1}{3} & 0 & \frac{2}{7} \\ \frac{2}{5} & \frac{4}{7} & \frac{2}{3} & \frac{2}{5} & \frac{2}{7} & 0 \end{bmatrix}$$

Fig. 1. PPI network and similarity matrix

3 Similarity Matrix Reflecting Attribute Information

As we know from biological knowledge, if the attribute information of the amino acid sequence's similarity between two proteins is high, it is more likely that they are interacted to accomplish a biological function. Therefore, extracting attribute information from protein sequence is the key to determine the prediction results.

3.1 Pseudo-Amino Acid Composition (PseAAC)

This study selects the more typical pseudo amino acid composition in recent years (PseAAC) [7] to extract protein attribute information. By the use of this method, a protein P can be expressed as the following form:

$$P = [p_1, p_2, \ldots, p_{20}, p_{20+1}, \ldots, p_{20+k}, \ldots, p_{20+\lambda}] \qquad (2)$$

This method constructs a $(20 + \lambda)$ dimension vector, where $p_1, p_2, \ldots, p_{20}$ are 20 amino acids composition. In addition, it also includes λ dimensions discrete data vector $p_{20+1}, p_{20+2}, \ldots, p_{20+\lambda}$ which consists of a series of sequence order related factors, representing the fusion of amino acids' hydrophobic index, hydrophilic index as well as the sequence order of molecular weight of the side chain and so on.

We use PseAAC to express the ith and j^{th} protein respectively: P^i and P^j. Here, P^i and P^j are both written as vectors similar to (2). Thus no matter P^i or P^j, its dimensions are the same. The form is written as follows:

$$\mathbf{P}^i = [p_1^i, p_2^i, \ldots, p_{20+\lambda}^i] \qquad (3)$$

$$\mathbf{P}^j = [p_1^j, p_2^j, \ldots, p_{20+\lambda}^j] \qquad (4)$$

3.2 The Selection of Attribute Information

In order to predict PPI and protein space structure better, some researchers are more inclined to denote the predicted protein by using more attribute information. But when the attribute information is selected more, the redundant information will increase so that the feature space dimension is too high. In order to solve the problem, Lin Zhi-ren et al. [8] developed a program named fselect.py which can score for all attribute information of proteins. After ranking all scores, researchers select the highest as the final attribute of proteins.

3.3 The Calculation of Attribute Information

This paper uses the Pearson correlation coefficient (Pearson correlation coefficient, the PCC) to calculate the similarity between two vectors. Due to the PCC can reflect two variables changing trend well, it is widely used to measure the similarity of the two vectors. The calculation of the PCC as shown in (5):

$$r(\mathbf{P}^i, \mathbf{P}^j) = \frac{N \sum_{k=1}^{N} p_k^i p_k^j - \sum_{k=1}^{N} p_k^i \sum_{k=1}^{N} p_k^j}{\sqrt{\left(N \sum_{k=1}^{N} (p_k^i)^2 - \left(\sum_{k=1}^{N} p_k^i\right)^2\right) \left(N \sum_{k=1}^{N} (p_k^j)^2 - \left(\sum_{k=1}^{N} p_k^j\right)^2\right)}} \qquad (5)$$

where $r(\mathbf{P}^i, \mathbf{P}^j)$ is the computing result of the two similarity, p_k^i represents the kth element of the vector $\mathbf{P}^i$; p_k^j denotes the kth element of the vector $\mathbf{P}^j$; N is the dimension of the vector $\mathbf{P}^i$ and vector $\mathbf{P}^j$. After calculating the attribute similarity between two proteins, we can get attribute similarity matrix $B = [B_{ij}]_{n \times n}$ Where element $B_{ij} = (p^i, p^j)$ is calculated by the Formula (5), denoting attribute similarity value between the ith and j^{th} proteins.

4 Algorithm

We put forward a new method, which respectively maps the two information to different spatial. The mapping spaces are as close as possible, in order to make the two information organically integration.

Therefore, we set A as topology information similarity matrix, B as attribute information similarity matrix, we map U, V making the next formula minimum:

$$L(U, V) = \|AU - BV\|_F^2 + \lambda_1 \|U\|_F^2 + \lambda_2 \|V\|_F^2 \tag{6}$$

Here U, V is a $n \times r$ matrix, $\|.\|_F$ is a Frobenius norm, the latter two items of Formula (6) are penalty terms which prevent over fit. The mapping process is shown in Fig. 2.

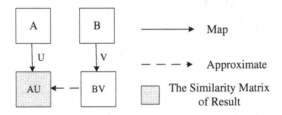

Fig. 2. Mapping the two similarity into different spaces

As we can see from Fig. 2, when $L(U,V)$ get the minimum, AU and BV will be very close. Thus, we can say that it does not favor one of the two, it can be regarded as the final result.

In order to obtain $\min_{u,v} L(U, V)$, we need to get partial derivative for the U and V of $L(U, V)$, and the specific process is as follows:

$$\frac{\partial L}{\partial U} = (A^T A + A A^T)U - 2V^T B^T A + 2\lambda_1 U$$

Considering A's symmetry, we get:

$$\frac{\partial L}{\partial U} = 2A^2 U - 2V^T B^T A + 2\lambda_1 U$$

Let $\frac{\partial L}{\partial U} = 0$, we get:

$$(A^2 + \lambda_1 I)U = V^T B^T A \tag{7}$$

Similarly, we get:

$$(B^2 + \lambda_2 I)V = B^T AU \tag{8}$$

Then, according to Formulas (7) and (8), we can use the EM algorithm to get U and V, the process is as follows:

Algorithm 1. EM-PPI

Input: Similarity matrix A,B;

The parameters of the corresponding λ_1, λ_2;

Nmax: Maximum iterations;

Output: Mapping matrix U, V;

Begin

    1. Take a set of initial value of V $V_{(0)}$; $i=1$;

    2. **Repeat**

    3. $U_{(i+1)} = (A^2 + \lambda_1 I)^{-1} V_{(i)}^T B^T A$

    4. $V_{(i+1)} = (B^2 + \lambda_2 I)^{-1} B^T AU_{(i)}$;

    5. $i=i+1$;

    6. **Until** convergence **or** $i>Nmax$.

End

The iteration formulas in the 3rd and 4th lines in Algorithm 1 can be obtained by Formulas (7) and (8) respectively.

Through the above analysis, we can get the framework of the algorithm, named PPIP - BSM (protein-protein interaction prediction-based on space mapping) which is as follows:

Algorithm 2. PPIP-BSM

Input: The vertex set V in graph G, the edge set E in graph G, parameter λ_k, $k=1,2$

Output: All the prediction links（PPI）: E'

Begin

    1. Calculating the topological similarity matrix according to formula(1);

    2. Given a protein, the protein is expressed in the form of formula(2);

    3. Using fselect.py to select the attribute information;

    4. Calculating attribute similarity matrix B according to the formula (5);

    5. Using EM - PPI algorithm to compute U, V which makes formula(6)minimum;

    6. Take AU for the final similarity matrix, and then predict the existence of links(PPI).

End

---

5   Experimental Results and Analysis

The experimental data based on the yeast PPI data which is issued in 1998 by Cho et al. [9].

The Number of Attributes Effects. We take the yeast protein interaction network [10] in the MIPS database (http://mips.helmholtz-muenchen.de) as the test data. After randomly removing 500 edges from the network, an incomplete network is formed. In this incomplete network, we predict the links with algorithm PPIP-BSM, and then compare the prediction results with the original standard network. The precision is defined as the number of links accurately predicted that are divided by the total number of links for prediction.

Table 1 shows the precision of the experimental results with the change of the number of attribute λ From Table 1 we can see that with the increase of λ, the accuracy is increasing as well, the average accuracy rate reached 71.87 %. When the number of attribute $\lambda = 37$, the prediction results has the highest accuracy rate which reaches to 76.51 %. At the same time, we use AAC instead of PseAAC for attribute extraction, and the results show that the average accuracy of the method is 69.65 % for the same data set. Obviously, the attribute extraction method PseAAC is more advantageous.

Table 1. The influence of the number of attributes on precision

Attribute number λ	Average accuracy (%)
11	65.90
24	70.34
37	76.51
50	74.72

The Influence of Different Levels of Network Integrity. Firstly we eliminated some edges randomly from the original yeast PPI network, and then keep 40 %, 50 %, 60 %, 70 %, 80 %, 90 % edges respectively to form some different incomplete networks. Figure 3 is the experimental results with the variation of observation points. As can be seen from Fig. 3, when 90 % of the network is known, the prediction accuracy is the highest; when only 40 % of the edges are known, the average accuracy of the prediction results can reach 68 %. From the above analysis results, we can conclude that our algorithm can effectively predict links.

The Influence of Different Weight Information. In the experiment, we compare our algorithm with CD-Dist [5] and weighted SRW [11]. It can be noted from the left of Fig. 4, when 90 % of the network is known, the algorithm prediction accuracy can reach 87 %, while SRW is 66 % and CD-Dist is 46 %.Similarily, we observe the AUC index of the predicted results in the case of different known links. AUC (area under the receiver operating characteristic curve) [12] is a measurement of the accuracy of the algorithm as a whole. It can be observed from the right of Fig. 4, compared with the other two algorithms, the AUC index of our algorithm is much better than others which

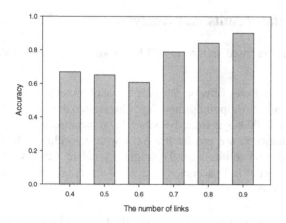

Fig. 3. The results vary with the number of known links

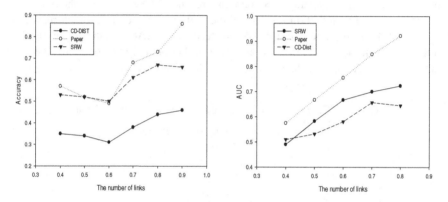

Fig. 4. The variation of the accuracy and AUC with the number of known links

shows that our algorithm can achieve higher accuracy. From the above analysis, we can learn that there are limitations by the use of single information in the link prediction, so it can achieve better results by combining the two kinds of weight information.

Independent Test Set Performance Evaluation. In order to further test the validity of this algorithm, we select the protein interaction data of the other 5 species from the DIP database, and the results are shown in Table 2. The results show that our algorithm can effectively make link prediction in PPI networks.

Table 2. Independent test set performance

Independent test set	Number of interacting proteins	Correct prediction results
D.melanogaster(fruit fly)	21975	18023(82.02 %)
E.coli	6954	5020(72.19 %)
H.saPiens(Human)	1439	1244(86.45 %)
H.Pylori	1420	987(69.51 %)
M.musculus(House mouse)	319	297(93.10 %)
Total average	32107	25571(79.64 %)

6 Conclusion

This paper put forward a new method for potential and hidden links prediction in PPI network based on spatial mapping. In this paper, according to topological structure and attribute information of two nodes in the network, we can predict whether there is a link between the two nodes by calculating their two similarities. In the aspect of balancing the two kinds of similarity, this paper considers the space mapping based method. The experimental results on yeast PPI network show that our proposed algorithm is effective.

Acknowledgements. This research was supported in part by the Chinese National Natural Science Foundation under grant Nos. 61379066, 61379064, 61472344, 61301220, 61402395, Natural Science Foundation of Jiangsu Province under contracts BK20130452, BK20151314 and BK20140492, and Natural Science Foundation of Education Department of Jiangsu Province under contract 12KJB520019, 13KJB520026.

References

1. Cannataro, M., Guzzi, P.H., Veltri, P.: Protein-to-protein interactions: technologies, databases, and algorithms. J. ACM Comput. Surv. **43**(1), 1–36 (2010)
2. Saito, R., Suzuki, H., Hayashizaki, Y.: Interaction generality, a measurement to assess the reliability of a protein-protein interaction. J. Nucleic Acids Res. **30**, 1163–1168 (2002)
3. Saito, R., Suzuki, H., Hayashizaki, Y.: Construction of reliable protein-protein interaction networks with a new interaetion generality measure. J. Bioinf. **19**, 756–763 (2003)
4. Chen, J., Hsu, W., Lee, M.L., et al.: Increasing confidence of protein interactomes using network topological metrics. J. Bioinf. **22**, 1998–2004 (2006)
5. Brun, C., Chevenet, F., Martin, D., Wojcik, J., Guénoche, A.: Bernard Jacq.: functional classification of proteins for the prediction of cellular function from a protein-protein interaction network. J Genome Biol. **5**, R6 (2003)
6. Chua, H.N., Sung, W.K., Wong, L.: Exploiting indirect neighbors and topological weight to predict protein function from protein-protein interactions. J. Bioinf. **22**, 1623–1630 (2006)
7. Chou, K.C.: Prediction of protein cellular attributes using pseudo-amino acid composition. J. Proteins Struct. Funct. Bioinf. **43**(3), 246–255 (2001)
8. Department of Computer Science & Information Engineering: http://www.csie.ntu.edu.tw/~cjlin/
9. Cho, R.J., Campbell, M.J., Winzeler, E.A., Steinmetz, L., Conway, A., Wodicka, L., Wolfsberg, T.G., Gabrielian, A.E., Landsman, D., Lockhart, D.J., Davis, R.W.: A genome-wide transcriptional analysis of the mitotic cell cycle. J Mol. Cell. **2**(1), 65–73 (1998)
10. Mewes, H.W., Frishman, D., Güldener, U., Mannhaupt, G., Mayer, K., Mokrejs, M., Morgenstern, B., Münsterkötter, M., Rudd, S., Weil, B.: MIPS: a database for genomes and protein sequences.J. Nucleic Acids Res. **30**(1), 31–34 (2002)
11. Backstrom, L., Leskovec, J.: Supervised random walks: predicting and recommending links in social networks. In: Proceedings of the Fourth ACM International Conference on Web Search and Data Mining, Hong Kong, pp. 635–644 (2011)
12. Hanley, J.A., McNeil, B.J.: The meaning and use of the area under a receiver operating characteristic(ROC)curve. J. Radiol. **143**, 29–36 (1982)

A Compatible Model of Unstructured Data for Cross-Media Retrieval in the Field of Tourism

Han Hu[1,2($\boxtimes$)], Xiaoyu Li[2($\boxtimes$)], Wenyi Wu[2], and Zhaoyi Liu[2]

[1] Chengdu Research Institute of UESTC, University of Electronic Science and Technology of China, Chengdu 610054, Sichuan, People's Republic of China
543160303@qq.com
[2] School of Information and Software Engineering, University of Electronic Science and Technology of China, Chengdu 610054, Sichuan, People's Republic of China
xiaoyu33521@163.com

Abstract. With the development of multimedia, the amount of unstructured data of multimedia is increasing, especially for the tourism. Meanwhile, it is one of the key issue to analysis large-scale unstructured data, which helps us to find the hidden relevance between redundant and different data. How to retrieve efficiently, and recommend accurately for cross-media retrieval is more and more important. This paper proposes a new data mode for cross-media retrieval - unstructured data compatible model, short for UDC model. The UDC model is constructed by its own metadata. All metadata are organized by a certain hierarchical relationship. Every metadata consists of three layers: the feature layer, the semantic layer and the compatibility layer. Furthermore, this paper presents retrieval and recommendation algorithms based on UDC model. The experiment results demonstrate that the retrieval engine based on UDC mode can be more effective for cross-media retrieval and recommendation.

Keywords: Tourism · Cross-media · Retrieval · Recommendation · Unstructured data model

1 Introduction

With touring is becoming one of people's lifestyle, comprehensively and quickly understanding a spot if suits ourselves is particularly important. However, as structured and unstructured data like videos, pictures increasing rapidly on the internet, in order to obtain satisfying results, users have to filter useless information repeatedly and switch diverse search engines for help. For traditional search engines, when users submit a retrieval request in a certain kind of media form, the feedback is mainly in the same form. The recommendation and the retrieval information are only connected in the literal concept, which often cannot make accurate recommendations to the users. For instance, when a user posts a retrieval request like "Travelling in Yunnan province" within traditional picture search engine, the engine only returns pictures and recommends like

H. Yin et al. (Eds.): IDEAL 2016, LNCS 9937, pp. 114–125, 2016.
DOI: 10.1007/978-3-319-46257-8_13

"travelling in Jiangxi province", "governor in Yunnan province". Generally, if the feedback includes not only pictures but also corresponding videos and texts, and the recommendation is not only based on the literal concept, it will be of more significance for users' further understand and choices. So how to carry out cross-media retrievals, associate with unstructured data, and recommend more effectively is a more and more valuable research direction in the field of tourism. This paper presents a new data mode–unstructured data compatible model (referred to as the UDC model). We carried out experiments, whose results proved that the UDC mode can effectively help search engines associate with the unstructured data, carry out cross-media retrieval and make more valuable recommendations.

2 Related Work

Traditional cross-media retrieval is based on text annotation. However the traditional method of text-based manual annotation for cross-media retrieval is not only laborious, time-consuming, but also subject-biased. To deal with the problem, the cross-media retrieval methods based on the content was proposed in the 1990 s [1]. The basic idea is to calculate the similarity between the retrieved object and the user's query by visual, auditory or 3D model geometric features.

With the development of content-based cross-media retrieval methods in twenty-first Century, researchers have been widely recognized that images, videos and other multimedia data with high dimensions is determined by finite freedom degree. Analyzing the geometric topology of the data can not only optimize the similarity between the data, but can also reduce the computational complexity greatly. This can be conducive for cross-media retrieval. So manifold learning theory is proposed [2]. From the beginning of 2008, with the increase of accompanying text with the image on the internet and the arrival of the Web2.0, it has been a hot topic in recent years to extract the accurate annotation words from the text and users' annotation information [3], which can reflect the multimedia data's semantic. After 2010, researchers begin to focus on the problems of large-scale image annotation, inconsistency label, label's extension and so on [4].

Make a summary of previous research, there are two main ways to deal with cross-media retrieval [5]:

- Extract the underlying features of multimedia, and map the underlying features into a correlation table. By maintaining the correlation table, calculating the distance of the features' value in the correlation table, we can realize cross-media retrieval. This kind of method has promoted the development of cross-media retrieval. However, with the explosive growth of data, the space occupied by the correlation table is larger and larger, and the resources consumed by maintaining correlation table are also increasing.
- By building an index for multimedia, and building a multimedia knowledge database [6]. The method can effectively solve the problems of the first method, which need to consume the resources to maintain the correlation table, and has been used in a number of areas. However, this method is not only aimed at the cross-media retrieval in tourism, so there is still the possibility of further optimization in the field

of tourism. Based on this method and take the characteristics in the field of tourism into account, the paper puts forward UDC model for cross-media retrieval in the field of tourism.

3 Unstructured Data Compatible Model

The UDC model is composed of its own unique metadata according to the hierarchical structure. In order to optimize the retrieval process, the organization between metadata has good verticality.

3.1 Metadata of UDC Model

The UDC model is composed of its own unique metadata. Unlike specific video files and image files, every metadata represents an abstract concept. The structure of the metadata is shown in Fig. 1. Every metadata is divided into three layers according to the dependency relationship, and every layer logically belongs to the upper layer. The first layer is the feature layer. The feature layer is composed of the feature vector set. The feature vector set is composed of underlying features which are extracted after preprocessing the multimedia data, (such as de-noising, duplicate content removal, context correlation, etc.). Every feature vector of the feature set describes one side of the semantics, for example "Tiananmen Square" can be characterized by the "Beijing", "square" and so on. The second layer is the semantic layer. On the basis of the first feature vector set, the semantic layer adopts the methods like structure analysis, the intelligent analysis conversion or others to draw out the annotation words, which can reflect the semantic of multimedia resources, and annotate the semantic in semantic layer. The third layer is the compatibility layer. The semantic extracted from the second layer is related to various types of media in the compatibility layer, so that the metadata can be compatible with other media types through its compatibility layer. For example a text file which has been annotated in second layer will associated with other metadata

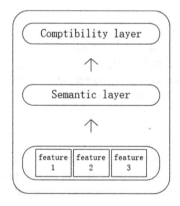

Fig. 1. Metadata structure diagram

with the same semantic. The metadata which is associated can be based on a video file, a picture file or other forms of media. When the retrieval is underway, the higher the matching accuracy between the first layer's feature set and the search request is, the higher their similarity is.

3.2   Metadata's Vertical Organization

With the explosive growth of the information on the internet, the vertical search engine has become the trend of the search engine's development. Vertical engines' search strategy is obviously different from the traditional one. The purpose of the traditional is to traverse the entire web network as quickly as possible, and to collect as many web pages as possible. As to the vertical search engines, its search scope is limited to a specific topic or special field, thus web pages that with enough correlation to the topic will be only selected in the search process. So, in order to search vertically better, the organization of metadata in UDC model is top-down and hierarchical, as shown in Fig. 2.

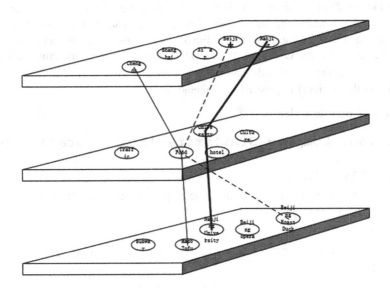

Fig. 2. The UDC model metadata's top-down organization

Every node in the figure represents a metadata, its name is the semantics annotated for metadata in its semantic layer. According to the particularity of retrieval in tourism, the nodes with more abstract semantics are always on the upper floor of the nodes with more specific semantics. For example, the node that represents 'food' is always located on the upper floor of the node which represents "Beijing roast duck". From the top to the bottom, on every retrieval path, the $(N + 1)^{th}$ node is conceptually subordinated to the Nth node, namely if the Nth node represents "Chengdu", the $(N + 1)^{th}$ node represents "bus station", the concept is expressed as "bus station in Chengdu".

4   Main Algorithms Based on UDC Model

The algorithms based on UDC model include two algorithms "similarity calculation algorithm" and "vertical backtracking algorithm".

4.1   The Algorithm of Similarity Calculation

This algorithm is improved on the similarity calculation algorithm [7] and be used to calculate the similarity between the node and the retrieval request.

Algorithm 1: Similarity Calculation Algorithm
It has been described previously that the feature layer of every metadata is a feature set consisting of N feature vectors. Suppose a user post a picture as the retrieval request, extract its feature set F,take F as the target feature set, and F contains n feature vectors. Now we will calculate the similarity between F and a metadata M.

- **Step 1:** Obtain M's feature set f in feature layer, f usually has more features than F. Compare f with F, remove its features that are not contained in F, conversely if F has features which are not contained in f, set them to 0 in f, and ensure every features in F has its counterpart in f. Finally f contains n feature vectors too.
- **Step 2:** Calculate the Euclidean distance $D_i(0 < i < n)$ between the same features in the target feature set F and feature set f, define $d = \frac{1}{n}\sum_{i=0}^{n} D_i$ as the similarity between the retrieval request and the metadata.

Algorithm 1 pseudo code:

```
Function: Compare_Similarity (F[n], metadata [N])
{
Initialization:
  F[n] = the feature set of posted pictures;
  Similarity=0;
Step 1:
  f(n)=descending_dimension (metadata[N],F[n]);
Step 2:
  For (i=0; i<f[i].lenghth; i++) {
    Di = (metadata[i]-f[i]) ²;
    distance+=Di;
}
  Similarity=1/n √distance;
  return Similarity;
}
```

4.2   Vertical Backtracking Algorithm Based on UDC Model

Suppose a user posts a picture as the retrieval request.

Algorithm 2: Vertical Backtracking Algorithm

- **Step 1:** Use feature extraction algorithms to extract the image feature set F, take F as the target feature set.
- **Step 2:** Assuming that the algorithm has been carried out to floor N (N = 1, 2...) in UDC. Use algorithm 1 to calculate the similarity between the target feature set F and every node's feature layer in floor N. If the similarity is less than the threshold value T, it will be discarded directly. Record and get the closest K nodes to F on floor N, take them as the result set of Nth floor R_n, and record their floor N. If the number of elements in R_n is 0, the query finishes, and goes to step 5.
- **Step 3:** Traverse the nodes in floor (N + 1) which are subordinated to the nodes in R_n, Use algorithm 1 to calculate the similarity between the target feature set F and every available node's feature layer in floor (N + 1). If the similarity is less than the threshold value T, it will be discarded directly. Record and get the closest K nodes to F among available nodes, take them as the result set of the (N + 1) floor R_{n+1}, and record their floor (N + 1). If the number of elements in the R_{n+1} is 0, the query is finished, and go to step 5.
- **Step 4:** Repeat step 3 until reach the bottom.
- **Step 5:** Sort all the nodes in R_i(i = 1, 2, 3...) according to the similarity, select the closest K nodes to F to form a set R.
- **Step 6:** In R_i(i = 1,2,3...), utilize the nodes' compatibility layer to obtain the nodes' all related nodes, and calculate their similarities with the target feature set F according to Algorithm 1. Rank their similarities, and choose the closest K nodes to form the association result set $R^{'}$.
- **Step 7:** Traverse the nodes' floor in R, obtain the most densely distributed floor M, back to the result set $R_{(m-1)}$, randomly select k nodes in floor M from the nodes which is subordinated to the nodes in $R_{(m-1)}$, take them as the recommended result set R_r.
- **Step 8:** Rank $R \cup R^{'}$, and return to the user as the retrieval result, at the same time return to the user R_r as the recommended result.

Algorithm 2 pseudo code:

```
Function: Vertical_Backtrack () {
Step 1:
    F[n]=picture's feature set ;
    Max_floor=the UDC model's max floor;
    Max_collected_floor=
the max floor of the nodes collected in step 2,3,4;
    Max_collected_node=
the total number of the nodes collected in step 2,3,4;
    T=the threshold of similarity;
    Rₙ=null;
Step 2:
    do{
     N=1;        // Start from the first floor
     for(i=1;i<floor[N].node_number;i++){
       similarity=Compare_Similarity(F[n],floor[N].node[i]);
       if(similarity>t){
         collection[floor[N].node[i]];
         }
       else{
         throw_away(floor[N].node[i]);
         }
       }
     Rₙ=collection.get_nearest_K_node();
     if(Rₙ.node_number==0){
         goto step 5; }
}
Step 3:
    N=N+1;        //enter the (N+1) floor
    floor[N]=Rₙ;
Step 4:
}while(N<=Max_floor)
Step 5:
    for(j=1;j< Max_collected_floor;j++){
      Rᵣₐₙₖₑ𝒹=rank(Rⱼ)
}
    R= Rᵣₐₙₖₑ𝒹.get_nearest_K_node();
Step 6:
    R'= Compare_Similarity(F[n],R.relatedNode.get_near-
est_K_node());
Step 7:
    m= get_thickest_floor(R);// obtain the most densely dis-
tributed floor in R
    backtrack(R₍ₘ₋₁₎);
    Rᵣ=R₍ₘ₋₁₎.get_random_K_node(m);
Step 8:
    return rank(R∪R')+Rᵣ;
}
```

As for the Vertical Backtracking Algorithm, one example could be illustrated for instance. When a user submits a text retrieval request for "Beijing Beijing roast duck", the retrieval process is:

- **Step 1:** Carry out a top-down retrieval as the path "Beijing→food→Beijing roast duck", and hit the relevant node "Beijing roast duck", obtain relevant text information.
- **Step 2:** Obtain the related other types of media nodes through node "Beijing roast duck", such as the pictures of "Beijing roast duck" or the introduction videos.
- **Step 3:** From the "Beijing roast duck" node back to the "food" node, get some nodes subordinated to the node "food" in concept, such as: "Beijing→food→The old Beijing Noodles with Soy Bean Paste", "Beijing→food→Donkey Burger" and take the nodes as the recommended result.
- **Step 4:** Rank and sort the results, return to the user.

5 Experiment Results Analysis

5.1 Implementation and Test of UDC Model

In order to verify and test the UDC model and algorithms, one system with B/S structure is designed and implemented, which composed of two parts: the internet crawlers and the cross-media retrieval engine based on the UDC model. As shown in Fig. 3, the program language of the system is Java in JDK 1.8, the crawlers is based on Heritrix which is an open source web crawler. The open source full text search engine Lucene is utilized to set up an index database. Every data file has its own logging file, every metadata of the UDC model consists of a data file and its corresponding logging file, the feature layer, semantic layer, compatibility layer's information, and the hierarchy of the metadata are all recorded in logging file. Tomcat is used for the web server, the front-end for Web is based on CSS, JavaScript, JSP and other website frameworks.

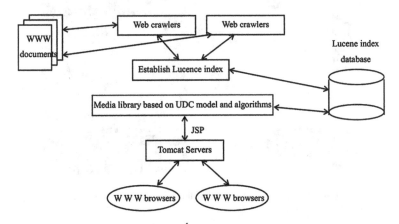

Fig. 3. The architecture of the experimental system

The hardware environment of the system is:

Table 1. Experimental hardware environment configuration table

OS	Windows 10
CPU	AMD FX8350
Memory	8 g
GPU	NVidia 660

It has been done that 200 web pages, 100 pictures,50 videos and 50 audios about the tourism in Chengdu are crawled, mainly in the official website of "Qunaer" and "Xiecheng" and with the help of the "Baidu" search engine. Place them into local database as the sample media library. Extract and annotate the underlying features for non-textual data. In the initial stage of the query, the threshold of the vertical backtracking algorithm is 0.5, namely T = 0.5.

5.2   Result Analysis

5.2.1   Cross-Media Retrieval

When the user inputs keywords "Chengdu take food" in Chinese "成都 冒菜", which is a famous food in Chengdu, in the retrieval engine based on the UDC model, the Vertical backtracking algorithm is called, then the engine returns multimedia information on the subject of "Chengdu take food" and make recommendations on the theme of "Chengdu food". In Fig. 4, the left is the return of the retrieval engine based on the UDC model, the right is the return of traditional retrieval engine as the comparison chart. According to the results, retrieval engine based on the UDC model can carry out efficient cross-media retrieval and make more effective recommendations.

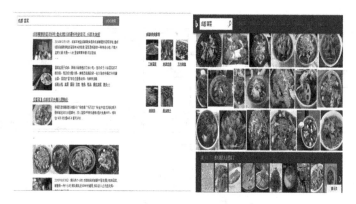

Fig. 4. The comparison chart of the UDC and the traditional

5.2.2   Recall Rates

The recall rate is one of the important criteria to measure the retrieval engine. Because the amount of data in the sample database is not large, manual validation is adopted to count the recall rate. This experiment has carried on 5 experiments, the query keywords are "Chengdu take food" (No. 1), "Chengdu quick hotel" (No. 2), "Chengdu LeShan Giant Buddha" (No. 3), "Chengdu Kam" (No. 4), "Sanxingdui" (No. 5). Figure 5 shows the statistical results. The recall rate of the 5 experiments are 0.70, 0.76, 0.80, 076 and 0.74, respectively. It shows that the retrieval engine based on this model can effectively return the relevant information that users want.

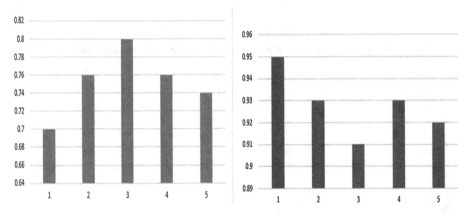

Fig. 5. The recall rate histogram   **Fig. 6.** The accuracy rate histogram

5.2.3   Accuracy Rate

The accuracy rate is also one of the important criteria to measure the retrieval engine too. The accuracy rate of the 5 tests is verified by the method of manual verification too. Figure 6 shows the statistical results. The accuracy of the 5 experiments, showed in Fig. 6, were: 0.95, 0.93, 0.91, 0.93, 0.92, respectively. To identify the gap between it and the traditional, we count and compare its accuracy with the traditional. The result is put on Fig. 7. From Figs. 6 and 7 we can infer that the search engine based on this model not only return what users want effectively, but also ensure a high accuracy.

5.2.4   Robustness Analysis

In this paper, the experiments are carried out on the basis that the threshold value of the vertical backtracking algorithm is $T = 0.5$, so it is necessary to verify the robustness of the algorithm when the threshold T changes. By analyzing the trend of the recall rate and the accuracy rate when the threshold change from 0 to 1, we can verify the robustness of the algorithm. As is shown in Fig. 8, by analyzing the robustness of the first query "Chengdu take food", we can see that if the threshold value T is not too large, more than 0.9, the algorithm is very stable, so the algorithm also has a very strong robustness in a reasonable range.

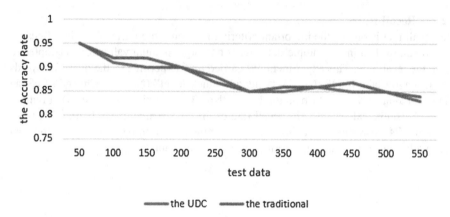

Fig. 7. The comparison chart of accuracy between the UDC and the traditional

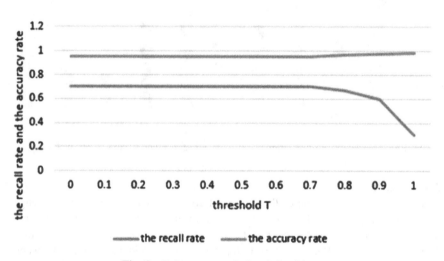

Fig. 8. Robustness analysis of algorithms

6 Conclusions

In order to effectively associate unstructured data and make a more accurate cross-media retrieval, this paper proposes a new data mode—unstructured data compatible model, short for UDC model, which has the function of cross-media retrieval and contributes to recommendations. The paper designs and realizes the retrieval engine based on the UDC model and proposes the vertical backtracking algorithm. Finally, the experiment result shows the model and its retrieval algorithms can be efficient in cross-media retrieval. It can also makes more valuable recommendations for users. However the model and algorithms of this paper still have some practical problems. For example, extracting files' features for UDC model's metadata, and organizing hierarchically used some mature frameworks and carry out manual sort when the data is small. Also every metadata must maintain a logging file when implement the system. The two problems above still need a better practical application to deal with when facing large amount of data. Future work will focus on how to automate metadata's organization and deal with the massive data.

Acknowledgement. This work is supported by the National Science Foundation of China (Grant Nos. 61502082), the Fundamental Research Funds for the Central Universities, ZYGX2014J065 and the Smart Cities Foundation of Sichuan (Grant Nos. RWS-CYHKF-04-2015003).

References

1. Foote, J.T.: Content-based retrieval of music and audio, multimedia storage and archiving system II. Proc of Spie **3229**, 138–147 (1997)
2. Saul, L.K., Weinberger, K.Q., Fei, S., et al.: Spectral methods for dimensionality reduction. Semisupervised Learn. **2006**, 1041–1048 (2006)
3. Wang, X.J., Zhang, L., Li, X., et al.: Annotating images by mining image search results. IEEE Trans. Pattern Anal. Mach. Intell. **30**(11), 1919–1932 (2008)
4. Wang, J.Z., Geman, D., Luo, J., et al.: Real-world image annotation and retrieval: an introduction to the special section. IEEE Trans. Softw. Eng. **30**(11), 1873–1876 (2008)
5. Tao, H.U., Gang-Shan, W.U., Ren, T.W., et al.: Ontology-based cross-media retrieval technique. Comput. Eng. **35**(8), 266–268 (2009)
6. Zhang, H., Zhuang, Y., Wu, F.: Cross-modal correlation learning for clustering on image-audio dataset. In: International Conference on Multimedia 2007, Augsburg, Germany, pp. 273–276, September 2007
7. Wu, F., Zhuang, Y.: Cross media analysis and retrieval on the web: theory and algorithm. J. Comput.-Aided Des. Comput. Graph. **22**(1), 1–9 (2010)

A Route Planning Method for Underwater Terrain Aided Positioning Based on Gray Wolf Optimization Algorithm

Jian Shen[1(✉)], Jing Shi[1], and Lu Xiong[2]

[1] Naval Academy of Armament, Beijing, China
sterrain@126.com
[2] Wuhan Ordnance N.C.O Academy, Wuhan, China

Abstract. Underwater terrain aided positioning technology is an effective method to improve the navigation accuracy of underwater vehicle. The route pre-planning on digit map can reduce matching time and improve matching accuracy for underwater terrain aided positioning. In this paper, a kind of route planning method for underwater terrain aided positioning based on gray wolf Optimization (GWO) algorithm is proposed. Firstly, the GWO algorithm was introduced. The objective functions and route planning method was researched combining with terrain matching problem. Secondly, the calculation formulas of underwater terrain entropy were introduced as well as the terrain information distribution. Thirdly, simulation parameters were set and the best planning route was get using GWO route planning method. Finally, the terrain matching simulation of ICCP was implemented along with the planned route which proved the feasibility of the planning method.

Keywords: Underwater terrain aided positioning · Route planning · Gray wolf optimization algorithm · Terrain entropy · ICCP

1 Introduction

Underwater terrain aided positioning technology is an effective method to improve the navigation accuracy of underwater vehicle which is currently a hot issue [1–4]. Terrain matching module is an important part of the underwater terrain aided positioning system whose performance depends on the terrain information of navigation routes [5, 6]. Unfortunately, underwater digital map cartography requires a large sum of finance to support a wide topographic survey which is also time-consuming, so a route pre-planning on digit map for underwater terrain matching is more feasible which improves the algorithm convergence speed and matching effect.

Generally there are five kinds of basic path planning methods for vehicle navigation. Voronoi diagram path planning method based on graph theory [7], A* planning method based on heuristic search algorithm [8], rapid random tree method (RRT) based on stochastic programming [9], artificial potential field method based on potential field theory [10]; genetic algorithm and particle swarm algorithm based on intelligent optimization [11].

© Springer International Publishing AG 2016
H. Yin et al. (Eds.): IDEAL 2016, LNCS 9937, pp. 126–133, 2016.
DOI: 10.1007/978-3-319-46257-8_14

Grey Wolves Optimization algorithm (GWO) was proposed by Mirjalili. It is a kind of swarm intelligence optimize method mimics the leadership hierarchy and hunting mechanism of grey wolves in nature. GWO algorithm has better global search ability and convergence performance compared to other Optimizations [12]. This work proposes a kind of route planning method for underwater terrain aided positioning based on GWO algorithm.

2 Route Planning Method Based on GWO Algorithm

2.1 GWO Algorithm

The GWO algorithm was inspired from prey of gray wolves. The algorithm is composed of 3 parts: social hierarchy, encircling and hunting prey, search and attacking [12].

(1) Social Hierarchy. GWO mimics the wolves' social hierarchy, the wolves of the top three fitness levels are labeled with α, β, and δ sequentially. The remaining wolves are marked with ω. The hunting action is initiated by α, β and δ.

(2) Encircling and Hunting Prey. Firstly, wolves must encircle the preys before hunting them, the mathematical model formulas are:

$$D = \left| C \cdot X_p(t) - X(t) \right| \tag{1}$$

$$X_p(t+1) = X_p(t) - A \cdot D \tag{2}$$

$$A = 2a \cdot r_1 - a \quad C = 2 \cdot r_2 \tag{3}$$

Where A and C is coefficient vectors. X_P is the is the position vector of the prey, X indicates the position vector of a grey wolf, r_1 and r_2 are random vectors of uniformly distributed in [0,1], a are linearly decreased from 2 to 0 over the course of iterations.

After the prey was trapped, α, β, and δ always occupy the top three attack positions, the other wolves adjustment their positions based on α, β, and δ. The hunt is usually guided by the alpha and the prey's positions are determined by α, β, and δ. The following formulas are proposed in this regard.

$$D_\alpha = \left| C_1 \cdot X_\alpha - X \right| \quad D_\beta = \left| C_2 \cdot X_\beta - X \right| \quad D_\delta = \left| C_3 \cdot X_\delta - X \right| \tag{4}$$

$$X_1 = X_\alpha - A_1 \cdot D_\alpha \quad X_2 = X_\beta - A_2 \cdot D_\beta \quad X_3 = X_\delta - A_3 \cdot D_\delta \tag{5}$$

$$X(t+1) = \frac{X_1 + X_2 + X_3}{3} \tag{6}$$

(3) Search and Attacking. A is a random value in the interval $[-2a, 2a]$, where a is decreased from 2 to 0 over the course of iterations. $|A<1|$ forces the wolves to attack towards the prey while $|A > 1|$ forces the grey wolves to diverge from the prey to hopefully find a fitter prey. C vector contains random values in $[0,2]$. This component provides random weights for prey in order to stochastically emphasize $(C > 1)$ or deemphasize $(C<1)$ the effect of prey in defining the distance in Eq. (1). This component is very helpful in case of local optima stagnation, especially in the final iterations [12].

2.2   Route Planning Method for Terrain Matching Based on GWO

At present, the common terrain matching algorithm is TERCOM, ICCP and nonlinear filtering matching algorithm. In this paper, as an example, the route planning method is designed for ICCP [13], the rest the algorithms are as the same method.

To ensure the accuracy of measurement data, underwater vehicles usually use the fixed-deep navigation mode. Thus, the problem belongs to the two-dimensional route planning issues.

Underwater terrain matching route planning should be complying with two constraints. The first is that the total distance of voyage should be as short as possible, and the second is that the terrain information surround the measurement position should be sufficient for terrain matching algorithm. Therefore, based on the above two conditions the fitness function is designed as $F_{fit} = F_R + F_T$, F_R is fitness function of route length and F_T is the fitness function of terrain information. Figure 1 is the illustrator of route planning course.

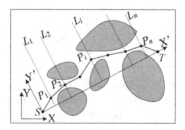

Fig. 1. Representation of a planned route for underwater vehicle

Where S and T are the starting and ending positions respectively and the gray areas are terrain flat regions. The planned route is the tactic connecting line of the point set $P = \{S, p_1, p_2 \cdots p_m, T\}$ [14]. For a legible expression, S and G are substituted by p_0 and p_{m+1}, fitness function of gray wolf can be expressed as follows:

$$F_R = \sum_{j=0}^{m-1} L_{p_j p_{j+1}} = \sum_{j=0}^{m-1} \sqrt{d + (y'_{i,j} - y'_{i,j+1})^2} \qquad (7)$$

Where $d = \left[\frac{L_{SG}}{m+1}\right]^2$ represents the shortest direct distance, $y'_{i,j}$ is the ordinate of ith in coordinate system of $X'SY'$. The boundary value $y_{i,j}^{\min}$ and $y_{i,j}^{\max}$ are the intersection points of set $\{l_1, l_2 \cdots l_m\}$ and coordinate axis. The coordinate transformation can be described as:

$$
\begin{bmatrix} x'_i \\ y'_i \end{bmatrix} = \begin{bmatrix} -\cos\theta & -\sin\theta \\ \sin\theta & -\cos\theta \end{bmatrix} \cdot \begin{bmatrix} x_s \\ y_s \end{bmatrix} + \begin{bmatrix} \cos\theta & \sin\theta \\ -\sin\theta & \cos\theta \end{bmatrix} \cdot \begin{bmatrix} x_i \\ y_i \end{bmatrix}
\tag{8}
$$

We also need to consider the terrain information surround each measurement point. In this article terrain entropy H_j is used to represents the terrain characteristics. The greater the entropy of the terrain is, the richer the terrain feature is. The specific formula will be introduced in the next chapter. So F_T can be expressed as follow:

$$
F_T = \sum_{j=1}^{m-1} H_j
\tag{9}
$$

As can be seen from F_R and F_T, the measure units are different as well as inconsistency, so we need to standardize them with the following formula:

$$
F_{fit} = \frac{F_R - F_{\min}}{F_{\max} - F_{\min}} + \frac{H_{\max} - H_T}{H_{\max} - H_{\min}}
\tag{10}
$$

Where $F_{\max}$ and $F_{\min}$ are maximum and minimum route fitness value of $y_{i,j}^{\max}$ and $y_{i,j}^{\min}$. $H_{\max}$ and $H_{\min}$ are the sum of terrain entropy values with the ranks of the top m and the lase m in digital map. Therefore, route planning based on GWO is using iterative method to find the coordinates set $Y' = (0, y'_1, y'_2 \cdots y'_m, 0)$, which minimize $F_{fit}(y)$, distributes in the parallel set $\{l_1, l_2 \cdots l_m\}$.

2.3   Steps of Method

The GWO route planning steps can be described as follows:

1. Calculate the terrain entropy of underwater digital map.
2. Calculate the θ and do coordinate transformation using Eq. (8).
3. Generate gray wolves $Y_i(i = 1, 2, \cdots n)$ on $\{l_1, l_2 \cdots l_m\}$ and initialize the GWO parameters a, A, C using Eq. (3).
4. Calculate fitness of each gray wolves using Eq. (10) and initialize the $Y_\alpha, Y_\beta, Y_\delta$ by using the top three gray wolves.
5. When iterations $I_{new} < I_{\max}$, Calculate the prey's position using Eqs. (5) and (6) and update the wolves position Y_i.
6. Update a, A and C using Eq. (3).
7. Calculate F_{fit} and get the newer $Y_\alpha, Y_\beta, Y_\delta$ and $I_{new} = I_{old} + 1$, Loop to Step 5 and repeat until a given maximum iteration number is attained.
8. Do coordinate inverse transformation from $X'SY'$ to XOY using Eq. (8).

3 Terrain Entropy of Underwater Digital Map

3.1 Terrain Entropy Definition

The conception of entropy was first proposed by the German physicist, Rudolf Clausius. It was used to represent the average distribution of any kind of energy in space. After that, C. Shannon, the initiator of information theory, theorized the entropy information measurement using probability statistical models. Nowadays, entropy is widely used in a large number of subject ambits as average information metrics. [16]

Digital maps usually adopt the DEM (digital elevation model) which are defined in the WGS-84 geodetic coordinate system. Therefore, terrain entropy can be defined as the following formula. [16]

$$H_{terrain} = -\sum_{i=1}^{M} p(i)lbp(i) \approx -\sum_{i=1}^{M} p(i)(p(i) - 1) \tag{11}$$

$$p(i) = \frac{|h(i)|}{\sum\limits_{i=1}^{M} |h(i)|} \tag{12}$$

Where $H_{terrain}$ represents terrain entropy, M is the total number of terrain data points in a water region. $h(i)$ is the depth data of each measuring points.

Terrain entropy is the metrics of underwater terrain information. Large entropy implies that the depths are well-distributed in the water region whose underwater terrain characters are significant. The region can supply sufficient terrain information for terrain matching algorithm; less entropy implies that the depths concentrate in some single values which imply that the water region is too flat to suit for the terrain matching algorithm.

3.2 Underwater Digital Map for Simulation

A kind of bottom topography reconstruction method based on the GeoTIFF format image was used to plot the underwater digital maps [15]. The seabed topography images were offered by international surveying and mapping website. Figure 2 is a GeoTIFF format seafloor topography images and it's three-dimensional reconstruction view provide by U.S. Geological Survey. The map covers a water extent of 7.1 × 7.8 km^2 and grid node distance is 50 m.

3.3 Terrain Entropy Calculation

The size of digital map is 710 × 780 after interpolation. We divide the DEM into small parts with every neighboring 10 × 10 blocks according to the reference [6]. The terrain entropy was calculated using Eqs. (11) and (12). Figure 3 is the terrain entropy distribution of the digital map.

Fig. 2. Seafloor topography image of GeoTIFF format and it's three-dimensional view

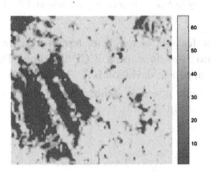

Fig. 3. Terrain entropy distribution

In Fig. 3, dark color parts indicate the low entropy regions which are poor in terrain information and unsuitable for terrain matching; light color parts indicate the large entropy regions which are suitable for terrain matching.

4 Simulation and Analyze

As it can be seen from Figs. 2 and 3, the terrain information in the left side region is scarcity. To this end, we choose the starting point S of the route in the left-down corner of the map and the target point G on the right, which is in order to test the optimize performance of the route planning method. The simulation parameters of iteration maximum number is 500 times, a are linearly decreased from 2 to 0 over the course of iterations, $A \in [-2a, +2a]$, C is the random number on $[0, 2]$. The Fig. 4 are the route planning results.

In Fig. 4, the planning route can effectively avoid the flat terrain. The constraints of maximization total entropy and shortest route length are taking into account. As can be seen from Fig. 4, after 32 iterations, the objective function is tending to convergence. It means that the rough planning method has better convergence speed and real-time feature, which can be used as a proposal method for engineering applications.

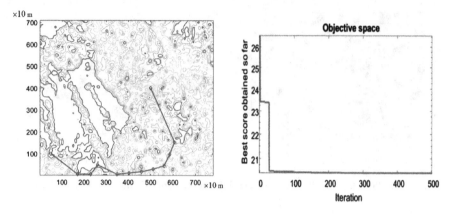

Fig. 4. Best route obtained by GWO route planning method and the iterative searching process

In order to prove the feasible of rough for terrain matching position, underwater terrain matching simulation was done using ICCP algorithm. The INS initial error in x and y direction are 200 m, course deviation is 3°, INS velocity directions variance are both 0.2 $(m/s)^2$, submersible speed is 4knot, measurement variance is 0.2 m^2, the simulation results shown in Fig. 5.

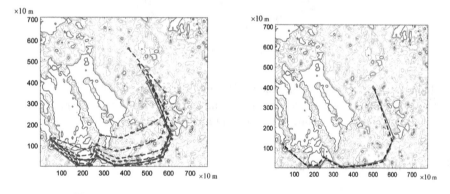

Fig. 5. Matching iterative process and matching result of ICCP

The simulation results show that ICCP can quickly converge to the points nearby actual measurement position of route, and the average error between matching positions and actual position is less than 16 m. A high resolution digital map will further improve the matching accuracy.

5 Conclusion

This paper presents a route planning method for underwater terrain matching position based on GWO algorithm. The method quickly locates the unsuitable regions to terrain matching in digital maps and plans the best matching position route. The route planning simulation proved to be feasible and establish the foundation for underwater terrain matching position.

References

1. Meduna, D., Rock, S., Mcewen, R.: Low-cost terrain relative navigation for long-range AUVs. In: OCEANS. MTS/IEEE, Quebec (2008)
2. Nygren, I.: Terrain navigation for underwater vehicles using the correlatormethod. IEEE J. Ocean. Eng. **29**(3), 906–915 (2004)
3. Mcphail, S.: Autosub6000: a deep diving long range AUV. J. Bionic Eng. **6**, 55–62 (2009)
4. Anonsen, K.B., Hagen, O.K.: An analysis of real-time terrain aided navigation results from a HUGIN AUV. In: OCEANS, Seattle (2010)
5. Tian, F.: Research on prior map data processing based terrain aided navigation methods for underwater vehicles. Harbin engineering university, Harbin (2007)
6. Feng, Q.: The research on new terrain elevation matching approaches and their applicability. National university of defense technology, Changsha (2004)
7. Pehlivanoglu, Y.V.: A new vibrational genetic algorithm enhanced with a Voronoi diagram for path planning of autonomous UAV. Aerosp. Sci. Technol. **16**(1), 47–55 (2012)
8. Yang, H.I., Zhao, Y.J.: Trajectory planning for autonomous aerospace vehicles amid known obstacles and conflicts. J. Guidance Control Dyn. **27**(6), 997–1008 (2004)
9. Liu, W., Zheng, Z., Cai, K.Y., et al.: QS-RRT basedmotion planning for unmanned aerial vehicles using quick and smooth convergence strategies. Sci. Chin. Inf. Sci. **42**(11), 1403–1422 (2012)
10. Jaradat, M.A.K., Garibeh, M.H., Feilat, E.A.: Autonomousmobile robot dynamic motion planning using hybrid fuzzy potential field. Soft. Comput. **16**(1), 153–164 (2011)
11. Zhu, W.: Real-time path planning system based on PSO for underwater vehicles. Harbin engineering university, Harbin (2008)
12. Seyedali, M., Seyed, M.M., Lewis, A.: Grey wolf optimization. Adv. Eng. Softw. **69**(7), 41–46 (2014)
13. Hongmei, Z., Zhao, J., Wang, A.: Research on simplified ICCP matching algorithm based on pre translation. Geomat. Inf. Sci. Wuhan Univ. **35**(12), 1432–1435 (2010)
14. Sun, B., Chen, W., Xi, Y.: Particle swarm optimization based global path planning for mobile robots. Control Decis. **20**(9), 1052–1055 (2005)
15. Shen, J., Yan, P., Zhang, J.: A method for reconstructing 3D seabed digital map based upon image processing. Comput. Simul. **26**(10), 90–93 (2009)
16. Zhang, H., Lu, Y.: Modeling method of massive terrain based on elevation entropy. Comput. Eng. Des. **27**(4), 4644–4647 (2006)

Feature Selection in Click-Through Rate Prediction Based on Gradient Boosting

Zheng Wang$^{(\boxtimes)}$, Qingsong Yu, Chaomin Shen, and Wenxin Hu

School of Computer Science and Software Engineering,
East China Normal University, Shanghai 200062, China
zenwan0553@gmail.com

Abstract. Click-Through Rate (CTR) prediction is one of the key techniques in computational advertising. At present, CTR prediction is commonly conducted by linear models combined with L_1 regularization, which is based on previous feature engineering including feature normalization and cross combination. In this case, the model cannot realize automatic feature learning. This paper uses the ensemble method for reference and proposes a feature selection algorithm based on gradient boosting. The algorithm employs the methods of Gradient Boosting Decision Tree (GBDT) and Logistic Regression (LR), and further conducts a positive analysis in the data set of kaggle-CTR prediction on display ads. The experimental result verifies the feasibility and validity of feature selection method. Moreover, it improves the performance of CTR prediction model, whose AUC value reaches 0.908.

Keywords: Gradient boosting · CTR · Logistic regression · Feature selection

1 Introduction

Click-Through Rate (CTR) prediction is the core technology in computational advertising, whose accuracy directly influences the advertising profits and user experience. The traditional CTR prediction model is mainly through Logistic Regression (LR) [1]. As a kind of linear model, LR can easily be implemented parallelly. Due to the limited learning capacity of linear models, however, large quantities of feature engineering are in need to enforce the non-linear learning capacity of LR. In real situations, few features can be directly applied to machine learning. Whether we can extract useful features in original Click-Through Log directly influences the model.

With the development of machine learning and the Internet, large quantities of online Click-Through Log can largely reflect users clicking and it can be used as a train set in learning supervision. How to extract effective features and build a train set is an important part of model leaning. In recent years, many researchers have explored CTR prediction,such as position-normalized [2], contextual analysis by combining relevance with click feedback [3,4] and probabilistic graphical model [5]. Many other scholars have proposed some methods

© Springer International Publishing AG 2016
H. Yin et al. (Eds.): IDEAL 2016, LNCS 9937, pp. 134–142, 2016.
DOI: 10.1007/978-3-319-46257-8_15

for optimizing CTR, including dealing with sparse data and adding personalized information [6,7]. Due to the characteristics of big data, numerous features, unbalanced categories and loud noises of the Click-Through Log, selecting effective features and shortening experimental periods are all urgent issues.

This paper employs boosting in the ensemble method [8]. We use base learner in the linear weighted combination, promoting weak learner be strong learner. Gradient Boosting Decision Tree (GBDT) is a realization of boosting. It is an iterative decision tree algorithm, and one decision tree is added at a time in the gradient of residual reduce. It is a non-linear model, having advantages of auto discovery of distinctive features and feature combinations. It solves the inability of learning class feature in linear model and helps to learn the relationship between feature data. We use the output characteristic of GBDT to conduct One-Hot-Encoding transformation, which is the input characteristic of LR Model. We conduct positive analysis in the data set of kaggle-CTR prediction on display ads, and testify the feasibility of feature selection based on Gradient Boosting.

2 Gradient Boosting Decision Tree

2.1 Principles of Gradient Boosting Decision Tree

GBDT was proposed by Friedman [9] in 1999, and it was first used in CTR prediction by Yahoo. It is a non-linear model, and realizes iterative learning of weak learner by boosting. It aims to solve decision function with least expected losses. As is shown in Fig. 1, it is linear combination of weak learners:

$$F^* = argminE_{x,y}[L(y, F(x))] \tag{1}$$

$$F(x; \rho_m, a_m) = \sum_{m=0}^{M} \rho_m h(x; a_m) \tag{2}$$

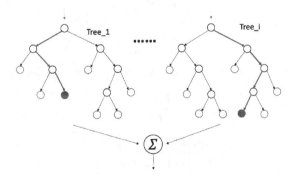

Fig. 1. Linear combination of weak learners

Algorithm 1. The Algorithm Framework of GBDT

1. $F^*(x) = argmin E_y(L(y, F(x))|x)$
2. $F_0(x) = f_0(x)$
3. for $m = 1$ to M :
4. $g_m(x) = -\frac{\partial E_y[L(y, F(x))|x]}{\partial F(x)}|F(x) = F_{m-1}(x)$ //descent direction
5. $\rho_m = argmin E_y[L(y, F_{m-1}(x) + \rho g_m(x)|x]$ //step size
6. $f_m(x) = \rho_m g_m(x)$
7. $F_m(x) = F_{m-1}(x) + \rho_m g_m(x)$
8. end for
9. $F^*(x) \approx F_M(x) = f_0(x) + \sum\limits_{m=1}^{M} \rho_m g_m(x)$

We can solve it by using the gradient descent method, and the algorithm framework is as the following pseudo-code.

The derivation based on LogLoss Function is as follows:

$$L(y, F) = log(1 + exp(-2yF)), y \in \{-1, 1\} \tag{3}$$

First, we need to initialize F_0 and set its partial derivative to be 0:

$$F_0 = argmin \sum_{i=1}^{N} L(y_i, F) \tag{4}$$

$$\frac{\partial \sum\limits_{i=1}^{N} L(y_i, F)}{\partial F} = 0, \tag{5}$$

$$\Rightarrow \sum_{i=1}^{N} \frac{exp\{-2y_i F\}(-2y_i)}{1 + exp\{-2y_i F\}} = 0 \tag{6}$$

$$\Rightarrow \sum_{i:y_i=1} \frac{-2exp\{-2F\}}{1 + exp\{-2F\}} + \sum_{i:y_i=-1} \frac{2exp\{2F\}}{1 + exp\{2F\}} = 0 \tag{7}$$

$$\Rightarrow F_0(x) = \frac{1}{2} log \frac{1 + \frac{1}{N}\sum\limits_{i=1}^{N} y_i}{1 - \frac{1}{N}\sum\limits_{i=1}^{N} y_i} = \frac{1}{2} log \frac{1 + \bar{y}}{1 - \bar{y}} \tag{8}$$

Then we need to estimate $g_m(x_i)$, and use decision tree for fitting:

$$g_m(x_i) = -\frac{\partial L(y_i, F)}{\partial F}|F = F_{m-1} \tag{9}$$

$$= \frac{2y_i exp\{-2y_i F_{m-1}(x_i)\}}{1 + exp\{-2y_i F_{m-1}(x_i)\}} \tag{10}$$

$$= \frac{2y_i}{(1 + exp\{2y_i F_{m-1}(x_i)\}} \tag{11}$$

Finally, we find approximate solution to learning rate using the Newton-Raphson method. In real situations, we often skip this step and use the shrinkage strategy [10], avoiding the model overfitting by setting learning rate of parameters. The shrinkage represents the learning rate of the model. Smaller shrinkage means lower learning rate and bigger shrinkage means higher learning rate. Generally, we often set shrinkage to be smaller.

$$f(r) = \sum_{x_i \in R_{jm}} log(1 + exp(-2y_i(F_{m-1}(x_i) + r))) \tag{12}$$

$$f'(r) = \sum_{x_i \in R_{jm}} \frac{-2y_i}{1 + exp(2y_i(F_{m-1}(x_i) + r))} \tag{13}$$

$$f''(r) = \sum_{x_i \in R_{jm}} \frac{-2y_i exp(2y_i(F_{m-1}(x_i) + r)}{[1 + exp(2y_i(F_{m-1}(x_i) + r))]^2} \tag{14}$$

$$\gamma_{jm} \approx \gamma_0 - f'(\gamma_0)/f''\gamma_0) = \frac{\sum_{x_i \in R_{jm}} \tilde{y}_i}{\sum_{x_i \in R_{jm}} |\tilde{y}_i|(2 - |\tilde{y}_i|)} \tag{15}$$

Compared with linear models, GBDT has strong advantages. It can process multiple types of features as well as collinearity simultaneously. It is insensitive to missing data, and there is no need for feature normalization. It can conduct automatic feature selection and can characterize importance of different features to enhance models interpretability. It is suitable for several missing functions and is fit for non-sparse data processing.

2.2 CTR Prediction Model Based on Gradient Boosting Decision Tree Feature Selection

CTR prediction is a typical prediction and LR in combination with L_1 regularization [11] is frequently used in sparse learning. The output of the model is the probability that some sample is labeled as a positive one. For a given user μ, a given webpage ρ, and a given advertisement α:

$$p = (c|\rho, \mu, \alpha) = \frac{1}{1 + exp(\sum_{i=1}^{d} \omega_i x_i(\rho, \mu, \alpha))} \tag{16}$$

The regular term is the L_1 regularization of parameter vector:

$$min\frac{1}{N} \sum_{i=1}^{N} L(y_i, \hat{y}_i) + \lambda \parallel \omega \parallel_1 \tag{17}$$

Among which: $x_i(\rho, \mu, \alpha)$ represents feature of dimension i based on GBDT selection; ω_i represents weight of feature i; $p = (\rho, \mu, \alpha)$ represents the probability that user μ would click advertisement α in webpage ρ; L is the loss function and $\hat{y}$ is the predicted value of the model.

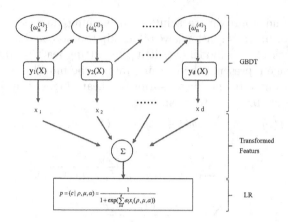

Fig. 2. Framework of CTR prediction mode (Colour figure online)

The framework of CTR prediction model is shown in Fig. 2. The red arrows represent the data flow, the blue ones represent the dependence of weight iteration, and the green ones represent of the weighted form of base classifiers.

Every base classifier $y_m(x)$ is trained based on a weighted form in the training data set. The iteration of weight $\omega_n^{(m)}$ depends on the performance of the previous base classifier $y_{m-1}(x)$. Every base classifier in GBDT is a decision tree. Input every sample x in the data set and after traversing all the trees, every x falls on the corresponding leaf node in the tree. Feature transformation of leaf nodes is conducted using One-Hot-Encoding, the output of which is used as input characteristic of LR model. Then CTR is solved using the LR linear model.

3   Experimental Results and Analysis

3.1   Data Description

The data set used in this experiment is from the Display Advertising Challenge provided by Kaggle in 2014. We chose 100,000 in the dac-samples, 75 % of which are used as train set and 25 % of which are used as the validation set. Both the train and test sets retain the distribution of source data and they cover users actions within seven days. Every line of the data corresponds to an advertising record. The first row of the record is an object variable, and clicking the advertisement is labeled as 1, otherwise as 0. The data includes 39 features, among which Row I1 to Row I13 are numerical characteristics and most of them are statistic. C1 to C26 are categorical characteristics, as they relate to users sensitive information, these features are desensitized. The experimental data are shown in Table 1.

Table 1. Experiment data

	Samples	Positive sample	Negative samples
Data set	100000	22663	77337
Train set	75000	16810	58190
Test set	25000	5853	19147

3.2   Experiment Design

The experiment is conducted from the following aspects:

(1) It mainly studies feature selection based on gradient boosting in CTR prediction. The model used is mainly GBDT+LR algorithm design, and is compared with hybrid models including LR, RT (Random Trees Embedding) and RF (Random Forest). It testifies that feature selection based on gradient boosting can promote CTR prediction.
(2) The model prediction mainly uses five indices including precision, recall, F1-score, Log-loss and AUC (Area Under ROC Curve) [12,13], among which F1-socre is to balance precision and recall in the model, Log-loss is to predict the fitting ability, and AUC is to predict the relative ranking ability.
(3) Major parameter design of the experiment: missing data are filled by the average. There are 500 trees in the GBDT model based on gradient boosting. The extreme depth is 5 and the learning rate is 0.1. L_2 regularization is used in the LR model with the regularization co-efficiency is 0.01, and the biggest iterations is 500. The parameters of trees in the RT and the RF models are the same as those in the GBDT model.

3.3   Result Analysis of the Experiment

After feature selection based on gradient boosting, dimensions of the data are promoted from 39 to 15241 dimensions after One Hot Encoding feature transformation, and every dimension represents one leaf node in a tree.

As shown in Fig. 3, judging relative ordering capability of models from AUC in the graph, we can find that GBDT+LR is the model with the best performance. It is worth mentioning that if feature selection based on RF model, then train LR model on these features, it can also realize good performance. The AUC of every model is shown in Table 2.

Table 2. AUC of every model

Model	GBDT+LR	RF+LR	RT+LR	GBDT	RF	LR
AUC	**0.908**	0.844	0.739	0.765	0.729	0.698

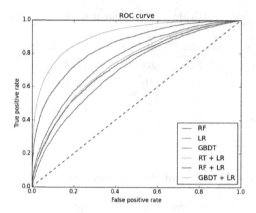

Fig. 3. Line char of ROC

Besides, we also study on the fitting ability of the models. We can see from Table 3 that we should choose GBDT rather than RF when creating trees. Though there are numerous trees in RF, however, it is not as good as GBDT. Moreover, we can find that the fitting ability of hybrid models are generally better than a single model. In addition, we also find that two numerical features and eight categorical features in the top 10 features. This is to explain that the ensemble decision tree [14] based on gradient boosting can not only distinguish distinctive features, but also conduct effective combinations of features, and the fitting ability is obviously more advantageous than creating trees using RF.

As shown in Table 4, we further investigate on precision, recall and F1-score. The experiment result shows that GBDT+LR hybrid model makes best performance in CTR prediction. We find that AUC, precision, recall and F1-score of LR model have poor performance. This is because in this experiment, large quantities of feature engineering has not been conducted in LR model. While the

Table 3. Logloss of every model

Model	GBDT+LR	RF+LR	RT+LR	GBDT	RF	LR
Log-loss	**0.315**	0.389	0.398	0.455	0.482	0.498

Table 4. Evaluation indices of the models

Model	GBDT+LR	RF+LR	RT+LR	GBDT	RF	LR
Precision	**0.87**	0.83	0.76	0.77	0.76	0.73
Recall	**0.87**	0.83	0.79	0.80	0.78	0.77
F1-score	**0.87**	0.81	0.75	0.77	0.71	0.69

good performance of GBDT+LR model shows that it is feasible to use feature output based on Gradient Boosting as the feature input of LR model.

4 Conclusion

CTR prediction is core technology in computational advertising. Accurate CTR prediction can not only increase the profits for advertisers, but also promote the satisfaction in user experience. This paper refers to boosting in ensemble method, and explains the learning process of gradient boosting, then it proposes the framework of feature selection based on gradient boosting and realized it using the GBDT+LR hybrid model. The experiment shows that the model can make good distinctions between all the features and can characterize importance of different features. The output of GBDT model can be directly used as the input of LR linear model and it can enhance the performance of CTR prediction.

This paper testifies the feasibility and correctness of the method proposed by conducting an empirical study. However, there is relationship of dependence between gradient boosting. Moreover, parallelization of large quantities of advertising data cannot be easily realized. This may be solved in further researches.

References

1. Chapelle, O., Manavoglu, E., Rosales, R.: Simple and scalable response prediction for display advertising. ACM Trans. Intell. Syst. Technol. **5**(4), 1–34 (2015)
2. Chen, Y., Yan, T.: Position-normalized click prediction in search advertising. In: Proceedings of the 18th ACM SIGKDD International Conference on Knowledge Discovery and Data Mining, pp. 795–803 (2012)
3. Chakrabarti, D., Agarwal, D., Josifovski, V.: Contextual advertising by combining relevance with click feedback. In: Proceedings of the 17th International Conference on World Wide Web, pp. 417–426 (2008)
4. Zhang, W., Jones, R.: Comparing click logs, editorial labels for training query rewriting. In: WWW 2007 Workshop on Query Log Analysis: Social and Technological Challenges (2007)
5. Yue, K., Wang, C., Zhu, Y.L., Liu, W.Y.: Click-through rate prediction of online advertisements based on probabilistic graphical model. J. East Chin. Normal Univ. **53**(3), 15–25 (2013)
6. Duchi, J., Jordan, M., McMahan, B.: Estimation, optimization, and parallelism when data is sparse. In: Advances in Neural Information Processing Systems, pp. 2832–2840 (2013)
7. Shen, S., Hu, B., Chen, W., Yang, Q.: Personalized click model through collaborative filtering. In: Proceedings of the Fifth ACM International Conference on Web Search and Data Mining, pp. 323–332 (2012)
8. Zhou, Z.-H., Methods, E.: Foundations and Algorithms, 1st edn. Chapman & Hall/CRC, New York (2012)
9. Friedman, J.: Greedy function approximation: a gradient boosting machine. Ann. Stat. **29**, 1189–1232 (2000)
10. Tibshirani, R.: Regression shrinkage and selection via the lasso. J. Roy. Stat. Soc. Ser. B **58**, 267–288 (1994)

11. Yuan, G.X., Ho, C.H., Lin, C.J.: An improved glmnet for l1-regularized logistic regression. J. Mach. Learn. Res. **13**, 1999–2030 (2012)
12. Fawcett, T.: ROC graphs: notes and practical considerations for researchers. Mach. Learn. **31**(1), 1–38 (2004)
13. Lobo, J.M., Jimnez-Valverde, A., Real, R.: AUC: a misleading measure of the performance of predictive distribution models. Glob. Ecol. Biogeogr. **17**(2), 145–151 (2008)
14. Gashler, M., Giraud-Carrier, C., Martinez, T.: Decision tree ensemble: small heterogeneous is better than large homogeneous. In: Seventh IEEE International Conference on Machine Learning and Applications, pp. 900–905 (2008)

Freight Vehicle Travel Time Prediction Using Sparse Gaussian Processes Regression with Trajectory Data

Xia Li$^{(\boxtimes)}$ and Ruibin Bai

University of Nottingham Ningbo China, Ningbo 315100, Zhejiang, China
XiaLi@nottingham.edu.cn

Abstract. Travel time prediction is important for freight transportation companies. Accurate travel time prediction can help these companies make better planning and task scheduling. For several reasons, most companies are not able to obtain traffic flow data from traffic management authorities, but a large amount of trajectory data were collected everyday which has not been fully utilised. In this study, we aim to fill this gap and performed travel time prediction for freight vehicles at individual level using sparse Gaussian processes regression (SGPR) models with trajectory data. The results show that the prediction performance can be gradually improved by adding more mean speed estimates of traveled distance from the first 5 min as the real-time information. The overall performances of SGPR models are very similar to full GP, supported vector regression (SVR) and artificial neural network (ANN) models. The computational complexity of SGPR models is $O(mn^2)$, and it does not require lengthy model fitting process as SVR and ANN. This makes GP models more practicable for real-world practice in large-scale transportation data analyses.

Keywords: Travel time prediction · Machine learning · Gaussian Processes · Sparse approximation · Trajectory data · Freight vehicle

1 Introduction

Travel time prediction is essential to transportation research. Accurate travel time prediction can: help to improve traffic management and control; help personal and business travellers make better decision on departure time, transportation mode and route selection; help public and freight transportation companies to make better planning and task scheduling. Travel time prediction is difficult because travel time fluctuates from time to time and can be influenced by many situational factors such as traffic flow conditions, weather and public events. The prediction can be even more difficult when it is performed at individual level due to the difficulties of modelling driving behaviours of each individual driver.

A large number of data-driven methods were proposed for traffic and travel time prediction in the last few decades. The data-driven methods can be classified into: parametric and non-parametric methods. The parametric method

© Springer International Publishing AG 2016
H. Yin et al. (Eds.): IDEAL 2016, LNCS 9937, pp. 143–153, 2016.
DOI: 10.1007/978-3-319-46257-8_16

includes linear regression models (e.g., [7,18,19]), Kalman filtering models (e.g., [2,8]), and autoregression integrated moving average (ARIMA) based models (e.g., [11,16,21]). In those models, the temporal (time-varying) or temporal-spatial (e.g., taking neighbourhood road links into account) relations between inputs and predictive outputs are required to be explicitly defined and parameters are carefully estimated. Specifying such predictive functions can be very difficult and parameters estimation varies from case to case.

When the dimension of model inputs becomes larger, it becomes very difficult to model the relations between the inputs and the observed targets exogenously. This led to the adoption of non-parametric methods with strong function approximation ability such as artificial neural network (ANN). Numerous types of ANN models were proposed for travel time prediction (e.g., [14,15]). Fitting ANN models is always a challenge work: network should be carefully configured to avoid overfitting; model training can take quite long time to avoid local minima. Also, the interpretation of the model results can be difficult.

Kernel method is another popular non-parametric method for travel time prediction. Wu et al. in [17] used a support vector regression (SVR) model to predict travel time on highways. Haworth et al. in [5] used ridge regression models to perform both offline and online predictions. Gaussian processes (GP) is a full Bayesian kernel method. It does not require lengthy model fitting as SVR and ANN models, and the model result can be interpreted in a probabilistic sense (thus prediction uncertainty can be reported). Current few related studies (see [6,20]) used full GP regression (GPR) models to predict traffic flow and travel time. The computational complexity of full GP is $O(n^3)$. The computation of predictive model can be very costly if the data size is large. Thus, in this study, we used sparse GPR (SGPR) models to predict travel time for freight vehicle at individual level, with features extracted and composed from the vehicles' temporally sparse trajectory data.

The remaining paper is organised as follows: it begins with a brief introduction of SGPR model. Section 3 introduces the experimental setup includes data preparation, feature selection, kernel selection, inducing set selection and experiment design. Next, the results are reported and discussed. At the end, conclusions and ideas of future research are outlined.

2   Preliminary. Sparse GPR (SGPR) Model

A regression problem has a dataset $\mathcal{D}$ consisting of n input vectors $\mathbf{X} = \{\mathbf{x}_i\}_{i=1}^n$ of D-dimension and corresponding targets $\mathbf{y} = \{y_i\}_{i=1}^n$. Our task is to find the latent function f that $y_i = f(\mathbf{x}_i) + \epsilon$, where ϵ is some i.d.d. Gaussian noise, $\epsilon \sim \mathcal{N}(0, \sigma_n^2)$. In GPR we assume the values of f are normal random variables. Let $\mathbf{f} = \{f(\mathbf{x}_i)\}_{i=1}^n$, we have the prior distribution over $\mathbf{f}$:

$$\mathbf{f} \sim \mathcal{N}(\mathbf{m}, \mathbf{K}) \tag{1}$$

where $\mathbf{m} = [m(\mathbf{x}_1), m(\mathbf{x}_2), ..., m(\mathbf{x}_n)]^T$ is the vector of mean functions and $\mathbf{K} = \mathbf{K}(\mathbf{X}, \mathbf{X})$ is the $n \times n$ matrix of covariance functions. In this study, we set $m(\mathbf{x}) =$

0, thus a GP is fully specified by its covariance matrix $\mathbf{K}$. Applying Bayes' theorem, we get the posterior distribution over $\mathbf{f}$:

$$p(\mathbf{f}|\mathbf{X}, \mathbf{y}) = \frac{p(\mathbf{y}|\mathbf{X}, \mathbf{f})p(\mathbf{f}|\mathbf{X})}{p(\mathbf{y}|\mathbf{X})} \tag{2}$$

where $p(\mathbf{y}|\mathbf{X}, \mathbf{f})$ is the likelihood and $p(\mathbf{y}|\mathbf{X}) = \prod_{i=1}^{n} p(y_i|\mathbf{x}_i, \theta)$ is the marginal likelihood. θ is the hyperparameter of covariance function $k(\mathbf{x}, \mathbf{x})$ in $\mathbf{K}$ which can be solved by maximising the log marginal likelihood [12]:

$$\log p(\mathbf{y}|\mathbf{X}) = \log[\mathcal{N}|0, \mathbf{K} + \sigma_n^2 \mathbf{I}] \tag{3}$$

$\mathbf{I}$ is the identical matrix of size n. To predict the latent value f_* for a test case $\mathbf{x}_*$, we can first compute the $n+1$ dimensional joint prior $p(f_*, \mathbf{f})$, then apply Bayes' theorem to get the joint posterior $p(f_*, \mathbf{f}|\mathcal{D}, \mathbf{x}_*)$. Marginalising $\mathbf{f}$, the predictive distribution can be obtained with mean and variance [10]:

$$\bar{f}_* = \mathbf{k}_*^T (\mathbf{K} + \sigma_n^2 \mathbf{I})^{-1} \mathbf{y}, \tag{4}$$

$$\mathbb{V}[f_*] = k(\mathbf{x}_*, \mathbf{x}_*) - \mathbf{k}_*^T (\mathbf{K} + \sigma_n^2 \mathbf{I})^{-1} \mathbf{k}_* \tag{5}$$

where $\mathbf{k}_* = \mathbf{k}(\mathbf{X}, \mathbf{x}_*)$. Equations (4) and (5) requires inverse of $n \times n$ covariance matrix $\mathbf{K}$ which reveals that the computational complexity is $\mathcal{O}(n^3)$. Making prediction can be very slow if n is more than few thousands.

To reduce the computational complexity of GPR model, various of *sparse* approximation methods were proposed and those GPR models are known as sparse GPR (SGPR) models. In SGPR models, an *inducing set* was introduced that contains m inducing inputs $\mathbf{X}_u$ and corresponding inducing targets $\mathbf{u} = \{u_i\}_{i=1}^{m}$. The joint prior can then be approximated by assuming that f_* and $\mathbf{f}$ are conditionally independent given $\mathbf{u}$ [9]:

$$p(f_*, \mathbf{f}) \simeq q(f_*, \mathbf{f}) = \int q(f_*|\mathbf{u})q(\mathbf{f}|\mathbf{u})p(\mathbf{u})d\mathbf{u} \tag{6}$$

Quionero-Candela et al. in [9] summarised that the main difference of different sparse approximation methods is their additional assumptions about $q(f_*|\mathbf{u})$ and $q(\mathbf{f}|\mathbf{u})$. This treatment reduces the computational complexity to $O(mn^2)$. Snelson, et al. in [12] proposed a method where $\mathbf{X}_u$ are not real inputs selected from training set but pseudo-inputs. $\mathbf{X}_u$ and the hyperparameters (θ, σ_n^2) are learnt using an approximation to the true log marginal likelihood in Eq. (3) [13]:

$$\log p(\mathbf{y}|\mathbf{X}) \simeq \log p(\mathbf{y}|\mathbf{X}, \mathbf{X}_u) = \log[\mathcal{N}(\mathbf{y}|0, \mathbf{Q} + \sigma_n^2 \mathbf{I})] \tag{7}$$

where $\mathbf{Q} = \text{diag}[\mathbf{K} - \mathbf{K}_{f,u}\mathbf{K}_{u,u}^{-1}\mathbf{K}_{u,f}] + \mathbf{K}_{f,u}\mathbf{K}_{u,u}^{-1}\mathbf{K}_{u,f}$, $\mathbf{K}_{f,u} = \mathbf{K}(\mathbf{X}, \mathbf{X}_u)$, $\mathbf{K}_{u,u} = \mathbf{K}(\mathbf{X}_u, \mathbf{X}_u)$ and $\mathbf{K}_{u,f} = \mathbf{K}(\mathbf{X}_u, \mathbf{X})$. Titsias in [13] proposed a variational inducing learning method to improve Snelson and Ghahramani's model: in stead of modifying the exact GP model, author tries to minimise a distance between the exact posterior GP and a variational approximation. Then $\mathbf{X}_u$ becomes variational parameters that can be jointly optimised with (θ, σ_n^2). Minimising the distance is equivalent to maximising a lower bound of regularised approximation to the true log marginal likelihood [13]:

$$L = \log[\mathcal{N}|\mathbf{0}, \mathbf{Q} + \sigma_n^2\mathbf{I})] - \frac{1}{2\sigma_n^2}\mathrm{Tr}(\tilde{\mathbf{K}}) \tag{8}$$

where $\mathbf{Q} = \mathbf{K_{f,u}K_{u,u}^{-1}K_{u,f}}$ and $\tilde{\mathbf{K}} = \mathbf{K} - \mathbf{K_{f,u}K_{u,u}^{-1}K_{u,f}}$. The regularisation term $\frac{1}{2\sigma_n^2}\mathrm{Tr}(\tilde{\mathbf{K}})$ is added to avoid overfitting. In this study, we adopted Titsias's method as the sparse treatment. For more details, please refer to [13].

3   Experimental Setup

3.1   Data

Three route segments were selected from the drayage network of Ningbo Port for this study. The trajectory data was collected from 16 March to 30 April, 2014. The trajectory reporting frequency is 30 s. Trips were first identified using geospatial fences predefined using Google Map and their travel times were estimated by estimating the time that passed over the start and end points of each route segment. The description of the three corresponding datasets are shown in Table 1.

Table 1. Datasets.

Dataset	Segment	Length (meters)	# of valid trips
R1	2nd Tong Dao (West→East)	7187	1143
R2	2nd Tong Dao (East→West)	5847	1643
R3	G329 (East→Wast)	7993	1320

3.2   Feature Selection

Three types of features are used in this study: basic, historical travel time and mean speed sequence. The basic features are mainly date-time relevant features as shown in Table 2. The historic travel time is the historical interval mean travel time. A look-up table was composed for each vehicle where each row contains an interval and its corresponding mean travel time calculated using training set only. In this study, we use a 15-min sampling window to collect trips for each interval. Two rules are applied in response to the situation when no trip is found at a given interval derived from the departure time of a given test case: (a) if no trip is found at this interval, the interval mean travel time of the trips from all the other vehicles is used; (b) if no trip is found at this interval given all vehicles, the mean travel time at the previous interval is used.

With the basic and historical travel time features, we can perform *pre-start* (before a trip starts) prediction for all the vehicles. Furthermore, we can add

Table 2. Basic information in a trip tuple.

Feature	Date type	Comment
Departure time	String	String format: 'yyyy-mm-dd h:m:s'
Day of week	Numeric	{1, 2, ..., 7}, 1: Monday, 7: Sunday
Month	Numeric	{1, 2, ..., 12}
Day in month	Numeric	{1, 2, ..., 31}
Day in year	Numeric	{1, 2, ..., 366}
Weekday	Binary	{0, 1}, 1: Yes, 0: No
Workday	Binary	{0, 1}, 1: Yes, 0: No
Public holiday	Binary	{0, 1}, 1: Yes, 0: No

real-time information that is extracted from the currently available trajectory data to perform *post-start* (after a trip starts but before it ends) prediction. Mean speed sequence is introduced as the real-time information. It is a sequence of mean speed estimates of traveled distance. Assume we want to perform post-start prediction at the Tth minute after trip starts, let the mean speed sequence be $S_T = \{\hat{v}_i | i = 1, 2, ..., k\}$ where $\hat{v}_i = d_i/t_i (t_i \leq T)$ is the ith mean speed estimate of the traveled distance d_i, t_i is the traveled time. Since the trajectory report frequency is 30 s, this allows us to estimate the mean speed of traveled distance every 30 s. Thus, k is equivalent to the number of trajectory reports received within T minutes.

The importance of the basic and historical travel time features was also examined using gradient boosting regression tree (GBRT) models where the importance is measured according to the frequency of co-occurrences of these features in all the splits of GBRT models. The results show that: departure time and historical interval mean travel time are relatively important features; month, weekday, workday and public holiday are relatively unimportant features. Month is less useful probably because of the relatively short time span of the data. The selected route segments are far away from the urban area and are less likely to be effected by in-stream and out-stream traffic flows of the urban area. This makes weekday, workday and public holiday less important in our study. Therefore, month, weekday, workday and public holiday are removed from the feature column in our experiments.

3.3 Kernel Selection

The covariance matrix $\mathbf{K}$ consists of $n \times n$ covariance functions, each of which is also known as a *kernel function*. A kernel function defines the similarity between two inputs. There are commonly used kernel functions such as squared exponential (SE), rational quadratic (RQ), periodic (PER) and linear (LIN). SE is a very popular kernel, which presents a smooth transition between two neighbouring inputs. RQ can be seen as an infinite sum of SE with multiple length-scales [10].

Early travel time analyses showed that there is no significant peak hours at daily or weekly level for all the three route segments. Also, no continued increasing and decreasing can be identified during the 46 days. Thus, we selected RQ as the kernel, to which automatic relevance determination (ARD) is applied. The ARD version RQ is expressed as follows:

$$RQ_{ard}(\mathbf{x}, \mathbf{x}') = \sigma^2 (1 + \frac{\|\mathbf{x} - \mathbf{x}'\|^T L^{-1} \|\mathbf{x} - \mathbf{x}'\|}{2\alpha})^{-\alpha} \tag{9}$$

where σ^2 is the variance w.r.t. the function's mean, α is the shape parameter, L is a diagonal matrix of length scale l_i^2 for each input dimension i.

3.4 Inducing Inputs Selection

Various of inducing set selection choices are discussed in [9]. Since we adopted Titsias's sparse treatment for GPR models, the inducing inputs are pseudo-inputs that are first randomly selected from training set and then jointly optimised with (θ, σ_n^2) using gradient based optimisation algorithm. Early study showed that the initial selection of inducing inputs can place an impact on the final result. Gal et al. in [3] initialised their inducing inputs using k-means. Arthur et al. in [1] proposed k-means++ which improves k-means by augmenting a randomised seeding technique. In this study, we used k-means++ to initialise the inducing inputs.

3.5 Experiment Design

The last 200 trips were used as the test sets where the rest were used to fit and train SGPR models for each route segment. Pre-start predictions were first performed using only basic and historical travel time features. Then post-start predictions were conducted with different mean speed sequence features. All the SGPR models were implemented using GPy - a Python Gaussian processes framework [4]. The limited memory bound-constrained optimisation method (L-BFGS-B) was adopted as the gradient optimisation method with the maximum iteration of 300 to jointly optimise ($\mathbf{X}_u$, θ, σ_n^2). SGPR models were also compared with full GPR, SVR and ANN models. The parameters of SVR and ANN models were tuned using grid search approach. The optimum parameter settings

Table 3. Parameter settings for SVR and ANN.

	ANN	SVR
R1	$N = 10$, $l = 0.01$, $m = 0.2$	Polynomial, $degree = 1$, $\epsilon = 0.125$, $C = 0.25$, $\gamma = 0.125$
R2	$N = 10$, $l = 0.01$, $m = 0.2$	RBF, $\epsilon = 0.25$, $C = 8.0$, $\gamma = 0.0078$
R3	$N = 14$, $l = 0.01$, $m = 0.2$	Polynomial, $degree = 1$, $\epsilon = 0.125$, $C = 0.25$, $\gamma = 0.031$

The network is a single hidden layer back-propagation ANN where the activation function is sigmoid, max epochs is 500. L - number of hidden layers, N - hidden neuron size, l - learning rate, m - momentum.

are shown in Table 3. Root mean squared error (RMSE) and mean absolute percentage error (MAPE) were used for the prediction performance evaluation.

4   Results

Figure 1 shows the RMSE and MAPE results of pre-start predictions using basic and historical travel time features with different inducing input sizes. Both RMSE and MAPE results show that: for R1, SGPR has the largest error with 5 inducing inputs; for R2, the difference is not easily identified among the results of different inducing input sizes; for R3, the prediction error experienced a rise-and-fall along with the increase of inducing input size. These observed results allow us to reduce the computational complexity of SGPR with the fewest possible inducing inputs. Therefore, we chose 20 inducing inputs for R1 and R2, and 5 inducing inputs for R3. For the sake of convenience, we use SGPR-5 and SGPR-20 to denote SGPR models with 5 and 20 inducing inputs respectively. Table 4 shows the comparison results of pre-start predictions between full GPR and SGPR models. Same kernel RQ_{ARD} is used for full GPR models. The SGPR models here refer to SGPR-20 (for R1 and R2) and SGPR-5 (for R3). The RMSE and MAPE (represented in 'mean±standard deviation') results show that the prediction performance of SGPR models is as good as full GPR models, but the inference time is significantly reduced.

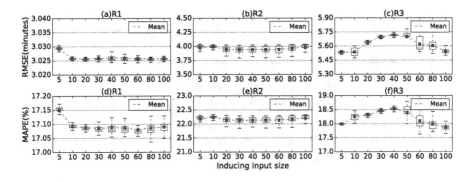

Fig. 1. RMSE and MAPE of pre-start predictions using different inducing input sizes.

To perform post-start predictions, mean speed sequence features were added as the real-time information. Figure 2 shows the RMSE and MAPE results of post-start predictions with mean speed sequences of the first T minutes where $T = \{0, 1, 2, 3, 4, 5\}$. When $T = 0$, no mean speed sequence feature is added, which is referred to as the pre-start prediction. From the figure, it can be seen that for R1 and R2, the prediction performance can be gradually improved by adding more mean speed estimates. Compared with pre-start predictions, the prediction error is reduced by about 2 % and 5 % in the case of R1 and

Table 4. RMSE and MAPE of pre-start predictions using full GPR and SGPR models.

	Full GPR (RMSE)	SGPR (RMSE)	Full GPR (MAPE)	SGPR (MAPE)
R1	3.026	3.026 ± 0.001	17.099 %	17.088 ± 0.015 %
R2	3.997	3.947 ± 0.047	22.240 %	22.146 ± 0.096 %
R3	5.473	5.534 ± 0.010	19.091 %	17.988 ± 0.045 %

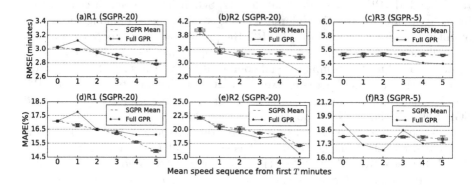

Fig. 2. RMSE and MAPE of post-start predictions using full GPR and SGPR models with mean speed sequences from the first T minutes. $T = \{0, 1, 2, 3, 4, 5\}$. When $T = 0$, no mean speed feature is added. This is equivalent to pre-start prediction.

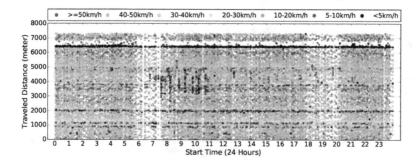

Fig. 3. Traveled distance versus departure time in a day in R3. The coloured dots are instant speed measures. (Colour figure online)

R2 respectively. In the case of R3, the mean speed sequence feature is not as helpful as in R1 and R2. Figure 3 shows the relation between traveled distance and departure time of trips in R3 where each data point is an instant speed measure. Several low instant speed measures (i.e., orange and red points) are observed in the range between 3000 and 5000 m during the morning time (i.e., 7–11 AM). This indicates that unusual and non-recurring events (e.g., bad signal settings, incident, road works) occurred that lessened the moving continuity of

Table 5. Comparison results of SGPR, SVR and ANN models.

Model	RMSE			MAPE		
	R1	R2	R3	R1	R2	R3
SGPR	<u>2.793</u>	3.287	5.549	<u>15.050 %</u>	<u>17.373 %</u>	18.078 %
SVR	2.909	3.716	<u>5.489</u>	16.186 %	20.333 %	<u>17.594 %</u>
ANN	3.474	<u>3.195</u>	5.933	17.601 %	19.506 %	18.074 %

some vehicles which can not be fully capture by the current features. One way to improve the prediction performance is to allow estimating additional travel times caused by unusual event based on real-time monitoring of traffic. This usually requires accessing real-time traffic flow data and expertise on traffic analyses which can be difficult. But then it still can be too late to make accurate prediction if an event occurs at a location close to the destination.

SGPR models were also compared with ANN and SVR models. The basic, historical travel time and mean speed sequence of the first 5 min are used for the comparative study. Table 5 shows the RMSE and MAPE results of those models. Since k-means++ produces different initial inducing inputs selection sometimes due to the randomised seeds, the results for SGPR models here are represented using the worse cases of 100 rounds predictions (i.e., the max values of RMSE and MAPE). The results show that in general the performance of SGPR models is very close to SVR and ANN models and at some cases it is slightly better. But SGPR models have two unique advantages: (a) they do not require lengthy model fitting processes as SVR and ANN models; (b) the model result is represented as a predictive distribution which is easier to interpret.

5   Conclusions

In this study, we performed travel time prediction using SGPR models for freight vehicles using their temporally sparse trajectory data. The test results show that the performance of SGPR models is as good as full GPR models in both pre-start and post-start predictions. The prediction performance can be gradually improved by adding more mean speed estimates of traveled distance from the first 5 min after trip starts. The comparative study results show that the performance of SGPR models is very close to SVR and ANN models and sometimes even slightly better. The inference time using SGPR models is much less compared with full GPR models. And unlike SVR and ANN models, SGPR models do not require lengthy model fitting processes and overfitting can be avoided.

Because SGPR is a full Bayesian framework, the model results can be interpreted in a probabilistic sense. In the next step, we will focus on predicting travel time uncertainty using SGPR models.

References

1. Arthur, D., Vassilvitskii, S.: k-means++: the advantages of careful seeding. In: Symposium on Discrete Algorithms (2007)
2. Chen, M., Chien, S.: Dynamic freeway travel-time prediction with probe vehicle data: link based versus path based. Transp. Res. Rec. J. Transp. Res. Board **1768**, 157–161 (2001)
3. Gal, Y., Der Wilk, M.V., Rasmussen, C.E.: Distributed variational inference in Sparse Gaussian Process Regression and latent variable models. Adv. Neural Inf. Process. Syst. **27**, 3257–3265 (2014)
4. The GPy authors: GPy: a Gaussian process framework in python (2012). http://github.com/SheffieldML/GPy
5. Haworth, J., Shawetaylor, J., Cheng, T., Wang, J.: Local online kernel ridge regression for forecasting of urban travel times. Transp. Res. Part C Emerg. Technol. **46**, 151–178 (2014)
6. Ide, T., Kato, S.: Travel-time prediction using Gaussian process regression: a trajectory-based approach. In: Proceedings of SIAM International Conference on Data Mining (2009)
7. Kwon, J., Coifman, B., Bickel, P.J.: Day-to-day travel-time trends and travel-time prediction from loop-detector data. Transp. Res. Rec. J. Transp. Res. Board **1717**, 120–129 (2000)
8. Nanthawichit, C., Nakatsuji, T., Suzuki, H.: Application of probe-vehicle data for real-time traffic-state estimation and short-term travel-time prediction on a freeway. Transp. Res. Rec. J. Transp. Res. Board **1855**, 49–59 (2003)
9. Quinonerocandela, J., Rasmussen, C.E.: A unifying view of sparse approximate Gaussian process regression. J. Mach. Learn. Res. **6**, 1939–1959 (2009)
10. Rasmussen, C.E., Williams, C.K.: Gaussian Processes for Machine Learning (Adaptive Computation and Machine Learning). The MIT Press, Massachusetts (2005)
11. Smith, B.L., Williams, B.M., Oswald, R.K.: Comparison of parametric and nonparametric models for traffic flow forecasting. Transp. Res. Part C Emerg. Technol. **10**(4), 303–321 (2002)
12. Snelson, E., Ghahramani, Z.: Sparse Gaussian processes using pseudo-inputs. Adv. Neural Inf. Process. Syst. **18**, 1257–1264 (2006)
13. Titsias, M.K.: Variational learning of inducing variables in sparse Gaussian processes. In: JMLR Workshop and Conference Proceedings, AISTATS, vol. 5, pp. 567–574 (2009)
14. Van Hinsbergen, C.P., Van Lint, J.W., Van Zuylen, H.J.: Bayesian committee of neural networks to predict travel times with confidence intervals. Transp. Res. Part C Emerg. Technol. **17**(5), 498–509 (2009)
15. Van Lint, J.W., Hoogendoorn, S.P., Van Zuylen, H.J.: Accurate freeway travel time prediction with state-space neural networks under missing data. Transp. Res. Part C Emerg. Technol. **13**(5–6), 347–369 (2005)
16. Williams, B.M., Hoel, L.A.: Modeling and forecasting vehicular traffic flow as a seasonal ARIMA process: theoretical basis and empirical results. J. Transp. Eng. **129**(6), 664–672 (2003)
17. Wu, C., Ho, J., Lee, D.T.: Travel-time prediction with support vector regression. IEEE Trans. Intell. Transp. Syst. **5**(4), 276–281 (2004)
18. Fei, X., Chung-Cheng, L., Liu, K.: A Bayesian dynamic linear model approach for real-time short-term freeway travel time prediction. Transp. Re. Part C Emerg. Technol. **19**(6), 1306–1318 (2011)

19. Zhang, X., Rice, J.A.: Short-term travel time prediction. Transp. Res. Part C Emerg. Technol. **11**(3–4), 187–210 (2003)
20. Xie, Y., Zhao, K., Sun, Y., Chen, D.: Gaussian Processes for short-term traffic volume forecasting. Transp. Res. Rec. J. Transp. Res. Board **2165**, 69–78 (2010)
21. Zhang, Y., Zhang, Y., Haghani, A.: A hybrid short-term traffic flow forecasting method based on spectral analysis and statistical volatility model. Transp. Res. Part C Emerg. Technol. **43**(Part 1), 65–78 (2014)

Study on the Detection of Cross-Site Scripting Vulnerabilities Based on Reverse Code Audit

Fen Yan[✉] and Tao Qiao

College of Information Engineering, Yangzhou University, Yangzhou 225127, China
yanfen@yzu.edu.cn, 1258858977@qq.com

Abstract. Cross-Site Scripting (XSS) is one of the most popular methods of current network attacks. The attackers mainly put malicious script into a web page through the vulnerabilities of the web application. This paper proposes an improved approach based on reverse code audit and static analysis to detect and extract the XSS vulnerabilities in the source code of the web application. In this paper, we give the theoretical definition and implementation algorithm related to this method. Also, our method can find the location of the vulnerability and the vulnerability of data source through the data link, so that testers and developers can fix vulnerabilities in Web applications immediately. Finally, the method is verified by experiment, which show that the method can not only effectively detect the potential XSS vulnerabilities in the code, but also significantly improve the detection efficiency of XSS vulnerabilities based on static analysis.

Keywords: Cross-Site scripting · Reverse code audit · Static analysis · Web application security

1 Introduction

In Recent years, with the rapid development of Internet technology, all kinds of security vulnerabilities in Web system presents a trend of more styles and more influence. Security reports pointed out that the reasons for these vulnerabilities are insufficient sanitation of inputs. According to the reports from OWASP [1] in 2013, Cross-Site Scripting (XSS) is one of the most destructive and influential web security vulnerabilities. The potential reasons for the existence of XSS is that the web applications quote the unsafe data in output operations without sufficient sanitation. According to its different characteristics and principles, XSS is mainly divided into three types [2]: Reflected XSS, Stored XSS and DOM Based XSS.

At present, the method of defending XSS vulnerability mainly includes two aspects. One is dynamic defense against XSS attacks in the operation, and the other is static analysis of the source code. By summarizing the previous work, this paper improves the XSS vulnerability detection method based on code audit. On the basis of source code analysis tool SOOT [3], we propose an improved approach based on reverse code audit and static analysis to detect and extract the XSS vulnerabilities in the source code of the web application.

© Springer International Publishing AG 2016
H. Yin et al. (Eds.): IDEAL 2016, LNCS 9937, pp. 154–163, 2016.
DOI: 10.1007/978-3-319-46257-8_17

Hydara [5] pointed out that development team should detect XSS vulnerabilities as completely as possible and make timely correction before the web application is on the line. Pixy [6] is the first relatively early open source static source scanning tool written by Java language. It can effectively detect the vulnerability in source code by tracking the dynamic data. However, the basis of this method is the data tag which is complicated. J. Dahse developed a static source analysis tool [8] which can detect a variety of vulnerabilities, including XSS vulnerabilities and SQL injection [7] vulnerabilities. The taint analysis [9] on the basis of information flow to determine the relationship between the vulnerability of the system and the user input node can effectively extract the XSS vulnerabilities in web application. In article [10], Shar proposed a method using code audit to detect XSS vulnerabilities. This method extracts the defense module in the source code so that the development team and the testers to verify whether these modules are sufficient against XSS attacks. It generates Control Flow Graph [11] through code audit, and then, it retrieves all the paths that may be passed from the input node to find all possible XSS vulnerabilities.

Through the research and improvement on the method of static analysis and code audit, this paper proposed a method based on reverse code audit to detect XSS vulnerabilities. The method can effectively ensure the detection effect, and greatly improve the detection efficiency.

2   Research Ideas

In the study of static analysis, the taint analysis treats each of the external data as the original tainted data. Once another data is generated from the tainted data, the data is marked as a stain data. As shown in Fig. 1, after the spread of a project, the original tainted data will form a huge tree data structure, with the original tainted data as the vertex. In general research methods, it needs to start from each original tainted data to traverse all routes, in order to determine the final flow of data. The complexity of this method in the detection of XSS vulnerabilities is very high.

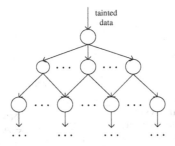

Fig. 1. Tainted data transmission diagram

Considering that the main cause of XSS is that the output node references outside data without sufficient sanitation. This paper defines all the output nodes as potential vulnerability node, then makes reverse scan from those nodes to determine whether the output node references a tainted data. As shown in Fig. 2, it starts scanning from node o1, and node o1 only pays attention to the relation of the node itself, that is node s3 and s5. This kind of analysis method greatly reduces the time needed to detect XSS vulnerabilities, and can improve the detection efficiency.

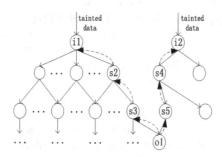

Fig. 2. Reverse data analysis diagram

The detection framework model we constructed is shown in Fig. 3, it mainly contains three modules: Firstly, it analyzes the source code to generate CFG [12] by prototype tool Soot [3]. The Control Flow Graph is a representation, using graph notation, of all paths that might be traversed through a program during its execution. Nowadays, the CFG is essential to many static analysis tools. Because the CFG can be more efficiently constructed directly from the program by scanning it for basic blocks [13], we use the CFG to track tainted data in this article. For example, the CFG in Fig. 5 is generated by the JSP code in Fig. 4. Secondly, on the basis of the CFG, we analyze the attributes of each node and the relationship among the nodes. Lastly, through the reverse scan of each output node, we extract the source of risk data which is used in output node.

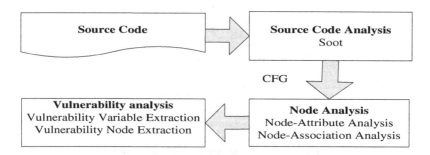

Fig. 3. The detection framework model

```
1:  String guestname=request.getParamter("guestname");
2:  Connection stmt=new Connection();
3:  if(guestname!=null&&guestname.length()<20){
4:    Stmt.Update("insert into guestbook values("+ guestname+")");
5:    String ww="welcome"+guestname.trim();
6:    out.println(ww);}
7:  ResultSet rs=stmt.Query("select * from guestbook");
8:  out.println("GuestList");
9:  while(rs.next()){
10: String name=rs.getString("guestname");
11: out.println(name);
12: If(name=="admim")
13: out.println("hello,admin!");}
```

Fig. 4. The JSP code of message book

3   Reverse Code Audit Method

3.1   Definition of Node Analysis

In Node Analysis module, we analyze each node in the CFG according to the following definition of rules.

Definition 1.   Data Source: In this paper, Data Source is the basis for judging whether the vulnerability exists. We divides the Data Source into two categories, one is the trusted data, and the other is the distrustful data.

trusted data: The constant string is the literal value, as the logical part of the program, they will not deliberately cause security vulnerabilities.

distrustful data: Distrustful data may come from different parts in the program. These parts may be client, document, database, session, etc.

Definition 2.   Node Class Attribute: In a CFG, each node has its special Node Class Attribute including input-node, judge-node, normal-node and out-node.

input-node: If a node's US(Used variable Set, Details are described below) contains distrustful data, then the node is called the input node.

judge-node: If a node is a judgment statement, such as if(…), while(…), for(…), etc., then the node is called judge-node.

out-node: If a node performs the HTML response operation, the node is called out-node. If the US of the node is not empty and all the variables in the node's US are not previously shown to be trusted, the node is also called potential vulnerability out-node.

normal-node: If a node only does the assignment, data processing and other operations, and it does not involve the above node division, the node is called normal-node.

Definition 3.   Node Variable Attribute: Each node in the CFG includes three variable attributes, respectively, DS, AS and US.

Define variable Set (DS): In a CFG, if a variable in the node is first defined, then the variable is called defined variable, and it is added to the US.

Assignment variable Set (AS): In a CFG, if the value of a variable in a node is been changed, the variable is called assignment variable, and it is added to the AS.

Used variable Set (US): In a CFG, if a variable in the node makes other variable's value changed, the variable is called used variable, and it is added to the US.

The result of node attribute analysis of CFG shown in Fig. 5 is shown in Table 1.

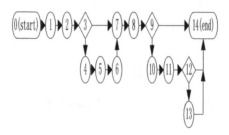

Fig. 5. The CFG of the JSP code shown in Fig. 4

Table 1. The table of node attribute analysis

Node ID	Attribute	DS	AS	US
0	start	Ø	Ø	Ø
1	input	{guestname}	{guestname}	Ø
2	normal	{stmt}	{stmt}	Ø
3	judge	Ø	Ø	{gusetname}
4	normal	Ø	Ø	{stmt,gusetname}
5	normal	{ww}	{ww}	{guestname}
6	out	Ø	Ø	{ww}
7	input	{rs}	{rs}	{stmt}
8	out	Ø	Ø	Ø
9	judge	Ø	Ø	{rs}
10	normal	{name}	{name}	{rs}
11	out	Ø	Ø	{name}
12	judge	Ø	Ø	{name}
13	out	Ø	Ø	Ø
14	end	Ø	Ø	Ø

Definition 4. Node Association Relationship: In node A and node B, if the US of node A and the AS of node B generate the intersection, that is US(A)∩AS(B) ≠ Ø, then the node B is associated with node A and this relationship is one-way, B—> A. For example, in node 1 and node 5, US(5)∩AS(1) ≠ Ø, node 1—> node 5.

Node Associated Path: Assuming that there are a number of nodes, node A,B,C,D..., if node B is associated with node A, node C is associated with node B, node D is associated with node C, and so on, then the Node Associated Path of A is {A,B,C,D...}.

Definition 5. Reverse scan path: In a CFG, the reverse path from the out-node to node O(start) is defined reverse scan path. The reverse scan paths of the CFG are {(6,5,4,3,2,1,start), (11,10,9,8,7,6,5,4,3,2,1,start), (11,10,9,8,7,3,2,1,start)}.

3.2   Node Analysis Algorithm

The premise of reverse path scan and analysis is the accurate node analysis of the CFG to determine the each node's DS, AS and US. In this section, Algorithm 1 "Node-Attribute Analysis Algorithm" gives a detailed description of node attribute analysis algorithm. The algorithm analyzes the attributes of each node In CFG according to the upper segment Definitions 2 and 3. According to the rules of the Definition 3, the variables will be divided into different sets. According to the rules of the Definition 2, each node's Node Class Attribute will be determined.

Algorithm 1. Node-Attribute Analysis Algorithm

```
Input: CFG
Output: Target.Attribute,Target.DS,Target.AS,Target.US
1: Attribute←normal, DS←∅,AS←∅,US←∅
2: for target_i∈CFG(0<i<CFG.length) do
3:   if target_i has variables that first defined
4:       target_i.DS←target_i.DS∪variables,end if
5:   if target_i has variables that values are changed
6:       target_i.AS←target_i.AS∪variables,end if
7:   if target_i has variables that change other variables'
     value
8:       target_i.US←target_i.US∪variables,end if
9:   if target_i.AS has variable submitted by user
10:      target_i.Attribute←input,end if
11:  if target_i performs a HTML response operation
12:      target_i.Attribute←out,end if
13:  if there are two branches that pass through target_i
14:      target_i.Attribute←judge,end if
15:Target←target_i.Attribute,taret_i.DS,target_i.AS,target_i.U
S
16: end for
17: return(Target.Attribute,Taget.DS,Target.AS,Target.US)
```

On the basis of Algorithm 1 and CFG, we look for a set of nodes association relationships according to the Definition 4. We use Algorithm 2 to analyze the relationship between the nodes and obtain the reverse scan path. Firstly, the algorithm makes Scan-node-Set empty. Then, the out-nodes which are needed to be scanned are added to Scan-node-Set. The nodes which are associated with the nodes in Scan-node-Set should be added to Scan-node-Set, and at the same time, the scanned nodes should been removed from Scan-node-Set., and so on, until the Scan-node-Set is empty. By Algorithm 2, we obtain associated path chain of the out-node. If there is input-node in associated path chain, then the chain is vulnerable.

Algorithm 2. Reversed Associated-Route Scan Algorithm

```
input: CFG,Target.Attribute,Target.DS,Target.AS,Target.US
output: Associated-Paths, PV-Variables
1:Associated-Paths←∅,PV-Variables←∅
2:for outi∈target(target.Attribute="out") do
3:   Path←∅,Scan-node-Set←∅,Associated-Path←∅
4:   if outi.US≠∅ do
5:   search the all Paths from outi to target0 through CFG
6:   Path←Path∪Paths,Scan-node-Set←Scan-node-Set∪outi,
     Associated-Path←Associated-Path∪outi
7:   for Pathn∈Path do
8:      while Scan-node-Set≠∅ do
9:        next-Scan-node-Set←∅
10:       scan-node=Scan-node-Set[0]
11:       next-Scan-node-Set←targetj(0≤j<scan-node.location)
12:       for nextnode∈next-Scan-node-Set
13:         if scan-node.US⊠nextnode.AS≠∅ do
14:         Associated-Path←Associated-Path∪nextnode,
15:         Scan-node-Set← Scan-node-Set∪nextnode
16:         end if, end for
17:         delete Scan-node-Set[0]
18:      end while, end for, end if
19: Associated-Paths←Associated-Paths∪Associated-Path
20: if node.Attribute=="input"&&node.AS≠∅ (node∈Associ
    ated-Path)
21: PV-Variables←PV-Variables∪node.AS, end if
22: end for
23: return(Associated-Paths,PV-Variables)
```

Next, we verify the vulnerable path to detect whether there are vulnerabilities in the path. If there are multiple input-nodes on the vulnerable path of the out-node A, we control the variables of the input-nodes. We can use data variation method [12] for constructing cross site scripting case in one input-node and use normal variables in other input-nodes. If it is proved that an input-node B has defects when we test, then it is said that input-node B is the vulnerability of the out-node A.

4   Experimental Results and Analysis

As shown in Fig. 1, the conventional algorithm of Taint Analysis is mainly to traverse all the tainted data propagation graph. Through the CFG and the initial input-node that needed to scan, the conventional algorithm of Taint Analysis uses Ant Algorithms [4] to find out all the associated paths and vulnerable routes of the input-node.

The reverse code audit algorithm proposed by this paper first uses Soot to generate CFG of the source code. Then the Algorithm 1 is used to analyze the Node Class

Attribute, DS, AS and US of the each node in the CFG. Next, the Algorithm 2 is used to get the Associated-Path. If there is an input-node in this path, this path is determined as a vulnerable path.

We use the conventional algorithm of Taint Analysis and the reverse code audit algorithm proposed in this paper to scan the JSP code shown in Fig. 4 separately. The result is shown in Table 2.

Table 2. The table of potential vulnerability scan results

Scanning method	Taint Analysis Algorithm		Reverse Code Audit Algorithm	
Initial scan node	1	7	6	11
Number of scanning paths	6	3	1	2
Associated Route	{(1,3), (1,4), (1,5,6)}	{(7,9), (7,10,11), (7,10,12)}	{(6,5,1)}	{(11,10,7,2)}
Vulnerable routes	{(1,5,6)}	{(7,10,11)}	{(6,5,1)}	{(11,10,7,2)}
Potential vulnerability	{(6:ww)}	{(11:name)}	{(6:ww)}	{(11:name)}
Source of vulnerability	{(1:guestname)}	{(7:rs)}	{(1:guestname)}	{(7:rs)}
Scanning time	6914.02μs		534.53μs	

It can be seen from the experiment result from Table 2 that the time efficiency of the proposed algorithm proposed by us is greatly improved compared with the original one. As in Taint Analysis Algorithm, all the data entered from the input-node are regraded as the source of the stain. In the process of data transmission, as long as the tainted data interact with any other data, other data are also considered dangerous. However, not all data can cause potential XSS risk. In the process of scanning, the paths generated by the data will increase with the increase scale of the source code of web application, thus the cost of scanning will increase also. In this paper, the algorithm is proposed to remove the redundant scan of data. It carries out the reverse scan from the output node using the idea of reverse scan. It only seeks nodes that have relationship with itself in order to locate the potential vulnerability node more quickly.

The test result shows that the risk of XSS is existing in JSP code shown in Fig. 4. Variable "guestname" of node 1 is the source vulnerability of node 6,and Variable "rs" of node 7 is the source vulnerability of node 11

Table 3. The table of the test to WebGoat 5.0

	Taint Analysis Algorithm	Reverse Code Audit Algorithm
Number of scanning paths	134	107
Number of XSS vulnerabilities	31	31
Scaiing time	78.47ms	27.34ms

In order to prove the effectiveness of the algorithm proposed by this paper, we used both Reverse Code Audit Algorithm and Taint Algorithm to test WebGoat [14]. Web-Goat is a Java Web Application full of loopholes that OWASP is responsible for it. We detected 12 JSP pages of the WebGoat which contain 855 JSP codes and the result is shown in Table 3.

Through the analysis of Table 3, we can know that Reverse Code Audit Algorithm can successfully detect the XSS vulnerabilities detected by Taint Analysis Algorithm. In addition, the Reverse Code Audit Algorithm does not need backtracking to find data branch, so the Reverse Code Audit Algorithm improves the efficiency of searching for XSS vulnerabilities.

5   Conclusion

From the point of view of code audit, we proposes reverse code audit algorithm for the source code. Using our method in this paper, we can find the vulnerabilities in the source code more quickly through the reverse analysis of the data stream. Finally, based on our method, we implement a source analysis tool shown in Fig. 3 to test the effectiveness and efficiency of the method. In the follow-up research work, we will consider making the tool apply to more complex frameworks, such as Struts. In addition, we will explore how to automatically fix the different vulnerabilities found.

References

1. Open Web application security project. OWASP top 10-2013. The Ten Most Critical Web Application Security Risks (2013). https://www.owasp.org/index.php/Top_10_2013
2. Wang, W., Li, J.: Web Application Security Threats and Prevention: Based on OWASP Top 10 and ESAPI, vol. 1. Electronic Industry Press, Beijing (2013)
3. Soot. Soot: a Java Optimization Framework. http://www.sable.mcgill.ca/soot/. Accessed 12 Feb. 2009
4. Dorigo, M., Caro, G.D., Gambardella, L.M.: Ant algorithms for discrete optimization. Artif. Life 5(2), 137–172 (1999)
5. Hydara, I., Sultan, A.B.M., Zulzalil, H., et al.: Current state of research on cross-site scripting (XSS)–a systematic literature review. Inf. Softw. Technol. 58, 170–186 (2014)
6. Jovanovic, N., Kruegel, C., Kirda, E.: Pixy: a static analysis tool for detecting web application vulnerabilities (short paper). In: 2006 IEEE Symposium on Security and Privacy, pp. 258–263 (2006)
7. Anley, C.: Advanced SQL injection in SQL server applications. Insight Security Research (2002)
8. Dahse, J.: A vulnerability scanner for different kinds of vulnerabilities. http://rips-scanner. sourceforge.net
9. Newsome, J., Song, D.: Dynamic taint analysis for automatic dedection, analysis, and signature generation of exploits on commodity software. In: Network and Distributed System Security Symposium (NDSS) (2005)
10. Shar, L.K., Tan, H.B.K.: Auditing the defense against cross site scripting in web applications. In: Proceedings of the 2010 International Conference on Security and Cryptography (SECRYPT), pp. 1–7. IEEE (2010)

11. Sinha, S., Harrold, M.J., Rothermel, G.: Interprocedural control dependence. ACM Trans. Softw. Eng. Methodol. **10**(2), 209–254 (2001)
12. Chen, J.F., Wang, Y.D., Zhang, Y.Q., et al.: Automatic generation of attack vectors for stored-XSS. J. Grad. Univ. Chin. Acad. Sci. **29**(6), 815–820 (2012)
13. Tarr, P.L., Wolf, A.L.: Engineering of Software: The Continuing Contributions of Leon J, p. 58. Osterweil. Springer, Heidelberg (2011). ISBN 978-3-642-19823
14. WebGoat, OWASP WebGoat Project. https://www.owasp.org/index.php/Category

A MapReduce-Based ELM for Regression in Big Data

B. Wu[1], T.H. Yan[1(✉)], X.S. Xu[1], B. He[2], and W.H. Li[3]

[1] School of Mechanical and Electrical Engineering,
China Jiliang University, Hangzhou 310018, China
{wu_bin0106, thyan, lionkingxxs}@163.com
[2] School of Information Science and Engineering College,
Ocean University of China, Qingdao 266100, China
bhe@ouc.edu.cn
[3] School of Mechanical, Materials and Mechatronic Engineering,
University of Wollongong, Wollongong, NSW, Australia
weihuali@uow.edu.au

Abstract. Regression is one of the most basic problems in machine learning. In big data era, for regression problem, extreme learning machine (ELM) can get better generalization performance and much fast training speed. However, the enlarging volume of dataset for training makes regression by ELM a challenging task, and it is hard to finish the training in a reasonable time or it will be out of memory. In this paper, through analyzing the theory of ELM, a MapReduce-Based ELM method is proposed. Under the MapReduce framework, ELM submodels are trained in every slave node parallelly. A combination method is designed to combine all the submodels as a complete model. The experiment results demonstrate that the MapReduce-Based ELM can efficient process big dataset on commodity hardware and it has a good performance on speedup under the cloud environment where the dataset is stored as data block in different machines.

Keywords: ELM · Regression · Machine learning · Mapreduce · Big data

1 Introduction

With the rapid development of information technology and the Internet of Things, we have entered the big data era. The volume, velocity, variety and value (4 V) have been the main features of big data [1]. The knowledge hides in big data is valuable to make decision in various fields. Machine learning has become one of the popular methods for knowledge discovery, which has been a focus in big data field. Extreme learning machine (ELM) as a novel machine learning algorithm for single-hidden layer feedforward networks (SLFNs) has been one of the most important algorithms for regression and classification [2–5]. Many variations of ELM have been proposed in practical applications, such as basic ELM [6, 7], kernel ELM [8, 9] and incremental ELM [10, 11], etc.

© Springer International Publishing AG 2016
H. Yin et al. (Eds.): IDEAL 2016, LNCS 9937, pp. 164–173, 2016.
DOI: 10.1007/978-3-319-46257-8_18

Unlike most practical implementations that all the parameters of networks need to be tuned, ELM as a single-hidden layer feedforward networks, both input weights and hidden layer biases are chosen randomly, which are not necessarily iteratively tuned [6]. And the output weights of ELM can be determined by calculating the Moore-Penrose generalized inverse of the hidden layer output matrix H analytically [12]. Compared with the traditional SLFNs, ELM can provide good generalization performance and fast learning speed.

However, efficiency and scalability are great challenges for ELM when the dataset is large scale. Big sample data together with vast of hidden node will lead to a huge matrix H, and calculating general inverse of H becomes so complex [13] that it may cause low efficiency and out of memory problem. Because of that, it drives the increasing research on parallel algorithm and distributed computing [14–16]. The MapReduce model, as a cloud computing framework, is proposed by Google for parallel computing in a distributed environment [17, 18], and Hadoop is one of its open source implementation [19]. The main idea of MapReduce is to allow the computing across the cluster including many machines and users can only focus on the data processing strategy. The MapReduce environment provides two primary functions, Map and Reduce, to design the parallel computing task. Besides, it provides fault tolerance and dynamic flexibility support, so it can be deployed on common hardware easily.

This paper is organized as follows. Section 2 introduces MapReduce framework. Section 3 gives out some important theories of ELM. ELM for regression based on MapReduce is proposed in Sects. 4 and 5 shows the experimental result. Finally, in Sect. 6, some conclusions are presented.

2 MapReduce Framework

As one of current cloud computing technology, MapReduce provides a programming model for big data processing. The processing is parallel and it can handle machine failures. Generally, one job contains two main phases, which are called Map and Reduce, respectively. As shown in Fig. 1, the input data is divided into large-volume sample blocks which are located in different nodes of Hadoop cluster, and the whole blocks can be stored by the Hadoop distribute file system (HDFS). In the Map phase, the input data is processed by the map function, and it generates some intermediate result as the input of the combine function or reduce function. In the reduce phase, the intermediate result from map function or combine function are processed by the reduce function. It should be noted that the combine function, as a small reduce function, is not must exist. In our method, it exists between the map function and reduce function. What users just need to do is to design the map function, reduce function and combine function if exists. It is need not to care about the communication between the functions and cluster.

In MapReduce, the basic data structure is the <key, value> pair. In the Map phase, for each <key1, value1> pair in blocks, the map function is invoked, and after processing by the map function, it generates some intermediate <key2, value2> pairs. In our study, the Combine phase is existence, so it generate some <key3, value3> pairs by processing the <key2, value2> pairs. In Reduce phase, the intermediate pairs <key3,

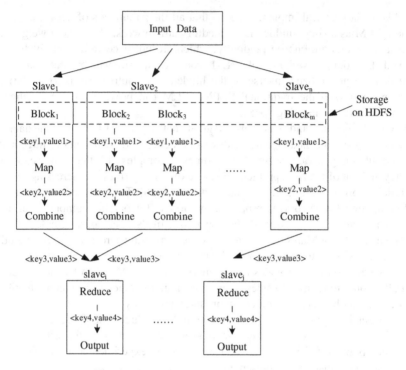

Fig. 1. Illustration of MapReduce framework

value3> are invoked by the function of reduce, and the final output <key4, value4> pairs are generated.

It should be noted that, the output <key, value> pair of map function, combine function or reduce function maybe different from the type of their input <key, value> pair. But, the type of the input of combine function should be the same with the type of the output of map function. Meanwhile, the type of the input of reduce function must be the same with the type of the output of combine function. The relationship of the type of <key, value> pairs is shown as follows:

$$\text{Map: input} <key1, value1>, \text{intermediate output} <key2, value2> \qquad (1)$$

$$\text{Combine: input} <key2, value2>, \text{intermediate output} <key3, value3> \qquad (2)$$

$$\text{Reduce: input} <key3, value3>, \text{final output} <key4, value4> \qquad (3)$$

Developers can implement their applications by using the three functions. And after that, MapReduce runtime distributes and executes the task automatically. Thus, the complexity of parallel programming is dramatically reduced.

3 ELM

ELM is proposed by Huang in his paper [20], and it is developed from the single-hidden layer feedforward neural networks (SLFNs) which is shown in Fig. 2. ELM randomly choses the input weight a_i and the hidden layer biases b_i, and then analytically determines the output weight β_i of SLFNs through generalized inverse operation of the hidden layer output matrix. The hidden layer of SLFNs need not be iteratively tuned, which is different from the common SLFNs. It can achieve better generalization performance than other conventional algorithms at an extremely fast learning speed, and it not only reach the smallest training error but also the smallest norm of the output weight.

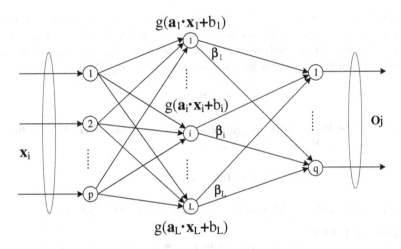

Fig. 2. Illustration of ELM

For N arbitrary distinct samples (x_i, t_i), where $x_i = \left[x_{i1} x_{i2} \cdots x_{ip}\right]^T \in R^p$ and $t_i = \left[t_{i1} t_{i2} \cdots t_{iq}\right]^T \in R^q$. SLFNs with L hidden nodes activation function $g(x)$ can be represented as

$$\sum_{i=1}^{L} \beta_i g_i(x_j) = \sum_{i=1}^{L} \beta_i g(a_i x_j + b_i) = o_j, \quad j = 1, 2, \cdots, N \tag{4}$$

where $a_i = \left[a_{i1} a_{i2} \cdots a_{ip}\right]^T$ is the weight vector connecting the ith hidden node and the input nodes, $\beta_i = \left[\beta_{i1} \beta_{i2} \cdots \beta_{iq}\right]^T$ is the weight vector connecting the ith hidden node and the output nodes, b_i is the bias if the ith hidden node, and $o_j = \left[o_{j1} o_{j2} \cdots o_{jq}\right]^T$ is the jth output vector of the SLFNs [6]. SLFNs with L hidden nodes can approximate these N samples with zero error means that there exist a_i, β_i and b_i such that

$$\sum_{i=1}^{L} \beta_i g(a_i x_j + b_i) = t_j, \quad j = 1, 2, \cdots, N \tag{5}$$

The above N equations can be written as

$$H\beta = T \tag{6}$$

where

$$H = \begin{bmatrix} g(a_1 \cdot x_1 + b_1) & \cdots & g(a_L \cdot x_1 + b_L) \\ \vdots & \cdots & \vdots \\ g(a_1 \cdot x_N + b_1) & \cdots & g(a_L \cdot x_N + b_L) \end{bmatrix}_{N \times L}, \; \beta = \begin{bmatrix} \beta_1^T \\ \vdots \\ \beta_L^T \end{bmatrix}_{L \times q}, \; \text{and} \; T = \begin{bmatrix} t_1^T \\ \vdots \\ t_N^T \end{bmatrix}_{N \times q}.$$

In most cases, the number of hidden node is much less than the number of training sample. H is a nonsquare matrix and there not exist a_i, β_i and b_i such that $H\beta = T$. But we can see the Eq. (5) as a multiple regression system, which β is the regression vector to be solved. The smallest norm least-squares solution of Eq. (5) is

$$\widehat{\beta} = H^+ T \tag{7}$$

where H^+ is the *Moore-Penrose* generalized inverse of matrix H [12]. So the estimation value matrix Y can be expressed by

$$Y = H\widehat{\beta} = HH^+ T \tag{8}$$

As discussed above, ELM algorithm can be described in Algorithm 1.

Algorithm 1. ELM

Given a training set $\aleph = \{(x_i, t_i) | x_i \in R^p, t_i \in R^q, i = 1, \cdots, N\}$, activation function $g(x)$, and hidden node number L.

Step 1: Randomly assign input weight a_i and bias b_i, $i = 1, \cdots, L$.

Step 2: Calculate the hidden layer output matrix H.

Step 3: Calculate the output weight $\widehat{\beta}$.

4   ELM for Regression Based on MapReduce

Under the MapReduce framework, each Hadoop node trains an ELM submodel with the local sample block which can be one or more. Training is moved to the nodes which stores sample blocks. So it costs much less I/O resources than collecting all the blocks together for training. After training, we propose a method to combine the ELM submodels by adopting the generalized inverse method to calculate the weights of each Hadoop node. When combined, the output weights β are recalculated with the combination weights w which is calculated by using the Moore-Penrose generalized inverse with additional small data set. The whole framework is shown in Fig. 1.

In Fig. 1, the whole framework contains two main parts: (1) Every data block is trained in Map and Combine phase. (2) Combine all the submodels in Reduce phase. It contains n slave nodes in cluster and the dataset is divided into m data blocks in HDFS. So m submodels will be trained in n slave nodes. Each Map function outputs the pair of <key2, value2>. The String of key2 contains the hidden nodes' input weight a and the bias b. Value2 contains the output of every hidden node. The String of key3 contains the hidden nodes' input weight a and the bias b, too. Value3 contains the output weight β for all hidden nodes. The pair of <key3, value3> will be executed by Reduce function and all the submodels will be combined in the Reduce phase. In this process, the combination weight w is calculated with one Reduce function by using another small dataset. The final output pair of <key4, value4> contains the input weight a, biases b, output weight β and the combine weight w.

In Fig. 2, we assume that the dataset of block$_i$ is x_i. Based on the algorithm in Sect. 3, we can get the value of β as

$$\beta = H^+ T_x \tag{9}$$

Then, another small dataset X is adopted to calculate the combination weight w. With X, the output of each ELM submodel can be represented as

$$o_i = \sum_{j=1}^{L} \beta_j g(a_j X + b_j), \quad i = 1, 2, \cdots, m \tag{10}$$

Assuming that $o = \left[o_1^T o_2^T \cdots o_m^T \right]^T$. The output of all blocks is represented as

$$O = w_1 o_1 + w_2 o_2 + \cdots + w_m o_m \tag{11}$$

where $ow = T_X$, so $w = o^+ T_X$.

It should be noted that, in our mode, only one Reduce function is used to calculate the combination weight w. Now, the whole model of MapReduce-Based ELM is established. The detail of MapReduce-Based ELM is shown in Algorithm 2.

Algorithm 2. MapReduce-Based ELM

Set the size of block as 8 M and the number of Reduce as 1. Assuming the big dataset for training is x and the small sample data for calculating the combination weight is X. The number of hidden node is L, and the activation function is $g(x)$.

Step 1: Randomly assign input weight and bias in every submodel.

Step 2: Calculate the hidden layer output matrix H in every submodel by using blocks located in different Hadoop node which is divided by the big dataset x.

Step 3: Output <key2, value2> pair.

Step 4: Calculate the output weight in every submodel by combine function.

Step 5: Output <key3, value3>.

Step 6: Calculate the output of each submodel by using the dataset X.

Step 7: Calculate the combination weight by using the output of every submodel.

Step 8: Output <key4, value4>.

5 Experiments

In this section, two experiments are designed to evaluate the performance of the MapReduce-Based ELM proposed in Sect. 4. The first experiment is designed to show the accuracy for regression and the second experiment is to display the performances with different size of cluster.

The experiments run in a cluster with some virtual machines on Linux Operation System, where each has 512 M memory and a 2.4 GHz core. The MapReduce framework is configured with Hadoop version 1.1.2 and JDK version 1.8.0_77. In our cluster environment, the size of each block is set as 8 M.

5.1 Experiments for Regression Accuracy

In order to illustrate regression accuracy of MapReduce-Based ELM, the datasets sinc, winequality-red, winequality-white and some others are used in experiments. All of them come from two famous machine learning repositories which are UCI and FCUP. The detail of datasets are shown in Table 1. Besides, the number of hidden node is assigned 20, the transfer function of every hidden node is "Sigmoid". In the above conditions, the regression accuracy on datasets is shown in Table 2.

Table 1. Datasets for regression

Name	Training data	Combining data	Testing data	Attributes
sinc	5000	1000	5000	2
winequality-red	1200	300	99	12
winequality-white	4000	700	198	12
airfoil self noise	1000	300	100	6
Yacht Hydrodynamics	200	50	50	7

In Table 2, the average root mean square error (RMSE) of training dataset and testing dataset are calculated. It is shown in Table 2 that the accuracy of MapReduce-Based ELM is as small as the classical ELM.

Table 2. Regression accuracy of ELM and MapReduce-Based ELM

Name	Size	ELM		MapReduce-base ELM	
		Training RMSE	Testing RMSE	Training RMSE	Testing RMSE
sinc	5000	0.1162	0.0062	0.1347	0.0238
winequality-red	1200	0.7826	0.6731	0.0683	0.5349
winequality-white	4000	0.9042	0.7720	0.9001	0.8316
airfoil self noise	1000	0.6703	0.7142	0.7204	0.8116
Yacht Hydrodynamics	200	0.7786	1.1770	0.8846	1.2112

5.2   Performance of MapReduce-Based ELM

In the experiment, five datasets which come from UCI are used for regression. The detail of datasets is shown in Table 3. All the datasets have large number of samples to train the submodels. A relatively small dataset is used to determine the combine weight. With different size of Hadoop cluster, by using the MapReduce-Based ELM, the different running time are shown in Table 4. From Table 4, we can find that the more slave has faster running time, which reflects positive correlation between training speed and cluster scale.

Table 3. Dataset for regression

Name	Training data	Combining data	Testing data	Attributes
sinc	3000000	5000	2000	2
ethylene_CO	4000000	170000	8000	19
ethylene_methane	4000000	170000	8000	19
Household power consumption	2000000	70000	5000	9
YearPredictionMSD	463715	30000	20000	90

Table 4. Runing time (s) of MapReduce-Based ELM

Name	Number of slave				
	1	3	5	7	9
sinc	15780.4	5352.6	5263.8	2778.6	2874.1
ethylene_CO	NaN	57894.3	35273.2	23314.4	19477.6
ethylene_methane	NaN	58344.5	34159.5	23112.2	18264.0
Household power consumption	35546.8	15234.9	8865.7	5477.4	5547.9
YearPredictionMSD	NaN	41327.4	23477.8	20874.1	15533.8

To display the effiency for MapReduce-Based ELM, the speedup concept is given as follow

$$speedup(m) = \frac{computing\ time\ on\ 1\ slave}{computing\ time\ on\ m\ slaves}. \tag{12}$$

Based on Table 4, the speedup can be gotten which is shown in Fig. 3. From Fig. 3, we can find that Hadoop cluster accelerates the training speed and it reflects positive correlation between training speed and cluster scale, too. But, when the cluster is large enough, the cluster can't further accelerates the training speed because of the number of block in each Hadoop node increasing more and more slow.

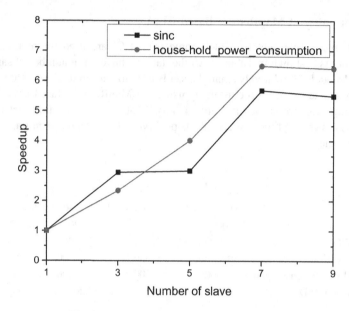

Fig. 3. Speedup of MapReduce-Based ELM

6   Conclusions

ELM algorithm has recently attracted a significant amount of research attention and it has been used to solve many regression problems. However, the data of large samples in practical applications makes the regression by ELM a challenging task. In this paper, by analyzing the theory of ELM, a MapReduce-Based ELM algorithm is proposed for big data learning. It trains submodels in each slave node and combines all the submodels after training with a small sample. In the cluster environment, experiments demonstrate that the proposed algorithm can learning from big data with high accuracy, and it has a good performance of speedup. In the future work, we will improve the combine function to achieve better performance.

Acknowledgment. This work is partially supported by the Natural Science Foundation of China & Key research and development program of China (51379198, 2016YFC0301404, 41176076, 31202036).

References

1. Lynch, C.: Big data: how do your data grow. Nature **455**, 28–29 (2008)
2. Ayerid, B., Grana, M.: Hyperspectral image nonlinear unmixing and reconstruction by ELM regression ensemble. Neurocomputing **174**, 299–309 (2016)
3. Qiu, S.S., Gao, L.P., Wang, J.: Classification and regression of ELM, LVQ and SVM for E-nose data of strawberry juice. J. Food Eng. **144**, 77–85 (2015)

4. Sa, J.J.D., Backes, A.R.: ELM based signature for texture classification. Pattern Recogn. **51**, 395–401 (2015)
5. Li, J.J., Wang, B.T., et al.: Probabilistic threshold query optimization based on threshold classification using ELM for uncertain data. Neurocomputing **174**, 211–219 (2016)
6. Huang, G.B., Zhu, Q.Y., Siew, C.K.: Extreme learning machine: theory and applications. Neurocomputing **70**, 489–501 (2006)
7. Cambria, E., Huang, G.B.: Extreme learning machine. IEEE Intell. Syst. **28**, 30–31 (2013)
8. Ma, C., Ouyang, J.H., et al.: A novel kernel extreme learning machine algorithm based on self-adaptive artificial bee colony optimisation strategy. Int. J. Syst. Sci. **47**, 1342–1357 (2016)
9. Deng, W.Y., Ong, Y.S., Zheng, Q.H.: A fast reduced kernel extreme learning machine. Neural Netw. **76**, 29–38 (2016)
10. Huang, G.B., Chen, L.: Enhanced random search based incremental extreme learning machine. Neurocomputing **71**, 3460–3468 (2008)
11. Huang, G.B., Li, M.B., et al.: Incremental extreme learning machine with fully complex hidden nodes. Neurocomputing **71**, pp. 576–583 (2008)
12. Lindstrom, A.: Generalized inverse of matrices and its applications. J. Oper. Res. Soc. **23**, 598 (1972)
13. Xin, J.C., Wang, Z.Q., et al.: Elastic extreme learning machine for big data classification. Neurocomputing **149**, 464–471 (2015)
14. Wang, B.T., Huang, S., et al.: Parallel online sequential extreme learning machine based on MapReduce. Neurocomputing **149**, 224–232 (2015)
15. Wang, X.L., Chen, Y.Y., et al.: Parallelized extreme learning machine ensemble based on min-max modular network. Neurocomputing **128**, 31–41 (2014)
16. He, Q., Zhuang, F., Li, J., Shi, Z.: Parallel implementation of classification algorithms based on MapReduce. In: Yu, J., Greco, S., Lingras, P., Wang, G., Skowron, A. (eds.) RSKT 2010. LNCS, vol. 6401, pp. 655–662. Springer, Heidelberg (2010)
17. Ghemawat, S., Gobioff, H., Leung, S.T.: The google file system. In: Proceedings of the 19th ACM Symposium on Operating Systems Principles, pp. 29–43 (2003)
18. Borthakur, D.: The Hadoop Distributed File System, Architecture and Design (2007)
19. Hadoop Official Website: http://hadoop.apache.org
20. Huang, G.B., Zhu, Q.Y., Siew, C.K.: Extreme learning machine: a new learning scheme of feedforward neural networks. In: Proceedings of the International Joint Conference on Neural Networks (IJCNN2004), vol. 2, pp. 985–990 (2004)

Very Deep Neural Network for Handwritten Digit Recognition

Yang Li[✉], Hang Li, Yulong Xu, Jiabao Wang, and Yafei Zhang

College of Command Information Systems, PLA University of Science and
Technology, Nanjing 210007, China
solarleeon@outlook.com

Abstract. Handwritten digit recognition is an important but challenging task. However, how to build an efficient artificial neural network architecture that can match human performance on the task of recognition of handwritten digit is still a difficult problem. In this paper, we proposed a new very deep neural network architecture for handwritten digit recognition. What is remarkable is that we did not depart from the classical convolutional neural networks architecture, but pushed it to the limit by substantially increasing the depth. By a carefully crafted design, we proposed two different basic building block and increase the depth of the network while keeping the computational budget constant. On the very competitive MNIST handwriting benchmark, our method achieve the best error rate ever reported on the original dataset ($0.47\% \pm 0.05\%$), without data distortion or model combination, demonstrating the superiority of our work.

Keywords: Neural network · Convolutional neural networks · Deep learning · Handwritten digit recognition

1 Introduction

Handwritten digit recognition is a promising subfield of object recognition with various applications. In the last ten years, automatic handwritten digit recognition capabilities have dramatically improved due to advances in deep learning and convolutional neural networks (CNNs). The performance of the new methods on the well-known MNIST dataset have reduce the recognition error rate from several percentage points to 1% [1], then down to 0.5% [2], then down to 0.23% [3].

However, almost all of the successful methods were trying to improve recognition accuracy in three different ways. The first method attempts to design a better architecture, which is better for handwritten digit recognition [4,5]. The second method improves the accuracy by enlarging the MNIST dataset. So far, the best results on MNIST were obtained by deforming training images, thus greatly increasing their number [3]. The third method combines several training models and makes decision by swarm intelligence [6].

© Springer International Publishing AG 2016
H. Yin et al. (Eds.): IDEAL 2016, LNCS 9937, pp. 174–182, 2016.
DOI: 10.1007/978-3-319-46257-8_19

To the best of our knowledge, data augmentation and model combination can improve almost all of the recognition methods, but they also lead to more training time. Even if one was able to train many different large networks, using them all at test time would be infeasible in applications where it is important to respond quickly. One discouraging news is that a lot of this progress is not a consequence of new ideas, algorithms and improved network architectures, but mainly just the result of a larger dataset and combination of several models.

In this paper, we will focus on a new efficient deep neural network architecture for handwritten digit recognition. The basic idea of this paper is taking inspiration and guidance from the theoretical work by Andrew Zisserman et al. [7] and Christian Szegedy et al. [8], who use smaller receptive window size and smaller stride of convolutional layer to build very deep CNNs for ILSVRC-2014. By a carefully crafted design, we increased the depth of the network while keeping the computational budget constant. It was demonstrated that the representation depth is beneficial for the classification accuracy, and that the state-of-the-art performance on the MNIST dataset can be achieved using a CNNs architecture with substantially increased depth. The benefits of the architecture are experimentally verified on the MNIST dataset without data augmentation and model combination, where it could reach comparable performance of the state-of-the-art approaches with less computation burden and shorter training time.

The rest of this paper is organized as follows. Section 2 describes the proposed architecture using CNNs module in details. Then experimental results and comparisons are shown in Sect. 3. Finally, the conclusion and future work are given in Sect. 4.

2   The Proposed Architecture

In this section, we elaborate how to build a hierarchical feature extraction and classification system with CNNs module. As illustrated in Fig. 1, the model architecture is mainly based on CNNs. We first briefly review CNNs, and then we depict the proposed mode in details.

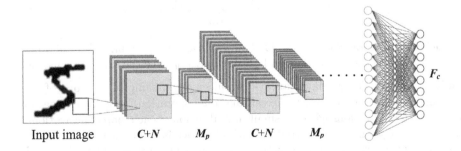

Input image $C+N$ M_p $C+N$ M_p

Fig. 1. CNNs feature extractor architecture.

2.1   A Brief Review of CNNs

Starting from LeCun [1], CNNs can be considered to be made up of two main parts. The first part typically had a standard structure stacked convolutional layers, which followed by contrast rectification layers (denoted as R) and average-pooling layers (denoted as A_p). The input of each layer is just the output of its previous layer. As a result, this forms a hierarchical feature extractor that maps the original input images into feature vectors. The second part are one or more fully-connected layers, which is a typical feed forward neural network trying to classify the extracted features vectors. In this paper, the proposed architecture is composed by 6 different component: convolutional layer (denoted as C), normalization layer (denoted as N), max-pooling layer (denoted as M_p), fully-connected layers (denoted as F_c), and we also adopt dropout (denoted as D_p) and Rectied Linear Units ($ReLU$) method in our network.

Convolutional Layer: In convolutional layer, each neuron is connected locally to its inputs of the previous layer, which functions like a 2D convolution with certain filter, then its activation could be computed as the result of a nonlinear transformation. In this paper, convolutional layer computes the convolution of the input image $x \in \mathrm{R}^{H \times W \times D}$ with a filter bank $f \in \mathrm{R}^{H' \times W' \times D \times D''}$ with D'' multi-dimensional. Formally, the output is $y \in \mathrm{R}^{H'' \times W'' \times D''}$ given by:

$$y_{i''j''d''} = b_{d''} + \sum_{i'=1}^{H'} \sum_{j'=1}^{W'} \sum_{d=1}^{D} f_{i'j'd} \times x_{i'+i''-1,j'+j''-1,d',d''} \qquad (1)$$

where H represents the input image height, W represents the input image width, D represents the number of channel, H' represents filter height, W' represents filter width, D'' represents the number of filter bank, H'' represents the output image height, W'' represents the output image width.

Normalization Layer: The local contrast normalization layer is inspired by computational neuroscience models. Normalization layer applied independently at each spatial location and groups of channels to get:

$$y_{ijk} = x_{ijk} (\kappa + \alpha \sum_{t \in G(k)} x_{ijt}^2)^{-\beta} \qquad (2)$$

For each output channel k, $G(k) \subset \{1, 2, ..., D\}$ is a corresponding subset of input channels. And the input image x and output image y have the same dimensions.

Pooling Layer: Pooling layer always produces down sampled versions of the input maps. This pooling involves executing some operation, typically average or max, over the activations within a small spatial region of each map of activations. Typically max pooling is preferred as it avoids cancellation of negative elements and prevents blurring of the activations and gradients throughout the network since the gradient is placed in a single location during back propagation. The max pooling operator computes the maximum response of each feature channel in a $H' \times W'$ patch, resulting in an output of size $y \in \mathrm{R}^{H'' \times W'' \times D}$:

$$y_{i''j''d} = \max_{1 \le i' \le H', 1 \le j' \le W'} x_{i'+i''-1, j'+j''-1, d} \tag{3}$$

ReLU: Typically the convolutional responses are passed through a non-linear activation function such as sigmoids, tanh, or *ReLU* [14] to produce activation maps. The *ReLU* can be compute as follows:

$$y_{ijd} = \max(0, x_{ijd}) \tag{4}$$

Dropout: The key idea of dropout is to randomly drop units (along with their connections) from the neural network during training. This method significantly reduces overfitting and gives major improvements over other regularization methods. The choice of which units to drop is random. In the simplest case, each unit is retained with a fixed probability p independent of other units, where p can be chosen using a validation set or can simply be set at 0.5, which seems to be close to optimal for a wide range of networks and tasks. For any layer l, $r^{(l)}$ is a vector of independent Bernoulli random variables each of which has probability p of being 1. This vector is sampled and multiplied element wise with the outputs of that layer $y^{(l)}$, to create the thinned outputs $\widetilde{y}^{(l)}$ [4].

$$r^{(l)} = Bernoulli(p) \tag{5}$$

$$\tilde{y}^{(l)} = r^{(l)} * y^{(l)} \tag{6}$$

Fully-Connected Layer: After multiple convolutional and pooling layers, a convolutional network typically has one or more fully connected neural net layers with weights W and biases b before the final classier. The entire network is trained with back-propagation of a supervised loss such as the cross entropy of a softmax classier output and the target labels y represented as a 1 of c vector, where c is the number of classes to discriminate.

$$y = -\sum_{ij} \left(x_{ijc} - \log \sum_{d=1}^{D} e^{x_{ijd}} \right) \tag{7}$$

2.2 Combining Modules into a Hierarchy Architecture

In order to design a new deep convolutional neural network architecture which is discriminative enough for handwritten digit recognition. Here we use the basic module to build up our model. However, different architectures can be produced by cascading the above-mentioned modules in various ways. According to [18], they point out the basic building block of convolutional networks are $C + A_p$ layer, $C + R + A_p$ layer, $C + R + N + A_p$ layer, $C + M_p$ layer. In this paper, we propose two different basic building block for our architecture.

$C+N+M_p$ **Layer:** This is the first basic building block of our convolutional networks. This block is composed of convolutional layer, normalization layer and max pooling down sampling layer. This block is just like human visual cortex which is used for feature extraction and nonlinear dimensionality reduction in our model.

C+N+ReLU Layer: This is second the basic building block of our convolutional networks, which compose of a convolutional layer followed by a normalization layer and a *ReLU* layer. This block is used for higher level feature extraction.

Table 1 shows the whole setting of our CNNs architecture. The input to our network is a fixed-size 28×28 gray image. The image is passed through a stack of different layers. Except for the first convolution filter, we use very small 3×3 receptive fields throughout the whole net. The max pooling layer here is non-overlapping and no rectification. Max-pooling is carried out over a 2×2 pixel window, with stride 2. It should be noticed that all weight convolutional layers are equipped with normalization. To the best of our knowledge, our CNNs architecture is the deepest model for handwritten digit recognition.

Table 1. CNNs architecture and parameters.

Layer	Type	Output size	Kernel size/Stride
1	Convolutional	$20 \times 24 \times 24$	$5 \times 5/1$
2	Normalization	$20 \times 24 \times 24$	—
3	Max pooling	$20 \times 12 \times 12$	$2 \times 2/2$
4	Convolutional	$40 \times 10 \times 10$	$3 \times 3/1$
5	Normalization	$40 \times 10 \times 10$	—
6	Max pooling	$40 \times 5 \times 5$	$2 \times 2/2$
7	Convolutional	$150 \times 3 \times 3$	$3 \times 3/1$
8	Normalization	$150 \times 3 \times 3$	—
9	ReLU	$150 \times 3 \times 3$	—
10	Convolutional	$150 \times 1 \times 1$	$3 \times 3/1$
11	Normalization	$150 \times 1 \times 1$	—
12	ReLU	$150 \times 1 \times 1$	—
13	Dropout(rate 0.4)	$150 \times 1 \times 1$	—
14	Convolutional	$150 \times 1 \times 1$	$1 \times 1/1$
15	Normalization	$150 \times 1 \times 1$	—
16	ReLU	$150 \times 1 \times 1$	—
17	Dropout (rate 0.1)	$150 \times 1 \times 1$	—
18	Fully connected	10	—

3   Experiments and Analysis

The experiments were run on the MNIST dataset. The MNIST dataset consists of handwritten digits 0–9 which are gray scale 28×28 pixel digit images. There are 60,000 training images and 10,000 testing images in total.

So far, the best results on MNIST were obtained by deforming training images, thus greatly increasing their number. This allows for training networks with many weights, making them insensitive to in-class variability. However, in this paper our network is not trained on numerous slightly deformed images, because we want to build a better learning model but not a simple model with seeing more data.

Our implementation is derived from the publicly available MatConvNet toolbox [9]. Matlab 2015a is used to conduct all the operations, running on a system with Intel Core i5-4690 CPU (3.50 GHz), 16 GB DDR3. Initial weights of the CNNs are drawn from a uniform random distribution in the range $[-0.01, 0.01]$. The training is carried out using mini-batch gradient descent (based on back-propagation) with momentum. The batch size was set to 100, momentum to 0.9. The training was regularized by weight decay (the L_2 penalty multiplier set to 5×10^{-4}) and dropout regularization for the two fully-connected layers (dropout ratio set to 0.4 and 0.1). The learning rate was initially set to 0.3, and then decreased by a factor of 3 when the validation set accuracy stopped improving.

The classification performance is evaluated using two measures: the top-1 and top-5 error. The former is a multi-class classification error, i.e. the ratio of incorrectly classified images; the latter is the main evaluation criterion used in the ILSVRC, and is computed as the ratio of images such that the ground-truth category is not within the top-5 categories. Figure 2 show one result of the training and testing accuracy on each epoch. From the left picture, we can see that the energy (training and testing loss) was decay dramatically with the training epoch. From the right picture, we can see that our method was learning fast which can get optimal performance 0.47 % error rate after 29 epoch iteration.

Finally, we compare our best-performing single network result with ten state-of-the-art methods. Table 2 shows the comparison of the recognition rate between the proposed architecture method and other recently reported results. All the methods in the experiment do not using data augmentation and model combination, our model secure the first place with 0.47 % ± 0.05 % test error rate. To the best of our knowledge, this is the best error rate ever reported on the original MNIST dataset, without distortions or model combination. The best previously reported error rate was 0.53 % [18]. In addition, without data augmentation, our method dramatically relieve the training procedure. In terms of training time, the proposed architecture method takes 34.84 min. It is much faster than [3], which needs to train up to 35 CNNs and costs 14 h even when the GPU parallelization is carried out.

To further understand the learnt model, we also draw the first convolutional layer of the learnt filters in Fig. 3. An intriguing pattern is observed in the filters of MNIST dataset. We can see both horizontal and vertical stripes, for these patterns attempt to capture the edges of the images.

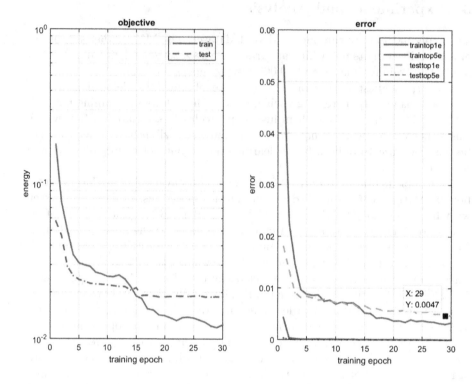

Fig. 2. Experimental results on MNIST dataset

Table 2. Test set misclassification rates for the best single model methods on the permutation invariant MNIST dataset.

Method	Test error
Srivastava et al. [4]	1.05 %
Salakhutdinov et al. [10]	0.95 %
Ranzato et al. [11]	0.60 %
Maxout NET [12]	0.94 %
Goodfellow et al. [13]	0.91 %
Deng et al. [14]	0.83 %
Rifai et al. [15]	0.81 %
Hinton et al. [16]	0.79 %
Zeiler et al. [17]	0.59 %
Jarrett et al. [18]	0.53 %
Our method	**0.42**% (0.47 % ± 0.05 %)

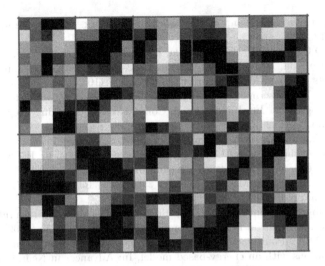

Fig. 3. The filters learned on MNIST dataset. There are 20 filter in the first stage and their size are 5×5

4 Conclusion and Future Work

In this paper, we propose a novel very deep network for handwritten digit recognition tasks. Unlike the shallow neural network used in many 1990s applications, ours are very deep. By a carefully crafted design, we propose two different basic building blocks and increase the depth of the network. The experiment result demonstrated that the representation depth is beneficial for the classification accuracy. Last and most importantly, the proposed model is a simple and computationally efficient approach for handwritten digit recognition tasks. On the very competitive MNIST handwriting benchmark, the proposed method achieve the best error rate ever reported on the original dataset, without distortions or model combination (0.47 % ± 0.05 %). In the future, we will further explore the potential representation ability of CNNs for various visual recognition tasks.

References

1. LeCun, Y., Bottou, L., Bengio, Y., Haffner, P.: Gradient-based learning applied to document recognition. Proc. IEEE **86**, 2278–2323 (1998)
2. Simard, P.Y., Steinkraus, D., Platt, J.C.: Best practices for convolutional neural networks applied to visual document analysis. In: 7th IEEE International Conference on Document Analysis and Recognition, pp. 958–963 (2003)
3. Ciresan, D., Meier, U., Schmidhuber, J.: Multi-column deep neural networks for image classification. In: IEEE Conference on Computer Vision and Pattern Recognition (CVPR), pp. 3642–3649 (2012)
4. Srivastava, N.: Improving neural networks with dropout. University of Toronto (2013)

5. Lin, M., Chen, Q., Yan, S.: Network in network. ArXiv preprint (2014). arXiv:1312.4400
6. Ciresan, D.C., Meier, U., Gambardella, L.M., Schmidhuber, J., Cire, D.C., Meier, U., Gambardella, L.M.: Handwritten digit recognition with a committee of deep neural nets on gpus. ArXiv preprint (2011). arXiv:1103.4487
7. Simonyan, K., Zisserman, A.: Very deep convolutional networks for large-scale image recognition. In: International Conference on Learning Representations (ICLR), pp. 1–14 (2015)
8. Szegedy, C., Liu, W., Jia, Y., Sermanet, P., Reed, S., Anguelov, D., Erhan, D., Vanhoucke, V., Rabinovich, A.: Going deeper with convolutions. In: IEEE Conference on Computer Vision and Pattern Recognition (CVPR), pp. 1–9 (2015)
9. Vedaldi, A., Lenc, K.: MatConvNet. In: 23th ACM International Conference on Multimedia, pp. 689–692 (2015)
10. Salakhutdinov, R., Hinton, G.E.: Deep Boltzmann machines. In: 12th International Conference on Artificial Intelligence and Statistics, pp. 448–455 (2009)
11. Ranzato, M.A., Poultney, C., Chopra, S., Lecun, Y.: Efficient learning of sparse representations with an energy-based model. In: Advances in Neural Information Processing Systems (NIPS), pp. 1137–1134 (2006)
12. Goodfellow, I.J., Warde-Farley, D., Mirza, M., Courville, A., Bengio, Y.: Maxout networks. ArXiv preprint (2013). arXiv:1302.4389
13. Goodfellow, I., Courville, A., Bengio, Y.: Joint training of deep Boltzmann machines for classification. In: International Conference on Learning Representations Workshops (ICLRW) (2013)
14. Deng, L., Yu, D.: Deep convex net: a scalable architecture for speech pattern classification. In: Proceedings of the Annual Conference of the International Speech Communication Association (INTERSPEECH), pp. 2285–2288 (2011)
15. Rifai, S., Dauphin, Y.: The manifold tangent classifier. In: Advances in Neural Information Processing Systems (NIPS), pp. 2294–2302 (2011)
16. Hinton, G.E., Srivastava, N., Krizhevsky, A., Sutskever, I., Salakhutdinov, R.R.: Improving neural networks by preventing co-adaptation of feature detectors. ArXiv preprint (2012). arXiv:1207.0580
17. Wan, L., Zeiler, M., Zhang, S., LeCun, Y., Fergus, R.: Regularization of neural networks using dropconnect. In: Proceedings of the International Conference on Machine Learning (ICML), pp. 1058–1066 (2013)
18. Jarrett, K., Kavukcuoglu, K., Ranzato, M., LeCun, Y.: What is the best multi-stage architecture for object recognition? In: Proceedings of the IEEE International Conference on Computer Vision (ICCV), pp. 2146–2153 (2009)

An FW-BF Based Approach on Elimination of Duplicated Web Pages

Leiming Ma and Zhengyou Xia[✉]

College of Computer Science and Technology,
Nanjing University of Aeronautics and Astronautics, Nanjing 211106, China
Mlm19910311@163.com, zhengyou_xia@nuaa.edu.cn

Abstract. With the blooming development of social network, Internet turns into the most widely information source. However, there are a large amount of duplicated web pages most of which are from being reprinted. Border et al. used to do an experiment on a collection of 30,000,000 HTML and text documents. It turned out that nearly 18 % of the pages are exactly the same and 41 % of the pages share 51 % similarity. These replicas of web pages has brought a major burden for the search engines and affecting the performance of the search engines badly. So elimination of duplicated web pages has become a very hot spot in information retrieval field in these years. In this paper, we have proposed a function word(FW) based approach which involves the concept of Bloom Filter(BF) to eliminate duplicated web pages without extracting the web main text. Our approach involves three separate stages. Stage 1 is to extract sample text according to function words feature in web pages. In stage 2, the feature code is extracted using function words. In stage 3, the duplicated web pages would be eliminated by similarity calculation of their BloomFilters.

Keywords: Duplicated web page elimination · Function Word · Bloom Filter · Feature code

1 Introduction

Duplicated web page elimination is originated from the text copy detection algorithm, but there is a clear distinction between the two [1]. Web page is weakly structured or semi-structured, where paragraphs and title information are vague as there is no obvious identification [12, 13]. The definition of duplicated web pages is different according to different applications. These related algorithms can be generally divided into three types:

(1) URL-based elimination algorithm has a basic assumption that the network resources with the same URL are the same. Effective hash function should be designed to convert URL strings into less memory usage hash value as the strings need great storage overhead, such as MD5, SHA-1 and so on [2]. In addition, Ding Zhenguo [3] proposed an URL-based algorithm using Bloom Filter to save space. However, these methods do not make full use of the structure of web page as there exists a webpage with several URL.

© Springer International Publishing AG 2016
H. Yin et al. (Eds.): IDEAL 2016, LNCS 9937, pp. 183–191, 2016.
DOI: 10.1007/978-3-319-46257-8_20

(2) Link-based elimination algorithm. Hyperlink is an important part of web page reflecting link relations and topic relevance between web pages, that is, if the similarity of any two pages' inbound and outbound link information is greater than a predetermined threshold, then the two pages thought to be repeated. However, a new page has less inbound link, and the outbound link is often determined by the site management personnel according to the requirement for the template or theme.

(3) Content-based elimination approach. Most of the existing methods use web content to identify the similarity as the two methods above is not very mature and often with inefficient results. Generally speaking, content-based method should firstly extract a set of features from the web page text. Compare the feature of this web page with those in the feature library. The web page is thought to be duplicated page if the similarity is greater than the threshold.

In this paper, An FW-BF based approach that integrates Bloom Filter with the function words' characteristic is designed to eliminate duplicated web pages. This is a novel content-based elimination approach as it doesn't need to extract the whole main text of web page, which means it is time-saving. Besides, the feature code used in Bloom Filter is made of the function words, which is also different with other methods. Function words are the word that have no real meaning but semantic or functional meaning that include adverb, preposition, conjunction, auxiliary word, interjection and mimetic word [4, 5].

The remainder of this paper is organized as follows. Sect. 2 presents the FW-BF based approach for eliminating duplicated web pages from UL library. In order to evaluate our approach, we conduct a few experiments on real world data sets, and the comparison results are presented in Sect. 3. Finally, we make a conclusion in Sect. 4.

2   FW-BF Based Approach

In this paper, we use the standard proposed by Yang. H [7] as that of this paper: if there are more than 80 % are the same in the use of words between the two, and the length of the difference is not more than 20 %, then the 2 documents are duplicated.

Our algorithm does not need to extract the main body of web page like other algorithms. We only do removal operation on the URL database, so the efficiency of the algorithm is greatly improved compared with others. This paper mainly consists of two tasks: extract feature code according to the characteristics of function words and web page elimination using Bloom filter.

To be convenient, we have the following definition:

Big paragraph: the paragraph having at least 3 function words.

Long paragraph: the number of the paragraph is not less than L. For example, the paragraph is long paragraph if the number is not less than 20 when L is 20.

2.1 Sample Text Extraction

The more information a paragraph carry, the more uncertainly the information receiver can exclude from the perspective of information science. Syntactically, the two pages are repeated if the long paragraphs in the two pages are almost the same. And the more words a paragraph has, the more function words it requires. Conversely, a paragraph has higher possibility to be long paragraphs if it has more function words. Therefore, we can judge whether two pages are duplicated by comparing their big paragraphs.

We do an experiment on the 12000 URL library to calculate the precision and recall whether t big paragraphs can instead of main text or not. The precision is 99.4 % when t is 3, which means most web pages can be judged to be duplicated or not according to their three big paragraphs. The recall is 86.4 % as there are some special cases that cannot be determined by big paragraphs. The precision is 90 % when t is 2 and the recall is 78.5 % when t is 4. So this paper takes 3 big paragraphs to represent the whole page.

2.2 Feature Code Extraction

Although whether two pages are similar can be judged by sample text, sample text often is very long and contains a lot of noise characters such as punctuation. Comparing the similarity between two sample texts needs a lot of time, so we need to extract feature code. A big paragraph with at least 3 function words has a 99.2 % probability to be a body paragraph of the main text [8]. Thus, the feature code can consist of the first Chinese character from the beginning, the ending, both sides of function words and the function words. And the basis is as follows:

(1) feature code is a part of the web page, and its length is short enough. It is not easy to be disturbed by noise information, and it can represent the sample text.
(2) the probability of two different sample text with same feature code is very small. In Chinese, there are 6700 commonly used Chinese characters, and function words selected in this paper are 52. Assume a sample text owns k function words, the probability that another sample text repetition is the same with it is $1/(52)^k(6700)^{3k+2}$, which can rarely happen.

2.3 BloomFilter Generation

A Bloom filter is a method for representing a set $A = \{a_1, a_2, ..., a_n\}$ of n elements (also called keys) to support membership queries [6]. The idea is to allocate a vector v of m bits, initially all set to 0, and then choose k independent hash functions each with range $(1,...,m)$. There is a certain probability that we are wrong. This is called a "false positive". The parameters k and m should be chosen such that the probability of a false positive is acceptable.

The salient feature of Bloom filters is that there is a clear tradeoff between m and the probability of a false positive. Observe that after inserting n keys into a table of size m, the probability that a particular bit is still 0 is exactly:

$$\left(1 - \frac{1}{m}\right)^{kn} \approx e^{-kn/m} \tag{1}$$

In order to facilitate the description, we make $p = e^{-kn/m}$. The value of all corresponding positions of the k hash values must be 1 when it is a false positive. We assume that the set of every position is an independent event, the probability of a false positive in this situation is:

$$f = (1 - p)^k = (1 - (1 - \frac{1}{m})^{kn})^k \approx (1 - e^{-kn/m}) \tag{2}$$

We can know from formula 2 that false positive rate of BF algorithm is determined by the size of the array, the number and random of hash function and the size of the data set. Michael Mitzenmacher [9] found that when

$$k = (\ln 2) \bullet (m/n) \tag{3}$$

The false positive rate is the lowest. At this point, $p = 1/2$, that is, the false positive rate is the lowest when the vector v is half empty, so the false positive rate f is:

$$f = (0.6285)^{\frac{m}{n}} \tag{4}$$

Jiao meng [10] fount that m should be at least 1.44 times than the minimum value. From formula4 we can know that $f = 0.0214$ when $m = 8n$ and the number of hash functions is 6, and $f = 0.0082$ when $m = 10n$ and the number of hash functions is 7.

We can ensure that the correct rate is above 97 % when $m = 8n$, and we can get k equals 6 according to formula 4.

2.4 Similarity Definition

In order to express the similarity degree of two web pages more precisely, we call the proportion of the same feature code the two pages share "similarity". Similarity of two web pages A and B is defined as follows according to the DSC algorithm:

$$s(A, B) = \frac{|F(A) \cap F(B)|}{|F(A) \cup F(B)|} \tag{5}$$

Where F(A) and F(B) are feature code set of A and B, and |A| is the size of set A.

The containment of A in B is defined as:

$$c(A, B) = \frac{|F(A) \cap F(B)|}{|M(A, B)|} \tag{6}$$

Hence the similarity is a number between 0 and 1, and it is always true that $r(A, A) = 1$. Similarly, the containment is a number between 0 and 1 and $c(A,B) = 1$ if $A \subset B$.

2.5 Evaluation Metrics

We adopt the metrics in the paper of Eduardo et al. [11] to assess our results, which include precision, recall and F-measure. A higher value of precision indicates fewer wrong classifications, while a higher value of recall indicates less false negatives. They are calculated as follows:

$$\Pr ecision = \frac{|bag(C) \cap bag(Rel(D))|}{|bag(C)|} \tag{7}$$

$$Recall = \frac{|bag(C) \cap bag(Rel(D))|}{|bag(Rel(D))|} \tag{8}$$

Where bag(C) denotes the bag of extracted duplicated web pages. |bag(C)| is the number of the extracted duplicated web pages. Rel(D) is the bag of correct duplicated web pages.

$$F - measure = \frac{2 \cdot precision \cdot recall}{precision + recall} \tag{9}$$

3 Experiment and Analysis

In this section, we describe a set of experiments conducted on a corpus of real news Web sites in order to evaluate the performance of FW-BF. The crossover-randomized experiment is done in Sect. 3.1. The time complexity of our method is analyzed in Sect. 3.2. Finally, Sect. 3.3 provides the summary of the performance evaluation of FW-BF.

3.1 Crossover-Randomized Experiment

The experiment is performed on 12000 URL library including 10000 corresponding different pages, 1000 corresponding pages of complete repetition and 1000 corresponding pages of partial repetition(two contained relation web pages are partial repetition in this paper), then we random fetched 6000 URL library including 5000 different pages, 500 complete repeated pages and 500 partial repeated pages as training set, and the other 6000 URL library is test set.

We do an experiment on the training set to obtain statistical length distribution. The result shows that 49.5 % of feature code lengths are over 80 and 50.5 % of that are less than 80. The average is about 80, so the length of feature code is set to be 80 according to the training set, that is, n = 80. So k = 6, m = 8n = 640.

To every URL, do the following:

Step1: Read the corresponding page source code row by row. We regard every row as an independent paragraph and the paragraph including text is text paragraph in this paper for convenience. Judge if the paragraph is big paragraph when it is a text paragraph. If it is a big paragraph, record it.

Step2: Extract the first three big paragraphs as sample text if there exist at least three big paragraphs in the source code. Otherwise, go to step3.

Step3: Extract the three paragraphs having the longest words as sample text if there doesn't exist three big paragraphs.

Step4: Convert every text sample to a character stream, remove the number, letter and natural punctuation marks.

Step5: Read the character stream obtained in Step4. Record the first Chinese character, Chinese character from both sides of function words and the last Chinese character (Fig. 1).

Step6: Feature code extraction. Each web page has different number of paragraphs, and each paragraph has different length, so the length of each corresponding feature code length is different. Some only has a dozen Chinese characters, while some are more than 100 characters. To ensure that the length of the feature code is not too long, the length of feature code is set to be n. And n equals 80 in this paper. There are two sample feature code from two web pages: (http://www.js.xinhuanet.com/2015-10/15/c_1116835291.htm and http://sports.sina.com.cn/china/womenfootballs/2016-04-25/doc-ifxrpvea1189043.shtml)

Step7: MD5 generation. To the feature code obtained in step5, use MD5 hash function to get the unique MD5 values of the URL. If there exists a same MD5 in the database, then the page is completely duplicated. Otherwise, go to the next step.

新读会活动在江赛从今月的征爱就是就是天学会名誉会长持了颁式慈阴是千之都漂报的千化的源化把爱富的幸力让生活在淮阴这片上的人有更加优美的生境更加捷的配务更加全的社社

(a) feature code of sample 1

新发了新队的集帅比尼目的是的是考赛的球均不在不在本运会女足比赛式在巴行最终分在三档的中典和南分在了在了同运会女足比赛日在巴行为运会进下教拉比法国

(b) feature code of sample 2

Fig. 1. Feature code of sample pages

To sample 1 and 2, their corresponding MD5 values are:
"f1b28ce190a03f2a2ff57291fa6f2daa",   "a76bd43b43ff0853ab8f862d3cac227f".
They are not completely duplicated.

Step8: BloomFilter generation. Generate BloomFilter according to the feature code obtained above.

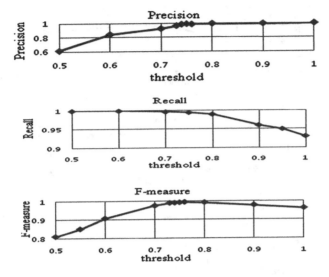

Fig. 2. Experiment results on URL training set

Step9: Threshold selection. Do an experiment on URL training set to select the threshold and the result is shown in Fig. 2.

Form the figure, we can get that F-score reaches 99.4 %, the maximum value, when the threshold value is 0.76. So the threshold value is selected as 0.76.

To samples, the similarity is 9.3 %, so the two sample pages is not duplicated.

Step10: Conduct the experiment on test URL set based on the selected threshold in Step9. When the similarity threshold is 0.76F-measure is 99.1 % and precision is 98.2 %.

Step11: Conduct the experiment on 10000 corresponding different pages(DP), 1000 corresponding pages of complete repetition(CRP) and 1000 corresponding pages of partial repetition(PRP) in test URL, and the result is shown in Table 1. Compared with Bloom Filter based and feature code based approaches, they are all good at accuracy, but our approach is time-saving. It is more obvious when dealing with big data.

Step12: To ensure accuracy and randomness, exchange the training set and test set. Carry out all steps above. The result is shown in Fig. 3 and Table 2.

Table 1. The runtime and accuracy of test set

Performance	FW-BF			BloomFilter			Feature code		
	CRP	PRP	DP	CRP	PRP	DP	CRP	PRP	DP
Avr Time(s)	3.191	6.614	31.173	3.191	9.098	39.146	16.012	15.739	150.401
F-score	1	98.60 %	99.30 %	1	99.20 %	99.30 %	1	93.40 %	97.70 %

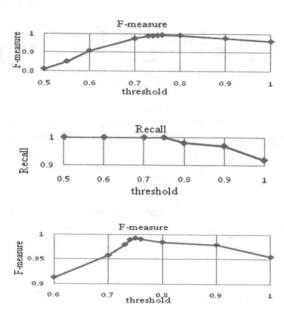

Fig. 3. Experiment results on exchanged test sets

The precision, recall and F-measure of total 6000 exchanged test sets are 98.6 %, 100 % and 99.3 % when the threshold is 0.75. From the results we can draw a conclusion that the FW-DTSS method is excellent in both efficiency and accuracy (Table 2).

Table 2. The runtime and accuracy of exchanged test set

Performance	FW-BF			BloomFilter			Feature code		
	CRP	PRP	DP	CRP	PRP	DP	CRP	PRP	DP
Avr Time(s)	3.542	7.003	33.912	3.8	10.171	42.207	18	19.174	179.176
F-score	1	99.00 %	99.20 %	1	99.10 %	99.20 %	1	95.70 %	91.00 %

4   Conclusion

In this paper, we study the FW-BF based duplicated web page elimination strategy, and prove the validity of the method through experiments. The FW-BF based method is a method without extracting main text from web pages, which is obviously time-saving. Experiment results show that the average runtime for our method is less than BloomFilter and Feature code based approaches. In addition to being efficient, our method can also achieve a satisfactory accuracy when compared with the others.

References

1. Weng, Y.: Research on NLP-based duplicated web pages detection algorithm. Beijing University of Posts and Telecommunications (2009)
2. Yang, H., et al.: Eliminated duplicate search web pages with Hash algorithm. Control Autom. **27**, 299–301 (2006)
3. Ding, Z., et al.: Research of large-scale URL filter based on Bloom filter. New Technol. Libr. Inf. Serv. **3**, 45–50 (2008)
4. Zhang, J., et al.: A study of the identification of authorship for Chinese texts. In: IEEE International Conference on Intelligence and Security Informatics, ISI 2008, pp. 263–264 (2008)
5. Ding, J., et al.: Existential state and presentation of Chinese style. Rhetor. Learn. **3**, 1–6 (2006)
6. Xu, N., et al.: BloomFilter based duplicated webpage elimination approach. Microcomput. Appl. **27**(3), 48–51 (2011)
7. Yang, H., Callan, J.: Near-duplicate detection for eRulemaking. In: National Conference on Digital Government Research. Digital Government Society of North America, pp. 78–86 (2005)
8. Ma, L., Xia, Z.: An FW-DTSS based approach for news page information extraction. In: Tan, Y., Shi, Y. (eds.) DMBD 2016. LNCS, vol. 9714, pp. 227–234. Springer, Heidelberg (2016). doi:10.1007/978-3-319-40973-3_22
9. Mitzenmacher, M.: Compressed Bloom filters. IEEE/ACM Trans. Netw. **10**(5), 604–612 (2002)
10. BloomFilter concepts and principles. http://pages.cs.wisc.edu/~cao/papers/summary-cache/node8.html
11. Laber, E.S., et al.: A fast and simple method for extracting relevant content from News web pages. In: Proceedings of CIKM, pp. 1685–1688 (2009)
12. Xia, Z., Bu, Z.: Community detection based on a semantic network. Knowl. Based Syst. **26**, 30–39 (2012)
13. Bu, Z., Xia, Z.: A last updating evolution model for online social networks. Phys. A Stat. Mech. Appl. **392**(9), 2240–2247 (2013)

Uncertain Frequent Itemsets Mining Algorithm on Data Streams with Constraints

Qun Yu[1,2(✉)], Ke-Ming Tang[2], Shi-Xi Tang[2], and Xin Lv[1]

[1] College of Computer and Information, Hohai University, Nanjing, China
yuqun_yctu@126.com, lvxin.gs@163.com
[2] School of Information Engineering, Yancheng Teachers University, Yancheng, China
tkmchina@126.com, tsxlyh@163.com

Abstract. Nowadays, many emerging applications in real-life can produce amount of uncertain data streams, while people are often interested in some aspects. To mine constrained frequent itemsets on uncertain data streams, this paper presents a method. First, determining the order of items in the transactions of data streams according to the properties of constraints; then, inserting items into the tree in order; finally, mining constrained frequent itemsets from the tree. Existing algorithms are compared with the proposed method and the performances are analyzed. Results indicate that the proposed method is effective and efficient, which mines constrained frequent itemsets when users request for the mining results and need no additional memory.

Keywords: Uncertain data · Data streams · Frequent itemsets · Constraint

1 Introduction

Many real-life applications produce a great deal of data streams, data in which contains certain degree of inherent uncertainty. The causes of uncertainty may include but are not limited to error of data collection equipment and error generated during transmission [1]. The uncertain data streams are growing concerned by academics and industry. However, people are more interest in some aspects not all of them. For example, the policemen keep eyes on the vehicles that exceed the limited speed. So it is important and meaningful to mine frequent itemsets with user-specified constraints in uncertain data streams. Although there are many methods of frequent itemsets mining on uncertain data, data stream and constraints mining, only few works on uncertain data streams with constraints. The properties of streams are not considered in the algorithms of constrained frequent itemsets mining on uncertain data, and those of constraints are also not considered in the algorithms on data streams. In this paper, an algorithm called CUSF-growth (Constrained Uncertain data Stream Frequent itemsets growth) is proposed for constrained frequent itemsets mining on uncertain data stream with consideration of the characteristic both of constrained uncertain data mining and streams. The proposed algorithm uses sliding window model and applies a "delayed" mode for mining frequent itemsets that satisfy the constraints of user-specified. Firstly, a depth-first CUSF-tree is created according to the order of items determined by the properties of constraints; then,

H. Yin et al. (Eds.): IDEAL 2016, LNCS 9937, pp. 192–201, 2016.
DOI: 10.1007/978-3-319-46257-8_21

frequent itemsets satisfied the constraints are mined from CUSF-tree. Based on the proposed approach, the mining process needs no additional storage and the mining operation are performed whenever users request for. The contributions of the proposed CUSF-growth algorithm are listed below:

(1) The proposed CUSF-growth algorithm can mine the frequent itemsets which satisfy users-specified constraints.
(2) For space complexity, the proposed CUSF-growth algorithm saves only one tree structure. For time complexity, the proposed algorithm adopts a "delayed" mining mode which mines constrained frequent itemsets whenever users need.

The remaining parts of this paper are organized as follows. Section 2 introduces related works. Section 3 defines the concepts used in this paper. Section 4 describes the details of our algorithms. Section 5 shows experimental results and the comparison with existing methods. Conclusions are presented in Sect. 6.

2   Related Work

The algorithms of frequent itemsets mining on uncertain data streams are originated from those of traditional static database, uncertain data and data streams. There are many methods for mining frequent itemsets on the above different types of data. Apriori Algorithm [2] and FP-growth algorithm [3] were proposed for the static data. The well-known data structure of FP-growth algorithm is FP-tree. Most of existing algorithms of frequent itemsets mining on uncertain data are extensions of the above two classic mining algorithms of deterministic data, such as U-Apriori algorithm [4] and UF-growth algorithm [5]. The tree-based data structure of UF-growth algorithm is UF-tree and it is derived from FP-tree of FP-growth algorithm. To reduce the size of UF-tree, many improved tree-based methods of frequent itemsets mining were proposed for uncertain data. That is, UFP-growth [6], CUFP-tree [7], CUF-tree [8], PUF-growth [9], PUF-tree* [10, 11].

FP-streaming algorithm [12] and algorithms in [13–16] were designed to mine frequent itemsets on data streams. On basis of FP-streaming and UF-growth algorithm, UF-streaming algorithm [17] and SUF-growth algorithm [17] were presented for mining frequent itemsets from uncertain data streams. Methods in [18–20] were also proposed for mining frequent itemsets from uncertain data streams, but they do not use tree-based method.

In real-world applications, although large amount of data is produced, whether they are static, uncertain or streams, people are more interest in some aspects not all of them. So it is meaningful to mine frequent itemsets which satisfied user-specified constraints. Ng et al. proposed a framework for mining frequent itemsets with constraints [21]. Users can use a lot of SQL format of constraints to guide the mining process. ACUF-growth algorithm [22] was proposed to mine constrained frequent itemsets on uncertain data. Algorithms in [23–26] were proposed to mining constrained frequent itemsets on distributed uncertain data. There are so many methods to mine constrained frequent itemsets on uncertain data that methods on uncertain data streams are rarely proposed.

Leung et al. proposed algorithms (UF-streaming[+], UF-streaming[*] and CUF-streaming) to mine constrained frequent itemsets from uncertain data stream [27]. All of three methods take an additional tree structure "UF-stream" to store frequent itemsets, which need extra storage space. In addition, these methods adopt an "immediate" mode to mine each batch of data in the window no matter when users request for mining results, which causes a lot of waste of computation.

3   Preliminaries

3.1   Uncertain Data

Each transaction of uncertain data contains items and their existential probabilities. The existential probability $P(x, t_i)$ is the possibility of item x occurring in transaction t_i. Define itemsets $IS = \{x_1, x_2, \ldots, x_n\}$, if an itemset X is the subset of IS, that is $X \subseteq IS$. There are d transactions in uncertain dataset D, that is $UDS = \{t_1, t_2, \ldots, t_d\}$, and each transaction t_i has a probability $P(x, t_i)$. We assume that the probabilities of each item are independent, then we can use expected support $\exp Sup(X)$ to express the expected support of itemset X in transaction t_i. According to literature [4], we can get the below expression:

$$\exp Sup(X) = \sum_{i=1}^{d} (\prod_{x \in X} P(x, t_i)) \tag{1}$$

With this setting, an itemset X is frequent if its expected support exceeds or equals to the user-specified support threshold min sup.

3.2   Constraints

Users are often interested in some aspects of massive information, so we can specify given constraints during mining frequent itemsets. It is pointed out in [21] that users can specify SQL-style constraints, such as aggregate constraints $C_1 \equiv \min(X.speed) \geq 80\ km/h$, $C_2 \equiv \max(X.price) \geq \$200$, $C_3 \equiv avg(X.temperature) \geq -2°C$ and $C_4 \equiv sum(X.ra\ inf\ ull) \geq 100\ mm$. To effectively mine constrained frequent itemsets, it is necessary to study on the properties of these constraints. The properties of constraints are classified as four kinds which were discussed in [18]: anti-monotone, monotone, convertible anti-monotone and convertible monotone.

4   The Proposed Algorithm

In this section, the proposed CUSF-growth algorithm is described and an example is used to illustrate the algorithm. Sliding window model is used in CUSF-growth algorithm and the process of CUSF-growth algorithm is as follows.

(1) Determining the order of items in transactions according to the properties of constraints given by users;

(2) Determining the order of tree root to tree leaves in CUSF-tree on the basis of order obtained by (1) and constructing CUSF-tree with the order;
(3) Mining frequent itemsets which satisfied the constraints.

4.1   Determine the Order of Items

According to the above four properties of constraints discussed in Sect. 3.2, we list the classification of constraints and the orders of items in Table 1 to facilitate the construction of CUSF-tree. For the following constraints, *attr* is the attribute of itemset X, θ represents $\geq$, $>$, $\leq$ or $<$, *const* is a constant, X^+ represents nonnegative number, X^- represents non-positive number.

Table 1. Classification of constraints and the corresponding orders of items

Properties	Form of constraints	Order of items in a transaction(order of tree leaves to root in CUSF-tree)
Anti-monotone	$max(X.attr)\theta const$ (θ is $\leq$ or $<$)	non-increasing
	$min(X.attr)\theta const$ (θ is $\geq$ or $>$)	non-descending
Monotone	$max(X.attr)\theta const$ (θ is $\geq$ or $>$)	non-increasing
	$min(X.attr)\theta const$ (θ is $\leq$ or $<$)	non-descending
Convertible anti-monotone	$avg(X.attr)\theta const$ (θ is $\leq$ or $<$)	non-descending
	$avg(X.attr)\theta const$ (θ is $\geq$ or $>$)	non-increasing
	$sum(X^-.attr)\theta const$ (θ is $\geq$ or $>$)	non-increasing
	$sum(X^+.attr)\theta const$ (θ is $\leq$ or $<$)	non-descending
Convertible monotone	$sum(X^-.attr)\theta const$ (θ is $\leq$ or $<$)	non-descending
	$sum(X^+.attr)\theta const$ (θ is $\geq$ or $>$)	non-increasing

4.2   Construct CUSF-Tree

After the order of tree leaves to root is determined, we will construct the CUSF-tree. Each node in CUSF-tree contains (a) item, (b) its expected support, (c) a list of occurrence count of such an expected support. In the lists, there are occurrence counts with different batches of data in the current sliding window. The steps of constructing a CUSF-tree are as follows.

Step 1: Insert items of the first transaction into CUSF-tree from tree root to leaves, the order of tree root to leaves corresponds with the reverse order of items determined by Table 1.

Step 2: When insert a new transaction, first check the root node whether its expected support is the same as that of the items in the new transaction. If they are same, go Step 2-1, otherwise go Step 2-2.

Step 2-1: merge them and add one to the occurrence count, then compare the children nodes. If the items and their expected support of children nodes are same, go Step 2-1-1, otherwise go Step 2-1-2.

Step 2-1-1: merge them and update the occurrence count successively.

Step 2-1-2: insert a new branch below the root node and insert the remainder of the transaction into the new branch in order.

Step 2-2: insert a new branch parallel to the existing root nodes and insert the remainder of the transaction into the branch in order.

Repeat the Step2 until the transactions in current window are inserted into the CUSF-tree.

When the new data arrive, the window slides. We need to delete the oldest batch and add the new one. The method of inserting new batch into CUSF-tree is the same as the above.

After CUSF-tree was built, the algorithm deletes the nodes whose values in the count lists are zero.

4.3 Mine Constrained Frequent Itemsets

Users can mine frequent itemsets from the newest CUSF-tree when they need. The steps are as follows.

Step 1: search for the valid items by traversing the path upwards (from tree leaves to tree root), delete the leaf nodes dissatisfied the constraints.

Step 2: CUSF-growth algorithm use "delayed" mode to recursively mining frequent itemsets from CUSF-tree using an appropriate min sup. First, compute the expected support of each node in the global CUSF-tree. Then, compare expected support of each node with min sup, if the former is bigger than or equal to the latter, the item is frequent. Finally, delete nodes which are infrequent items in CUSF-tree.

Step 3: form projected database by traversing the path upwards (from tree leaves to tree root), decide whether the itemsets containing two, three and more items are frequent. For the extensions of valid and invalid items, whether they need to be further checked constraints can refer to Sect. 4.1.

To get a better understanding of our proposed CUSF-growth algorithm, let us consider Example 4.1.

Let min sup = 1.0, the use-specified constraint is $max(X.price) < \$60$ and the window size $w = 2$. The steps of mining constrained frequent itemsets are as follows.

Table 2. Uncertain data streams

Batch	Transaction	Items and their probabilities
B_1	T_1	{a:0.8, b:0.7, c:0.8, d:1.0, e:0.5}
	T_2	{a:0.8, b:0.7, c:0.8, e:0.9, f:0.1}
	T_3	{a:0.7, c:0.6, d:1.0}
B_2	T_4	{a:0.8, b:0.7, d:1.0}
	T_5	{a:0.7, f:0.3}
	T_6	{a:0.8, b:0.7, c:0.8, d:1.0, e:0.2}
B_3	T_7	{a:0.8, b:0.7, c:0.8, e:0.3, f:0.1}
	T_8	{a:0.7, c:0.6, d:0.9, e:0.4}
	T_9	{a:0.8, b:0.7, d:1.0, f:0.1}

(1) Determine the order of items in transactions. The order is (e, c, f, d, a, b) (Table 2).
(2) Construct CUSF-tree. Figure 1 is the CUSF-tree.

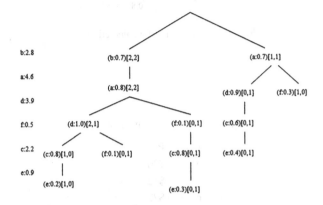

Fig. 1. Global CUSF-tree for the second and third batches

(3) Search for frequent itemsets satisfied constraint in CUSF-tree.
(a) Look for valid items satisfied constraint. In Fig. 1, we remove nodes e, f and c.
(b) Find the frequent single items, delete infrequent items. In Fig. 1, we compute the expected support of every single node. Then combine with step (a), we can get CUSF-tree in Fig. 2
(c) Mine all constrained frequent itemsets from Fig. 2. First according to step (a) and (b), we know that $\{d\}$ is a valid item and $\{a\}$, $\{b\}$, $\{d\}$ are frequent items. Then the algorithm extends $\{d\}$, generates the UF-tree for $\{d\}$ and its projected database, which is shown in Fig. 3. From Fig. 3, we get $\{a, d\}$, $\{b, d\}$ and $\{b, a, d\}$ which are constrained frequent itemsets. Besides $\{d\}$, $\{a\}$ is also a valid item, then extends $\{a\}$, generates the UF-tree for $\{a\}$ and its projected database, which is shown in Fig. 4. From Fig. 4, we get $\{b, a\}$ which we need. As a result, our proposed CUSF-growth algorithm reports all the true frequent itemsets which satisfied the constraint in the second and third batches of the uncertain data stream at time T'. They are $\{a\}$, $\{b, a\}$, $\{b\}$, $\{d\}$, $\{a, d\}$, $\{b, d\}$ and $\{b, a, d\}$ with their corresponding expected supports of 4.6, 2.24, 2.8, 3.9, 3.03, 2.1, 1.68 (Table 3).

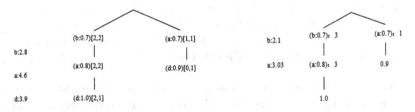

Fig. 2. CUSF-tree after deleting tree leaves   **Fig. 3.** UF-tree for $\{d\}$–projected DB
dissatisfied constraint and infrequent nodes

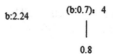

Fig. 4. UF-tree for {a}–projected DB

Table 3. Attributes information

Items	Price
a	$30
b	$12
c	$72
d	$45
e	$84
f	$65

5 Experimental Analysis

In order to evaluate the performance of our algorithm, we used different datasets and compared the memory usage and runtime of CUSF-growth with CUF-streaming algorithm, as CUF-streaming algorithm is better than the other two algorithms in [27]. All experiments are performed on a PC with 2.8 GHz CPU and 2G main memory. Two algorithms are implemented using Microsoft Visual C++. The testing datasets Mushroom [28] and T10I4D100K [28] are used. Mushroom contains 8124 pieces of records and the number of items in T10I4D100K is 100K. Random existential probabilities from the range (0, 1.0] were assigned to the items in the datasets. In our experiments, we set the window size as $5K$. We compare two algorithms with variable min sup, the constraints C_1, C_2, C_3 and C_4 discussed above. Which are shown in Figs. 5 and 6 are the experimental results of dataset Mushroom. The trends on dataset T10I4D100K are consistent with those of Mushroom.

The runtime and memory usage of CUSF-growth algorithm is compared with that of CUF-streaming respectively. In order to save space, we only list the compared results with constraint C_4. The trends of runtime and memory usage of two algorithms with C_1, C_2 and C_3 are similar to those of C_4. Based on quantitative comparison, compared with CUF-streaming algorithm, the proposed algorithm saves 40 % runtime and 60 % memory usage.

Theoretically, the runtime and memory usage of CUSF-growth should be better than those of CUF-streaming. Since it uses "delayed" mode, CUSF-growth algorithm mines constrained frequent itemsets whenever the user requests for. It mines in the current window and uses only one tree structure. As a result, a lot of unnecessary computation and storage has been saved. However, when the size of sliding window is much bigger,

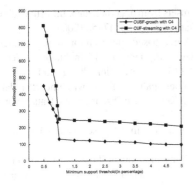
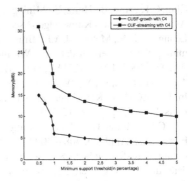

Fig. 5. Runtime vs. support threshold   **Fig. 6.** Memory usage vs. support threshold

the memory consumption of CUF-streaming is less than CUSF-growth algorithm. The reason is that the items contained in CUSF-tree become much more, it needs more memory to store CUSF-tree.

6 Conclusions

An algorithm of CUSF-growth is proposed for mining constrained frequent itemsets on uncertain data streams. Based on the sliding window model, the algorithm analyzes the properties of constraints and determines the order of items in transactions; it constructs a CUSF-tree by the order and mines frequent and constrained itemsets from CUSF-tree. The CUSF-growth algorithm is implemented and compared with CUF-streaming algorithm in terms of running time and memory consumption. Our experimental results on both datasets show that CUSF-growth is more effective and efficient. We will further extend the algorithm to the application of high-dimensional data.

Acknowledgements. This work is supported partially by the National Natural Science Foundation of China under grant No. 61379064, No. 11472238, No. 61300122 and No. 61402394, Special Fund for Public Welfare Industry of the Ministry of Water Resources of China under grant No. 201501007, NSF-China and Guangdong Province Joint Project under grant No. U1301252, and the National Twelfth Five-Year Key Technology Research and Development Program of the Ministry of Science and Technology of China under grant No. HNKJ13-H17-04.

References

1. Aggarwal, C.C., Yu, P.S.: A framework for clustering uncertain data streams. In: 24th IEEE International Conference on Data Engineering, pp. 150–159 (2008)
2. Agrawal, R., Srikant, R.: Fast algorithm for mining association rules. In: 20th International Conference on Very Large Data Bases, pp. 487–499 (1994)
3. Han, J., Pei, J., Yin, Y.: Mining frequent patterns without candidate generation. In: ACM SIGMOD International Conference on Management of Data, pp. 1–12 (2000)

4. Chui, C.K., Kao, B., Hung, E.: Mining frequent itemsets from uncertain data. In: 11th Pacific-Asia Conference on Knowledge Discovery and Data Mining, pp. 47–58 (2007)
5. Leung, C.K.-S., Mateo, M.A.F., Brajczuk, D.A.: A tree-based approach for frequent pattern mining from uncertain data. In: Washio, T., Suzuki, E., Ting, K.M., Inokuchi, A. (eds.) PAKDD 2008. LNCS (LNAI), vol. 5012, pp. 653–661. Springer, Heidelberg (2008)
6. Aggarwal, C.C., Li, Y., Wang, J.Y., Wang, J.: Frequent pattern mining with uncertain data. In: 15th ACM SIGKDD International Conference on Knowledge Discovery and Data Mining, Paris, France, pp. 29–38 (2009)
7. Lin, C.W., Hong, T.P.: A new mining approach for uncertain databases using CUFP trees. Expert Syst. Appl. **39**, 4084–4093 (2012)
8. Leung, C.K.-S., Tanbeer, S.K.: Fast tree-based mining of frequent itemsets from uncertain data. In: Lee, S.-g., Peng, Z., Zhou, X., Moon, Y.-S., Unland, R., Yoo, J. (eds.) DASFAA 2012, Part I. LNCS, vol. 7238, pp. 272–287. Springer, Heidelberg (2012)
9. Leung, C.K.-S., Tanbeer, S.K.: PUF-tree: a compact tree structure for frequent pattern mining of uncertain data. In: Pei, J., Tseng, V.S., Cao, L., Motoda, H., Xu, G. (eds.) PAKDD 2013, Part I. LNCS, vol. 7818, pp. 13–25. Springer, Heidelberg (2013)
10. MacKinnon, R.K., Leung, C.K.S., Tanbeer, S.K.: A scalable data analytics algorithm for mining frequent patterns from uncertain data. In: 18th Pacific-Asia Conference on Knowledge Discovery and Data Mining (PAKDD), Tainan, Taiwan, 13–16 May, pp. 101–416 (2014)
11. Cuzzocrea, A., Leung, C.K.S., Mackinnon, R.K.: Approximation to expected support of frequent itemsets in mining probabilistic sets of uncertain data. In: 19th International Conference on Knowledge-Based and Intelligent Information & Engineering Systems, Singapore, 07–09 September, pp. 613–622 (2015)
12. Giannella, C., Han, J., Pei, J., Yan, X., Yu, P.S.: Mining frequent patterns in data streams at multiple time granularities. In: Kargupta, H., Joshi, A., Sivakumar, D., Yesha, Y. (eds.) Data Mining: Next Generation Challenges and Future Directions, pp. 191–212. AAAI/MIT Press, Massachusetts (2004)
13. Nori, F., Deypir, M., Sadreddini, M.H.: A sliding window based algorithm for frequent closed itemsets mining over data streams. J. Syst. Softw. **86**, 615–623 (2013)
14. Shin, S.J., Lee, D.S., Lee, W.S.: CP-tree: an adaptive synopsis structure for compressing frequent itemsets over online data streams. Inf. Sci. **278**, 559–576 (2014)
15. Calders, T., Dexters, N., Gillis, J.J.M., Goethals, B.: Mining frequent itemsets in a stream. Inf. Syst. **39**, 233–255 (2014)
16. Troiano, L.G., Scibelli, G.: Mining frequent itemsets in data streams within a time horizon. Data Knowl. Eng. **89**, 21–37 (2014)
17. Leung, C.K.S., Hao, B.: Mining of frequent itemsets from streams of uncertain data. In: IEEE International Conference on Data Engineering, pp. 1663–1670 (2009)
18. Han, D.H., Carrier, C.G., Li, S.R.: Efficient mining of high-speed uncertain data streams. Appl. Intell. **43**, 773–785 (2015)
19. Akbarinia, R., Masseglia, F.: Fast and exact mining of probabilistic data streams. In: Blockeel, H., Kersting, K., Nijssen, S., Železný, F. (eds.) ECML PKDD 2013, Part I. LNCS, vol. 8188, pp. 493–508. Springer, Heidelberg (2013)
20. HewaNadungodage, C., Xia, Y.N., Lee, J.J., Tu, Y.C.: Hyper-structure mining of frequent patterns in uncertain data streams. Knowl. Inf. Syst. **37**, 219–244 (2013)
21. Ng, R.T., Lakshmanan, L.V.S., Han, J., Pang, A.: Exploratory mining and pruning optimizations of constrained associations rules. In: ACM SIGMOD International Conference on Management of Data, pp. 13–24 (1998)

22. Leung, C.K.S., Brajczuk, D.A.: Efficient algorithms for mining constrained frequent patterns from uncertain data. In: ACM SIGKDD Workshop on Knowledge Discovery from Uncertain Data, pp. 9–18 (2009)
23. Cuzzocrea, A., Leung, C.K.: Distributed mining of constrained frequent sets from uncertain data. In: Xiang, Y., Cuzzocrea, A., Hobbs, M., Zhou, W. (eds.) ICA3PP. LNCS, vol. 7016, pp. 40–53. Springer, Heidelberg (2011)
24. Jiang, F., Leung, C.K.-S., MacKinnon, R.K.: BigSAM: mining interesting patterns from probabilistic databases of uncertain big data. In: Peng, W.-C., Wang, H., Bailey, J., Tseng, V.S., Ho, T.B., Zhou, Z.-H., Chen, A.L. (eds.) PAKDD 2014 Workshops. LNCS, vol. 8643, pp. 780–792. Springer, Heidelberg (2014)
25. Leung, C.K.S., MacKinnon, R.K., Jiang, F.: Distributed uncertain data mining for frequent patterns satisfying anti-monotonic constraints. In: 28th International Conference on Advanced Information Networking and Applications Workshops, Victoria BC, 13–16 May, pp. 1–6 (2014)
26. Cuzzocrea, A., Leung, C.K.S., MacKinnon, R.K.: Mining constrained frequent itemsets from distributed uncertain data. Future Gener. Comput. Syst. **37**, 117–126 (2014)
27. Leung, C.K.S., Hao, B., Jiang, F.: Constrained frequent itemsets mining from uncertain data streams. In: International Conference on Data Engineering-workshops, pp. 120–127 (2010)
28. Dataset. http://fimi.cs.helsinki.fi/

Ubiquitous Robot: A New Paradigm for Intelligence

Tiantian Zhang[1], Bo Yuan[1(✉)], Tao Meng[1], Yinghao Ren[1],
Houde Liu[2], and Xueqian Wang[2]

[1] Intelligent Computing Lab, Division of Informatics, Graduate School at Shenzhen,
Tsinghua University, Shenzhen 518055, People's Republic of China
2573546543@qq.com, yuanb@sz.tsinghua.edu.cn,
zdhmengtao@163.com, 790241114@qq.com
[2] Shenzhen Laboratory of Space Robotics and Telescience, Graduate School at Shenzhen,
Tsinghua University, Shenzhen 518055, People's Republic of China
{liu.hd,wang.xq}@sz.tsinghua.edu.cn

Abstract. This paper presents a systematic review of the recent development in ubiquitous robotics, with a focus on its implication on intelligence. Ubiquitous robot (Ubibot), the third generation robot, is characterized by its unprecedented power to sense environments and provide sophisticated services autonomously. Based on the architectures of existing Ubibot projects, we describe a general framework containing the perception, intelligence and execution modules as well as the middleware layer used to integrate the three modules to make them collaborate seamlessly. Two representative projects are introduced to exemplify the state-of-the-art progress in Ubibot, along with a brief discussion of various underlying communication techniques. In the perspective of intelligence, we point out that Ubibot opens up new horizons for investigating and applying popular AI techniques such as computer vision and pattern recognition, compared to traditional robots. Furthermore, a list of new challenging research topics is identified that deserve full consideration in the future to make Ubibot more robust, effective and adaptive.

Keywords: Ubiquitous robot · Middleware · Intelligence · Communication

1 Introduction

Ubiquitous robot (Ubibot) refers to the third generation robots following industrial robots (first generation) and personal robots (second generation) [1]. Ubibot is built upon classical robotics and the concept of ubiquitous computing, which was first proposed by researchers in the Robot Intelligence Technology (RIT) Lab., KAIST in 2002 [2]. Naturally, it is expected that robots can perform a variety of tasks in the increasingly complicated environment [3]. However, developing a standalone robot that can provide all desired services can be extremely expensive and time consuming. Instead, it is reasonable to combine robots with limited abilities via the ubiquitous network containing sensors and portable devices to provide a wide range of services promptly at much lower cost [4]. In this ubiquitous space (U-space), people can gain access to required services anytime, anywhere and in various ways.

© Springer International Publishing AG 2016
H. Yin et al. (Eds.): IDEAL 2016, LNCS 9937, pp. 202–211, 2016.
DOI: 10.1007/978-3-319-46257-8_22

Each personal robot is typically a standalone platform with a computer and wired network, providing "one person, one robot, and user-commanded" services [5]. This class of robots is characterized by the spatially localized entity, which prevents them from serving humans better [1]. By contrast, Ubibot removes the necessity of the conventional notion of a single platform by defining three components: intelligence, perception and execution. In practice, these three components are implemented by software robots, sensors and devices, and mobile robots, respectively [6]. This multi-robots platform breaks the spatial limitation of traditional robots and can provide users with seamless context-aware services timely via wireless ubiquitous network whenever the environment changes, providing "one person, many robots, seamless and context-aware" services [5].

Ubibot is closely related to networked robots (the two terms are sometimes used interchangeably by researchers) as the components of Ubibot need to exchange data via the network frequently. With the help of networks, the sensing and operating ranges of robots can be extended, allowing them to communicate with each other over long distances to coordinate their activity [7]. The key difference is that Ubibot is a collection of robot components, instead of fully functional robots. Also, the communication among components is likely to be much more complex and intensive as they need to collaborate extensively to finish a specific task (just imaging the level of collaboration required among eyes, brain and legs for human to walk properly).

Another related concept is cloud robot that employs cloud technologies such as cloud computing, cloud storage and other Internet technologies to enhance the capability of single-robot platforms [8]. Cloud robots can benefit from the powerful computing and storage resources (e.g., knowledge base) in the cloud and share information with other robots and smart agents. By contrast, Ubibot focuses more on the collaboration of various components, although the intelligence module can also invoke cloud technologies when necessary.

In 1985, Brady defined robotics as "the intelligent connection of perception to action" [9]. Traditional robots are often designed for a single scene and tasks are statically pre-programmed during the production process. As a result, they are unable to provide adaptive services as well as operate in uncertain and dynamically changing environments. Featuring context-awareness, self-learning and higher-level intelligence, Ubibot can sense changes in the environment continuously via advanced communication technologies and make proper decisions/actions [10]. In other words, Ubibot is an ensemble of mini-robots with specific functions distributed in the space and, when strategically combined together, they can be regarded as a single powerful virtual robot. More importantly, since Ubibot always exists as a collection of interacting components, there are many challenging issues that require intelligent solutions, including the optimal utilization of hardware/software resources and the reconfiguration of components in the event of partial system failure.

Section 2 presents a general architecture of Ubibot and two representative case studies. Section 3 introduces the underlying communication technologies used in Ubibot while Sect. 4 gives an in-depth analysis of the implication of Ubibot on intelligence. This paper is concluded in Sect. 5 with some discussions and prospects for future Ubibot systems.

2   Ubiquitous Robot Systems

2.1   General Architecture

As an emerging field of robotics, the notion of Ubibot was created by Jong-Hwan Kim et al. [11]. Afterwards, Bong Keun Kim et al. developed the ubiquitous function services [12] and provided a simulation framework for Ubibot and sensors using RT-Middleware [13] to better solve the ubiquitous localization and mapping problems. Saffiotti et al. presented the vision of an Ecology of Physically Embedded Intelligent System (PEIS-Ecology) and developed a technique for autonomous self-configuration and reconfiguration as well as a robotics middleware structure to cope with highly heterogeneous systems [14]. Yong-Guk Ha et al. investigated the problem that the robot system needs to be interoperable with sensors and devices in its dynamic service environments automatically [15]. Furthermore, Yukihiro Nakamura et al. introduced a framework for the network robot platform (NWR-PF) and a service allocation method for heterogeneous distributed robots [4].

Despite of the diversity of Ubibot in applications and functionalities, Fig. 1 gives the common architecture of Ubibot systems. According to this architecture, Ubibot is divided into three modules: intelligence, perception, and execution modules. The intelligence module refers to software applications with self-learning ability that can interact with human and environment. Due to the essence of software, they can also move anywhere in the ubiquitous network. The perception module contains hardware facilities such as sensors, cameras and other smart devices embedded in the environment. When combined with processors, they can gather and process video, voice, and position data

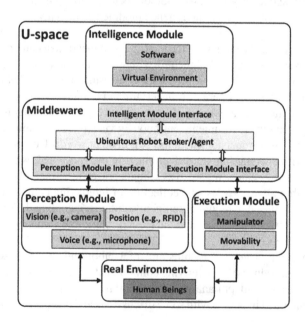

Fig. 1. The general architecture of ubiquitous robot systems

from the environment and provide valuable information to other robot components. The execution module resembles the hands and feet of traditional robots, which can move physically and provide various services. Furthermore, the mobile components can also be regarded as a complementary data acquisition platform, gathering additional information about the U-space by exploring the physical environment. Meanwhile, Fig. 1 contains a component named middleware. This is an independent system or service program for source sharing and communication within and among robot components via a variety of network interfaces and protocols. The middleware structure usually contains three interfaces to link corresponding modules and one broker to enable the system to make an offer of service irrespective of the operating structure, position and type of interface [1].

2.2 URC (Ubiquitous Robotics Companion) Project

The URC project is a conceptual vision of ubiquitous service robots proposed by KAIST in 2003 to provide users with required services, anytime and anywhere in ubiquitous computing environments [16]. To realize URC with robot systems automatically interoperable with sensors and devices, this project designed and implemented a Semantic-based Ubiquitous Robotic Space (SemanticURS), which enables automated integration of networked robots into ubiquitous computing environment, exploiting Semantic Web Services and AI-based planning technologies. According to robots' functions and roles in the U-space, Ubibot was divided into three categories: Software Robot (Sobot), Embedded Robot (Embot), and Mobile Robot (Mobot), corresponding to the three modules in Fig. 1. These three classes of components were implemented separately and connected via network and middleware.

For the middleware, researchers proposed a multi-layered architecture of ubiquitous robot system for integrated services [17]. It consists of five layers: software agent layer, Sobot management layer, context provider/task schedule layer, device management layer and physical layer. In this architecture, Embots and Mobots in physical layer directly operate within the ubiquitous environment while Sobots in software agent layer are transferred into various Mobots services. The proposed middleware was implemented and simulated with virtual sensors in the virtual environment.

2.3 NRS (Network Robot System) Project

The NRS project is a five-year research plan on Ubibot proposed by MIAC (Ministry of Internal Affairs and Communications, Japan) in 2004, which aimed at establishing fundamental Ubibot technologies that can greatly improve the ability of recognition and communication, compared with the single-robot platform. The goal of this project was to establish a multi-robots service platform in commercial district. A structured platform of environmental information was built for detecting people's locations and actions by installing cameras, laser range finders and RFID tags in the urban pedestrian street, Osaka. With this platform, robots can acquire information about people's locations and actions, in order to provide them with route guidance and targeted information about surrounding stores.

Similarly, researchers divided Ubibot into virtual-type robots, inconspicuous-type robots, and visible-type robots, corresponding to the three modules in Fig. 1. To integrate heterogeneous distributed robots, a network robot platform (NWR-PF) was introduced with three layers: connection units, area management gateway, and robot-user interaction database [4]. Firstly, the connection units use device-dependent protocols to obtain data from robots and then generate the status of users and robots in the 4Ws (When, Who, Where, What) format. After receiving the information from connection units, the area management gateway determines the most suitable combination of scenario and robots and controls service execution. The robot-user interaction database stores the information of the users, robots and services submitted from the connection units. Meanwhile, another simulation framework for Ubibot using RT-Middleware was also developed [13]. In this scheme, sensors and robots are implemented as RT components and combined by RT-Middleware. Since RT components have common interface module, they can be reused to lower the cost of development.

3   Ubiquitous Communication

In this section, various communication techniques are briefly introduced, which can be adopted in Ubibot to enable collaboration and interaction among different Ubibots (mainly wired communications), as shown in Fig. 2(a) or different modules within the same Ubibot (mainly wireless communications), as shown in Fig. 2(b).

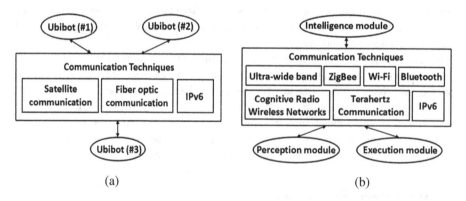

(a)   (b)

Fig. 2. Ubiquitous communication: (a) among Ubibots; (b) within Ubibot

Fiber optic communication plays an important role in telecommunication infrastructure for broadband networks. It can provide enormous and unsurpassed transmission bandwidth with negligible latency, and is widely used for long distance and high data rate transmission in telecommunication networks [18]. Satellite communication is also an alternative, which can be used to connect sparsely distributed Ubibots.

Short-range wireless communications are dominated by four protocols: Bluetooth, ultra-wide band (UWB), ZigBee, and Wi-Fi, corresponding to IEEE 802.15.11, 802.15.3, 802.15.4, and 802.11a/b/g standards, respectively [19]. Bluetooth is used for wireless personal area network (WPAN) and can be used in perception module for transmitting data to intelligence module. UWB, on the one hand, can be used for indoor short-range high-speed wireless communication [20]. ZigBee is a low data rate WPAN within 10 to 100 m. It features low power-consumption, multi-hop, and reliable mesh networking, making it appealing to battery powered devices. Wi-Fi can be used in the general situation to transmit data among various modules.

There are also many next-generation communication techniques that can be applied in Ubibot, such as cognitive radio wireless networks [21], terahertz communication technology [22, 23], and quantum communication [24]. Different from today's wireless networks regulated by the spectrum assignment policy, cognitive radio wireless networks can exploit existing wireless spectrum opportunistically. So, it can provide high bandwidth to mobile users via heterogeneous wireless architectures and dynamic spectrum access techniques [21]. Terahertz radio (0.1–10 THz) is characterized by high data transfer rate, good directivity and high transmittance [23]. Terahertz communication is envisioned as a key technology to satisfy the increasing demand for high speed wireless communication as it can alleviate the spectrum scarcity and capacity limitations of current wireless systems. It enables new applications in classical networking domains as well as in novel nanoscale communication paradigms [25]. Quantum communication combines classical communication and quantum mechanics and includes quantum teleportation, quantum superdense coding and quantum cryptography, ensuring highly secure communications.

4 Ubiquitous Intelligence

4.1 New Opportunities

Ubibot features unprecedented flexibility, versatility and robustness compared to traditional robots. For example, a standalone robot often has very restricted field of vision, limited by the single camera mounted on it. Its mobility is also limited, making it difficult to provide services in a wide area promptly. What is even worse is that a single component (e.g., camera) failure can result in the total loss of functionality of the robot. By contrast, Ubibot contains multiple sensing devices of different types, which are distributed in the U-space, providing a full range of sensing capabilities (Fig. 3). Also, multiple mobile components can be scheduled to provide services with minimum latency. Furthermore, due to the existence of components of similar functionality, Ubibot is intrinsically fault-tolerant. As a result, Ubibot has much better capability in sensing and interacting with the environment as well as providing a variety of services. Through reconfiguration, Ubibot can also handle component failures without significantly compromising its functionality. This feature is particularly important for military applications and aerospace industry where robots often need to work in severe and hostile environments.

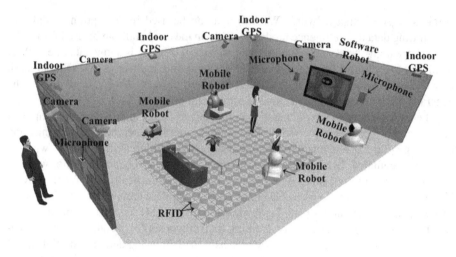

Fig. 3. An example of the Ubibot system. It shows the three key modules: perception (cameras, microphone, indoor GPS, RFID), intelligence (software robot) and execution (mobile robots).

Ubibot also provides potentially much wider opportunities for machine intelligence. For example, computer vision is one of the key techniques in robotics. It is essential for a robot to accurately recognize the surrounding environment to avoid obstacles on its path or grab an object. To provide more personalized services, it is desirable that the robot can identify different users using face recognition techniques. With the help of multiple cameras, Ubibot can acquire richer image/video information, which can dramatically increase the accuracy of recognition. For example, given multiple images of the same object from different angles, a super-resolution image can be constructed with more details [26]. Meanwhile, high-quality 3D reconstruction also becomes feasible, which can greatly help robots understand the shape of objects. With the RFID tags located on the floor, Ubibot can precisely identify the location of human, even in a totally dark situation. In the housekeeping scenario, Ubibot can easily find a misplaced item or provide crucial fall detection service for elderly people. By contrast, due to the limitation of sensing, traditional robots may not be able to be aware of the emergency and raise the alarm in a timely manner.

4.2 New Challenges

As a brand new class of robots, Ubibot itself creates a series of unique technical challenges that needs to be properly addressed to make Ubibot work efficiently and reliably. A list of possible research topics is as follows:

- *Optimal Planning Problem.* For example, in Ubibot, video cameras may be fixed on the ceiling in an indoor setting but can be also mounted on a UAV (Unmanned Aerial Vehicle) or an AGV (Automatic Guided Vehicle) for extended surveillance range. Due to the limited onboard power supply and tight restriction on the data acquisition latency, Ubibot needs to optimize the route and schedule of UAVs or

AGVs to fulfill the above objectives. This problem is related to TSP with Neighborhoods (TSPN) [27] with the additional factor that multiple agents can collaborate to finish the data collection. Similarly, given multiple users to be serviced/ tasks to be fulfilled, there is also a need of intelligent algorithms for scheduling one or more mobile components.

- *Load Balancing Problem.* Since Ubibot can continuously receive information from the environment, the amount of streaming data to be processed is significantly greater than in traditional robots. In the presence of multiple intelligence components, there is a requirement for properly balancing the workload on each component so that all data can be processed with minimum delay. Furthermore, certain tasks may be passed onto external computing facilities in the cloud for better processing capability. There are several factors that need to be taken into account to ensure efficiency, such as current workload level, task type, security requirement, timeliness, communication overhead, data parallelism vs. task parallelism, priority and granularity.

- *Reconfiguration Problem.* One of the most distinctive features of Ubibot is its capability of reconfiguration to cover component failures or extend system capacity. Components should be self-aware and self-organizing, requiring little if any human intervention. For example, newly added components should automatically be configured and added into Ubibot, following the "plug-and-play" paradigm. When one or more components become inoperative, other components should adjust their technical parameters (e.g., surveillance range) to maintain coverage and capability. Note that in military applications, some components may be deliberately designed as disposable devices, such as the warhead for intercepting missiles or satellites. With the reconfiguration feature, most other components (e.g., sensing and intelligence) in Ubibot can be reused, reducing system cost dramatically.

- *Self-Adaptive Component Problem.* Once deployed in the ocean, on the land or in the orbital space, Ubibot should require minimum service from human. In face of the unforeseen environment changes, components in Ubibot should be self-adaptive to some extent and update their functionality to better suit the new operational environment or compensate for faults. This attractive feature can be achieved using the idea of evolvable hardware [28], which applies EAs (Evolutionary Algorithms) to evolve a population of candidate circuits and implements the solution in FPGA. This is particularly appealing when there is no sufficient prior knowledge on circuit design and the specification only states the desired behavior of the component.

- *Ubiquitous Modeling Problem.* It is essential for a robot to build a systematic model of its working environment. In general, the U-space can be seen as a sophisticated data acquisition and processing platform and Ubibot can build comprehensive models using the diverse data acquired from various sensors. For example, Ubibot can monitor and record the movement of human users and conduct trajectory mining to better understand their behavior patterns, in order to provide more personalized services. Meanwhile, Ubibot can analyze the historical data regarding the occurrence of key tasks and the amount of processing time required, to obtain insightful guidance for maximizing the utilization of resources. A key issue is that the data is produced by various devices and is likely to be stored separately. Consequently, distributed modeling/mining techniques are required to create a unified view of the data.

5 Conclusion

As the third generation robot, ubiquitous robot has been developed following the paradigm shift of robotics and ubiquitous computing, which can provide services anytime and anywhere. Based on various existing Ubibot projects, we present a general architecture of Ubibot consisting of three modules: intelligence, perception, and execution, which correspond to software robots, sensors and devices, and mobile robots, respectively. A middleware layer is used to integrate the three modules to make them collaborate seamlessly. Meanwhile, two representative Ubibot projects (URC and NRS) are introduced to demonstrate their principles. Furthermore, typical communication techniques are discussed to highlight their unique characteristics. We point out that, Ubibot brings more opportunities for current AI techniques and significantly extends their applicability in robotics. More importantly, Ubibot comes with new technical challenges that deserve in-depth investigation and competent techniques are to be developed to ensure its efficiency, flexibility and robustness.

Acknowledgement. This work was supported by the 863 Program (No. 2015AAXX46201), Natural Science Foundation of Guangdong Province (No. 2015A030313881) and Research Foundation of Shenzhen (No. JCYJ20140509172959962).

References

1. Kim, J.H., Lee, K.H., Kim, Y.D., Kuppuswamy, N.S., Jo, J.: Ubiquitous robot: a new paradigm for integrated services. In: Proceedings of 2007 IEEE International Conference on Robotics and Automation, pp. 2853–2858 (2007)
2. Kim, Y.D., Kim, Y.J., Kim, J.H., Lim, J.R.: Implementation of artificial creature based on interactive learning. In: Proceedings of the FIRA Robot World Congress, pp. 369–373 (2002)
3. Ubiquitous Robotics. http://www-arailab.sys.es.osaka-u.ac.jp/research/Integrating-Ambi ent-Intelligence/e_index.html
4. Nakamura, Y., Machino, T., Motegi, M., Iwata, Y., Miyamoto, T., Iwaki, S., Muto, S., Shimokura, K.I.: Framework and service allocation for network robot platform and execution of interdependent services. Robot. Auton. Syst. **56**(10), 831–843 (2008)
5. Kim, J.H.: Ubiquitous robot. In: Reusch, B. (ed.) Computational Intelligence, Theory and Applications. Advances in Soft Computing, vol. 33, pp. 451–459. Springer, Heidelberg (2005)
6. Kim, J.H.: Ubiquitous robot: recent progress and development. In: Proceedings of SICE-ICASE International Joint Conference, pp. I-25–I-30. IEEE (2006)
7. Sanfeliu, A., Hagita, N., Saffiotti, A.: Network robot systems. Robot. Auton. Syst. **56**(10), 793–797 (2008)
8. Cloud Robotics. https://en.wikipedia.org/wiki/Cloud_robotics
9. Brady, M.: Artificial intelligence and robotics. Artif. Intell. **26**(1), 79–121 (1985)
10. Gigras, Y., Gupta, K.: Ambient intelligence in ubiquitous robotics. Int. J. Comput. Sci. Inf. Technol. **2**(4), 1438–1440 (2011)
11. Kim, J.H., Kim, Y.D., Lee, K.H.: The third generation of robotics: ubiquitous robot. In: Proceedings of the 2nd International Conference on Autonomous Robots and Agents (2004)

12. Kim, B.K., Tomokuni, N., Ohara, K., Tanikawa, T., Ohba, K., Hirai, S.: Ubiquitous localization and mapping for robots with ambient intelligence. In: Proceedings of 2006 IEEE/RSJ International Conference on Intelligent Robots and Systems, pp. 4809–4814. IEEE (2006)

13. Do, H.M., Kim, B.K., Kim, Y.S., Lee, J.H., Ohara, K., Sugawara, T., Tomizawa, T., Liang, X., Tanikawa, T., Ohba, K.: Development of simulation framework for ubiquitous robots using RT-middleware. In: Proceedings of International Conference on Control, Automation and Systems, pp. 2483–2486. IEEE (2007)

14. Saffiotti, A., Broxvall, M., Gritti, M., LeBlanc, K., Lundh, R., Rashid, J., Seo, B.S., Cho, Y.J.: The PEIS-ecology project: vision and results. In: Proceedings of IEEE/RSJ International Conference on Intelligent Robots and Systems, pp. 2329–2335. IEEE (2008)

15. Ha, Y.G., Sohn, J.C., Cho, Y.J., Yoon, H.: A robotic service framework supporting automated integration of ubiquitous sensors and devices. Inf. Sci. **177**(3), 657–679 (2007)

16. Ha, Y.G., Sohn, J.C., Cho, Y.J., Yoon, H.: Towards a ubiquitous robotic companion: design and implementation of ubiquitous robotic service framework. ETRI J. **27**(6), 666–676 (2005)

17. Jeong, I.B., Kim, J.H.: Multi-layered architecture of middleware for ubiquitous robot. In: Proceedings of 2008 IEEE International Conference on Systems, Man and Cybernetics. pp. 3479–3484. IEEE (2008)

18. Idachaba, I., Ike, D.U., Hope, O.: Future trends in fiber optics communication. In: Proceedings of the World Congress on Engineering, vol. 1 (2014)

19. Lee, J.S., Su, Y.W., Shen, C.C.: A comparative study of wireless protocols: Bluetooth, UWB, ZigBee, and Wi-Fi. In: Proceedings of the 33rd Annual Conference of the IEEE Industrial Electronics Society, pp. 46–51 (2007)

20. Porcino, D., Hirt, W.: Ultra-wideband radio technology: potential and challenges ahead. Commun. Mag. **41**(7), 66–74 (2003)

21. Akyildiz, I.A., Lee, W.Y., Vuran, M.C., Mohanty, S.: Next generation/dynamic spectrum access/cognitive radio wireless networks: a survey. Comput. Netw. **50**(13), 2127–2159 (2006)

22. Koening, S., Lopez-Diaz, D., Antes, J., Boes, F., Henneberger, R., Leuther, A., Tessmann, A., Schmogrow, R., Hillerkuss, D., Palmer, R., Zwick, T., Koos, C., Freude, W., Ambacher, O., Leuthold, J.: Wireless sub-THz communication system with high data rate. Nat. Photonics **7**(12), 977–981 (2013)

23. Yao, J., Chi, N., Yang, P., Cui, H., Wang, J., Li, J., Xu, D., Ding, X.: Study and outlook of terahertz communication technology. Chin. J. Lasers **36**(9), 2213–2233 (2009)

24. Duan, L.M., Lukin, M.D., Cirac, J.I., Zoller, P.: Long-distance quantum communication with atomic ensembles and linear optics. Nature **414**(6862), 413–418 (2001)

25. Akyildiz, I.F., Jornet, J.M., Han, C.: Terahertz band: next frontier for wireless communications. Phys. Commun. **12**, 16–32 (2014)

26. Park, S.C., Park, M.K., Kang, M.G.: Super-resolution image reconstruction: a technical overview. IEEE Signal Process. Mag. **20**(3), 21–36 (2003)

27. Arkin, E.M., Hassin, R.: Approximation algorithms for the geometric covering salesman problem. Discrete Appl. Math. **55**, 197–218 (1994)

28. Yao, X., Higuchi, T.: Promises and challenges of evolvable hardware. IEEE Trans. Syst. Man Cybern. Part C **29**(1), 87–97 (1999)

A Method of Discriminative Features Extraction for Restricted Boltzmann Machines

Song Guo, Changjun Zhou$^{(\boxtimes)}$, Bin Wang, and Shihua Zhou

Key Laboratory of Advanced Design and Intelligent Computing,
Dalian University, Ministry of Education, Dalian, China
zhou-chang231@163.com

Abstract. The Restricted Boltzmann Machine (RBM) is a kind of stochastic neural network. It can be used as basic building blocks to form deep architectures. Since Hinton solved the problem of computational inefficiency by using a so called greedy layer-wise unsupervised pre-training algorithm, much more attention is focused on deep learning and achieved significant success in areas of speech recognition, object recognition, natural language processing, etc. In addition to initializing deep networks, RBMs can also be used to learn features from the raw data. In this paper, we proposed a method to learn much better discriminative features for RBMs based on using a novel objective function. We test our idea on MNIST handwritten digit dataset. In our experiments, the features learnt by RBM were further fed to a multinomial logistic regression and results show that our objective function could result in much higher accuracy ratio of classification.

Keywords: Restricted Boltzmann machines · Deep learning · Discriminative feature · Objective function

1 Introduction

Deep feed-forward neural networks often refer to neural networks with multiple hidden layers. Deep networks aim at learning features hierarchies with higher level of features composed of lower level features so as to learn much abstract representation. As discussed in [1] since the objective function is highly non-convex that traditional gradient-based training (e.g. back propagation) of deep neural networks often falls into bad local optima. What's more, during the training procedure of back propagation algorithm, the gradient becomes smaller and smaller when it comes to the formal layer so that it's hard to learn from the raw data efficiently and this phenomenon is what so-called diffusion of gradients [19]. Since 2006, Hinton proposed a greedy layer-wise unsupervised pre-training algorithm [7] to address this issue, deep learning method attracted a lot of attention because it's successful application in objective recognition, speech recognition [11], image analyse, natural language processing [15], etc. Deep architectures are often formed by stacking multiple Restricted Boltzmann Machines (RBMs)

© Springer International Publishing AG 2016
H. Yin et al. (Eds.): IDEAL 2016, LNCS 9937, pp. 212–219, 2016.
DOI: 10.1007/978-3-319-46257-8_23

together such as deep autoencoder [5], deep Boltzmann Machine (DBM) [14], etc. RBM is a special type of Markov Random Field and it is a probability generative model to model the probability distribution of the training data. Hence, it is important to have an efficient algorithms to train RBM.

Most researches on RBM concentrate on its training algorithm. Since the log-likelihood gradient with respect to the parameters contains one intractable term, different algorithms use different methods to approximate that term, such as contrastive divergence (CD) [6], average contrastive divergence [12], Persistent CD (PCD) [17], Fast Persistent CD [18], parallel tempering (PT) [4], multiple tempering [2]. Some researcher study on self-contained classifier of RBM such as classification RBM [8]. Some variants model based on RBM such as mcRBM [13] could handle continuous data and sparse RBM [10] to learn sparse over-complete representation. In this paper, we use RBM as a stand-alone feature extractor, the activation probabilities of its hidden units are regarded as features learnt from the raw data, and further the features are fed to a multinomial logistic regression classifier. And the accuracy ratio of classification on the test data is used to measure the efficiency of our method.

This paper is organised as follows. In Sect. 2, we describe the RBM and its training algorithm in details. In Sect. 3, our method is introduced. In Sect. 4, experiments and results are described. Finally, in Sect. 5, we conclude this paper and make a discussion on the direction of future research.

2   Restricted Boltzmann Machines

The Restricted Boltzmann Machine (RBM) [16] is a stochastic neural network with one visible layer and one hidden layer. Suppose that the RBM consists of m visible units $v = (v_1, \ldots, v_m)$ and n hidden units $h = (h_1, \ldots, h_n)$. The connections between visible layer and hidden layer are fully connected while there are no connections between visible to visible and hidden to hidden units. We suppose that each unit takes binary values, that is either 0 or 1.

The RBM is an energy-based model [9], and its probability distribution is defined through an energy function. The energy function is defined as follows:

$$E(v, h; \theta) = -\sum_{i=1}^{m} v_i b_i - \sum_{j=1}^{n} h_j c_j - \sum_{i=1}^{m} \sum_{j=1}^{n} v_i w_{ij} h_j \qquad (1)$$

where W_{ij} is the connecting weight between i-th visible unit and j-th hidden unit. b and c are bias vectors corresponding to visible layer and hidden layer, respectively. We use $\theta = \{W, b, c\}$ denotes the parameters of the model.

The convention from the energy function to probability distribution is through the gibbs distribution. So the probability distribution over the visible and hidden units is defined as follows:

$$p(v, h; \theta) = \frac{e^{-E(v, h; \theta)}}{Z} \qquad (2)$$

where $Z = \sum_v \sum_h e^{-E(v, h; \theta)}$, the normalizing constant.

The probability that the model assigns to a observed data v is defined as the marginal over h:

$$p(v;\theta) = \sum_h p(v,h;\theta) = \frac{\sum_h e^{-E(v,h;\theta)}}{Z} \qquad (3)$$

The conditional probability over hidden units given visible units are independent and conditional probability over visible units given hidden units are independent too. The conditional probability formulas are defined as follows:

$$p(h_j = 1|v;\theta) = \sigma(c_j + \sum_{i=1}^{m} v_i W_{ij}) \qquad (4)$$

$$p(v_i = 1|h;\theta) = \sigma(b_i + \sum_{j=1}^{n} h_j W_{ij}) \qquad (5)$$

where $\sigma(x) = 1/(1 + \exp(-x))$, the sigmoid activation function.

Traditionally, maximum likelihood estimation (MLE) is used to learn the parameters of RBM. From the point of view of energy-based model, the loss function corresponds to negative log-likelihood on the training dataset. We use one training sample for simplicity and then learning corresponds to the following optimization problem:

$$minimize_\theta - \log p(v;\theta) \qquad (6)$$

Using stochastic gradient descent, the learning formulas become:

$$W_{ij} := W_{ij} + \eta(\langle v_i h_j \rangle_d - \langle v_i h_j \rangle_m) \qquad (7)$$

$$b_i := b_i + \eta(\langle v_i \rangle_d - \langle v_i \rangle_m) \qquad (8)$$

$$c_j := c_j + \eta(\langle h_j \rangle_d - \langle h_j \rangle_m) \qquad (9)$$

where we use notation $\langle \cdot \rangle_d$ represents expectation under the distribution defined by the data. While $\langle \cdot \rangle_m$ denotes the expectation under the distribution defined by the model. η is the learning rate. Commonly, they are called the positive phase gradient and the negative phase gradient, respectively.

The positive phase gradient can be computed exactly, which only need to compute the conditional probability using Eq. (4) while the visible units are clamped to a training sample. But the negative phase gradient in some sense is incomputable as it needs to calculate the expectation value under the model distribution which need to sum over $2^{(m+n)}$ configurations of the visible and hidden units. Different training algorithms for RBM differs in approximating the negative phase gradient. The CD [6] learning algorithm was the first efficient method proposed to approximate the negative phase gradient. Instead of following the exact gradient, CD-n approximates the gradient by drawing from a Markov chain which performs only n steps of transition operations. The starting states of the Markov chain is the training sample other than a random states to speed up convergence. Meanwhile, the transfer operator refers to gibbs sampling. The learning formula for weight becomes:

$$W_{ij} := W_{ij} + \eta(\langle v_i h_j \rangle_d - \langle v_i h_j \rangle_{p_n}) \tag{10}$$

where n denotes the number of gibbs sampling steps.

3 A Novel Objective Function and Its Training Algorithm

3.1 The Novel Objective Function

Traditionally, the loss function is negative log-likelihood function when training RBM.

$$loss(\boldsymbol{v}; \theta) = -\log p(\boldsymbol{v}; \theta) = F(\boldsymbol{v}; \theta) + \log \sum_{\tilde{\boldsymbol{v}}} e^{-F(\tilde{\boldsymbol{v}};\theta)} \tag{11}$$

while $F(\boldsymbol{v}; \theta)$ is the free energy function and is defined below:

$$F(\boldsymbol{v}; \theta) = -\log \sum_{h} e^{(-E(\boldsymbol{v},h;\theta))} \tag{12}$$

From the Eq. (11), we can conclude that minimizing the loss function corresponds to decrease the free energy value of a given data and increase the free energy value for all configurations of visible units.

As training the RBM in an unsupervised way with no label associated information involved during the training there is no guarantee that the feature learnt is useful for discriminative task. While energy-based model aims at giving lower energy value to observed data and higher value to unobserved data. During experiments, we observed that the energy value of training samples are mostly below zero while randomly generated samples are mostly above zero. Some researchers proposed to use the ratio of the probability of the training data over the probability of the data whose probability should be low [3] and they believe that the ratio is as high as possible when training RBM as an generative model. But, when it comes to considering RBM as a feature extractor, whether the free energy value is as high as possible is not clear. We wonder whether it is workable by constraining the free energy of unobserved variables from being too high, but their values are still higher than observed variables. Based on this idea, the novel objective function is proposed which consists of two parts: the old negative log-likelihood function, an penalty term. The extra penalty term penalizes big free energy values. It is formalized as follows:

$$loss'(\boldsymbol{v}; \theta) = loss(\boldsymbol{v}; \theta) + \frac{\lambda}{2} F(\tilde{\boldsymbol{v}}; \theta)^2$$

$$= F(\boldsymbol{v}; \theta) + \log \sum_{\tilde{\boldsymbol{v}}} e^{-F(\tilde{\boldsymbol{v}};\theta)} + \frac{\lambda}{2} F(\hat{\boldsymbol{v}}; \theta)^2 \tag{13}$$

where λ is the penalty coefficient to constraint the parameter space. And $\hat{\boldsymbol{v}}$ is the data point near the training data $\boldsymbol{v}$.

3.2   Training Algorithms

For training traditional negative log-likelihood objective function, CD and other variants algorithms are used to draw a sample from the model distribution. The update rules for parameters only differ in the penalty term and the data point $\hat{v}$ near the training data v could be obtained through CD, or other CD-like algorithms. Since the nature behind CD is that the energy values of the points near the training data are pushed up. Our penalty term is used to limit that energy value from going too high.

Until now, we get the update rules using stochastic gradient descent:

$$W_{ij} := W_{ij} + \eta(p(h_j = 1|v;\theta)v_i - p(h_j = 1|\tilde{v})\tilde{v}_i + \lambda \cdot f \cdot (h_j = 1|\hat{v})\hat{v}_i) \quad (14)$$

$$b_i := b_i + \eta(v_i - \tilde{v}_i + \lambda \cdot f \cdot \hat{v}_i) \quad (15)$$

$$c_j := c_j + \eta(p(h_j = 1|v;\theta) - p(h_j = 1|\tilde{v};\theta) + \lambda \cdot f \cdot p(h_j = 1|\hat{v};\theta)) \quad (16)$$

where $\hat{v}$ is a sample drawn the model distribution which can be obtained through CD, CD-n, etc. In our experiments below, $\hat{v}$ and $\tilde{v}$ takes the same value. η is the learning rate. And f is the free energy value which is defined in Eq. (12). Note that the update rules shown above only differs in one term compared with Eqs. (7)~(9), from the point of view of computation complexity that with an overhead of computing the free energy value we could learn much beneficial features.

4   Experiments and Results

To verify that our proposed objective function could learn much excellent features, several experiments were implemented and the corresponding results were described below.

Data Set: All experiments were performed on the MNIST handwritten digit dataset. The MNIST dataset consists of 70000 gray-scale images. Each image has 28×28 pixels which ranges from 0 to 255. The dataset has been split into a training set with 60000 images and a testing set with the remaining 10000 images. We scale the pixel densities of each image between 0 and 1 and which are interpreted as probabilities.

Evaluation Metrics: We used RBM as a feature extractor. The activation probabilities of the hidden units are regarded as the features learnt by the model. Further, the features were fed to a multinomial logistic regression classifier, and the accuracy ratio of classification on test dataset is regarded as the evaluation metric.

Parameter Settings: In the following experiments, the RBM consists of 784 visible units and 500 hidden units and stochastic gradient descent with mini-batches of 100 samples used. For comparison, no regularization or momentum method is involved. A fixed learning rate was used.

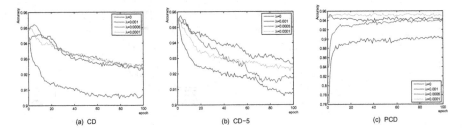

Fig. 1. The accuracy ratio on test dataset against the training epoch. Learning rate was set to 0.1. Plots of (a),(b),(c) correspond to training algorithms CD, CD-5, PCD respectively. Note that $\lambda = 0$ means that training the RBM using negative log-likelihood objective function.

Two groups of experiments having been performed, the only difference between each group is in the value of the learning rate and penalty coefficient. In the first group experiments, learning rate was set to 0.1 and $\lambda \in \{1e^{-3}, 5 \times 1e^{-4}, 1e^{-4}\}$. To acquire a sample drawn from the model distribution, CD, CD-5 and PCD were used. The accuracy ratio of classification on test dataset against the iteration number was depicted in Fig. 1.

From Fig. 1, we observe that when CD was adopted the accuracy ratio improves nearly 2 % and the influence of penalty coefficient becomes smaller and smaller along with the training. Obviously, When using CD-5 algorithm, the difference among different penalty coefficient becomes important. But three different settings of penalty coefficient all achieved much higher accuracy ratio. The rightmost one which trained by PCD shows that too bigger penalty coefficient could result in much lower accuracy ratio, only one smallest penalty coefficient could lead to the much higher accuracy ratio. For the experience of the first group experiments we concluded that learning rate set to 0.1 might be too big so that the curve drop significantly along with the training but our method still works well and achieved impressive improvement. Since PCD algorithm draw a sample from the model distribution from the persistent Markov chain, it is supposed that the model changes slightly after each parameter update and there is a connection between the parameter learning rate and mixing rate of Markov chain. For learning rate equals 0.1, the parameter learning rate grows much faster than the mixing rate so that the sample drawn from the model is much farther from the exact model distribution. We believe that based on this idea, the PCD behaves not very well, but experiments below show that when the learning rate was set reasonable our objective function works much better.

In the second group of experiments, the only difference is that learning rate was set to 0.01 and $\lambda \in \{1e^{-2}, 1e^{-3}\}$. CD, CD-5, PCD were adopted as usual and the results were depicted in Fig. 2.

From Fig. 2, we observe that our objective function achieved much higher accuracy ratio and improved by at least 1 % among all three training algorithms. Having a look at the leftmost one, after reaching the top point the blue curve

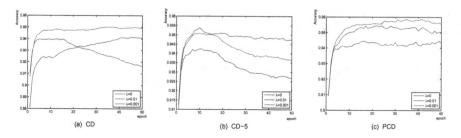

Fig. 2. The accuracy ratio on test dataset against the training epoch. Learning rate was set to 0.01. Plots of (a),(b),(c) correspond to training algorithms CD, CD-5, PCD respectively. Note that $\lambda = 0$ means that training the RBM using negative log-likelihood objective function. (Colour fig online)

changes slightly and gradually decrease, while the red one after reaching its top point, it changes slightly along with the training epoch which means that it is much stable. The same phenomenon could be also be observed when it comes to CD-5 algorithms. In this group experiments, PCD works as desired. We observe that the PCD favors in small penalty coefficient for no matter in the first group of experiments or the second one the smallest penalty coefficient achieved the highest accuracy ratio.

5   Conclusion and Future Work

In this paper, we attempted to learn much better features of RBM from raw data through minimizing a novel objective function. Our idea concentrates on limiting the free energy value of the data points near training data from being too high. An extra term is added to the negative log-likelihood objective function. And experimental results show that the accuracy ratio of classification on test dataset was improved. Although the proposed method works well, we are not clear of the theoretical reason behind the experimental results. And this could be regarded as the direction for future research.

Acknowledgment. This work is supported by the National Natural Science Foundation of China (Nos.61425002, 61572093, 61402066, 61402067, 31370778, 61370005), Program for Changjiang Scholars and Innovative Research Team in University (No. IRT_15R07), the Program for Liaoning Innovative Research Team in University(No. LT2015002), the Basic Research Program of the Key Lab in Liaoning Province Educational Department (Nos. LZ2014049, LZ2015004), Natural Science Foundation of Liaoning Province (No.2014020132), Scientific Research Fund of Liaoning Provincial Education (Nos. L2015015, L2014499), Liaoning BaiQianWan Talents Program (No.2013921007), and the Program for Liaoning Key Lab of Intelligent Information Processing and Network Technology in University.

References

1. Bengio, Y.: Learning deep architectures for AI. Found. Trends Mach. Learn. **2**(1), 1–127 (2009)
2. Brakel, P., Dieleman, S., Schrauwen, B.: Training restricted Boltzmann machines with multi-tempering: harnessing parallelization. In: Villa, A.E.P., Duch, W., Érdi, P., Masulli, F., Palm, G. (eds.) ICANN 2012. LNCS, vol. 7553, pp. 92–99. Springer, Heidelberg (2012). doi:10.1007/978-3-642-33266-1_12
3. Buchaca, D., Romero, E., Mazzanti, F., Delgado, J.: Stopping criteria in contrastive divergence: alternatives to the reconstruction error (2013). arXiv preprint arXiv:1312.6062
4. Desjardins, G., Courville, A.C., Bengio, Y., Vincent, P., Delalleau, O.: Tempered markov chain Monte Carlo for training of restricted Boltzmann machines. In: Proceedings of the Thirteenth International Conference on Artificial Intelligence and Statistics (AISTATS 2010), vol. 9, pp. 145–152 (2010)
5. Hinton, G.E., Salakhutdinov, R.R.: Reducing the dimensionality of data with neural networks. Science **313**(5786), 504–507 (2006)
6. Hinton, G.E.: Training products of experts by minimizing contrastive divergence. Neural Comput. **14**(8), 1771–1800 (2002)
7. Hinton, G.E., Osindero, S., Teh, Y.W.: A fast learning algorithm for deep belief nets. Neural Comput. **18**(7), 1527–1554 (2006)
8. Larochelle, H., Bengio, Y.: Classification using discriminative restricted Boltzmann machines. In: Proceedings of the 25th International Conference on Machine Learning, ICML 2008, pp. 536–543. ACM, New York (2008)
9. LeCun, Y., Chopra, S., Hadsell, R., Ranzato, M., Huang, F.: A tutorial on energy-based learning. In: Predicting Structured Data. MIT Press (2006)
10. Lee, H., Ekanadham, C., Ng, A.Y.: Sparse deep belief net model for visual area v2. In: Advances in Neural Information Processing Systems 20, pp. 873–880. Curran Associates, Inc. (2008)
11. Deng, L., Yu, D.: Deep convex network: a scalable architecture for speech pattern classification. In: International Speech Communication Association, August 2011
12. Ma, X., Wang, X.: Average contrastive divergence for training restricted Boltzmann machines. Entropy **18**(1), 35 (2016)
13. Ranzato, M., Hinton, G.E.: Modeling pixel means and covariances using factorized third-order Boltzmann machines. In: 2010 IEEE Conference on Computer Vision and Pattern Recognition (CVPR), pp. 2551–2558, June 2010
14. Salakhutdinov, R., Hinton, G.E.: Deep Boltzmann machines. J. Mach. Learn. Res. **5**(2), 1967–2006 (2009)
15. Sarikaya, R., Hinto, G.E., Deoras, A.: Application of deep belief networks for natural language understanding. IEEE/ACM Trans. Audio, Speech Lang. Process. **22**(4), 778–784 (2014)
16. Smolensky, P.: Information processing in dynamical systems: foundations of harmony theory. In: Parallel Distributed Processing: Explorations in the Microstructure of Cognition, vol. 1, pp. 194–281. MIT Press, Cambridge (1986)
17. Tieleman, T.: Training restricted Boltzmann machines using approximations to the likelihood gradient. In: Proceedings of the 25th International Conference on Machine Learning, ICML 2008, pp. 1064–1071. ACM, New York (2008)
18. Tieleman, T., Hinton, G.: Using fast weights to improve persistent contrastive divergence. In: Proceedings of the 26th Annual International Conference on Machine Learning, ICML 2009, pp. 1033–1040. ACM, New York (2009)
19. Stanford University: Difficulty of training deep architectures. http://ufldl.stanford.edu/wiki/index.php/Deep_Networks:_Overview

An Approach to Design Growing Echo State Networks

Li Fan-jun[1(✉)] and Li Ying[2]

[1] School of Mathematical Science,
University of Jinan, Shandong 250022, China
ss_lifj@ujn.edu.cn
[2] School of Science, Qilu University of Technology,
Jinan 250353, Shandong, China
lzhbb07@sina.com

Abstract. Echo State Networks (ESNs) have attracted wide attention for their superior performance in time series prediction. However, it is difficult to design an ESN to match with the given application. In this paper, an approach is proposed to design growing echo state networks. The basic idea of the proposed method is to design a growing reservoir with multiple sub-reservoirs by adding hidden units to the network group by group. First, several subservoirs are synchronously constructed by using the singular value decomposition. Then, every subservoir is evaluated and the best one is selected to be added to the network. Finally, two time series are used to validate the proposed approach.

Keywords: Echo state network · Reservoir · Sub-reservoir · Growing · Recurrent neural network

1 Introduction

Theoretically, recurrent neural networks (RNNs) are powerful tools to model temporal correlations between the input and output sequences, and can approximate any open dynamical systems with an arbitrary accuracy [1]. Nonetheless, several factors still hinder the wide application of RNNs. There are few learning rules and most suffer from slow convergence rates, thus limiting their applicability [2–4]. As special RNNs, echo state networks (ESNs) have attracted a lot of attention in recent decades for their superior performance in time series predictions [5, 6]. In ESNs, only the weights feeding into the output units are calculated by solving a simple linear regression problem, whereas all the other weights remain unchanged [7]. This simple and effective training approach makes ESNs to perform well when some stability conditions are satisfied [8, 9]. ESNs offer an intuitive methodology for using the temporal processing power of RNNs without the hassle of training them, and have been successfully applied in many fields [10–13].

The performance of ESNs mainly depends on a large number of randomly and sparsely connected hidden units called reservoir [7, 8]. There have been many attempts to find more efficient reservoir schemes to improve the performance of ESNs [14–16]. For example, the simple cycle reservoir (SCR) has been developed, which is often

H. Yin et al. (Eds.): IDEAL 2016, LNCS 9937, pp. 220–230, 2016.
DOI: 10.1007/978-3-319-46257-8_24

sufficient for obtaining performance comparable with random ones [17]. Reservoirs with biological properties (small-world and scale-free) have been constructed by applying the complex network theory [18]. Decoupled ESN (DESN) with multiple sub-reservoirs has been proposed to reduce the coupling effects between the units within a single reservoir [19]. However, it is difficult for above schemes to determine their reservoir size. An Growing Echo State Network (GESN) with multiple sub-reservoirs has been proposed to design self-organizing ESNs to match with the given application, which can determine their reservoir size and sparsity automatically [20]. During our recent study, it is found that some of the hidden units in such networks may play a very minor role in the network output and thus may eventually increase the network complexity.

As a result of the research noted above, an approach is proposed to design growing ESNs for improving the performance of ESNs. First, several subservoirs are synchronously constructed with random singular values (smaller than 1) and orthogonal matrices by using the singular value decomposition (SVD) method. Then, a novel index function is derived to evaluate the contribution of subservoirs to the network, and the best one is added to the network. When some termination criterion is satisfied, the growth of the network is stopped. Finally, Mackey–Glass time series and multiple superimposed oscillator (MSO) problem are used to validate the proposed approach. Simulation results show that the ESN constructed by our method has better prediction performance.

The remainder of this paper is organized as follows. Section 2 briefly reviews the original ESN. The details of the proposed method are described in Sect. 3. Section 4 presents the simulation results which show the superior performance of the ESN constructed by our method. Finally, Sect. 5 summarizes the main conclusions.

2 Brief of ESNs

An ESN without output feedback connections, as shown in Fig. 1, is a recurrent neural network with K input units providing external stimulations $u(n) \in R^K$, N internal units possessing internal states $x(n) \in R^N$, and L output units generating the output signals $y(n) \in R^L$. The connections between the input units and the internal units are given by an $N \times K$ weight matrix W^{in}, connections between the internal units are collected in an $N \times N$ weight matrix W, connections feeding into output units are given in a $L \times N$ weight matrix W^{out}. The internal states $x(n)$ and output signals $y(n)$ can be calculated as follows,

$$x(n) = f(Wx(n-1) + W^{in}u(n)) \tag{1}$$

$$y(n) = f^{out}(W^{out}X(n)) \tag{2}$$

Where $X(n) = (x(n)^T, u(n)^T)^T$ is the concatenation of the internal states and input vectors, f and f^{out} are the activation functions of internal units and output units respectively. After initial washout period, the internal states and input vectors are collected in the matrix H, and the corresponding target outputs are given in matrix T. If f^{out} is identity function, the output weight W^{out} can be calculated as follows,

$$W^{out} = ((H^T H)^{-1} H^T T)^T \qquad (3)$$

An important element for the ESN to work is that the reservoir should have the echo state property (ESP) or stability. As long as the largest singular value of W ($\sigma(W)$) is less than 1, ESNs can possess the ESP. For practical purpose, the reservoir weight matrix W is scaled by a scaling parameter α to satisfy the necessary condition for the ESP: the spectral radius of W ($\rho(W)$) is less than 1.

The main steps of the original ESN (without output feedback connections) can be summarized as follows,

Step 1: Procure a reservoir with the necessary condition (W, W^{in}, α);
Step 2: Run the reservoir using the training input $\mathbf{u}(n)$, then collect the internal state $x(n)$ after initial washout period;
Step 3: Calculate the output weight W^{out} by using Eq. (3);
Step 4: Test the trained ESN on unseen data.

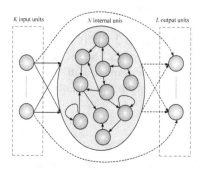

Fig. 1. Basic structure of an ESN without output feedback connections (Dashed arrows: connections to be trained)

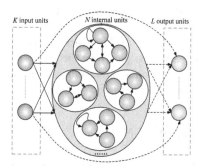

Fig. 2. Basic structure of an GESN without output feedback connections (Dashed arrows: connections to be trained)

3 Approach to Design Growing ESNs

3.1 Method to Generate Sub-Reservoir Weight Matrices

The method to generate sub-reservoir weight matrices is similar with that in GESN. The basic idea of GESN is to design a growing reservoir by analyzing the linear algebraic properties of the reservoir weight matrix. As shown in Fig. 2, the growing reservoir consists of sub-reservoirs that are added to the network one by one until a stopping criterion is satisfied. The sub-reservoir weight matrices are constructed by the SVD method with the predefined singular values as follows [20],

Step 1: Randomly generate a diagonal matrix $S_i = diag(\sigma_1, \sigma_2, \cdots, \sigma_{n_i})$ where $0 < \sigma_j < 1(j = 1, 2, \cdots, n_i)$;

Step 2: Randomly generate two orthogonal matrices $U_i = (u_{kj})_{n_i \times n_i}$ and $V_i = (v_{kj})_{n_i \times n_i}$ where $-1 < u_{kj}, v_{kj} < 1(k, j = 1, 2, \cdots, n_i)$;

Step 3: Generate sub-reservoir weight matrix $\varDelta W_i = U_i S_i V_i$.

According to SVD, $\varDelta W_i$ and S_i have the same singular values which are smaller than 1.

3.2   Index Function to Evaluate Sub-Reservoirs

Theorem 1: Given M training samples, an sub-reservoir with n_i hidden units, sigmoidal activation function, the internal state matrix $H_{M \times n_i} = (\,h_1 \quad h_2 \quad \cdots \quad h_{n_i}\,)^1$, the corresponding training error vector E_{i-1} after i-1 sub-reservoirs have been added, define the internal state space as $V_i = span(h_1, h_2, \cdots, h_{n_i})$, the distance between E_{i-1} and V_i as $d(E_{i-1}, V_i)$. Then we have

$$d^2(E_{i-1}, V_i) = \frac{G(h_1, h_2, \cdots, h_{n_i}, E_{i-1})}{G(h_1, h_2, \cdots, h_{n_i})} \tag{4}$$

where $G(h_1, h_2, \cdots, h_{n_i})$ is the gramian of $h_1, h_2, \cdots, h_{n_i}$ as follows

$$G(h_1, h_2, \cdots, h_{n_i}) = \begin{vmatrix} \langle h_1, h_1 \rangle & \langle h_1, h_2 \rangle & \cdots & \langle h_1, h_{n_i} \rangle \\ \langle h_2, h_1 \rangle & \langle h_2, h_2 \rangle & \cdots & \langle h_2, h_{n_i} \rangle \\ \vdots & \vdots & \vdots & \vdots \\ \langle h_{n_i}, h_1 \rangle & \langle h_{n_i}, h_2 \rangle & \cdots & \langle h_{n_i}, h_{n_i} \rangle \end{vmatrix} \tag{5}$$

Proof: Let e_{i-1} be the orthogonal projection of E_{i-1} onto V_i, then we have

$$\langle E_{i-1} - e_{i-1}, h_j \rangle = 0, j = 1, 2, \cdots, n_i \tag{6}$$

and

$$G(h_1, h_2, \cdots, h_{n_i}, e_{i-1}) = 0 \tag{7}$$

According to (7), we have

---

[1] According to reference [20], the column vectors of H can be made full-rank with probability one.

$$
\begin{aligned}
G(h_1, h_2, \cdots, h_{n_i}, E_{i-1}) &= G(h_1, h_2, \cdots, h_{n_i}, (E_{i-1} - e_{i-1}) + e_{i-1}) \\
&= G(h_1, h_2, \cdots, h_{n_i}, (E_{i-1} - e_{i-1})) + G(h_1, h_2, \cdots, h_{n_i}, e_{i-1}) \\
&= \begin{vmatrix}
\langle h_1, h_1 \rangle & \langle h_1, h_2 \rangle & \cdots & \langle h_1, h_{n_i} \rangle & \langle h_1, E_{i-1} - e_{i-1} \rangle \\
\langle h_2, h_1 \rangle & \langle h_2, h_2 \rangle & \cdots & \langle h_2, h_{n_i} \rangle & \langle h_2, E_{i-1} - e_{i-1} \rangle \\
\vdots & \vdots & \cdots & \vdots & \vdots \\
\langle h_{n_i}, h_1 \rangle & \langle h_{n_i}, h_2 \rangle & \cdots & \langle h_{n_i}, h_{n_i} \rangle & \langle h_{n_i}, E_{i-1} - e_{i-1} \rangle \\
\langle E_{i-1} - e_{i-1}, h_1 \rangle & \langle E_{i-1} - e_{i-1}, h_2 \rangle & \cdots & \langle E_{i-1} - e_{i-1}, h_{n_i} \rangle & \langle E_{i-1} - e_{i-1}, E_{i-1} - e_{i-1} \rangle
\end{vmatrix}
\end{aligned}
\tag{8}
$$

According to (6), we have

$$
G(h_1, h_2, \cdots, h_{n_i}, E_{i-1}) =
\begin{vmatrix}
\langle h_1, h_1 \rangle & \langle h_1, h_2 \rangle & \cdots & \langle h_1, h_{n_i} \rangle & 0 \\
\langle h_2, h_1 \rangle & \langle h_2, h_2 \rangle & \cdots & \langle h_2, h_{n_i} \rangle & 0 \\
\vdots & \vdots & \cdots & \vdots & \vdots \\
\langle h_{n_i}, h_1 \rangle & \langle h_{n_i}, h_2 \rangle & \cdots & \langle h_{n_i}, h_{n_i} \rangle & 0 \\
0 & 0 & \cdots & 0 & \langle E_{i-1} - e_{i-1}, E_{i-1} - e_{i-1} \rangle
\end{vmatrix}
\tag{9}
$$

$$
= \langle E_{i-1} - e_{i-1}, E_{i-1} - e_{i-1} \rangle G(h_1, h_2, \cdots, h_{n_i})
$$

Hench, we have

$$
d^2(E_{i-1}, V_i) = \langle E_{i-1} - e_i, E_{i-1} - e_i \rangle = \frac{G(h_1, h_2, \cdots, h_{n_i}, E_{i-1})}{G(h_1, h_2, \cdots, h_{n_i})}
\tag{10}
$$

The smaller distance between E_{i-1} and V_i means more contributions of the ith sub-reservoir to the network, hence index function is defined as (11) to evaluate sub-reservoirs.

$$
d(E_{i-1}, V_i) = \sqrt{\frac{G(h_1, h_2, \cdots, h_{n_i}, E_{i-1})}{G(h_1, h_2, \cdots, h_{n_i})}}
\tag{11}
$$

3.3   Steps to Construct Growing ESNs

Given a training sequence $(u(n); y(n))$ $n = 0, 1, \cdots, n_{\max}$ where the input signals come from a compact set, the internal activation function (sigmoid function) and the output activation function (identity function). The main steps of the proposed approach to design growing ESNs (without output feedbacks) can be described by the flowchart as shown in Fig. 3, which are explained further as follows:

Step 1: Generate J sub-reservoirs by using SVD as done in Sect. 3.1;
Step 2: Calculate the distance between every sub-reservoir and the training error vector by using Eq. (11) to evaluate the contribution of sub-reservoirs to the network;

Step 3: Select the sub-reservoir with the largest contribution (the smallest distance) to be added to the network, which leads to a growing internal weight matrix as Eq. (12);

Step 4: Calculate the output weight matrix W^{out} by using Eq. (3);

Step 5: Calculate training error vector and validation error;

Step 6: If some terminate condition is satisfied, the process of growing is stopped, or go to step 1.

$$W = \begin{pmatrix} \Delta W_1 & & & \\ & \Delta W_2 & & \\ & & \Delta W_3 & \\ & & & \ddots \end{pmatrix} \quad (12)$$

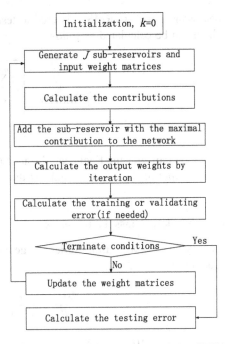

Fig. 3. Flowchart of approach to design GESN

In this paper, the l-step reduction of validation error, caused by the successive l sub-reservoir, is selected as the terminate criterion, which can be calculated as follows,

$$Error^l_{validate} = Error_{validate}(i - l) - Error_{validate}(i) \quad (13)$$

where $i > l$ and $Error_{validate}(i)$ is the validation error after the ith sub-reservoir is added to the network. When $Error_{validate}(i)$ is not more than zero or the size of reservoir exceeds the allowable maximum, the growth of the network is stopped.

4 Performance Evaluation

In this section, all simulations are carried out in MATLAB 2010a environment, running on a Pentium 4 with 3.4 GHZ CPU. Several ESNs with different reservoirs[2], including the original ESN (O-ESN) [8], SCR [17], DESN [19] and GESN [20], are used to test performance of the proposed approach. Each data set is divided into parts for training, validating, and testing, the lengths of which are denoted by L_{train}, $L_{validate}$, and L_{test}, respectively. The initial washout period is denoted by $L_{washout}$. The normalized root

---

[2] Size, sparsity and spectral radius of reservoirs are optimized by the conventional grid search method.

mean square error (NRMSE) is used to test the performances of ESNs in this section, which can be calculated as follows,

$$NRMSE = \sqrt{\frac{\sum_{i=n_{min}+1}^{n_{max}} \|\mathbf{t}(i) - \mathbf{y}(i)\|^2}{\sum_{i=n_{min}+1}^{n_{max}} \|\mathbf{t}(i) - \tilde{t}\|^2}} \tag{14}$$

where $\|\bullet\|$ indicates the Euclidean distance (or norm), $t(i)$ is the target ouput, $y(i)$ is the actual output of the network, and $\tilde{t}$ is the average over the whole target outputs.

4.1   Mackey–Glass Time Series

Mackey-Glass (MG) time series is generated by the following differential delay equation,

$$\frac{dx(t)}{dt} = \frac{ax(t-\tau)}{1+x^n(t-\tau)} + bx(t) \tag{15}$$

Where $n = 10$, $a = 0.2$, $b = -0.1$, $\tau = 17$. 10000 points are generated using the fourth-order Runge–Kutta method with random initialization, and split into three parts with $L_{train} = 3000$, $L_{validate} = 3000$, $L_{test} = 4000$ and $L_{washout} = 1000$.

One hundred independent simulations of MG are teacher forced on the trained network for 2000 steps, and then the network remains free running for 2000 steps. The network's prediction outputs are compared with the target outputs of MGS after 84 steps to obtain a testing $NRMSE_{84}$, which is calculated as follows,

$$NRMSE_{84} = (\sum_{i=1}^{100} \frac{(t_i(2084) - y_i(2084))^2}{100\sigma^2})^{1/2} \tag{16}$$

where $t_i(2084)$ is the target output, $y_i(2084)$ is the prediction output of the trained network, and σ^2 is the variance of the MG signal.

The detailed simulation results and the selected parameters are listed in Table 1. It can be seen that the proposed approach obtains the best prediction performance with the smallest reservoir size. Although slower than GESN, the proposed approach learns much faster than O-ESN and DESN. Meanwhile, the largest singular values of the proposed approach and GESN are smaller than 1, which means that they can guarantee the ESP without posterior scaling of the weights. Figure 4 shows the first 1500 steps free running outputs of the trained networks with the continuation of the original series. It can be seen from Fig. 4 that the prediction outputs for all algorithms start to deviate perceptibly from the original series after about 1200 steps, while the proposed approach performs better. The training error curve and the contributions of sub-reservoirs are shown in Fig. 5, which illustrates the effectiveness of the derived index function.

Table 1. The selected parameters and simulation results for Mackey-Glass time series

Approach	Training time	Testing NRMSE$_{84}$	Reservoir size	Spectral radius	The largest singular value	Sparsity
Proposed	1.8593e + 03	**3.35e-05**	**905**	0.9568	0.9887	0.0053
GESN	**1.0475e + 03**	4.18e-05	945	0.9483	0.9868	0.0055
O-ESN	2.8154e + 03	5.24e-05	1000	0.8500	1.6920	0.0350
DESN	2.7711e + 03	1.50e-04	1000	0.8000	1.5571	0.0250
SCR	3.5541e + 02	5.39e-04	1000	0.8000	0.8000	0.0010

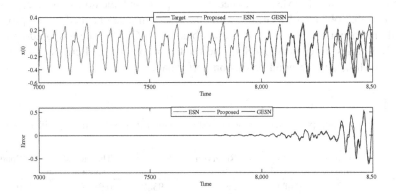

Fig. 4. Prediction results for Mackey-Glass time series

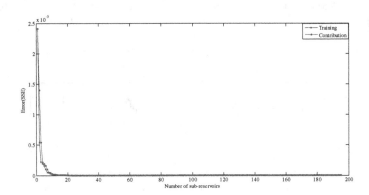

Fig. 5. Training error for Mackey-Glass time series

4.2 MSO Problem

In contrast to MG time series prediction, the MSO problem is a challenge for ESNs, even though the function consists of only two sine waves. In this simulation, the MSO problem with five sine waves is derived from the following equation

$$y(t) = \sum_{i=1}^{5} \sin(\alpha_i t) \tag{17}$$

where $\alpha_1 = 0.2$, $\alpha_2 = 0.311$, $\alpha_3 = 0.42$, $\alpha_4 = 0.51$, $\alpha_5 = 0.63$. 1800 points are generated by using Eq. (17) and split into three parts with $L_{train} = 1000$, $L_{validate} = 500$, $L_{test} = 300$ and $L_{washout} = 100$.

For all algorithms, 100 independent simulations have been conducted and the averaged results are listed in Table 2. It can be seen that the proposed approach obtains the parsimonious reservoir architecture with the best prediction performance. For example, the testing NRMSE obtained by the proposed method only accounts for 63.63 % of that obtained by GESN, while the reservoir size obtained by the proposed method only accounts for 83.33 % of that obtained by GESN. The proposed approach learns 16.9 times faster than O-ESN. The prediction outputs and errors are shown in Fig. 6, which means the good prediction performance.

Table 2. The selected parameters and averaged simulation results for MSO problem with five sine waves

Approach	Training time	Testing NRMSE	Reservoir size	Spectral radius	The largest singular value	Sparsity
Proposed	7.2122	**0.0035**	**25**	0.9380	0.9869	0.1923
GESN	**1.7624**	0.0055	30	0.9589	0.9855	0.1667
O-ESN	121.9974	0.0732	80	0.9000	1.6931	0.2000
DESN	85.3838	0.1413	90	0.7500	1.2515	0.8000
SCR	8.5149	0.0529	40	0.7500	0.7500	0.0250

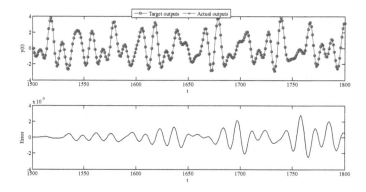

Fig. 6. Prediction results for Mackey-Glass time series

5   Conclusions

In this paper, an approach is proposed to design growing ESNs with multiple sub-reservoirs. The basic idea of the proposed method is to design a growing reservoir by adding hidden units to the network group by group. First, the index function is derived to evaluate the contribution of sub-reservoirs. Then, several subservoirs are synchronously constructed by using the singular value decomposition. Finally, every subservoir is evaluated by the derived index function, and the best one is selected to be added to the network. Two benchmark time series are used to validate the proposed approach. Simulation results show that the ESNs constructed by the proposed approach have smaller reservoir size and better prediction performance.

References

1. Schafer, A.M., Zimmermann, H.G.: Recurrent neural networks are universal approximators. Int. J. Neural Syst. **17**(04), 253–263 (2007)
2. Jaeger, H.: Tutorial on training recurrent neural networks, covering BPTT, RTRL, EKF and echo state network approach. Tech. report 159, German National Research Center for Information Technology, St. Augustin, Germany (2002)
3. Doya, K.: Recurrent networks: learning algorithms. In: Arbib, M.A. (ed.) The Handbook of Brain Theory and Neural Networks, pp. 955–960. MIT Press, Cambridge (2003)
4. Sutskever, I.: Training recurrent neural networks. Dissertation, University of Toronto (2013)
5. Li, D.C., Han, M., Wang, J.: Chaotic time series prediction based on a novel robust echo state network. IEEE Trans. Neural Netw. Learn. Syst. **23**(5), 787–799 (2012)
6. Zhang, B., Miller, D.J., Wang, Y.: Nonlinear system modeling with random matrices: echo state networks revisited. IEEE Trans. Neural Netw. Learn. Syst. **23**(1), 175–182 (2012)
7. Jaeger, H., Haas, H.: Harnessing nonlinearity: predicting chaotic systems and saving energy in wireless communication. Science **304**(5667), 78–80 (2004)
8. Jaeger, H.: The echo state approach to analyzing and training recurrent neural networks. Technical report 148, German National Research Center for Information Technology, St. Augustin, Germany (2001)
9. Rao, J.S.: Optimization. In: Rao, J.S. (ed.) History of Rotating Machinery Dynamics. HMMS, vol. 20, pp. 341–351. Springer, Heidelberg (2011)
10. Pan, Y.P., Wang, J.: Model predictive control of unknown nonlinear dynamical systems based on recurrent neural networks. IEEE Trans. Ind. Electron. **59**(8), 3089–3101 (2012)
11. Skowronski, M.D., Harris, J.G.: Automatic speech recognition using a predictive echo state network classifier. Neural Netw. **20**(3), 414–423 (2007)
12. Xia, Y., Jelfs, B., Van Hulle, M.M., et al.: An augmented echo state network for nonlinear adaptive filtering of complex noncircular signals. IEEE Trans. Neural Netw. **22**(1), 74–83 (2011)
13. Shi, Z.W., Han, M.: Support vector echo-state machine for chaotic time-series prediction. IEEE Trans. Neural Netw. **18**(2), 359–372 (2007)
14. Lukosevicius, M., Jaeger, H., Schrauwen, B.: Reservoir computing trends. KI-Knstliche Intelligenz **26**(4), 365–371 (2012)
15. LukosEviclus, M., Jaeger, H.: Reservoir computing approaches to recurrent neural network training. Comput. Sci. Rev. **3**(3), 127–149 (2009)

16. Strauss, T., Wustlich, W., Labahn, R.: Design strategies for weight matrices of echo state networks. Neural Comput. **24**(12), 3246–3276 (2012)
17. Rodan, A., Tino, P.: Minimum complexity echo state network. IEEE Trans. Neural Netw. **22** (1), 131–144 (2011)
18. Deng, Z., Zhang, Y.: Collective behavior of a small-world recurrent neural system with scale-free distribution. IEEE Trans. Neural Netw. **18**(5), 1364–1375 (2007)
19. Xue, Y., Yang, L., Haykin, S.: Decoupled echo state networks with lateral inhibition. Neural Netw. **20**(3), 365–376 (2007)
20. Qiao, J., Li, F., Han, H., et al.: Growing echo-state network with multiple subreservoirs. IEEE Trans. Neural Netw. Learn. Syst. **PP**(99), 1–14 (2016)

A Density-Based Clustering Algorithm with Sampling for Travel Behavior Analysis

Wang Tang, Dechang Pi$^{(\boxtimes)}$, and Yun He

College of Computer Science and Technology,
Nanjing University of Aeronautics and Astronautics, 29 Jiangjun Road,
Nanjing 211106, Jiangsu, People's Republic of China
{46015316,1561850387}@qq.com, dc.pi@nuaa.edu.cn

Abstract. In order to get the characteristics of travel behavior and reduce the cost of experiment, this paper presents a travel behavior clustering algorithm, sampling-based DBSCAN (SB-DBSCAN), which uses a sampling technique based on density-based spatial clustering of applications with noise (DBSCAN). By introducing sampling technique, SB-DBSCAN can obtain the approximate distribution of original data set. The improved algorithm solves the problem that DBSCAN needs large capacity memory when analyzing massive data. The effects of sampling rate and neighborhood radius on the efficiency of SB-DBSCAN are analyzed by experiments. After comparing SB-DBSCAN with DBSCAN, the result shows that SB-DBSCAN can maintain the effect of clustering, with a better scalability and less time cost simultaneously. Finally, SB-DBSCAN is used to analyze the travel behavior, which provides decision support for urban planning and construction.

Keywords: Data mining · Cluster analysis · Travel behavior · Sampling

1 Introduction

Travel behavior theory [1] is one of the most important theories in the field of traffic management and control. Travel behavior research plays an important role in the urban traffic planning and construction.

Hess et al. [2] derived a travel-time coefficient with mathematical method, and studied the time value of urban residential travel behavior. Based on the survey data of residents in Shanghai suburbs, Robert and Jennifer [3] analyzed the influence of suburban rail construction on the travel behavior and residential employment opportunities. Raquel et al. [4] used RP/SP data to analyze the behaviors of the suburban travelers choosing the means of transport, and calculated the time value of the travelers. Ma et al. [5] and Kieu et al. [6] used the traditional DBSCAN algorithm to mine spatial and temporal travel patterns from AFC data.

Compared to other methods, DBSCAN is more flexible to find the travel behavior pattern. Because DBSCAN can discover clusters of arbitrary shape as higher-density data points are more likely to be grouped into a cluster. However, for large-scale spatial databases, DBSCAN is expensive as it requires large volume of memory support due to

© Springer International Publishing AG 2016
H. Yin et al. (Eds.): IDEAL 2016, LNCS 9937, pp. 231–239, 2016.
DOI: 10.1007/978-3-319-46257-8_25

its operations over the entire database. So in this paper, we propose a sampling-based DBSCAN (SB-DBSCAN), which can reduce the operation time effectively in the large-scale spatial databases. And then, this algorithm is applied on the analysis of travel behavior. According to the analysis results, we will get the high traffic areas and the time period of congestion, which can promote the planning of urban public traffic route network and ensure rational use of resources.

The rest paper is organized as follows. In Sect. 2, some related definitions are introduced, and the process of DBSCAN is given. An improved algorithm named SB-DBSACN is proposed in Sect. 3. The analysis of the experimental results is provided in Sect. 4. Finally, the Sect. 5 offers concluding remarks.

2   The Classical DBSCAN Algorithm

2.1   Related Definitions

The classical DBSCAN algorithm is based on the following definitions.

Definition 1 (*Eps*-neighborhood): A user-specified parameter *Eps*, which is used to specify the radius of a neighborhood. Given a data set, D, if set P represents the point set in the *Eps*-neighborhood of point q, then:

$$P = \{p, p \in D \wedge d(p, q) \leq Eps\}$$

Definition 2 (core object): Core object is the object which *Eps*-neighborhood contains at least *MinPts* objects.

Definition 3 (directly density-reachable): Given a set S, if object q is within the *Eps*-neighborhood of object p, and p is a core object, then we say that q is directly density-reachable from p.

Definition 4 (density-reachable): Given a set S and a bunch of sample points $p_1, p_2, \ldots, p_n$ and $p = p_1$, $q = p_n$. If object p_i is directly density-reachable from object p_{i-1}, then we say that q is density-reachable from p.

Definition 5 (density-connected): Given a set S, for each point $o \in S$, if object p is density-reachable from object o, and object q is density-reachable from o, then we say that p and q are density-connected.

From the above definition, we can divide all the points in the set into three categories: core point, border point and noise point.

Definition 6 (core point): A point is a core point if the number of points within its *Eps*-neighborhood exceeds the certain threshold *MinPts*. In Fig. 1, point A is a core point.

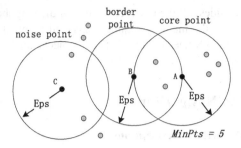

Fig. 1. Core, border and noise point

Definition 7 (border point): A border point is not a core point, but falls within the *Eps*-neighborhood of a core point. In Fig. 1 point *B* is a border point.

Definition 8 (noise point): A noise point is the point that is neither a core point nor a border point. In Fig. 1, point *C* is a noise point.

2.2 DBSCAN Algorithm Descriptions

The general idea of DBSCAN is, for each point p_i in data set *S*. If p_i is a core point, we can find the *Eps*-neighborhood, and create a new cluster for p_i and the point within the *Eps*-neighborhood of p_i. And then use these points within the *Eps*-neighborhood of p_i to expand their cluster until a complete cluster is found. We can also find the other clusters in the same way. In the end, the point that doesn't belong to any cluster is noise point. The process of DBSCAN is as follows:

Algorithm 1: DBSCAN

Input: data set *S*, the radius of a neighborhood *Eps*, the density threshold *MinPts*

Output: A set of cluster $C_j (1 \le j \le n)$, where *n* is the number of clusters

01. *for*$(i=1; i \le num; i++)$// *num*: the number of point in the data set *S*

02.   *if*(p_i is a core object)

03.   find all density-reachable point of p_i in the *Eps*-neighborhood of p_i

04. *for*$(j=1; j \le m; j++)$// *m*: the number of core objects in the data set *S*

05.   find all directly density-reachable point of the core object p_i, then obtain the set of density-connected point.

06. the point that doesn't belong to any cluster is noise point

---

In order to obtain more effective clustering results, DBSCAN uses the structure which called R* tree. Before clustering, DBSCAN creates an R* tree for all data [7]. Firstly, DBSCAN calculates the distance between the point and its *k* nearest neighbor point. Then, sort the distance in ascending order, and draw the *k-dist* diagram on the basis of the results. Users should choose a suitable *Eps* according to the *k-dist* diagram. Both creating the R* tree and drawing the *k-dist* diagram are time consuming, especially for a large database. Besides, all the untreated points in the neighborhood will be used to expand the cluster. So, the memory requirements will be great. Meanwhile, in the clustering process, DBSCAN needs to check all the objects, so the time complexity of DBSCAB is very high.

3   The SB-DBSACN Algorithm

3.1   SB-DBSCAN Algorithm Descriptions

Commonly, Sampling used to improve the efficiency of clustering when clustering large data sets. In order to improve the scalability and the time efficiency of DBSCAN in large database, we introduce a sampling technique to DBSCAN, and propose an improved clustering algorithm named SB-DBSACN. An approximate distribution of the original data set is obtained by sampling. With the increasing of the sampling rate, this approximate distribution will be much more similar to the original data set. In the research of data mining, sampling can reduce the size of data set, so that many data mining algorithms can be applied to large data sets and data streams. Besides, the correct and representative sampling can ensure the efficiency of data mining algorithms. According to the selected probability of each data point, the sampling method is divided into uniform sampling and biased sampling. Uniform sampling means that the data items are selected with the same probability, while biased sampling is opposite. Uniform sampling is divided into two types: Bernoulli sampling and Reservoir sampling [8]. In this paper, we use Reservoir sampling to sample data-set, because it can get the uniform sampling results. The process of SB-DBSACN is shown in Algorithm 2.

Algorithm 2: SB-DBSACN

Input: data set S, the radius of a neighborhood *Eps*, the density threshold *MinPts* and sampling rate *Per*

Output: A set of cluster $C_j (1 \leq j \leq n)$, where n is the number of clusters

01.   Sampling in S according to *per*, and obtain the S'.

02.   Create two R^* trees for S and S'.

03.   Clustering S' with algorithm 1.

04.   *for(i=1;i $\leq$ num;i++)* //*num*: the number of points in sample data set S'

05.   if(p_i is a core object AND $pi \subseteq S'$)

06.   find all density-reachable points of p_i in the *Eps*-neighborhood of p_i in the original dataset S.

07.   *for(j=1;j $\leq$ m;j++)*// m : the number of core objects sample the data set S'

08.   find all directly density-reachable points of the core object p_i, then obtain the set of density-connected, mark all points in the set by the cluster label of the p_i.

09.   deal with the problem of missing point and cluster splitting caused by data sampling in clustering process.

Compared with DBSCAN, SB-DBSACN has an advantage: The process of clustering analysis to the sampled data and the process of marking un-sampled data are synchronous, which means that SB-DBSACN is more efficient than the other algorithms.

In spite of the advantage mentioned above, there is no denying that the sampled data can't represent the original data set completely. The size of the sampling data set and the uniformity of the sampling distribution will have much effect on the final clustering results: missing point and cluster splitting. Missing point means that few objects are density-reachable from the point. If these points are not sampled, it will be possible to become a noise point. But in fact, these points can become an independent cluster. Besides, in the original database, some clusters may contain two or more parts, and these parts are simply constituted by one or a few density-reachable points. When these points are not been sampled, the cluster will be split into multiple clusters, which is called cluster splitting.

For the problems mentioned above, there are some ways to solve them. The missing points can be solved in this way: First, look for the neighborhood of the missing points whether there is such a point that has been marked by a cluster. If such a point does not exist, the missing point is a real noise point. Otherwise, the missing point belongs to the cluster. For the cluster splitting, look for the neighborhood of the boundary points of a cluster, if it contains a core point which has been marked by another cluster, then merge these two clusters.

3.2 Complexity Analysis

As we known, the complexity of the DBSCAN algorithm using a R^* tree is $O(nlogn)$, where n is the size of dataset. Because of the sampling, the SB-DBSCAN algorithm has the extra complexity $O(n)$, but the size of dataset is became small, so the complexity of the SB-DBSCAN algorithm is

$$O(m \log m) + O(n)$$

where m is the size of the dataset after sampling, which proves us that the time complexity of SB-DBSCAN is much less than SBSCAN.

4   Experiments and Analysis

This section uses SB-DBSACN to analyze the travel behavior. Software environment: Intel(R) Core(TM) i5-3470 3.2GHZ CPU, 4G main memory, 1T hard disk and Microsoft Windows7 system. Experiment platform: Eclipse. Language: Java.

4.1   The Experimental Data Source

The experimental data used in this study comes from Nanjing Citizen Card Co., Ltd, which covers a city citizen-card data from 2015-06-05 00:00:00 to 2015-07-06 00:00:00, 10 million data records in total. The data is stored in a data table named by the day.

As Table 1, the main structure of the data contains 30 attributes. For clustering more effectively, we summarize the data to a data table, and use 3 attributes: GetOnTime, GetOnStation and GetOnLine.

Table 1. Public bus card behavior data sheet

Number	Attribute name	Type	Remark	Description
1	Record ID	Num		
...	...	...	...	
18	GetOnTime	Time	yyyymmddhhmmss	The time of people getting on bus
19	GetOnStation	Num		The place of people getting on bus
20	GetOnLine	Num		The bus line that people sit on
...	...	...	...	

4.2 Parameter Determination of SB-DBSACN

Although the sampling technique can reduce the size of the data set and improve the efficiency of the algorithm, the results may be distorted. The experiment in this section is to find a sampling rate S_r, which can not only preserve the prototype of the cluster, but also improve the time efficiency of clustering. Firstly, this algorithm clusters the dataset by DBSCAN (*Eps* is 10.0). This experiment finds that there are 58 clusters in the original dataset and the experimental data set is not equally distributed.

Table 2 shows that the clustering results and clustering time of the improved SB-DBSACN algorithm under different sampling rates S_r.

Table 2. Comparison of clustering quality and efficiency between SB-DBSACN and DBSCAN under different sampling rates

Algorithm category	DBSCAN	SB-DBSACN			
		$S_r = 50\%$	$S_r = 30\%$	$S_r = 20\%$	$S_r = 10\%$
Cluster	58	56	55	53	50
Noise point	0	1708	2589	3590	6802
Missing point	0	2623	3692	8927	12539
Time cost(s)	312	172	118	80	45

From the Table 2, it can be seen that with the decreasing of sampling rate S_r, the running time of SB-DBSACN is getting shorter, and the efficiency is getting higher. It also can be found that the number of clusters gradually reduces. It shows that those clusters which contain few data points are missing because of the sampling. However, the clustering result is acceptable, because only 3 clusters are missing when the sampling rate is 30 %.

According to the results of Table 2, we set 30 % as the best S_r, and here we will discuss the extension of the data volumes. The comparison of the performance between DBSCAN and SB-DBSACN is shown in Fig. 2.

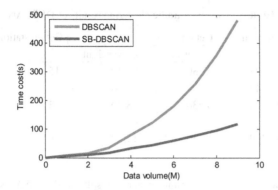

Fig. 2. The comparison of the performance between DBSCAN and SB-DBSACN

From Fig. 2, it can be seen that the time of the two algorithms is almost linear increasing when the data volume increases from 1 million to 3 million. But with the amount of data increasing, the running time of DBSCAN grows faster than SB-DBSACN, which shows that SB-DBSACN algorithm is better than DBSCAN.

Then, we study the effect of the radius of *Eps*-neighborhood on the efficiency of the algorithm. Based on previous results, we set the sampling rate S_r to 30 %. The efficiency of SB-DBSACN in different radius of *Eps*-neighborhood is shown in Fig. 3.

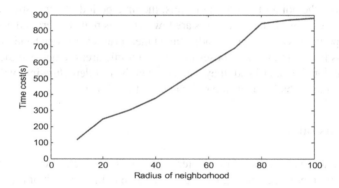

Fig. 3. The efficiency of the SB-DBSCAN algorithm in different radius of *Eps*-neighborhood

From Fig. 3, it can draw a conclusion that when neighborhood radius is in the range between 10 and 80, the time cost of the algorithm increases with the raising of the radius. Obviously, the efficiency reduces gradually. However, due to the limit of the number of points which are contained in the *Eps*-neighborhood, the time cost will not increase when it exceeds a certain value.

4.3 Travel Behavior Analysis

In this part, DBSCAN and SB-DBSACN are used to analyze the travel behavior, and the clustering results are shown in Tables 3 and 4.

Table 3. Customer traffic clustering results of DBSCAN

No.	Data point	Get on time		Get on station	Get on line
		Upper limit	Lower limit		
1	290423	07:30	08:30	15	50
2	271045	17:30	18:30	7	1
3	264576	07:30	08:30	13	136
...	...	...	...	...	...
460	409	14:30	15:30	6	25

Table 4. Customer traffic clustering results of SB-DBSACN

No.	Data point	Get on time		Get on station	Get on line
		Upper limit	Lower limit		
1	79856	07:30	08:30	13	136
2	75647	07:30	08:30	15	50
3	70345	17:30	18:30	7	1
...	...	...	...	...	...
402	120	09:30	10:30	12	34

From Tables 3 and 4, We can see that SB-DBSACN can find the main clusters. In addition, from the point of clustering results, the time period of congestion is the rush hour. At off-peak times, the data points are few, which is realistic. From the results, the relevant departments can take some adjustment measures. On the one hand, they can set up more bus lines and put more buses in the high traffic areas to achieve the objective of diversion. On the other hand, they can lengthen the bus departure intervals properly at off-peak times, which can ensure rational use of resources.

5   Conclusion

Travel behavior analysis based on clustering is a new research direction, and it has a wide application prospect and a great reference value in the research of urban planning and infrastructure construction. This paper presented a travel behavior clustering algorithm, SB-DBSACN. The influence of the sampling rate on clustering quality and efficiency were analyzed. Then, we studied the extension of SB-DBSACN and the result showed that SB-DBSACN is better than DBSCAN. Finally, we analyzed travel behavior by SB-DBSACN, and provided decision supports for urban construction.

Acknowledgments. The research work is supported by National Natural Science Foundation of China (U1433116) and Foundation of Graduate Innovation Center in NUAA (kfjj20151602).

References

1. Xiaolu, Z.: Understanding spatiotemporal patterns of biking behavior by analyzing massive bike sharing data in Chicago. PLoS ONE **10**(10), e0137922 (2015)
2. Stephane, H., Michel, B., Polak, J.W.: Estimation of value of travel-time savings using mixed logit models. Transp. Res. Part A **39**, 221–236 (2005)
3. Cervero, R., Day, J.: Suburbanization and transitoriented development in China. Transp. Policy **15**(5), 315–323 (2008)
4. Espino, R., de Dios Ortuzar, J., Concepción, R.: Understanding suburban travel demand: flexible modeling with reveabled and stated choice data. Transp. Res. Part A **41**(10), 899–912 (2007)
5. Ma, X., Wu, Y.-J., Wang, Y., et al.: Mining smart card data for transit riders' travel patterns. Transp. Res. Part C **36**, 1–12 (2013)
6. Kieu, L.-M., Ashish, B., Edward, C.: Passenger segmentation using smart card data. IEEE Trans. Intell. Transp. Syst. **6**, 1537–1548 (2015)
7. Ester, M., Kriegel, H.P., Sander, J., et al.: A density-based algorithm for discovering clusters in large spatial databases. In: 2nd International Conference on Knowledge Discovery and Data Mining, pp. 226–231 (2008)
8. Levy, P.S., Lemeshow, S.: Sampling of Populations: Methods and Applications, pp. 349–360. Wiley, New York (2013)

Key Course Selection for Academic Early Warning Based on Gaussian Processes

Min Yin, Jing Zhao, and Shiliang Sun[✉]

Department of Computer Science and Technology, East China Normal University,
500 Dongchuan Road, Shanghai 200241, People's Republic of China
jzhao2011@gmail.com, slsun@cs.ecnu.edu.cn

Abstract. Academic early warning (AEW) is very popular in many colleges and universities, which is to warn students who have very poor grades. The warning strategies are often made according to some simple statistical methods. The existing AEW system can only warn students, and it does not make any other analysis for academic data, such as the importance of courses. It is significant to discover useful information implicit in data by some machine learning methods, since the hidden information is probably ignored by the simple statistical methods. In this paper, we use the Gaussian process regression (GPR) model to select key courses which should be paid more attention to. Specifically, an automatic relevance determination (ARD) kernel is employed in the GPR model. The length-scales in the ARD kernel as hyperparameters can be learned through the model selection procedure. The importance of different courses can be measured by these corresponding length-scales. We conduct experiments on real-world data. The experimental results show that our approaches can make reasonable analysis for academic data.

Keywords: Academic early warning · Key course selection · Gaussian process regression · Automatic relevance determination kernel

1 Introduction

Many colleges and universities are working on academic early warning (AEW) which can give prompt warnings to the students who have poor test scores. AEW is among the recent computational education problems discussed in Sun [8]. Effective warning strategies will promote the students' learning or improve their learning methods. In most colleges and universities, the existing warning strategies are made according to some simple statistical methods. For example, the AEW system will send warning letters to the students who have more than ten failed credits in a semester. However, this simple statistical method can only discover the explicit information appears in the data. Using machine learning methods to model the data can excavate some hidden but useful information

---

M. Yin and J. Zhao—The authors contributed equally to this work.

© Springer International Publishing AG 2016
H. Yin et al. (Eds.): IDEAL 2016, LNCS 9937, pp. 240–247, 2016.
DOI: 10.1007/978-3-319-46257-8_26

implicit in the data. Further, the existing AEW system can only make some warning specific to students [3]. It does not make any analysis about the courses. Some useful advice on the courses will help the students to study purposefully and effectively. Therefore, selecting key courses by machine learning methods is significant and has a practical value.

Key course selection can be regarded as feature selection in the machine learning area. There are several popular methods for feature selection. For example, Lasso regression [1,10,12] and $\mathcal{L}_1$-norm support vector regression [5,13] both use $\mathcal{L}_1$-norm regularization to achieve sparse feature selection. The selection results from the above two methods are deterministic. Some probabilistic models based on the Bayesian framework can provide uncertainty estimates for features. The importance of the features can be presented by the weight ratios of each features. The features with much higher weight ratios are more likely to be the key features. In allusion to the key course selection in the AEW system, given the weight ratio of each course, users can filter the key courses depending on the practical demands.

When it comes to probabilistic models, the Gaussian process regression (GPR) model is a typical probabilistic model [4]. The GPR model provides a flexible framework for probabilistic regression and classification. It is widely used to solve the nonlinear regression problems attributed to its elegant formulation [2,4]. Several improved approaches for the GPR model are successively put forward, such as sparse Gaussian process [11,14] and mixtures of Gaussian processes [7,9]. Besides the elegant nonlinear modeling form, the flexibility of choosing kernel functions is another attractive feature of GPR models. Some special data characteristics can be captured by the particular kernel functions. For example, the automatic relevance determination (ARD) kernel is able to capture the importance of different features. The ARD kernel has been used successfully for removing irrelevant features [4]. The hyperparameters introduced by kernel functions can be adaptively learned by model selection methods.

In this paper, we use a GPR model with an ARD kernel to select key courses which should be paid more attention to. The length-scales as hyperparameters in the ARD kernel can be learned through the model selection procedure. The importance of different courses can be measured by these length-scales. We conduct experiments on real-world data. Due to the practical situation that different students sometimes choose different courses, the collected data need to be reconstructed. After reconstructing the data by the nearest neighbor data-filling algorithm, we use the GPR model with an ARD kernel to model the reconstructed data and select the key courses.

2 Gaussian Process Regression Model

In this section, we will introduce the GPR model from the function-space view, and analyze the model selection methods for the GPR model.

2.1   Gaussian Process Regression Model

A Gaussian process is a collection of random variables, any finite number of which have a joint Gaussian distribution. From the perspective of function space, Gaussian process can be seen as a distribution of function. The characteristic of Gaussian process is determined by mean function and covariance function. Define mean function $m(\mathbf{x})$ and the covariance function $\kappa(\mathbf{x}, \mathbf{x}')$ of a Gaussian process $f(\mathbf{x})$ as

$$m(\mathbf{x}) = \mathbb{E}[f(\mathbf{x})], \tag{1}$$
$$\kappa(\mathbf{x}, \mathbf{x}') = \mathbb{E}[(f(\mathbf{x}) - m(\mathbf{x}))(f(\mathbf{x}') - m(\mathbf{x}'))]. \tag{2}$$

The Gaussian process $f(\mathbf{x})$ can be written as

$$f(\mathbf{x}) \sim \mathcal{GP}(m(\mathbf{x}), \kappa(\mathbf{x}, \mathbf{x}')). \tag{3}$$

In most cases, we can only get access to noisy versions thereof $y = f(\mathbf{x}) + \epsilon$. Assuming additive independent identically distributed Gaussian noise ϵ with variance σ_n^2, the joint distribution of the observed values $\mathbf{y}$ and the test outputs $\mathbf{f}_*$ is

$$\begin{bmatrix} \mathbf{y} \\ \mathbf{f}_* \end{bmatrix} \sim \mathcal{N} \left(0, \begin{bmatrix} K(X, X) + \sigma_n^2 I & K(X, X_*) \\ K(X_*, X) & K(X_*, X_*) \end{bmatrix} \right), \tag{4}$$

where K is the covariance matrix calculated by the kernel function $\kappa(\mathbf{x}, \mathbf{x}')$.

2.2   Model Selection

The GPR model can use different covariance functions. Squared exponential covariance function is a common covariance function. Furthermore, by using the ARD squared exponential kernel, the model selection procedure allows us to automatically infer the importance of the input features without introducing explicit regularization. For the purpose of doing automatic model selection of the dimensionality of latent space, the kernel can be chosen to follow the ARD squared exponential form:

$$\kappa(\mathbf{x}, \mathbf{x}') = \sigma_f^2 \exp\{-\frac{1}{2} \sum_{d=1}^{D} \frac{1}{\ell_d^2}(x_d - x_d')^2\}, \tag{5}$$

where ℓ_d is the length-scale of the covariance and σ_f^2 is the signal variance. σ_n^2 mentioned in (4) is the noise variance. In general we call these free parameters hyperparameters. We use symbol $\boldsymbol{\theta}$ to denote the hyperparameters in the Gaussian process regression model, i.e., $\boldsymbol{\theta} = \{\{\ell_d^2\}, \sigma_f^2, \sigma_n^2\}$. Such a covariance function implements automatic relevance determination, since the inverse of the length-scale determines how relevant an input feature is. We will introduce two kinds of model selection methods for the GPR model. One is the Type II maximum likelihood and the other one is maximizing a posteriori [4,6].

Type II Maximum Likelihood. In Type II Maximum Likelihood (ML-II), one needs to calculate the negative logarithmic marginal likelihood of the samples, $L(\boldsymbol{\theta}) = -\log p(\mathbf{y}|X, \boldsymbol{\theta})$, and then calculate the partial derivatives of $L(\boldsymbol{\theta})$ with respect to $\boldsymbol{\theta}$ [4]. Through the model selection procedure, the hyperparameters in the ARD kernel which represent the importance of the features can be automatically determined. In our practical problems, the key courses can be chosen according to the ratios of the magnitudes of the inverse length-scales. We denote these ratios as the weight ratios of courses.

Maximizing a Posteriori. In maximizing a posteriori (MAP), one needs to compute the posterior of the hyperparameters which is expressed as

$$p(\boldsymbol{\theta}|\mathbf{y}, X, \mathcal{H}_p) = \frac{p(\mathbf{y}|X, \boldsymbol{\theta}, \mathcal{H}_p)p(\boldsymbol{\theta}|\mathcal{H}_p)}{p(\mathbf{y}|X, \mathcal{H}_p)}, \tag{6}$$

where $p(\boldsymbol{\theta}|\mathcal{H}_p)$ is the prior for the hyperparameters named as hyper prior. $\mathcal{H}_p$ represents the parameters in the hyper prior distribution, which can be set by hand according to actual situations.

In our experimental settings, we use Gamma hyper prior for the inverse length-scales, and use Gaussian hyper prior for both the logarithmic signal variance and logarithmic noise variance. The Gamma hyper prior for the inverse length-scale $\frac{1}{\ell_d}$ is expressed as

$$\frac{1}{\ell_d} \sim \text{Gamma}(\alpha, \lambda), \tag{7}$$

where the expectation and variance of the Gamma hyper prior are

$$\mathbb{E}(\frac{1}{\ell_d}) = \frac{\alpha}{\lambda}, \quad \mathbb{V}(\frac{1}{\ell_d}) = \frac{\alpha}{\lambda^2}. \tag{8}$$

The Gaussian hyper prior of the logarithmic signal and noise variance are expressed as

$$\log(\sigma_f^2) \sim \mathcal{N}(\mu_0, \sigma_0), \quad \log(\sigma_n^2) \sim \mathcal{N}(\mu_1, \sigma_1). \tag{9}$$

Given the above Gamma hyper prior and Gaussian hyper prior, $\mathcal{H}_p$ represents the parameters $\{\alpha, \lambda, \mu_0, \sigma_0, \mu_1, \sigma_1\}$. The hyperparameters can be learned through maximizing the posterior distribution in (6) using some gradient based optimization algorithms. As α and λ control the expectation and variance of the Gamma hyper prior for the inverse length-scales, we can obtain the inverse length-scales $\{\ell_d\}$ with different characteristics through adjusting the settings of α and λ. For example, the difference between the inverse length-scales $\{\ell_d\}$ will be larger if the variance of the Gamma prior is larger.

3   Data Collection and Reconstruction

We collect the students' scores from the department of computer science and technology in a certain university. The data are collected from two grades which are Grade 2010 and Grade 2011. In each grade, there are two classes which are pedagogical class and regular class. In each class, the numbers of students are different which are 47, 51, 23 and 52, respectively. We separate the data from each class into seven groups with each one corresponding to one semester. As different students are likely to choose different courses, the course numbers are different in each group of data. Therefore, we have to employ data-filling methods to reconstruct the data.

When reconstructing the data, the nearest neighbor (NN) is a common data-filling method which is to find the most appropriate data for the missing value. In NN, for the missing score of every course from every student, the score from the most similar student is used to fill the missing score. The similarity is measured by the distance of two students' scores for the chosen courses. Since different students often choose different courses, the simple Euclidean distance are inappropriate for measuring the distance between two samples. For fairness, we compute the averaged distance per course.

4   Experiments

4.1   Experimental Settings

We use two kinds of GPR model selection methods which are ML-II and MAP for learning the importance of courses. The simple statistical method is used as reference. Particularly in the MAP model selection method, we assume two different Gamma hyper prior distributions which have small and large variances respectively for the inverse length-scales. We denote these two MAP methods as MAP1 and MAP2. The Gamma hyper priors are set to Gamma(0.1, 0.1) and Gamma(0.1, 0.02). The Gaussian hyper priors for $\log \sigma_f^2$ and $\log \sigma_n^2$ are set to $\mathcal{N}(-1, 1)$ and $\mathcal{N}(-3, 3)$. For both the two GPR model selection methods, the maximum iteration numbers for optimization are set to 1000.

4.2   Experimental Results and Analysis

We demonstrate and analyze our experimental results by taking an example from a certain semester. The analysis of the data from other semesters is similar. Particularly, we analyze the selected courses by different GPR model selection methods as well as the simple statistical method. We plot the details of the key courses from the two classes (pedagogical class in Fig. 1 and regular class in Fig. 2) in the first semester in Grade 2010.

From Fig. 1, we find that the four methods all regard "Introduction to Computer Science and Practice (ICSP)" as the most critical course. This is convictive because "ICSP" contains the basic operations on computer. But beyond that, most results from the simple statistical method only show the appearance instead of the hidden characteristics in the data. For example, it shows that "C" is prone to fail, but it weakens the importance of "Advanced Mathematics (AM)". Differently, MAP1 and MAP2 both put more weights on "AM". ML-II selects the

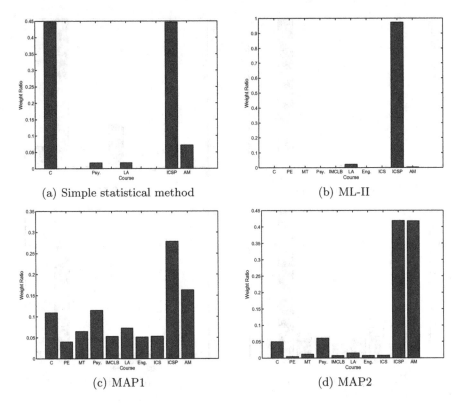

(a) Simple statistical method   (b) ML-II

(c) MAP1   (d) MAP2

Fig. 1. Key course selection results for the pedagogical class in the first semester in Grade 2010.

"Linear Algebra (LA)" and "Advanced Mathematics (AM)" as the additional key courses. Through further analysis of the course characteristics, we know that "AM" is the foundation course for computer related courses. "AM" is recognized as a challenging and important course, especially for pedagogical students who have relative poor mathematical basis. After the analysis of these key courses, it is recommended to focus more on the education of mathematical courses for pedagogical students in this semester.

From Fig. 2, we find that the four methods all regard "ICSP" as the most critical course. This is consistent with the conclusion from the pedagogical class. However, except for "ICSP", the simple statistical method shows that there is few failed courses for regular class. ML-II selects two additional courses, "Linear Algebra (LA)" and "AM", as key courses while MAP1 and MAP2 treat the other courses almost the same except "ICSP". Such phenomenon implies that the students in the regular class have better mathematical foundation compared with pedagogical students. It is recommended to balance attentions to every course on the basis of laying emphasis on "ICSP" for regular students in this semester.

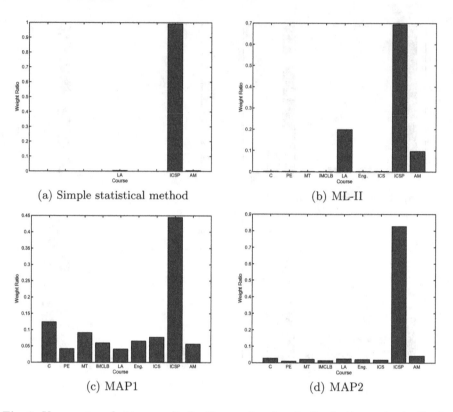

Fig. 2. Key course selection results for the regular class in the first semester in Grade 2010.

Comparing MAP1 and MAP2, we find that the two MAP methods with two different Gamma hyper priors actually obtain two trends of weight ratios. The Gamma hyper prior with larger variance tends to infer the length-scales with wide difference. The introduction of hyper priors can overcome the over-fitting to some extent and bring convenience to adjust the magnitude of the difference between the weights.

5   Conclusion and Future Work

We have selected key courses from the academic data by GPR model selection methods. From the experimental results on the real-world data, we conclude that the MAP model selection method based on GPR is a reasonable and flexible method for key course selection. GPR model selection methods can discover the hidden information about courses. Combining the key courses selected by the GPR model selection methods and those selected by the simple statistical method, the ultimately selected courses are very useful no matter for students or educational administrators. From the students' point of view, it can warn the students to pay more attention to some specific courses. From the educational administrators' point of view, it can help to evaluate the students according to the scores of some key courses and then make proper educational policies.

Key course selection is an important task for AEW and it is only at the primary stage of work on AEW. In the future work, we will discuss how to warn students to pay more attention to some specific courses in the next semester by their performance in the current semester.

Acknowledgments. The first two authors Min Yin and Jing Zhao are joint first authors. The corresponding author Shiliang Sun would like to thank support by NSFC Project 61370175.

References

1. Efron, B., Johnstone, I., Hastie, T., Tibshirani, R.: Least angle regression. Ann. Stat. **32**, 407–499 (2002)
2. Ghahramani, Z.: Probabilistic machine learning and artificial intelligence. Nature **521**, 452–459 (2015)
3. Kuzilek, J., Hlosta, M., Herrmannova, D., Zdrahal, Z., Wolff, A.: OU analyse: analysing at-risk students at the open university. Learn. Anal. Rev. **LAK15-1**, 1–16 (2015)
4. Rasmussen, C.E., Williams, C.K.I.: Gaussian Process for Machine Learning. MIT Press, Cambridge (2006)
5. Shawe-Taylor, J., Sun, S.: A review of optimization methodologies in support vector machines. Neurocomputing **74**, 3609–3618 (2011)
6. Shi, J., Choi, T.: Gaussian Process Regression Analysis for Functional Data. CRC Press, Boca Raton (2011)
7. Sun, S.: Infinite mixtures of multivariate Gaussian processes. In: Proceedings of the International Conference on Machine Learning and Cybernetics, pp. 1011–1016 (2013)
8. Sun, S.: Computational education science and ten research directions. Commun. Chin. Assoc. Artif. Intell. **9**, 15–16 (2015)
9. Sun, S., Xu, X.: Variational inference for infinite mixtures of Gaussian processes with applications to traffic flow prediction. IEEE Trans. Intell. Transp. Syst. **12**, 466–475 (2011)
10. Tibshirani, R.: Regression shrinkage and selection via the lasso. J. Roy. Stat. Soc. B **58**, 267–288 (1996)
11. Titsias, M.K.: Variational learning of inducing variables in sparse Gaussian processes. In: Proceedings of the 12th International Conference on Artificial Intelligence and Statistics, pp. 567–574 (2009)
12. Vidaurre, D., Bielza, C., Larrañaga, P.: A survey of L1 regression. Int. Stat. Rev. **81**, 361–387 (2013)
13. Zhang, Q., Hu, X., Zhang, B.: Comparison of L1-norm SVR and sparse coding algorithms for linear regression. IEEE Trans. Neural Netw. Learn. Syst. **26**, 1828–1833 (2015)
14. Zhu, J., Sun, S.: Single-task and multitask sparse Gaussian processes. In: Proceedings of the International Conference on Machine Learning and Cybernetics, pp. 1033–1038 (2013)

Privacy-Preserving Scalar Product Computation in Cloud Environments Under Multiple Keys

Hong Rong[✉], Huimei Wang, Kun Huang, Jian Liu, and Ming Xian

State Key Laboratory of Complex Electromagnetic Environment
Effects on Electronics and Information System, National University of Defense
Technology, Changsha, China
r.hong_nudt@hotmail.com, {freshcdwhm,khuang_123}@163.com,
ljabc730@gmail.com, qwertmingx@tom.com

Abstract. With the advent of big data era, clients lack of computational resources tend to outsource their data and mining tasks to resourceful cloud service providers. Generally, the outsourced data contributed by multiple clients should be encrypted under multiple keys for privacy and security concerns. Unfortunately, existing secure outsourcing protocols are either restricted to a single key or quite inefficient due to frequent client interactions, making the deployment far from practical. In this paper, we focus on addressing these outsourced problems over encrypted data under multiple keys, and propose an efficient Outsourced Privacy-Preserving Scalar Product (OPPSP) protocol. Theoretical analysis shows that the proposed solution preserves data confidentiality of all participating users in the semi-honest model with negligible computation and communication costs. Experimental evaluation also demonstrates its practicability and efficiency.

Keywords: Big data · Cloud computing · Privacy-preserving data mining · Scalar product computation

1 Introduction

With rapid growth of data captured nowadays, outsourcing both data and data mining tasks to resourceful cloud service providers becomes a natural option. Though outsourcing has advantages for reducing computation and management costs, it is critical to protect sensitive raw data and valuable mining results for security concerns. To prevent potential risks of privacy leakage, data owners tend to encrypt their data before outsourcing. Therefore, we need secure protocols to allow servers to process encrypted data coming from multiple users.

In this paper, we focus on a useful building block called privacy-preserving scalar product in the cloud environment, which securely computes the sum of the products of corresponding values from two vectors owned by different users. It has been widely utilized as a fundamental operation among such field of data mining as, similar document detection [1], association rule mining [2], etc.

© Springer International Publishing AG 2016
H. Yin et al. (Eds.): IDEAL 2016, LNCS 9937, pp. 248–258, 2016.
DOI: 10.1007/978-3-319-46257-8_27

This topic has been intensively studied since the introduction of privacy-preserving data mining, and many protocols have been proposed in the past decades. Most research assumes a distributed model where the computation is carried out interactively between different participating parties who hold their own plaintext data. These methods are designed based on Secure Multiparty Computation (SMC) [2] techniques. But SMC is not secure for outsourced model in that the final results will be revealed to all parties. Fully Homomorphic Encryption (FHE) can be used to conduct arbitrary functions on condition of a single key pair. F. Liu *et al.* [3,4] designed a protocol for outsourced scalar product by using BGN FHE cryptosystem. However, their scheme not only requires the participating clients online, but also contributes to high communication overhead. Besides, the current implementations of FHE schemes are still inefficient, thus impractical in real life application.

Main Contributions. The problem of privacy-preserving computation over encrypted data under multiple keys was studied only recently [5,6]. Nevertheless, their methods either incur significant ciphertext transformation cost, or have restrictions on server counts and message size. To the best of our knowledge, there's no existing work that addresses OPPSP (Outsourced Privacy-Preserving Scalar Product) problem under multiple users setting. In this paper, we propose an efficient secure protocol–OPPSP* that allow users to encrypt data with their own keys while the cloud servers can perform scalar product computation over the encrypted inputs. The main contributions of this work are three-fold:

- Our solution does not demand users to participate in outsourced computation except for encrypting their data locally with their own keys. In order to process the ciphertexts under multiple keys, we propose a series of privacy-preserving primitives by leveraging proxy re-encryption technique to transform ciphertexts into ones under a unified key. Those primitives can ensure the cloud servers conduct any arithmetic operations over ciphertexts correctly and securely without revealing any private information about the inputs, intermediate results and outputs.
- The proposed OPPSP* protocol is based on multiplicatively homomorphic property. Its secure addition subprotocol is also optimized and only requires two servers to complete addition with minimal interactions.
- We demonstrate our protocols' correctness, security and efficiency both theoretically and empirically compared with previous works.

The rest of the paper is organized as follows. Our system model and threat model are described in Sect. 2. In Sect. 3, we briefly introduce proxy re-encryption techniques. The design of privacy-preserving building blocks under multiple keys and the corresponding OPPSP protocol are presented in Sect. 4. Then, we evaluate the performance of our schemes in Sect. 5. Finally, we summarize the paper and outline future work in Sect. 6.

2   Problem Statement

In this section, we formally describe our system model and threat model, and we aim at designing a correct, secure and efficient protocol under this model.

2.1   System Model

There are m users $U_1, U_2, ..., U_m$ in our system model, who hold their own plaintext vectors $v_1, v_2, ..., v_m$ along with their respective public/private key pairs, denoted as $pk_1/sk_1, pk_2/sk_2, ..., pk_m/sk_m$. The cloud environment, composed of two servers, namely C_1 and C_2, is capable of offering large-scale data storage and computation services. As depicted in Fig. 1, U_i ($i \in [1, m]$) can upload its encrypted vector $\mathsf{Enc}_{pk_i}(v_i)$ under pk_i to C_1, while evaluation over the scalar product function $f(v_i, v_j) = v_i \cdot v_j$ ($i, j \in [1, m]$ and $i \neq j$) is achieved by the two servers. More specifically, U_i can initiate a secure scalar product query with respect to U_j's vector to C_1. Only after receiving an approval from U_j will C_1 begin computing $f(v_i, v_j)$ with C_2 through a set of interactive operations on the ciphertexts or blinded data. Finally, the encrypted result is returned to U_i under its public key pk_i.

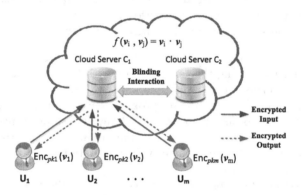

Fig. 1. System model

Our system model is appropriate and applicable for the following two reasons. On one hand, owing to the concerns about privacy leakage in the cloud, it is essential for users to encrypt their data before outsourcing. Besides, the encryptions are conducted by using their own keys, which reduces the risks of key leakage and message interception. On the other hand, two non-colluding servers to perform privacy-preserving computations are commonly used to eliminate users' interactions [7], and previous work has proven that a non-interactive solution is impossible to implement [8]. Furthermore, two servers are chosen from different cloud providers, generally driven by different business model and competing relationship, thus lowering down the chances of collusion attacks.

2.2 Threat Model

As Fig. 1 shows, our threat model primarily includes two cloud servers and multiple users.

(1) Servers: The two cloud servers are assumed to be semi-honest, which means that each server will follow the protocol, but may try to analyze user's inputs, intermediate results, as well as outputs in order to infer sensitive information. In addition, there is no collusion between the two servers or between the servers and the users.

(2) Users: The users are also assumed to be semi-honest, but non-colluding. They can cooperate with other users to submit queries for scalar product to the cloud server, meanwhile they may attempt to gather others' private information. Besides, their online periods are relatively short and non-deterministic.

3 Preliminaries

Proxy Re-Encryption (PRE) is a useful primitive introduced by Blaze, Bleumer and Strauss [9]. In a PRE system, a proxy is given a re-encryption key $rk_{i \to j}$ so that it can transform a ciphertext under public key pk_i into a ciphertext of the same plaintext under another user's public key pk_j. The proxy, however, learns nothing in terms of the corresponding plaintext. In this work, we use the classic bidirectional PRE scheme in [9], which is secure against chosen-plaintext attacks (CPA) and consists of the following five algorithms [10]:

- KeyGen$(\mathbb{G}, p, g) \to \{pk_i, sk_i\}$: Let $\mathbb{G}$ be a multiplicative cyclic group of an order of p, and g be a generator of $\mathbb{G}$. U_i uses this key generation algorithm to generate a key pair $sk_i = a \in Z_p^*$ and $pk_i = g^a \in \mathbb{G}$.
- ReKeyGen$(sk_i, sk_j) \to \{rk_{i \leftrightarrow j}\}$: The re-encryption key generation algorithm takes two private keys sk_i and sk_j as inputs, and outputs a re-encryption key $rk_{i \leftrightarrow j} = sk_j/sk_i \in \mathbb{Z}_p^*$. Here, it is required that $i \neq j$ in that there's no point to re-encrypt oneself's ciphertext.
- Enc$(pk_i, b) \to \{CT_i\}$: The encryption algorithm takes a public key pk_i and a message $b \in \mathcal{M}$ as inputs. It outputs a ciphertext $CT_i = (b \cdot g^r, pk_i^r)$ under pk_i. Here, $\mathcal{M}$ denotes the message space, and r is a random number generated from $\mathbb{Z}_p^*$.
- ReEnc$(rk_{i \leftrightarrow j}, CT_i) \to \{CT_j\}$: The re-encryption algorithm takes re-encryption key $rk_{i \leftrightarrow j}$ and an original ciphertext CT_i as inputs, and outputs a transformed ciphertext $CT_j = (b \cdot g^r, (pk_i^r)^{rk_{i \leftrightarrow j}})$ under pk_j.
- Dec$(sk_i, CT_i) \to \{m\}$: The decryption algorithm takes a private key sk_i and an original or converted ciphertext CT_i under public key pk_i as inputs. It outputs a plaintext message $b \leftarrow bg^r/((pk_i^r)^{sk_i^{-1}})$.

4 Privacy-Preserving Scalar Product Under Multiple Keys

In this section, we first give an overview of the outsourced privacy-preserving scalar product protocol that leverages multiplicatively homomorphic property of bidirectional PRE scheme. Then, we present the design details as well as correctness, security and complexity analysis.

4.1 Overview

The basic idea of our OPPSP* scheme is to utilize the bidirectional ciphertext conversion properties of PRE. Initially, C_1 runs a setup operation and distributes the system public parameters, based on which cloud users and C_2 generates their respective public and private key pairs. After this, both servers and users jointly generate the re-encryption keys. Once users' encrypted vectors under their keys are uploaded to the cloud, the server C_1 transforms all the ciphertexts into encryptions under C_2's key pk_u. Then, scalar product can feasibly be performed through our proposed cryptographic protocols. The final results under pk_u should be converted back to the ciphertexts under pk_i. The outsourced computation part of scalar product is conducted with no interactions of users whatsoever. Finally, U_i retrieves the encrypted output of C_1 and decrypts it locally with its respective sk_i in order to obtain the result.

4.2 Constructions of OPPSP*

OPPSP* is constructed based on PRE scheme [9] with multiplicatively homomorphic property. In other words, the cloud server C_1 can evaluate multiplication over encrypted data without interactions with C_2. By contrast, addition over ciphertexts still requires such interactions between the servers. The processing procedures of OPPSP* can be divided into three privacy-preserving primitive subprotocols as building blocks under the threat model. Next, we discuss the subprotocols as well as complete protocol in detail.

The Key Initialization (KI) Protocol: At first, the server C_1 runs a setup process that initializes the ElGamal cryptosystem and distributes system parameters to other participants. Each parties generates its key pairs by KeyGen($\mathbb{G}, p, g$), including users' key pairs $\{(pk_i, sk_i)|i = 1, 2, ..., m\}$ and cloud server C_2's unified key (pk_u, sk_u). C_1 computes n re-encryption keys for each user through secure interactions. The overall communication is protected by secure subprotocol like SSL. The complete steps can be found in [6].

The Secure Multiplication (SM) Protocol: Given that C_1 holds private inputs $\mathsf{Enc}_{pk_u}(a)$ and $\mathsf{Enc}_{pk_u}(b)$ while C_2 holds the secret key sk_u, the goal of this protocol is compute the encryption of multiplication of a and b, i.e., $\mathsf{Enc}_{pk_u}(a \cdot b)$ as output to C_1. Since the ElGamal encryption is multiplicatively homomorphic, the multiplication over the two ciphertexts can be performed by C_1 independently as follows, where r_a, r_b are random numbers in $\mathbb{Z}_p^*$.

$$\begin{aligned} \mathsf{Enc}_{pk_u}(a \cdot b) &= \mathsf{Enc}_{pk_u}(a) \times \mathsf{Enc}_{pk_u}(b) \\ &= ((a \cdot b) \cdot g^{r_a + r_b}, g^{(r_a + r_b) \cdot sk_u}), \end{aligned} \tag{1}$$

The Secure Addition (SA) Protocol: Assume that the server C_1 holds private input $\mathsf{Enc}_{pk_u}(a)$ and $\mathsf{Enc}_{pk_u}(b)$ while server C_2 holds the secret key sk_u. The goal of this protocol is to compute the encrypted addition of a and b, i.e., $\mathsf{Enc}_{pk_u}(a + b)$ as output to C_1. Nevertheless, it has been proven that the two non-colluding servers setting will disclose private information about input data.

Algorithm 1. $\mathsf{SA}(\mathsf{Enc}_{pk_u}(a), \mathsf{Enc}_{pk_u}(b)) \rightarrow \mathsf{Enc}_{pk_u}(a + b)$

Require: C_1 has $\mathsf{Enc}_{pk_u}(a)$ and $\mathsf{Enc}_{pk_u}(b)$; C_2 has sk_u.

C_1:
  (a) Generate four random numbers $\alpha, \beta, r_1, r_2 \in_R \mathbb{G}$;
  (b) Compute four encrypted elements in a vector, denoted by L_i ($i = 1, 2, ..., 4$):
    – $\mathsf{Enc}_{pk_u}(r_1\beta a) \leftarrow \mathsf{Blind}(\mathsf{Enc}_{pk_u}(a), r_1\beta)$;
    – $\mathsf{Enc}_{pk_u}(r_1\alpha) \leftarrow \mathsf{Enc}(pk_u, r_1\alpha)$;
    – $\mathsf{Enc}_{pk_u}(r_2\beta b) \leftarrow \mathsf{Blind}(\mathsf{Enc}_{pk_u}(b), r_2\beta)$;
    – $\mathsf{Enc}_{pk_u}(-r_2\alpha) \leftarrow \mathsf{Enc}(pk_u, -r_2\alpha)$;
    – $\Gamma \leftarrow \{L_i | i = 1, 2, ..., 4\}$ and $I \leftarrow \{1, 2, ..., 4\}$;
  (c) Generate a random permutation function π;
  (d) $\Gamma' \leftarrow \pi(\Gamma)$ and $I' \leftarrow \pi(I)$;
  (e) Send Γ' above to C_2;
C_2:
  (a) Receive encrypted results Γ' from C_1;
  (b) Decryption: $L_i' \leftarrow \mathsf{Dec}(sk_u, \Gamma'[i])$, for $1 \leq i \leq 4$;
  (c) **for** $i = 1$ to $|\Gamma'| - 1$ **do:**
    – **for** $j = i + 1$ to $|\Gamma|$ **do:**
      $*$ $S_k \leftarrow L_i' + L_j'$; $k \leftarrow k + 1$; // Initial $k \leftarrow 1$
  (d) Encryption: $S_i' \leftarrow \mathsf{Enc}(pk_u, S_i)$, for $1 \leq i \leq k$;
  (e) Send encrypted set S' to C_1;
C_1:
  (a) Receive encrypted set S' from C_2;
  (b) Compute $index \leftarrow \mathsf{FindSumIndex}(I', \{1, 2\})$; $\gamma_1 \leftarrow \mathsf{Blind}(S'[index], r_1^{-1})$;
  (c) Compute $index \leftarrow \mathsf{FindSumIndex}(I', \{3, 4\})$; $\gamma_2 \leftarrow \mathsf{Blind}(S'[index], r_2^{-1})$;
  (d) Send γ_1, γ_2 to C_2;
C_2:
  (a) Receive γ_1, γ_2 from C_1;
  (b) Decryption: $\gamma_i' \leftarrow \mathsf{Dec}(sk_u, \gamma)$, for $i = 1, 2$;
  (c) Compute $\lambda \leftarrow \gamma_1' + \gamma_2'$; $\lambda' \leftarrow \mathsf{Enc}(pk_u, \lambda)$;
  (d) Send λ' to C_1;
C_1:
  (a) Receive λ' from C_2;
  (b) Compute $\mathsf{Enc}_{pk_u}(a + b) \leftarrow \mathsf{Blind}(\lambda', \beta^{-1})$;

Algorithm 2. FindSumIndex($I, \{a, b\}$) $\rightarrow index$

Require: C_1 has permuted index set I and original subscript set $\{a, b\}$.

C_1:
  (a) Initialize $index \leftarrow 0$;
  (b) **for** $i = 1$ to $\|I\| - 1$ **do:**
    – **for** $j = i + 1$ to $\|I\|$ **do:**
      * **if** $(I(i) = a \wedge I(j) = b) \vee (I(i) = b \wedge I(j) = a)$ **then:**
        $index \leftarrow (i - 1)\|I\| - i(i - 1)/2 + j - i$;
        break;

The blinding factors can be removed by computing $b_1/b_2 \leftarrow rb_1/rb_2$, where the randomized rb_1 and rb_2 are known to C_2. By this way, C_2 is able to distinguish the plaintexts b_1 and b_2. Therefore, a third cloud server is introduced to address this problem [6]. But this additional cost may be unacceptable for those clients whose budget is constrained and inadequate. In this paper, we still consider the two non-colluding server model while lower down the opportunities that semi-honest server can recover true fractional relationship of inputs to preserve data privacy. The overall steps of SA are presented in Algorithm 1. $\mathsf{Blind}(C, r)$ is an

---

Algorithm 3. OPPSP*($\{V_1, V_2, ..., V_m\}, \kappa$) $\rightarrow \{S_{i,j} | i \neq j \wedge i, j \in [1, m]\}$

Require: U_i ($i \in [1, m]$) holds a private vector V_i; C_1 has security parameter κ.
  {Secure Data Outsourcing Phase}
$C_1, C_2, \{U_i | i = 1, 2, ..., m\}$:
  (a) All entities run KI protocol cooperatively to obtain their own keys respectively;
  (b) U_i, **for** $i = 1$ to m **do:**
    – $CV_i \leftarrow \{\mathsf{Enc}_{pk_i}(v_{i,l}) | l = 1, 2, ..., n\}$ under pk_i;
    – Upload CV_i to server C_1;
  {Scalar Product Query Phase}
  (a) U_i sends a scalar product query regarding U_j's vector V_j to C_1;
  (b) **if** U_j disapproves the query **then:** C_1 terminates this protocol;
  {Outsourced Computation Phase}
C_1 and C_2:
  (a) Compute CV_i^u and CV_j^u using its corresponding re-encryption keys;
  (b) **for** $l = 1$ to n **do:**
    – Compute multiplication of the corresponding elements of CV_i^u and CV_j^u by $\mathsf{Enc}_{pk_u}(v_{i,l} \cdot v_{j,l}) \leftarrow \mathsf{SM}(\mathsf{Enc}_{pk_u}(v_{i,l}), \mathsf{Enc}_{pk_u}(v_{j,l}))$, where $CV_i^u = \{\mathsf{Enc}_{pk_u}(v_{i,l}) | l = 1, 2, ..., n\}$ and $CV_j^u = \{\mathsf{Enc}_{pk_u}(v_{j,l}) | l = 1, 2, ..., n\}$;
  (c) Initialize $\mathsf{Enc}_{pk_u}(S_{i,j}) \leftarrow \mathsf{Enc}_{pk_u}(v_{i,1} \cdot v_{j,1})$;
  (d) **for** $l = 2$ to n **do:**
    – Compute $\mathsf{Enc}_{pk_u}(S_{i,j}) \leftarrow \mathsf{SA}(\mathsf{Enc}_{pk_u}(S_{i,j}), \mathsf{Enc}_{pk_u}(v_{i,l} \cdot v_{j,l}))$;
  (e) $\mathsf{Enc}_{pk_i}(S_{i,j}) \leftarrow \mathsf{ReEnc}(\mathsf{Enc}_{pk_u}(S_{i,j}), rk_{i \leftarrow u}^{-1})$, and Send it to U_i;
  {Result Retrieval Phase}
U_i:
  (a) Receive the encrypted result from C_1;
  (b) Compute $S_{i,j} \leftarrow \mathsf{Dec}(sk_i, \mathsf{Enc}_{pk_i}(S_{i,j}))$;

operation that randomizes c_1 of ciphertext C by multiplying random value r so that the plaintext b is blinded by r, where $C = (c_1, c_2)$, $c_1 = bg^{r'}$, $r, b \in \mathbb{G}$, and $r' \in Z_p^*$.

The Proposed OPPSP* Protocol: On the basis of the building blocks described above, we present the OPPSP* protocol in cloud environments under multiple keys. This protocol handles scalar product queries delivered from any two users in a privacy-preserving manner while ensures their ability to retrieve data at any moment from cloud storage. More specifically, the total process of OPPSP* comprises of four phases, that is, Secure Data Outsourcing Phase, Scalar Product Query Phase, Outsourced Computation Phase, and Result Retrieval Phase, the major steps of which are shown in Algorithm 3.

4.3 Correctness, Security and Complexity Analysis of OPPSP*

Correctness of OPPSP*. It is apparent that KI and SM schemes are correct due to the exactness and homomorphic property of the underlying cryptosystem. Then, we show the correctness of SA protocol in the following.

In our model, C_1 has $\mathsf{Enc}_{pk_u}(a)$ and $\mathsf{Enc}_{pk_u}(b)$ while C_2 has sk_u. Recall that C_1 generates four random numbers and computes encrypted values which are $\mathsf{Enc}_{pk_u}(r_1\beta a)$, $\mathsf{Enc}_{pk_u}(r_1\alpha)$, $\mathsf{Enc}_{pk_u}(r_2\beta b)$, $\mathsf{Enc}_{pk_u}(-r_2\alpha)$. The order of these ciphertexts are changed by permutation function π. Then C_2 decrypts those and calculates a vector composed of sums between any two elements. The sums are later encrypted under pk_u by C_2, among which C_1 finds two encrypted values, namely, $\mathsf{Enc}_{pk_u}(r_1\beta a + r_1\alpha)$ and $\mathsf{Enc}_{pk_u}(r_2\beta b - r_2\alpha)$ by FindSumIndex function as shown in Algorithm 2. By blinding with r_1^{-1} and r_2^{-1}, C_1 obtains $\mathsf{Enc}_{pk_u}(\beta a + \alpha)$ and $\mathsf{Enc}_{pk_u}(\beta b - \alpha)$. After that, C_2 decrypts these two and gets $\beta a, \beta b$. $\mathsf{Enc}_{pk_u}(\beta a + \beta b)$ is computed through encrypted addition over βa and βb. Observe that the blinding value β can be removed by multiplying β^{-1} on ciphertext of $\beta a + \beta b$, yielding the desired output $\mathsf{Enc}_{pk_u}(a + b)$.

Security of OPPSP*. During OPPSP* protocol, it is difficult for cloud servers C_1 and C_2 to distinguish the inputs, intermediate results and final outcome of oursourced scalar product computation over the ciphertext under multiple users' keys with high probability as long as ElGamal cryptosystem is semantically secure, and blinding factors are completely randomly selected. Moreover, the security can be enhanced by inserting more encrypted random numbers in Γ. The proof part is omitted due to space limitation.

Complexity of OPPSP*. Let Exp, Mult denote an operation of exponentiation, multiplication, respectively. During the Outsourcing Computation Phase, SM needs $2\mathsf{Mult}$ operations for C_1 while SA needs $4\mathsf{Exp} + 11\mathsf{Mult}$ and $20\mathsf{Exp} + 13\mathsf{Mult}$ operations for C_1 and C_2, respectively. The overall computation cost for cloud servers is $(26n - 23)\mathsf{Exp} + (26n - 24)\mathsf{Mult}$ and communication cost is $26n\|\mathbb{G}\|$. We don't compare our work with Vitamin$^+$ [6], since it has limitation on the message size and needs to solve discrete logarithm hard problem. Compared to Vitamin* in [6] with $(26n - 23)\mathsf{Exp} + (34n - 32)\mathsf{Mult}$ computation cost, our scheme requires less computation time and no extra third server. Additionally, OPPSP* can be optimized by by moving the computation of random ciphertexts to offline preparation.

5   Experimental Results

In this section, we evaluate the performance of our schemes for outsourced scalar product computation under multiple keys in cloud environments and compare our work with similar methods.

The experiments of OPPSP* are performed on two non-colluding servers while similar work Vitamin* [6] requires three non-colluding servers. The servers have identical configurations that are Intel Xeon E5-2620 @ 2.10 GHz with 8 GB RAM running CentOS 6.5. We implement all protocols in C++.

We generate a synthetic dataset consisting of 100000 uniformly distributed random vectors. The encryption key size is chosen to be 1024 bits in all our experiments. We evaluate the performance of our protocol based on the parameters of the vector dimension size (n) and the number of noise (k). Each user's encrypted vector is outsourced to simulated servers for scalar product calculation. The results presented in the following are averaged over ten data samples.

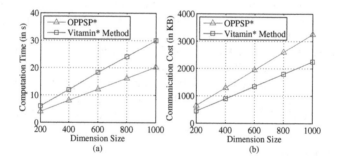

Fig. 2. Performance of OPPSP* for varying size of vector dimension

First, we assess the entire time for cloud servers varying with dimension size. From Fig. 2(a), it can be obviously observed that our OPPSP* outperforms Vitamin*. When $n = 1000$, the computation time of optimized OPPSP* is 20.26 s whereas that of Vitamin* is 30.01 s, saving around 32.49 % computation time. Moreover, this figure also demonstrates that our methods have nice scalability on account of linear growth of computation time. Figure 2(b) shows that the OPPSP* consumes the larger communication costs, but the latency is relatively small compared to computation time. For example, when $n = 600$, OPPSP* only incurs 1.95 ms latency for 1 Gbps bandwidth.

Figure 3 shows the computation time and communication costs grow almost linearly with number of random noises inserted in Γ of SA protocol. Because the

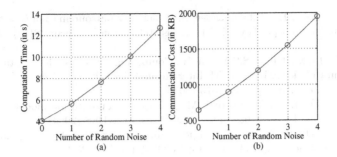

Fig. 3. Performance of OPPSP* for varying number of random noises ($n = 200$)

more noises there are, the more data need to be processed by C_2. Thereby, it's essential to make a tradeoff between security and efficiency. For example, when $k = 3$ and $n = 200$, the probability of guessing correctly can be as low as $1/210$, while the computation time is 10.02 s with 1550 KB costs. This kind of cost is acceptable considering the guess difficulty.

6 Conclusion

In this paper, we focused on the outsourced scalar product computation and proposed a novel privacy-preserving protocol based on the bidirectional proxy re-encryption scheme to address this issue. With a prevalent security model of two semi-honest but non-colluding servers in cloud environments, our methods do not require any user interaction during the outsourcing period. Theoretical analysis show that the proposed protocols can compute the results correctly and efficiently while ensuring the confidentiality of users' data and results. We also highlight the practicability of our protocols by performing experiments under different parameter settings with similar works. As future work, we plan to investigate more secure and efficient OPPSP protocols to deal with the malicious adversarial challenges.

References

1. Murugesan, M., Jiang, W., Clifton, C., Si, L., Vaidy, J.: Efficient privacy-preserving similar document detection. VLDB J. **19**(4), 457–475 (2010)
2. Vaidya, J., Clifton, C.: Privacy-preserving association rule mining in vertically partitioned data. In: ACM KDD 2002, pp. 1639–1644 (2002)
3. Liu, F., Ng, W.K., Zhang, W.: Encrypted scalar product protocol for outsourced data mining. In: IEEE International Conference on Cloud Computing, pp. 336–343 (2014)
4. Liu, F., Ng, W.K., Zhang, W.: Secure scalar product for big-data in MapReduce. In: The 1st IEEE International Conference on Big Data Computing Service and Applications, pp. 120–129 (2015)

5. Peter, A., Tews, E., Katzenbeisser, S.: Efficiently outsourcing multiparty compu- .
   tation under multiple keys. IEEE Trans. Inf. Forensics Secur. 8(12), 2046–2058
   (2013)
6. Wang, B., Li, M., Chow, S.S.M., Li, H.: Computing encrypted cloud data efficiently
   under multiple keys. In: The 4th International Workshop on Security and Privacy
   in Cloud Computing, pp. 504–513 (2013)
7. Chow, S.S.M., Lee, J.H., Strauss, M.: Two-party computation model for privacy-
   preserving queries over distributed databases. In: NDSS (2009)
8. Van, M.D., Juels, A.: On the impossibility of cryptography alone for privacy-
   preserving cloud computing. In: HotSec 2010, pp. 1–8 (2010)
9. Blaze, M., Bleumer, G., Strauss, M.: Divertible protocols and atomic proxy cryptography. In: Nyberg, K. (ed.) EUROCRYPT 1998. LNCS, vol. 1403, pp. 127–144.
   Springer, Heidelberg (1998). doi:10.1007/BFb0054122
10. Weng, J., Deng, R.H., Liu, S., Chen, K.: Chosen-ciphertext secure bidirectional
   proxy re-encryption schemes without pairings. Inf. Sci. 180, 5077–5089 (2010)

Semi-discriminative Multiview Canonical Correlation Analysis for Recognition

Yun-Hao Yuan[1], Yun Li[1(✉)], Hong-Kun Ji[2], Chong-Guang Ren[3], Xiao-Bo Shen[2], and Quan-Sen Sun[2]

[1] School of Information Engineering,
Yangzhou University, Yangzhou 225127, China
liyun@yzu.edu.cn
[2] School of Computer Science and Engineering,
Nanjing University of Science and Technology, Nanjing 210094, China
[3] School of Computer Science and Technology,
Shandong University of Technology, Zibo 255000, China

Abstract. Different with typical supervised canonical correlation methods where intraclass and interclass information of samples are exploited at the same time for classification tasks, in this paper, we follow the principle of Occam's razor and thus propose a new supervised multiview dimensionality reduction method for image recognition, called semi-discriminative multiview canonical correlations (SemiDMCCs), which takes partial class information into account but generates discriminative low-dimensional projections. Experimental results on benchmark databases show the more effectiveness of the proposed method, in contrast to existing feature reduction methods.

Keywords: Pattern recognition · Canonical correlation · Multiview data learning · Discriminative learning

1 Introduction

In numerous practical applications in pattern recognition and computer vision, there are large volumes of multiple view data derived from the same objects. Because of the information diversity and complementary nature of different views, multiview learning has recently attracted more and more attentions. Currently, a great deal of work has been done on multi-view learning, including multiview clustering [1–3], multiview classification [4, 5], and multiview dimensionality reduction (MDR) [6, 7]. This paper focuses on MDR for recognition purpose, which aims to simultaneously learn multiple meaningful low-dimensional representations from multiple views.

Since multiview data are derived from the same objects, there are naturally close *relationships* between them. In this sense, canonical correlation analysis (CCA) [8], which was designed to extract the correlations between two multidimensional random variables or data sets, is very suitable for two-view dimensionality reduction or feature extraction tasks. However, it should be noted that CCA essentially is a two-view feature extractor and thus only applicable to two-view scenario. To solve this issue, multiview CCA (MCCA) [9, 10] was proposed to deal with more (than two) view

© Springer International Publishing AG 2016
H. Yin et al. (Eds.): IDEAL 2016, LNCS 9937, pp. 259–266, 2016.
DOI: 10.1007/978-3-319-46257-8_28

cases, where view-specific canonical projection directions, one for each view, are found by maximizing the total correlations among all the views.

However, MCCA does not take discriminant information of training samples into account and thus is an unsupervised method. This means that it may be unable to yield *good* low-dimensional projections for real-world classification tasks. In order to improve the discriminative power of MCCA, Su et al. [11] presented a multiset discriminant canonical correlation method, called multiple principal angle (MPA), where within-class subspaces possess minimal principal angles, while between-class subspaces have maximal principal angles. Experimental results show the extracted features by MPA are powerful for visual recognition. Yuan et al. [12] presented graph regularized multiset canonical correlations which explicitly considers both discriminative and intrinsic geometrical structure information hidden in multiple view data. Moreover, Shen et al. [13] proposed a unified graph embedding-based MCCA framework for MDR, where three specific supervised methods were presented to demonstrate the effectiveness of the framework.

In this paper, we propose a novel supervised multiview dimensionality reduction method for image recognition tasks, called semi-discriminative multiview canonical correlations (SemiDMCCs). Different with typical supervised canonical correlation methods (see, for example, [14]) where both intraclass and interclass information are simultaneously exploited for classification, our SemiDMCC does not consider the aforementioned all supervised information hidden in multiview data, while only using the intraclass information of samples. This is because, according to the principle of Occam's razor [15], a good algorithm should be *simple* and *economical*. Thus, we attempt to build such an algorithm (i.e., SemiDMCC) using less supervised information. Experimental results show SemiDMCC is promising and able to yield more discriminative low-dimensional projections, in contrast with existing methods.

2   Multiview CCA

Assume that m-view training samples derived from the same n objects are given as $\{X^{(i)} \in \mathbb{R}^{p_i \times n}\}_{i=1}^m$, where $X^{(i)} = (x_1^{(i)}, x_2^{(i)}, \cdots, x_n^{(i)})$ is the sample matrix of the ith view that contains n observations in its columns and p_i denotes the dimensionality of sample vectors, $i = 1, 2, \cdots, m$. The objective of MCCA is to find one set of canonical projection directions, $\{\alpha^{(i)} \in \mathbb{R}^{p_i}\}_{i=1}^m$, which maximize the total correlations of the low-dimensional projections $\{\alpha^{(i)T} X^{(i)}\}_{i=1}^m$. Specifically, directions $\alpha^{(1)}, \alpha^{(2)}, \cdots, \alpha^{(m)}$ are obtained by the following optimization problem:

$$
\begin{aligned}
\max_{\alpha^{(1)}, \alpha^{(2)}, \cdots, \alpha^{(m)}} & \sum_{i=1}^m \sum_{j=1}^m \alpha^{(i)T} X^{(i)} X^{(j)T} \alpha^{(j)} \\
s.t. \quad & \alpha^{(i)T} X^{(i)} X^{(i)T} \alpha^{(i)} = 1, i = 1, 2, \cdots, m
\end{aligned}
\tag{1}
$$

where each $X^{(i)}$ is supposed to be centered, i.e., $\sum_{j=1}^n x_j^{(i)} = 0$, $i = 1, 2, \cdots, m$.

By the Lagrange multipliers technique, the solution vectors $\{\alpha^{(i)}\}_{i=1}^{m}$ of the problem in (1) reduce to the solution of the following multivariate eigenvalue problem (MEP) [16]:

$$\Phi\alpha = \Delta D\alpha, \tag{2}$$

where $\Phi \in \mathbb{R}^{p \times p}$ is a block matrix with (i, j)th block entry as $X^{(i)}X^{(j)^{T}}$, $D \in \mathbb{R}^{p \times p}$ is a block diagonal matrix with ith diagonal block as $X^{(i)}X^{(i)^{T}}$, $p = \sum_{i=1}^{m} p_i$, and $\Delta = diag(\lambda_1 I_{p_1}, \lambda_2 I_{p_2}, \cdots, \lambda_m I_{p_m})$ with $\{\lambda_i \in \mathbb{R}\}_{i=1}^{m}$ as multivariate eigenvalues and $I_{p_i} \in \mathbb{R}^{p_i \times p_i}$ as the identity matrix, $i, j = 1, 2, \cdots, m$.

3 Semi-discriminative MCCA

3.1 View-Specific Intraclass Scatter

Suppose the samples of m views of the same n objects for a total of c classes are given as $\{X^{(i)} = (X_1^{(i)}, X_2^{(i)}, \cdots, X_c^{(i)}) \in \mathbb{R}^{p_i \times n}\}_{i=1}^{m}$, where $X_j^{(i)} = (x_{j1}^{(i)}, x_{j2}^{(i)}, \cdots, x_{jn_j}^{(i)}) \in \mathbb{R}^{p_i \times n_j}$ describes the sample matrix from the ith view of the jth class, whose columns are composed of the observations, p_i denotes the dimensionality of sample vectors, and n_j is the number of samples in the jth class and $\sum_{j=1}^{c} n_j = n$, $i = 1, 2, \cdots, m$ and $j = 1, 2, \cdots, c$.

For c classes of n training samples $\{x_{j1}^{(i)}, x_{j2}^{(i)}, \cdots, x_{jn_j}^{(i)}\}_{j=1}^{c}$ from view i, we get their images $\{y_{j1}^{(i)}, y_{j2}^{(i)}, \cdots, y_{jn_j}^{(i)}\}_{j=1}^{c}$ after the projections onto the projection axis $\alpha^{(i)}$. The class-specific scatter is defined by

$$J_{wj}^{(i)} = \frac{1}{n_j} \sum_{k=1}^{n_j} (y_{jk}^{(i)} - \bar{y}_j^{(i)})(y_{jk}^{(i)} - \bar{y}_j^{(i)})^{T}, \tag{3}$$

where $\bar{y}_j^{(i)}$ is the mean of the projected training samples from the ith view of jth class, $i = 1, 2, \cdots, m$ and $j = 1, 2, \cdots, c$.

According to (3), the view-specific intraclass scatter can thus be characterized by

$$\begin{aligned}
J_w^{(i)} &= \sum_{j=1}^{c} \frac{n_j}{n} J_{wj}^{(i)} = \frac{1}{n} \sum_{j=1}^{c} \sum_{k=1}^{n_j} (y_{jk}^{(i)} - \bar{y}_j^{(i)})(y_{jk}^{(i)} - \bar{y}_j^{(i)})^{T} \\
&= \frac{1}{n} \sum_{j=1}^{c} \sum_{k=1}^{n_j} \alpha^{(i)T} (x_{jk}^{(i)} - \frac{1}{n_j} \sum_{k=1}^{n_j} x_{jk}^{(i)})(x_{jk}^{(i)} - \frac{1}{n_j} \sum_{k=1}^{n_j} x_{jk}^{(i)})^{T} \alpha^{(i)} \\
&= \alpha^{(i)T} \left[\frac{1}{n} \sum_{j=1}^{c} \sum_{k=1}^{n_j} (x_{jk}^{(i)} - \frac{1}{n_j} \sum_{k=1}^{n_j} x_{jk}^{(i)})(x_{jk}^{(i)} - \frac{1}{n_j} \sum_{k=1}^{n_j} x_{jk}^{(i)})^{T} \right] \alpha^{(i)} \\
&= \alpha^{(i)T} S_w^{(i)} \alpha^{(i)}
\end{aligned} \tag{4}$$

where $S_w^{(i)} = \frac{1}{n} \sum_{j=1}^{c} \sum_{k=1}^{n_j} (x_{jk}^{(i)} - \frac{1}{n_j} \sum_{k=1}^{n_j} x_{jk}^{(i)})(x_{jk}^{(i)} - \frac{1}{n_j} \sum_{k=1}^{n_j} x_{jk}^{(i)})^T$ is referred to as within-class

scatter matrix that is positive semidefinite. In terms of (4), it is easy to characterize the total intraclass scatter of all views as

$$J_w = \sum_{i=1}^{m} J_w^{(i)} = \sum_{i=1}^{m} \alpha^{(i)T} S_w^{(i)} \alpha^{(i)} . \tag{5}$$

Clearly, we need to minimize the total intraclass scatter J_w for making intraclass sample points as close as possible.

3.2   Model and Solution

To facilitate classification, we expect that the proposed method can find a projection which not only captures the correlation information of multiple views, but also makes the close intraclass samples closer together. To achieve this goal, we build the following multi-objective optimization model:

$$\begin{cases} \max_{\alpha^{(1)}, \alpha^{(2)}, \cdots, \alpha^{(m)}} \sum_{i=1}^{m} \sum_{j=1}^{m} \alpha^{(i)T} S_{ij} \alpha^{(j)} \\ \min_{\alpha^{(1)} \neq 0, \cdots, \alpha^{(m)} \neq 0} J_w = \sum_{i=1}^{m} \alpha^{(i)T} S_w^{(i)} \alpha^{(i)} \end{cases} \tag{6}$$

with the constraints $\alpha^{(i)T} S_{ii} \alpha^{(i)} = 1$, $i = 1, 2, \cdots, m$, where $S_{ij}(i \neq j)$ is the between-view covariance matrix of views i and j, and S_{ii} is the within-view covariance matrix of view i, $i, j = 1, 2, \cdots, m$. As we consider *partial* class information hidden in training samples rather than full class information, the proposed method is called semi-discriminative multiview canonical correlations (SemiDMCCs).

Using the *evaluation function technique* [17], we can transform the two-objective optimization problem in (6) into the following single objective problem:

$$\max_{\alpha^{(1)}, \alpha^{(2)}, \cdots, \alpha^{(m)}} \sum_{i=1}^{m} \sum_{j=1}^{m} \alpha^{(i)T} S_{ij} \alpha^{(j)} - \delta \sum_{i=1}^{m} \alpha^{(i)T} S_w^{(i)} \alpha^{(i)} \tag{7}$$

$$s.t. \quad \alpha^{(i)T} S_{ii} \alpha^{(i)} = 1, i = 1, 2, \cdots, m$$

where $\delta \geq 0$ is a regularization parameter that controls the smoothness and balance between the correlation and scatter. It is obvious that SemiDMCC reduces to MCCA when $\delta = 0$. That is, MCCA can be viewed as a special case of SemiDMCC.

With Lagrange multiplier technique for the problem in (7), we get the following Lagrange function $\mathcal{L}$:

$$\mathcal{L} = \sum_{i=1}^{m} \sum_{j=1}^{m} \alpha^{(i)T} S_{ij} \alpha^{(j)} - \delta \sum_{i=1}^{m} \alpha^{(i)T} S_w^{(i)} \alpha^{(i)} - \sum_{i=1}^{m} \lambda_i (\alpha^{(i)T} S_{ii} \alpha^{(i)} - 1), \qquad (8)$$

where $\{\lambda_i \in \mathbb{R}\}_{i=1}^{m}$ are the Lagrange multipliers. Let $\partial \mathcal{L}/\partial \alpha^{(i)} = 0$, we have

$$\sum_{j=1}^{m} S_{ij} \alpha^{(j)} - \delta S_w^{(i)} \alpha^{(i)} - \lambda_i S_{ii} \alpha^{(i)} = 0, \ i = 1, 2, \cdots, m. \qquad (9)$$

It follows that

$$(S - \delta S_w) \alpha = \Delta S_D \alpha, \qquad (10)$$

where S is a block matrix with the (i, j)th block element as $S_{ij} \in \mathbb{R}^{p_i \times p_j}$, S_w, Δ, and S_D are respectively block-diagonal matrices, i.e., $S_w = diag(S_w^{(1)}, S_w^{(2)}, \cdots, S_w^{(m)})$, $S_D = diag(S_{11}, S_{22}, \cdots, S_{mm})$, and $\alpha^T = (\alpha^{(1)T}, \alpha^{(2)T}, \cdots, \alpha^{(m)T})$.

The problem in (10) is essentially a MEP [16], which is very difficult and has no analytical solution. Thus, in this paper, we relax the problem in (10) via setting $\lambda_1 = \lambda_2 = \cdots = \lambda_m = \lambda$ and reformulate it as

$$(S - \delta S_w) \alpha = \lambda S_D \alpha. \qquad (11)$$

We select the eigenvectors $\{\alpha_i^T = (\alpha_i^{(1)T}, \alpha_i^{(2)T}, \cdots, \alpha_i^{(m)T})\}_{i=1}^{d}$ corresponding to the top d eigenvalues of (11) as projection matrices $\{Q_i = (\alpha_1^{(i)}, \alpha_2^{(i)}, \cdots, \alpha_d^{(i)})\}_{i=1}^{m}$ for m views. Once projection matrices are obtained, for any one multiview sample $x \in \mathbb{R}^p$, the reduced sample can be described as:

$$\hat{x} = (Q_1^T, Q_2^T, \cdots, Q_m^T) x, \qquad (12)$$

which is used to represent the original sample x for subsequent recognition tasks.

4   Experiments

To validate the effectiveness of our proposed SemiDMCC, we perform experiments using the popular image databases. Furthermore, we compare SemiDMCC with local Fisher discriminant analysis (LFDA) [18], MCCA, and the baseline method PCA [19]. Note that the parameter δ in SemiDMCC is empirically set to 1 for avoiding the exhaustive search, i.e., $\delta = 1$. In our all experiments, we perform the Coiflets, Daubechies, and Symlets orthonormal wavelet transforms to extract three groups of low-frequency feature vectors from original images, as used in [12]. After that, the K-L transform is used to separately reduce their dimensions to 150. The resulting three-set vectors are employed for the performance evaluation.

4.1   Experiment on the Yale-B Database

The Yale-B+Extended Yale-B database consists of two widely used datasets. That is, one is the Yale-B dataset, and the other is the extended Yale-B dataset. The Yale-B dataset contains 5760 single-light-source images of 10 individuals, each under 576 viewing conditions (9 poses × 64 illumination conditions). The extended Yale-B dataset includes 16,128 images of 28 individuals with the same condition and data format as in the Yale-B dataset. In our experiment, we employ a combinatorial subset (still referred to as Yale-B without confusion) from these two datasets, as used in [20], which contains 2414 images with size 32 × 32 of 38 individuals and each has around 64 near frontal images under different illuminations.

In this experiment, 25 and 30 images per individual are randomly selected to form a training set, respectively, while the rest are used for testing. Ten independent tests are run to evaluate the performance of PCA, LFDA, MCCA, and SemiDMCC under the nearest neighbor (NN) classifier with cosine distance metric. Table 1 shows the average recognition results of each method. As can be seen, our SemiDMCC method obviously outperforms PCA, MCCA, and the state-of-the-art method LFDA, whether the number of training samples per person is 25 or 30.

Table 1. Average recognition rates (%) of PCA, LFDA, MCCA, and our SemiDMCC with different training sample sizes and corresponding standard deviations on the Yale-B database.

#/person	PCA	LFDA	MCCA	SemiDMCC
25	65.2 ± 0.7	91.8 ± 1.2	91.8 ± 0.7	**94.2 ± 0.9**
30	68.0 ± 0.9	94.4 ± 0.5	92.9 ± 1.0	**95.4 ± 0.7**

4.2   Experiment on the ETH-80 Database

The ETH-80 database [21] contains 80 objects from 8 different categories (i.e., apple, pear, tomato, cow, dog, horse, car, and cup). In each category, there are 10 objects and each object is represented by 41 images from viewpoints spaced equally over the upper viewing hemisphere, at distances of 22.5°−26°. In our experiments, each image is resized to 64 × 64 pixels for computational efficiency. All color images are converted to grayscale images.

In this experiment, three objects per category are randomly selected for training, while the remaining seven objects are used for testing. As a result, the number of training samples and testing samples is, respectively, 8 × 3 × 41 = 984 and 8 × 7 × 41 = 2296. Table 2 shows the average recognition results across 10 runs of

Table 2. Average recognition rates (%) of PCA, LFDA, MCCA, and our SemiDMCC and the corresponding standard deviations (Std) on the ETH-80 database.

Method	PCA	LFDA	MCCA	SemiDMCC
Accuracy	70.4	59.2	72.9	**74.3**
Std	3.4	1.8	2.2	**2.1**

PCA, LFDA, MCCA, and SemiDMCC under the cosine NN classifier. From Table 2, it can be seen that our proposed SemiDMCC significantly outperforms other methods. In addition, LFDA achieves the worst results in all the methods.

5   Conclusion

In this paper, we present a novel supervised MDR method for recognition tasks, called SemiDMCC, which considers partial class information rather than the full. The proposed SemiDMCC method is applied to face and object image recognition. The experimental results on the Yale-B and ETH-80 databases show the effectiveness of our proposed method and better recognition performance, in contrast with existing dimensionality reduction methods. A future attractive direction is to design its kernel or locality extensions for capturing the nonlinear relationships among multiple view data.

Acknowledgments. This work is supported by the National Natural Science Foundation of China under Grant Nos. 61402203, 61273251, 61170120, Natural Science Foundation of Jiangsu Province of China under Grant No. BK20161338, and the Fundamental Research Funds for the Central Universities under Grant No. JUSRP11458. Moreover, it is also supported by the Program for New Century Excellent Talents in University under Grant No. NCET-12-0881.

References

1. Bickel, S., Scheffer, T.: Multi-view clustering. In: Proceedings of the IEEE International Conference on Data Mining, pp. 19–26 (2004)
2. Zhou, D., Burges, C.J.: Spectral clustering and transductive learning with multiple views. In: Proceedings of the 24th International Conference on Machine Learning, pp. 1159–1166 (2007)
3. Li, S.-Y., Jiang, Y., Zhou, Z.-H.: Partial multi-view clustering. In: Proceedings of the 28th AAAI Conference on Artificial Intelligence, pp. 1968–1974 (2014)
4. Zien, A., Ong, C.S.: Multiclass multiple kernel learning. In: Proceedings of the 24th International Conference on Machine Learning, pp. 1191–1198 (2007)
5. Qian, Q., Chen, S., Zhou, X.: Multi-view classification with cross-view must-link and cannot-link side information. Knowl.-Based Syst. **54**, 137–146 (2013)
6. Xia, T., Tao, D., Mei, T., Zhang, Y.: Multiview spectral embedding. IEEE Trans. Syst. Man Cybern. B Cybern. **40**, 1438–1446 (2010)
7. Han, Y., Wu, F., Tao, D., Shao, J., Zhuang, Y., Jiang, J.: Sparse unsupervised dimensionality reduction for multiple view data. IEEE Trans. CSVT **22**, 1485–1496 (2012)
8. Hotelling, H.: Relations between two sets of variates. Biometrika **28**, 321–377 (1936)
9. Kettenring, J.R.: Canonical analysis of several sets of variables. Biometrika **58**, 433–451 (1971)
10. Rupnik, J., Shawe-Taylor, J.: Multi-view canonical correlation analysis. In: SiKDD (2010). http://ailab.ijs.si/dunja/SiKDD2010/Papers/Rupnik_Final.pdf
11. Su, Y., Fu, Y., Gao, X., Tian, Q.: Discriminant learning through multiple principal angles for visual recognition. IEEE Trans. Image Process. **21**, 1381–1390 (2012)

12. Yuan, Y.-H., Sun, Q.-S.: Graph regularized multiset canonical correlations with applications to joint feature extraction. Pattern Recogn. **47**, 3907–3919 (2014)
13. Shen, X., Sun, Q., Yuan, Y.: A unified multiset canonical correlation analysis framework based on graph embedding for multiple feature extraction. Neurocomputing **148**, 397–408 (2015)
14. Kim, T.-K., Kittler, J., Cipolla, R.: Discriminative learning and recognition of image set class using canonical correlations. IEEE Trans. PAMI **29**, 1005–1018 (2007)
15. Blumer, A., Ehrenfeucht, A., Haussler, D., Warmuth, M.K.: Occam's razor. Inf. Process. Lett. **24**, 377–380 (1987)
16. Chu, M.T., Watterson, J.L.: On a multivariate eigenvalue problem: I. algebraic theory and power method. SIAM J. Sci. Comput. **14**, 1089–1106 (1993)
17. Wei, Q.-L., Wang, R.-S., Xu, B., Wang, J.-Y., Bai, W.-L.: Mathematical Programming and Optimization Design. National Defense Industry Press, Beijing (1984)
18. Sugiyama, M.: Dimensionality reduction of multimodal labeled data by local fisher discriminant analysis. J. Mach. Learn. Res. **8**, 1027–1061 (2007)
19. Martinez, A.M., Kak, A.C.: PCA versus LDA. IEEE Trans. PAMI **23**, 228–233 (2001)
20. Fu, Y., Yan, S., Huang, T.S.: Correlation metric for generalized feature extraction. IEEE Trans. PAMI **30**, 2229–2235 (2008)
21. Leibe, B., Schiele, B.: Analyzing appearance and contour based methods for object categorization. In: Proceedings of IEEE Conference on Computer Vision and Pattern Recognition, pp. 409–415 (2003)

Gray-Coded Clonal Selection Algorithm for Optimization Problem

Hongwei Dai[✉] and Yu Yang

School of Computer Engineering, Huaihai Institute of Technology,
Lianyungang 222005, Jiangsu, China
hwdai@hhit.edu.cn

Abstract. Clonal Selection Algorithm (CSA), inspired by the clonal selection theory, has gained much attention and wide applications. In most common forms, the CSAs use a binary representation of variables, and the emulated immune operators, mutation, proliferation, selection, for example, are made to act on it. However, the binary representation often suffers from the so-called Hamming Cliff problem. In order to overcome this problem, a Gray-coded CSA is presented and used to solve optimization problems. The algorithm is applied to numerous bench-mark problems of numerical optimization problems and the computational results show effectiveness of the proposed algorithm.

Keywords: Clonal selection algorithm · Gray code · Optimization problem · Simulation

1 Introduction

Bio-inspired computation, as a major subset of natural computation, use of biology as inspiration for solving computational problems and the use of the natural world experiences to solve real world problems. Unlike the classical optimization approaches, which emphasize accurate and exact computation at the cost of spending much more computation time, the bio-inspired computation has better convergence speeds to the optimal or near optimal results.

Over the last few decades, much research has been done on the bio-inspired or nature-inspired computation. Different artificial algorithms have been designed and developed. Genetic Algorithm (GA), Particle Swarm Optimization (PSO), Ant Colony Optimization (ACO), Evolution Algorithm (EA), Gravitational Search Algorithm (GSA) are developed and used widely [4,6,8,11]. More recently however, a new biological immune system inspired computational paradigm - Artificial Immune Systems (AIS) has been receiving more attention [13]. AIS exploits the immune system's characteristics of learning and memory to develop adaptive systems capable of performing a wide range of tasks in various engineering applications. Numerous algorithms such as Negative Selection Algorithms, [10], Immune Network Theory-based model have been proposed and successfully applied to a wide range of problems [2,9,14]. In particular, Clonal Selection

© Springer International Publishing AG 2016
H. Yin et al. (Eds.): IDEAL 2016, LNCS 9937, pp. 267–273, 2016.
DOI: 10.1007/978-3-319-46257-8_29

Algorithms (CSA) [1] which based on the clonal selection principle proposed by Burnet [3] has received a rapid increasing interest and has been verified as a powerful approach to optimization problems.

In their most common forms, the above algorithms use binary strings to represent variables. One then mutates the strings by flipping one or more bits. It is known that there can be problems with this approach if the standard binary encoding is used: there exist so-called Hamming cliff problem points that are neighbours according to the topology of the space, but are not neighbours when considered as binary strings. In other words, a very large perturbation in binary space would cause only a small change in integer space. For example, the decoded value of the binary string 01 111 111 is 127 in the integer space, while the string 10 000 000, which is at a large Hamming distance apart, since all of its corresponding bits are different from the previous one, decodes as 128, the next integer. In such a situation, the usual binary representation is often pushed to the limits of its efficacy. In order to overcome the Hamming cliff problem, a Gray-coded clonal selection algorithm (GCCSA) is proposed. In this new method, all neighbours in the original space are also neighbours as strings. We applied our algorithm on a number of functions and compared the present results with our earlier investigations.

2   Clonal Selection Principles [5, 7]

Immune system, a highly evolved, parallel and distributed adaptive system, protects our body against bacteria and viruses etc. Lymphocytes are small leukocytes that possess a major responsibility in immune system. B cells, one of the most important lymphocytes, work mainly by secreting substances called antibody (Ab) as a response to antigen. The number of antibodies contained in immune system is known to be much inferior to the number of possible antigens, making the diversity and individual binding capability the most important properties to be exhibited by the antibody repertoire.

Clonal selection theory, proposed by Burnet, explains the essential properties which contain sufficient diversity, discrimination of self and non-self and long-lasting immunologic memory, [3]. This theory interprets the response of lymphocytes in the face of an antigenic stimulus. When an animal is exposed to an antigen, some subpopulation of its bone marrow derived cells (B lymphocytes) can recognize the antigen with a certain affinity (degree of match), the B lymphocytes will be stimulated to proliferate (divide) and eventually mature into terminal (non-dividing) antibody secreting cells, called plasma cells. Proliferation of the B lymphocytes is a mitotic process whereby the cells divide themselves, creating a set of clones identical to the parent cell. The proliferation rate is directly proportional to the affinity level, i.e. the higher affinity levels of B lymphocytes, the more of them will be readily selected for cloning and cloned in larger numbers.

3   Gray-Coded Clonal Selection Algorithm

According to what described above, a Gray-coded CSA is proposed. This new algorithm, based on Gray coding method, includes initialization, affinity evaluation, clonal operator, adaptive affinity maturation, and clonal selection. These above procedures are iterated until a pre-specified termination criterion G_{max} is satisfied. The process can be explained as follows:

Step 1. Randomly create an initial pool of m antibodies $(A_1, A_2, ..., A_m)$. Antibody is defined as as a N bit binary string:

$$A_i = \{a_i^0, a_i^1, ..., a_i^N\} \tag{1}$$

where, $a_i^j = 0 \; or \; 1, i = 1, 2, ...m; j = 1, 2, ..., N$.

In this paper, the proposed algorithm is used to solve the following function optimization problem:

$$\begin{cases} min \; f(X_1, ..., X_r) \\ s.t. \quad X_i \in [minv, maxv], i = 1, 2, ..., r \end{cases} \tag{2}$$

where $[minv, maxv] \subset R$; f is a real-valued continuous function; r is the number (dimension) of optimization variables.

Step 2. Compute the affinity of all antibodies $(Af(A_1), Af(A_2), ..., Af(A_m))$ and then sort them in a descending order, $Af(.)$ is the function to compute the affinity.

Step 3. Select the n $(n \leq m)$ best elite antibodies according to their affinities from the m original antibodies.

Step 4. Place each of the n selected elites in n separate and distinct pools $(EP_1, EP_2, ..., EP_n)$.

Step 5. Convert elite antibody to their Gray equivalent. The leftmost bit in the binary unchanged to the same location in the Gray representation. The resultant of an XOR (Exclusive OR) operation between the next bit in the binary and its left-hand side neighbor fills its corresponding position in the Gray string, and the XOR operation continues till all the bit locations are filled up.

Step 6. Clone the elites in each elite pool with a rate proportional to its fitness. The amount of clone generated for these antibodies is given by Eq.(3):

$$p_i = round(\frac{(n - i)}{n} \times M) \tag{3}$$

where i is the ordinal number of the elite pools, M is a multiplying factor which determines the scope of the clone and $round(.)$ is the operator that rounds its argument towards the closest integer.

Step 7. Subject the clones in each pool through maturation process. Then convert these Gray-coded antibodies to binary ones.

Step 8. Compute the affinity and select the fittest individual Bi ($A(Bi) = argmax(A(EP_{i,1}, ..., EP_{i,p_i}))$, $i = 1, 2, ..., n$) in each elite pool from amongst its mutated clones.

Step 9. Update the parent antibodies in each elite pool with the fittest individual of the clones.

Step 10. The process will be terminated when the iteration number reaches a pre-specified maximal generation number G_{max}. Otherwise, it returns to Step 3.

4   Simulations

4.1   Parameters Setting and Benchmark Functions

The meaning and value of parameters of GCCSA are listed in Table 1.

Table 1. Parameter setting of GCCSA for the experiments on the numerical optimization problems

Number of initial populations	m	150
Number of selected populations	n	100
Clone multiplying factor	M	50
Maximum number of mutation points	PM	10
Termination condition	G_{max}*: see Table 3	

*: the maximum number of generation G_{max}

In this paper, six numerical optimization functions named *Sphere, Ackley, Griewank, Rastrigin, Schwefel,* and *Rosenbrock* are used to evaluate the effectiveness of GCCSA. Table 2 presents the detailed information of these functions.

In simulation, the number of the binary-bits for the six functions are set as 18, 18, 21, 17, 22, and 18 bits (per variable), respectively. The number of the dimension for all the six functions is 30.

4.2   Experimental Results

For the comparison purpose, the termination condition with the maximum number of generations is used, since the experiments referred from [12] were tested with a fixed number of generations. Classical evolutionary programming (CEP), fast EP (FEP) [12], CSA, receptor editing based CSA (RECSA) are compared with the proposed algorithm.

Simulation results are shown in Table 3. For functions *Sphere, Ackley,* and *Griewank* which are relatively simple functions compared with other ones, binary-coded CSA and Gray-coded CSA have almost the same results which are better than the results generated by CEP and FEP. For *Rastrigin* function, the algorithm RECSA outweighs other algorithms. In the term of average results,

Table 2. Test functions

Function	Mathematic expression	Initial range	f_{min}	X_{min}
Sphere	$f(X) = \sum_{i=1}^{N} x_i^2$	$[-100.0, 100.0]$	0.0	$(0,0,...,0)$
Ackley	$f(X) = -20\exp(-0.2\sqrt{\frac{1}{N}\sum_{i=1}^{N} x_i^2})$ $-\exp(\frac{1}{N}\sum_{i=1}^{N}\cos(2\pi x_i)) + 20 + e$	$[-32.0, 32.0]$	0.0	$(0,0,...,0)$
Griewank	$f(X)$ $= \frac{1}{4000}\sum_{i=1}^{N} x_i^2 - \prod_{i=1}^{N}\cos(\frac{x_i}{\sqrt{i}}) + 1$	$[-600.0, 600.0]$	0.0	$(0,0,...,0)$
Rastrigin	$f(X)$ $= 10N + \sum_{i=1}^{N}(x_i^2 - 10\cos(2\pi x_i))$	$[-5.12, 5.12]$	0.0	$(0,0,...,0)$
Schwefel	$f(X)$ $= 418.9829N - \sum_{i=1}^{N} x_i\sin(\sqrt{|x_i|})$	$[-500.0, 500.0]$	0.0	$(-420.9687,$ $-420.9687,...,$ $-420.9687)$
Rosenbrock	$f(X) =$ $\sum_{i=1}^{N-1}(100(x_{i+1} - x_i^2)^2 + (x_i - 1)^2)$	$[-30.0, 30.0]$	0.0	$(1,1,...,1)$

Table 3. Experimental results of the six numerical optimization functions

		CEP	FEP	CSA	RECSA	GCCSA
Sphere t =1500	b.	2.2×10^{-4}	5.7×10^{-4}	4.4×10^{-6}	4.4×10^{-6}	1.5×10^{-5}
	m.	-	-	4.5×10^{-6}	6.2×10^{-6}	2.2×10^{-5}
	w.	-	-	5.5×10^{-6}	9.0×10^{-6}	3.1×10^{-5}
	δ	5.9×10^{-4}	1.3×10^{-4}	3.9×10^{-7}	1.3×10^{-6}	6.2×10^{-6}
Ackley t =1500	b.	9.2	2.8	4.9×10^{-4}	5.5×10^{-4}	8.3×10^{-4}
	m.	-	-	5.4×10^{-4}	6.2×10^{-4}	1.0×10^{-3}
	w.	-	-	6.6×10^{-4}	6.6×10^{-4}	1.3×10^{-3}
	δ	1.8×10^{-2}	2.1×10^{-3}	5.6×10^{-5}	4.3×10^{-5}	1.6×10^{-4}
Griewank t =2000	b.	8.6×10^{-2}	1.6×10^{-2}	1.6×10^{-7}	1.8×10^{-7}	6.8×10^{-7}
	m.	-	-	1.8×10^{-7}	2.0×10^{-7}	1.3×10^{-6}
	w.	-	-	2.0×10^{-7}	2.5×10^{-7}	2.5×10^{-6}
	δ	0.12	2.2×10^{-2}	1.0×10^{-8}	2.0×10^{-8}	5.8×10^{-7}
Rastrigin t =5000	b.	89.0	4.6×10^{-2}	9.4	9.1×10^{-6}	1.1×10^{1}
	m.	-	-	12.5	9.6×10^{-6}	1.4×10^{1}
	w.	-	-	15.8	1.2×10^{-5}	1.8×10^{1}
	δ	23.1	1.2×10^{-2}	2.2	1.0×10^{-6}	2.5
Schwefel t =9000	b.	4652.3	14.987	103.8	0.5	3.8×10^{-4}
	m.	-	-	161.9	1.1	9.5×10^{1}
	w.	-	-	249.7	1.4	2.4×10^{2}
	δ	634.5	52.6	46.8	0.2	9.3×10^{1}
Rosenbrock t =20000	b.	6.17	5.06	24.3	24.3	1.1×10^{-4}
	m.	-	-	25.8	25.8	3.9×10^{-2}
	w.	-	-	27.6	27.1	1.6×10^{-1}
	δ	13.61	5.87	1.0	1.1	5.0×10^{-1}

* b., m., w. δ, and t represent the best, the mean, the worst, the standard deviation, and the maximum number of generations respectively

GCCSA performs the same ability as other methods for solving *Schwefel* function. However, the best solution of GCCSA is much better than others.

In mathematical optimization, the *Rosenbrock* function is a non-convex function used as a performance test problem for optimization algorithms. The global minimum is inside a long, narrow, parabolic shaped flat valley. To find the valley is trivial. To converge to the global minimum, however, is difficult. It is worthwhile to mention that GCCSA performs very well in optimizing *Rosenbrock* function than all other algorithms.

Figure 1 shows the convergence process of CSA and GCCSA for six functions. The vertical axis of each figure shows the average function value, and the horizontal axis represents the number of generations. From these figures, except the *Rastrigin* function, it can be easily confirmed that the traditional binary-coded CSA is easy to be trapped in local optima so as to be premature convergence, whereas Gray-coded CSA are able to escape the local optima.

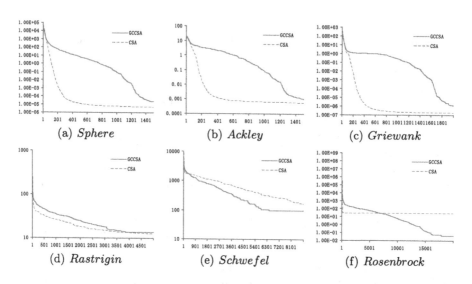

Fig. 1. Convergence process of CSA and GCCSA for six test functions

5   Conclusions

Like other algorithms, clonal selection algorithms (CSA) use a binary representation of variables, and the emulated immune operators, mutation, proliferation, selection, for example, are made to act on it. However, these binary-coded algorithms have not effectively been applied for global optimizations of continuous variable in nonlinear models. The Hamming distance between the two closest integers in binary code is very large. These methods often suffer from the so-called Hamming Cliff problem. To overcome these difficulties relating to binary encoding for continuous variable optimizations, a Gray-coded CSA (GCCSA) is presented.

The algorithm is applied to numerous bench-mark problems of numerical optimization problems and the computational results show effectiveness of the proposed algorithm.

Acknowledgements. This work was supported by the Prospective Joint Research of University-Industry Cooperation of Jiangsu (No. BY2015248, BY2016056-02), the Six Talent Peaks Project of Jiangsu (No. XXRJ-013), Lianyungang Science and Technology Project (No. CG1413, CG1501), and the Natural Science Foundation of Huaihai Institute of Technology (No. z2015005, z2015012).

References

1. Afaneh, S., Zitar, R.A., Alhamami, A.: Virus detection using clonal selection algorithm with genetic algorithm (VDC algorithm). Appl. Soft Comput. **13**(1), 239–246 (2013)
2. Bayar, N., Darmoul, S., Hajri-Gabouj, S., Pierreval, H.: Fault detection, diagnosis and recovery using artificial immune systems: a review. Eng. Appl. Artif. Intell. **46**(Part A), 43–57 (2015)
3. Burnet, F.M.: The Clonal Selection Theory of Acquired Immunity. Cambridge Press, New York (1959)
4. Gao, S., Vairappan, C., Wang, Y., Cao, Q., Tang, Z.: Gravitational search algorithm combined with chaos for unconstrained numerical optimization. Appl. Math. Comput. **231**, 48–62 (2014)
5. Goldsby, R.A., Kindt, T.J., Osborne, B.A., Kuby, J.: Immunology. W. H. Freeman Co., New York (2002)
6. Inkaya, T., Kayaligil, S., Evin Özdemirel, N.: Ant colony optimization based clustering methodology. Appl. Soft Comput. **28**, 301–311 (2015)
7. U.S. Department of Health, Human Services National Institutes of Health: Understanding The Immune System - How It Works. NIH Publication (2003)
8. Ryu, T., Kanemaru, T., Kataoka, S., Arihama, K., Yoshitake, A., Arakawa, D., Ando, J.: Optimization of energy saving device combined with a propeller using real-coded genetic algorithm. Int. J. Naval Archit. Ocean Eng. **6**(2), 406–417 (2014)
9. Seresht, N.A., Azmi, R.: MAIS-IDS: a distributed intrusion detection system using multi-agent AIS approach. Eng. Appl. Artif. Intell. **35**, 286–298 (2014)
10. Stibor, T.: Foundations of r-contiguous matching in negative selection for anomaly detection. Nat. Comput. **8**(3), 613–641 (2008)
11. Wang, J., Liao, J., Zhou, Y., Cai, Y.: Differential evolution enhanced with multiobjective sorting-based mutation operators. IEEE Trans. Syst. Man. Cybern. **44**(12), 2792–2805 (2014)
12. Yao, X., Liu, Y., Lin, G.: Evolutionary programming made faster. IEEE Trans. Evol. Comput. **3**(2), 82–102 (1999)
13. Zhang, Y.D., Wang, S.H., Wu, L.N., Huo, Y.K.: Artificial immune system for protein folding model. JCIT J. Convergence Inf. Technol. **6**(1), 55–61 (2011)
14. Zuo, X.Q., Fan, Y.S.: A chaos search immune algorithm with its application to neuro-fuzzy controller design. Chaos Solitons Fractals **30**(1), 94–109 (2006)

Improved PCA-BP Face Recognition
Based on Construction of Virtual Sample

Jing Lin$^{(\boxtimes)}$, Xisheng Wu, and Xudong Yan

College of Internet of Things Engineering, Jiangnan University, Wuxi 214122, China
1203126534@qq.com

Abstract. According to the lack of training samples and the shortage of the traditional PCA and BP algorithm, this paper proposes an improved PCA-BP face recognition algorithm based on the construction of virtual samples. Firstly, the algorithm uses original training samples to generate virtual samples by mirror and rotation transformation, and then all of the original samples and virtual samples are used for the improved PCA algorithm to extract the feature vectors. Moreover, the extracted feature vectors are input to the improved BP neural network for training, and finally combined K-Nearest Neighbor algorithm comprehensive discriminant classification, the generalization ability of face image recognition is enhanced. The experimental result on ORL, YALE, FERET and AR face database proves that the proposed algorithm can predict the possible changes of the samples in a certain extent, and the recognition rate is improved obviously.

Keywords: Virtual samples · PCA algorithm · BP neural network · K-Nearest neighbor algorithm · Face recognition

1 Introduction

Face recognition has always been an important research direction in the field of pattern recognition, because it has a good application prospect in the military, government, finance, electronic commerce and other industries [1–3]. In the process of face image acquisition, due to the changes of illumination, facial expression, position, posture and the influence of equipment, the collected images can not accurately reflect all the information of human face [4–6]. Therefore, the methods of constructing virtual samples have been proposed one after another [7–11], such as using optical flow method, noise, geometric transformation and weighted average method to generate a new virtual sample, but the number of virtual samples generated by these methods is not enough to reflect all the information of the face. So it is necessary to generate more adequate virtual samples according to the existing human face images.

J. Lin—The National Natural Science Foundation of China (61373055), Industry-Academic Collaboration Innovation fund project of Jiangsu Province (BY2013015-35).

H. Yin et al. (Eds.): IDEAL 2016, LNCS 9937, pp. 274–284, 2016.
DOI: 10.1007/978-3-319-46257-8_30

Although the traditional BP algorithm can update the weights through continuous learning to improve the recognition rate [12–14], but the training process is easy to fall into local minimum, and the BP neural network which is easily affected by the training shock has slow convergence speed [15–17]. When extracting feature vectors, traditional PCA algorithm did not take the contribution of different components into account, and the illumination transformation would affect the first 3 principal components, thus affecting the final recognition results.

For this reason, this paper proposes a face recognition algorithm based on the construction of virtual samples and the improved BP neural network and PCA method. Firstly the virtual samples are constructed through the mirror and rotating method, and then all of the original images and virtual samples are used for the improved weighted PCA algorithm to extract feature vectors. Then the extracted feature vectors are input to the improved BP neural network for training, and finally combined K-Nearest neighbor algorithm to classify test images. To a certain extent, it reflects the possible changes of the samples, and the recognition accuracy is improved.

2   Principle of Face Recognition Algorithm

2.1   PCA Feature Extraction

PCA algorithm is also called principal component analysis algorithm, it is a data processing method that extracts the feature vectors by constructing covariance matrix, so as to achieve the purpose of reducing the dimension, compression and extraction of the original complex data. Because of its simple computation and easy realization, it is widely applied in various fields of pattern recognition and data mining [18].

The basic idea of PCA algorithm is that the complex and multivariate raw data are processed by Karhunen-Loeve Transform to find out the most representative elements of the face and form feature face space, so the interference of irrelevant attributes to the recognition result is reduced. Then all the images are projected onto the obtained characteristic subspace to get the projection coefficients of different faces, and these coefficients can represent the main features of the face image. Finally, the appropriate classifier for face classification and recognition is used to achieve the purpose of compression and dimension reduction of data, grasp the essence of the problem, simplify the data and reduce the amount of computation.

2.2   Basic Principle of BP Algorithm

BP network algorithm [19] is a multilayer feed-forward neural network algorithm which uses the error between the output and the expected value as the adjustment signal for the back propagation. Robert Hecht-Nielsen found the 3-layer BP network that has the hidden layer can approximate any continuous function well, and has strong adaptive ability and good classification effect [20].

The error exists between the output of the BP algorithm and the expected value. If the error exceeds a certain value, the error will be used as an adjustment signal for back propagation, so in the continuous inverse propagation process we can adjust the weight value and threshold value according to the desired error value in order to obtain ideal output result. At present, there are two kinds of algorithms to improve the performance of BP neural network: additive momentum method and elastic gradient descent method.

2.3   K-Nearest Neighbor Algorithm

K-Nearest Neighbor classification algorithm is a mature method in theory and also one of the most simple machine learning algorithms. The idea of this algorithm is that if most of k similar samples in the feature space which are similar to one sample belong to a certain category, then the sample also belong to this category.

3   Improved PCA-BP Algorithm Based on Virtual Sample Construction

3.1   Virtual Sample Generation

In the process of practical application, the face images may be affected by position, illumination, pose and expression, so that the training samples can not accurately represent all the characteristics of the target face. Therefore, before the PCA feature extraction, the images should be processed first so that the training samples can better reflect the impact of a variety of circumstances. By the combination of mirror and rotation transformation, the obtained virtual sample is still the same target, and it can include the change information in the posture and position of the sample.

As we all know, the collected face image is generally positive, and the structure of the human face is symmetrical. Therefore, the collected images can be horizontally mirrored by the central axis to obtain virtual samples. The obtained virtual sample is symmetrical with the original image, and it can be regarded as the image that the face rotates a certain angle around the vertical axis, so it can reduce the effect of horizontal rotation of the head on the recognition result. The processing results are shown in Fig. 1.

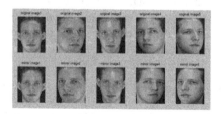

Fig. 1. Part of the original images and their horizontal mirroring virtual samples

In the process of face image acquisition, may be due to the acquisition equipment is not fixed and the movement of the target, the face image skews and is not vertical. Therefore we can make the image rotate 10 degrees clockwise and counterclockwise, and the obtained image can be seen as a human face rotating a certain angle around the horizontal axis so as to reduce the impact of head tilt on the recognition, and the processing results are shown in Fig. 2.

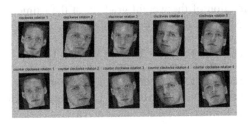

Fig. 2. Part of the virtual samples of original images clockwise and counter clockwise rotation

3.2 Improved PCA Algorithm

In the feature extraction process of the traditional PCA algorithm, no consideration is given to the effects of different components on the recognition results and the contributions of all components are calculated with equal weight method, this caused the components of small contribution effect the final recognition results. Face images will inevitably be influenced by the change of environment in the collection process, and the first three feature vectors of PCA feature extraction can be easily affected by the change of light, so the improved algorithm proposed in this paper gives the first three eigenvectors with adaptive weights after the feature extraction by using the method of least squares linear fitting so as to reduce the impact of illumination changes on the recognition results. Because of taking the contribution of different components into account, this method improves the recognition rate.

3.3 BP Neural Network Improved by the Method of Weight Adjustment

Traditional BP network algorithm uses gradient descent method to study in the process of training, and its convergence is not very good and its training time is too long. Further more, it will be affected by the training shocks, and easy to fall into local minimum points. As we all know, although the additional momentum method can prevent the BP neural network from falling into local minimum point in the training process, its learning rate is so slow that caused long network training time. On the other hand, the elastic gradient descent method reduces the training time of BP neural network and converges more quickly, but it is easy to be trapped into local minima. In this paper, these two improved methods are combined with innovation, and a new algorithm of BP network weights adjustment is formed. Among them, the weight adjustment formula of t + 1 time is:

$$W(t+1) = \begin{cases} W(t) - sign(\dfrac{\partial E(t)}{\partial W})\Delta t & \dfrac{\partial E(t)}{\partial W} \neq 0 \\ 0 & \dfrac{\partial E(t)}{\partial W} = 0 \end{cases} \tag{1}$$

The $\Delta W(t)$ is weight variation, t is the number of training, mc is the momentum coefficient and Δt is the update value.

(1) If the change direction of the weight gradient at the t time and the t−1 time are consistent, then Δt increase. The formula of Δt is:

$$\Delta t = \alpha \times (1 - mc) \times \Delta(t - 1) + mc \times \Delta(t - 1), \; mc \in (0, 1) \tag{2}$$

(2) If the change direction of the weight gradient at the t time and the t−1 time is opposite, then Δt reduce, and the weight W is adjusted according to a time-varying probability P. The formula of P is:

$$P = \beta/((1 + \beta)(e^{0.2t} + 1)), \; P \in (0, 1) \tag{3}$$

The improved weight adjustment method not only reduces the convergence time, but also reduces the possibility of the BP network to be trapped in local minimum points during the training process. At this point, the formula of Δt is:

$$\Delta t = \beta \times (1 - a) \times \Delta(t - 1) \times sign(\frac{\partial E(t)}{\partial W}) + a \times \Delta(t - 1), \; 0 < \beta < 1, \; a \in (0, \frac{\beta}{1 + \beta}) \tag{4}$$

(3) In addition, the weight adjustment direction is according to the probability 1−P, then the formula of Δt is:

$$\Delta t = \beta \times (1 - mc) \times \Delta(t - 1) + mc \times \Delta(t - 1), \; mc \in (0, 1) \tag{5}$$

Under the other conditions, Δt remains unchanged. Therefore, the improved algorithm proposed in this paper reduces the effect of network shaking on the convergence speed, shortens the training time, and minimizes the possibility of the network falling into local minima in a certain extent, so it significantly improves the performance of BP network.

3.4   General Description of Improved Algorithm

According to the above analysis of the algorithm principle, the specific steps of the algorithm in this paper can be summarized as follows:

(1) Virtual samples generation
   We can read the images in the four face databases respectively, and take parts of the images as the training samples and construct the same number of virtual samples by the mirror method. Then the original samples and the obtained mirror samples are used for the rotation method, so the number of training samples changed from 1 to 4.
(2) Improved PCA algorithm to extract feature vectors
   The virtual samples obtained from the step (1) and the original training samples are all used for the weighted PCA algorithm proposed in this paper to extract the main features of the human face.

Average face:

$$u = \frac{1}{N} \sum_{j=1}^{N} x_j \tag{6}$$

Deviation matrix:

$$S_r = \sum_{j=1}^{N} (x_j - u)(x_i - u)^T = AA^T, A = [x_1 - ux_2 - u \cdots x_N - u] \tag{7}$$

$$p = \min_k \left\{ \frac{\sum\limits_{i=1}^{k} l_i}{\sum\limits_{i=1}^{r} l_i} \geq 0.9, k \leq r \right\} \tag{8}$$

The characteristic value is arranged according to the size. According to the formula (8), the eigenvectors which are relative to the first p sorted characteristic values are extracted to form a feature face space, to achieve the purpose of dimensionality reduction, and then the first 3 components extracted from the feature vectors are added adaptive weights.

The normalized feature vector of the $A^T A$ is:

$$u_i = \frac{1}{\sqrt{l_i}} A v_i (i = 1, 2, \cdots, p) \tag{9}$$

So the feature face space is:

$$U = [u_1, u_2, \cdots, u_p] \tag{10}$$

(3) The projection coefficients of training and testing images which have been normalized are used as the input of the BP network for training. The network weights are calculated by using the weight adjustment method proposed in the paper.

(4) BP neural network method and k-Nearest Neighbor algorithm are combined to classify face images. For each input test sample x_j, the output of the BP network is t, given a threshold value $\beta(\beta < 0.5)$, if $\max\{t(i,1)\} > \beta$, the category is $k = \max\{t(i, 1)\}$; if $\max\{t(i,1)\} < \beta$, the corresponding input x_j is classified by k-Nearest Neighbor algorithm. Finally, by calculation the correct recognition rate can be obtained.

4   Experiment and the Result Analysis

In windows 7 system (Intel®Core™ i5-4590 CPU @ 3.30 GHz processor and 8 GRAM) and Matlab 2012 environment, we choose ORL, FERET, AR and Yale four face databases images for face recognition experiments. Firstly reading images in the databases, take the AR face database for example, there are 120 people and 14 original images per person. The first 7 images are used for training and the last 7 images are used for testing, then the 840 training samples are used to obtain other 840 virtual samples through mirror method, and then through the method of rotating the 1680 virtual samples can be obtained, so the total number of training samples changed from the original 840 to 3360. Consequently, according to the method of this paper, the experiments are carried out in three aspects.

4.1   Adaptive Weight Coefficients of the First Three Principal Components Optimization

It can be seen From Table 1 that by the least square linear fitting method, the first three characteristic components of the feature vectors which are extracted from PCA algorithm are added adaptive weights of less than 1. By comparing the data in four different face databases, we can see that the adaptive weight coefficients in different face databases are different, and the adaptive weight coefficients of first three different principal components in the same face database are also different. So it can be seen that the effects of different components on face recognition are different, and it also verifies the necessity of the weighted improved PCA algorithm.

Table 1. Different weight coefficients of first 3 principal components in 4 different databases

Face database	First component	Second component	Third component
ORL	0.42	0.40	0.45
FERET	0.39	0.34	0.41
AR	0.08	0.23	0.39
YALE	0.40	0.36	0.32

4.2   Comparison of the Proposed Algorithm and Other Algorithms

The comparative experiments between the proposed algorithm and the other mainstream face recognition methods based on the whole face image were carried out on the ORL, FERET, AR and Yale four person face image databases. The recognition rate of various classification algorithms in the different face databases is shown in Table 2.

Among them, the literature [11] proposes a method that constructs virtual samples by weighted average method and then identifies human faces in the classification framework based on sparse representation. In literature [12], the advantages of principal component analysis (PCA) and linear discriminant analysis (LDA) are fully fused together. The literature [6] proposed a method to recognize face images across pose based on orthogonal discriminant vector (ODV) and validate the feasibility of modeling the face images of different individuals on different linear manifolds. In literature [4], the fusion of two useful descriptors, i.e., the Zernike moments (ZMs) and the local binary pattern (LBP)/local ternary pattern (LTP), has been proposed to generate superior results.

Table 2. Recognition rate of various classification algorithms under different face databases

Algorithm	ORL	FERET	AR	YALE
Original PCA algorithm	0.77	0.81	0.79	0.82
"Literature [11]" algorithm	0.95	0.77	0.69	–
"Literature [12]" algorithm	0.94	0.53	0.55	–
"Literature [6]" algorithm	–	0.94	0.88	0.92
"Literature [4]" algorithm	0.94	0.62	–	0.80
Our algorithms	0.96	0.87	0.91	0.94

It can be seen from Table 2 that the face recognition algorithm proposed in this paper which is based on constructing virtual samples and the improved PCA-BP algorithm predicts the possible changes of samples to a certain extent, and has a positive impact on the classification recognition rate, so it generally improves the image recognition rate and has better effect. But in the FERET and AR face database, the data in Table 2 is the result without the rotating virtual samples, so on the FERET and AR face database, the effect is less than the literature [6].

In order to observe the test results more intuitively and easily, we have designed the graphical human-computer interaction interface based on the face recognition algorithm proposed in this paper. As shown in Fig. 3, the face recognition experiment is carried out on the ORL face database.

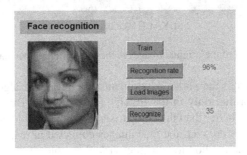

Fig. 3. Human-computer interaction interface of face recognition

4.3 The Applicability of the Rotation Method to Construct Virtual Samples

From Table 2 and the above analysis, it can be seen that the algorithm proposed in this paper has better recognition effect on the ORL and YALE face database, but the recognition rate of the FERET and AR face database is different. The details are shown in Table 3.

Table 3. Recognition rate of the proposed algorithm in different human face database and different training samples

Sample type	ORL	FERET	AR	YALE
Original sample	0.87	0.84	0.883	0.84
Add mirror sample	0.895	0.87	0.91	0.92
Add mirror + rotation sample	0.96	0.81	0.82	0.94

As can be seen from Table 3, the proposed algorithm has a very good effect in the ORL and YALE face database, but in the FERET and AR face database, the proposed algorithm is better in the case of no rotating samples.

By observing the characteristics of the images in the four face databases (shown in Fig. 4), we can find that the images of the FERET and AR face databases are all straight and not inclined, so the rotation method is not applicable to these two face databases. But the images of the ORL and YALE face databases are often inclined, so the rotation method has a good effect for the recognition.

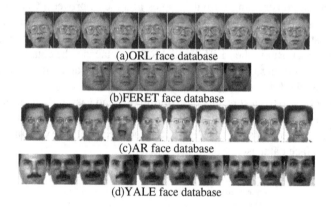

(a)ORL face database

(b)FERET face database

(c)AR face database

(d)YALE face database

Fig. 4. Part of the original images in the four different face databases

As can be seen from the above experiments, the proposed algorithm is less than "Literature [6]" algorithm only in the FERET face database. But in other cases, the recognition results are better than the comparison literature methods (including "Literature [6]" algorithm). Moreover, the face images collected in real life could not be all straight, so the rotation method to construct virtual samples is still a very meaningful work.

5   Concluding Remarks

In this paper, an adaptive weighted PCA face recognition algorithm based on improved BP neural network is proposed under the environment of constructing virtual samples. Firstly, the virtual samples are obtained from the original training samples by mirror and rotation transformation, which can reduce the impact of head rotation and tilt. Then all of the original training samples and constructed virtual samples are used for the improved PCA algorithm to find out the representative feature vectors and then according to the extracted image data, we can determine the parameters of the BP network, and finally use the improved BP neural network and K-Nearest Neighbor algorithm to comprehensively identify the face images.

Simulation experiments show that through the recognition in ORL, Yale, FERET and AR face databases, the algorithm proposed in this paper predicts the possible variations of samples in a certain extent and significantly improves the recognition rate. In addition, more improvements are needed to study in order to carry out the experiments in real-time environment.

References

1. Chellappa, R., Wilson, C.L., Sirohey, S.: Human and machine recognition offaces: a survey. Proc. IEEE **83**(5), 705–740 (1995)
2. Zhao, W., Chellappa, R., Rosenfeld, A., et al.: Face recognition: a literature survey. ACM Comput. Surv. **35**(4), 399–458 (2003)
3. Jie, Z., Chun-yu, L., Chang-shui, Z., et al.: A survey of automatic human face recognition. Acta Electronica Sin. **4**(4), 102–106 (2000)
4. Guan-jie, F., Wan-pei, C., Cai-kou, C., et al.: Combination of WPCA and WLDA for face recognition. Radio Commun. Technol. **39**(5), 89–92 (2013)
5. Wen-fan, S.: The Study of Improves PCA and BP Neural Network Face Recognition. Yanshan University, Qinhuangdao (2013)
6. Ying-ying, Z., Kai-fa, D.: Second wavelet transform and PCA face recognition algorithm based on improved BP neural network. Inf. Technol. **2015**(6), 8–11 (2015)
7. Singh, C., Mittal, N., Walia, E.: Complementary feature sets for optimal face recognition. Eurasip J. Image Video Process. **2014**(1), 1–18 (2014)
8. Kautkar, S.N., Atkinson, G.A., Smith, M.L.: Face recognition in 2D and 2.5D using ridgelets and photometric stereo. Pattern Recogn. **45**(9), 3317–3327 (2012)
9. Wang, J., You, J., Li, Q., et al.: Orthogonal discriminant vector for face recognition across pose. Pattern Recogn. **45**(12), 4069–4079 (2012)
10. Shen, Yu., Jian-wu, D., Xin, F., et al.: Fusion of infrared and visible images based on tetrolet transform. Spectro. Spectral Anal. **33**(6), 1506–1511 (2013)
11. Krommweh, J.: Tetrolet transform: a new adaptive haar wavelet algorithm for sparse image representation. J. Visual Commun. Image Represent. **21**(4), 364–374 (2010)
12. Xiang, Y., Han-lin, Q.: Image fusion based on tetrolet transform. J. Optoelectronics laser **24**(8), 1629–1633 (2013)
13. Hong-jin, S.: Image multiresolution decomposition and reconstruction based on haar wavelet. Coal Technol. **29**(11), 157–159 (2010)
14. Zi, L., Xiao-ning, S., Zhen-min, T.: Integrating original images and its virtual samples for face recognition. Comput. Sci. **42**(5), 289–294 (2015)
15. Li-qiang, Yu., Dao-jun, Ma.: Neural network for speaker recognition of PCA technology. Comput. Eng. Appl. **46**(19), 211–213 (2010)
16. Xian-feng, C., Zhi-bing, S., Ying-kai, Z.: Application of resilient BP network to the recognition of English letter with noise. Comput. Simul. **22**(9), 153–155 (2005)
17. Li Zhi-qing, F., Xiu-fen, F.: Performance study on three kinds of improved BP algorithm based on principal component analysis. Comput. Eng. **37**(21), 108–110 (2011)

18. Jose, C.A.: Fast on-line algorithm for PCA and its convergence characteristics. IEEE Trans. Neural Netw. **4**(2), 299–307 (2000)
19. Hecht-Nielsen, R.: Theory of the back propagation neural network. In: Proceedings of the International Joint Conference on Neural Networks 1989, pp. 593–605 (1989)
20. Xu Yi-shen, G., Ji-hua, T.Z., et al.: Handwritten character recognition based on improved BP neural network. Commun. Technol. **5**(44), 106–109 (2011)

A Duplication Task Scheduling Algorithm in Cloud Environments

Min Ruan, Yun Li$^{(\boxtimes)}$, and Yinjuan Zhang

College of Information Engineering, Yangzhou University, Yangzhou, China
liyun@yzu.edu.cn

Abstract. In this paper, we propose an improved adaptive heuristic algorithm with duplication (D-IAHA). It turns the workflow into complex directed acyclic graph (DAG) in cloud environments, and then modifies the improved adaptive heuristic algorithm (IAHA) considering duplication. Specifically, D-IAHA repeats important predecessor tasks in the free time slots of the processors, in order to avoid long communication cost between tasks. Meanwhile, elimination of redundant tasks is taken into account. The experimental results show that the proposed method can achieve good performance, significantly obtain the response quickly moreover optimize makespan, load balancing on resources and failure rate of tasks.

Keywords: Cloud environments · Duplication task scheduling · DAG

1 Introduction

Task scheduling is a vital component for cloud computing, which aims at adopting appropriate strategy to schedule tasks for available resources satisfying Quality of Service (QoS) requirements. This problem is NP-complete in the general case, as well as some restricted cases. Loads of algorithms are proposed to challenge this puzzle, such as Heterogeneous Earliest Finish Time (HEFT) [1] and Critical Path [2]. Most of them are often used in simple model and can not reflect real parallel system accurately. According to such issue, some meta heuristic algorithms are proposed, like Partial swarm optimization (PSO) [3], Simulated Annealing (SA) [4], and Genetic Algorithm (GA) [5].

The proposed algorithm, improved adaptive heuristic algorithm with duplication (D-IAHA), modifies the improved adaptive heuristic algorithm (IAHA) [6] considering duplication. It utilizes free time slot of processors to repeat the current task's parent in order to reduce communication cost between tasks. In this way, the start time of current task's immediate child can be advanced, and then the whole scheduling time can be shortened. In addition, the elimination of redundant tasks is taken into account as well, decreasing the computational load of the processors. Since task's start time is restricted by the arrival time of its parents. There are two ways to solve this issue:

(1) Repeat current task's important immediate predecessor in suitable free time slot to reduce data transmission cost between tasks;
(2) Schedule the child task to the processor where its immediate parent is located if possible, thus to avoid the communication cost.

© Springer International Publishing AG 2016
H. Yin et al. (Eds.): IDEAL 2016, LNCS 9937, pp. 285–292, 2016.
DOI: 10.1007/978-3-319-46257-8_31

However, the second method will lead to a situation that too many tasks are on the same processor obviously, and then the load of processor is too heavy, finally advanced entire completion time will not be available. Therefore D-IAHA adopts the first method, here are the two duplicate rules it follows:

1. Repeat the current task's immediate parent with largest degree preferentially. As this task will affect more child tasks' start time when its degree is larger. So repeat a task like this, not only the start time of current task can be advanced, the start time of the rest child tasks belonged to the same parent will be advanced accordingly.
2. Repeat the most important immediate parent of the current task whose data arrives at the latest. Why we choose this task is that the start time of current task is restricted by it. Hence, repeating this task can make the start time advanced.

2 Improved Adaptive Heuristic Algorithm (IAHA)

The IAHA algorithm improves HSGA algorithm [7] in two aspects: task prioritization, crossover and mutation operation.

2.1 Task Prioritization

In this phase, IAHA mainly contributes on solving the discordance between priorities of tasks and the topology sequence of DAG after sort stage. Based on this problem, IAHA adopts the following sort method to calculate task's weight:

$$
T_{priority}(v_i) = w(v_i) + \sum_{d(v_j)=\alpha}^{\beta} \left(d(v_j) \times w(e_{i,j}) \right), \quad where
$$

$$
\alpha = \beta - \left\lfloor \beta/2 \right\rfloor, \quad v_j \in T_{succ}(v_i)
$$

(1)

Where $T_{succ}(v_i)$ is a set of child tasks of task v_i, β is the most depth value of successors with the maximum-length sequence. In this way, it will ensure the reasonableness of task execution, and optimize the entire completion time as well.

2.2 Crossover and Mutation Operations

In IAHA, adaptive crossover and mutation rate are adopted in order to avoid premature and reduce of search efficiency due to the determined and unchanged rate. The evaluation index Δ of population's premature length is in literature [8].
Crossover rate is:

$$
P_c = 1/(1 + \exp(-k_1\Delta))
$$

(2)

Mutation rate is:

$$
P_m = 1 - 1/(1 + \exp(-k_2\Delta))
$$

(3)

The crossover and mutation result is influenced by P_c and P_m. In the evolutionary process, if Δ becomes larger, the ability of developing best individual will strengthen; if Δ becomes smaller, the ability of producing new individuals will strengthen.

3 Proposed Algorithm

On the basis of IAHA, we selectively repeat tasks which have been executed on other processors in the free slot of processors to reduce the communication cost between processors. However, it is puzzle that how to choose the appropriate duplicate tasks.

3.1 Duplicating Phase

At this stage, first, tasks are scheduled to processors according to IAHA, and then, repeat the scheduled parent task in task sequence on the processor where its child task is executed on in the free time slot based on the duplicate rules in the Sect. 1. It should be noted that the performance of current processor may not be the best-suited one for the parent task, this means, it will not achieve the fastest processing speed. According to this, D-IAHA takes computing power and other properties of candidate processor into consideration before scheduling.

Since task v_i is scheduled to processor P_j, typically define this time as the task's arrival time $T_{avail}(v_i, P_j)$. However, the real start time is not the task's arrival time, because whether the processor is free and whether all data are conveyed should be considered.

We can see that the free time slot plays a vital role in D-IAHA, its generation relays on the constraints of priority in DAG and communication delay, which makes the advanced arrival of data unavailable. Definition 1 presents the idea of how to judge whether a suitable free time slot exists.

Definition 1. [9] Given a set of n tasks $\{v_1, v_2, \cdots, v_n\}$ scheduled on a processor P_k, a free time slot (gap) G_r (between t_r and t_{r+1}), is suitable for task v_i if,

$$G_r^F - \max\{DAT(v_i, P_k), G_r^S\} \geq w_{i,k},$$

$$where \quad G_0^S = 0, \quad G_0^F = T_{st}(v_1, P_k), \quad G_n^S = T_{ft}(v_n, P_k), \quad and \quad G_n^F = \infty \tag{4}$$

G_r^S and G_r^F refer to the free time slot's start and finish time respectively. $w_{i,k}$ is the execution time of task v_i on the current processor P_k. $T_{st}(v_i, P_k)$ refers to the start execution time of v_i. Assuming no suitable free time slot exists, then schedule v_i to P_k, so that its start time depends on the processor's idle time. Figure 1 shows it.

Definition 2. Define the most important immediate parent (MIIP) task as the immediate parent of task v_i whose data arrives at the latest, represented by M_i. Notice that it hinders the start time of v_i. Then, we denote the data arrival time of MIIP task v_i as:

$$DAT(v_i, P_k) = \max_{v_j \in pred(v_i)} \{\min\{T_{ft}(v_j, P_k), T_{ft}(v_j, P_l)\} + c_{ji}\} \tag{5}$$

Here, $pred(v_i)$ addresses to the set of predecessor tasks of v_i, $T_{ft}(v_j, P_k)$ refers the finish time in the case that predecessor task v_j of v_i and v_i itself are executed on the

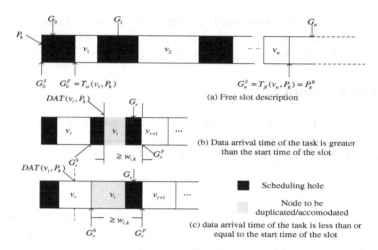

Fig. 1. Free time slot and duplicate task

same processor, here c_{ji} is 0(c_{ji} is the communication cost from v_j to v_i); $T_{ft}(v_j, P_l)$ shows that v_j and v_i are executed on different processors, and now c_{ji} is described in literature [6].

The start execution time of v_i on the current processor is:

$$T_{st}(v_i, P_k) = \max\left\{DAT(M_i, P_k),\ \min\{P_k^R, G_r^S\}\right\} \qquad (6)$$

Where G_r^S refers to the start time of the first free time slot which can accommodate v_i, P_k^R describes the free time on P_k. $DAT(M_i, P_k)$ is the data arrival time of MIIP task M_i of v_i.

The finish time and the entire completion time of v_i are described respectively:

$$T_{ft}(v_i, P_k) = T_{st}(v_i, P_k) + w_{i,k} \qquad (7)$$

$$makespan = \max\{T_{ft}(v_i, P_k)\} \qquad 0 \le i \le n \ \ and \ \ 0 \le k \le m \qquad (8)$$

In order to avoid redundant duplication, we only repeat the parent task with important data output. Each time after repeating v_i, update the finish time of it. If the entire completion time (*makespan*) has no change after repeating, we can delete it. Repeat this operation until no MIIP task, parent task with largest degree, or free time slot exists. And then check all tasks, make sure there is no redundant duplication.

3.2   Task Scheduling Phase

We state the detail about duplication process by now. In order to have an overall concept, Fig. 2 presents the whole task scheduling process in cloud environments. Where, A is a task sequence used to store tasks needed to be repeated; B is a task

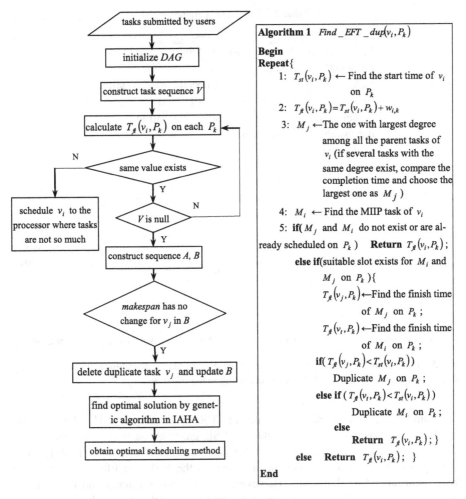

Fig. 2. Task scheduling flowchart

sequence as well, it is used to store the already repeated task from A. Algorithm 1 shows the illustration on duplication.

4 Simulation and Result Analysis

To test the performance of D-IAHA, we compare it with HEFT and IAHA in CloudSim simulation platform: HEFT uses the strategy of repeating tasks in free time slot of processor; D-IAHA improves IAHA considering duplication. The simulation parameters are listed in Table 1. The performance parameters NSL and Efficiency are described in literature [10]. Additionally, CCR [11] is the communication to computation ratio. The value of CCR changes in D-IAHA.

Table 1. Simulation parameters

Parameter	Value
Number of tasks in application	30–100
The number of resources	30
Task length	1200000–7200000 (MIPS)
Failure rates of tasks	0.0001–0.001
Resources speed	500–1000 (MIPS)
Bandwidth between resources	10–100 (mbps)
CCR value	0.25

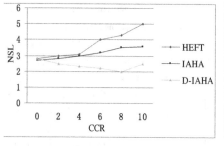

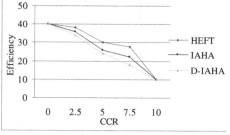

(a)NSL with increasing value of CCR   (b)Efficiency with increasing value of CCR

Fig. 3. The performance achieved by the proposed D-IAHA algorithm and the competing algorithms includes HEFT, IAHA under increasing value of CCR.

From Fig. 3, with the increasing value of CCR, the makespan of HEFT and IAHA is affected increasingly, then the value of NSL increases as well. Nevertheless, as D-IAHA exchanges computation cost for communication cost by repeating task, makespan will not have a significant increase, so the NSL and Efficiency value are lower than other two algorithms. However, if CCR becomes larger and larger, communication between tasks is frequent, there will be a hard issue to choose duplicate tasks. As a result, the makespan will rebound. But for all that, the performance of D-IAHA is better.

In Fig. 4 (a), since IAHA and D-IAHA adopt the method that schedule tasks to an efficient processor and consider about load of resources during crossover and mutation phase, the load balancing rates of these two methods are less than that of HEFT. However, as D-IAHA exchange computation cost for communication cost to reduce the makespan, the load balancing rate on each node is comparatively higher than IAHA, but it still reduces the load balancing rate of resources compared with other algorithms. From Fig. 4 (b), we can see that the failure frequency of IAHA and D-IAHA is less than that of HEFT, this is because HEFT takes no consideration about error probability of tasks. On the contrary, in IAHA and D-IAHA, they consider it in the part of whether or not to choose mutated resources. Consequently, the failure frequency when task is executed on processors has been severely depleted.

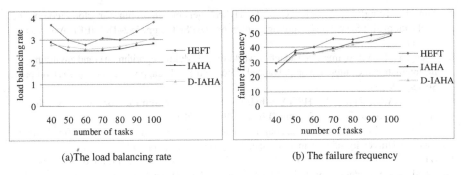

(a) The load balancing rate (b) The failure frequency

Fig. 4. The values of load balancing rate and failure frequency under different numbers of tasks achieved by D-IAHA, comparing with HEFT and IAHA

5 Conclusion

In this paper, we deal with the tasks in DAG by pre-merging them, then divide tasks into task packages according to the partition strategy which mainly aims to reduce communication cost between tasks, and minimize entire completion time finally. We consider about the load of processors during partition phase as well. After this, the proposed D-IAHA algorithm is used to utilize the free time slot of processors to repeat tasks, and reduce the entire completion time of task even further. The experimental results show that the proposed method can achieve visible performance gain.

Acknowledgements. This research was supported in part by the Chinese National Natural Science Foundation under grant No. 61402396, 61402203 and 61379066, Natural Science Foundation of Jiangsu Province under contract BK20161338, The high-level talent project of "Six talent peaks" of Jiangsu Province under contract 2012-WLW-024, Joint innovation fund project of industry, education and research of Jiangsu Province (prospective joint research) under contract BY2013063-10 and the talent project of "Green Yangzhou and golden phoenix" under contract 2013-50.

References

1. Tang, X., Li, K., Liao, G., Li, R.: List scheduling with duplication for heterogeneous computing systems. J. Parallel Distrib. Comput. **70**, 323–329 (2010)
2. N'Takpé, T., Suter, F.: Critical path and area based scheduling of parallel task graphs on heterogeneous platforms. In: Proceedings of the 12th International Conference on Parallel and Distributed Systems (ICPADS 2006), pp. 3–10. IEEE Computer Society (2006)
3. Pandey, S.: Scheduling and management of data intensive application workflows in grid and cloud computing environments. J. Doctoral thesis, Department of Computer Science and Software Engineering, The University of Melbourne, Australia (2010)
4. Kalashnikov, A.V., Kostenko, V.A.: A parallel algorithm of simulated annealing for multiprocessor scheduling. J. Comput. Syst. Sci. Int. **47**(3), 455–463 (2008)

5. Zhu, K., Song, H., Liu, L., Gao, J.: Hybrid genetic algorithm for cloud computing applications. In: Services Computing Conference (APSCC), pp. 182–187. IEEE Asia-Pacific (2011)
6. Zhang, Y., Li, Y.: An improved adaptive workflow scheduling algorithm in cloud environments. In: International Conference on Advanced Cloud and Big Data, pp. 112–116. IEEE Computer Society (2015)
7. Delavar, A.G., Aryan, Y.: HSGA: a hybrid heuristic algorithm for workflow scheduling in cloud systems. J. Cluster Comput. **17**(1), 129–137 (2014)
8. Xin, L.: The improvement of the adaptive genetic algorithm and its application (in Chinese). NanJing University of Information Science & Technology, pp. 32–34 (2008)
9. Bansal, S., Kumar, P., Singh, K.: Dealing with heterogeneity through limited duplication for scheduling precedence constrained task graphs. J. Parallel Distrib. Comput. **65**(4), 479–491 (2005)
10. Ali, J., Khan, R.Z.: Optimal partitioning strategy with duplication (OTPSD) in parallel computing environments. J. Int. J. Comput. Distrib. Syst. **4**(1), 7–15 (2013)
11. Mezmaz, M., Melab, N., Kessaci, Y., Lee, Y.C., Talbi, E.-G., Zomaya, A.Y., Tuyttens, D.: A parallel bi-objective hybrid meta heuristic for energy-aware scheduling for cloud computing systems. J. Parallel Distrib. Comput. **71**(11), 1497–1508 (2011)

Discussion of Graph Reachability Query with Keyword and Distance Constraint

Wen Juping[1,2(✉)]

[1] School of Electronic and Information Engineering,
Foshan University, Foshan 528000, Guangdong, China
wenjuping@sohu.com
[2] Foshan City Computer Society, Foshan 528000, Guangdong, China

Abstract. Focusing on the problem of graph reachability query with keyword and distance constraint, a method of graph reachability query based on reference node embedding was proposed. Firstly, a very small part of representative reference nodes were selected from all nodes, shortest path distance between all nodes and these reference nodes were previously calculated. Next distance range was obtained based on the triangle inequality relation. Lastly according to the distance constraint and keyword in query condition, a conclusion of reachability could be drawn quickly. Comparative tests were done on data of social network and road network. Compared with Dijkstra algorithm, the proportion of drawing a reachability conclusion directly of the proposed algorithm is 84.6 % in the New York Road network, while 66.6 % in the Digital Bibliography & Library Project (DBLP) network, moreover, the running time are both shortened greatly, which is reduced by 87.4 % and 77.3 % respectively. The testing experimental data demonstrates that the computational complexity of online queries is reduced by using a small index cost through the method proposed in this paper, which is a good solution to graph reachability query with keyword and distance constraint suitable for weighted graph as well as unweighted graph, with wider value in application.

Keywords: K-hop reachability query · Graph reachability query with keyword and distance constraint · Reference node embedding · Triangle inequality relation · Node attribute

1 Introduction

The problem of how to efficiently answer reachability queries has attracted a lot of interest lately [1]. Nowadays, there is a new research direction called k-hop reachability query problem [2]. It answers the question whether there exists a path from s to t, whose number of steps is no more than k. Today, many real-world networks emerging nowadays have labels or textual contents on the nodes. For instance, in a road network, a location may have labels such as "hospital", "McDonald's", "gas station", etc. In a social network, a person may have information include name, hobbies and skills, etc. Therefore, the reachability query with distance constraint is usually attached a conditional restriction, the route must go through some special point with some attribute.

© Springer International Publishing AG 2016
H. Yin et al. (Eds.): IDEAL 2016, LNCS 9937, pp. 293–301, 2016.
DOI: 10.1007/978-3-319-46257-8_32

For example, we query whether we can travel from place A to place B or not, during the route there is a gas station within a certain distance.

In this paper, we discuss the issue of graph reachability query with keyword and distance constraint. In a network G modeled as an graph, each node is attached with zero or some keywords, and each edge is assigned with a weight measuring its length. Give a query pair nodes a and b in G, the distance value k and a keyword λ, the question we study is denoted as the form of $Q = (a,b,\lambda,k)$, which represents whether node a can reach node b within k value distance, the route must go through a node which has λ attribute.

2 Related Work

Reachability is a fundamental problem on large-scale networks which has been studied extensively [1]. There are also several extensions to the classic reachability problem: (1) reachability in uncertain graphs where the existence of an edge is given by a probability [3]; (2) reachability with constraints such as edges on the path or nodes must have certain labels [4, 5]; (3) a new type of reachability query in weighted undirected graphs, the answers to which are meaningful only if the edge or node weight is also captured in the reported path, in many real-world applications [6]. Issue of the reachability query with distance constraint is discussed in literature [3], which is suitable for uncertain graphs. Different from the literature [3], in this paper, we discuss the reachability query with distance constraint based on certain graphs.

Sometimes it is conditional on the reachability query, and it needs to ensure that the path passes through special nodes. In fact, these special nodes are nodes which have some attribute. For example, in social networks, sometimes we need to query whether or not a person can make acquaintance with another person within some certain steps, through some important person. We define this type of query as a query with a keyword, similar with keyword search discussed in literature [2], the issue discussed in which is that given a query node q in G and a keyword λ, a k top nearest keyword query seeks the nearest k nodes to q which contain λ attribute.

3 Graph Reachability Query with Distance Constraint

Given a graph $G = (V, E, w)$, where V denotes the set of nodes, E denotes the set of edges, $w: E \rightarrow R$ denotes function of edge weight, which indicates the weight mapped from a edge into a real number field. $w(u,v)$ denotes a positive number, which presents the distance from node u to node v, and the value of $w(u,v)$ is 1, when a graph is an unweighted graph.

3.1 Algorithm of Graph Reachability Query with Distance Constraint

Let's see an example, which presents the principle of the algorithm:

The query is whether node a can reach node b or not within the distance of k value. Firstly, we compute the distance range's upper and lower value between node a and node b, the values of which are *upper* and *lower* respectively calculated by an algorithm designed in this paper. Secondly, k is compared with *upper* and *lower*, if $k <$ *lower* or $\geq$ *upper*, the reachability conclusion can be drawn directly, otherwise, the exact shortest path distance between a and b must be calculated online, which is used to be compared with k to draw the reachability conclusion. The query $Q = (a,b,k)$ is implemented by the *Algorithm 1*, as the following shows:

```
input:    a graph G=(V,E,w), two query nodes: a and b, and
distance value k
output:   a Boolean indicator whether a->ₖb
1. range(a,b,&upper,&lower)/*This function is implemented
by algorithm 2 */
2. if (k<lower)
3.      return false;
4. if (k≥upper)
5.      return true;
6. if (k≥lower&&k<upper)
7.      {dist=shortestpathdistance(a,b);
8.         if (dist≤k)
9.            return true;
10.        else
11.           return false;}
```

3.2   Solution of the Shortest Path Distance Range

The idea of reference nodes embedding is introduced into the discussion of graph reachability query in this paper which is used to determine the distance range. The idea of reference nodes embedding comes from literature [7], which is used for the shortest path distance estimation.

We define the selected reference nodes set $S = \{l_1, l_2, ..., l_d\} \subseteq V$, for each $l_i \in V$ in set S, a n-dimensional vector is pre-computed, which is used to denote the shortest path distance between the reference node l_i and all nodes, shown as Eq. (1):

$$\vec{D}(l_i) = \langle o(l_i, v_1), o(l_i, v_2), \ldots, o(l_i, v_j), \ldots, o(l_i, v_n) \rangle (1 \leq i \leq d, 1 \leq j \leq n) \quad (1)$$

Suppose $o(a,b)$ is the distance of query point pair (a,b), according to distance pre-stored and triangle inequality relation, the upper value and lower value of $o(a,b)$ for each reference node can be obtained. Then the maximum value is selected from all the lower values, the minimum value is selected from all the upper values, which form the distance range of $o(a,b)$, as the following Eqs. (2) and (3) show:

$$o(a,b) \geq \max\{|o(l_i,a) - o(l_i,b)|\}(1 \leq i \leq d, l_i \in S) \tag{2}$$

$$o(a,b) \leq \min\{o(l_i,a) + o(l_i,b)\}(1 \leq i \leq d, l_i \in S) \tag{3}$$

The algorithm of the shortest path distance range is defined as *Algorithm 2*, shown as the following:

```
input:    a graph G=(V,E,w), two query nodes: a and b
output:   upper, lower
1. Compute a vertex set S={l₁,l₂,...,l_d}⊆V/*computation of
reference nodes set S*/
2. for each l_i∈S do
3.     Pre-compute vector D(l_i)=<o(l_i,v₁), o(l_i,v₂),...,
o(l_i,v_j),... o(l_i,v_n)>(1≤i≤d,1≤j≤n)
4. for each l_i∈S do
5.     Compute |o(l_i,a)-o(l_i,b)|
6.     Compute o(l_i,a)+o(l_i,b)
7. lower:=max{|o(l_i,a)-o(l_i,b)|}(l_i  S)
8. upper:=min{ o(l_i,a)+o(l_i,b)} (l_i∈S)
9. return upper,lower
```

The first line in *Algorithm 2* is corresponding to looking for the set of reference nodes, which is very important for the determination of distance range. When the strategy of selecting reference nodes is relatively optimizational, the distance range computed by triangle inequality relation will be narrower. Heuristic algorithm based on degrees is used in this paper.

4 Graph Reachability Query with Keyword and Distance Constraint

Suppose there are some special nodes having m attribute values, such as $\lambda_1, \lambda_2...\lambda_m$ in a graph $G = (V,E,w)$, each of special nodes has one of attribute values. We define a set $\wedge = \{\lambda_1, \lambda_2...\lambda_m\}$. For each attribute value, there are deferent number of special nodes, that is $t_1, t_2,.....,t_m$. The problem of graph reachability query with keyword and distance constraint is changed into the query $Q = (a,b,k,\lambda)$.

4.1 Method for Determining the Shortest Path Distance Range with Keyword

Similar with the theory described in Sect. 3.2, d n-dimensional vectors should be pre-computed, shown as Eq. (1).

Let's define $X_\lambda = \{x_{\lambda 1}, x_{\lambda 2}..., x_{\lambda t}\}$ as the set of special nodes with λ attribute value, then the distance range of the path through each special node is acquired by the equations below:

$$\max\{|o(a, l_i) - o(l_i, x_{\lambda j})|\} \leq o(a, x_{\lambda j}) \leq \min\{o(a, l_i) + o(l_i, x_{\lambda j})\}(1 \leq i \leq d, 1 \leq j \leq t, l_i \in S) \tag{4}$$

$$\max\{|o(x_{\lambda j}, l_i) - o(l_i, b)|\} \leq o(x_{\lambda j}, b) \leq \min\{o(x_{\lambda j}, l_i) + o(l_i, b)\}(1 \leq i \leq d, 1 \leq j \leq t, l_i \in S) \tag{5}$$

The algorithm of the shortest path distance range is defined as *Algorithm 3* which corresponds to function *range(a,b, λ,&upper,&lower)*, shown as following:

```
input:   a graph G=(V,E,w), two query nodes: a and b, at-
tribute value λ
output:  upper, lower
1. Compute a vertex set S={l₁,l₂,…,l_d}⊆V/*computation of
reference nodes set S*/
2. for each lᵢ∈S do
3.     Pre-compute vector D(lᵢ)=<o(lᵢ,v₁), o(lᵢ,v₂),...,
o(lᵢ,vⱼ),... o(lᵢ,vₙ)>(1  i  d,1  j  n)
4. for j=1 to t do
5.    for each lᵢ∈S do
6.        Compute o(a, lᵢ)+o(lᵢ, x_λj)+ o(x_λj, lᵢ)+o(lᵢ, b)
7.        Compute |o(a, lᵢ)-o(lᵢ, x_λj)|+|o(x_λj, lᵢ)-o(lᵢ, b)|
8.     templower[j]:=min{ o(a, lᵢ)+o(lᵢ, x_λj)+ o(x_λj,
lᵢ)+o(lᵢ, b)} (lᵢ  S)
9.     tempupper[j]:=max{|o(a, lᵢ)-o(lᵢ, x_λj)|+|o(x_λj, lᵢ)-
o(lᵢ, b)|} (lᵢ∈S)
10.lower:=min{templower[j]}( 1⩽j⩽t)
11.upper:=max{tempupper[j]}( 1⩽j⩽t)
12.return lower, upper
```

4.2   Algorithm of Graph Reachability Query with Keyword and Distance Constraint

The query $Q = (a,b,k,\lambda)$ is defined as a function, which is implemented by the algorithm similar with *Algorithm 1*, in which *range(a,b,&upper,&lower)* is replaced with *range(a,b, λ,&upper,&lower)*.

5   Experiment

The experimental data used in this paper are derived from two real graphs: the DBLP (Digital Bibliography & Library Project) data set and the New York road network data set. Among them, DBLP is an unweighted graph, in which each node represents an author, each edge represents the relationship of coauthor, New York road network is a weighted graph, in which the weight of each edge reflects the distance between two nodes.

Experiment is done from two aspects to test by using the two kinds of data sets.

(1)   The reachability judgment of query point pair is made by two algorithms. One is Dijkstra algorithm which calculates the exact distance, the other is algorithm of graph reachability query with keyword and distance constraint which is proposed in this paper. The running time of the two algorithms is tested to see which is shorter.
(2)   The ratio of the reachability conclusion given directly is calculated aiming at the algorithm proposed in this paper.

5.1   Time Test

Aiming at the two data sets, a graph is formed including 2000 nodes respectively. For comparability, the same query data is used by the two algorithms of which one is Dijkstra algorithm, the other is the algorithm proposed in this paper. All point pairs distribution maps of the two graph is shown as Figs. 1 and 2.

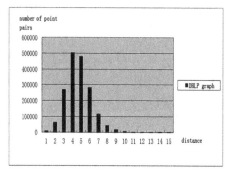
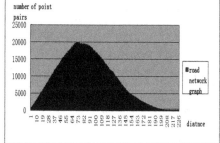

Fig. 1. Point pairs distribution map in DBLP

Fig. 2. Point pairs distribution map in road network

For each graph, we do data collection and comparison. The running time is collected from the Dijkstra algorithm and the algorithm proposed in this paper respectively, the data of which is shown in the following chart, where n represents the number nodes in the graph, *number* represents the number of query point pairs, d represents the number of reference nodes, t_z represents the average running time for each query point pair using Dijkstra algorithm, t_k represents the time of algorithm proposed in this paper.

From the above table data, we can find whether in DBLP graph or road network graph its running time is greatly shortened using the algorithm in this paper and the effect on the road network is more obvious. Compared with the shortest path algorithm, the running time is decreased by 87.4 % and 77.3 % respectively, in DBLP graph and road network graph.

Next, the trend of the above running time with the increase of the number of reference nodes is expressed directly with the chart of Fig. 3. We can find that the running time using the algorithm proposed in this paper become shorter with the increase of the number of the reference nodes. The reason is that the range distance is more accurate as the number of reference nodes is increasing, therefore, the running time becomes shorter and shorter. While after the running time reaches the minimum value, it becomes increasing trend, the reason is that the expense of calculating the distance range becomes bigger with the number of reference nodes further increasing. Combined with the change trend of data from Table 1 and Fig. 3, we can see the running time reaches minimum value when d is 20 or 40 respectively in DBLP graph and road network graph.

5.2 Ratio Test

In this part of experiment, we statistics the ratio of number of point pairs giving the reachability directly to number of all point pairs, which value is represented by *percentage*, and the value changes with the number of reference nodes(d), as shown in the Table 2.

Table 1. Query running time form

Data	n	Number	$t_z(s)$	Algorithm proposed in this paper	
				d	$t_k(s)$
DBLP	2000	500	0.213385	5	0.062968
				10	0.055438
				20	0.042468
				30	0.042562
				40	0.043344
				50	0.04425
Road network graph	2000	500	0.15812	5	0.02378
				10	0.02175
				20	0.021156
				30	0.0205
				40	0.015782
				50	0.016312

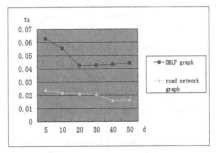

Fig. 3. Query running time chart

From the data above, we can find *percentage* is low for DBLP graph, the highest is 66.6 %, the lowest 45.4 %, while the highest is 84.6 %, the lowest is 71.2 % in the road network graph. The reason lies in the analysis of the two data sets belonging to different type of data. DBLP data is social network of small radius, in accordance with the six degrees of separation theory. The distance difference of point pair is not big, therefore more accurate distance range can be obtained only through increasing the number of reference nodes to tighten the upper and lower value. Because road network belongs to a weighted graph, the span between two nodes is larger, the ratio is higher with the same number of reference nodes in road network.

Next, *percentage* is presented in the form of chart, as shown in Fig. 4. From the curve in the chart, we can find *percentage* increases with the increase of the number of reference nodes.

Table 2. Ratio form

Data	n	Number	Algorithm proposed in this paper	
			d	Percentage
DBLP	2000	500	5	45.40 %
			10	52.40 %
			20	65.20 %
			30	66.00 %
			40	66.40 %
			50	66.60 %
Road network	2000	500	5	71.20 %
			10	73.60 %
			20	75.40 %
			30	77.20 %
			40	84.20 %
			50	84.60 %

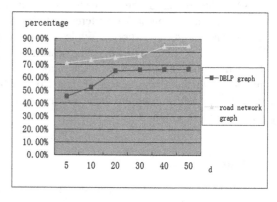

Fig. 4. Ratio chart

6 Conclusion

An algorithm to solve the problem of reachability query with keyword and distance constraint is presented in this paper. The algorithm is suitable both for the weighted graph and for unweighted graph. So the algorithm has more extensive and practical application value. The innovation of this paper lies in introducing the reference node embedding mechanism into the graph query research. The focus of the work in the future is to study the selection strategy of reference nodes, and to consider how to solve the problem of reachability query which's query condition includes the condition constraint that the path must pass through special edges.

Acknowledgements. This work is supported by Foundation for Distinguished Young Talents in Higher Education of Guangdong, China (2014KQNCX184).

References

1. Van Schaik, S.J., De Moor, O.: A memory efficient reachability data structure through bit vector compression. In: Proceedings of the 2011 ACM SIGMOD International Conference on Management of Data, pp. 913–924. ACM press, New York (2011)
2. Qiao, M., Qin, L., Cheng, H., Xu Yu, J., Tian, W.T.: Top-K nearest keyword search on large graphs. Proc. VLDB Endowment 6(10), 901–912 (2013)
3. Jin, R., Liu, L., Ding, B., Wang, H.: Distance-constraint reachability computation in uncertain graphs. PVLDB 4(9), 551–562 (2011)
4. Jin, R., Hong, H., Wang, H., Ruan, N., Xiang, Y.: Computing label-constraint reachability in graph databases. In: 2010 SIGMOD Conference, pp. 123–134. ACM press, New York (2010)
5. Xu, K., Zou, L., Xu Yu. J.: Answering Label-Constraint Reachability in large graphs. In: CIKM 2011, pp. 1595–1600. ACM press, New York (2011)
6. Qiao, M., Cheng, H., Qin, L., Xu Yu, J.: Computing weight constraint reachability in large networks. VLDB 22(3), 275–294 (2013)
7. Potamias, M., Bonchi, F., Castillo, C., Gionis, A.: Fast shortest path distance estimation in large networks. In: Proceedings of the 18th ACM Conference on Information and Knowledge Management, pp. 867–876. ACM press, New York (2009)

Supervised Isometric Mapping Based Classification Algorithm

Ping He$^{(\boxtimes)}$, Tianyu Jing, Xiaohua Xu$^{(\boxtimes)}$, Lei Zhang, and Huihui Lin

Department of Computer Science, Yangzhou University,
Yangzhou 225009, China
angeletx@gmail.com, arterx@gmail.com

Abstract. In this paper, we propose a novel supervised classification algorithm named Supervised Isometric Mapping Based classification Algorithm (SIMBA). The main idea of SIMBA is to integrate the supervised information into the well-known ISOmetric MAPping (ISOMAP) manifold learning algorithm and classify the transformed data in a low-dimensional feature space. By virtue of the integrated supervised information, the manifold mapping becomes more discriminative, thus the classification performance can be improved. SIMBA can deal with complex high-dimensional data lying on an intrinsically low-dimensional manifold, but only has one free parameter, which is the number of nearest neighbors. Sufficient experiment results demonstrate that SIMBA shows higher classification accuracy on real-world datasets than the state-of-the-art support vector machine classifier.

Keywords: Multi-class classification · Dimension reduction · Manifold learning

1 Introduction

Classification plays an important role in the research areas of machine learning and data mining. There has been a large number of state-of-the-art classification algorithms proposed by far, among which the kernel methods achieve great success. However, people often need to deal with very high dimensional data in various real-world application scenarios. For instance, in the automated medical analysis, potential patients are usually required to undergo different tests, including blood tests, medical imaging, neuropsychological tests and etc., while the produced high dimensional test results are used for disease diagnosis. If we directly apply the kernel classifiers to such data, it will cause the curse of dimensionality [1]. A straightforward solution is to reduce the data dimension before the application of classification algorithms. It not only avoids the curse of dimensionality, but also is more beneficial for visualization, removal of noise, reduction of space and time cost clustering, and even improvement of classification performance [2].

Up to now, many methods have been proposed for dimension reduction. In general, the existing dimension reduction methods can be classified into two categories, linear methods and nonlinear methods. Principal Component Analysis (PCA) [3], Linear Discriminate Analysis (LDA) [4] and Multidimensional Scaling (MDS) [5] are some of

© Springer International Publishing AG 2016
H. Yin et al. (Eds.): IDEAL 2016, LNCS 9937, pp. 302–309, 2016.
DOI: 10.1007/978-3-319-46257-8_33

the most popular linear dimension reduction methods. Compared with linear dimension reduction methods, nonlinear dimension reduction algorithms can handle more complex real-world problems. Among them, most approaches are developed by extending linear methods through kernel trick, such as kernel PCA and kernel LDA. However, they may still suffer from the curse of dimensionality when the dimension of the original feature space is very high. Unlike the kernel methods that transform original data into a higher-dimension feature space, manifold learning [6], a promising branch of nonlinear dimension reduction, assumes that the data of interest lie on a low-dimensional embedded non-linear manifold within the original feature space. Under this manifold assumption, many approaches have been developed, including Local Linear Embedding (LLE) [7], Laplacian Eigenmaps (LE) [1], ISOmetric MAPping (ISOMAP) [8] and etc. LLE supposes that the neighboring points are locally linear, thus maximally preserves the locally linear structures of the nearest neighborhood when mapping into a low-dimensional subspace. Similarly, LE also preserves the local structure by maintaining the neighborhood distance instead of local linear reconstruction weight. Different from LLE and LE, ISOMAP is a global manifold learning algorithm whose optimization target is to preserve the pairwise distance of all the data as much as possible. Therefore, it helps to reveal the global structure of the underlying manifold.

Since the traditional manifold learning algorithms are unsupervised, in recent years researchers have made a great endeavor to develop supervised dimension reduction methods to preserve discriminative information while embedding high-dimensional data into low-dimensional feature space. The supervised version of LLE is Supervised LLE (SLLE) [9]. It integrates the label information into the neighborhood determination by redefining a distance that increases the between-class scatter. Local Sensitive Discriminant Analysis (LSDA) [10] constructs both inter-class graph and intra-class graph to characterize the discriminant and geometrical structure of data manifold, and then determines the linear transformation matrix by preserving the combined local neighborhood information. SLLE and LSDA both focus on local manifold structure and fail to take into account the global structure like ISOMAP.

In this paper, we propose a novel supervised manifold learning algorithm, named Supervised Isometric Mapping Based classification Algorithm (SIMBA). Compared with the traditional ISOMAP that only aims at dimension reduction and data visualization, SIMBA is developed for classification by adapting the distance matrix in a block-wise manner so that not only the discriminative information is enforced on the data belonging to different classes, but also the intrinsic manifold structure of the within-class data can be preserved. In addition, SIMBA also derives the optimal mapping of the out-of-sample test data that can minimize the difference of their distance to all the training data on the low-dimensional embedding from that on the original manifold. After that, SIMBA predicts the class labels of the mapped test data using linear support vector machine classifier. The extensive experimental results on the real-world datasets demonstrate the superiority of our proposed approach in comparison with Supervised Local Linear Embedding, and Local Sensitive Discriminant Analysis.

The rest of this paper is organized as follows: Sect. 2 briefly reviews the unsupervised ISOMAP algorithm. In Sect. 3, the procedure of SIMBA is described in details. The experimental results on the real-world datasets are discussed in Sect. 4. In the end, Sect. 5 concludes the whole paper.

2 Background

ISOmetric MAPping (ISOMAP) is a global geometric framework for nonlinear dimensionality reduction. Its approach establishes on classical MDS but explores to preserve the original similarity of the data in lower embedding, as obtained in the geodesic manifold distances which indicate the real distance better in the manifold between all pairwise data points.

Suppose there are a set of n data points $X = \{x_i\}_{i=1}^n \subset \mathbb{R}^m$ that denotes the original input space. The procedures of ISOMAP can be summarized as follows:

Step 1. Determine the neighborhood of each data point on the manifold by means of K-nearest neighbors method or fixed-radius ϵ method to construct a weighted undirected graph G. If two arbitrary data points $x_i, x_j \in X$ are neighbors, the weighted edge between them $d_X = (x_i, x_j)$ represents their neighborhood relationships.

Step 2. Let $d_g(x_i, x_j) = d_X(x_i, x_j)$ if x_i and x_j are neighbors and $D_G = \{d_g(x_i, x_j)\}$ is the matrix of geodesic distance between all the pairwise data points. We can compute d_g by the shortest path distance in the graph G using Floyd-Warshall algorithm.

$$d_g(x_i, x_j) = \min\{d_g(x_i, x_j), d_g(x_i, x_p) + d_g(x_k, x_j)\} \tag{1}$$

Step 3. Seek the low-dimensional embedding Y by applying the classical MDS method to the results of step2. Minimize the cost function:

$$E = \| \tau(D_G) - \tau(D_Y) \|_{L^2} \tag{2}$$

where $D_Y = \{d_Y(y_i, y_j)\} = \{\| y_i - y_j \|\}$. D_Y is the matrix of Euclidean distance between any pair of data points on the low-dimensional embedding. The function τ converts distances to inner products, which uniquely characterize the geometry of the data in a form that supports efficient optimization. The global minimum of Eq. (2) is top d eigenvectors of the matrix $\tau(D_G)$ [8].

3 Supervised Isometric Mapping Based Classification Algorithm

In this section, we introduce our Supervised Isometric Mapping Based classification Algorithm (SIMBA). It is composed of four steps, including integration of supervised information, embedding of low-dimensional manifold, extensions of out-of-sample data and label prediction of test data, each of which will be described in detail later.

Let $\mathcal{X} = \{x_1.x_x, \ldots, x_N\}$ denote the high-dimensional input dataset with $x_i \in \mathbb{R}^D$, $X_S = [x_1^T x_2^T \ldots x_m^T]^T$ denote the training data, $X_T = [x_{m+1}^T \ldots x_n^T]^T$ denote the test data. Suppose $\mathcal{Z}$ is the low-dimensional embedding dataset of $\mathcal{X}$ with $z_i \in \mathbb{R}^d (d < < D)$, Z_S is the embedding of X_S, while X_T is the corresponding embedding of X_T. Besides, $D_G = \{d_g(x_i, x_j)\}$ is the distance matrix, where $d_g(x_i, x_j)$ is the geodesic distance between data points x_i and point x_j.

3.1 Integration of Supervised Information

SIMBA integrates the supervised label information into the classical ISOMAP algorithm by directly enforcing them on the data points belonging to different classes and
preserving the intrinsic manifold structure of the within-class data simultaneously, so as
to obtain a discriminative low-dimensional embedding of the original input data. The
specific formula of incorporating the supervised information is as follows,

$$\tilde{d}_{ij} = \begin{cases} d_g(x_i, x_j) & \text{if } x_i \text{ and } x_j \text{ are in the same class} \\ \max\limits_{\substack{\forall p,q,y(p)=y(i) \\ y(q)=y(j)}} d_g(x_p, x_q) & \text{otherwise} \end{cases} \tag{3}$$

where $\tilde{d}_{ij}$ is the adapted geodesic distance between x_i and x_j, $\tilde{d}_{ij} = d_g(x_i, x_j)$ if and only
if x_i and x_j belong to the same class, otherwise $\tilde{d}_{ij}$ is set the maximal distance between
the two classes that x_i and x_j respectively belong to.

It can be seem from Eq. (3) that the within-class geometric structure is preserved
because the within-class distances are kept untouched, while the inter-class geometric
structure is updated with a block-wise structure since the between-class distances are
set the maximal distance between the corresponding two different classes. The reason
for choosing the block-wise distance adaptation method is that it not only maximizes
the inter-class discrepancy, but also relies on no free parameter.

3.2 Embedding of Low-Dimensional Manifold

To embed the training data onto a low-dimensional manifold, SIMBA takes advantage
of the traditional multidimensional scaling algorithm. Let $D_S = \{\tilde{d}_{ij}^2\}$ be the supervised
squared geodesic distance matrix, SIMBA first computes the τ function by computing
the doublely-centered squared distance matrix,

$$\tau(D_S) = -\frac{HD_SH}{2} \tag{4}$$

where $H = I - \frac{1}{m}\mathbf{1}\mathbf{1}^T$ is the centralizing matrix and I is the identity matrix. Next,
SIMBA performs eigenvalue decomposition on the symmetric $\tau(D_S)$, i.e. $\tau(D_S) = U^T\Lambda U$.

The low-dimensional embedding of the training data then can be obtained,

$$Z = \sqrt{\Lambda_d}U_d \tag{5}$$

where $\Lambda_d = \text{diag}(\lambda_1, \lambda_2, \ldots, \lambda_d)$ is the diagonal matrix composed of the largest
$d << D$ eigenvalues and $U_d = (u_1, u_2, \ldots, u_d)$ is the column matrix composed of the
corresponding eigenvectors.

3.3 Extensions of Out-of-Sample Data

Due to the lack of supervised information, the out-of-sample data cannot be directly embedded into the low-dimensional manifold like the training data. To solve this problem, we reuse the cost function of ISOMAP (Eq. (2)), but only replace D_G with the distance matrix between the training data X_S and the test data X_T on the original high-dimensional manifold, and replace D_Y with distance matrix between Z_S and Z_T on the embedded low-dimensional manifold respectively. Accordingly, we also redefine their inner product conversion function τ in a different way.

$$\tau_g(D(X_S, x_t)) = -\frac{1}{2}(S(X_S, x_t) - E_x[S(x, x_t)] - E_{x'}[S(x_S, x')] + E_{x,x'}[S(x, x')]) \quad (6)$$

where $x_t \in \mathcal{X}_T$ is an arbitrary out-of-sample test data, $S(a, b) = d^2(a, b)$. Equation (6) essentially computes the doubly-centered squared geodesic distance between X_S and x_t on the original manifold. As to the inner product conversion function for the embedding manifold, it is defined in a much simpler way,

$$\tau_z(D(Z_S, z_t)) = Z_S \Lambda_d z_t^T \quad (7)$$

where $z_t \in \mathcal{Z}_T$ is the embedding of x_t. Equation (7) actually computes the weighted inner product of the mapped training data Z_S and the mapped test data z_t on the low-dimensional embedding, while the weight matrix is the same as the diagonal eigenvalue matrix indicating the importance of each selected eigenvector.

By rewriting the cost function in the matrix form and enforcing the two measurements to stay the same as much as possible, SIMBA derives its out-of-sample extension formula.

$$\begin{aligned} Z_T &= \arg\min \sum_{i \in T} \| \tau_g(D(X_S, x_i)) - \tau_z(D(Z_S, z_i)) \|^2 \\ &= \tau_g(D(X_S, X_T))^T Z_S \Lambda^{-1} \end{aligned} \quad (8)$$

It can be found that the above out-of-sample extension shares the same form as that derived for the five unsupervised learning algorithms using Nyström approximation method.

3.4 Label Prediction of Test Data

In the end, SIMBA adopts linear SVM to estimate the decision boundary on the embedded manifold, where the optimization target is

$$\begin{aligned} &\min_{w, b} \quad \frac{1}{2} \| w \|^2 \\ &\text{s.t.} \quad y_i((w \cdot z_i) + b) \geq 1, i = 1, \cdots, m \end{aligned} \quad (9)$$

whose equivalent dual form is

$$\min_{\alpha} \; \frac{1}{2} \sum_{i=1}^{m} \sum_{j=1}^{m} y_i y_j (z_i \cdot z_j) \alpha_i \alpha_j - \sum_{j=1}^{m} \alpha_j$$

$$\text{s.t.} \quad \sum_{i=1}^{m} y_i \alpha_i = 0, \quad \alpha_i \geq 0, \; i = 1, \cdots, m \tag{10}$$

After obtaining the solution to Eq. (10), the label of the test data can be predicted as follows.

$$y = \text{sgn}(\sum_{i=1}^{m} y_i \alpha_i^* (z_i \cdot z_t) + b^*) \tag{11}$$

4 Experiments

In this section, we evaluate the performance of SIMBA by applying it on several real-world medical datasets, including liver disorder, diabetes, acute bacterial meningitis, breast cancer, muscular dystrophy, mammographic mass, heart disease and very low birth weight infant datasets, in comparison with other famous manifold classification methods.

First, we show the procedures of SIMBA on liver disorder dataset, which contains 345 instances and seven attributes for the diagnosis of the alcoholic liver disease (ALD). Figure 1(a) depicts the training data in the first two dimensions of the original feature space. The red dots represent normal samples, while blue dots represent disordered samples. They are considerably hard to separate from each other. Figure 1(b) illustrates the mapped training data after the supervised isometric mapping of SIMBA. We find that the two classes of training data are clearly separated and a single line is built as a classifier for it. To extend the classifier to the out-of-sample test data, we first depict the test data in the first two dimensions of the original feature space. Then we apply the out-of-sample-extension formula of SIMBA to the test data. The low-dimensional embedding of the test data is shown in Fig. 1(d). It shows that not only the training data but also the test data can be reallocated with intra-class data nearby and inter-class data faraway. Moreover, the simple classifier built for the training data can classify the test data very well after the unsupervised out-of-sample extension mapping of SIMBA.

Furthermore, we compare the mean classification accuracy of SIMBA with another two state-of-the-art manifold classification algorithms SLLE and LSDA in a 10-fold cross validation. Table 1 summarizes the comparison result, which proves that SIMBA performs the best among the three on real-world medical diagnosis datasets.

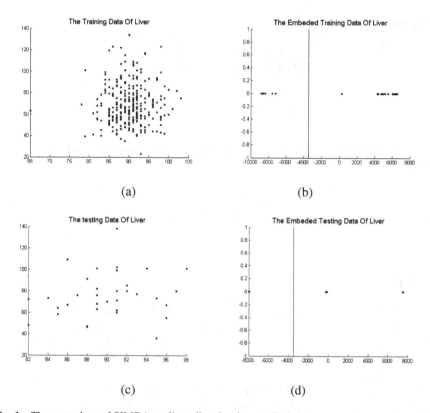

Fig. 1. The procedure of SIMBA on liver disorder dataset. Red dots represent normal samples, blue dots represent disordered samples (Color figure online)

Table 1. Comparison of classification accuracy among SLLE, LSDA and SIMBA

Datasets	SLLE	LSDA	SIMBA
Liver	83.23 %	86.25 %	**92.3 %**
Diabetes	73.96 %	79.41 %	**84.5 %**
Meningitis	72.98 %	73.42 %	**77.97 %**
Breast Cancer	71.55 %	74.98 %	**78.34 %**
Dystrophy	83.72 %	84.58 %	**90.48 %**
Mammographic Mass	70.22 %	71.35 %	**76.29 %**
VLBW Infant	74.67 %	76.52 %	**80.88 %**

5   Conclusions

In this paper, we show a supervised classification algorithm SIMBA, short for Supervised Isometric Mapping Based classification Algorithm. It begins with supervised dimension reduction and ends with classification in a reduced dimensional feature space. During the supervised dimension reduction procedure, SIMBA reallocate the data belonging to the same class nearby and separate the data from different classes faraway. The approach proposed is theoretically extended to the out-of-sample test data to ensure its generalization. In our experiments, SIMBA is both visually illustrated and compared with another two state-of-the-art manifold classification algorithms on real-world medical datasets. The promising experimental results demonstrate the superiority of our method on automated medical diagnosis.

Acknowledgement. This research was supported in part by the Chinese National Natural Science Foundation under Grant nos. 61402395, 61472343 and 61379066, Natural Science Foundation of Jiangsu Province under contracts BK20151314, BK20140492 and BK20130452, Natural Science Foundation of Education Department of Jiangsu Province under contract 13KJB520026, and the New Century Talent Project of Yangzhou University.

References

1. Belkin, M., Niyogi, P.: Laplacian Eigenmaps for dimensionality reduction and data representation. Neural Comput. **15**(6), 1373–1396 (2003)
2. Orsenigo, C., Vercellis, C.: Kernel ridge regression for out-of-sample mapping in supervised manifold learning. Expert Syst. Appl. **39**(9), 7757–7762 (2012)
3. Jolliffe I T.: Principal Compon. Anal. **87**(100), pp. 41–64 (2010). Springer, Berlin
4. Nasrabadi, N.M., Bishop, C.M.: Pattern Recognition and Machine Learning, pp. 186–189. Academic Press
5. Groenen, P.J.F., Velden, M.: Multidimensional scaling. Econometric Inst. Res. Pap. **46**(2), 1050–1057 (2014)
6. Camastra, F., Vinciarelli, A.: Feature extraction methods and manifold learning methods. In: Camastra, F., Vinciarelli, A. (eds.) Machine Learning for Audio, Image and Video Analysis, pp. 341–386. Springer, London (2015)
7. Roweis, S.T., Saul, L.K.: Nonlinear dimensionality reduction by locally linear embedding. Science **290**(5500), 2323–2326 (2000)
8. Tenenbaum, J.B., Desilva, V., Langford, D.C.: A global geometric, for nonlinear dimensionality reduction. Science **290**(5500), 2319–2323 (2000)
9. Kouropteva, O., Okun, O., Pietikäinen, M.: Supervised locally linear embedding algorithm for pattern recognition. In: Perales, F.J., Campilho, A.J.C., de la Blanca, N.P., Sanfeliu, A. (eds.) IbPRIA 2003. LNCS, vol. 2652, pp. 386–394. Springer, Heidelberg (2003)
10. Liu, X.M., Deng, S.G., Yin, J.W., et al.: Locality sensitive discriminant analysis based on matrix representation. J. Zhejiang Univ. **2**, 290–296 (2009)

Successive Ray Refinement and Its Application to Coordinate Descent for Lasso

Jun Liu[1]([⊠]), Zheng Zhao[2], Ruiwen Zhang[1], and Yan Xu[1]

[1] SAS Institute Inc., 500 SAS Campus Dr, Cary, NC 27513, USA
{jun.liu,ruiwen.zhang,yan.xu}@sas.com
[2] Google Inc., 1600 Amphitheater Pkwy, Mountain View, CA 95014, USA
alzhao@google.com

Abstract. Coordinate descent is one of the most popular approaches for solving Lasso and its extensions due to its simplicity and efficiency. When applying coordinate descent to solving Lasso, we update one coordinate at a time while fixing the remaining coordinates. Such an update, which is usually easy to compute, greedily decreases the objective function value. In this paper, we aim to improve its computational efficiency by reducing the number of coordinate descent iterations. To this end, we propose a novel technique called Successive Ray Refinement (SRR). SRR makes use of the following ray continuation property on the successive iterations: for a particular coordinate, the value obtained in the next iteration almost always lies on a ray that starts at its previous iteration and passes through the current iteration. Motivated by this ray-continuation property, we propose that coordinate descent be performed not directly on the previous iteration but on a refined search point that has the following properties: on one hand, it lies on a ray that starts at a history solution and passes through the previous iteration, and on the other hand, it achieves the minimum objective function value among all the points on the ray. We propose a scheme for defining the search point and show that the refined search point can be efficiently obtained. Empirical results for real and synthetic data sets show that the proposed SRR can significantly reduce the number of coordinate descent iterations, especially for small Lasso regularization parameters.

1 Introduction

Lasso [11] is an effective technique for analyzing high-dimensional data. It has been applied successfully in various areas, such as machine learning, signal processing, image processing, medical imaging, and so on. Let $X = [\mathbf{x}_1, \mathbf{x}_2, \ldots, \mathbf{x}_p] \in \mathcal{R}^{n \times p}$ denote the data matrix composed of n samples with p variables, and let $\mathbf{y} \in \mathcal{R}^{n \times 1}$ be the response vector. In Lasso, we compute the $\boldsymbol{\beta}$ that optimizes

$$\min_{\boldsymbol{\beta}} f(\boldsymbol{\beta}) = \frac{1}{2}\|X\boldsymbol{\beta} - \mathbf{y}\|_2^2 + \lambda\|\boldsymbol{\beta}\|_1, \tag{1}$$

Z. Zhao—This work was done when Zheng Zhao was with SAS Institute Inc.

© Springer International Publishing AG 2016
H. Yin et al. (Eds.): IDEAL 2016, LNCS 9937, pp. 310–320, 2016.
DOI: 10.1007/978-3-319-46257-8_34

where the first term measures the discrepancy between the prediction and the response and the second term controls the sparsity of β with ℓ_1 regularization. The regularization parameter λ is nonnegative, and a larger λ usually leads to a sparser solution.

Researchers have developed many approaches for solving Lasso in Eq. (1). Least Angle Regression (LARS) [3] is one of the most well-known homotopy approaches for Lasso. LARS adds or drops one variable at a time, generating a piecewise linear solution path for Lasso. Unlike LARS, other approaches usually solve Eq. (1) according to some prespecified regularization parameters. These methods include the coordinate descent method [4,15], the gradient descent method [1,14], the interior-point method [6], the stochastic method [10], and so on. Among these approaches, coordinate descent is one of the most popular approaches due to its simplicity and efficiency. When applying coordinate descent to Lasso, we update one coordinate at a time while fixing the remaining coordinates. This type of update, which is easy to compute, can effectively decrease the objective function value in a greedy way.

To improve the efficiency of optimizing the Lasso problem in Eq. (1), the screening technique has been extensively studied in [5,8,9,12,13,16]. Screening (1) identifies and removes the variables that have zero entries in the solution β and (2) solves Eq. (1) by using only the kept variables. When one is able to discard the variables that have zero entries in the final solution β and identify the signs of the nonzero entries, the Lasso problem in Eq. (1) becomes a standard quadratic programming problem. However, it is usually very hard to identify all the zero entries, especially when the regularization parameter is small. In addition, the computational cost of Lasso usually increases as the regularization parameter decreases. The computational cost increase motivates us to come up with an approach that can accelerate the computation of Lasso for small regularization parameters.

In this paper, we aim to improve the computational efficiency of coordinate descent by reducing its iterations. To this end, we propose a novel technique called Successive Ray Refinement (SRR). Our proposed SRR is motivated by an interesting ray-continuation property on the coordinate descent iterations: for a given coordinate, the value obtained in the next iteration almost always lies on a ray that starts at its previous iteration and passes through the current iteration. Figure 1 illustrates the ray-continuation property by using the data specified in Sect. 2. Motivated by this ray-continuation property, we propose that coordinate descent be performed not directly on the previous iteration but on a refined search point that has the following properties: on one hand, the search point lies on a ray that starts at a history solution and passes through the previous iteration, and on the other hand, the search point achieves the minimum objective function value among all the points on the ray. We propose a scheme for defining the search point, and we show that the refined search point can be efficiently computed. Experimental results on both synthetic and real data sets demonstrate that the proposed SRR can greatly accelerate the convergence of coordinate descent for Lasso, especially when the regularization parameter is small.

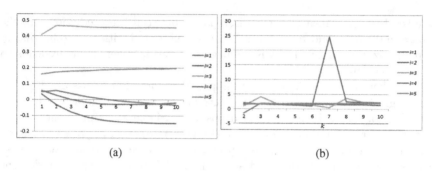

(a)   (b)

Fig. 1. Illustration of the iterations of coordinate descent. For both plots, the x-axis corresponds to the iteration number k. The y-axis of plot (a) denotes β_i^k, the value of the ith coordinate in the kth iteration. The y-axis of plot (b) denotes α_i^k, which is computed using the equation $\beta_i^{k+1} = \alpha_i^k \beta_i^{k-1} + (1 - \alpha_i^k)\beta_i^k$. Ray-continuation property: for a given coordinate i, the value obtained in the next iteration denoted by β_i^{k+1} almost always lies on a ray that starts at its previous iteration, β_i^{k-1}, and passes through the current iteration, α_i^k.

Organization. The rest of this paper is organized as follows. We introduce the traditional coordinate descent for Lasso and present the ray-continuation property that motivates this paper in Sect. 2, propose the SRR technique in Sect. 3, and discuss the efficient computation of the refinement factor that is used in SRR in Sect. 4. We report experimental results on both synthetic and real data sets in Sect. 5, and we conclude this paper in Sect. 6. Throughout this paper, we assume that X does not contain a zero column; that is, $\|\mathbf{x}_i\|_2 \neq 0, \forall i$.

2   Coordinate Descent for Lasso

In this section, we first review the coordinate descent method for solving Lasso, and then analyze the adjacent iterations to motivate the proposed SRR technique.

Let β_i^k denote the ith element of $\boldsymbol{\beta}$, which is obtained at the kth iteration of coordinate descent. In coordinate descent, we compute β_i^k while fixing $\beta_j = \beta_j^k, 1 \leq j < i$, and $\beta_j = \beta_j^{k-1}, i < j \leq p$. Specifically, β_i^k is computed as the minimizer to the following univariate optimization problem:

$$\beta_i^k = \arg\min_\beta f([\beta_1^k, \ldots, \beta_{i-1}^k, \beta, \beta_{i+1}^{k-1}, \ldots, \beta_p^{k-1}]^T).$$

It can be computed in a closed form as:

$$\beta_i^k = \frac{S(\mathbf{x}_i^T \mathbf{y} - \sum_{j<i} \mathbf{x}_i^T \mathbf{x}_j \beta_j^k - \sum_{j>i} \mathbf{x}_i^T \mathbf{x}_j \beta_j^{k-1}, \lambda)}{\|\mathbf{x}_i\|_2^2}, \tag{2}$$

where $S(\cdot, \cdot)$ is the shrinkage function

$$S(x, \lambda) = \begin{cases} x - \lambda & x > \lambda \\ x + \lambda & x < -\lambda \\ 0 & |x| \leq \lambda. \end{cases} \tag{3}$$

Let

$$\mathbf{r}_i^k = \mathbf{y} - X[\beta_1^k, \ldots, \beta_{i-1}^k, \beta_i^k, \beta_{i+1}^{k-1}, \ldots, \beta_p^{k-1}]^T \qquad (4)$$

denote the residual obtained after updating β_i^{k-1} to β_i^k. With Eq. (4), we can rewrite Eq. (2) as

$$\beta_i^k = S(\beta_i^{k-1} + \frac{\mathbf{x}_i^T \mathbf{r}_{i-1}^k}{\|\mathbf{x}_i\|_2^2}, \frac{\lambda}{\|\mathbf{x}_i\|_2^2}). \qquad (5)$$

In addition, with the updated β_i^k, we can update the residual from $\mathbf{r}_{i-1}^k$ to $\mathbf{r}_i^k$ as

$$\mathbf{r}_i^k = \mathbf{r}_{i-1}^k + \mathbf{x}_i(\beta_i^{k-1} - \beta_i^k). \qquad (6)$$

We demonstrate the coordinate descent algorithm using the following randomly generated X and $\mathbf{y}$:

$$X = \begin{bmatrix} -0.204708 & 0.478943 & -0.519439 & -0.555730 & 1.965781 \\ 1.393406 & 0.092908 & 0.281746 & 0.769023 & 1.246435 \\ 1.007189 & -1.296221 & 0.274992 & 0.228913 & 1.352917 \\ 0.886429 & -2.001637 & -0.371843 & 1.669025 & -0.438570 \\ -0.539741 & 0.476985 & 3.248944 & -1.021228 & -0.577087 \end{bmatrix}, \qquad (7)$$

$$\mathbf{y} = [0.124121, 0.302614, 0.523772, 0.000940, 1.343810]^T. \qquad (8)$$

We show the iterations of coordinate descent for Lasso with $\lambda = 0$ in Fig. 1 (a). It can be observed from the results in Fig. 1 (a) that we can obtain an approximate solution with a small objective function value within a few iterations. However, achieving a solution with high precision takes quite a few iterations for this example. More interestingly, for a particular coordinate, the value obtained in the next iteration almost always lies on a ray that starts at its previous iteration and passes through the current iteration. To show this, we compute α_i^k that satisfies the following equation:

$$\beta_i^{k+1} = \alpha_i^k \beta_i^{k-1} + (1 - \alpha_i^k)\beta_i^k. \qquad (9)$$

Figure 1 (b) show the values of α_i^k for different iterations. It can be observed that the values of α_i^k are almost always positive except α_1^2 for this example. In addition, most of the values of α_i^k are larger than 1. We tried quite a few synthetic data and observed a similar phenomenon.

For a particular iteration number k, if $\alpha_i^k = \alpha, \forall i$, we can easily achieve $\beta^{k+1} = \alpha\beta^{k-1} + (1 - \alpha)\beta^k$ without needing to perform any coordinate descent iteration. This motivated us to come up with the successive ray refinement technique to be discussed in the next section.

3   Successive Ray Refinement

In the proposed SRR technique, we make use of the ray-continuation property shown in Fig. 1. Our idea is as follows: To obtain β^{k+1}, we perform coordinate descent based on a refined search point $\mathbf{s}^k$ rather than on its previous solution $\boldsymbol{\beta}^k$. We propose setting the refined search point as:

$$\mathbf{s}^k = (1 - \alpha^k)\mathbf{h}^k + \alpha^k \boldsymbol{\beta}^k, \tag{10}$$

where $\mathbf{h}^k$ is a properly chosen history solution, $\boldsymbol{\beta}^k$ is the current solution, and α^k is an optimal refinement factor that optimizes the following univariate optimization problem:

$$\min_{\alpha}\{g(\alpha) = f((1 - \alpha)\mathbf{h}^k + \alpha\boldsymbol{\beta}^k)\}. \tag{11}$$

The setting of $\mathbf{h}^k$ to one of the history solutions is based on the following two considerations. First, we aim to use the ray-continuation property to reduce the number of iterations. Second, we need to ensure that the univariate optimization problem in Eq. (11) can be efficiently computed. We discuss the computation of Eq. (11) in Sect. 4.

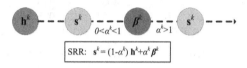

Fig. 2. The proposed SRR technique. The search point $\mathbf{s}^k$ lies on the ray that starts from a properly chosen history solution $\mathbf{h}^k$ and passes through the current solution $\boldsymbol{\beta}^k$, and meanwhile it achieves the minimum objective function value among all the points on the ray, optimizing Eq. (11).

Figure 2 illustrates the proposed SRR technique. When $\alpha^k = 1$, we have $\mathbf{s}^k = \boldsymbol{\beta}^k$; that is, the refined search point becomes the current solution $\boldsymbol{\beta}^k$. When $\alpha^k = 0$, we have $\mathbf{s}^k = \mathbf{h}^k$; that is, the refined search point becomes the specified history solution $\mathbf{h}^k$. However, our next theorem shows that $\mathbf{s}^k \neq \mathbf{h}^k$ because α^k is always positive. In other words, the search point always lies on a ray that starts with the history point $\mathbf{h}^k$ and passes through the current solution $\boldsymbol{\beta}^k$.

Theorem 1. *Assume that the history point* $\mathbf{h}^k$ *satisfies*

$$f(\mathbf{h}^k) > f(\boldsymbol{\beta}^k). \tag{12}$$

Then, α^k *that minimizes Eq. (11) is positive. In addition, if* $X\mathbf{h}^k \neq X\boldsymbol{\beta}^k$, α^k *is unique.*

Proof. It is easy to verify that $g(\alpha)$ is convex. Therefore, α^k that minimizes Eq. (11) has at least one solution. Equation (12) leads to

$$g(1) < g(0). \tag{13}$$

Therefore, the global refinement factor $\alpha^k \neq 0$. Next, we show that α^k cannot be negative.

If $\alpha^k < 0$, due to the convexity of $g(\alpha)$, we have

$$g((1-\theta)\alpha^k + \theta) \leq (1-\theta)g(\alpha^k) + \theta g(1), \forall \theta \in [0,1]. \tag{14}$$

Setting $\theta = \frac{\alpha^k}{\alpha^k - 1}$, we have

$$g(0) \leq \frac{-1}{\alpha^k - 1}g(\alpha^k) + \frac{\alpha^k}{\alpha^k - 1}g(1), \forall \theta \in [0,1]. \tag{15}$$

Making use of Eq. (13), we have $g(1) < g(\alpha^k)$. This contradicts the fact that α^k minimizes Eq. (11). Therefore, α^k is always positive.

If $X\mathbf{h}^k \neq X\beta^k$, $g(\alpha)$ is strongly convex and thus α^k is unique. This ends the proof of this theorem. □

For coordinate descent, the condition in Eq. (12) always holds, because the objective function value keeps decreasing. The selection of an appropriate $\mathbf{h}^k$ is key to the success of the proposed SRR, and the following theorem says that if $\mathbf{h}^k$ is good enough, the refined search solution $\mathbf{s}^k$ is an optimal solution to Eq. (1).

Theorem 2. *Let β^* be an optimal solution to Eq. (1). If*

$$\beta^* - \mathbf{h}^k = \gamma(\beta^k - \mathbf{h}^k), \tag{16}$$

for some positive γ, $\mathbf{s}^k$ achieved by SRR in Eq. (10) satisfies $f(\mathbf{s}^k) = f(\beta^*)$.

Proof. When setting $\alpha^k = \gamma$, we have $\mathbf{s}^k = \beta^*$ under the assumption in Eq. (16). Therefore, with the SRR technique, we can obtain a refined solution $\mathbf{s}^k$ that is an optimal solution to Eq. (1). □

In the following, we discuss a scheme for choosing the history solution $\mathbf{h}^k$. Specificially, we set

$$\mathbf{h}^k = \beta^{k-1}. \tag{17}$$

Figure 3 demonstrates this scheme. Since the generated points follow a triangle structure, we call this scheme the Successive Ray Refinement Triangle (SRRT). We name the resulting method as CD+SRRT, where CD stands for the traditional coordinate descent.

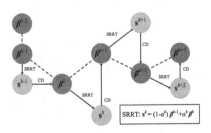

Fig. 3. Illustration of the proposed SRRT technique. $\mathbf{s}^k$ is the refined search point and β^{k+1} is the point obtained by applying coordinate descent (CD) based on the refined search point $\mathbf{s}^k$. In this illustration, it is assumed that the optimal refinement factor α^k is larger than 1. When $\alpha^k \in (0,1)$, $\mathbf{s}^k$ lies between β^{k-1} and β^k.

4   Efficient Refinement Factor Computation

In this section, we discuss how to efficiently compute the refinement factor α^k in Eq. (11). The function $g(\alpha)$ can be written as:

$$
\begin{aligned}
g(\alpha) &= \frac{1}{2} \left\| X((1-\alpha)\mathbf{h}^k + \alpha\boldsymbol{\beta}^k) - \mathbf{y} \right\|_2^2 + \lambda \|(1-\alpha)\mathbf{h}^k + \alpha\boldsymbol{\beta}^k\|_1 \\
&= \frac{1}{2} \left\| \mathbf{r}_h^k - \alpha(\mathbf{r}_h^k - \mathbf{r}^k) \right\|_2^2 + \lambda \|\mathbf{h}^k - \alpha(\mathbf{h}^k - \boldsymbol{\beta}^k)\|_1,
\end{aligned}
\tag{18}
$$

where $\mathbf{r}_h^k = \mathbf{y} - X\mathbf{h}^k$ and $\mathbf{r}^k = \mathbf{y} - X\boldsymbol{\beta}^k$ are the residuals that correspond to $\mathbf{h}^k$ and $\boldsymbol{\beta}^k$, respectively. Before the convergence, we have $\mathbf{r}_h^k \neq \mathbf{r}^k$. Therefore, $g(\alpha)$ is strongly convex in α, and α^k, the minimizer to Eq. (11), is unique.

When $\lambda = 0$, Eq. (11) has a nice closed form solution,

$$
\alpha^k = \frac{\langle \mathbf{r}_h^k, \mathbf{r}_h^k - \mathbf{r}^k \rangle}{\|\mathbf{r}_h^k - \mathbf{r}^k\|_2^2}.
\tag{19}
$$

Next, we discuss the case $\lambda > 0$. The subgradient of $g(\alpha)$ with regard to α can be computed as

$$
\partial g(\alpha) = \alpha \|\mathbf{r}_h^k - \mathbf{r}^k\|_2^2 - \langle \mathbf{r}_h^k, \mathbf{r}_h^k - \mathbf{r}^k \rangle + \lambda \sum_{i=1}^p (\beta_i^k - h_i)\mathrm{SGN}(h_i - \alpha(h_i - \beta_i^k)).
\tag{20}
$$

Computing α^k is a root-finding problem. According to Theorem 1, we have $\alpha^k > 0$. Next, we consider only $\alpha > 0$ for $\partial g(\alpha)$. We consider the following three cases:

1. If $h_i = 0$, we have

$$
(\beta_i^k - h_i)\mathrm{SGN}(h_i - \alpha(h_i - \beta_i^k)) = \{|\beta_i^k|\}.
$$

2. If $h_i(\beta_i^k - h_i) > 0$, we have

$$
(\beta_i^k - h_i)\mathrm{SGN}(h_i - \alpha(h_i - \beta_i^k)) = \{|\beta_i^k - h_i|\}.
$$

3. If $h_i(\beta_i^k - h_i) < 0$, we let

$$
w_i = \frac{h_i}{h_i - \beta_i^k},
\tag{21}
$$

and we have

$$
(\beta_i^k - h_i)\mathrm{SGN}(h_i - \alpha(h_i - \beta_i^k)) = \begin{cases} \{-|\beta_i^k - h_i|\} & \alpha \in (0, w_i) \\ \{|\beta_i^k - h_i|\} & \alpha \in (w_i, +\infty) \\ |\beta_i^k - h_i|\{[-1,1]\} & \alpha = w_i. \end{cases}
\tag{22}
$$

For the first two cases, the set $\mathrm{SGN}(h_i - \alpha(h_i - \beta_i^k))$ is deterministic. For the third case, $\mathrm{SGN}(h_i - \alpha(h_i - \beta_i^k))$ is deterministic when $\alpha \neq w_i$. Define

$$
\Omega(\mathbf{h}^k, \boldsymbol{\beta}^k) = \{i : h_i(\beta_i^k - h_i) < 0\}.
\tag{23}
$$

Figure 4 illustrates the function $\partial g(\alpha), \alpha > 0$. It can be observed that $\partial g(\alpha)$ is a piecewise linear function. If $\Omega(\mathbf{h}^k, \boldsymbol{\beta}^k)$ is empty, $\partial g(\alpha)$ is continuous; otherwise, $\partial g(\alpha)$ is not continuous at $\alpha = w_i, i \in \Omega(\mathbf{h}^k, \boldsymbol{\beta}^k)$.

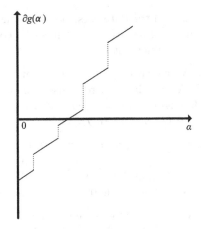

Fig. 4. Illustration of $\partial g(\alpha)$. When $\lambda > 0$, it is a non-continuous piecewise linear monotonically increasing function. The intersection between $\partial g(\alpha)$ and the horizontal axis gives α^k, the solution to Eq. (11).

4.1   An Algorithm Based on Sorting

To compute the refinement factor, one approach is to sort w_i as follows:

First, we sort $w_i, i \in \Omega(\mathbf{h}^k, \boldsymbol{\beta}^k)$, and assume $w_{i_0} \leq w_{i_1} \leq \cdots \leq w_{i_{|\Omega(\mathbf{h}^k, \boldsymbol{\beta}^k)|}}$.

Second, for $j = 1, 2, \ldots, |\Omega(\mathbf{h}^k, \boldsymbol{\beta}^k)|$, we evaluate $\partial g(\alpha)$ at $\alpha = w_{i_j}$ with the following three cases:

1. If $0 \in \partial g(w_{i_j})$, we have $\alpha^k = w_{i_j}$ and terminate the search.
2. If an element in $\partial g(w_{i_j})$ is positive, α^k lies in the piecewise line starting $\alpha = w_{i_{j-1}}$ and ending $\alpha = w_{i_j}$, and it can be analytically computed.
3. If all elements in $\partial g(w_{i_j})$ are negative, we set $j = j + 1$ and continue the search.

Finally, if all elements in $\partial g(w_{i_j})$ are negative when $j = |\Omega(\mathbf{h}^k, \boldsymbol{\beta}^k)|$, α^k lies on the piecewise line that starts at $\alpha = w_{i_j}$. Thus, α^k can be analytically computed.

With a careful implementation, the naive approach can be completed in $O(p + m \log(m))$, where $m = |\Omega(\mathbf{h}^k, \boldsymbol{\beta}^k)|$. In Lasso, the solution is usually sparse, and thus m is much smaller than p, the number of variables. In addition, with a similar implementation as in [7], we can develop an improved bisection approach that has a time complexity of $O(p)$. Note that the cost for computing the refinement factor is much less than each coordinate iteration that costs $O(np)$ operations. Thus, we only report the number of iterations in the experiments.

5   Experiments

In this section, we report experimental results for synthetic and real data sets, studying the number of iterations of CD and CD+SRRT for solving Lasso.

We begin with the discussion of the data sets, then give the experimental setup, and finally report the number of iterations consumed by coordinate descent and the proposed successive ray refinement.

Synthetic Data Sets. We generate the synthetic data as follows. The entries in the $n \times p$ design matrix X and the $n \times 1$ response $\mathbf{y}$ are drawn from a Gaussian distribution. We try the following three settings of n and p: (1) $n = 500, p = 1000$, (2) $n = 1000, p = 1000$, and (3) $n = 1000, p = 500$.

Real Data Sets. We make use of the following three real data sets provided in [2]: leukemia, colon, and gisette. The leukemia data set has $n = 38$ samples and $p = 7129$ variables. The colon data set has $n = 62$ samples and $p = 2000$ variables. The gisette data set has $n = 6000$ samples and $p = 5000$ variables.

Experimental Settings. For the value of the regularization parameter, we try $\lambda = r\|X^T\mathbf{y}\|_\infty$, where $r = 0.5, 0.1, 0.05, 0.01$. For the synthetic data sets, the reported results are averaged over 10 runs. For a particular regularization parameter, we first run CD until $\|\beta^k - \beta^{k-1}\|_2 \leq 10^{-6}$, and then run CD+SRRT until the obtained objective function value is less than or equal to the one obtained by CD.

Results. Table 1 shows the results for the synthetic and real data sets. We can see that when the solution is very sparse (for example, $\lambda = 0.5\|X^T\mathbf{y}\|_\infty$), the proposed CD+SRRT consumes comparable number of iterations to the traditional CD. The reason is that the optimal refinement factor computed by SRR in Eq. (11) is equal to or close to 1, and thus CD+SRRT is very close to the traditional CD. Note that a regularization parameter $\lambda = 0.5\|X^T\mathbf{y}\|_\infty$ is usually too large for practical applications because it selects too few variables, and we

Table 1. Performance for the synthetic data sets and the real data sets. The results for the synthetic data sets are averaged over 10 runs. The sparsity is defined as the number of zeros in the solution divided by the total number of variables p.

Synthetic data					Real data				
data set	λ	CD	CD+SRRT	sparsity	data set	λ	CD	CD+SRRT	sparsity
$n = 500$	0.5	10.0	9.2	0.9395	leukemia	0.5	122	84	0.9982
	0.1	151.7	59.5	0.6406		0.1	155	103	0.9964
$p = 1000$	0.05	463.2	109.1	0.5763		0.05	254	127	0.9961
	0.01	4132.7	326.1	0.5146		0.01	2053	343	0.9948
$n = 1000$	0.5	7.9	7.7	0.9397	colon	0.5	31	24	0.9975
	0.1	54.9	29.0	0.473		0.1	157	78	0.9840
$p = 1000$	0.05	125.4	47.7	0.3126		0.05	308	118	0.9775
	0.01	748.0	128.9	0.118		0.01	2766	375	0.9715
$n = 1000$	0.5	7.9	7.8	0.8856	gisette	0.5	26	11	0.9994
	0.1	26.3	17.1	0.3102		0.1	180	108	0.9932
$p = 500$	0.05	35.5	20.2	0.1678		0.05	823	432	0.9876
	0.01	47.9	25.1	0.0382		0.01	4621	1368	0.8340

usually need to try a smaller $\lambda = r\|X^T\mathbf{y}\|_\infty$ for example, $r = 0.01$. It can be observed that the proposed CD+SRRT requires much fewer iterations, especially for smaller regularization parameters.

6 Conclusion

In this paper, we propose a novel technique called successive ray refinement. Our proposed SRR is motivated by an interesting ray-continuation property on the coordinate descent iterations: for a particular coordinate, the value obtained in the next iteration almost always lies on a ray that starts at its previous iteration and passes through the current iteration. We propose a scheme for SRR and apply them to solving Lasso with coordinate descent. Empirical results for real and synthetic data sets show that the proposed SRR can significantly reduce the number of coordinate descent iterations, especially when the regularization parameter is small.

It is interesting to study the convergence rate of CD+SRR. We focus on a least squares loss function in (1), and we plan to apply the SRR technique to solving the generalized linear models. We compute the refinement factor as an optimal solution to Eq. (11), and we plan to obtain the refinement factor as an approximate solution, especially in the case of generalized linear models.

References

1. Beck, A., Teboulle, M.: A fast iterative shrinkage-thresholding algorithm for linear inverse problems. SIAM J. Imaging Sci. **2**(1), 183–202 (2009)
2. Chang, C.C., Lin, C.J.: LIBSVM: a library for support vector machines. ACM Trans. Intell. Syst. Technol. **2**, 27:1–27:27 (2011)
3. Efron, B., Hastie, T., Johnstone, I., Tibshirani, R.: Least angle regression. Ann. Stat. **32**, 407–499 (2004)
4. Friedman, J.H., Hastie, T., Tibshirani, R.: Regularization paths for generalized linear models via coordinate descent. J. Stat. Softw. **33**(1), 1–22 (2010)
5. Ghaoui, L., Viallon, V., Rabbani, T.: Safe feature elimination in sparse supervised learning. Pac. J. Optim. **8**, 667–698 (2012)
6. Koh, K., Kim, S., Boyd, S.: An interior-point method for large-scale l1-regularized logistic regression. J. Mach. Learn. Res. **8**, 1519–1555 (2007)
7. Liu, J., Ye, J.: Efficient Euclidean projections in linear time. In: International Conference on Machine Learning (2009)
8. Liu, J., Zhao, Z., Wang, J., Ye, J.: Safe screening with variational inequalities and its application to lasso. In: International Conference on Machine Learning (2014)
9. Ogawa, K., Suzuki, Y., Takeuchi, I.: Safe screening of non-support vectors in path-wise SVM computation. In: International Conference on Machine Learning (2013)
10. Shalev-Shwartz, S., Tewari, A.: Stochastic methods for ℓ_1 regularized loss minimization. In: Proceedings of the 26th International Conference on Machine Learning (2009)
11. Tibshirani, R.: Regression shrinkage and selection via the lasso. J. Roy. Stat. Soc. Ser. B **58**, 267–288 (1996)

12. Tibshirani, R., Bien, J., Friedman, J.H., Hastie, T., Simon, N., Taylor, J., Tibshirani, R.J.: Strong rules for discarding predictors in lasso-type problems. J. Roy. Stat. Soc. Ser. B **74**, 245–266 (2012)
13. Wang, J., Lin, B., Gong, P., Wonka, P., Ye, J.: Lasso screening rules via dual polytope projection. In: Advances in Neural Information Processing Systems (2013)
14. Wright, S.J., Nowak, R.D., Figueiredo, M.A.T.: Sparse reconstruction by separable approximation. IEEE Trans. Sign. Process. **57**(7), 2479–2493 (2009)
15. Yuan, G.X., Ho, C.H., Lin, C.J.: An improved glmnet for l1-regularized logistic regression. J. Mach. Learn. Res. **13**, 1999–2030 (2012)
16. Zhen, J.X., Hao, X., Peter, J.R.: Learning sparse representations of high dimensional data on large scale dictionaries. In: Advances in neural information processing systems (2011)

Multi-view Subspace Clustering via a Global Low-Rank Affinity Matrix

Lei Qi[1], Yinghuan Shi[1(✉)], Huihui Wang[1], Wanqi Yang[1,2], and Yang Gao[1]

[1] State Key Laboratory for Novel Software Technology,
Nanjing University, Nanjing, China
syh@nju.edu.cn
[2] School of Computer Science and Technology,
Nanjing Normal University, Nanjing, China

Abstract. Subspace clustering is a technique which aims to find the underlying low-dimensional subspace in a high-dimensional data space. Since the multi-view data exists generally and it can effectively improve the performance of the learning task in real-world applications, multi-view subspace clustering has gained lots of attention in recent years. In this paper, to further improve the clustering performance of multi-view subspace clustering, we propose a novel subspace clustering method based on a global low-rank affinity matrix. In our method, we introduce a global affinity matrix, and use a sparse term to fit the difference between the global affinity matrix and local affinity matrices. Meanwhile, our method explores the global consistent information from different views and simultaneously guarantees the global affinity matrix for segmentation is low-rank. The objective function can be solved efficiently by the inexact augmented Lagrange multipliers (ALM) optimization method. Experiments results on two public real face datasets demonstrate that our method can improve the clustering performance against with the state-of-the-art methods.

Keywords: Multi-view · Subspace clustering · Global low-rank affinity matrix

1 Introduction

In recent years, with the advent of the era of big data, the amount of data in our life is growing explosively. Meanwhile, the dimension and intrinsic structure of data become higher and more complex, respectively. Conventional clustering techniques, such as k-means, do not efficiently deal with the data with a union of subspaces. Subspace clustering is one of popular clustering methods which clusters the samples belonging to the same subspace into the same cluster. Furthermore, Subspace clustering has been proved effectively and applied in many applications, such as face clustering [1], motion segmentation [2], image segmentation [3], etc.

© Springer International Publishing AG 2016
H. Yin et al. (Eds.): IDEAL 2016, LNCS 9937, pp. 321–331, 2016.
DOI: 10.1007/978-3-319-46257-8_35

Recently, subspace clustering has been extended to the multi-view domain for machine learning tasks. Multi-view data can provide consistent, diversiform and relatively complementary information from different views, and thus it can efficiently improve the performance than the single-view in the same learning task [4]. In fact, multi-view data exists widely in numerous real-world applications, such as a video can be described by it's image, text and voice information. In this paper, aiming at the multi-view subspace clustering problem, we propose a novel method. In particular, we introduce a global low-rank affinity matrix which is shared in all views. It has been proved that a low-rank model may output an approximate block diagonal affinity matrix and efficiently improves the clustering performance in [5]. In our method, we use a sparse term to fit the difference between the affinity matrix in each view (local affinity matrix) and the global affinity matrix to explore the consistent information from different views. Experiments results demonstrate that our method improve the clustering performance against with the state-of-the-art methods on two real face datasets, Extended YaleB and ORL database.

The remainder of the paper is organized as follows. In Sect. 2, we provide the mainly related work about multi-view subspace clustering. Section 3 presents our multi-view subspace clustering method and the relevant optimization procedure in detail. The experimental results evaluated on two real-world datasets are reported in Sect. 4. Finally, Sect. 5 gives some concluding remarks.

2   Related Work

Subspace clustering has been concerned widely in numerous research fields, such as computer vision, pattern recognition and machine learning. Existing subspace clustering methods can be divided into four categories [6], which include algebraic methods, iterative methods, statistical methods and spectral clustering-based methods. Among of these methods, the first three methods need to obtain the dimension of each subspace. However, computing the dimension of the subspace is a very complex problem. In spectral clustering-based methods, the dimension of the subspace does not need to be calculated. It only requires an assumption that any one sample can represented by the linear combination of these samples in the same subspace. Therefore, subspace clustering method based on spectral clustering has gained a lot of attention in recent years. These methods generally include two steps. First, according to the above-mentioned assumption, an affinity matrix (similarity matrix) can be computed by subspace clustering methods. Second, the result of clustering is obtained by using the conventional spectral clustering method for the affinity matrix.

Multi-view subspace clustering can effectively exploit the information between different views and improve the clustering performance. Cheng *et al.* proposed a collaborative subspace clustering framework for image segmentation in [3]. In this work, multiple different types of image features were extracted as the multi-view data. For each view, an affinity matrix was learned and then the matrix was expanded as a vector. Finally, these vectors were combined into a new matrix with

$\ell_{2,1}$-norm constraint which made each affinity matrix has more zero elements and ensured the corresponding element consistent. Cao *et al.* proposed a diversity-induced multi-view subspace clustering (DiMSC) method which aimed at exploring the complementary information among multi-view features in [7]. DiMSC extended the existing subspace clustering into the multi-view domain, and utilized the Hilbert Schmidt Independence Criterion (HSIC) as a diversity term to explore the complementarity of multi-view representations. Therefore, it can make the learned affinity matrices in different views complementary. These above multi-view subspace clustering methods were based on the idea that the final affinity matrix can be accumulated by these similarity matrices in each view. Although these methods can enhance the connection between different views, they may destroy the structure of the affinity matrix, such as the block diagonal property. Further, Gao *et al.* proposed a unified optimization framework which can simultaneously learn the local affinity matrix and the spectral clustering result in [8]. However, it required the strict initialization, which made it infeasible to obtain the optimal solution.

3   The Proposed Method

3.1   Notations

In this paper, multi-view data is denoted by $\{X^v\}_{v=1}^s$, $X^v \in \mathbb{R}^{d^v \times m}$, where s and m are the number of views and samples, respectively. d^v is the dimension of sample in view v. $Z^v \in \mathbb{R}^{m \times m}$ is the affinity matrix and $E^v \in \mathbb{R}^{d^v \times m}$ is the loss in view v. Suppose that R is a matrix, $\|R\|_* = trace(\sqrt{R^T R}) = \sum_i \sigma_i$, is the nuclear norm, where σ_i is a singular value of matrix R. $\|R\|_{2,1}$ is the sum of columns R_js with the ℓ_2 norm, which can be formulated as $\|R\|_{2,1} = \sum_j \|R_j\|_2$. $\|R\|_F$ is Frobenius norm or the Hilbert-Schmidt norm, $\|R\|_F = \sqrt{\sum_i \sum_j |R_{i,j}|^2} = \sqrt{trace(R^T R)} = \sqrt{\sum_i \sigma_i^2}$. $\|R\|_1 = \sum_i \sum_j |R_{i,j}|$, is the sum of absolute values of all elements.

3.2   Formulation

To solve the multi-view subspace clustering task, we propose a novel multi-view subspace clustering method via a global low-rank affinity matrix. In particular, we introduce a global affinity matrix, and use a sparse term to force the affinity matrix of each view adequately close to the global affinity matrix for multi-view subspace clustering. Our method explores the consistent information from different views and guarantees the global affinity matrix for segmentation is low-rank. Figure 1 shows the framework of our method. The global affinity matrix guarantees the local affinity matrices have a high correlation, thus the information in the different view can be mutually complemented. In subspace clustering, the Low-Rank Representation (LRR) is an effective method and has been famously applied for single view clustering [5]. Meanwhile based on the assumption that

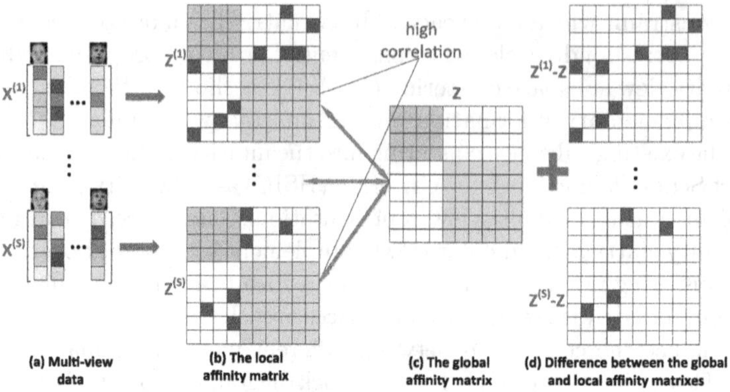

(a) Multi-view data   (b) The local affinity matrix   (c) The global affinity matrix   (d) Difference between the global and local affinity matrixes

Fig. 1. The framework of our method. With the multi-view input (a), our method dependently learns the affinity matrix (b) of each view. The global affinity matrix in (c) is used to explore the global consistent information from all single views. (d) is the difference between the global affinity matrix and affinity matrices of different views.

the underlying data should be low-rank and the noise in the given data should be sparse, Robust Principal Component Analysis (RPCA) can effectively restores the underlying low-rank data from the given data. In the subspace clustering problem, a desired matrix Z should be block diagonal. It has been proved that a low-rank affinity matrix can satisfy the block diagonal characteristic [5]. In this paper, we assume that there is sparse error between the global affinity matrix and local affinity matrices. Inspired by LRR and RPCA, our multi-view subspace clustering method can be mathematically formulated as follows:

$$\min_{Z,Z^v,E^v} \sum_{v=1}^{s} \|E^v\|_{2,1} + \sum_{v=1}^{s} \lambda \|Z^v - Z\|_1 + \alpha \|Z\|_*$$
$$s.t \ X^v = X^v Z^v + E^v \tag{1}$$

where the first term is the difference intra-view, the second term is the difference between the local and global similarity matrix. The final term denotes the shared affinity matrix with the low-rank constraint. In order to facilitate the variable optimization, we introduce auxiliary matrices including $F^v \in \mathbb{R}^{m \times m}$ and $Q \in \mathbb{R}^{m \times m}$ in the objective function with additional equality constraints. Therefore, the original problem in Eq. (1) can be reformed as:

$$\min_{Z,Z^v,E^v,F^v,Q} \sum_{v=1}^{s} (\|E^v\|_{2,1} + \lambda \|F^v\|_1) + \alpha \|Q\|_*$$
$$s.t \ X^v = X^v Z^v + E^v,$$
$$F^v = Z^v - Z, Q = Z \tag{2}$$

Basically, the augmented Lagrangian methods for the clustering problem in Eq. (2) can be mathematically formulated as:

$$\min_{Z,Z^v,E^v,F^v,Q} \sum_{v=1}^{s} (\|E^v\|_{2,1} + \lambda \|F^v\|_1) + \alpha \|Q\|_*$$
$$+ \sum_{v=1}^{s} (\langle V^v, E^v - X^v + X^v Z^v \rangle + \frac{\mu}{2} \|E^v - X^v + X^v Z^v\|_F^2$$
$$+ \langle W^v, F^v - Z^v + Z \rangle + \frac{\mu}{2} \|F^v - Z^v + Z\|_F^2)$$
$$+ \langle P, Q - Z \rangle + \frac{\mu}{2} \|Q - Z\|_F^2 \tag{3}$$

where $\mu > 0$ is a penalty parameter, and V^v, W^v and $P \in \mathbb{R}^{m \times m}$ are Lagrange multipliers.

3.3 Optimization Procedures of Our Method

In this subsection, we adopt an efficient augmented Lagrange multipliers (ALM) method [9], to solve the multi-variable optimization in our proposed clustering method. Equation (3) is a non-convex function in terms of multi-variable, however, the objective function with respect to only one variable is a convex function, which can be effectively solved by the conventional optimization methods. In the following, a more detailed description of optimization procedures will be presented.

Update E^v: It optimizes E^v while fixing all the other variables, and thus the optimization problem in Eq. (3) with respect to E^v can be rewritten as follows:

$$\min_{E^v} \sum_{v=1}^{s} \|E^v\|_{2,1} + \sum_{v=1}^{s} (\langle V^v, E^v - X^v + X^v Z^v \rangle + \frac{\mu}{2} \|E^v - X^v + X^v Z^v\|_F^2) \tag{4}$$

For simplification, the linear and quadratic terms can be combined by scaling V^v in Eq. (4). Thus, the objective with respect to E^v can be rewritten in a slightly scaled form as follows:

$$\min_{E^v} \frac{1}{\mu} \|E^v\|_{2,1} + \frac{1}{2} \|E^v - M^v\|_F^2,$$
$$M^v = X^v - X^v Z^v - V^v / \mu \tag{5}$$

The scaled form in Eq. (5) is clearly equivalent to the augmented Lagrangian in Eq. (4). Furthermore, the augmented Lagrangian is more convenient to solve the original problems, and it has a closed solution [10].

Update Z^v: Fixing Z, Q, E^v and F^v, the updating of Z^v can be denoted as:

$$\min_{Z^v} \sum_{v=1}^{s} (\langle V^v, E^v - X^v + X^v Z^v \rangle + \frac{\mu}{2} \|E^v - X^v + X^v Z^v\|_F^2$$
$$+ \langle W^v, F^v - Z^v + Z \rangle + \frac{\mu}{2} \|F^v - Z^v + Z\|_F^2) \tag{6}$$

Note that computing the derivative of Eq. (6) with respect to Z^v and setting it to zero, we can obtain

$$Z^v = (\mu I + \mu X^{v^T} X^v)^{-1} (W^v + \mu(F^v + Z) - X^{v^T} V^v - \mu X^{v^T} (E^v - X^v)) \quad (7)$$

Update Z: Similar with the updating of Z^v, the variable Z can be updated as follows:

$$Z = (\mu s I + \mu I)^{-1} (P + \mu Q - \sum_{v=1}^{s} (W^v + \mu(F^v - Z^v))) \quad (8)$$

Update F^v: It optimizes F^v while fixing other variables, and thus the optimization problem of F^v can be rewritten as follows:

$$\min_{F^v} \sum_{v=1}^{s} \lambda \|F^v\|_1 + \sum_{v=1}^{s} (\langle W^v, F^v - Z^v + Z \rangle + \frac{\mu}{2} \|F^v - Z^v + Z\|_F^2) \quad (9)$$

By the same way in Eq. (4) the objective with respect to F^v can be rewritten in a slightly scaled form as follows:

$$\min_{F^v} \frac{\lambda}{\mu} \|F^v\|_1 + +\frac{1}{2} \|F^v - N^v\|_F^2, \quad (10)$$
$$N^v = Z^v - Z - W^v/\mu$$

It has been proved that the objective function of F^v has a closed solution [9].

Update Q: By fixing all variables except to Q, the updating process of Q can be mathematically formulated as:

$$\min_{Q} \alpha \|Q\|_* + \langle P, Q - Z \rangle + \frac{\mu}{2} \|Q - Z\|_F^2 \quad (11)$$

Similar with the updating of F^v, the variable Q can be simply rewritten as:

$$\min_{Q} \frac{\lambda}{\mu} \|Q\|_* + \frac{1}{2} \|Q - (Z - P/\mu)\|_F^2 \quad (12)$$

Similar to Eq. (10), it also has a closed solution [9].

In this paper, the ALM iterations give the following convergence guarantee. For any $\mu > 0$, the stop criteria of our algorithm are listed as:

$$F^v - Z^v + Z \to 0, Q - Z \to 0, E^v - X^v - X^v Z^v \to 0. \quad (13)$$

These stop criteria meet these conditions to guarantee the objective function in problem (2) converges to the optimal objective function of problem (1). Typically, the specific process of our method is described in Algorithm 1.

Algorithm 1. Our method for multi-view subspace clustering by the inexact ALM

Input: Data matrices $\{X^v\}_{v=1}^s$, parameters λ and α.
Initialize: $Z = Q = Z^v = F^v = W^v = P = 0$, X, and $E^v = V^v = 0$, $\mu = 10^{-6}$, $\rho = 1.1$, $max_\mu = 10^{10}$.
while not converge **do**

1. Fix the other variables, update E^v by solving Eq.(5).
2. Z^v can be updated by the following formula,

$$Z^v = (\mu I + \mu X^{v^T} X^v)^{-1}(W^v + \mu(F^v + Z) - X^{v^T} V^v - \mu X^{v^T}(E^v - X^v))$$

3. The updating of Z is donated as,

$$Z = (\mu s I + \mu I)^{-1}(P + \mu Q - \sum_{v=1}^s (W^v + \mu(F^v - Z^v)))$$

4. Fix the others, F^v can be optimized by Eq.(10).
5. Fix the others and update Q by solving Eq.(12).
6. Update the multipliers

$$W^v = W^v + \mu(F^v - Z^v + Z)$$

$$V^v = V^v + \mu(X^v - X^v Z^v - E^v),$$

$$P = P + \mu(Q - Z)$$

7. Update the parameter μ by $\mu = \min(\rho\mu, max_\mu)$.
8. Check the stop criteria,

$$F^v - Z^v + Z \to 0, Q - Z \to 0, E^v - X^v - X^v Z^v \to 0.$$

end while
Output: Z.

---

4   Experimental Results

4.1   DataSets and Experimental Settings

The two public real face datasets are adopted for clustering performance evaluation in this paper:

Extended Yale B[1]: The dataset consists of 192×168 pixel cropped face images of 38 subjects, where there are 64 frontal face images for each subject acquired under various lighting conditions.

ORL[2]: The dataset contains 92×112 pixel images of 40 distinct subjects, and each subject has 10 different images. For some subjects, the images are taken at different times, varying the lighting, facial expressions and facial details. All the

---

[1] http://vision.ucsd.edu/~leekc/ExtYaleDatabase/ExtYaleB.html.
[2] http://www.cl.cam.ac.uk/research/dtg/attarchive/facedatabase.html.

images are taken against a dark homogeneous background with the subjects in an upright, frontal position.

In all experiments, two different features extracted by the real and imaginary component of the Gabor filter are considered as two different views. For the Extended YaleB and ORL datasets, we choose 5 and 10 subjects as the experimental datasets, respectively. Moreover, we use PCA to reduce the dimension of original feature with reserving 99 % energy in these datasets. To evaluate the clustering performance, we adopt **normalized mutual information (NMI), accuracy (ACC), F-score, Precision and Recall** as evaluation metrics. For all these metrics, the higher value indicates a better clustering quality. To improve the reliability of the experimental results, we run all methods 10 times and report the average performance and standard derivation in our experiments. For parameter settings, we choose the optimal parameters of the corresponding method for fair comparison.

4.2   Comparison with Other Clustering Methods

To validate the superiority of our method, we compare it with six methods, i.e., **K-means, SSC** [11], **LRR** [5], **LSR** [12], **MSC** [8] and **MLAP** [3] in the experiments. In particular, SSC, LRR and LSR are single view subspace clustering methods, while MSC and MLAP are multi-view subspace clustering methods. In these methods with the exception of MSC, the corresponding affinity matrix Z in each method is generated, and normalized spectral clustering is used to cut the affinity matrix. MSC combines the spectral clustering and affinity matrix learning into a unified optimization framework which can simultaneously learn the affinity matrix and the spectral clustering, and then it directly outputs the result of spectral clustering. In the experimental report, *_odd and *_even mean that the single view clustering methods are adopted for dealing with the imaginary and real datasets extracted from the experimental datasets, respectively.

The results of our method comparing with these baseline methods on dataset Extended YaleB are shown in Table 1. In these methods, the highest mean on the evaluation metrics are in bold. Table 1 shows that our proposed method can obtain the best clustering results than the others on the Extended YaleB dataset. Therefore, our method is suitable to solve the subspace clustering problem. The traditional clustering method known as K-means has a big gap in clustering results. The main reason is that face images varying with illumination contain multiple subspace structures, and thus K-means is infeasible for this kind data in real practice. Therefore, in the specific problem, the subspace clustering method is more effective than the clustering method based on the traditional distance metric.

Furthermore, comparing all the clustering methods on dataset ORL in Table 2, our method performs better than other methods in terms of almost all evaluation metrics except Recall. According to the mean of standard derivations of experiment results on the two datasets in Tables 1 and 2, our proposed method has a more stable performance than others. Although MSC is a multiview subspace clustering method, it's clustering performance is worst than the

Table 1. Clustering results of comparative methods on Extended Yale B dataset with 5 subjects.

	ACC	NMI	Precision	Recall	F-score
k-means_odd	0.302(±0.038)	0.144(±0.072)	0.254(±0.035)	0.274(±0.029)	0.264(±0.032)
k-means_even	0.392(±0.044)	0.254(±0.056)	0.300(±0.027)	0.333(±0.033)	0.315(±0.029)
LRR_odd	0.69(±0.081)	0.536(±0.074)	0.506(±0.056)	0.587(±0.073)	0.543(±0.062)
LRR_even	0.768(±0.059)	0.603(±0.067)	0.574(±0.061)	0.649(±0.067)	0.609(±0.062)
LSR_odd	0.6(±0.051)	0.475(±0.051)	0.467(±0.049)	0.527(±0.052)	0.495(±0.048)
LSR_even	0.601(±0.06)	0.472(±0.054)	0.462(±0.049)	0.524(±0.053)	0.491(±0.049)
SSC_odd	0.722(±0.045)	0.567(±0.056)	0.439(±0.066)	0.654(±0.034)	0.524(±0.057)
SSC_even	0.746(±0.038)	0.592(±0.044)	0.468(±0.053)	0.661(±0.02)	0.547(±0.043)
MSC	0.565(±0.104)	0.395(±0.118)	0.306(±0.062)	0.666(±0.034)	0.414(±0.057)
MLAP	0.729(±0.094)	0.577(±0.093)	0.548(±0.09)	0.625(±0.092)	0.583(±0.09)
Our method	**0.836**(±0.032)	**0.708**(±0.041)	**0.636**(±0.069)	**0.746**(±0.036)	**0.686**(±0.055)

Table 2. Clustering results of comparative methods on ORL-face dataset with 10 subjects.

	ACC	NMI	Precision	Recall	F-score
LRR_odd	0.661(±0.089)	0.683(±0.076)	0.439(±0.097)	0.61(±0.091)	0.509(±0.095)
LRR_even	0.516(±0.066)	0.553(±0.073)	0.29(±0.068)	0.566(±0.064)	0.381(±0.071)
LSR_odd	0.62(±0.066)	0.63(±0.069)	0.35(±0.071)	0.621(±0.096)	0.445(±0.079)
LSR_even	0.491(±0.071)	0.501(±0.065)	0.462(±0.059)	0.518(±0.078)	0.341(±0.06)
SSC_odd	0.518(±0.107)	0.588(±0.1)	0.312(±0.096)	0.674(±0.098)	0.42(±0.098)
SSC_even	0.351(±0.055)	0.342(±0.082)	0.173(±0.046)	0.371(±0.092)	0.234(±0.056)
MSC	0.329(±0.083)	0.352(±0.111)	0.146(±0.049)	**0.757**(±0.049)	0.24(±0.059)
MLAP	0.668(±0.034)	0.693(±0.04)	0.459(±0.039)	0.638(±0.075)	0.533(±0.045)
Our method	**0.685**(±0.034)	**0.707**(±0.046)	**0.481**(±0.053)	0.65 (±0.079)	**0.552**(±0.059)

other comparative methods. The reason may be that MSC is sensitive for initialization. In all the experiments, the initialization of variables in MSC are to be set as the optimal clustering results of other methods. Contrary to MSC, our method is insensitive for initialization and all the variables can be initialized arbitrarily.

In this experiment, we also validate the convergence speed of our proposed method on Extended YaleB and ORL datasets. The convergence curve of the objective function is showed as Fig. 2. With the increasing of the number of iterations, the objective function value gradually reduces. The convergence curve on Extended YaleB in Fig. 2 (a) gradually decreases until converges, while the convergence curve on ORL in Fig. 2 (b) fluctuates sharply in the iterations and it also converges finally. The main reason may be that the experimental dataset Extended YaleB has more samples than ORL, and thus it can obtain a more stable result in each iteration. In our method, the stop criteria guarantee that the objective function can converge to the optimal solution.

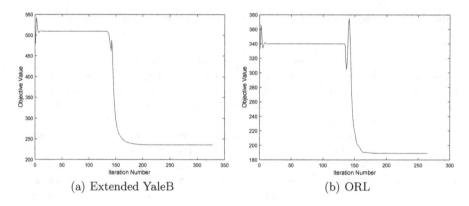

(a) Extended YaleB   (b) ORL

Fig. 2. The convergence curves of our method on the two datasets.

5   Conclusion

In this paper, to effectively solve the multi-view subspace clustering problem, we propose a novel subspace clustering method via a global low-rank affinity matrix. Inspired by LRR and RPCA, we introduce an global affinity matrix with the low-rank constraint, and use a sparse term to fit the difference between the global and local affinity matrices in our method. Then, the clustering problem can be adequately solved by the augmented Lagrange multipliers (ALM) optimization method. In the experiments, two public real face image datasets are utilized for performance evaluation. The experimental results show that our method can improve the clustering performance compared with the state-of-the-art methods. Therefore, our proposed method is suitable to deal with the multi-view data for clustering.

Acknowledgements. The work was supported by NSFC (U1435214, 61305068), Jiangsu Nature Science Foundation (JSNSF) (BK20130581), and the Open Project Program of State Key Laboratory for Novel Software Technology (KFKT2016B16).

References

1. Zhou, C., Zhang, C., Fu, H., Wang, R., Cao, X.: Multi-cue augmented face clustering. In: Proceedings of the 23rd Annual ACM Conference on Multimedia Conference, MM 2015, Brisbane, Australia, 26–30 October 2015, pp. 1095–1098 (2015)
2. Elhamifar, E., Vidal, R.: Sparse subspace clustering. In: 2009 IEEE Computer Society Conference on Computer Vision and Pattern Recognition (CVPR 2009), 20–25 June 2009, Miami, Florida, USA, pp. 2790–2797 (2009)
3. Cheng, B., Liu, G., Wang, J., Huang, Z., Yan, S.: Multi-task low-rank affinity pursuit for image segmentation. In: IEEE International Conference on Computer Vision, ICCV 2011, Barcelona, Spain, 6–13 November 2011, pp. 2439–2446 (2011)
4. Xu, C., Tao, D., Xu, C.: A survey on multi-view learning (2013). arXiv preprint. arXiv:1304.5634

5. Liu, G., Lin, Z., Yan, S., Sun, J., Yu, Y., Ma, Y.: Robust recovery of subspace structures by low-rank representation. IEEE Trans. Pattern Anal. Mach. Intell. **35**(1), 171–184 (2013)
6. Vidal, R.: A tutorial on subspace clustering. IEEE Signal Process. Mag. **28**(2), 52–68 (2010)
7. Cao, X., Zhang, C., Fu, H., Liu, S., Zhang, H.: Diversity-induced multi-view subspace clustering. In: IEEE Conference on Computer Vision and Pattern Recognition, CVPR 2015, Boston, MA, USA, 7–12 June 2015, pp. 586–594 (2015)
8. Gao, H., Nie, F., Li, X., Huang, H.: Multi-view subspace clustering. In: 2015 IEEE International Conference on Computer Vision, ICCV 2015, Santiago, Chile, 7–13 December 2015, pp. 4238–4246 (2015)
9. Lin, Z., Chen, M., Ma, Y.: The augmented lagrange multiplier method for exact recovery of corrupted low-rank matrices (2010). arXiv preprint. arXiv:1009.5055
10. Liu, G., Lin, Z., Yu, Y.: Robust subspace segmentation by low-rank representation. In: Proceedings of the 27th International Conference on Machine Learning (ICML-10), 21–24 June 2010, Haifa, Israel, pp. 663–670 (2010)
11. Elhamifar, E., Vidal, R.: Sparse subspace clustering: algorithm, theory, and applications. IEEE Trans. Pattern Anal. Mach. Intell. **35**(11), 2765–2781 (2013)
12. Lu, C., Min, H., Zhao, Z., Zhu, L., Huang, D., Yan, S.: Robust and efficient subspace segmentation via least squares regression. In: Computer Vision - ECCV 2012–12th European Conference on Computer Vision, Florence, Italy, 7–13 October 2012, Part VII, pp. 347–360 (2012)

An Integrated Method for Road Network Centerline Detection from Multispectral Imagery

Yuefan Du, Jie Li, and Ying Wang$^{(\boxtimes)}$

VIPSL Lab, School of Electronic Engineering, Xidian University, Xi'an, China
yuefandu@stu.xidian.edu.cn,
leejie@mail.xidian.edu.cn, yingwang@xidian.edu.cn

Abstract. Automatic road network detection from multispectral imagery is an effective and economic way to obtain the road and related information. This paper presents an integrated method to extract road network centerline from multispectral imagery. It includes four main steps. First, support vector machine (SVM) is used to analyze spectral information to classify the imagery into road and non-road regions. Then, shape feature, morphological top-hat transform and multiple directional filters are cascaded to reduce the misclassification. Based on these procedures, morphological thinning algorithm and Hough transform are introduced to detect the road centerline which will be regarded as road primitives. Finally, a novel road tracking method is developed to refine and improve the whole road network. The proposed method is verified on a multispectral images acquired from SPOT-6 satellite and QuickBird data sets.

Keywords: Multispectral imagery · Road network detection · Spectral classification · Multiple directional filters · Kernel-based density estimation · Road tracking

1 Introduction

Automatic road network detection from satellite remote sensing image is a challenging and difficult problem. Road as a man-made object whose information is of importance in cartography, urban planning, traffic management, and industrial development. Since manual road information extraction from satellite image is too boring and time consuming, research for developing automatic methods has been of much interest in recent years. An overview of road detection methods in this area can be seen in [1–3] and a test of automatic road detection methods results can be seen in [4]. In the literature, there are several methods proposed to deal with the detection of roads from satellite images [5–10].

The segmentation-based method (often using a classifier) first segment the imagery into regions, followed by a rule to refine extracted road networks. Shi and Zhu [5] proposed a detection model that first segmented imagery into a binary map by simple thresholding and then followed by the line segment match rule and morphological processing. Song and Civco [11] used the support vector machine (SVM) to classify the

© Springer International Publishing AG 2016
H. Yin et al. (Eds.): IDEAL 2016, LNCS 9937, pp. 332–341, 2016.
DOI: 10.1007/978-3-319-46257-8_36

imagery and introduced a two step-based method to detect road network. The straight-line-based method often uses line, edge and ridge detectors to extract potential road points. Then road points are connected to generate road network. Cem and Beril [12] first using canny edge detectors to extract edge pixels as primitives. The mathematical morphology methods often employ segmentation method to discriminate the road regions from their surroundings. Katartzis et al. [13] applied local analysis with morphological filters and line tracking to detect road network in their work. Conditional morphological techniques are used to significantly improve the segmentation results in [14]. The active contour methods intend to minimum energy function which combines intensity and gradient information of images to extract the edge of the road. Mayer et al. [15] using snakes method to extract road from aerial images based on the multi-scale detection of roads in combination with geometry constrained edge extraction. A family of quadratic snakes [16] which are able to split, merge, and disappear as necessary, is used to extract disconnected road networks and enclosed regions.

However, observed in high-resolution multispectral satellite images, it is difficult to model all the variants in different types of phenomena (spectral reflectance, shape, contrast, shadow, and occlusion) observed in roads. Therefore, how to incorporate them into a single processing module is a very complex problem. Most of the existing road extraction methods for multispectral imagery rely on an automated and reliable classification of road surfaces [11, 17]. Unfortunately, because of other objects similar to road existing, the classification accuracy of roads is far from satisfactory whether supervised or unsupervised classification method is used.

In this paper, we proposed a novel approach to detect road network centerlines from multispectral image. We first use SVM to implement spectral classification, which can segments image into road and non-road regions. Top-hat transform and multiple directional filters are then employed to remove the false road pixels. After that, the thinning algorithm and Hough transform are combined to obtain more accuracy road centerline as road primitives. Finally, we develop a novel road tracking method which introduces the searching window idea to refine and improve the whole road network.

The remainder of this paper is organized as follows: Sect. 2 describes the proposed methodology in detail. Experimental results are given in Sect. 3. Finally, conclusions are drawn in Sect. 4.

2   Methodology

The proposed road network centerline detection method is presented in this section, whose framework is shown in Fig. 1. The details of the aforementioned steps will be given in detail in the following.

2.1   Spectral Classification

Employing the spectral information provided by multispectral image, the spectral classification step is to segment the imagery into two groups: road group and non-road group. The road group will be used as the initial road segments. The digital number

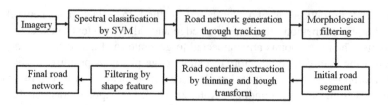

Fig. 1. Flowchart of the proposed method

(DN) vectors in multispectral image is regarded as a variable v, $v \in R^d$, where d denotes the number of bands/channels in multispectral image. In this paper, we use SVM to classify the multispectral image, and the result is defined as:

$$S(v) = \begin{cases} 1, & \text{if } v \text{ is classified as road} \\ 0, & \text{otherwise} \end{cases} \tag{1}$$

The original multispectral image becomes a binary image $I(x, y)$ after SVM spectral classification.

2.2   Road Segment Extraction

Due to the phenomenon that different objects reflect the similar spectral features in multispectral image, other land-cover types, such as buildings, parking lots, and bare soil, tend to be misclassified as roads. To overcome this problem, some typical road properties should be studied and used to remove the misclassified groups.

Generally, the misclassified pixels can be divided into groups that connected and disconnect with road regions.

As we knew, roads have unique geometric property, which is quite different from other land-cover features. Roads are always elongated with small changes of curvature. Hence, shape information can also be used to filter false road segments. In this paper, second-order moments are used to measure road shape feature. We first use connected component labeling [18] to extract the disjoint components from classification result. Then the major and minor axis lengths of each component are computed using normalized second central moments [19] which are given as follows:

$$\mu_{20} = M_{20} - \bar{x}M_{10} \quad \mu_{02} = M_{02} - \bar{y}M_{01} \tag{2}$$

where

$$\bar{x} = \frac{M_{10}}{M_{00}} \quad \bar{y} = \frac{M_{01}}{M_{00}} \quad M_{pq} = \sum_x \sum_y x^i y^i I(x, y) \tag{3}$$

The ratio of major axis length to minor one of each component is computed as:

$$E = \mu_{20}/\mu_{02} \tag{4}$$

Components with E value less than the threshold T_E are usually considered as non-road structures and hence removed. The steps of the algorithm, depicting this filtering module, are given in Algorithm 1. We denoted the image filtered with shape feature by $I_{shape}(x, y)$.

Algorithm 1. Steps of filtering roads by shape feature

1. Label the connected components.
2. Compute eccentricity (E) of each connected component.
3. For each component
    if ($E \leq T_E$) then
        delete that component
    end if

Most of misclassified region disconnected with roads will be removed to some extent by shape feature. However, false positive ones, especially those connected with roads, still exist, and further refinement procedure is necessary to improve the reliability of road extraction.

In fact, roads possess a limited maximum width w in satellite image, while, other spectrally similar objects, such as parking lots, buildings, are usually wider than roads. As we knew, morphological opening operation can weed out the convex shape smaller than structure element SE_w and cut slender connection between areas in binary image. Therefore, the region wider than limited maximum width w can be preserved. After a morphological top-hat transform, that convex shape larger than structure element will be removed through top-hat transform. The top-hat transform can be defined as follows:

$$I_{top} = I_{shape} - (I_{shape} \ominus SE_w) \oplus SE_w \tag{5}$$

Then the multiple directional filters are used to remove the false positive road regions which closed connected with roads. At first, we construct the multiple directional linear structure element SE_{L,α_i} as follows:

$$SE_{L,\alpha_i} = \begin{cases} y_i = x_i \tan(\alpha_i) & |\alpha_i| < 45° \\ x_i = y_i \cot(\alpha_i) & 45° < |\alpha_i| < 90° \end{cases} \tag{6}$$

$$\begin{aligned} x_i &= 0, \pm 1, \ldots \pm (L-1)cos(\alpha_i)/2 \\ y_i &= 0, \pm 1, \ldots \pm (L-1)sin(\alpha_i)/2 \end{aligned} \tag{7}$$

$$\alpha_i = i \times 3° \quad i \in [-30, 30] \tag{8}$$

where i denotes the ith linear structure element, α_i and L denote the direction angle and length of the ith linear structure element respectively. Then the morphological opening operation with structure element as SE_{L,α_i} is used to image. The multiple directional filters can be then defined as follows:

$$I_{multi-line} = \bigcup_{i=-30}^{30} I_{top} \circ SE_{L,\alpha_i} \tag{9}$$

Overall, the misclassified regions which closed connected with road regions could be removed through morphological top-hat transform and multiple directional filters.

2.3   Road Centerline Extraction

After road segment extraction, morphological thinning operation [20] is a commonly used method to extract road centerlines. By using the morphological top-hat transform and multiple directional filters, misclassified regions can be removed, but some real road segments are improperly removed simultaneously. Hough transform [21] is also a popular method to extract road centerlines, which can detect straight line overcome the noise. Hence, it is used to link the cutting off linear structure of road segments. Then the final road centerlines are obtained by combining thinning results with hough transform detected results through logical "OR" operation. This fusion operation can make up part of lost road owing to the road segment extraction process and make the detection results more accurate. The fusion results obey the following rule:

$$I_{fusion}(x, y) = \begin{cases} 1, & if\ I_{thin}(x,y) = 1\ or\ I_{hough}(x,y) = 1 \\ 0, & otherwise \end{cases} \tag{10}$$

2.4   Road Centerline Tracking

Although the basic road segments can be gathered through the above processing, sometimes, the road is still incomplete due to occlusion and road information loss. Therefore, road centerline extracted in the previous section need further tracking step. The tracking procedure can be divided into three stages: sampling, decision and fusion. The tracking process will stop when the decision stage can not found any road pixels or the number of iteration reach a prefixed value. Three stages will be detailed in following.

Sampling Stage: In this stage, connected component analysis is used to find the endpoints (having just one neighbor) of each road segment. The angle of road segment endpoint's tangent line θ can denote the direction of roads. The tangent line of endpoint can be represented by the connecting line between the endpoint and the point 5 pixels away approximately. It is first used in [14] to estimate the road centerlines according to road edge pixels as:

$$N(x,y) = \frac{1}{\kappa}\exp(-(\frac{x^2}{a^2} + \frac{y^2}{b^2})) \quad a > b \tag{11}$$

where κ is the normalizing constant, the scale of a and b control the size of searching window for road endpoint. Now the endpoints of road segments can denote by $(x_i, y_i, \theta_i, a, b)$, where (x_i, y_i) are the location of endpoint and θ_i denotes the tangential direction of endpoint. Then the spatial probability density function of ith endpoint become

$$N_i(x,y) = \frac{1}{\kappa}\exp(-(\frac{(x'\cos(\theta_i) - y'\sin(\theta_i))^2}{a^2} + \frac{(y'\cos(\theta_i) + x'\sin(\theta_i))^2}{b^2})) \quad a > b \tag{12}$$

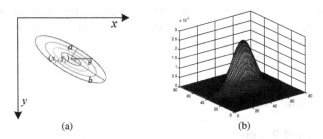

(a)   (b)

Fig. 2. The spatial probability density function. (a) The contour line of elongated Gaussian kernel function. (b) The p. d. f. of elongated Gaussian kernel function.

where

$$x' = x - x_i \quad y' = y - y_i \tag{13}$$

The spatial *p. d. f.* is presented in Fig. 2 intuitively. The red represents very high possibility, and blue represents low possibility.

Decision Stage: The pixels located in searching window will be judged whether they belong to roads or not. Employing the mind of maximum likelihood classification (MLC) in remote sensing, we obtain the posterior probability p_{road} that a pixel belongs to road. The final probability for one pixel belongs to road can be expressed as:

$$p_{road} = N(v \mid \mu, \sum) \tag{14}$$

$$p = N_i(x,y) * p_{road} \tag{15}$$

where v denotes the DN vector, μ and $\sum$ denote the mean vector and covariance matrix of road pixels. A probability threshold T_p is set to decide whether the pixel belongs to road or not.

Fusion Stage: The road centerline in sampling stage will be combined with decision result by using logical "OR" operation and the thinning algorithm is used to extract the centerline of fusion result.

The tracking process will be repeated when the decision stage can found road pixels or the number of iteration does not reach the prefixed value. The procedure of tracking process is given in Algorithm 2.

Algorithm 2. Procedure of road centerline tracking

Repeat
  (1) Determine the searching window and searching region for each road endpoint.
  (2) Judge pixels located in searching window whether belong to road or not according to Eq. (11)-(14).
  (3) Combine the sampling image and decision image.
Until the decision image are all-zero image or the number of iteration is reached.

In sampling stage, the proposed algorithm combines the information of road's direction and the searching window idea which is popular in video target detection. As for decision stage, we make use of kernel density estimate and posterior probability to thoroughly analyze whether the pixels belong to road or not.

3 Experimental Analysis

3.1 Experimental Results

The first test image acquired by SPOT-6 satellite with size of 501×501 pixels is acquired on December 05, 2009, which covers Xianyang urban road network. The experimental results of this test image are presented in Fig. 3.

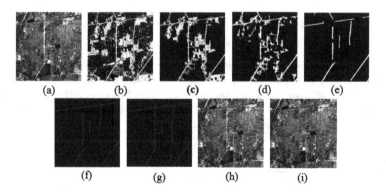

Fig. 3. (a) Original image. (b) Classification result. (c) Shape feature filtering result. (d) Morphological top-hat transform. (e) Multiple directional filtering result. (f) Thinning algorithm and Hough transform fusion result. (g) Final tracking result. (h) Road network and original image superposition. (i) The hand-drawn road reference map.

The second test image acquired by SPOT-6 with size of 501×501 pixels and acquired on December 05, 2009, which covers Xianyang rural road network. The experimental results of second test image are presented in Fig. 4.

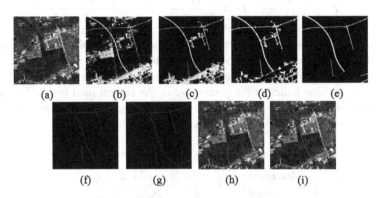

Fig. 4. (a) Original image. (b) Classification result. (c) Shape feature filtering result. (d) Morphological top-hat transform. (e) Multiple directional filtering result. (f) Thinning algorithm and Hough transform fusion result. (g) Final tracking result. (h) Road network and original image superposition. (i) The hand-drawn road reference map.

The third test image acquired by QuickBird remote sensed imageries is 512×512 pixels and available at http://www.cse.iitm.ac.in/~sdas/vplab/satellite.html. The experimental results are presented in Fig. 5.

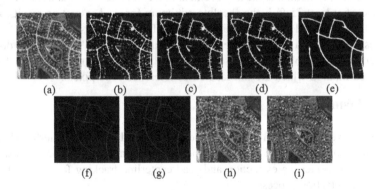

Fig. 5. (a) Original image. (b) Classification result. (c) Shape feature filtering result. (d) Morphological top-hat transform. (e) Multiple directional filtering result. (f) Thinning algorithm and hough transform fusion result. (g) Final tracking result. (h) Road network and original image superposition. (i) The hand-drawn road reference map.

3.2   Evaluation Criteria

In quantifying the performance values, we follow the criteria used by [21] as:

$$E_1 = \frac{TP}{TP+FN} \quad E_2 = \frac{TP}{TP+FP} \quad E_3 = \frac{TP}{TP+FP+FN} \qquad (16)$$

where E_1, E_2 and E_3 denote completeness, correctness, and quality, respectively, and TP, FN, and FP represent true positive, false negative, and false positive, respectively. The completeness value indicates the percentage of ground truth road pixels detected. The correctness value indicates the percentage of the correct road pixels. Finally, the quality value indicates the goodness of the result. Table 1 shows the performance of proposed method using these three criteria.

Table 1. Setting Word's margins.

Experiment		$E_1(\%)$	$E_2(\%)$	$E_3(\%)$
1	Initial detection	62.68	78.52	71.51
	After tracking	86.78	83.84	83.65
2	Initial detection	73.56	80.24	82.28
	After tracking	84.49	86.82	86.92
3	Initial detection	57.03	72.45	70.87
	After tracking	67.84	70.40	74.30

Mayer et al. [4] claim that the achievement of a completeness of at least 60 % and a correctness of at least 75 % is the absolute minimum for road extraction results, before they can be considered useful in practice. From Table 1, it is seen that the first and second test image could achieve these goals. The third test image's correctness does not achieve goals because of the high false positives. Some small narrow roads are detected by proposed method which are not labeled as reference road pixels.

4   Conclusion

In this paper, we proposed an integrated method for road network centerline detection from multispectral imagery. This method includes spectral classification, road segment extraction, road centerline extraction and road centerline tracking four main stages to complete the detection process.

The experimental results have been presented on six multispectral images to evaluate the proposed method, which shows that the proposed method achieves a good performance for road network detection.

References

1. Mena, J.B.: State of the art on automatic road extraction for GIS update: a novel classification. Pattern Recogn. Lett. **24**(16), 3037–3058 (2003)

2. Fortier, M.F.A., Ziou, D., Armenakis, C., Wang, S.: Survey of work on road extraction in aerial and satellite images. Technical report 241, University of Sherbrooke, Sherbrooke, QC, Canada, vol. 24, no. 16, pp. 3037–3058 (2003)
3. Das, S., Mirnalinee, T.T., Varghese, K.: Use of salient features for the design of a multistage framework to extract roads from high-resolution multispectral satellite images. IEEE Trans. Geosci. Remote Sens. **49**(10), 3906–3931 (2011)
4. Mayer, H., Hinz, S., Bacher, U., Baltsavias, E.: A test of automatic road extraction approaches. Int. Arch. Photogram. Remote Sens. Spatial Inf. Sci. **36**(3), 209–214 (2006)
5. Shi, W., Zhu, C.: The line segment match method for extracting road network from high-resolution satellite images. IEEE Trans. Geosci. Remote Sens. **40**(2), 511–514 (2002)
6. Zhu, C., Shi, W., Pesaresi, M., Liu, L., Chen, X., King, B.: The recognition of road network from high-resolution satellite remotely sensed data using image morphological characteristics. Int. J. Remote Sens. **26**(24), 5493–5508 (2005)
7. Hu, X., Tao, C.V.: A reliable and fast ribbon road detector using profile analysis and model based verification. Int. J. Remote Sens. **26**(5), 887–902 (2005)
8. Long, H., Zhao, Z.: Urban road extraction from high-resolution optical satellite images. Int. J. Remote Sens. **26**(22), 4907–4921 (2005)
9. Yang, J., Wang, R.S.: Classified road detection from satellite images based on perceptual organization. Int. J. Remote Sens. **28**(20), 4653–4669 (2007)
10. Huang, X., Zhang, L.: Road centerline extraction from high-resolution imagery based on multiscale structural features and support vector machines. Int. J. Remote Sens. **30**(8), 1977–1987 (2009)
11. Song, M.J., Civco, D.: Road extraction using SVM and image segmentation. Photogram. Eng. Remote Sens. **70**(12), 1365–1371 (2004)
12. Unsalan, C., Sirmacek, B.: Road network detection using probabilistic and graph theoretical methods. IEEE Trans. Geosci. Remote Sens. **50**(11), 4441–4453 (2012)
13. Guo, D., Weeks, A., Klee, H.: Robust approach for suburban road segmentation in high-resolution aerial images. Int. J. Remote Sens. **28**(2), 307–318 (2007)
14. Laptev, I., Mayer, H., Lindeberg, T., Eckstein, W., Steger, C., Baumgartner, A.: Automatic extraction of roads from aerial images based on scale space and snakes. Mach. Vis. Appl. **12**(1), 23–31 (2000)
15. Marikhu, R., Dailey, M.N., Makhanov, S., Honda, K.: A family of quadratic snakes for road extraction. In: Yagi, Y., Kang, S.B., Kweon, I.S., Zha, H. (eds.) ACCV 2007, Part I. LNCS, vol. 4843, pp. 85–94. Springer, Heidelberg (2007)
16. Doucette, P., Agouris, P., Stefanidis, A., Musavi, M.: Selforganized clustering for road extraction in classified imagery. ISPRS J. Photogram. Remote Sens. **55**(5/6), 347–358 (2001)
17. Haralick, R.M., Shapiro, L.G.: Computer and Robot Vision. Addison-Wesley, Reading (1992)
18. Mukundan, R., Ramakrishnan, K.R.: Moment Functions in Image Analysis Theory and Applications. World Scientific, Singapore (1998)
19. Soille, P.: Morphological Image Analysis: Principles and Applications. Springer, New York (1998)
20. Poullis, C., You, S.: Delineation and geometric modeling of road networks. ISPRS J. Photogram. Remote Sens. **65**(2), 165–181 (2010)
21. Peng, T., Jermyn, I.H., Prinet, V., Zerubia, J.: Incorporating generic and specific prior knowledge in a multiscale phase field model for road extraction from VHR images. IEEE J. Sel. Topics Appl. Earth Observ. Remote Sens. **1**(2), 139–146 (2008)

Clustering Evolutionary Data
with an r-Dominance Based Multi-objective
Evolutionary Algorithm

Wenhao Gao, Wenjian Luo$^{(\boxtimes)}$, Chenyang Bu, Li Ni,
and Daofu Zhang

Anhui Province Key Laboratory of Software Engineering in Computing
and Communication, School of Computer Science and Technology,
University of Science and Technology of China, Hefei 230027, Anhui, China
{gao1991,bucy1991,nlcs,dfl01}@mail.ustc.edu.cn,
wjluo@ustc.edu.cn

Abstract. Clustering evolutionary data (or called *evolutionary clustering*) has received an enormous amount of attention in recent years. A recent framework (called *temporal smoothness*) considers that the clustering result should depend mainly on the current data while simultaneously not deviate too much from previous ones. In this paper, evolutionary data is clustered by a multi-objective evolutionary algorithm based on *r-dominance*, and the corresponding algorithm is named *rEvoC*. The *rEvoC* considers the previous clustering result (or historical data) as the reference point. We propose three strategies to define the reference point and to calculate the distance between a reference point and an individual. Based on the reference point and the *r-dominance* relation, the search could be guided into the region, in which a solution not only could cluster the current data well, but also does not shift two much from the previous one. Additionally, the *rEvoC* adopts one step *k-means* as a local search operator to accelerate the evolutionary search. Experimental results on two different data sets are given. The experimental results demonstrate that, the *rEvoC* achieves better performance than the corresponding static clustering algorithm and the evolutionary *k-means* algorithm.

Keywords: Evolutionary data · Clustering · Reference point · r-dominance

1 Introduction

Evolutionary data is ubiquitous in the real-world, such as daily news, blogs, social network data, and capital market trading data. The key challenge to learn from evolutionary data is that both the distribution and the underlying concept of the evolutionary data may change over time. Such changes usually include: (1) short-term variation due to the noise; (2) long-term trend due to the underlying concept drift [1].

For most traditional methods, the data to be clustered are seen as a static whole set (called static clustering problems). However, this kind of method may not perform well for clustering evolutionary data, because the temporal information is omitted. In order to cluster evolutionary data (called *evolutionary clustering*), Chakrabarti et al. [1] proposed

H. Yin et al. (Eds.): IDEAL 2016, LNCS 9937, pp. 342–352, 2016.
DOI: 10.1007/978-3-319-46257-8_37

the *temporal smoothness* framework, which considers that: on the one hand, the current clustering result should depend mainly on the current data futures; on the other hand, the clustering result should not deviate too much from the recent past time steps. The framework divides the objective function into two parts: *snapshot quality* (*Sq*), measuring the clustering quality of the current snapshot data, and *history quality* (*Hq*), verifying how similar the current clustering is to the previous one. It is worth pointing out that the *evolutionary clustering* is different from evolutionary algorithms for clustering. The word '*evolutionary*' in *evolutionary clustering* refers to evolutionary data.

For static clustering problems, Multi-Objective Evolutionary Algorithms (MOEAs) have been widely used (e.g. [2, 3]). MOEAs simultaneously optimize of multiple objectives and provide a set of non-dominant solutions, which is one of the reasons to utilize MOEAs for clustering [4]. However, in the real-world applications, because the final decision is always a unique solution, the decision maker is interested in a part, but not all, of Pareto-optimal front. Ben Said et al. [5] expressed the decision maker's preference as a reference point and introduced a new variant of Pareto dominance relation, called *r-dominance*, which could be in various MOEAs. The main idea of the *r-dominance* is to sort Pareto-equivalent solutions according to their distances to the reference point. Hence, the *r-dominance* relation is able to guide the search into the part of the Pareto-optimal region, in which the decision maker is interested.

Inspired by the concept of the reference point and the *r-dominance* relation proposed in [5], we deal with the *evolutionary clustering* problem by a multi-objective evolutionary algorithm based on *r-dominance*. Specifically, the previous clustering result (or historical data) is considered as a reference point, such that the search can be guided to the region, in which a solution not only could cluster the current data well, but also does not shift two much from the previous one. In this way, the clustering result obtained at each time step can avoid some interference from the short-term noise and concept drift, and can reflect a reasonable partition for the current data.

The rest of this paper is organize d as follows. Section 2 introduces the backgrounds, including the *evolutionary clustering* and the *r-dominance* based multi-objective evolutionary algorithm. Section 3 describes the proposed algorithm in detail. Experimental results on a synthetic dataset and a real-world dataset are presented in Sect. 4. Finally, Sect. 5 concludes the paper.

2 Related Works

2.1 Evolutionary Clustering

Evolutionary clustering has attracted significant attention in recent years. A typical work was done by Chakrabarti et al. [1] and they proposed a general framework named *temporal smoothness*. Two classic clustering algorithm (*k*-means and agglomerative hierarchical clustering) were studied within their framework. Based on the similar framework, Chi et al. [6] extended static spectral clustering to evolutionary clustering and proposed two algorithms (PCQ and PCM). The goal of the framework is to optimize two objectives, i.e., *snapshot quality (Sq)* and *history quality (Hq)*, through the following cost function:

$$C_{total} = \alpha \cdot Sq + (1 - \alpha) \cdot Hq \tag{1}$$

where Sq measures how well the clustering found represents the data at the current time; Hq is the measure of history quality; and $\alpha \in [0, 1]$.

2.2 Evolutionary Clustering Based on Multi-objective Evolutionary Algorithms

As far as we know, existing works based on multi-objective evolutionary algorithm for *evolutionary clustering* are to optimize the two objectives, i.e., *snapshot quality* (e.g. modularity) and *history quality* (e.g. *NMI*).

Typically, to discover dynamic community structure, Folino and Pizzuti [7, 8] formulated the detection of community structure with *temporal smoothness* as a multi-objective optimization problem and proposed an evolutionary multi-objective algorithm for dynamic community discovery. The algorithm was named *DYNMOGA*, which optimizes *modularity* and *normalized mutual information (NMI)* simultaneously. Based on the similar idea in [7, 8], some extended works have been done. Ma et al. [9] employed the decomposition based multi-objective evolutionary algorithm (i.e. *MOEA/D*) to reveal dynamic community structure. Zhou et al. [10] proposed a multi-objective biogeography based optimization algorithm with decomposition to deal with dynamic community detection. Chen et al. [11] proposed a multi-objective evolutionary algorithm for dynamic community discovery, which optimized *modularity density* and *NMI* at the same time.

To address attributed data, Ma et al. [12] adopted *MOEA/D* to cluster evolutionary data and the new algorithm is named *EKM-MOEA/D*. The result of *EKM-MOEA/D* is a set of Pareto-optimal solutions and how to get a better unique solution, that satisfies the decision maker's preference, has not be studied yet.

2.3 The *r-dominance* Based Multi-objective Evolutionary Algorithms

The work in this paper is inspired by the algorithm proposed in [5]. In [5], a variant of the Pareto-optimal relation, named *r-dominance*, is proposed. The *r-dominance* has the ability to build a strict partial order between Pareto-equivalent solutions, such that the search can be led to the part of the Pareto-optimal region, where the decision maker is interested.

The definition of the *r-dominance* is described as follows [5]: Given a population P and a reference point g, an individual u is said to *r-dominance* an individual $v(u <_r v)$, if one of the following conditions is satisfied:

(1) $u < v$.;
(2) $u \nless v$, $v \nless u$, and $Dist(u, v, g) < - \delta$, where $\delta \in [0, 1]$ and

$$Dist(u, v, g) = \frac{Dist(u, g) - Dist(v, g)}{Dist_{max} - Dist_{min}} \tag{2}$$

$$Dist_{max} = max_{z \in P} Dist(z, g) \tag{3}$$

$$Dist_{min} = min_{z \in P} Dist(z, g) \tag{4}$$

$$\delta = 1 - \frac{1 - \delta_{user}}{G} * gen \tag{5}$$

where $Dist(z, g)$ is the distance between the individual z and the reference point g; G is the maximum evolutionary generation; gen is the current generation; and $\delta_{user} \in [0, 1]$. The first condition means u dominates v according to the *Pareto dominance* definition. The second condition means u and v are *Pareto-equivalent*, but u is closer to g than v. Additionally, the adaptive setup of δ (decreasing from 1 to δ_{user}) enables the individuals gradually move toward the region close to the reference point, and avoid premature convergence.

3 Algorithm Descriptions

The evolutionary data with T time steps is usually denoted as a sequence of snapshots $X = \{X^1, \ldots, X^t, \ldots, X^T\}$, where X^t denotes the d-dimensional data points at time step t, i.e., $X^t = \{x_1^t, x_2^t, \ldots, x_n^t\}; \forall i, x_i^t \in \Re^d$. The obtained clustering result for data set X^t is denoted as C^t, where $C^t = \{c_1^t, \ldots, c_i^t, \ldots, c_K^t\}$; c_i^t is the i-th cluster; and K is the number of clusters.

3.1 Chromosome Representation and Operators

The coordinates of all cluster centers are encoded into a real-coded chromosome as a candidate solution [3, 13]. And a chromosome is denoted by a vector of length $K * d$, where K is the cluster number and d is the dimension of the data. For example, in the two-dimensional space, the chromosome

2.0	5.5	0.5	2.5	5.0	10.0

encodes three cluster centers, i.e., (2.0, 5.5), (0.5, 2.5), and (5.0, 10.0).

Binary tournament selection, single-point crossover and polynomial mutation [14] are used. And the one step *k-means* operator [15] is adopted for fine-tuning partitions found by the genetic operators. Note that for the single-point crossover operator, the crossover points should fall between two cluster centers to avoid disruption of cluster centers.

3.2   Two Objective Functions

In this paper, we do not adopt the *temporal smoothness* framework in [1], but minimize two objectives to cluster current data, i.e., the overall cluster deviation (*Dev*) [2, 15] and the Davies-Bouldin (*DB*) index [16]. The *Dev* (Eq. 6) is computed as the summed distances between each data point and its corresponding center, and the *DB* (Eq. 7) is the ratio of the measure *within–cluster scatter* (Eq. 8) to the measure *between–cluster separation* (Eq. 9) [4]. Other objective functions used in the traditional multi-objective clustering will be investigated in the future.

The form of *Dev* is

$$Dev(C) = \sum_{c_i^t \in C^t} \sum_{x^t \in c_i^t} D^2(\mu_i^t, x^t) \tag{6}$$

where μ_i^t is the center of cluster c_i^t, and $D(.,.)$ is the Euclidean distance.

The form of *DB* index is

$$DB = \frac{1}{K} \sum_{i=1}^{K} R_i \tag{7}$$

where

$$R_i = \max_{j,j \neq i} \left\{ \frac{S_i + S_j}{d_{i,j}} \right\} \tag{8}$$

$$S_i = \frac{1}{|c_i^t|} \sum_{x^t \in c_i^t} D^2(\mu_i^t, x^t) \tag{9}$$

$$d_{i,j} = D^2(\mu_i^t, \mu_j^t) \tag{10}$$

S_i is the scatter within the i-th cluster; $|c_i^t|$ is the number of elements in the cluster c_i^t; and $d_{i,j}$, defined as the distance between the center of cluster c_i^t and the center of cluster c_j^t, is the measure of between-cluster separation.

3.3   The Reference Points

In this subsection, we introduce three strategies to define the reference point g and to calculate the distance between g and an individual z, i.e., $Dist(z, g)$. The three strategies, i.e., *rEvoC-NMI*, *rEvoC-dist*, and *rEvoC-ct*, are detailed below, where C^t is the clustering result decoded by the individual z.

(1)   *rEvoC-NMI*

For *rEvoC-NMI*, the previous clustering result C^{t-1} is considered as the reference point g, and the *NMI* metric [17] is used to measure the distance. The form of the distance function *Dist* is given as follows.

$$Dist(z, g) = 1 - NMI(C^t, C^{t-1}) \tag{11}$$

the *NMI* metric is defined as follows, where the *NMI* score is between 0 and 1; and the greater the *NMI* is, the more similar the two clusterings are.

$$NMI = \frac{\sum_{i=1}^{K} \sum_{j=1}^{K} n_{i,j} \log(\frac{n \cdot n_{i,j}}{n_i \cdot n_j})}{\sqrt{\sum_{i=1}^{K} n_i \log \frac{n_i}{n}} \sqrt{\sum_{j=1}^{K} n_j \log \frac{n_j}{n}}} \tag{12}$$

n_i and n_j are the numbers of data points in cluster i and cluster j, respectively; and $n_{i,j}$ denotes the number of data points in both cluster i and in cluster j.

(2) *rEvoC-dist*

For *rEvoC-dist*, the set of previous cluster centers at time $t - 1$ is considered as the reference point g, and the summed distance between the centers in z and previous corresponding ones [1] is used to compute the distance. Let μ_i^t denote the i-th cluster center of the individual z, and $\mu_{f(i)}^{t-1}$ denotes the corresponding cluster center at time step $t - 1$. Then *Dist* is defined as the following.

$$Dist(z, g) = \sum_i D^2(\mu_i^t, \mu_{f(i)}^{t-1}) \tag{13}$$

(3) *rEvoC-ct*

For *rEvoC-ct*, the historical data at time $t - 1$ is considered as the reference point. To penalize the current individuals that do not fit well with the historical data, the temporal cost defined in [6] is adopted as the *Dist*. The form is given as follows.

$$Dist(z, g) = \sum_{l=1}^{K} \sum_{i \in c_l^t} D^2(x_i^{t-1}, \eta_l^{t-1}) \tag{14}$$

where x_i^{t-1} is the i-th data point at the previous time step and $\eta_l^{t-1} = \sum_{j \in c_l^t} \frac{x_j^{t-1}}{|c_l^t|}$.

3.4   Description of the Algorithm Framework

At the first time step, the static version of the proposed algorithm without considering the reference point (called *rEvoC-static*) is used to cluster the dataset because of no historical data. The result of *rEvoC-static* is a set of Pareto-optimal individuals and the membership function in [10, 18] is used to select an individual as the final decision.

The sorting process at Step 9 in Algorithm 1 is detailed as follows. First, conduct the sorting procedure of NSGA-II [19] and classify R_{gen} into several fronts. Next, compare any two individuals u and v at each front based on the *r-dominance* relation, i.e., if $Dist(u, v, gen) < -\delta$, then $u <_r v$.

Because the *r-dominance* relation is adopted to sort Pareto-equivalent individuals and the parameter δ_{user} in Eq. (5) is set to 0, the final solution at Step 13 is a unique individual in the Pareto-optimal front that is closed to the reference point. In this way, we do not need another measure to select a final solution in the Pareto-optimal front.

Algorithm 1. The framework of *rEvoC*

Input:	A sequence of snapshot data $X = \{X^1, X^2, ..., X^T\}$
Output:	The clustering result for each time step, i.e., $C^t = \{c_1^t, ..., c_i^t, ..., c_K^t\}, t = 1, ..., T$

1:　　Conduct *rEvoC-static* to obtain the initial clustering result C^1;
2:　　**for** $t = 2$ to T
3:　　　　Generate an initial population P_0 consisting of NP random individuals; and set $gen = 0$;
4:　　　　**while** $gen <= G$ **do**
5:　　　　　　$Q_{gen} \leftarrow$ Conducting selection, crossover and mutation on P_{gen};
6:　　　　　　$R_{gen} \leftarrow P_{gen} \cup Q_{gen}$;
7:　　　　　　Decode each individual in R_{gen}, conduct the local search operator (i.e., one step *k-means*), and then assign each data point to its nearest cluster center;
8:　　　　　　Evaluate the values of *Dev* and *DB* index;
9:　　　　　　Sort all the individuals according to the *r-dominance* relation and then assign each individual a rank;
10:　　　　　Select NP individuals based on both the rank and their crowding distances to form a new population P_{gen+1};
11:　　　　　　$gen = gen + 1$;
12:　　　　**end while**
13:　　　　Decode the final individual to obtain the clustering result C^t.
14:　　**end for**

4　Experiments

The proposed *rEvoC* are tested both on synthetic and real-world data sets, compared with *rEvoC-static* and evolutionary *k-means* clustering (*EKM*) [1]. The *NMI* (Eq. 11) is adopted to measure the similarity between the clustering result and the ground truth. All experiments are conducted 30 times independently and the average results are given. For convenience, in the tables, *rEvoC-NMI, rEvoC-dist, rEvoC-ct* and *rEvoC-static* are abbreviated as r1, r2, r3, r4, respectively.

The parameter settings are as follows. The population size $NP = 50$, the maximum generation $G = 39$, crossover rate $= 0.8$, mutation rate $= 1/K * d$, the distribution index in polynomial mutation is 20, $\delta_{user} = 0$. And parameter α in *EKM* is set to 0.7.

4.1   Synthetic Dataset

The data generation method in [20] is adopted. For each time step, 800 samples are obtained from a mixture of two 2-D Gaussian distributions, the first with mean (3, 3) and the second with mean (−3, −3). The covariance matrixes of both Gaussians are [1, 0; 0, 1]. The mixture proportion is initially set to 1:1, i.e. each cluster with 400 samples. From time step 1 to 10, the second cluster moves towards the first by (0.4, 0.4) at each time. To simulate the change in cluster membership, the mixture proportion is switched to 5:3 and 3:1 at times 11 and 12, respectively. From time step 13 onward, both the means and the mixture proportion are kept stationary. The scatter plots in three typical times are shown in Fig. 1.

The results on the synthetic dataset are shown in Table 1. The results show that *rEvoC-NMI* and *rEvoC-ct*, outperform *rEvoC-static* and *EKM* in most cases. And the *rEvoC-dist* has similar results with *rEvoC-static*.

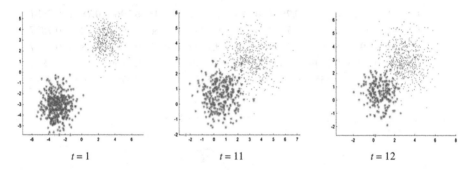

$t = 1$ $t = 11$ $t = 12$

Fig. 1. The synthetic dataset at three typical time steps

Table 1. Experimental results on the synthetic data set

	1	2	3	4	5	6	7	8	9	10
r1	**1.0000**	**1.0000**	**1.0000**	**0.9874**	**0.9874**	**1.0000**	**0.9546**	**0.9192**	**0.8359**	**0.7541**
r2	**1.0000**	**1.0000**	**1.0000**	**0.9874**	**0.9874**	**1.0000**	**0.9546**	**0.9192**	**0.8359**	**0.7541**
r3	**1.0000**	**1.0000**	**1.0000**	**0.9874**	**0.9874**	**1.0000**	**0.9546**	**0.9192**	**0.8359**	**0.7541**
r4	**1.0000**	**1.0000**	**1.0000**	**0.9874**	**0.9874**	**1.0000**	**0.9546**	**0.9192**	**0.8359**	0.7489
EKM	**1.0000**	**1.0000**	**1.0000**	**0.9874**	0.9681	0.9437	0.8348	0.7210	0.5304	0.3919
	11	12	13	14	15	16	17	18	19	20
r1	**0.7026**	**0.7757**	0.7335	0.7050	**0.7266**	0.7190	**0.6967**	**0.7190**	**0.7792**	**0.6779**
r2	0.6953	0.7314	0.6745	0.6149	0.6841	0.7043	0.6838	0.7020	0.6802	0.6544
r3	0.7022	0.7748	**0.7337**	**0.7051**	0.7252	**0.7214**	0.6954	0.7170	0.7785	0.6635
r4	0.6960	0.7515	0.6691	0.6260	0.6987	0.7122	0.6798	0.6910	0.7142	0.6509
EKM	0.4510	0.5514	0.5852	0.5203	0.5797	0.5077	0.6026	0.5578	0.5663	0.5591

4.2   Real-World Dataset

The *pendigits* dataset from the UCI dataset is adopted. The original dataset has 10992 data instances with 16 dimensions. The dataset has 10 classes, labeled as 0–9, respectively. To simulate 10 time steps, each class is averagely divided into 10 parts randomly.

The experimental results on the *pendigits* dataset are presented in Table 2 and Fig. 2. Table 2 shows *EKM* is far worse than other approaches in this dataset. Figure 2 shows *rEvoC-NMI* and *rEvoC-ct* outperform *rEvoC-static* for almost all the times (except the first time), and that *rEvoC-dist* also has higher *NMI* than *rEvoC-static* in most cases. The results of EKM is not shown in Fig. 2 because it is obviously the worse one.

Table 2. Experimental results on the pendigits data set

	1	2	3	4	5	6	7	8	9	10
r1	**0.7749**	**0.7945**	0.7929	**0.7904**	**0.7894**	0.7811	0.7943	**0.7966**	**0.7871**	**0.7981**
r2	**0.7749**	0.7774	0.7811	0.7892	0.7837	0.7751	0.7866	0.7931	0.7808	0.7827
r3	**0.7749**	0.7925	**0.7944**	0.7899	0.7884	**0.7820**	**0.7944**	0.7956	0.7866	0.7977
r4	**0.7749**	0.7860	0.7749	0.7809	0.7844	0.7719	0.7840	0.7869	0.7740	0.7858
EKM	0.7211	0.6627	0.5555	0.5442	0.5586	0.5338	0.5589	0.5706	0.5675	0.5742

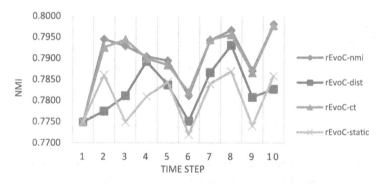

Fig. 2. Comparison of rEvoC-ct and rEvoC-static on the pendigits data set

5   Conclusions

In this paper, inspired by the work in [5], we deal with the *evolutionary clustering* problem from the view of reference point. Three forms of the reference point and their corresponding distance functions are introduced. Based on the reference point and the *r-dominance* relation, an evolutionary clustering algorithm (i.e. *rEvoC*) is proposed. Experimental results demonstrate that, compared with typical algorithms, the proposed *rEvoC* algorithm is suitable to cluster evolutionary data and could achieve better performance.

Acknowledgements. This work is partly supported by Anhui Provincial Natural Science Foundation (No. 1408085MKL07) and National Natural Science Foundation of China (No. 61573327).

References

1. Chakrabarti, D., Kumar, R., Tomkins, A.: Evolutionary clustering. In: The 12th ACM SIGKDD International Conference on Knowledge Discovery and Data Mining, pp. 554–560 (2006)
2. Handl, J., Knowles, J.: An evolutionary approach to multiobjective clustering. IEEE Trans. Evol. Comput. **11**(1), 56–76 (2007)
3. Ripon, K.S.N., et al.: Multi-objective evolutionary clustering using variable-length real jumping genes genetic algorithm. In: The 18th International Conference on Pattern Recognition, pp. 1200–1203 (2006)
4. Mukhopadhyay, A., Maulik, U., Bandyopadhyay, S.: A survey of multiobjective evolutionary clustering. ACM Comput. Surv. (CSUR) **47**(4), 61 (2015)
5. Ben Said, L., Bechikh, S., Ghédira, K.: The r-dominance: a new dominance relation for interactive evolutionary multicriteria decision making. IEEE Trans. Evol. Comput. **14**(5), 801–818 (2010)
6. Chi, Y., et al.: Evolutionary spectral clustering by incorporating temporal smoothness. In: The 13th ACM SIGKDD International Conference on Knowledge Discovery and Data Mining, pp. 153–162 (2007)
7. Folino, F., Pizzuti, C.: A multiobjective and evolutionary clustering method for dynamic networks. In: The International Conference on Advances in Social Networks Analysis and Mining, pp. 256–263 (2010)
8. Folino, F., Pizzuti, C.: An evolutionary multiobjective approach for community discovery in dynamic networks. IEEE Trans. Knowl. Data Eng. **26**(8), 1838–1852 (2014)
9. Ma, J., et al.: Decomposition-based multiobjective evolutionary algorithm for community detection in dynamic social networks. Sci. World J. **2014**, 1–22 (2014)
10. Zhou, X., et al.: Multiobjective biogeography based optimization algorithm with decomposition for community detection in dynamic networks. Physica A **436**, 430–442 (2015)
11. Chen, G., Wang, Y., Wei, J.: A new multiobjective evolutionary algorithm for community detection in dynamic complex networks. Math. Probl. Eng. **2013**, 1–7 (2013)
12. Ma, J., et al.: Spatio-temporal data evolutionary clustering based on MOEA/D. In: The 13th Annual Conference Companion on Genetic and Evolutionary Computation, pp. 85–86 (2011)
13. Chen, G., Luo, W., Zhu, T.: Evolutionary clustering with differential evolution. In: The 2014 IEEE Congress on Evolutionary Computation (CEC), pp. 1382–1389 (2014)
14. Deb, K., Goyal, M.: A combined genetic adaptive search (GeneAS) for engineering design. Comput. Sci. Inform. **26**, 30–45 (1996)
15. Bandyopadhyay, S., Maulik, U.: An evolutionary technique based on K-means algorithm for optimal clustering in RN. Inf. Sci. **146**(1), 221–237 (2002)
16. Davies, D.L., Bouldin, D.W.: A cluster separation measure. IEEE Trans. Pattern Anal. Mach. Intell. PAMI **1**(2), 224–227 (1979)
17. Chen, W.-Y., et al.: Parallel spectral clustering in distributed systems. IEEE Trans. Pattern Anal. Mach. Intell. **33**(3), 568–586 (2011)

18. Agrawal, S., Panigrahi, B., Tiwari, M.K.: Multiobjective particle swarm algorithm with fuzzy clustering for electrical power dispatch. IEEE Trans. Evol. Comput. **12**(5), 529–541 (2008)
19. Deb, K., et al.: A fast and elitist multiobjective genetic algorithm: NSGA-II. IEEE Trans. Evol. Comput. **6**(2), 182–197 (2002)
20. Xu, K.S., Kliger, M., Hero III, A.O.: Adaptive evolutionary clustering. Data Min. Knowl. Disc. **28**(2), 304–336 (2014)

Sparse Non-negative Matrix Factorization with Generalized Kullback-Leibler Divergence

Jingwei Chen[1], Yong Feng[1], Yang Liu[2(✉)], Bing Tang[3], and Wenyuan Wu[1]

[1] Chongqing Key Laboratory of Automated Reasoning and Cognition,
Chongqing Institute of Green and Intelligent Technology, CAS,
Chongqing 400714, China
{chenjingwei,yongfeng,wuwenyuan}@cigit.ac.cn
[2] College of Information Science and Engineering, Chongqing Jiaotong University,
Chongqing 400074, China
ly1246@qq.com
[3] School of Computer Science and Engineering,
Hunan University of Science and Technology, Xiangtan 411201, China
btang@hnust.edu.cn

Abstract. Non-negative Matrix Factorization (NMF), especially with sparseness constraints, plays a critically important role in data engineering and machine learning. Hoyer (2004) presented an algorithm to compute NMF with exact sparseness constraints. The exact sparseness constraints depends on a projection operator. In the present work, we first give a very simple counterexample, for which the projection operator of the Hoyer (2004) algorithm fails. After analysing the reason geometrically, we fix this bug by adding some random terms and show that the fixed one works correctly. Based on the fixed projection operator, we propose another sparse NMF algorithm aiming at optimizing the generalized Kullback-Leibler divergence, hence named SNMF-GKLD. Experimental results show that SNMF-GKLD not only has similar effects with Hoyer (2004) on the same data sets, but is also efficient.

Keywords: Non-negative Matrix Factorization · Projection operator · Generalized Kullback-Leibler divergence

1 Introduction

Since Lee and Seung's *Nature* paper [11], Non-negative Matrix Factorization (NMF) has been extensively studied and has a great deal of applications in science and engineering. In contrast to Principal Component Analysis [9] and Independent Component Analysis [8], NMF is strictly required that the entries of both resulting matrices are non-negative. Such a constraint is very meaningful in many applications, in which the data representation is purely additive, for instance, the parts-based representation of face image data from CBCL database.

Given a non-negative matrix $V \in \mathbb{R}^{m \times n}$ and a positive integer $r < \min\{m, n\}$, the goal of NMF is to find non-negative matrices $W \in \mathbb{R}^{m \times r}$ and $H \in \mathbb{R}^{r \times n}$

© Springer International Publishing AG 2016
H. Yin et al. (Eds.): IDEAL 2016, LNCS 9937, pp. 353–360, 2016.
DOI: 10.1007/978-3-319-46257-8_38

minimizing the function $f(W, H) = \|V - WH\|_F^2$, or the function $d(W, H) = \sum_{i,j} (V_{i,j} \log (V_{i,j}/(WH)_{i,j}) - V_{i,j} + (WH)_{i,j})$, where $\| \cdot \|_F$ is the Frobenius norm of a matrix. We call $f(W, H)$ the *Square of Euclidean Distance* (SED) and $d(W, H)$ the *Generalized Kullback-Leibler Divergence* (GKLD). The product WH is called an *NMF* for V. Note that, in most cases, WH is not equal to V. The parameter r is problem-dependent, and is set by users. Usually, r is chosen to satisfy $r \ll \min\{m, n\}$ such that WH can be thought of as a compressed form of the original data. Lee and Seung [12] presented two NMF algorithms based on multiplicative formulae whose objective functions are SED and GKLD, respectively. The two NMF algorithms can be seen as the basic ones, on which many other NMF algorithms are based. In 2004, Hoyer [6] introduced the sparseness definition for any non-zero n-dimensional vector $\boldsymbol{x}$, i.e., Sparseness($\boldsymbol{x}$) = $(\sqrt{n} - \|\boldsymbol{x}\|_1/\|\boldsymbol{x}\|_2) / (\sqrt{n} - 1)$, and proposed an algorithm to perform NMF with sparseness constraints (NMFSC). NMFSC adopts Lee and Seung's multiplicative formulae to optimize SED and uses a non-linear projection operator to control the sparseness of W and H.

In the present paper, we first revisit the Hoyer's projection operator [6, p. 1463]. In fact, for a kind of examples, the projection operation may fail. Geometrically, the failure case corresponds to that a direction vector of the projection operator is $\boldsymbol{0}$, so that the operator can not decide how to process further. We modify the operator a little to fix this bug and prove its correctness in Sect. 2. In Sect. 3, we propose a sparse NMF algorithm based on the fixed Hoyer's projection operator, named SNMF-GKLD, whose objective function is GKLD. We show experimentally that SNMF-GKLD is efficient and has similar effects with NMFSC in Sect. 4.

1.1   Related Work

Here, we only focus on algorithms to compute sparse NMF. We refer to [1, 4, 7] and references therein for general NMF discussion.

Li et al. proposed a local NMF algorithm [13], in which the key idea is to limit the columns of W orthogonal to each other, which makes W sparse, but H may be far from sparse. Hoyer [5] combined sparse coding and NMF. Liu et al. [15] gave a similar algorithm, but with SED replaced by GKLD. Liu and Zheng [14] presented an ℓ_p-NMF algorithms, which uses GKLD as its objective function and limits $\|W_i\|_p = 1$. For larger p, ℓ_p-NMF gives sparser representations. Hoyer [6] adopted SED as the objective function to propose an NMF algorithm (NMFSC) with exact sparseness constraints. The exact constrain on sparseness depends on a nonlinear projection that may fail for some cases. We fix the little bug in this paper. Stadlthanner et al. extended NMFSC [17] with exact sparseness constraints on W_i and H_i and disscussed the uniqueness of Hoyer's projection operator. Cichocki et al. [3] presented an NMF algorithm which uses $\|V - WH\|_F^2 + \alpha J_1(W) + \beta J_2(H)$ as its objective function. Their algorithms are modified from the basic ones and easy to implement, however, they may diverge. Pascual-Montano et al. [16] proposed the nsNMF algorithm and showed that it balances the sparseness and the capability of representing the original data. Kim and Park [10] gave SNMF/L and SNMF/R. Tong et al. [18] proposed an NMF algorithm which combines SVD initialization technique from [2] and the extended NMFSC from [17]. Although a large number of sparse NMF algorithms have been proposed, there seems to be no sparse NMF algorithm in literature combining Hoyer's projection operator and the multiplicative formula for GKLD. In this paper, we explore this combination.

2   The Hoyer's Projection Operator Revisited

As indicated in [6] by the author, Hoyer's NMFSC algorithm is essentially the multiplication iteration for the gradient descent algorithm with a projection operator which enforces the desired degree of sparseness. Actually, the projection operator solves the following problem: given any vector x, find the closest (in the euclidean sense) non-negative vector s with a given ℓ_1-norm L_1 and a given ℓ_2-norm L_2. This operator is naturally used to control the sparseness exactly. We recall this operator as in Algorithm 1.

Algorithm 1. (The projection operator in [6]).

Input: A vector $x \in \mathbb{R}^n$, norm conditions L_1 and L_2.
Output: A closest non-negative s to x with $\|s\|_i = L_i$, $i = 1, 2$.
1: Set $s := x + (L_1 - \|x\|_1)/ne$ with $e = (1, 1, \cdots, 1)^T \in \mathbb{R}^n$. Set $m := (L_1/n)e$.
2: Set $s := m + \alpha(s - m)$ with $\alpha > 0$ such that $\|s\|_2 = L_2$.
3: **if** there exists j with $s_j < 0$ **then**
4:     Set $s_j := 0$. Remove j-th coordinate of x.
5:     Decrease dimension $n := n - 1$.
6:     **goto** 1.
7: **end if**

The projection algorithm starts from orthogonally projecting the given vector x onto the hyperplane $\sum_{i=1}^n s_i = L_1$. Next, within this hyperplane, it projects to the closest point on the joint constraint hypersphere. Namely, computing a point s satisfying $\sum_{i=1}^n s_i = L_1$ and $\|s\|_2 = L_2$ simultaneously. This is done by, step 2, moving radially outward from the center of the sphere (the center is given by the point where all components have equal values). If the result is completely non-negative, we have arrived at our destination. If not, then we have $\|s\|_1 > L_1$, and hence those components that attained negative values must be fixed to zero (step 4) and the new point must be projected onto the hyperplane $\sum_{i=1}^n s_i = L_1$ again (step 5 and 6), until the algorithm converges. The above iteration terminates after at most n iterations since at each iteration, the algorithm either terminates, or at least one component is set to zero and removed.

2.1   A Counterexample

It is ingenious to design the projection algorithm. Further, Stadlthanner et al. [17] proved the uniqueness of the projection operator. However, there exists a case, for which the projection algorithm may fail. The case happens when $s = m$ before step 2, i.e., $\forall i \notin Z$, all components s_i's are equal. Geometrically, in this case, we can not moving from m, the center of the joint constraint hypersphere, to the closest point. Instead, the algorithm will return m, however, $\|m\|_2$ is not be L_2 in general. For this case, the projection algorithm fails. Here is a simple counterexample for which Hoyer's Matlab implementation does not work:

```
>> s =[-1,-1]'
>> projfunc(s, 3, 5.598076212, 1)
```

In the code, the parameter 3 is the ℓ_1-norm and 5.598076212 is the square of the ℓ_2-norm. For this example, projfunc falls into an endless loop.

Note that this kind of examples does not contradict with the uniqueness in [17, Theorem 1], because it has been already indicated that the exception set for uniqueness has Lebesgue measure 0. In fact, the set of counterexamples pointed out here has exact Lebesgue measure 0.

2.2 Bug Fixing

Here is a modification to fix the above bug. The basic idea is the following: if $s = m$ before step 3, we re-choose s such that $\sum_{i=1}^n s_i = L_1$ and that $s - m \neq 0$. In particular, one can insert "If $s = m$, then randomly choose s_i such that $\sum_i s_i = L_1$. Repeat this step until $s \neq m$" to the location between step 1 and 2.

Proposition 1. *Algorithm 1 with the above modification correctly computes a closest non-negative s to x with $\|s\|_i = L_i$, $i = 1, 2$.*

Proof. As indicated in Sect. 2.1, $s = m$ means that all components s_i's are equal. Then the orthogonal projection of s onto $\sum_{i=1}^n s_i = L_1$ is exactly m, the center of the joint constraint hypersphere. This means that the distances between $m (= s)$ and each intersection point of the sum and the ℓ_2-norm constraints are equal, so do the distances between x and each intersection point. Thus, in the constraint hypersphere we can move from m along any direction (i.e., either $\alpha \geq 0$ or $\alpha < 0$) to the closest point. The correctness follows.

3   Sparse NMF with GKLD

We now present a sparse NMF algorithm with the generated Kullback-Leibler divergence $d(W, H)$ as its objective function. We call the algorithm SNMF-GKLD, which can be seen as a result of combing the corresponding multiplicative iterations from [12] and the modified Hoyer's projection operator in Sect. 2.

SNMF-GKLD has a similar structure with Hoyer's NMFSC algorithm [6], however, besides the modified projection operator, it is different from NMFSC in at least the following two aspects. Firstly, we use GKLD as our objective function, while NMFSC uses SED. The two objective functions are well-known as basic objective functions for NMF. It is natural and necessary to investigate the performance of GKLD plus Hoyer's projection operator. Secondly, NMFSC uses the additive version of gradient descent algorithm with automatically chosen stepsizes to make the objective function decrease. SNMF-GKLD gives up this step, since according to a large number of experimental observations, it seems that the objective function $d(W, H)$ always decreases after each iteration, if only one sparseness of W and H is constrained. Unfortunately, this observation does not hold when both sparseness of W and H are constrained. As a consequence, SNMF-GKLD only allows to constrain the sparseness of one factor, i.e., either W or H. An advantage is that SNMF-GKLD has a more practical efficiency than NMFSC for large size data, as the experiments will show in the next section.

Algorithm 2. (SNMF-GKLD).

Input: A non-negative matrix V of size $m \times n$; $r \in \mathbb{Z}$; $0 \le \gamma_W \le 1$ or $0 \le \gamma_H \le 1$.
Output: A matrix W of size $m \times r$ and a matrix H of size $r \times n$ minimizing $d(W, H)$.
1: Initialize W and H as random non-negative matrices.
2: If sparseness constraints on W apply, then projection each column W to be non-negative, have unchanged ℓ_2-norm, but ℓ_1-norm set to achieve desired sparseness.
3: If sparseness constraints on H apply, then projection each row H to be non-negative, have unit ℓ_2-norm and ℓ_1-norm set to achieve desired sparseness.
4: **while** converge or stop **do**
5: $\quad W_{i,a} := W_{i,a} \dfrac{\sum_\mu H_{a,\mu} V_{i,\mu}/(WH)_{i,\mu}}{\sum_\nu H_{a,\nu}}.$
6: $\quad$ **if** sparseness constraints on W apply **then**
7: $\quad\quad$ Project each column W to be non-negative, having unchanged ℓ_2-norm, but set ℓ_1-norm to achieve desired sparseness.
8: $\quad$ **end if**
9: $\quad H_{a,\mu} := H_{a,\mu} \dfrac{\sum_i W_{i,a} V_{i,\mu}/(WH)_{i,\mu}}{\sum_k W_{k,a}}.$
10: $\quad$ **if** sparseness constraints on H apply **then**
11: $\quad\quad$ Project each row H to be non-negative, having unit ℓ_2-norm and set ℓ_1-norm to achieve desired sparseness.
12: $\quad$ **end if**
13: **end while**

---

4   Experiments

We now report some experimental results, which show that SNMF-GKLD has almost the same capability as NMFSC, but is more practical for "big data". The data sets we use are CBCL, ORL and ON/OFF filtered natural image database that is included in Hoyer's NMF software package. These data sets were also used in [6] to test NMFSC. We run all experiments in Matlab® R2015b on a Win 10 PC with Intel® Core™ i5-4300U CPU and 8 GB memory. In addition, we fix the number of iterations at 300 for all experiments.

For the CBCL data, some resulting bases are shown in Fig. 1, in which the parameters are taken from [6, Fig. 3]. i.e., for (a) $\gamma_W = 0.8$, for (b) $\gamma_H = 0.8$ and for (c), $\gamma_W = 0.2$. As NMFSC, setting a high sparseness value for W results in a local representation, and global features can be learned by setting a low sparseness value for W or a high sparseness for H. Figure 2 shows bases learned by

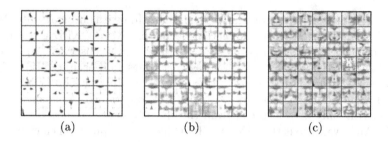

(a)   (b)   (c)

Fig. 1. Features learned from the CBCL database by SNMF-GKLD

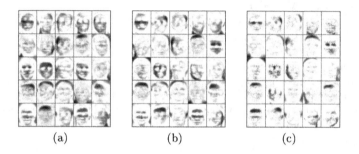

(a)   (b)   (c)

Fig. 2. Features learned from the ORL database by SNMF-GKLD

SNMF-GKLD for various sparseness settings, where sparseness levels were set to (a) $\gamma_W = 0.4$, (b) $\gamma_W = 0.5$ and (c) $\gamma_W = 0.6$. The representation switches from global to local with sparseness increasing. Figure 3 shows that SNMF-GKLD is also able to learn oriented features. According to these experiments, SNMF-GKLD has similar effects as NMFSC (comparing with [6, Figs. 3, 4, 5]).

According to our experiments, for CBCL and ORL data, SNMF-GKLD has a similar efficiency with NMFSC, however, for ON/OFF-filtered natural images, SNMF-GKLD costs only about a half time of NMFSC. More specifically, for learning the features in Fig. 3, SNMF-GKLD uses 87.5 s while NMFSC uses 155.9 s. For another example, SNMF-GKLD uses 85.9 s when the sparseness of H was fixed at 0.8, while NMFSC uses 155 s. There may be two reasons for this phenomenon. Firstly, ON/OFF-filtered natural images have larger size than that of CBCL and ORL data. Secondly, NMFSC has to re-choose the stepsize to make the objective function decrease, but SNMF-GKLD omits this step thanks to the observation obtained by Fig. 4.

Figure 4 shows the evolution of the objective function $d(W, H)$ for all experiments above. It shows that the objective function decreases after each iteration in SNMF-GKLD. This can be seen as an experimental evidence for convergence. However, we lack a mathematical convergence proof for the moment.

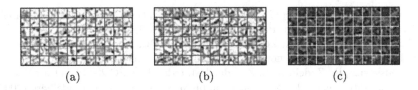

(a)   (b)   (c)

Fig. 3. Basis vectors learned from ON/OFF-filtered images by SNMF-GKLD

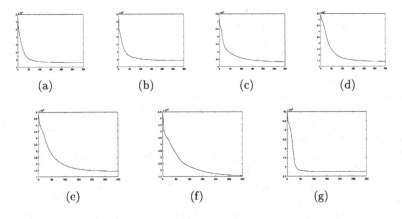

(a)   (b)   (c)   (d)

(e)   (f)   (g)

Fig. 4. The evolution of $d(W, H)$ for all experiments in Figs. 1, 2 and 3

5   Conclusion and Discussion

In the present paper, we analyse and fix a bug of the key part of Hoyer's NMFSC algorithm, the projection operator, and prove the fixed version is correct. Combining the fixed projector operation with the generalized Kullback-Leibler divergence objective function, we propose a sparse NMF algorithm, SNMF-GKLD. Experiments shows that SNMF-GKLD has almost the same capability as NMFSC and that it is efficient.

SNMF-GKLD can control the sparseness exactly, but, unfortunately, the sparseness can be constrained only on one factor, i.e., W or H. How to extend SNMF-GKLD such that it can be used to control the sparseness of W and H simultaneously is an open problem. In addition, it would be very interesting to explore the theoretical convergence of SNMF-GKLD.

Acknowledgments. This work was partially supported by NSFC (11471307, 11501540, 61572024), CAS "Light of West China" Program (2014), NSF of Hunan Province (2015JJ3071) and Chongqing Research Program (cstc2015jcyjys40001).

References

1. Berry, M.W., Browne, M., Langville, A.N., Pauca, V.P., Plemmons, R.J.: Algorithms and applications for approximate nonnegative matrix factorization. Comput. Stat. Data Anal. **52**(1), 155–173 (2007)

2. Boutsidis, C., Gallopoulos, E.: SVD based initialization: a head start for nonnegative matrix factorization. Pattern Recogn. **41**(4), 1350–1362 (2008)
3. Cichocki, A., Amari, S.I., Zdunek, R., Kompass, R., Hori, G., He, Z.: Extended SMART algorithms for non-negative matrix factorization. In: Rutkowski, L., Tadeusiewicz, R., Zadeh, L.A., Żurada, J.M. (eds.) ICAISC 2006. Lecture Notes in Artificial Intelligence (LNAI), vol. 4029, pp. 548–562. Springer, Heidelberg (2006). doi:10.1007/11785231_58
4. Cichocki, A., Zdunek, R., Phan, A.H., Amari, S.I.: Nonnegative Matrix and Tensor Factorizations: Applications to Exploratory Multi-way Data Analysis and Blind Source Separation. John Wiley & Sons Ltd., Chichester (2009)
5. Hoyer, P.O.: Non-negative sparse coding. In: Bourlard, H., Adali, T., Bengio, S., Larsen, J., Douglas, S. (eds.) NNSP 2012, pp. 557–565. IEEE, New York (2002)
6. Hoyer, P.O.: Non-negative matrix factorization with sparseness constraints. J. Mach. Learn. Res. **5**, 1457–1469 (2004)
7. Huang, Z., Zhou, A., Zhang, G.: Non-negative matrix factorization: a short survey on methods and applications. In: Li, K., Li, J., Liu, Y., Castiglione, A. (eds.) ISICA 2015. CCIS, vol. 575, pp. 331–340. Springer, Heidelberg (2012). doi:10.1007/978-3-642-34289-9_37
8. Hyvärinen, A., Karhunen, J., Oja, E.: Independent Component Analysis. John Wiley & Sons, New York (2001)
9. Jolliffe, I.T.: Principal Component Analysis, 2nd edn. Springer, New York (2002)
10. Kim, H., Park, H.: Sparse non-negative matrix factorizations via alternating non-negativity-constrained least squares for microarray data analysis. Bioinformatics **23**(12), 1495–1502 (2007)
11. Lee, D.D., Seung, H.S.: Learning the parts of objects by non-negative matrix factorization. Nature **401**(6755), 788–791 (1999)
12. Lee, D.D., Seung, H.S.: Algorithms for non-negative matrix factorization. In: Leen, T.K., Dietterich, T.G., Tresp, V. (eds.) NIPS*2000, pp. 556–562. MIT Press, Cambridge (2001)
13. Li, S.Z., Hou, X., Zhang, H., Cheng, Q.: Learning spatially localized, parts-based representation. In: CVPR 2001, vol. 1, pp. 207–212. IEEE, Los Alamitos (2001)
14. Liu, W., Zheng, N.: Learning sparse features for classification by mixture models. Pattern Recogn. Lett. **25**(2), 155–161 (2004)
15. Liu, W., Zheng, N., Lu, X.: Non-negative matrix factorization for visual coding. In: ICASSP 2003, vol. 3, pp. 293–296. IEEE, Piscataway (2003)
16. Pascual-Montano, A., Carazo, J.M., Kochi, K., Lehmann, D., Pascual-Marqui, R.D.: Nonsmooth nonnegative matrix factorization (nsNMF). IEEE Trans. Pattern Anal. Mach. Intell. **28**(3), 403–415 (2006)
17. Stadlthanner, K., Theis, F.J., Puntonet, C.G., Lang, E.W.: Extended sparse nonnegative matrix factorization. In: Cabestany, J., Prieto, A., Sandoval, F. (eds.) IWANN 2005. LNCS, vol. 3512, pp. 249–256. Springer, Heidelberg (2005). doi:10.1007/11494669_31
18. Tong, M., Guo, J., Tao, S., Wu, Y.: Independent detection and self-recovery video authentication mechanism using extended NMF with different sparseness constraints. Multimedia Tools Appl. **75**(13), 8045–8069 (2016)

An Efficient Auction Mechanism Toward Heterogeneous Spectrum Allocation

Haiyan Qin[1], Xin Li[1], Yonglong Zhang[1], and Bin Li[1,2(✉)]

[1] College of Information Engineering, Yangzhou University,
Yangzhou, China
1309200650@163.com, 846065272@qq.com, {ylzhang,lb}@yzu.edu.cn
[2] State Key Laboratory for Novel Software Technology,
Nanjing University, Nanjing, China

Abstract. It's widely recognized that auction is an efficient method to allocate spectrum resource. However, due to exaggerated price asked in the primary market, secondary users with limited budget cannot access to benefits in such auction. In our paper, we consider the scenario that spectrum holder releases heterogeneous channels to secondary users. Therefore, we propose an Efficient Auction Mechanism Toward Heterogenous Spectrum Allocation, dubbed EATHER, where channels are allocated as a greedy mode based on 'bid density'. Our auction scheme gives sufficient consideration to heterogeneity of channel which is one of the challenges in spectrum auction. We show analytically that EATHER has polynomial time complexity. More precisely, EATHER is efficient. Our analysis demonstrates EATHER achieves truthfulness, individual rationality and budget balance. The simulation evaluates the performance of EATHER minutely in teams of buyer satisfaction ratio, channel utilization ratio and social welfare.

1 Introduction

As an efficient resource allocation method, auction has been introduced to various fields such as cooperative communication [5], bandwidth [6], electricity [7] and cloud resource [8]. The dramatic application in wireless promotes competition in spectrum resource nowadays. Therefore, auction is applied in spectrum allocation [1] which allows mobile phone operators supply an abundance of new services [2]. In the primary spectrum market, specialized management institutions, such as the FCC (Federal Communications Commission) has held several auctions to allocate spectrum resource. Three carriers, Telefonica, Deutsche Telecom and Vodafone Germany obtained twenty blocks channel and priced up to 5.081 billion euros in Germany multi-frequency spectrum auction which held in June 2015 [3]. Thereout, we observe that only several biggest communication companies can afford the fee in the primary spectrum market, while some small wireless service providers cannot benefit from it. Therefore, we bring double auction into the secondary spectrum market so as to redistribute spectrum.

In secondary market [4], the Spectrum Holder (SH) releases some idle channels to Secondary Users (SUs) for the sake of monetary award. SUs are the users

© Springer International Publishing AG 2016
H. Yin et al. (Eds.): IDEAL 2016, LNCS 9937, pp. 361–370, 2016.
DOI: 10.1007/978-3-319-46257-8_39

who manage small wireless application. Likewise, the SU makes diverse requests for heterogeneous channels and pays minority fee if he wins. Without loss of generality, a trustworthy agent works as the auctioneer and decides the auction results. The problem of spectrum redistribution between multiple channels and several SUs can be modeled as a single-round multi-item double auction.

However, designing available auction mechanism for spectrum redistribution has several challenges. One major challenge is heterogeneity of spectrum. Other than isomorphic items, buyers are enthusiastic to present preference for several channels which makes allocation troublesome. Fundamental essence of auction brings another challenge, truthfulness (please refer to Definition 1 for definition). In a truthful auction, players are enforced to report their true bid or ask. In other word, his optimal strategy is his true bid or ask. The last challenge which has to be addressed is time complexity. Most existing work allocate channels viably while sacrificing time complexity.

In this paper, we propose EATHER, an Efficient Auction Mechanism Toward HEteRogeneous Spectrum Allocation to address the above-mentioned challenges. We present a novel allocation mechanism on the basis of 'bid density' with polynomial time complexity. The members in Daniel Grosu group make great contributions on 'bid density' [9,10]. Meanwhile, we fully consider the heterogeneity of the channel and import weight to measure the channel. Moveover, EATHER guarantees truthfulness, individual rationality and budget balance.

2 Preliminaries and Problem Formulation

2.1 Auction Model

As shown in Fig. 1, we consider the scenario as an open market of secondary spectrum where n secondary users (SUs) purchase channels from spectrum holder (SH). In the auction, there is one spectrum holder in the model who holds m idle heterogeneous channels to sell. The SH is seller, SUs are buyers and channels are goods. We assume that each SU presents various demands to channels but would like to buy one channel ultimately and each channel can be assigned to at most one buyer. We also assume that the seller is trustworthy whose reserve price is truthful. For simplicity, the SH plays the role of the auctioneer. In the procedure of auction, buyers submit their bid to the auctioneer in a sealed way. In other words, the SU has no possibility to obtain information of other participants'. In addition, we assume that bidders do not collude with each other to improve theirs utility. The SH executes the auction to decide the auction results, including one-to-one map between buyers and channels and the payment collected from the buyers. We now characterize the entities in the auction.

Secondary User: Let $U = \{u_1, \ldots, u_n\}$ denote the set of SUs. The valuation of u_i for c_j is v_i^j and $V_i = \{v_i^1, \ldots, v_i^m\}$ is valuation vector for all channels. Correspondingly, $B_i = \{b_i^1, \ldots, b_i^m\}$ is the bid vector for all channels where b_i^j is u_i's bid for c_j. $B = \{B_1, \ldots, B_n\}$ denotes the bid matrix of all buyers. If u_i is in the set of winning buyers W_u, u_i can rent $c_{\sigma(i)}$ for a period of time. σ is

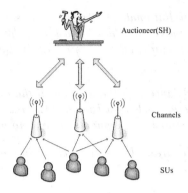

Fig. 1. An model of secondary spectrum allocation

map function here. In the meantime, u_i pays $p_i^{\sigma(i)}$ for renting $c_{\sigma(i)}$. The payment vector of all buyers is denoted by $P = \{p_1^{\sigma(1)}, \ldots, p_n^{\sigma(n)}\}$.

Spectrum Holder: There are a set of heterogeneous channels $C = \{c_1, \ldots, c_m\}$ where $c_j \in C$ are provided by the same SH. For the sake of interests, each $c_j \in C$ sets up his reverse price rp_j which is the lowest price c_j would like to tolerate. Channels in C are heterogeneous in teams of bandwidth, frequencies, maximum allowed transmission powers and so on. Therefore, we introduce weight w_j as a measure of the heterogeneity of c_j. $A = \{< rp_1, w_1 >, \ldots, < rp_m, w_m >\}$ is a two-dimension ask of all channels. In addition, the SH who acts as the auctioneer determines the results of the auction, including W_c, W_u, P and σ. W_c and W_u indicates the set of winning channels and users respectively.

Therefore, the utility of $u_i \in W_u$ is the difference between valuation $v_i^{\sigma(i)}$ and payment to SH $p_i^{\sigma(i)}$. If u_i is a loser, the utility is zero.

$$U_i^b = \begin{cases} v_i^{\sigma(i)} - p_i^{\sigma(i)} & \text{if } u_i \in W_u, \\ 0 & \text{otherwise.} \end{cases} \tag{1}$$

Similarly, the utility of seller is the sum of the difference between the charge from SU $p_{-\sigma(j)}^j$ and reserve price rp_j, $c_j \in W_c$.

$$U^s = \sum_{c_j \in W_c} (p_{\sigma^{-1}(j)}^j - rp_j) \tag{2}$$

2.2 Definition of Concepts

Definition 1 *(Truthfulness). Truthfulness means that no player can improve his utility by misreporting his bid or ask regardless of other participants' bid or ask.*

For players, their dominant strategy [11] is their true valuation irrespective of other players' strategy. In that way, players have no worry about manipulation in the market.

Definition 2 *(Individual Rationality). An auction is individual rationality if the winning seller is charged more than his ask and the winning buyer is paid less than his bid. In other words, all players in the auction gain a non-negative utility.*

Definition 3 *(Budget Balance). In the auction, if the auctioneer gains a non-negative revenue, the auction is budget balance. The revenue is the fee collected from all winning buyers minus the payments charged to all winning sellers.*

Definition 4 *(Social Welfare). The social welfare in our auction is the sum of winning buyers' valuation:*

$$SW = \sum_{i=1}^{n} \sum_{j=1}^{m} v_i^j x_{ij} \tag{3}$$

Subject to:

$$x_{ij} = \{0, 1\}, \quad \forall i, \forall j \tag{4}$$

$$\sum_{j=1}^{m} x_{ij} = \{0, 1\}, \forall i; \quad \sum_{i=1}^{n} x_{ij} = \{0, 1\}, \forall j \tag{5}$$

In Eq. (4), if $x_{ij} = 1$, it means u_i acquires c_j successfully. On the contrary, $x_{ij} = 0$ indicates c_j isn't assigned to u_i. In Eq. (5), $\sum_{j=1}^{m} x_{ij} = \{0, 1\}, \forall i$ denotes that each buyer would like to purchase at most one channel. Similarly, $\sum_{i=1}^{n} x_{ij} = \{0, 1\}, \forall j$ signifies each $c_j \in C$ can be assigned to at most one buyer.

According to [12], truthfulness, individual rationality, budget balance and system efficiency cannot be achieved in any double auction at the same time. System efficiency stands for maximizing social welfare in our paper. Our scheme, EATHER looses system efficiency so as to ensure the other three properties.

3   Auction Design

We describe the specific auction design which consists of two stages, Winner-Determination and Clearing-Pricing. In addition, we reveal that our scheme satisfies three traditional properties and has a polynomial time complexity.

3.1   Auction Procedure

Winner-Determination. We present a greedy mode based on bid density to screen winning buyers. After gathering offer from buyers and seller such as U, B, C and A, the auctioneer conducts the auction process. Among A, weight w_j of $c_j \in C$ is public for buyers. In such way, it's convenient for buyers to express diversified preferences for channels. We first sum up all bids for c_j in $sumb_j$ as

---

Algorithm 1. Winner-Determination(C, A, U, B)

1: $W_u \leftarrow \emptyset, W_c \leftarrow \emptyset$
2: **for** $j = 1, \ldots, m$ **do**
3:     $sumb_j \leftarrow \sum_{i=1}^{n} b_i^j$
4: **end for**
5: **for** $i = 1, \ldots, n$ **do**
6:     **for** $j = 1, \ldots, m$ **do**
7:         $\gamma_i^j \leftarrow b_i^j / sumb_j$
8:     **end for**
9: **end for**
10: **for** $i = 1, \ldots, n$ **do**
11:     $\delta_i \leftarrow \sum_{j=1}^{m} \left(\gamma_i^j w_j \right)$
12: **end for**
13: Sort U to get an ordered list
    $U' = \{u_{i_1}, \ldots, u_{i_n}\}$ such that $\delta_{i_1} \geq \ldots \geq \delta_{i_n}$
14: **for** $i = i_1, \ldots, i_n$ **do**
15:     Sort B_i to get an ordered list
    $B_i' = \{b_i^{j_1}, \ldots, b_i^{j_m}\}$ such that $\gamma_i^{j_1} \geq \ldots \geq \gamma_i^{j_m}$
16:     **for** $j = j_1, \ldots, j_m$ **do**
17:         **if** $b_i^j \geq rp_j$ and $c_j \notin W_c$ **then**
18:             $W_u \leftarrow W_u \cup \{u_i\}, W_c \leftarrow W_c \cup \{c_j\}, \sigma(i) = j$
19:         **end if**
20:     **end for**
21: **end for**
22: **return** W_c, W_u, σ

---

the base of each channel. Unlike [9,10], 'bid density' γ_i^j in our paper is calculated as the proportion of b_i^j in $sumb_j$, i.e.,

$$\gamma_i^j = b_i^j / sumb_j \qquad \forall u_i \in U, \quad \forall c_j \in C \tag{6}$$

In order to compare priorities of buyers, we figure up δ_i as a measure. The u_i in the ahead of the sequence of $\{\delta_i\}$ takes precedence to assign channel. We prefer to distribute unallocated channel c_j whose γ_i^j is bigger in the vector of $\{\gamma_i^j\}$. Then c_j can be assigned to u_i if b_i^j is greater than rp_j so as to ensure the utility of c_j. The detailed mechanism is shown in Algorithm 1.

Clearing-Pricing. In this stage, we figure out the fee the winning buyer should pay for the matched channel which is decided in Winner-Determination stage. The payment of $u_i \in W_u$ is greater than the reserve price of matched channel so as to ensure budget balance. On the same time, the payment should be no more than buyer's bid in order to ensure individual rationality. Algorithm 2 shows the pricing schedule. When u_i is removed from U, $c_{\sigma(i)}$ may be assigned to another buyer u_k. The clearing price is divided into two condition refer to $b_k^{\sigma(i)}$. If $b_i^{\sigma(i)} \geq b_k^{\sigma(i)}$, p_i^j is the bid of u_k for $c_{\sigma(i)}$. Otherwise, p_i^j is the reserve price of c_j. Similarly, if $c_{\sigma(i)}$ hasn't been assigned to another channel when u_i is removed from the U, u_i should pay rp_j for c_j.

Algorithm 2. Clearing-Price(W_c, W_u, σ)

1: **for each** $u_i \in W_u$ **do**
2:   $W_u', W_c' \leftarrow$ Winner-Determination(C, RP, U_{-i}, B_{-i})
   // $U_{-i} = \{u_1, \ldots, u_{i-1}, u_{i+1} \ldots, u_n\}$,
         $B_{-i} = \{B_1, \ldots, B_{i-1}, B_{i+1}, \ldots, B_n\}$
3:   **if** $c_{\sigma(i)} \in W_c'$ s.t. $c_{\sigma(i)}$ allocated to u_k, $u_k \in W_u'$ **then**
4:     **if** $b_i^{\sigma(i)} \geq b_k^{\sigma(i)}$ **then**
5:       $p_i^j = b_k^{\sigma(i)}$
6:     **else**
7:       $p_i^j = rp_{\sigma(i)}$
8:     **end if**
9:   **end if**
10:   **if** $c_{\sigma(i)} \notin W_c'$ **then**
11:     $p_i^j = rp_{\sigma(i)}$
12:   **end if**
13: **end for**
14: **return** W_c, W_u, P

For Winner-Determination stage, the time complexity depends on Line 14–21 in Algorithm 1. The sort algorithm can apply quick sort, bubble sort or selection sort. No matter which sort mechanism adopted, $T \in \{mlogm, m^2\}$ is polynomial. The loop of this part is $n \cdot m$. Hence the time complexity of Algorithm 1 is $O(n \cdot T \cdot m)$. In Pricing stage, we go through the winning buyers and re-execute the Winner-Determination. Therefore, the time complexity of Algorithm 2 is $O(l \cdot n \cdot T \cdot m)$. l is the amount of winning channel-buyer pairs here. Generally speaking, the time complexity of EATHER is $O(l \cdot n \cdot T \cdot m)$ which is polynomial.

3.2 Analysis

In this section, we prove that our auction mechanism achieves truthfulness individual rationality and budget balance.

Theorem 1. *EATHER achieves truthfulness.*

proof: As mentioned in Definition 1, truthfulness is that the player cannot benefit from manipulating his bid or ask no matter what strategy other players choose. Specially, for buyers, if $b_i^{\sigma(i)} \neq v_i^{\sigma(i)}$ with B_{-i} fixed, $\tilde{U}_i^u \leq U_i^u$, where $\tilde{U}_i^u$ and U_i^u is the utility when u_i bids $b_i^{\sigma(i)}$ and $v_i^{\sigma(i)}$. We discuss the truthfulness of buyers from four cases:

(1) *Case 1:* If u_i loses no matter he bids $b_i^{\sigma(i)}$ or $v_i^{\sigma(i)}$, then $\tilde{U}_i^u = U_i^u = 0$.
(2) *Case 2:* u_i wins if bidding $v_i^{\sigma(i)}$, but loses while bidding $b_i^{\sigma(i)}$. Therefore, it is clearly $U_i^u \geq \tilde{U}_i^u = 0$ in this case.
(3) *Case 3:* If u_i loses when bidding $v_i^{\sigma(i)}$ while winning by bidding $b_i^{\sigma(i)}$. u_i changes his bid on $c_{\sigma(i)}$ with other bids fixed, hence u_i is assigned to $c_{\sigma(i)}$ when bidding untruthfully. We have no idea whether $c_{\sigma(i)}$ is allocated to

another buyer u_k when u_i loses by bidding $v_i^{\sigma(i)}$, so the proof in this case can be divided into two condition:

- If $c_{\sigma(i)}$ is assigned to u_k when u_i bids $v_i^{\sigma(i)}$. Therefore, u_i should enlarge his $v_i^{\sigma(i)}$ until bigger than $b_k^{\sigma(i)}$ so that $c_{\sigma(i)}$ can be preferentially allocated to u_i instead of u_k. In other words, $b_i^{\sigma(i)} > b_k^{\sigma(i)} \geq v_i^{\sigma(i)}$. According to our pricing schedule, $p_i^{\sigma(i)} = b_k^{\sigma(i)}$ because $c_{\sigma(i)}$ will be allocated to u_k when u_i is removed from U. We can get $\tilde{U}_i^u = p_i^{\sigma(i)} - v_i^{\sigma(i)} = b_k^{\sigma(i)} - v_i^{\sigma(i)} < U_i^u = 0$.

- If $c_{\sigma(i)}$ hasn't been assigned to any buyer when u_i bids $v_i^{\sigma(i)}$. When it's turn to allocate u_i, we will compare $b_i^{\sigma(i)}$ with $rp_{\sigma(i)}$. $c_{\sigma(i)}$ hasn't been assigned to any buyer when u_i bids $v_i^{\sigma(i)}$ but assigned to u_i when bidding $b_i^{\sigma(i)}$, so that $b_i^{\sigma(i)} > rp_{\sigma(i)} > v_i^{\sigma(i)}$. We know $p_i^{\sigma(i)} = rp_{\sigma(i)}$ on account of our pricing schedule. To sum up, $\tilde{U}_i^u = p_i^{\sigma(i)} - v_i^{\sigma(i)} = rp_{\sigma(i)} - v_i^{\sigma(i)} < U_i^u = 0$.

(4) *Case 4:* u_i wins when bidding $v_i^{\sigma(i)}$ or $b_i^{\sigma(i)}$, so that $\tilde{U}_i^u = U_i^u$.

The truthfulness of buyers is proved above. We have assumed that the reserve price of channels is true, i.e., EATHER is true for sellers. Taken together, EATHER achieves truthfulness.   ∎

Theorem 2. *EATHER achieves individual rationality.*

proof: If c_j will be sold to $u_{\sigma^{-1}(j)}$, $p_{\sigma^{-1}(j)}^j$ can be distinguished two condition:

- If c_j is assigned to u_k when $u_{\sigma^{-1}(j)}$ is removed from U, we can get $p_{\sigma^{-1}(j)}^j = b_k^j$ and $b_k^j \geq rp_j$. Therefore, $p_{\sigma^{-1}(j)}^j - rp_j = b_k^j - rp_j \geq 0$.
- If c_j hasn't been assigned to any buyer when u_i bids v_i^j, we can get $p_{\sigma^{-1}(j)}^j = rp_j$. Therefore, $p_{\sigma^{-1}(j)}^j - rp_j = 0$.

According to Eq. (5), U^s is the sum of all $p_{\sigma^{-1}(j)}^j - rp_j$, so that $U^s \geq 0$. Therefore, the individual rationality of sellers holds. The truthfulness of buyers can be proof in the same way, hence we omit here.   ∎

Theorem 3. *EATHER achieves budget balance.*

proof: In EATHER, the seller plays the role of auctioneer, hence the fee collected from the winning buyers is the revenue of the auction which is non-negative. In other words, EATHER achieves budget balance.   ∎

4   Simulation

In this section, we simulate EATHER and evaluate its performance from three metrics.

4.1 Evaluation Setup

We simulate a SH provides a set of channels to numbers of SUs who ask for channel. The reserve price of channels is uniformly distributed over $(2, V_{max}]$ where V_{max} can be within the range of $[4, 10]$. The buyers' bids are randomly selected in the interval $[0, V_{max}]$. In order to embody the competition to channel, the lowest reserve price is 2 here. For simplicity, we assume $w_i < w_j$ if $i < j$ here. All the results on performance are averaged over 1000 runs.

The metrics evaluated in the simulation are as follows:

- Buyer satisfaction ratio: Buyer satisfaction ratio is the proportion of the total buyers who are the winners.
- Channel utilization ratio: Channel utilization ratio is the ratio of the total channels which are assigned to buyers.
- Social welfare: As mentioned in Definition 4, the social welfare in our auction is the sum of winning buyers' valuation.

Buyer satisfaction ratio and channel utilization ratio are as a measure of the system performance of the auction. Social welfare reflects the economic benefit this spectrum auction brings in.

4.2 Impact on System Performance

Figure 2 shows the performance of EATHER when the number of channels and buyers is changing. The number of buyers is varied between 10 and 100 when there are 20, 40 and 60 channels. The V_{max} is fixed at 6.

Figure 2(a) illustrates buyer satisfaction ratio reaches its maximum at the beginning but decreases when the number of buyers extends. When there are 60 channels, buyer satisfaction ratio almost attains 1 which means each buyer can be allocated to one channel. The reason for the decline in buyer satisfaction ratio is that more buyers in the auction leads to more intense competition. Buyer satisfaction ratio ascends with the growth of channels, because larger amount of channels means more buyers' request satisfied. In Fig. 2(b), we can see that channel utilization ratio almost linearly increases until the turning point reaches. Channel utilization ratio grows at first, this is due to limitation on the number of buyers brings about a portion of channels have no chance to be allocated. Once the number of buyers saturated, channel utilization ratio stays in a stable state. Figure 2(c) indicates social welfare grows with the scale of buyers enlarged, but the speed of growth slows down and gradually enters a stable state. This is because there will be more successful trades as the number of buyers increasing, i.e., more winners. We notice that social welfare is higher when there are larger channels which means more successful trades.

In Fig. 3, we fix the number of buyers at 20 and vary the number of channels from 5 to 40 with increment of 5. Here we obverse the performance of EATHER with V_{max} varied from 4 to 10 as contrasts.

Figure 3(a) shows that buyer satisfaction ratio almost linearly increases with the growth of channels. Gradually, the speed of growth slows down because

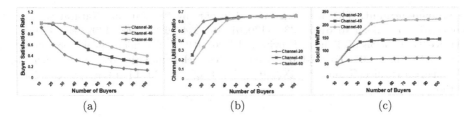

Fig. 2. The performance when the number of channels varies

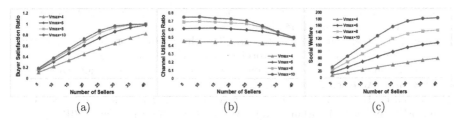

Fig. 3. The performance when V_{max} varies

the number of buyers is fixed in 20 which means no more effective trades. In Fig. 3(b), we observe channel utilization ratio decreases as the scale of sellers extends. Again, the number of buyers limits the growth of channel utilization ratio. The turning point is around the 25 channels no matter which V_{max} is. It means the number of channels around 25 is the transaction saturation point to 20 buyers. It is shown in Fig. 3(c) that social welfare almost linearly goes up but the growth is slowing as the number of channels increasing. Similarly, the number of buyers limits the successful trades as well social welfare. Furthermore, in Fig. 3, we observe the performance of EATHER is better with higher V_{max} under the condition of the same number of channels. The higher V_{max}, the value of bids is looser which means the expected value of buyers' valuation is higher. Because of that, the probability that bid of buyers greater than the reserve price is higher which brings about more trades. For this reason, buyer satisfaction ratio, channel utilization ratio and social welfare are greater when V_{max} is bigger.

5   Conclusion and Future Work

In this paper, we have designed EATHER to address the problem of spectrum allocation in the secondary market. In a novel method, the buyer ahead in the sequence $\{\delta_i\}$ has priority to be assigned to channel. Considering efficiency of the auction, we allocate channels greedily in view of 'bid density'. Our theory analysis demonstrates that EATHER possesses good properties such as truthfulness, individual rationality and budget balance though sacrificing social welfare. Extensive experiment exposes detailed performance of EATHER when the number of channels and V_{max} varies respectively. As for future work, EATHER can be applied to group-buying or approach the problem in online spectrum allocation.

Acknowledgment. This work was funded in part by the National Natural Science Foundation of China (No. 61070133, 61170201, 61472344, 61402396, 61402234); Natural Science Foundation of Jiangsu Province (BK20150460); Six talent peaks project in Jiangsu Province (2011-DZXX-032); Scientific Research Foundation of Graduate School of Jiangsu Province(Grant No. CXZZ16_1889). Professor Bin Li is the corresponding author.

References

1. Ting, S.: Auction, an effective means for radio frequency spectrum management. World Telecommun. **6**, 41–44 (1999)
2. Grimm, V., Riedel, F., Wolfstetter, E.: The third generation (UMTS) spectrum auction in Germany. In: Cesifo Working Paper, vol. 48, no. 10 (2001)
3. Germany multi-frequency spectrum auction. http://www.srrc.org.cn/NewsShow 12887.aspx. Accessed on 20 May 2016
4. Zhu, Y., Li, B., Li, Z.: Truthful spectrum auction design for secondary networks. In: IEEE INFOCOM, Orlando, vol. 131, no. 5, pp. 873–881, March 2012
5. Yang, D., Fang, X., Xue, G.: Truthful auction for cooperative communications. In: Proceedings of the ACM Mobihoc, Paris, France, pp. 89–98 (2011)
6. Zheng, Z., Gui, Y., Wu, F., Chen, G.: STAR: strategy-proof double auctions for multi-cloud, multi-tenant bandwidth reservation. IEEE Trans. Comput. **64**(7), 2071–2083 (2015)
7. Zhang, L., Li, Z., Wu, C.: Randomized auction design for electricity markets between grids and microgrids. ACM Sigmetrics Perform. Eval. Rev. **42**(1), 99–110 (2014)
8. Bonacquisto, P., Di Modica, G., Petralia, G., Tomarchio, O.: A strategy to optimize resource allocation in auction-based cloud markets. In: IEEE International Conference on Services Computing, pp. 339–346 (2014)
9. Nejad, M.M., Mashayekhy, L., Grosu, D.: Family of truthful Greedy mechanisms for dynamic virtual machine provisioning and allocation in clouds. IEEE Trans. Parallel Distrib. Syst. **26**(2), 594–603 (2015)
10. Zaman, S., Grosu, D.: Combinatorial auction-based allocation of virtual machine instances in clouds. J. Parallel Distrib. Comput. **73**(4), 495–508 (2010)
11. Osborne, M.J., Rubenstein, A.: A Course in Game Theory. MIT Press, Cambridge (1994)
12. Myerson, R.B., Satterthwaite, M.A.: Efficient mechanisms for bilateral trading. J. Econ. Theor. **29**(2), 265–281 (1983)

An Improved Recommender Model by Joint Learning of Both Similarity and Latent Feature Space

Yunxiang Tao and Ming Yang$^{(\boxtimes)}$

School of Computer Science and Technology, Nanjing Normal University,
Nanjing 210023, China
taoyunxiang@hotmail.com, myang@njnu.edu.cn

Abstract. The matrix factorization recommender system based on manifold regularization, taking into account the similarity of local neighbors and manifold structure, can improve the quality of a recommendation system. However, the similarity between samples may not be accurate due to the sparsity of the data or the incompleteness of the tag information. Therefore, we propose a new model called SI-GMF (Similarity-learning-based Improved Graph Regularized matrix Factorization) by embedding the new similarity measure strategy in GMF (Graph Regularized matrix Factorization) framework, and induce three new matrix factorization algorithms (SI-GMF_1, SI-GMF_2, SI-GMF_3) based on three initial similarities by employing three different similarity measures. The solutions to the newly developed algorithms can be effectively obtained by SGD method. The experimental results show that the newly designed algorithms significantly improve the accuracy of a recommender system.

Keywords: Matrix factorization · Manifold regularization · Similarity · Recommendation system

1 Introduction

Recommendation system help users find the key point in their own concerns from the mass of information. It plays an important role in people's daily lives.

Currently, the existing recommendation techniques include content-based recommendation [1], collaborative filtering [2] and so on. Collaborative filtering can be divided into the neighborhood-based method [3] and matrix factorization method [4]. In neighborhood-based collaborative filtering, similarity measure plays an important role, while matrix factorization holds low space complexity and high prediction accuracy. So, how to effectively embed the similarity into matrix factorization based collaborative filtering framework is the focus of this paper. In this paper, we propose a new model called SI-GMF by embedding the new similarity measure strategy in GMF framework. In SI-GMF, based on alternatively updating and stochastic gradient descent method, the latent feature space can be effectively obtained by simultaneously learning latent feature space and similarities.

© Springer International Publishing AG 2016
H. Yin et al. (Eds.): IDEAL 2016, LNCS 9937, pp. 371–378, 2016.
DOI: 10.1007/978-3-319-46257-8_40

The rest of this paper is organized as follows. In Sect. 2, we present some of the most relevant works. Section 3 presents our method. The results of experimental analysis are presented in Sect. 4, followed by the conclusion and future work in Sect. 5.

2 Related Works

So far, there are lots of works on matrix factorization algorithm. In this section, we only present relevant works on two classic approaches: recommendations based on basic matrix factorization and recommendations based on bias matrix factorization.

2.1 Basic Matrix Factorization

Rating prediction problem is the core of the recommendation system. The user's rating behavior can be expressed as a rating matrix R. Since a lot of elements in this matrix is missing, what we want to do is to predict these missing values.

From the view of matrix factorization, we decompose the rating matrix R into the product of two low dimensional matrices: $R = P^T Q$. $P \in R^{f \times m}$ and $Q \in R^{f \times n}$ are the two matrices after dimensionality reduction. The objective function is as follows:

$$\min \sum_{(u,i) \in Tain} \left(r_{ui} - p_u^T q_i\right)^2 + \lambda\left(||p_u||^2 + ||q_i||^2\right) \tag{1}$$

The model is named as Basic MF [5]. Here, λ is the regularization parameter, r_{ui} is the rating by user u on item i, p_u is user-factors vector of user u and q_i is item-factors vector of item i. The minimization process is generally achieved by stochastic gradient descent [6].

2.2 Bias Matrix Factorization

In some practical situations, a rating system has some inherent properties that have nothing to do with the item, users have some attributes that are not related to the items, items also get some attributes that are not related to users. Researchers put forward a matrix factorization algorithm which called Bias MF [7], the model is as follows:

$$\min \sum_{(u,i) \in Tain} \left(r_{ui} - u - b_u - b_i - p_u^T q_i\right)^2 + \lambda\left(||p_u||^2 + ||q_i||^2 + b_u^2 + b_i^2\right) \tag{2}$$

Compared to Basic MF, Bias MF adds the global average of all the rating records u, user bias item b_u, item bias b_i. b_u indicates that the user's rating habits are not related to the factors while b_i indicates that the rating of the item is not related to the users. The model needs to learn the parameters b_u, b_i, p_u, q_i. The minimization process can also be achieved by stochastic gradient descent.

3 Our Model: SI-GMF

In this section, we present our model. First, we introduce GMF model, and then describe the global framework SI-GMF. Next, we give the solution to SI-GMF. Finally, we use three different methods to obtain the initial similarity, which represent different precision, three recommendation algorithms are induced by SI-GMF model.

3.1 GMF

LPP (Locality Preserving Projection) [8] is a linear dimensionality reduction method. In LPP, the samples are similar while their projections are similar too. Hu et al. proposed a non negative matrix factorization model based on Manifold Regularization [9]. Lo et al. combined the collaborative filtering algorithm with the manifold regularization for the evaluation of Web service quality [10, 11]. In Refs. [10, 11], following the idea of LPP, if items are similar, then the item-factors vector after matrix factorization should also be similar. Using this locality strategy, by embedding similarity manifold regularization into biasMF, GMF model is obtained as follows.

$$
\begin{aligned}
\min \sum_{(u,i)\in Tain} & (r_{ui} - u - b_u - b_i - p_u^T q_i)^2 \\
& + \lambda_1 (\|p_u\|^2 + \|q_i\|^2 + b_u^2 + b_i^2) + \lambda_2 \sum_{j \in N(i)} S_{ij} \|q_i - q_j\|^2
\end{aligned}
\tag{3}
$$

In the model, S_{ij} is the similarity in the original space between item i and item j, $N(i)$ represents the set of item i's neighbors. Here, if s_{ij} is larger, the value of the distance between q_i and q_j should be small. This constraint is designed to minimize the difference of factors between the items and its nearest neighbors.

3.2 SI-GMF Model

Imprecise similarity may lead to a bad recommendation result. The method we proposed can tune similarity adaptively. S^0 is denoted as the initial similarity matrix. The initial value of the i-th and j-th S_{ij}^0 in S^0 may not be entirely accurate, which need to be adaptively dynamic adjusted. So, in this paper, we design a new model called SI-GMF, which can simultaneously learn latent feature space and similarities, corresponding formula as follows:

$$
\begin{aligned}
\min \sum_{(u,i)\in Tain} & (r_{ui} - u - b_u - b_i - p_u^T q_i)^2 + \lambda_1 (\|p_u\|^2 + \|q_i\|^2 + b_u^2 + b_i^2) \\
& + \lambda_2 \sum_{j \in N(i)} S_{ij} \|q_i - q_j\|^2 + \lambda_3 \sum_{j \in N(i)} \|s_{ij} - S_{ij}^0\|^2
\end{aligned}
\tag{4}
$$

3.3 Solution to SI-GMF

The parameters are alternatively updated using stochastic gradient descent. Note $e_{ui} = r_{ui} - u - b_u - b_i - p_u^T q_i$, γ is the learning rate, we can update parameters by the following rules:

$$b_u \leftarrow b_u + \gamma(e_{ui} - \lambda_1 b_u) \tag{5}$$

$$b_i \leftarrow b_i + \gamma(e_{ui} - \lambda_1 b_i) \tag{6}$$

$$p_u \leftarrow p_u + \gamma(e_{ui} \cdot q_i - \lambda_1 p_u) \tag{7}$$

Fix S_{ij}, update q_i and fix q_i, q_j, update S_{ij}, the formulas are as follows:

$$q_i \leftarrow q_i + \gamma \left(e_{ui} \cdot p_u - \lambda_1 q_i - \lambda_2 \sum_{j \in N(i)} S_{ij}(q_i - q_j) \right) \tag{8}$$

$$S_{ij} \leftarrow S_{ij} + \gamma \left(-\frac{1}{2} \lambda_2 \|q_i - q_j\|^2 - \lambda_3 (S_{ij} - S_{ij}^0) \right) \tag{9}$$

It is worth noting that the model itself is reduced to BiasMF when the parameter λ_3 is equal to 0. S_{ij} is no longer handled as a parameter and the solution to the model should follow the rules of BiasMF.

3.4 Three Recommendation Algorithms Induced by SI-GMF Model

In order to evaluate the influence of the initial similarity on SI-GMF model, three different similarity measures are employed as follows:

$$S_{ij} = \cos(l_i, l_j) = \frac{l_i \cdot l_j}{\|l_i\| \cdot \|l_j\|} \tag{10}$$

$$S_{ij} = \frac{1}{1 + e^{-p_{ij}}} \tag{11}$$

$$S_{ij} = \frac{\sum_{t=1}^{K} (p_t^{(i)} \times p_t^{(j)})}{\sqrt{\sum_{t=1}^{K} (p_t^{(i)})^2} \sqrt{\sum_{t=1}^{K} (p_t^{(j)})^2}} \tag{12}$$

In Eq. (10), the similarity between item i and item j is obtained based on the expert tags. l_i, l_j represent tag vectors. Cosine value between tag vectors is expressed as the similarity between items. The initial similarity obtained in this approach is roughly accurate.

In Eq. (11), similarity is obtained based on ratings. Pearson correlation coefficient $p_{ij} \in [-1, 1]$ is used to calculate the correlation between item i and item j, then p_{ij} is scaled between 0 and 1 using Eq. (11). Due to the small size of the data set and the sparsity of the ratings, the initial similarity is also roughly accurate.

The similarity in Eq. (12) is obtained on the topic space. $p_t^{(i)}$ represents the probability that item i belongs to topic t. K is the number of topics. In the experiment, we preprocess the MovieLens Tag Genome data set at first, find those records that their correlations are greater than 0.5, that is to find the tag words which can represent the attributes of the movie best. After the preprocessing, the record of each row is obtained as a "text" concept, the entire data set as a "text collection". Here, K is set to 100. The LDA (Latent Dirichlet Allocation) model [12] is used for obtaining the topic probability of the text. The initial similarity obtained in this way is relatively accurate.

Three recommendation algorithms are induced by SI-GMF model. When the initial similarity in SI-GMF model is obtained by Eq. (10), the algorithm is named as SI-GMF_1, analogously, when the initial similarities are obtained by Eq. (11) or Eq. (12), the corresponding algorithms are named as SI-GMF_2 and SI-GMF_3 respectively.

Algorithm 1. The algorithm of SI-GMF_1

1: The initial similarity is obtained by Eq.(10).
2: Initialize p_u, q_i, b_u, b_i (u=1...m, i=1...n).
3:    for t=1,2,...T do   /* T is the number of iterations*/
4:        for each user u and item i in the training set do
5:            Update the parameters $p_u, q_i, b_u, b_i, S_{ij}$ via Eqs. (5-9).
6:        Decrease the learning rate via $\gamma = \gamma \times 0.95$

Descriptions of SI-GMF_2 algorithm and SI-GMF_3 algorithm are similar to SI-GMF_1, the difference between them is the selection of initial similarity.

4 Experiments

4.1 Description of the Datasets

The 100 K MovieLens data set is used for our experiments. The data set contains 1000000 rating records from 943 users on 1682 films. Each record is a triple (u, i, r) which represents user u has rated item i and r is the rating. The rating is an integer from 1 to 5. The data set also includes tag information and demographic information.

Tag Genome MovieLens data set is used for obtaining the initial similarity. The data set contains relevant information between 1128 tags and 9734 films. The data format for the Tag Genome data set is: <MovieID> <TagID> <Relevance>. Relevance is the degree of correlation between the tag and the film, its value ranges from 0 to1, the lager the value is, the stronger the relevance between Tag and Movie is.

Mean Absolute Error (MAE) is used as the evaluation criterion. r_{ui} is the actual rating of the user i on the item u, r'_{ui} is a predictive rating achieved by the recommendation algorithm, T is the length of the test set, then MAE is defined as follows:

$$MAE = \frac{\sum_{u,i \in T} |r_{ui} - r'_{ui}|}{|T|} \tag{13}$$

4.2   Experimental Results and Analysis

The influence of λ_3 on the performance. λ_3 in formula (4) is the parameter which controls the offset range of the initial similarity and the updated similarity. The influence of parameter λ_3 on the prediction results are tested. Here, the latent semantic space dimension is fixed to 10, the number of the nearest neighbor of items is equal to 10, $\lambda_1 = 0.003$, $\lambda_2 = 0.001$, learning rate γ is set to 0.005. Experimental results on the test set are shown in Fig. 1, which describes the MAEs of SI-GMF_1, SI-GMF_2 and SI-GMF_3 under different λ_3:

As shown in Fig. 1., SI-GMF_1 and SI-GMF_3 obtain the minimum value of MAE when λ_3 is equal to 0.001, while SI-GMF_2 achieves the minimum value of MAE when λ_3 is equal to 0.003. If the value of λ_3 is too big or too small, the similarity may deviate from the true similarity and leads to a bad recommendation result.

Performance comparison. We fix adjustable parameter λ_3 for each algorithm and compare them with basic MF [5], Bias MF [7], GMF [10] model respectively.

For SI-GMF_1 and SI-GMF_2, let $\lambda_3 = 0.001$, for SI-GMF_3, let $\lambda_3 = 0.003$. In order to compare the performance of several algorithms intuitively, the experimental results are made into the line charts and bar charts. Figure 2 shows the MAE varies with the number of latent semantic space dimension grows and the performance comparison of four algorithms in the same latent semantic space dimension:

As shown above, the MAE of different algorithms all decrease as the value of latent semantic space dimension F becomes large. Meanwhile, the algorithms we proposed in

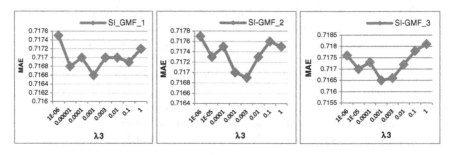

Fig. 1. Effect of λ_3 on MAE of SI-GMF_1, SI-GMF_2, SI-GMF_3 algorithms.

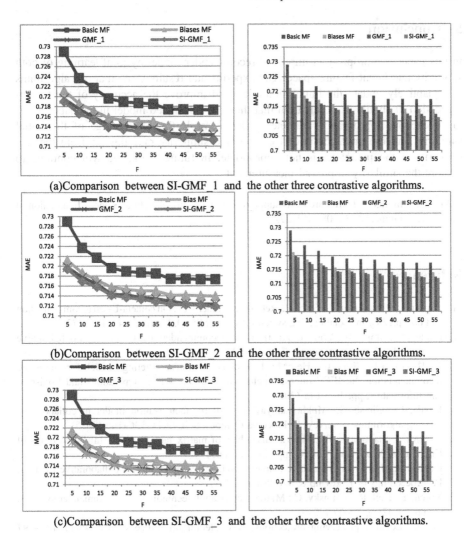

(a)Comparison between SI-GMF_1 and the other three contrastive algorithms.

(b)Comparison between SI-GMF_2 and the other three contrastive algorithms.

(c)Comparison between SI-GMF_3 and the other three contrastive algorithms.

Fig. 2. Line charts show the MAE value varies with the number of latent semantic space dimension grows; Bar charts show the comparison of MAE values in the same latent semantic space dimension.

this paper, SI-GMF_1, SI-GMF_2 and SI-GMF_3 all perform better than their contrastive algorithms on prediction accuracy in the same latent semantic space dimension.

From the three groups of experiments above we may see that when using different similarity metrics as the initial similarity, especially the initial similarity is not accurate, out method in the paper can improve the recommendation accuracy.

5 Conclusion

In this paper, we propose an improved recommender model by joint learning of both similarity and latent feature space. The experimental results show that our algorithms can adaptively update the initial similarity, and improve the accuracy of the recommendation. It should be pointed out that the newly designed model in the paper needs to choose the appropriate parameter which controls the offset range of the initial similarity and the updated similarity, since the similarity may deviate from the real similarity if the parameter is not appropriate. In future work, we will continue to explore the similarity improved measurement.

Acknowledgments. The research was supported by the National Natural Science Foundation of China under Grants 61432008, 61272222.

References

1. Puglisi, S., Parra-Arnau, J., Forné, J., Rebollo-Monedero, D.: On content-based recommendation and user privacy in social-tagging systems. Comput. Stand. Interfaces **41**, 17–27 (2015)
2. Linden, G., Smith, B., York, J.: Amazon.com recommendations: item-to-item collaborative filtering. IEEE Internet Comput. **7**(1), 76–80 (2003)
3. Ji, H., Li, J.F., Ren, C., He, M.: Hybrid collaborative filtering model for improved recommendation. In: IEEE International Conference on Service Operations and Logistics, and Informatics, pp. 142–145 (2013)
4. Bokde, D., Girase, S., Mukhopadhyay, D.: Matrix factorization model in collaborative filtering algorithms: a survey ✰. Procedia Comput. Sci. **49**(1), 136–146 (2015)
5. Wu, M.: Collaborative filtering via ensembles of matrix factorizations. In: Proceedings of Kdd Cup & Workshop, vol. 30, pp. 29–38 (2007)
6. Paterek, A.: Improving regularized singular value decomposition for collaborative filtering. In: Proceedings of Kdd Cup & Workshop (2007)
7. Koren, Y., Bell, R., Volinsky, C.: Matrix factorization techniques for recommender systems. Computer **42**(8), 30–37 (2009)
8. He, X.: Locality preserving projections. Adv. Neural Inf. Process. Syst. **45**(1), 186–197 (2005)
9. Hu, W., Choi, K.S., Wang, P., Jiang, Y., Wang, S.: Convex nonnegative matrix factorization with manifold regularization. Neural Netw. Official J. Int. Neural Netw. Soc. **63C**(1), 94–103 (2014)
10. Wei, L., Yin, J., Deng, S., Li, Y.: Collaborative web service QoS prediction with location-based regularization. In: IEEE International Conference on Web Services, pp. 464–471. IEEE (2012)
11. Yin, J., Wei, L., Deng, S., Li, Y., Wu, Z., Xiong, N.: Colbar: a collaborative location-based regularization framework for QoS prediction. Inf. Sci. **265**(5), 68–84 (2014)
12. Blei, D.M., Ng, A.Y., Jordan, M.I.: Latent Dirichlet allocation. J. Mach. Learn. Res. **3**, 993–1022 (2003)

Diagnosis Support for Orphan Diseases: A Case Study Using a Classifier Fusion Method

Xiaowei Kortum[1](✉), Lorenz Grigull[2], Urs Muecke[2], Werner Lechner[3],
and Frank Klawonn[1,4]

[1] Department of Computer Science, Ostfalia University of Applied Sciences,
Salzdahlumer Str. 46/48, 38302 Wolfenbuettel, Germany
{x.kortum,f.klawonn}@ostfalia.de
[2] Department of Paediatric Haematology and Oncology,
Medical University Hannover, Carl-Neuberg Str. 1, 30625 Hannover, Germany
{grigull.lorenz,urs.muecke}@mh-hannover.de
[3] Improved Medical Diagnostics IMD GmbH, Ostfeldstr. 25,
30559 Hannover, Germany
werner.lechner@improvedmedicaldiagnostics.com
[4] Helmholtz Center for Infection Research, Inhoffenstrasse 7,
38124 Braunschweig, Germany
Frank.Klawonn@helmholtz-hzi.de

Abstract. Orphan diseases post a particular problem to medical expert
and data analysts, because of the lack of data resources and sometimes
missing effective treatment. In order to shorten the diagnosing time for
rare diseases, we have gathered qualitative and quantitative data through
clinical observations, interviews and questionnaires of patients who suffer from rare diseases. From the perspective of data analysis, a pattern
recognition system based on an ensemble of classifiers was trained to
support the diagnosis of rare diseases. Our study shows that the combination of multiple classifiers has better performance in disease prediction
than any individual classifier. Furthermore, a testing method was used
to better understand the importance and influences of particular symptoms for certain diseases, and to determine how reliable or sensitive the
recognition of diseases through small adjustments in the given answers
of a patient.

Keywords: Fusion classifier · Multiple classifier system · Machine
learning · Pattern recognition

1   Introduction

There is no generalizable definition of an orphan or rare disease. However, most
countries define a rare disease as a disease that affects no more than one to ten
in 10,000 people.

For medical doctors (MDs), diagnosing an orphan disease is often a problem. Even though reliable tests for the disease might be available, MDs cannot

© Springer International Publishing AG 2016
H. Yin et al. (Eds.): IDEAL 2016, LNCS 9937, pp. 379–385, 2016.
DOI: 10.1007/978-3-319-46257-8_41

be aware of all possible orphan diseases, therefore sometimes they do not even consider the possibility of a specific type of rare disease. The time from the incipient symptoms until the final diagnosis of an orphan disease can sometimes take years or even decades. During this time, patients often invest a lot of time and money but still suffer from a bad and slowly worsening health status as well as wrong diagnoses and wrong treatments, even unnecessary surgery. For researchers, whether their central focus is about analysing the mechanism, disorders' pedigrees or clinical exploration of targeted therapies, the research on etiology of rare diseases always requires to obtain a sufficient number of samples. Therefore, collecting data and using statistical methods to understand the characteristics of rare diseases is a tedious and difficult task.

In order to shorten the long odyssey of patients with a rare disease, we established a collaboration research project with medical experts, created a questionnaire-based classification system to provide data support for assisting MDs to enhance their diagnosis. The design of the questionnaires is based on interviews, investigations and observations of patients that have been already diagnosed and reports from experienced MDs. The collection of questions is closely related to patients daily life but with strong correlation to typical early stage symptoms and pathogenesis of rare diseases. We streamlined and optimized the amount of questions, aiming to find questions that are representative for some specific disease symptoms. After distributing these questionnaires to relevant organizations and cooperating partners in Germany, we have collected more than 1,000 valid records from different categories of diseases and the responses are continuously increasing.

To discover implied knowledge on rare diseases, the main emphasis is to identify patients' early symptoms and potential issues by analysing the collected data and recognizing the answer patterns of different disease groups. We are also interested to figure out which questions are most useful to distinguish between different diseases to better understand the characteristics of the diseases and the predictions made by the classifier. The designed pattern recognition system and related methodology are introduced in the following sections.

2   Machine Learning Background

Jain et al. [4] discussed pattern recognition systems in detail and summarized some well-known machine learning algorithms and the way to select appropriate classifiers. Ho et al. [3] approved that because of the complementary recognition methods and flexibility in dynamic adaptation, a suitable combination of different classifiers makes the performance of multiple classifier systems more robust and accurate than individual classifiers.

Since the data we collected are all subordinate to their predefined disease groups, a supervised classification task with an ensemble of classifiers has been established. Support vector machines (SVM) have a good out-of-sample generalization performance. As a non-linear, non-parametric classification technique, SVM can work flexible with an appropriate kernel and can theoretically avoid over-fitting [2]. Linear discriminant analysis (LDA) attempts to discover the inherent structure in the data based on the assumption of normally distributed explanatory variables. Predictions are mainly related to estimate the mean and variance of training data. Logistic regression (LR) has no assumptions on the distribution of the features but requests classes are linearly separable. As a widely used classifier, LR is relatively robust, and the output of LR directly yields a probability for each category [7]. Random forests (RF), as a fast and non-parametric ensemble algorithm, are based on a set of decision trees using different (random) subsets of the data and features. Therefore RF are less prone to over-fitting than ordinary decision trees [5].

Multiple classifier systems have proven to often outperform single classifiers and have been widely used in medical studies. For instance, Sboner et al. [8] use multiple classifier systems to diagnose skin cancer and Ma et al. [6] use a fusion method to find signs for lung diseases in CT images.

3   The Process of Building the Diagnosis Support System

Figure 1 illustrates the process for the evaluation of our fusion classifier. The designed system operates in two modes: training (learning) mode and testing (classification) mode.

Assume a questionnaire that contains d types of diagnoses (d classes), each question is designed to represent a symptom or feature of the diseases. Thus if we have k questions in a diagnostic questionnaire, there will be k features or measurements for each diagnosis pattern, which corresponds to a k-dimensional feature vector.

The input to the recognition system is a set of raw data that includes n samples. Under the leave-one-out cross-validation (LOOCV), the data set is split into two parts. LOOCV removes one questionnaire sample from the data set and sets it as testing data. The remaining $(n-1)$ samples are then used for

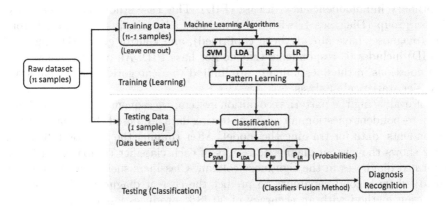

Fig. 1. Evaluation of the fusion classifier

training different models which then predict the class (diagnosis) for the sample that was left out. The role of LOOCV is to evaluate how well a corresponding model would perform on new data [1]. For training the classification model, it will have n different combinations of $(n-1)$ sample partitions and the model will be tested on each sample that has been left out. In this way, one obtains predictions of the classifiers for each sample where the sample for which the prediction is made was not involved in the training process.

In the training mode, we apply four classifiers SVM, LDA, RF and LR as the basis of the ensemble of classifiers to propose diagnoses based on the answer patterns in questionnaires. Each classifier has been trained independently with the same training set, and generates its own diagnosis prediction based on its learning results.

In the testing mode, the trained classifiers assign their classification pattern to the testing data. When we use K classifiers C_i ($i = 1, 2, ..., K$; $K = 4$ in our case), Eq. (1) calculates the average of classifiers' probabilities $P_{average}$ of each diagnosis group d (the values are in the range of $[0, 1]$).

$$P_{average}(C_i) = \frac{1}{K} \sum_{i=1}^{K} P_i(C_i) \tag{1}$$

Thus, each diagnosis d will obtain the value of $P_{average}^{(d)}$ as an indicator. We then choose the diagnosis with highest $P_{average}$ as final diagnostic result of the testing sample. By evaluating the compatibility and accuracy of individual classifiers, the fusion method takes advantages of each classifier. If a classifier is more certain about a diagnosis, it will gain a higher weight in the fusion process.

4   System Evaluation with Real-World Data

In order to test on a real-world problem, we applied the above mentioned approach to a data set with 126 respondents from a questionnaire that only focuses on primary immunodeficiency disease (PID). This case study has two groups: disease group (Diagnose 1, with PID) includes 63 patients who suffered from PID (diagnoses have already been confirmed), and control group (Diagnose 2, no PID) includes 63 respondents that do not have PID. All possible answers of such questions in the questionnaire are limited to a categorical or ordinal scale for easier statistical analysis.

For evaluating the pattern recognition system, we sequentially extracted one row of respondent questionnaire data as testing data, and used the remaining 125 respondents' data for training the model. After running the test for 126 times, Fig. 2 shows the system diagnostic results of each classifier compared with the confirmed diagnoses in the rightmost column. Classifiers such as SVM and RF already show a good capability to predict the correct diagnosis. Nevertheless, the fusion method with an accuracy of 90.48 % produces less errors than each single classifier.

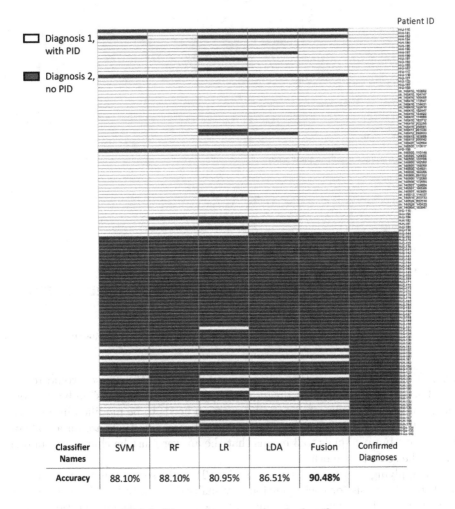

Patient ID

Diagnosis 1,
with PID

Diagnosis 2,
no PID

Classifier Names	SVM	RF	LR	LDA	Fusion	Confirmed Diagnoses
Accuracy	88.10%	88.10%	80.95%	86.51%	**90.48%**	

Fig. 2. The performance of each classifier

5   Understanding the Importance of Single Questions

The single classifiers and even more the fusion classifier function as black boxes
and the predictions – although they are quite reliable – are not comprehensible
or traceable. Nevertheless, one is interested in which questions are more and
which questions are less important, perhaps even unnecessary for making the
prediction. For this purpose, we consider each patient, modify the answer for
each single question and check whether changing the answer leads to a different
prediction. A question for which changing the answer does not affect the predicted diagnosis for all patients is not essential for the classification. In contrast,
a question for which a modification of the answer changes the prediction for all
or most of the patients is crucial for the diagnosis. Figure 3(1) shows a patient of

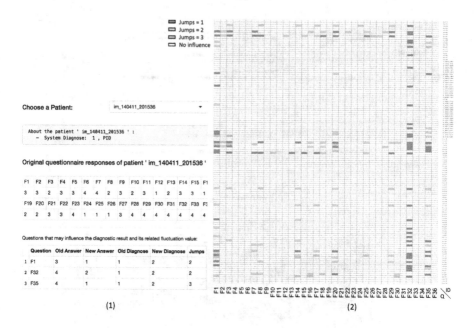

Fig. 3. Influence of a single question to the predicted diagnosis

whom there are three question where a modification of the answer would result in a different prediction of the diagnosis. The last column "Jumps" indicates how much the answer of the corresponding question needs to be modified in order to change the predicted diagnosis. The answers of most questions are on a scale of 1 to 4. So "Jump" equals to 3 means that the answer must be changed from one end of the scale to the other.

By applying this to all patients in the PID data set, the influences on the predicted diagnosis for each question is visualised in Fig. 3(2). The colored (or grey) boxes indicate where the change of the answer to a single question of the corresponding patient alters the predicted diagnosis. In the vertical direction, it can be seen that Question 32 is extremely sensitive, only one jump modification of the answers on this question leads to a change of the predicted diagnosis for most patients (85.7 %). The columns with entire white mean no matter how the single question changes its answer, it will not harm the final system diagnosis (Question 4, Question 19 etc.), which we call insensitive questions. In the horizontal direction, it can be observed that a lot of question boxes in several rows are colored. This means that the predicted diagnosis of such patients could easily change if one answer would be altered.

6   Conclusions and Future Work

Our approach to visualise when a change of an answer to a question leads to a different predicted diagnosis, helps MDs to better understand the importance of a single question in the questionnaire and also to see for which patients the classifier's diagnosis is robust against or sensitive to alterations of question answers.

In the future, our method will be trained and advanced with more questionnaire data according to rare disease categories. Meanwhile we will focus on the orientation of diseases by significant combination of symptom groups. We hope our research can provide modeling and reference methods for further research that focuses on rare diseases, promote rare diseases research and improve the level of diagnoses and treatment for patients.

References

1. Arlot, S., Celisse, A.: A survey of cross-validation procedures for model selection. Dev. Appl. Stat. **4**, 40–79 (2000)
2. Auria, L., Moro, R.A.: Support vector machines (SVM) as a technique for solvency analysis, pp. 1–16. Discussion Papers of Deutsches Institute of Wirtschaftsforschung, Berlin (2008)
3. Ho, T.K., Hull, J.J., Srihari, S.N.: Decision combination in multiple classifier systems. IEEE Trans. Pattern Anal. Mach. Intell. **16**(1), 66–75 (1994)
4. Jain, A.K., Duin, R.P., Mao, J.: Statistical pattern recognition: a review. IEEE Trans. Pattern Anal. Mach. Intell. **22**(1), 4–37 (2000)
5. Liaw, A., Wiener, M.: Classification and regression by random forest. R News **2**(3), 18–22 (2002)
6. Ma, L., Liu, X., Song, L., Zhou, C., Zhao, X., Zhao, Y.: A new classifier fusion method based on historical and on-line classification reliability for recognizing common CT imaging signs of lung diseases. Comput. Med. Imag. Graph. **40**, 39–48 (2015)
7. Pohar, M., Blas, M., Turk, S.: Comparison of logistic regression and linear discriminant analysis: a simulation study. Metodoloski zvezki **1**(1), 143 (2004)
8. Sboner, A., Eccher, C., Blanzieri, E., Bauer, P., Cristofolini, M., Zumiani, G., Forti, S.: A multiple classifier system for early melanoma diagnosis. Artif. Intell. Med. **27**(1), 29–44 (2003)

Fractional-Order Multiview Discriminant Analysis

Yun-Hao Yuan[1], Yun Li[1(✉)], Xiao-Bo Shen[2], Chong-Guang Ren[3], and Chao-Fei Li[4]

[1] School of Information Engineering, Yangzhou University,
Yangzhou 225127, China
liyun@yzu.edu.cn
[2] School of Computer Science and Engineering,
Nanjing University of Science and Technology,
Nanjing 210094, China
[3] School of Computer Science and Technology,
Shandong University of Technology, Zibo 255000, China
[4] Department of Computer Science and Technology,
Jiangnan University, Wuxi 214122, China

Abstract. Multi-view discriminant analysis (MvDA) is a powerful method for dimensionality reduction. However, intraclass and interclass sample scatter matrices in MvDA will deviate from true ones due to noise or limited training samples. To reduce the negative effect of the bias, in this paper we propose a novel method for learning multi-view low-dimensional representations, called fractional-order multi-view discriminant analysis (FMDA), which is based on fractional-order dispersion matrices built by sample spectrum reconstruction. Moreover, MvDA can be viewed as a special case of FMDA. A series of experiments show FMDA is effective and overall outperforms the state-of-the-art method MvDA.

Keywords: Image recognition · Multi-view discriminant analysis · Discriminative learning · Multi-view dimensionality reduction

1 Introduction

Linear discriminant analysis (LDA) [1] is a classical but effective statistical method for dimensionality reduction and feature extraction. LDA is essentially a supervised learning algorithm, which aims to search for the optimal linear transformation such that the within-class scatter of samples is minimized and simultaneously, the between-class scatter is maximized in the transformed low-dimensional space, thus achieving maximum discrimination. The optimal transformation can be obtained by maximizing the ratio of between-class scatter to within-class scatter, which leads to a generalized eigenvalue problem for solution. LDA has been successfully applied to many real-world applications such as face recognition [1, 2], document classification [3], image retrieval [4], and so on.

© Springer International Publishing AG 2016
H. Yin et al. (Eds.): IDEAL 2016, LNCS 9937, pp. 386–394, 2016.
DOI: 10.1007/978-3-319-46257-8_42

In many practical classification applications, the same objects are usually depicted at multiple different viewpoints. Such data are often referred to as multiple view data [5], which are informatively complementary to each other, thus helpful to improve the performance of feature extraction and dimensionality reduction methods. Since LDA is a single view based dimensionality reduction method in essence, it is not suitable to learn discriminative low-dimensional features from multi-view high-dimensional data. More recently, to solve this issue, Kan et al. [6, 7] proposed a multi-view discriminant analysis (MvDA) approach, where a discriminant common space is found in a non-pairwise manner by maximizing interclass variations and at the same time minimizing intraclass variations from both within-view and between-view. Experimental results on various face recognition show MvDA is more powerful compared with existing feature learning methods.

However, there is an important issue that the state-of-the-art method MvDA must face. That is, when training samples are small-scale or noisy, the within-class and between-class scatter matrices in MvDA will deviate from the true ones. This means that it is difficult for MvDA to generate a good projection suitable for classification in this situation. Concerning the deviation problem, Yuan et al. [8, 9] have already noticed it in canonical correlation based methods and proposed a fractional-order embedding strategy to correct the spectrums of sample covariance matrices, which has been proved to be effective for canonical correlation learning methods.

Motivated by recent progress in the aforementioned research [6–9], in this paper we propose a novel multi-view dimensionality reduction or feature extraction algorithm, called fractional-order multi-view discriminant analysis (FMDA), which is based on fractional-order dispersion matrices built by spectrum reconstruction. In addition, our FMDA can subsume MvDA as a special case. Also, we give an important property of MvDA. The proposed FMDA method is applied to handwritten digit recognition. Extensive experiments on different view combinations show FMDA is very effective for multi-view low-dimensional representations and overall outperforms the up-to-date state-of-the-art method MvDA.

2 Review on MvDA

Let $\mathcal{X}^{(i)} = \{x_{lk}^{(i)} | l = 1, 2, \cdots, c, k = 1, 2, \cdots, n_l^{(i)}\}$ be the sample set from the i-th view, $i = 1, 2, \cdots, m$, where $x_{lk}^{(i)} \in \mathfrak{R}^{d_i}$ is the k-th sample of the l-th class in i-th view, d_i is the dimensionality of samples, c is the number of classes, $n_l^{(i)}$ is the number of samples of the l-th class in i-th view, and m is the number of views. MvDA aims to search for m linear transformations, i.e., $W_1, W_2, \cdots, W_m$, such that the trace ratio defined as

$$J(W_1, W_2, \cdots, W_m) = \frac{Tr(W^T D W)}{Tr(W^T S W)}, \tag{1}$$

is maximized, where $Tr(\cdot)$ denotes the matrix trace, $W^T = [W_1^T, W_2^T, \cdots, W_m^T]$, $D \in \Re^{\tilde{d} \times \tilde{d}}$ $(\tilde{d} = \sum_{i=1}^{m} d_i)$ is a block matrix whose (i, j)th block entry is

$$D_{ij} = \left(\sum_{l=1}^{c} \frac{n_l^{(i)} n_l^{(j)}}{n_l} \mu_l^{(i)} \mu_l^{(j)T} \right) - \frac{1}{n} \left(\sum_{l=1}^{c} n_l^{(i)} \mu_l^{(i)} \right) \left(\sum_{l=1}^{c} n_l^{(j)} \mu_l^{(j)} \right)^T$$

with $n_l = \sum_{i=1}^{m} n_l^{(i)}$ as the number of samples of l-th class in all views, $n = \sum_{l=1}^{c} n_l$ as the number of samples from all classes and all views, and $\mu_l^{(i)} = \left(1 / n_l^{(i)} \right) \sum_{k=1}^{n_l^{(i)}} x_{lk}^{(i)}$ as the mean of all the samples of l-th class in i-th view, and $S \in \Re^{\tilde{d} \times \tilde{d}}$ is a block matrix with (i, j)th block entry as

$$S_{ij} = \begin{cases} \sum_{l=1}^{c} \left(\sum_{k=1}^{n_l^{(i)}} x_{lk}^{(i)} x_{lk}^{(i)T} - \left(n_l^{(i)} n_l^{(i)} / n_l \right) \mu_l^{(i)} \mu_l^{(i)T} \right), & i = j \\ -\sum_{l=1}^{c} \left(n_l^{(i)} n_l^{(j)} / n_l \right) \mu_l^{(i)} \mu_l^{(j)T}, & \text{else} \end{cases}$$

$i, j = 1, 2, \cdots, m$.

Due to the difficulty of maximizing the trace ratio in (1), the criterion used by MvDA is relaxed into the following form of ratio trace:

$$(W_1^*, W_2^*, \cdots, W_m^*) = \arg \max_{W_1, W_2, \cdots, W_m} Tr \left(\frac{W^T D W}{W^T S W} \right), \tag{2}$$

which can be solved analytically by generalized eigenvalue decomposition.

3 Approach

In this section, we first build fractional-order dispersion matrices of D and S through spectrum reconstruction, and then propose a new multi-view dimensionality reduction method for recognition purpose.

3.1 Characterize Fractional-Order Matrix

Let A represent one of matrices D and S. According to singular value decomposition (SVD) [10], it is easy that A can be decomposed as

$$A = U \Lambda V^T, \quad \Lambda = diag(\sigma_1, \sigma_2, \cdots, \sigma_r) \tag{3}$$

where U and V are two rotating matrices of matrix A such that $U^T U = V^T V = I_r$, $I_r \in \Re^{r \times r}$ is the identity matrix, $\sigma_1 \geq \sigma_2 \geq \cdots \geq \sigma_r > 0$ are the singular values in descending order, and $r = rank(A)$. Now, let us define fractional-order dispersion matrix of A based on (3), as follows:

$$A^g = U\Lambda^g V^T, \quad \Lambda^g = diag(\sigma_1^g, \sigma_2^g, \cdots, \sigma_r^g) \tag{4}$$

where $g \in \{\alpha, \beta\}$ and α and β are two fractions satisfying $0 \le \alpha, \beta \le 1$. Note that when $A = D$, $g = \alpha$ and when $A = S$, $g = \beta$.

Clearly, the fractional-order decomposition of matrix A in (4) can subsume its SVD in (3) as a special case. This implies that the decomposition in (4) is a more general case, in contrast to traditional SVD. As shown in [8], the fractional-order technique is very effective to reduce the bias between sample and true dispersion matrices. Motivated by that, in this paper we introduce this idea and thus build an improved version of MvDA for multi-view dimensionality reduction and classification tasks.

3.2 Formulation with Fractional Order

With the fractional-order dispersion matrices D^α and S^β as described in (4), our FMDA method aims to seek for a set of optimal projection matrices $W_1, W_2, \cdots, W_m$ which maximize the following generalized Rayleigh quotient:

$$J_f(W_1, W_2, \cdots, W_m) = \frac{Tr(W^T D^\alpha W)}{Tr(W^T S^\beta W)}. \tag{5}$$

It is obvious the criterion in (5) can reduce to the one in (1) when $\alpha = \beta = 1$. This means that the MvDA method can be incorporated into our FMDA method.

Since the objective in (5) is in the form of trace ratio, which suggests there is no existence of a closed-form solution according to [11], we relax it into the following form of ratio trace:

$$W^* = \arg \max_{W_1, W_2, \cdots, W_m} Tr\left(\left(W^T S^\beta W\right)^{-1} W^T D^\alpha W\right), \tag{6}$$

which can be directly solved with the generalized eigenvalue decomposition method:

$$D^\alpha W = S^\beta W \Delta, \tag{7}$$

where $\Delta \in \Re^{d \times d}$ is the diagonal matrix of the first d largest eigenvalues of D^α w.r.t S^β, and $W \in \Re^{\tilde{d} \times d}$ consists of the corresponding d generalized eigenvectors, $d \le rank(D^\alpha)$. In addition, for fractional-order dispersion matrix D^α, there is an important proposition, as follows:

Proposition 1. *The rank of D^α is at most $c - 1$.*

Proof. According to (3) and (4), it is easy to get $rank(D^\alpha) = rank(D)$. Let us set $\mu_i^T = [n_i^{(1)} \mu_i^{(1)T}, n_i^{(2)} \mu_i^{(2)T}, \cdots, n_i^{(m)} \mu_i^{(m)T}] \in \Re^{\tilde{d}}$. Then, it follows that

$$D = [D_{ij}]_{\tilde{d}\times\tilde{d}} = \left[\left(\sum_{l=1}^{c}\frac{n_l^{(i)}n_l^{(j)}}{n_l}\mu_l^{(i)}\mu_l^{(j)T}\right) - \frac{1}{n}\left(\sum_{l=1}^{c}n_l^{(i)}\mu_l^{(i)}\right)\left(\sum_{l=1}^{c}n_l^{(j)}\mu_l^{(j)}\right)^T\right]_{\tilde{d}\times\tilde{d}}$$

$$= \sum_{l=1}^{c}\frac{1}{n_l}\mu_l\mu_l^T - \frac{1}{n}\sum_{l=1}^{c}\sum_{k=1}^{c}\mu_l\mu_k^T = \sum_{l=1}^{c}\frac{1}{n_l}\mu_l\mu_l^T - \sum_{l=1}^{c}\mu_l\sum_{k=1}^{c}\frac{1}{n}\mu_k^T$$

$$= \sum_{l=1}^{c}\mu_l\left(\frac{1}{n_l}\mu_l^T - \sum_{k=1}^{c}\frac{1}{n}\mu_k^T\right) = \sum_{l=1}^{c}\mu_l\left(\frac{1}{n_l}\mu_l - \sum_{k=1}^{c}\frac{1}{n}\mu_k\right)^T = B\tilde{B}^T$$

where $\tilde{B} = [\tilde{\mu}_1, \tilde{\mu}_2, \cdots, \tilde{\mu}_c] \in \Re^{\tilde{d}\times c}$ with $\tilde{\mu}_l = (1/n_l)\mu_l - (1/n)\sum_{k=1}^{c}\mu_k$, $l = 1, 2, \cdots, c$, and $B = [\mu_1, \mu_2, \cdots, \mu_c] \in \Re^{\tilde{d}\times c}$. Note that $c \ll \tilde{d}$ holds in practical applications. Thus, we have

$$rank(D) = rank(B\tilde{B}^T) \leq \min\{rank(B), rank(\tilde{B})\}.$$

Since columns $\tilde{\mu}_1, \tilde{\mu}_2, \cdots, \tilde{\mu}_c$ of $\tilde{B}$ are linearly dependent, it is straightforward to show $rank(\tilde{B}) \leq c - 1$. Together with $rank(B) \leq c$, we get:

$$rank(D^\alpha) = rank(D) \leq c - 1.$$

As a result, Proposition 1 is true.   □

Corollary 1. *The rank of D in MvDA is at most c − 1.*

It is obvious that Corollary 1 is true according to the foregoing Proof. From Proposition 1 and Corollary 1, we can see that both D and D^α have at most $c - 1$ nonzero generalized eigenvalues w.r.t S and S^β, respectively. This means MvDA and FMDA can obtain at most $c - 1$ projection directions for each view. It should be pointed out that, to the best of our knowledge, this conclusion for MvDA is new.

4   Experiments

To test the performance of our proposed FMDA, a series of experiments are carried out using the famous multiple feature dataset (MFD)[1] [12] in UCI. To reveal effectiveness, we compare our FMDA with the state-of-the-art method MvDA[2]. In all the experiments, fractional-order parameters α and β in FMDA are, respectively, selected from $\{0.1, 0.2, \cdots, 1\}$. Moreover, the classification is performed by the nearest neighbor (NN) classifier with cosine distance measure in all the experiments. Note that, for

---

[1] http://archive.ics.uci.edu/ml/datasets/Multiple+Features.

[2] Matlab code available at http://vipl.ict.ac.cn/resources/codes.

MvDA and FMDA, if S and S^β are singular, we use the following strategy to regularize them:

$$S \leftarrow S + \kappa I \text{ and } S^\beta \leftarrow S^\beta + \kappa I,$$

where I is the identity matrix and $\kappa = 0.001$ in this paper.

4.1 Multiple Feature Dataset

The multiple feature dataset (MFD), which is widely used to test multi-view learning methods, is employed in our experiment. The digit dataset includes 10 classes of handwritten numerals (i.e., "0"–"9") extracted from a collection of Dutch utility maps. 200 samples per class (2,000 samples in total) are available in the form of 30×48 binary images. These numerals are represented in terms of six feature sets (views), as shown in Table 1.

Table 1. Six feature sets of handwritten numerals in MFD

Pix: 240-dimension pixel averages feature in 2×3 windows;
Fac: 216-dimension profile correlations feature;
Fou: 76-dimension Fourier coefficients of the character shapes feature;
Kar: 64-dimension Karhunen-Loève coefficients feature;
Zer: 47-dimension Zernike moments feature;
Mor: 6-dimension morphological feature.

4.2 Experiment Using Two-View Features

In this test, we choose any two different feature sets as two views. Thus, there are 15 different feature combinations in total. For each, the first 10 samples per class are chosen for training, while the remaining 190 samples are used for testing. Thus, the number of training and testing samples is 100 and 1900, respectively. Table 2 lists the recognition results of MvDA and FMDA under cosine NN classifier. As seen, FMDA performs better than state-of-the-art MvDA on 14 cases, while MvDA outperforms

Table 2. Recognition accuracy (%) of MvDA and FMDA with two-view features on MFD.

Views	Pix-Fac	Pix-Fou	Pix-Kar	Pix-Zer	Pix-Mor
MvDA	91.21	69.37	83.58	82.90	70.53
FMDA	**91.84**	**92.63**	**89.32**	**90.21**	**85.90**
Views	Fac-Fou	Fac-Kar	Fac-Zer	Fac-Mor	Fou-Kar
MvDA	76.95	90.11	91.47	**91.84**	65.47
FMDA	**88.79**	**91.74**	**91.63**	87.68	**92.79**
Views	Fou-Zer	Fou-Mor	Kar-Zer	Kar-Mor	Zer-Mor
MvDA	57.47	48.90	79.47	68.32	72.79
FMDA	**80.42**	**74.00**	**90.84**	**87.53**	**74.74**

FMDA only on one case. Also, MvDA performs much worse than FMDA on many cases such as Pix-Fou, Pix-Mor, Fou-Zer, and so on. Thus, on the whole, our FMDA method is more discriminative in contrast with MvDA.

4.3   Experiment Using Three-View Features

In this test, we select arbitrary three different feature sets as three views. Thus, there are a total of 20 feature combinations. For each combination, the first 10 samples per class are chosen for training, while the remaining 190 samples are used for testing. Table 3 summarizes the recognition results of MvDA and FMDA with cosine NN classifier. As can be seen, our FMDA performs better than state-of-the-art MvDA on 17 cases, while MvDA outperforms FMDA only on three cases.

Table 3.   Recognition accuracy (%) of MvDA and FMDA with three-view features on MFD.

Views	Pix-Fac-Fou	Pix-Fac-Kar	Pix-Fac-Zer	Pix-Fac-Mor	Pix-Fou-Kar
MvDA	82.90	89.68	92.05	**91.05**	81.42
FMDA	**93.95**	**91.32**	**92.79**	87.42	**92.00**
Views	Pix-Fou-Zer	Pix-Fou-Mor	Pix-Kar-Zer	Pix-Kar-Mor	Pix-Zer-Mor
MvDA	78.84	66.32	83.68	72.74	83.74
FMDA	**93.32**	**90.84**	**90.95**	**88.37**	**86.16**
Views	Fac-Fou-Kar	Fac-Fou-Zer	Fac-Fou-Mor	Fac-Kar-Zer	Fac-Kar-Mor
MvDA	85.37	82.84	74.95	91.42	**90.16**
FMDA	**93.58**	**89.79**	**87.42**	**93.00**	89.42
Views	Fac-Zer-Mor	Fou-Kar-Zer	Fou-Kar-Mor	Fou-Zer-Mor	Kar-Zer-Mor
MvDA	**90.90**	73.84	61.74	67.16	80.84
FMDA	86.95	**93.11**	**90.42**	**77.53**	**87.26**

4.4   Experiment Using Four-View Features

In this test, we select any four different feature sets as four views. Thus, there are a total of 15 different feature combinations. Each combination uses the first 10 samples per class for training and the remaining 190 samples for testing. Table 4 shows the recognition results of MvDA and FMDA under cosine NN classifier. As we can see, our proposed FMDA method outperforms the state-of-the-art method MvDA again on most cases. In addition, although MvDA performs better than FMDA on Pix-Fac-Zer-Mor and Fac-Kar-Zer-Mor combinations, FMDA achieves comparable results with MvDA. These conclusions are overall consistent with those drawn from the first two experiments.

In addition, from Tables 2 and 4, we can find that using more feature sets (views) is not of necessity to yield better recognition results. For instance, MvDA and FMDA respectively achieve better results on Pix-Fac combination than those on Pix-Fac-Kar and Pix-Fac-Kar-Mor combinations. This implies that not all of features can be used for multi-view feature learning and recognition tasks.

Table 4. Recognition accuracy (%) of MvDA and FMDA with four-view features on MFD.

Views	Pix-Fac-Fou-Kar	Pix-Fac-Fou-Zer	Pix-Fac-Fou-Mor	Pix-Fac-Kar-Zer	Pix-Fac-Kar-Mor
MvDA	86.79	88.68	81.79	89.84	89.53
FMDA	**93.58**	**93.84**	**93.11**	**92.37**	**89.63**
Views	Pix-Fac-Zer-Mor	Pix-Fou-Kar-Zer	Pix-Fou-Kar-Mor	Pix-Fou-Zer-Mor	Pix-Kar-Zer-Mor
MvDA	**91.16**	84.11	79.68	75.11	82.42
FMDA	88.11	**93.37**	**90.68**	**91.42**	**87.84**
Views	Fac-Fou-Kar-Zer	Fac-Fou-Kar-Mor	Fac-Fou-Zer-Mor	Fac-Kar-Zer-Mor	Fou-Kar-Zer-Mor
MvDA	88.37	84.05	80.90	**90.58**	76.00
FMDA	**93.95**	**92.16**	**87.74**	89.11	**91.58**

5 Conclusion

In this paper, we propose a novel multi-view learning method called FMDA for multi-view dimensionality reduction and classification tasks. FMDA is based on fractional-order dispersion matrices that are built by fractional spectrum modeling. Our FMDA method has many good properties. For example, it is a more general method due to MvDA being a special case, and it can obtain more discriminative low-dimensional projections than the state-of-the-art MvDA, which is demonstrated by a series of experiments on visual recognition. Moreover, our experimental results show that using more features does not necessarily yield better recognition performance. Thus, we should make a feature selection before multi-view learning, which will be our next work in future.

Acknowledgments. This work is supported by the National Natural Science Foundation of China under Grant Nos. 61402203, 61273251, 61170120, Natural Science Foundation of Jiangsu Province of China under Grant No. BK20161338, and the Fundamental Research Funds for the Central Universities under Grant No. JUSRP11458. Moreover, it is also supported by the Program for New Century Excellent Talents in University under Grant No. NCET-12-0881.

References

1. Belhumeur, P.N., Hespanha, J.P., Kriegman, D.J.: Eigenfaces vs fisherfaces: recognition using class specific linear projection. IEEE Trans. PAMI **19**(7), 711–720 (1997)
2. Liu, C., Wechsler, H.: Gabor feature based classification using the enhanced fisher linear discriminant model for face recognition. IEEE Trans. Image Process. **11**(4), 467–476 (2002)
3. Torkkola, K.: Linear discriminant analysis in document classification. In: IEEE ICDM Workshop on Text Mining, pp. 800–806 (2001)
4. Swets, D.L., Weng, J.J.: Using discriminant eigenfeatures for image retrieval. IEEE Trans. PAMI **18**(8), 831–836 (1996)
5. Long, B., Philip, S.Y., Zhang, Z.: A general model for multiple view unsupervised learning. In: Proceedings of the 2008 SIAM International Conference on Data Mining, pp. 822–833 (2008)

6. Kan, M., Shan, S., Zhang, H., Lao, S., Chen, X.: Multi-view discriminant analysis. In: Fitzgibbon, A., Lazebnik, S., Perona, P., Sato, Y., Schmid, C. (eds.) ECCV 2012, Part I. LNCS, vol. 7572, pp. 808–821. Springer, Heidelberg (2012)
7. Kan, M., Shan, S., Zhang, H., Lao, S., Chen, X.: Multi-view discriminant analysis. IEEE Trans. PAMI **38**(1), 188–194 (2016)
8. Yuan, Y.-H., Sun, Q.-S., Ge, H.-W.: Fractional-order embedding canonical correlation analysis and its applications to multi-view dimensionality reduction and recognition. Pattern Recogn. **47**, 1411–1424 (2014)
9. Yuan, Y.-H., Sun, Q.-S.: Fractional-order embedding multiset canonical correlations with applications to multi-feature fusion and recognition. Neurocomputing **122**, 229–238 (2013)
10. Golub, G.H., Van Loan, C.F.: Matrix Computations. The Johns Hopkins University Press, Baltimore (1995)
11. Wang, H., Yan, S., Xu, D., Tang, X., Huang, T.: Trace ratio vs ratio trace for dimensionality reduction. In: CVPR, pp. 1–8 (2007)
12. Jain, A.K., Duin, R.P.W., Mao, J.: Statistical pattern recognition: a review. IEEE Trans. PAMI **22**(1), 4–37 (2000)

Person Re-identification Using Cascade Filter

Xinyu Wang[1,2(✉)], Hua Yang[1,2], Ji Zhu[1,2], Lin Chen[1,2], and Junhao Huang[1,2]

[1] Institue of Image Communication and Network Engineering
of Shanghai Jiao Tong University, Shanghai 200240, China
xinyuwang@sjtu.edu.cn
[2] Shanghai Key Lab of Digital Media Processing and Transmission,
Shanghai 200240, China

Abstract. Feature fusion is proved to be very effective in the problem of person re-identification. Commonly the fusion feature can performs better than single features. In this paper, we address how to take advantages of different ranking results yield by using different features. We propose a cascade filter framework to alleviate the influence of the error determination when the ranking results of different features are not consistent. This method can make full use of the information provided by features. Extensive experiments on publicly available datasets show that the proposed method achieve favorable performance in terms of accuracy and efficiency.

Keywords: Person re-identification · Feature fusion · Cascade filter

1 Introduction

In the field of video surveillance, person re-identification is a very important issue. The aim of person re-identification is to identify the same pedestrian from different cameras by computing the similarity of two images. Although many researchers have devoted great effort to person re-identification, it still remains a challenging issue due to large variations of view, illumination, occlusion, and etc.

In order to solve these challenging problems, researchers have make effort on two aspects. One is to design a more reliable feature to represent images and another is metric learning. Our work focus on the former one by proposing a framework to better carry out feature fusion.

Many methods adopt feature fusion to form a reliable representation. For example, SDALF [1] achieved satisfactory results by forming a new representation with MSCR [2], WCH and RHSP. eBiCov [3] method combined wHSV, MSCR, as well as BiCov [3], which also achieved good performance. Other methods that evaluate the effectiveness of features are also proposed to improve the performance of re-identification. In these methods, an effective feature will be given a high weight when fusion with other features and the feature that is not effective enough will be given a low weight. Chunxiao Liu et al. [4] proposed a method to adjust the global weight of different features, by using a random

© Springer International Publishing AG 2016
H. Yin et al. (Eds.): IDEAL 2016, LNCS 9937, pp. 395–404, 2016.
DOI: 10.1007/978-3-319-46257-8_43

forest based method to evaluate the effectiveness of different features. And this method also improved the accuracy of re-identification. Liang Zheng et al. [5] used a non supervised method to measure the effectiveness of the features based on the attenuation of the score curve, and achieved better results.

The aim of feature fusion is to get a more discriminative representation by relying on the characteristics of different features. The method based on a single feature usually can not achieve a very satisfactory performance. For example, the method use texture feature can recognizes part of the pedestrian images well, but for other pedestrian images, texture feature may can not help us to find the right person. The method using color feature will face the same problem. So it is not very reliable to rely on a simple feature for person re-identification. However, using the method of feature fusion, we can get more accurate recognition accuracy by relying on several features. But if a feature has a very bad performance, it will reduce the accuracy when fusion with other features [4].

How to effectively fuse different features to achieve a satisfactory performance for re-identification is what we discussed in this paper. Given a query image. When two features are used to search for the similar pedestrian images in a database, these two features will yield different ranking results. Generally, there are two cases. The first one is the two features have the same ranking for an image in dataset. In this case, these two features will confirm each other's results and there is a great possibility the ranking for this image is reasonable. The second case is the two features have a great different ranking for an image. In this case, the judgment of one feature is correct and another is wrong. If the traditional fusion method is used, the correct judgment of one feature may be lost, which can decrease the accuracy of the method. However, we find that the correct gallery image will hardly be the last in the ranking queue when using a reasonable feature. So if a image from the dataset is ranked at the end by a feature, there is a great probability that this image is not the matched image of the query. Based on above observation, we design a re-identification framework with a cascade filter, which can perform well not only on the accuracy but also on the efficiency.

The contributions of this work are summarized as follows:

– We propose a re-identification framework with a cascade filter. When perform feature fusion and the features have great different performances, the feature that judge an image to be a wrong matched image will be trusted, which will alleviate the influence of the error determination when the performances of different features are not consistent.
– We use a cascade filter framework and this filter framework can improve the efficiency of re-identification.

The rest of the paper is organized as follows. In Sect. 2, the details of the proposed framework with cascade filter is described. Section 3 introduces the features that are used in the algorithm. Section 4 shows the experimental results of proposed method and gives analysis and comparison with existing algorithms. Finally, the conclusions are summarized in Sect. 5.

2 Person Re-identificaton Framework Using Cascade Filter

When evaluating the similarity between a probe pedestrian image and images in the gallery dataset, different kinds of features often give different similarity ranking results. In general, there are two possible situations using two kinds of features. One is that both features give the consistent ranking result for an image in the gallery dataset, which indicates there is a great possibility that the ranking result is reasonable. The other situation is the two features give a great different ranking for an image. One feature gives this image a high ranking, while another gives it a low ranking. There is always a correct judgement between these two rankings. In this case, we need to decide which ranking result is more credible. Traditional feature fusion methods try to make a trade off between different ranking results, which often lead to an averaged ranking score. It may lead to that the correct judgment of one feature may be lost.

However, we find that if a gallery image is given a very low ranking by one feature, there is a great probability that this image is not the right-matched image of the query. We choose 316 pairs of pedestrian images in VIPeR dataset. Three features SDC [1], CH [6], MSCR [7] are used for similarity ranking of 316 pairs of pedestrian images respectively. When comparing the similarity between a probe pedestrian image and images in the gallery, a rank list can be gotten. We calculate the proportion of the targets' right-matched images that distribute in different rankings and plot it in the Fig. 2. The horizontal coordinate indicates different rankings and the longitudinal coordinate shows the proportion.

According to Fig. 2, we find that the curves show a 'L' shape and the proportion will drop sharply when the ranking is low, which means that there is a large probability that the pedestrian image with a very low ranking is not the correct matching of target. Based on this observation, we design a re-identification framework with a cascade filter. By filtering the pedestrian images that with a very low ranking from gallery dataset, we can alleviate the influence of error determination when the performances of different features are not consistent. Figure 3 shows an example that the candidate images are given a very low ranking by the SDC are identified as the most similar images of the probe pedestrian image by MSCR. In this case, if we use the traditional method to make a trade off between different ranking results, we may lose the right ranking result of SDC. Our method can reduce the error decision caused by this situation. In addition, this filtering mechanism can also improve the efficiency of the method by reducing the computation cost. Because we decrease the number of the image samples, which can reduce the computation of feature extraction and matching.

Figure 1 is the person re-identificaton framework with cascade filter. In this framework, we used three features, SDC [1], CH [6], MSCR [7]. When constructing the framework, we always choose the feature that perform better to filter firstly. This is because if a feature performs better, the probability that pedestrian image with a very low ranking is not the right-matched image of a probe will be larger. This can been seen in Fig. 2. But if the performances of two features are equivalent, we should firstly use the feature which need less computation.

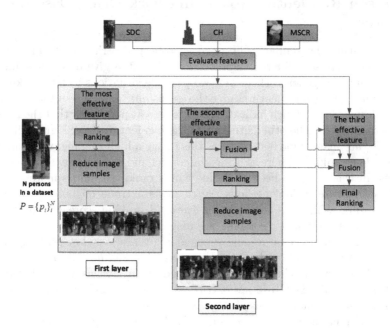

Fig. 1. Person re-identificaton framework with cascade filter.

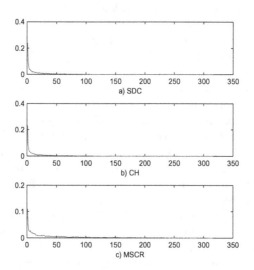

Fig. 2. The proportion of the targets' right-matched images that distribute in different rankings in VIPeR. (a) SDC, (b) CH, (c) MSCR.

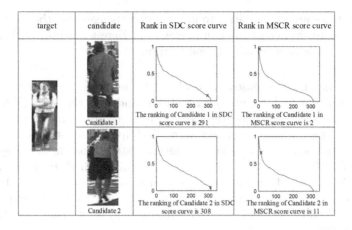

Fig. 3. Different ranking results yield by using different features.

This is because putting the feature with more computation in next layer will reduce the complexity of our method, since the image samples will be less in next layer. When filtering, we remove the images with low ranking from the gallery dataset. At the first layer, we firstly use the most effective feature to get a rank, then we remove k percent of pedestrian images from gallery dataset that appear at the end of the rank list. The rest of the pedestrian images in gallery dataset are sent into the next layer for feature extraction and recognition of pedestrians. In the second layer, the first feature and second feature are combined with the same weight to get a fusion representation and get a rank, then we also remove k percent of pedestrians images with very low ranking from the gallery dataset. The choice of k relates to the accuracy and complexity of the algorithm. The computation is less when k is larger. But when k is larger, the probability that the pedestrian image appears at last k being the unright one of the probe pedestrian is lower, which will reduce the accuracy of the algorithm. Considering the accuracy and complexity, we choose k = 10 here. And then we use the third feature to compare the similarity of images. These three feature are weighted equally when fused together to get the final ranking.

3   Feature Extraction

Due to the function of single feature can not be strong enough to distinguish every subtle differences of pedestrian images, fusing different type of features can improve the reliability of the feature. The fusion feature can better distinguish pedestrian images. In this paper, we use the SDC feature which is based on local texture and feature MSCR based on stable color region. We also use the color feature CH, based on Bag-of-Words (BoW) model. Compared to other features, color feature is not sensitive to view changes and pose changes. However, the accuracy rate of the color feature will be affected by the illumination changes.

Texture features are not sensitive to illumination changes, but the view changes and pose changes will cause some problems to it. Here we combine SDC, MSCR and CH to form a more reliable fusion feature.

SDC. SDC is a method based on local texture features proposed by RuiZhao et al. The image is divided into several blocks, and then the SIFT [8] feature is extracted from the blocks, and the LAB [9] features are extracted to combined with SIFT to form a more reliable feature. To improve the performance of person re-identification, human salience is incorporated in patch matching to find reliable and discriminative matched patches in this method. Here in our paper we use SVM salience detection methods. Just as the method used in [1], we get the $Score_{SDC}(p_i, q_j)$ represents the similarity between the two images. p_i is an image from P and P is image set from camera A. q_j is an image from Q and Q is image set from camera B.

CH. CH feature describes the color characteristics. Color histogram is widely used in person re-identification. The color histogram has strong robustness for pose changes and view changes, while the color histogram is very sensitive to changes of illumination. In order to reduce the influence of illumination changes, this paper uses the HS feature. We also employ the Bag-of-Words representation. The codebook size is set to 350. Local features are extracted by dense sampling: 44 image patches with step of 4. The final descriptor is 5600-dim for each image. HS histogram is used for each image patch. The similarity between the two images is represented by $Score_{CH}(p_i, q_j)$.

MSCR. The MSCR detects a set of blob regions. The MSCR operator detects a set of blob regions by looking at successive steps of an agglomerative clustering of image pixels. The detected regions are then described by their area, centroid, second moment matrix and average RGB color, forming 9-dimensional patterns. As mentioned in [2], we get the $d_{MSCR}(p_i, q_j)$, which represents the distance between the two vectors. We transform the distance to similarity by using Eq. (1):

$$Score_{MSCR}(p_i, q_j) = \min_{p_i \subset P, q_j \subset Q} (d_{MSCR}(p_i, q_j)) / d_{MSCR}(p_i, q_j). \quad (1)$$

4   Experiments

This experiment is conducted on the Window7 platform, Intel i7, 3.4 GHz, 8 GB RAM. We evaluate our approach on the public VIPeR dataset and ETHZ1 dataset. The experimental results are shown in the form of CMC curve. The matching rate in the table is defined as Eq. (2). In the equation, p_i is an image from P and $P = \{p_i | 1 \leq i \leq N\}$ is image set from camera A. q_i is an image from Q and $Q = \{q_i | 1 \leq i \leq N\}$ is image set from camera B. $\{p_i, q_i\}$ represents pedestrian image pair. The rank represents the ranking of image similarity. The ε represents unit step function, defined as equation Eq. (3).

$$\eta = \frac{\sum_{p_i \epsilon P, 1 \leq i \leq N} \varepsilon(r - rank(q_i)))}{N} \times 100\% \quad (2)$$

$$\varepsilon(t) = \begin{cases} 1 & t \geq 0 \\ 0 & t < 0 \end{cases} \qquad (3)$$

Dataset. The VIPeR dataset consists of images of pedestrians from two different camera views. This dataset contains 632 pairs of pedestrian images. The VIPeR dataset includes images with both view and illumination changes. These images are cropped and scaled to 128×48 pixels. This is one of the most challenging datasets for re-identification. The ETHZ1 dataset for appearance-based modeling was captured from moving cameras. It contains 83 people and the sizes the person images of are from 13*30 to 72*144. This camera setup provides a range of variations in people appearance with significant changes in pose, scale, occlusion and illumination. Figure 4 shows some examples of these two datasets.

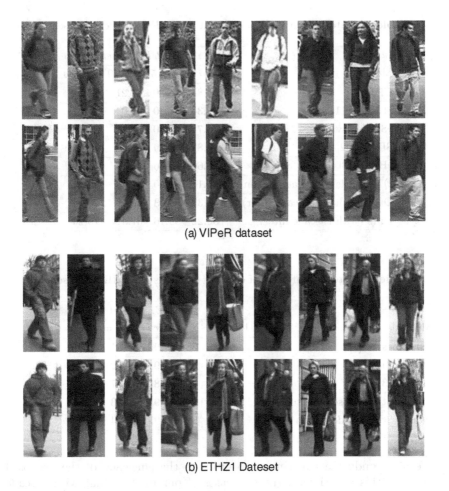

(a) VIPeR dataset

(b) ETHZ1 Dateset

Fig. 4. Example images of different datasets used in our evaluation. The columns denote image pairs examples of the same person. (a) VIPeR, (b) ETHZ1.

In this paper, three kinds of features are used to conduct re-identification and experiments in the VIPeR dataset and ETHZ1 dataset are done. When conducting experiments in VIPeR dataset, we randomly select 316 pairs of pedestrian images from the dataset. When conducting experiments in ETHZ1 dataset, we evaluate our method in the single-shot case. Experimental results are shown in Tables 1, 2 and Fig. 5. From the experimental results, we can conclude that the fusion feature significantly improves the accuracy of the re-identification, proving the effectiveness of feature fusion. At the same time, we compare the accuracy of the framework with the cascade filter and framework with non-filtering, and the framework with non-filtering which can be considered as a special case of our method where k = 0. We can see that the accuracy rate has been improved after the cascade filter. This is because we eliminate situation showed in Fig. 3, which will contributes to the performance.

Table 1. Matching rate (%) of different methods on VIPeR dataset.

Methods	VIPeR dataset			
	$r = 1$	$r = 5$	$r = 10$	$r = 20$
CPS [10]	21.84	44.00	57.21	71.00
eLDFV [11]	22.34	47.00	60.04	71.00
eSDC_ocsvm [1]	26.74	50.70	62.37	76.36
SDALF [2]	19.87	38.89	49.37	65.73
eBiCov [3]	20.66	20.66	56.18	68.00
Query-Adaptive Late Fusion [5]	30.17	51.60	62.44	73.81
Ours (without cascade filter)	31.32	55.03	66.96	78.64
Ours (with cascade filter)	30.22	55.16	67.31	79.24

Table 2. Matching rate (%) of different methods on ETHZ1 dataset.

Methods	ETHZ1 dataset						
	$r = 1$	$r = 2$	$r = 3$	$r = 4$	$r = 5$	$r = 6$	$r=7$
eSDC_ocsvm [1]	80	85	88	90	91	92	93
eLDFV [11]	83	87	90	91	92	93	94
eBiCov [3]	74	80	83	85	87	88	89
Ours (without cascade filter)	80	87	90	91	93	94	94
Ours (with cascade filter)	81	87	90	92	94	94	95

We also conduct a experiment to evaluate the efficiency of the proposed strategy. In this experiment, the two cases of our method that with cascade filter and without cascade filter are conducted on 83 pairwise people images from ETHZ1 dataset. From Table 3 we find our method with cascade filter can

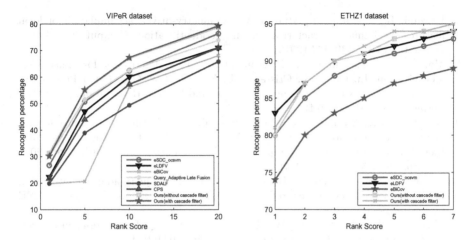

Fig. 5. Matching rate (%) of different methods on VIPeR dataset and ETHZ1 dataset.

Table 3. Time cost results of comparing framework with cascade filter and framework with non-filtering.

Methods	Our method (with cascade filter)	Our method (without cascade filter)
Cost time (ms)	760735	836006

increase computation efficiency. The efficiency improvement is mainly caused by the reduce the number of image samples, which will lead to the time consumption of feature extraction and similarity calculation be reduced.

5   Conclusions

In the field of video surveillance, person re-identification is a very important issue. In this paper, we propose a cascade filter framework to alleviate the influence of the error determination when the ranking results of different features are not consistent. This method can make full use of the information provided by features. This method can both improve the accuracy and efficiency. The experimental results demonstrate the effectiveness of our approach.

Acknowledgement. This work was supported in Science and Technology Commission of Shanghai Municipality (STCSM, Grant Nos. 15DZ1207403).

References

1. Zhao, R., Ouyang, W., Wang, X.: Unsupervised salience learning for person re-identification. In: Proceedings of the IEEE International Conference on Computer Vision and Pattern Recognition (2013)

2. Bazzani, L., Cristani, M., Murino, V.: Symmetry-driven accumulation of local features for human characterization and re-identification. Comput. Vis. Image Underst. **117**(2), 130–144 (2013)
3. Ma, B., Su, Y., Jurie, F.: Discriminative image descriptors for person re-identification. In: Gong, S., Cristani, M., Yan, S., Loy, C.C. (eds.) Person Re-Identification. Advances in Computer Vision and Pattern Recognition, pp. 23–42. Springer, New York (2014)
4. Liu, C., Gong, S., Loy, C.C., Lin, X.: Evaluating feature importance for re-identification. In: Gong, S., Cristani, M., Yan, S., Loy, C.C. (eds.) Person Re-Identification. Advances in Computer Vision and Pattern Recognition, pp. 203–228. Springer, New York (2014)
5. Zheng, L., Wang, S., Tian, L., He, F., Liu, Z., Tian, Q.: Query-adaptive late fusion for image search and person re-identification. In: CVPR, pp. 1741–1750. IEEE Computer Society (2015)
6. Zheng, L., Shen, L., Tian, L., Wang, S., Bu, J., Tian, Q.: Person re-identification meets image search (2015). arXiv preprint arXiv:1502.02171
7. Forssn, P.E.: Maximally stable colour regions for recognition and matching. In: IEEE Conference on Computer Vision and Pattern Recognition (2007)
8. Liu, C., Yuen, J., Torralba, A.: Sift flow: dense correspondence across scenes and its applications. TPAMI **33**, 978–994 (2011)
9. Wang, X., Doretto, G., Sebastian, T., Rittscher, J., Tu, P.: Shape and appearance context modeling. In: Proceedings of the IEEE International Conference on Computer Vision (2007)
10. Cheng, D., Cristani, M., Stoppa, M., Bazzani, L., Murino, V.: Custom pictorial structures for re-identification. In: BMVC (2011)
11. Ma, B., Su, Y., Jurie, F.: Local descriptors encoded by fisher vectors for person re-identification. In: Fusiello, A., Murino, V., Cucchiara, R. (eds.) ECCV 2012. LNCS, pp. 413–422. Springer, Heidelberg (2012). doi:10.1007/978-3-642-33863-2_41

Grouping Parallel Bayesian Network Structure Learning Algorithm Based on Variable Ordering

Xiaolong Qi[1,2], Yinhuan Shi[1(✉)], Hao Wang[1], and Yang Gao[1]

[1] State Key Laboratory for Novel Software Technology, Nanjing University,
Nanjing 210023, China
qxl_0712@sina.com, {syh,gaoy}@nju.edu.cn,
wanghao.hku@gmail.com
[2] Department of Electronics and Information Engineering,
Yili Normal University, Yining 835000, China

Abstract. Given an ordering constraint of n random variables and the maximum in-degree u for any variable, the search space of Bayesian network structure reduces from $n!2^{\frac{n(n-1)}{2}}$ to the $2^{\frac{n(n-1)}{2}}$. Even so, with the increase of the number of variables, the requirement of time cannot be tolerated. In this paper, we present a parallel Bayesian network structure learning algorithm based on variable ordering. The algorithm includes three components: 1. variable grouping: it completes variable partition; 2. group learning: it completes independently construct of sub-Bayesian network; 3. Between the groups learning: it asynchronously combines sub-Bayesian network in order to get the full Bayesian networks. We theoretically analyzed that time complexity of our algorithm is $O(mu^2nr)$, where u that is number of parent, In the worst case, $u = n$, complexity of the algorithm is $O(mn^3r)$. The empirical results present in term of time complexity grouping parallel algorithm has significance compared with the traditional algorithm.

Keywords: Bayesian network · Structure learning · Parallel learning

1 Introduction

Bayesian network learning is the obtained process of Bayesian network by analyzing the data, which includes the parameter learning and structure learning. Structure learning is primarily considered. Score-based learning, constraint-based learning, and hybrid methods are the top three main forms of learning problem. Scoring function was used in Score-based learning methods to assess quality of the Bayesian network structure and the one with highest score was likely to have high priority [7, 12]. If the dataset is limited in size and variables, score-based learning methods may be quiet suitable for problem solving because it's mainly considering the procedure as combinatorial optimization. However, this model may encounter a bottleneck with large datasets. Further details of this model will be discussed in the next section. Statistical analysis is used in constraint-based learning methods to construct a Bayesian network

© Springer International Publishing AG 2016
H. Yin et al. (Eds.): IDEAL 2016, LNCS 9937, pp. 405–415, 2016.
DOI: 10.1007/978-3-319-46257-8_44

with the best interpretation of independence relations through conditional independence relation identify [4, 10, 22]. As it model averaging methods are, constraint-based
methods are applicable with large datasets. Meanwhile, lacking of accuracy, insufficient data and noisy data will be greatly affect the result. So, it may not as precisely as
score-based learning methods dealing with small sets of data. There were both
advantages and limitations for the former two methods, while Hybrid methods were
evolved from the convergence of the integration and mix used of those two methods.
A very instructive action is firstly to construct a skeleton graph (use constraint-based
learning), then draw a subgraph of the skeleton (use score-based learning) [19, 21].
Bayesian model averaging methods are beyond our preference for we are concentrating
on the feature of model selection rather than edges [9, 11, 12].

Score-based structure learning will be further discussed here. $2^{\frac{n(n-1)}{2}}$ directed acyclic
graphs (DAGs) will be listed as for n variables, when the number of variable grows, the
size of the solution space will be enlarged exponentially. This might help us to
understand why score-based structure learning has been considered to be NP-hard [6].
Even though we tried to reduce the search space by constrains like setting of variable
orderings and in-degree, hardness still bothered us.

Given variable ordering, there are $2^{\frac{n(n-1)}{2}}$ directed acyclic graphs that are consistent
with it. Obviously, with the increase of the number of variables, it makes sense to find
the optimal structure for time reasonable. In this paper, we proposed grouping method
which partition variable in dependency between variables, based on the fact that given
a variable ordering, the parent node of any variable must be the predecessors of it in the
order. After partition, we proposed algorithm that all grouping independently and in
parallel generate sub-Bayesian network and every group directly combination to
generate full Bayesian network. Given u as the number of parent, we theoretically
analyzed the time complexity of the algorithm is $O(mu^2nr)$. In the worst case, u = n,
complexity of the algorithm is $O(mn^3r)$. The empirical results present in term of time
complexity grouping parallel algorithm has significance.

The remainder of the paper is structured as follows. Section 2 reviews the problem
of learning optimal Bayesian networks and other related work. Section 3 introduces the
grouping parallel algorithm. Section 4 theoretically analyzes that time complexity of
our method (algorithm) and presents empirical results for evaluating our algorithm
against existing approaches. Finally, Sect. 5 offers a conclusion.

2   Preliminary Knowledge

In this paper, suppose another set of variables of X, values over X was represented by a
vector noted as D_i, each data point consists a new dataset D = $\{D_1, D_2 ..., D_M\}$. When
we talk about leaning a Bayesian network under this situation, we are actually talking
about finding the fittest structure for D. In order to achieve that we have to assume that
there is a finite number of possible values for each variable and no missing values for
any data point.

2.1   Score-Based Learning

The score-based learning method has two major elements a scoring function Score(.) and a search strategy. The former is used as quality accessing towards the network structure; the latter is developed to find the optimal strategy. In order to find the optimal network effectively, researchers are making scoring functions satisfy local decomposability:

$$\text{Score}(G) = \sum_{x_i \in X} Score(x_i, \mathrm{P}_a(x_i)) \tag{1}$$

In general, the scoring functions can assign a very high probability to the network that best fits the data. Finally, the optimization criterion can find the optimal network that represents observation. In this paper, we use BD scoring function to find a Bayesian network with maximum BD score. Let n denotes the number of variable, q_i denotes case numbers of parent, r_i denotes numbers of states of variable. N_{ijk} be $x_i = $ k, case of the parent node is the number of the jth case. Let $N_{ij} = \sum_{k=1}^{r_i} N_{ijk}$. BD is defined as follows [7].

$$\max_{B_s}[(B_s, D)] = C \prod_{i=1}^{n} max \left[\prod_{j=1}^{q_i} \frac{(r_i - 1)!}{(N_{ij} + \_r_i - 1)!} \prod_{k=1}^{r_i} N_{ijk}! \right] \tag{2}$$

Approximate Search Strategies. A major difficulty of BN structure learning is super-exponential number of search space in the large number of variables. Given n variables, there are $n!$ Orderings and $2^{\frac{n(n-1)}{2}}$ structures under each ordering. So, early studies focused on the approximate algorithm [3, 12]. At present, the most commonly used methods are K2 algorithm, greedy hill climbing, stochastic search, etc.

K2 algorithm used a variable ordering ρ and a positive integer u to restrict search space [7]. It begins with an edgeless graph which only includes all nodes. According to variable orderings, K2 investigated individually variable in ρ to determine its parent nodes, then added relevant edge to networks. Its time complexity is $O(mu^2nr)$, in the worst case, u = n, is $O(mn^4r)$. Obvious, with the increase in the number of variables, the requirements of time cannot be tolerated.

A well-known disadvantage of Hill climbing method is that it is easy to fall into local optimum without finding global optimum. Extensions to this approach include random restart hill climbing. In addition, the other heuristic search in combination optimize can also be used. Such as tabu search with random restarts [13], simulated annealing [2], genetic algorithm [14, 15] etc. Of course, these methods also restrict a positive integer u as number of node of parent.

Exact Search Strategies. Based on a fact that Bayesian networks have at least one leaf node, researchers proposed a number of exact learning methods that are able to find the global optimum. More representatives, that is the introduction of dynamic programming to learn Bayesian network structure. Dynamic programming algorithm works as follows: first, it finds optimal structure for every variable; then, it selects a node from remainder n-1 variable as current sub-structure optimal leaf which guarantees that sub-structure score is max after added leaf. This process is repeated until an optimal network is found for X. Its time complexity is $O(n2^n)$ and space complexity is $O(n2^n)$ [18, 20]. In order to solve the problem of time and space complexity of the algorithm, researchers have proposed a parallel algorithm based on the algorithm.

Tamada et al.'s [16] show a parallel algorithm for globally optimal structure learning based on Ott et al.'s sequential algorithm. Its idea is to use superset to compute the optimal orderings for all subsets of the same size in parallel. Due to a large number of redundant computations, time complexity and space complexity of the algorithm are $O(n^{\sigma+1}2^n)$. Olga et al.'s [5, 17] presented a different parallel algorithm. Firstly, the algorithm decompose n-dimensional (n − D) hypercube into 2^{n-k} k − D hypercube. Then it assigns 2^k processor to k − D hypercube. Finally, it used pipeline to manage the execution of k − D hypercube. Its time complexity is $O\left(n^2 2^{n-k}\right)$.

In addition, there are other exact algorithms such as A* search [24], Integer linear programming [1, 8] branch and bound algorithm. The methods are suitable for small datasets, in other words, up to deal with dozens of variables. It will be unable to find solutions when encountering a large number of variables. Due to time or space running out, there may be no solutions.

The approximation algorithm can be extended to large datasets but the quality of the solution is unpredictable. On the contrary, exact algorithm is able to find the global optimum but can't extend to large datasets. So, a compromise approach is meaningful. K2 algorithm complies with this requirement. Given variable ordering, K2 algorithm guarantees to find that the optimal structure is consistent with given ordering. So, it is meaningful for us to promote the performance of the K2 algorithm when the orderings of variable is given.

3   Parallel Learning Bayesian Networks

3.1   Grouping Conditions

This paper proposes a method different from the above. Our search space is not 2^n unordered subsets of X but is $2^{\frac{n(n-1)}{2}}$ structure space. We Propose the idea of grouping parallelism to complete BN structure learning. The main idea is that variables are grouped in domain and the size of group is mutative. Then we assign processor to each group. These processors independently and in parallel to generate optimal sub-Bayesian Networks for group. So, the algorithm must satisfy the following conditions:

(1) The group is complete. The relationship between variables is not destroyed by group. In simple terms, itself exists dependence of two variables, since there is no assigned to the same group result in this dependence relationship has not been found.

(2) The cycle can't appear when a combination of computing results for each group. There are two cases. The first case is acyclic, and this is what we want. The other one is cycle. We can't simply refer addition, deletion, and reversal to remove cycle. To do so, the final Bayesian network may be not optimal.

For this purpose, we first present grouping parallel Bayesian network structure learning algorithm based on variable ordering. The advantage of the method is that the parent's choice of each variable is independently done in parallel; it means that n optimal sub-Bayesian network was generated. Finally, we need not to detect cycle when combining sub-Bayesian network directly in order to get the full Bayesian Networks.

3.2   Grouping Parallel Learning Method

Definition 1: We say that X is ordering set, when $\forall x_i \in X, x_j \in Pa_{x_i} \subseteq X$ and $\forall x_j$ precede x_i, where Pa_{x_i} denotes the parent of variable x_i.

We found a good property from Definition 1. Given variable ordering, parent of any variable is ahead of it. This property ensures that our grouping parallel algorithm satisfies the above two conditions. Our method (Grp algorithm) consists of two main elements: variable group and parallel Learning. First, we introduce variable group. Given an ordering set order $(X) = x_{i+1}, x_{i-2}, \ldots, x_i, \ldots x_n$, $\Pi(x_i)$ denotes index of variable x_i in order(X). According to the ordered set, we grouped variables.

$$Group_i = \{x_j | \Pi(x_j) \leq \Pi(x_j)\} \qquad (3)$$

Proposition 1: $\Pi(x_i) = 1$ and $P_a(x_i) = \emptyset$, when $|Group_i| = 0$.

The proposition implies that x_i is first element in order(X). The parent of remaining variables may include x_i. Be further, the optimal Bayesian Networks on $Group_i$ is a single variable x_i.

Proposition 2: Group is complete based on variable ordering.

Proof: By the formula (3) indicates a set of variables X been divided into n groups, where $x_i \in$ order(X) and $Group_i = \{x_{i+1}, x_{i-2}, \ldots, x_i\}$. In other words, each group contains only the variable and its preceding element in order(X). According to the fact that the parent node of each variable is located in front of the variable given order, the group is complete based on variable ordering.

Proposition 3: The group parallelization BN structure learning based on the variable ordering ensures that sub-BN structure combinations learning between groups will be acyclic.

Proof: Let $x_i \in Group_i$, $j \in [i+1, k], x_j \in Group_k$, i, j denotes index of variable in ordering. By the Eq. (3), $Group_i \subset Group_k$, By Proposition 2, $\forall x_i$ is x'_j nodes of parent, on the contrary, it is not true. Therefore, learning between groups will be acyclic.

The grouping in our algorithm is used to group the variables into groups according to the ordered set in Eq. (3), and the sound is guaranteed by Proposition 2. Specifically, we use an array *order*[] to store the prior order of X. Supposing variable $x_i, x_j \in order[]$.

i and j denote the indies of x_i, x_j, respectively. For each index i, if $j < i$, then variables x_i, x_j are grouped into an identical group. Finally, the pseudo-code of grouping is presented in Algorithm 1.

Algorithm 1: Grouping

input: order[]=$[x_{i+1}, x_{i-2}, ..., x_i, ... x_n]$, group(i)=Φ

output: group(i)

---

1. For each $x_i \in order[]$ do
2.   For j=1 to i do
3.   group(i)=group(i) ∪ x_j
4.   end for
5. end for

For parallel learning, our work includes group learning and learning between groups. Firstly, we introduce group learning. The main idea is that assigns a processor to each group and the processor's task is to search for the optimal parent of variable. Specifically, within each group, we use Eq. (2) to examine each variable in group for determining parent nodes of variable x_i, and then add the parent node which has high score to $P_a(x_i)$. The number of parent of variable x_i at most is $|Group_i \setminus (x_i)|$. So, in worst case, this score function is called at most $|Group_i \setminus (x_i)|$. We give formal description that search parent of variable x_i and pseudo-code of group learning in Algorithm 2.

$$P_a(x_i)_{x_i \in X} = max_{P_a(x_i) \subseteq group(i) \setminus \{x_i\}} Score(x_i, P_a(x_i)) \tag{4}$$

Algorithm 2: processor i find optimal parent of x_i

Input: group(i), D: dataset u: in-degree of variable

Output: $P_a(x_i)$

---

1. $P_a(x_i) = Φ$;
2. $t_1 = Score(x_i, P_a(x_i)|D)$
3. OK=true
4. While(OK and $|P_a(x_i)| < u$)
5. $k = max_{1 \leq k \leq i, x_k \notin P_a(x_i)} Score(x_i, P_a(x_i) \cup \{x_k\}|D)$
6. $t_2 = Score(x_i, P_a(x_i) \cup \{x_k\}|D)$
7.   if($t_2 > t_1$)
8. $t_1 = t_2$
9. $P_a(x_i) = P_a(x_i) \cup \{x_k\}$
10.   else
11   OK=false
12. End while
13. Return $P_a(x_i)$

The main task of the group is to find optimal parent of all the variables. In other words, it finds the optimal Bayesian network structure for each group. Learning between groups is used to find full Bayesian network structure for X. We use an adjacent matrix A[i,j] to store dependence between variable x_i and parent node x_j. If $x_j \in P_a(x_i)$, then set A[i,j] = 1, until $P_a(x_i) = \phi$. Its pseudo-code as follows.

Algorithm3: Learning Between groups
input:$P_a(x_i)$, X = { $x_1, x_2, ..., x_n$}
output:A[i,j]// adjacent matrix
1. For i=1 to n do
2.   While(true)
3.   if($P_a(x_i) \neq \phi$)
4. $x_j \in P_a(x_i)$
5.   A[i,order(x_j)]=1 // order(x_j)denotes the index of
6.   variable x_j in X = { $x_1, x_2, ..., x_n$}
7.   else
8.   Break;
9.   End if
10.   End while
11. End for

4 Complexity Analysis and Experimental Results

Now we analyze the time complexity of our algorithm. Time complexity of grouping algorithm is mainly due to two "for" statements namely $O(n^2)$. Time complexity of group learning: because variable x_i has at most n-1predecessors in ordering, execution of line 5 is at most n-1 time. The "while" statement requires time $O(u)$, requirements of time of formula Eq. (2) is $O(mur)$. The execution of line 5 requires time $O(munr)$. Combining these results, the overall complexity of the group learning is $O(munr)$ $O(u) = O(mu^2nr)$. In the worst case, u = n, and complexity of the group learning is $O(mn^3r)$. Complexity of the learning between groups consists of two parts: the while statement requires time $O(|P_a(x_i)|)$ and the "for" statement requires time $O(n)$. The overall complexity of the learning between groups is $O(n|P_a(x_i)|)$. So, the overall complexity of our algorithm is $O(mn^3r) + O(n^2) + O(n|P_a(x_i)|) = O(mn^3r)$. Fig. (1)

This section shows the experimental evaluation of the proposed algorithm. In the experiment, along with changes of the number of nodes and samples in the domain, respectively, from the large sample size, a small number of nodes; small sample size, the nodes of appropriate scale; large sample size, large-scale nodes estimated performance of the algorithm. For the experiments, we used the [7] proposed discrete model scoring functions BD meanwhile using artificial datasets and uci datasets. Eventually, through comparison with classic algorithms, this paper shows the algorithm can scale well to handle large-scale problems. (Fig. 2)

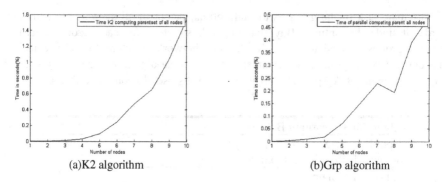

(a)K2 algorithm   (b)Grp algorithm

Fig. 1. Shows require time for both algorithm to calculate the parent of variables when dataset with 3000 samples, 10 nodes, and in-degree of variable is u = 4.

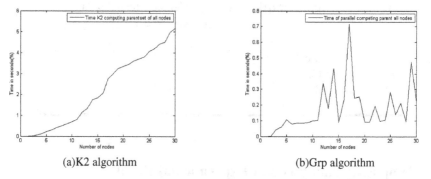

(a)K2 algorithm   (b)Grp algorithm

Fig. 2. Shows requirements time for both algorithm to calculate the parent of variables and when dataset with194 samples, 30 nodes, in-degree of variable is u = 6 and u = n−1.

Experiments above show that our algorithm exceeds to the classic K2 algorithm regardless of the degree of the nodes whether are given for any datasets. Mainly due to the overall complexity of K2 is $O(mn^2u^2r)$ when in-degree of variable is u, but the overall complexity of our algorithm is $O(mu^2nr)$. In the worst case, u = n, the overall complexity of K2 is $O(mn^4r)$, but the overall complexity of our algorithm (Grp) is $O(mn^3r)$. It is noteworthy that in the worst case, the time complexity of our algorithm (Grp) is less than in the best case, the time complexity of the K2, when $u > \sqrt{n-1}$. This is confirmed why the algorithm requires less time, when u = n-1 than the K2 when u = 6. (Fig. 3)

In our experiment, we only made the comparison with the K2 algorithm because the quality of the solution of approximation algorithm is unknown and exact algorithm is of poor scalability. But, given priori ordering, faced with any size of datasets, K2 is able to find optimal solution. (Fig. 4)

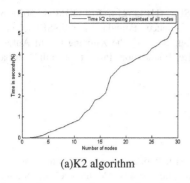 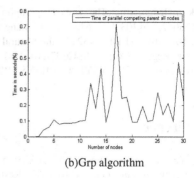

(a)K2 algorithm   (b)Grp algorithm

Fig. 3. Shows requirements time for both algorithm to calculate the parent of variables and when dataset with194 samples, 30 nodes, in-degree of variable is u = n−1.

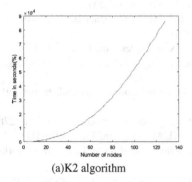 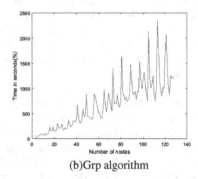

(a)K2 algorithm   (b)Grp algorithm

Fig. 4. Shows requirement time for both algorithm to calculate the parent of variables, when dataset with 6000 samples, 128 nodes and in-degree of variable is u = 1.

5   Conclusion

Based on the fact that we are able to find the optimal Bayesian Network that is consistent with given ordering, we proposed grouping parallel Bayesian network structure learning algorithm. According to the orderings, all variables are divided into different groups; each assigned a compute node, compute nodes simultaneously calculating the optimal parent of the corresponding variable in each group. The produce of optimal parent of corresponding variables also means that the optimal Bayesian Networks of every group is generated. Finally, learning between groups is used to find full Bayesian network structure for X. the comparison experiment shows the effectiveness of the algorithm. Validity and efficiency is theoretically analyzed. The problem is the strict precondition that group must based on variable ordering. An opening assumption of a priori ordering is under our future consideration.

Acknowledgments. The work was supported by NSFC (U1435214, 61432008, 61503178, 61175042, 61403208, 61321491), Jiangsu Nature Science Foundation (JSNSF) (Grant No. BK20130581,DE2015213), the Natural Science Foundation of the Xinjiang Uygur Autonomous Region (No: 201442137-26), and the Science Foundation Project of Yili Normal University (No: 2015YSYB30, 2014YSYB03).

References

1. Ye, S., Cai, H., Sun, R.: An algorithm for Bayesian networks structure learning based on simulated annealing with MDL Restriction. In: International Conference on Natural Computation, pp. 72–76. IEEE Computer Society (2008)
2. Bouckaert, R.R.: Properties of Bayesian belief network learning algorithms. In: Uncertainty Proceedings, pp. 102–109 (2013)
3. Cheng, J., Greiner, R., Kelly, J., et al.: Learning Bayesian networks from data: an information-theory based approach. Artif. Intell. **137**(1–2), 43–90 (2002)
4. Chen, Y., Tian, J., Nikolova, O., et al.: A parallel algorithm for exact Bayesian structure discovery in Bayesian networks. Eprint Arxiv (2014)
5. Chickering, D.M.: Learning Bayesian networks is NP-complete. In: Fisher, D., Lenz, H. J. (eds.) Learning from Data: Artificial Intelligence and Statistics V, pp. 121–130. Springer, New York (1996)
6. Cooper, G.F., Herskovits, E.: A Bayesian method for the induction of probabilistic networks from data. Mach. Learn. **9**(4), 309–347 (1992)
7. Cussens, J.: Bayesian network learning with cutting planes. In: Proceedings of the 27th Conference on Uncertainty in Artificial intelligence (UAI 2011), pp. 153–160 (2011)
8. Dash, D., Cooper, G.F.: Model averaging for prediction with discrete Bayesian networks. J. Mach. Learn. Res. **5**(5), 1177–1203 (2004)
9. de Campos, L.M.D., Huete, J.F.: A new approach for learning belief networks using inde-pendence criteria. Int. J. Approx. Reason. **24**(1), 11–37 (2000)
10. Friedman, N., Koller, D.: Being Bayesian about network structure. a Bayesian approach to structure discovery in Bayesian networks. Mach. Learn. **50**(1), 95–125 (2001)
11. Heckerman, D.: A tutorial on learning with Bayesian networks. In: Jordon, M.I. (ed.) Learning in Graphical Models. LNS, pp. 33–82. MIT Press, Cambridge (2000)
12. Glover, F.: Tabu search: a tutorial. Interfaces **20**(4), 74–94 (1990)
13. Hsu, W.H., Guo, H., Perry, B.B., et al.: A permutation genetic algorithm for variable ordering in learning Bayesian networks from data. In: Proceedings of the Genetic and Evolutionary Computation Conference, pp. 383–390. Morgan Kaufmann Publishers Inc. (2010)
14. Larranaga, P., Kuijpers, C.M.H., Murga, R.H., et al.: Learning Bayesian network structures by searching for the best ordering with genetic algorithms. IEEE Trans. Syst. Man Cybern. Part A Syst. Hum. **26**(4), 487–493 (1996)
15. Tamada, Y., Imoto, S., Miyano, S.: Parallel algorithm for learning optimal Bayesian network structure. J. Mach. Learn. Res. **12**, 2437–2459 (2011)
16. Nikolova, O., Zola, J., Aluru, S.: Parallel globally optimal structure learning of Bayesian networks. J. Parallel Distrib. Comput. **73**, 1039–1048 (2013)
17. Ott, S.: Finding optimal models for small gene networks. In: Pacific Symposium on Biocomputing, vol. 9, pp. 557–567 (2004)
18. Perrier, E., Imoto, S., Miyano, S.: Finding optimal Bayesian network given a super-structure. J. Mach. Learn. Res. **9**, 2251–2286 (2008)

19. Silander, T., Myllymaki, P.: A simple approach for finding the globally optimal Bayesian network structure. In: Proceedings of the 22nd Annual Conference on Uncertainty in Artificial Intelligence (UAI-06), pp. 445–452. AUAI Press (2006)
20. Tsamardinos, I., Brown, L.E., Aliferis, C.F.: The max-min hill-climbing Bayesian network structure learning algorithm. Mach. Learn. **65**(1), 31–78 (2006)
21. Xie, X., Geng, Z.: A recursive method for structural learning of directed acyclic graphs. J. Mach. Learn. Res. **9**(3), 459–483 (2008)
22. Yuan, C., Malone, B.: Learning optimal Bayesian Networks: a shortest path perspective. J. Artif. Intell. Res. **48**, 23–65 (2013)

One-Class Models for Continuous Authentication Based on Keystroke Dynamics

Maria Kazachuk, Alexander Kovalchuk, Igor Mashechkin, Igor Orpanen, Mikhail Petrovskiy[✉], Ivan Popov, and Roman Zakliakov

Computer Science Department, Lomonosov Moscow State University, MSU, Vorobjovy Gory, Moscow 119899, Russia
{kazachuk,kovalchuk,orpanen,ivan,zakliakov}@mlab.cs.msu.su,
{mash,michael}@cs.msu.su

Abstract. In this paper we discuss an applied problem of continuous user authentication based on keystroke dynamics. It is important for a user model to discover new intruders. That means we don't have the keystroke samples of such intruders on the training phase. It leads us to the necessity of using one-class models. In the paper we review some popular feature extraction, preprocessing and one-class classification methods for this problem. We propose a new approach to reduce dimensionality of a feature space based on two-sample Kolmogorov-Smirnov test and investigate how the quantile-based discretization technique can improve the one-class models' performance. We present two algorithms, which have not been used for keystroke dynamics before: Fuzzy kernel-based classifier and Random Forest Regression classifier. We conduct experimental evaluation of the proposed approach.

Keywords: Keystroke dynamics · User authentication · Kolmogorov-Smirnov test · Quantile discretization · Fuzzy classification · Random forest regression classification

1 Introduction

Nowadays computer systems play a very important role in people's life. These systems are used to store, search and process information. Therefore, it is essential to employ a high level of protection against unauthorized access on these systems.

One of the best known security mechanisms is authentication. It allows a system to confirm that the user is who he claims to be. The reliability of this process depends on the information used for user authentication. Authentication employs several factors, which are usually one of the following: knowledge, ownership, physiological biometrics and behavioral biometrics. Using the knowledge and the ownership factors for authentication has a major drawback, that information can be lost, stolen or divulged.

Physiological biometric images, such as fingerprints, retina, the geometry of the face and hands, are given to people by birth and can not be changed by

© Springer International Publishing AG 2016
H. Yin et al. (Eds.): IDEAL 2016, LNCS 9937, pp. 416–425, 2016.
DOI: 10.1007/978-3-319-46257-8_45

the will of its owner. Their use for authentication features are highly reliable, but they need additional equipment, and the theft of data samples can cause irreparable damage to the security.

Behavioral biometric samples include those human characteristics and traits that appear when the user performs a certain set of actions. Such data systems include voice, handwriting and gait authentication. Behavioral characteristics can be easily changed by the will of the owner, and it is almost impossible to betray your secret behavioral pattern. Today these systems are widely developed, but their quality is inferior to other authentication systems.

Use of continuous authentication systems allows us to detect intruders when they try to interact with a machine. It's possible that an intruder isn't known, so the construction of a multi-class model, that recognizes a specific user from a closed set, can cause an impostor to remain undetected. Therefore, it is necessary to research the methods that build one-class models for each legitimate user. When performing continuous authentication, model of the legitimate user is compared to the behavior of the current user. Legitimacy of the current user is determined based on this comparison.

This paper features an applied problem of continuous user authentication based on keystroke dynamics, given an assumption of inavailability of illegitimate users' data. This paper is structured into the following sections. Section 2 describes existing keystroke dynamics research. Section 3 describes our approach to enhance keystroke dynamics with better accuracy. Section 4 includes experiments. Section 5 ends this paper with final conclusions on keystroke dynamics.

2   Survey

User's keyboard interaction can be described as key press and release timing information. To collect this data OS tools [6,11–13,15] and web browser tools [1,4] can be used.

Collected data is split into windows, which contain events to be processed into a single feature vector. We have determined several events, which force a new window to be created: exceeding the maximum size of a window or the maximum pause between windows, active process change. In [15] the authors suggest to ignore windows with the number of events below the threshold.

Existing feature extraction approaches include analysis of single key presses and consecutive key presses, called n-graphs. The most frequently used n-graphs are digraphs ($n = 2$) and trigraphs ($n = 3$).

Hold time ($t_i^{up} - t_i^{down}$) and latency ($t_{i+1}^{down} - t_i^{up}$) are calculated for every key (Fig. 1). For n-graphs a subset of the following features is used: $\{(t_{k+n-1}^{up} - t_k^{down}),$ $(t_{k+n-1}^{down} - t_k^{up}), (t_{k+n-1}^{down} - t_k^{down}), (t_{k+n-1}^{up} - t_k^{up})\}$, here k is an index of n-graph in the current window [2–5,11–13,15,16].

Each key and each n-graph produces one or more features in the feature vector, which are equal to the mean or variance of the corresponding statistic (hold time, latency time). Features of keys and key sequences, which are not present in a window, are filled with zeros.

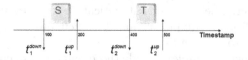

Fig. 1. Key presses and releases

The constructed feature space is high-dimensional, but not all of the extracted features have a significant impact on the classification results. The most common techniques for dimensionality reduction used in related problems are principal component analysis (PCA) [3] and random search approaches such as genetic algorithm (GA) [6,13], particle swarm optimization (PSO) [6,13] and gravitational search algorithm (GSA) [5]. However, our experiments showed that the use of PCA in most cases causes even worse accuracy of the classification, which can be explained by the presence of non-linear correlation between variables. Random search feature selection algorithms are rather computationally difficult.

Some classification algorithms might work considerably better, if features are standardized. This is done by subtracting feature's expected value Ex and dividing the difference by its standard deviation $\sqrt{Dx}$. Both expected value and standard deviation are determined on the training set.

The most commonly used classification algorithms in the field of arbitrary text keystroke dynamics are: metric (one-class KNN [16]) and probabilistic (one-class SVM [1,16], Gaussian mixture models [4], Bayes networks [2]) methods. The best results are achieved in the reviewed papers by using the one-class KNN and one-class SVM classifiers.

3   Proposed Approach

Our approach uses the most popular methods to define a feature space, suggests some new ways of feature selection and preprocessing, and introduces several new classifiers, which have not been used in this task before.

3.1   Feature Space

To construct a feature space, which will describe user's interaction with the keyboard, we propose to combine the analysis of single keys and digraphs. The sequence of collected events is split into windows. Window size is required to be between the fixed minimum and maximum size. If a user is inactive for a specified maximum pause between sequential events, then a window split is forced. These parameters are chosen experimentally.

In order to reduce the dimension of the feature space, we propose to keep only the most frequent digraphs and keys, where frequencies are determined on the training set.

For each of the remaining keys we propose to calculate hold time; for each of the remaining digraphs — the up-up and down-up latency. Also, we divide

Fig. 2. The proposed 12 key groups

keys into 12 groups (Fig. 2), for each of them hold times are calculated. Some special features, used in this work, are frequency of key presses, and the ratio of the number of left Shift key presses to the number of combined left and right Shift key presses.

3.2 Feature Preprocessing

After feature extraction we examine several feature preprocessing methods: dimensionality reduction of the feature space through selection of the most stable features, and quantile-based feature discretization. To reduce dimensionality of the extracted feature space we select only those features, which have insignificant variance over time. To achieve this we propose two-sample Kolmogorov-Smirnov test. It is used to evaluate a hypothesis, that two samples have the same distribution. The first sample contains unaveraged feature values for each window and the second sample contains congregated unaveraged values for all windows in the training set.

Let $F_{1,n}$ be the distribution function of a feature in the current window, where n is the number of occurrences of the feature in the current window. Let $F_{2,m}$ be the distribution function of the feature for all windows in the training set, where m is the number of occurrences of the feature in the training set. The main goal is to find such λ, that the following inequality is true:

$$\sqrt{\frac{nm}{n+m}}D_{n,m} \geq \lambda, \tag{1}$$

where $D_{n,m} = sup_x |F_{1,n}(x) - F_{2,m}(x)|$.

Next we use the quantile table of the Kolmogorov distribution to find a p-value, that corresponds to λ. This p-value represents the probability of rejecting the hypothesis, that these two samples have the same distribution.

To accumulate p-values across all windows we suggest Fisher's method, which combines all p-values into single test statistic:

$$\chi_{2k}^2 \approx -2 \sum_{i=1}^{k} \ln(p_i), \tag{2}$$

where k — number of p-values, p_i — i-th p-value.

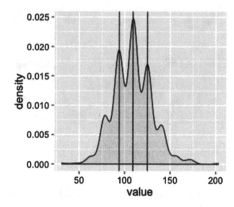

Fig. 3. Quantile-based discretization on 4 intervals

Subsequently the chi-squared distribution table is used to find a corresponding p-value with $2k$ degrees of freedom. The resulting p-value may be used to perform feature selection. We have studied selection of N features with the highest p-values and selection of all features, which exceed threshold p-value. After performing feature selection all unaveraged features are substituted with their mean value.

One of the most popular feature preprocessing techniques used for multimodal distributions is the quantile discretization. This approach has not been applied to the keystroke data before. For each feature in the training set we calculate k quantiles of order $\frac{1}{k}, \frac{2}{k}, ..., \frac{k-1}{k}$, which form several intervals (bins): $(-\infty, \frac{1}{k}), (\frac{1}{k}, \frac{2}{k}), ..., (\frac{k-1}{k}, +\infty)$. Then each feature value is replaced with the corresponding bin index, which it falls into.

Unlike the approach in [11], where it is stated, that digraph hold time has normal distribution, we assume most of the features, associated with key presses have multimodal distribution (Fig. 3). Quantile discretization method takes this into account, and subsequent experiments confirm our hypothesis.

3.3   Building User Model

As mentioned before, we consider only one-class classification methods. Fuzzy Kernel-based Method for Outliers Detection and feature values estimator based on Random Forest Regression were introduced for our problem.

We have developed Fuzzy kernel-based method [14] earlier for the task of outliers mining for intrusion detection. It inherits the ideas of SVM, but instead of looking for a crisp sphere in the RKHS feature space, we suggest to search for a fuzzy sphere including all RKHS data images. This problem can be viewed as calculating single fuzzy cluster in the RKHS feature space using possibilistic fuzzy clustering approach. In this case the fuzzy membership can be described as a measure of "typicalness" of data instances. Keystrokes with low "typicalness" are considered outliers. Changing the threshold does not lead to the recalculation of the models, as it was done for SVM and kernelized distance-based algorithms. Mathematically, the problem is to find $\min\limits_{U,a,\eta} J(U,a,\eta)$:

$$J(U, a, \eta) = \sum_{i=1}^{N} u_i^m (\phi(x_i) - a)^2 - \eta \sum_{i=1}^{N} (1 - u_i)^m, \tag{3}$$

where a is a center of the fuzzy cluster in the RKHS feature space; N is a number of instances in the initial feature space X; U is a membership vector, where $u_i \in [0, 1]$ is membership of the image $\phi(x_i)$ and besides the "typicalness" of datum x_i; m is fuzzyfier and η – parameter, that controls the size of cluster.

After the minimization of the functional, for each training sample $x_i \in X$ the measure of its "typicalness" u_i is calculated. The calculation of this value for a new item x is done as follows:

$$u(x) = \left[1 + \left(\frac{\sum_{j=1}^{N} u_j^m \sum_{i=1}^{N} u_i^m K(x_i, x_j)}{\eta \left(\sum_{i=1}^{N} u_i^m \right)^2} - 2 \frac{\sum_{i=1}^{N} u_i^m K(x, x_i)}{\eta \sum_{i=1}^{N} u_i^m} + \frac{K(x,x)}{\eta} \right)^{\frac{1}{m-1}} \right]^{-1}, \tag{4}$$

where N is a number of records in training set, $K(x,y)$ is a kernel for x and y.

Another algorithm that we have developed is a one-class classifier, based on the Random Forest Regression [8] used for the approximation of values of all features. The degree of normality of the data is calculated in accordance to how well it was approximated by the model. The idea is similar to Replicator Neural Networks [9], but instead of neural nets the Random Forest is used as an approximator. Regression trees are built as follows. Suppose, that we have p inputs and N observations with response $(x_i, y_i); x_i = (x_{i1}, \ldots, x_{ip}); i = \overline{1, N}$. The algorithm chooses splitting variables, split points and trees topology and we get M regions $R_1, \ldots, R_M$. We model a response (where c_m is a constant in each region, its best approximation is $\hat{c}_m = average(y_i \mid x_i \in R_m)$):

$$f(x) = \sum_{m=1}^{M} c_m I(x \in R_m). \tag{5}$$

When we use Random Forest for regression, we build an ensemble of trees $\{T(x; \theta_b)\}_1^B$, using the subset of the training data by recursively splitting the nodes on the best split-point among a subset of m random features until the minimum node size is reached. θ_B characterizes the b-th tree in terms of split variables, cutpoints at each node and terminal-node values. After this step, the random forest predictor is:

$$\hat{f}_{rf}^B(x) = \frac{1}{B} \sum_{b=1}^{B} T(x; \theta_b). \tag{6}$$

A one-class classifier can be built by creating a set of p regressors, each for a separate variable, where other variables are used as predictor values:

$$\hat{f}^B_{x_i}(x) = \frac{1}{B} \sum_{b=1}^{B} T(x_1, \ldots, x_{i-1}, x_{i+1}, \ldots, x_p; \theta_b). \tag{7}$$

When we get a new observation, it can be estimated using the built predictors:

$$(x_1, \ldots, x_p) \rightarrow (\hat{f}^B_{x_1}, \ldots, \hat{f}^B_{x_p}). \tag{8}$$

The resulting decision function is defined as a reconstruction error:

$$DF(x) = \frac{1}{n} \sum_{i=1}^{n} (x_i - \hat{f}^B_{x_i})^2. \tag{9}$$

Note that all features must be standardized. The main advantage of this algorithm is the ability to detect non-linear correlation between features and to ignore irrelevant features.

4   Experiments

The most suitable dataset we have found is the Villani keystroke public dataset [12,15], which consists of 144 users, who were instructed to respond to open-ended essay questions and produced 1345 samples overall. The following data was collected for each user: platform (desktop or laptop), gender, age group, handedness and awareness of data collection. In our work only the following data was used: keycode, keystroke press and release times. We only use 53 of 144 users because they provide 20 or more feature vectors when using parameters discussed below. For each of the 53 users half of their data was used as the training set and other half of the data was combined with the data of all the remaining users to be the test set.

Area under the ROC-curve (AUC) was used to evaluate classification results. The AUC is equal to the probability that a classifier will rank a randomly chosen positive instance higher than a randomly chosen negative one [7]. The distinguishing feature of the AUC is the invariance with respect to the proportion of positive and negative samples in the test set. We have used average AUC among all users to evaluate final results.

The essential part of conducting experiments is to select optimized parameters for the feature extraction, feature preprocessing and classification algorithms. We have conducted preliminary experiments on a dataset collected by ourselves to choose some hyperparameters. 20 users had been working for 10 days, 1 h per day, on 3 different configurations of computers doing their usual work: programming, text and presentation preparation, Internet browsing. For feature extraction algorithms optimal parameters were: minimum window size — 300 events; maximum window size — 500 events; maximum pause between events

— 40 s; amount of the most popular keys — 50; amount of the most popular digraphs — 100. Next we have chosen optimal parameters for the feature pre-processing algorithms. Feature selection based on stability depends either on the amount of the most stable features or on the threshold level of significance. Best results in our experiments were achieved by using the level of significance more than 0.1. Quantiles discretization has only one parameter: number of quantiles, which has the default value of 10.

Subsequent task is to choose optimal parameters for classification algorithms, which we have done for each of the following classifiers:

1. One-class KNN [16] is based on evaluating distances between samples in the training set. Threshold distance, which determines the maximum distance of an original class element, is chosen as the distance to the neighbor with index $\lfloor p * N \rfloor$ in the sorted array of k neighbors, where $p \in [0, 1]$.
2. One-class SVM classifier [1,16] is based on the construction of a hypersphere, which encompasses most of the user's data. In the course of classification elements, which fall inside this hypersphere are considered to belong to the original class. The parameters of the algorithm are: kernel type, kernel coefficient γ, an upper bound on the fraction of training errors and a lower bound of the fraction of support vectors ν.
3. Kernel principal component analysis (KPCA) [10] extends standard PCA to non-linear data distributions, which can also be used for one-class classification. The parameters of the algorithm are: kernel type, kernel coefficient γ, number of principal components n.
4. Replicator Neural Network (RNN) [9] is a forward propagation neural network with the same number of input and output neurons, which reconstructs

Table 1. Optimal values for the classification algorithms

	Parameter	Standardization	Selection using stability	Quantile discretization
KNN	k	4		
SVM	$kernel$	radial basis function (RBF)		
	γ	$1/N_{features}$		
	ν	0.1		
KPCA	kernel	radial basis function (RBF)		
	γ	$1/max\|x_i - x_j\|$, where x_i, x_j are in training set, $i \neq j$		
	n	10	10	5
RNN	$hidden$	three hidden layers (16 – 4 – 16)		
	$iterations$	500		
Fuzzy	$kernel$	radial basis function (RBF)		
	γ	$1/max\|x_i - x_j\|$, where x_i, x_j are in training set, $i \neq j$		
	m	1.5		
RFR	$nTrees$	3	3	20
	$minSize$	3	10	10

Table 2. Results on the Villani dataset

	Standardization	Selection using stability	Quantile discretization
KNN	0,5392	0,5596	0,6656
SVM	0,7665	**0,7933**	0,8861
KPCA	0,7770	0,7898	0,9081
RNN	0,7641	0,7918	0,8833
Fuzzy	0,7737	0,7888	**0,9122**
RFR	**0,7805**	0,7898	0,8991

original class elements better than outliers. The parameters of the algorithm are: a number of hidden layers of the neural network, a number of neurons on hidden layers, a number of iterations.

5. Fuzzy classification method is described in the previous section. The parameters of the algorithm are: kernel type, kernel coefficient γ, affiliation level's decrease rate m based on the distance to the center of the cluster.

6. Random Forest Regressor (RFR) classification method is described in the previous section. The parameters of the algorithm are: number of trees ($nTrees$) and minimum tree's leaf size ($minSize$).

Optimal parameters were selected on our dataset (Table 1), and then each of the classification algorithms with obtained hyperparameters was applied to Villani dataset (Table 2).

5 Conclusion

In this paper, the task of continuous user authentication using keystroke dynamics was considered. We have proposed a method to reduce dimensionality of a feature space based on two-sample Kolmogorov-Smirnov test and a quantile-based discretization technique to preprocess features with multimodal distribution. We have introduced two one-class classifiers, which haven't been applied to this problem before. We have conducted experiments on a benchmark dataset, which have confirmed that stability-based feature selection improves quality of authentication when using SVM, KNN, Fuzzy, KPCA and RNN classifiers, but almost no effect on RFR classifier. Furthermore, quantile-based discretization improves the results of all classifiers.

Acknowledgments. The research is financially supported by the Ministry of Education and Science of the Russian Federation (the subsidy agreement #14.604.21.0056, unique project identifier RFMEFI60414X0056).

References

1. Al Solami, E., Boyd, C., Clark, A., Ahmed, I.: User-representative feature selection for keystroke dynamics. In: 2011 5th International Conference on Network and System Security (NSS), pp. 229–233. IEEE (2011)

2. Alsultan, A., Warwick, K.: Keystroke dynamics authentication: a survey of free-text methods. Int. J. Comput. Sci. Issues **10**(4), 1–10 (2013)
3. Bailey, K.O., Okolica, J.S., Peterson, G.L.: User identification and authentication using multi-modal behavioral biometrics. Comput. Secur. **43**, 77–89 (2014)
4. Ceker, H., Upadhyaya, S.: Enhanced recognition of keystroke dynamics using Gaussian mixture models. In: Military Communications Conference, MILCOM 2015-2015 IEEE, pp. 1305–1310. IEEE (2015)
5. Chandrasekar, V., Akila, M., Maheswari, T.: Gravitional search optimization for the user authentication in biometrics. Middle-East J. Sci. Res. **23**(8), 1626–1631 (2015)
6. Everitt, R.A., McOwan, P.W.: Java-based internet biometric authentication system. IEEE Trans. Pattern Anal. Mach. Intell. **9**, 1166–1172 (2003)
7. Fawcett, T.: An introduction to ROC analysis. Pattern Recogn. Lett. **27**(8), 861–874 (2006)
8. Hastie, T., Tibshirani, R., Friedman, J., Franklin, J.: The elements of statistical learning: data mining, inference and prediction. Math. Intelligencer **27**(2), 83–85 (2005)
9. Hawkins, S., He, H., Williams, G., Baxter, R.: Outlier detection using replicator neural networks. In: Kambayashi, Y., Winiwarter, W., Arikawa, M. (eds.) DaWaK 2002. LNCS, vol. 2454, pp. 170–180. Springer, Heidelberg (2002). doi:10.1007/3-540-46145-0_17
10. Hoffmann, H.: Kernel PCA for novelty detection. Pattern Recogn. **40**(3), 863–874 (2007)
11. Kang, P., Cho, S.: Keystroke dynamics-based user authentication using long and free text strings from various input devices. Inf. Sci. **308**, 72–93 (2015)
12. Monaco, J.V., Bakelman, N., Cha, S.H., Tappert, C.C.: Developing a keystroke biometric system for continual authentication of computer users. In: 2012 European Intelligence and Security Informatics Conference (EISIC), pp. 210–216. IEEE (2012)
13. Namin, A.S.: Cyberspace security use keystroke dynamics. Ph.D. thesis, Texas Tech University (2015)
14. Petrovskiy, M.: A fuzzy kernel-based method for real-time network intrusion detection. In: Böhme, T., Heyer, G., Unger, H. (eds.) IICS 2003. LNCS, vol. 2877, pp. 189–200. Springer, Heidelberg (2003). doi:10.1007/978-3-540-39884-4_16
15. Tappert, C.C., Cha, S., Villani, M., Zack, R.S.: Keystroke biometric identification and authentication on long-text input. Int. J. Inf. Secur. Priv. (IJISP) **4**, 32–60 (2010)
16. Teh, P.S., Teoh, A.B.J., Yue, S.: A survey of keystroke dynamics biometrics. Sci. World J., 1–24 (2013). doi:10.1155/2013/408280

Incremental Nonnegative Matrix Factorization Based on Matrix Sketching and k-means Clustering

Chenyu Zhang, Hao Wang[✉], Shangdong Yang, and Yang Gao

State Key Laboratory for Novel Software Technology,
Nanjing University, Nanjing 210023, China
nju_zcy@163.com, wanghao.hku@gmail.com,
yangshangdong007@gmail.com, gaoy@nju.edu.cn

Abstract. Along with the information increase on the Internet, there is a pressing need for online and real-time recommendation in commercial applications. This kind of recommendation attains results by combining both users' historical data and their current behaviors. Traditional recommendation algorithms have high computational complexity and thus their reactions are usually delayed when dealing with large historical data. In this paper, we investigate the essential need of online and real-time processing in modern applications. In particular, to provide users with better online experience, this paper proposes an incremental recommendation algorithm to reduce the computational complexity and reaction time. The proposed algorithm can be considered as an online version of nonnegative matrix factorization. This paper uses matrix sketching and k-means clustering to deal with cold-start users and existing users respectively and experiments show that the proposed algorithm can outperform its competitors.

Keywords: Recommender system · Incremental recommendation · Matrix factorization · Matrix sketching · k-means clustering

1 Introduction

Conventional recommender systems are usually off-line, i.e., they train recommendation models based on historical data and recommend items to users regardless of their current behaviors [2]. Such offline recommendations may not be accurate enough in scenarios where it is crucial to capture users' real-time interests. For example, a user may enjoy energetic music while running but prefer soft music before sleeping. It's not a good practice to recommend energetic music based on historical data while the user is actually going to sleep. Therefore, *online and real-time recommendation* has recently attracted much attention in both academia and industry [5, 7, 12]. While users are browsing the web, such online recommenders should take users' real-time behaviors into consideration and update any underlying model to make recency-sensitive recommendations.

There have been extensive studies on recommender systems. To our understanding, existing recommendation methods can be classified into three categories. The first category includes *content-based methods*, which aim at modeling users with predefined

© Springer International Publishing AG 2016
H. Yin et al. (Eds.): IDEAL 2016, LNCS 9937, pp. 426–435, 2016.
DOI: 10.1007/978-3-319-46257-8_46

features [12]. These methods need many predefined features to describe both users and items, however in applications it is hard to get such sufficient content data. The second category is *collaborative filtering* (CF), which uses known user ratings of items to predict user preferences in item selection [4]. CF methods are usually based on a rating matrix which consists of users' ratings to items. Each row of the matrix stands for each user and each column for each item. Every non-zero entry represents a *preference* of some user over some item. In some applications the preferences can be explicit *ratings* (e.g., 5 for "very like" and 1 for "very dislike"), and in some other applications the preferences are numeric records of users' behaviors (e.g., durations and frequencies of visits). Matrix factorization (MF) is often applied for CF problems, which factorizes the rating matrix into two smaller matrices describing user features and item features respectively. Nonnegative matrix factorization (NMF) [13] is one MF method that generates models with only nonnegative values, which makes values more explicable. Finally, the third category is hybrid methods which combine multiple algorithms to get the recommendation lists.

This paper comes up with a new method to perform *incremental* NMF. It is known that NMF is a common recommendation method but it can only handle batch data and takes a long time to train the model when there is a large amount of data. To do the incremental update, we distinguish two different scenarios: The first scenario is about *cold-start* users, i.e., when the incoming data are due to a user who has no history in the system. In this case, we should insert a new row into the matrix; The second scenario is about existing users, in which case we should modify the corresponding row of the matrix. We propose two different incremental methods to deal with the two scenarios respectively.

The rest of the paper is organized as follows. Section 2 introduces related work. Section 3 summarizes our approach. Then, Sect. 4 describes our algorithm in detail. After that, Sect. 5 introduces the experimental results and analyzes the performance of our approach. Finally, Sect. 6 draws a conclusion for this paper.

2 Related Work

2.1 Recommender System

The existing recommendation methods are usually divided into three categories: *content-based methods*, CF, and hybrid methods.

Content-based methods use the personal preferences of a new user [8]. This kind of methods models users with predefined features and calculates the similarity between user preferences and item features, then it can recommend items that appeal to the target user. These methods do not take other users' preferences into consideration so they need not to analyze relations among users. The key of *content-based methods* is to define features for items, which also brings the problem that it's hard to describe items' features and get users' preferences sometimes.

CF contains two main types of recommendation: one based on item and the other based on user. Item-based CF recommends items that are similar to what the target user is interested in. User-based CF recommends what users who share similar interests with the target user are interested in. The similarity between users or items usually can be measured by Euclidean distance or cosine similarity. There are various matrix factorization based techniques in CF. [13] proposes NMF which trains models with only nonnegative features, and this method makes values more explicable.

Hybrid methods aim at combining multiple algorithms to get final results. These methods can overcome some shortcomings of single algorithm, such as sparsity of rating matrix. [9] classifies diverse combination types into several basic groups: weighted hybrid which sets different weights for results generated by each algorithm; switching hybrid which chooses distinct algorithms for different application scenarios; mixed hybrid which mixes results of several algorithms and presents to the users; cascade hybrid which uses one algorithm to generate recommendation candidates and another to determine items' order.

2.2   Incremental Recommendation

MF has been widely used in recommender systems and numerous different MF variants have been already published. Most of the methods can only handle stationary data and take a long time to train the model when the number of users and items grows up to a high order of magnitude. However, in practical application scenarios, data usually reaches the system in the form of stream which means the rating matrix is dynamic. If we retrain the model every time when new data reaches, it will take high computational complexity and time complexity.

Recently, several works come up with methods to update model in an incremental way. Some works modify a small part of feature matrix at a time, for example, [10] just updates one feature matrix while keeping the other invariant. These methods usually ignore the influence of data on one matrix and focus on the other matrix to speed up updating. [1] proposes a method via feature space re-learning strategy, it re-learns the feature space via auxiliary feature learning and matrix sketching strategies. Our approach for cold-start users is based on the method in [1], but we introduce the threshold of feature size and update both matrices to further accelerate updating.

3   Our Approach

3.1   Scheme

Our approach is based on NMF. We use NMF to deal with original rating matrix, then we use different incremental methods to deal with new data records from cold-start users and existing users separately. As shown in Fig. 1, the rating matrix consists of users' ratings on items. In some applications, users don't rate on items directly according their preferences and the website can only get records of user visit. In this case, systems usually set ratings based on numeric records of users' behaviors (e.g., durations or frequencies of visits). In our experiment, we set the rating matrix as binary matrix, 1 for that the user has visited the web page of the item and 0 for that the user has not visited the web page of the item.

	Item$_1$	...	Item$_k$	...	Item$_m$
User$_1$	R_{12}	...	R_{1k}	...	R_{1m}
...	...	...	...	...	...
User$_k$	R_{k1}	...	R_{kk}	...	R_{km}
...	...	...	...	...	...
User$_n$	R_{n1}	...	R_{nk}	...	R_{nm}

Fig. 1. Rating matrix

For data records from cold-start users, we use auxiliary features to refine the original model. The auxiliary features help better model new users' preferences but also bring the problem that the feature size will increase infinitely. To solve this problem, we set a threshold for the feature size and use matrix sketching on item feature matrix to reduce the feature size when it reaches the threshold. Matrix sketching can help shrink the matrix to a smaller size and retain the similarity between items.

For data records from existing users, we use k-means clustering to find the user set in which users share similar interests with the target user, then we just need to refresh the sub-matrix consisting of the user set. K-means is one of the most popular and efficient method for clustering. We can use it to locate similar user sets in a rapid and robust way [11]. This method will dramatically enhance the efficiency of factorization.

3.2 Problem Definition

In recommender systems, we use rating matrix $R_0 \in \mathbf{R}^{n \times m}$ to represent original historical data. n represents the number of users and m represents the number of items. Usually we train model on R_0 to get original decomposing matrix P_0 and Q_0. Here $P_0 \in \mathbf{R}^{n \times k}$ represents the user feature matrix and $Q_0 \in \mathbf{R}^{k \times m}$ represents the item feature matrix, k represents the feature size. We can predict the rate for user u on item i by using $p_u q_i$, where p_u represents the row of user u in P and q_i represents the column of item i in Q.

For new data belonging to a cold-start user, we should insert a new row into the rating matrix. In this paper, we use $R_1 \in \mathbf{R}^{c \times m}$ to represent the matrix that contains these new rows. Here, we use c to represent the number of cold-start users. For new data belonging to existing users, we should modify the corresponding rows of the rating matrix. In this paper, we use $R_2 \in \mathbf{R}^{d \times m}$ to represent the matrix that contains these rows of target users.

The goal of this paper is to update the model on the matrix which includes both original historical data and new data in an efficient way.

4 Specific Algorithm

4.1 Nonnegative Matrix Factorization

The goal of NMF is to get matrix P and Q so that $R \approx PQ$ and we can get the prediction for those items that has not been rated yet. We try to minimize the difference between the estimated value and the real one. Besides, we use regular terms to avoid over-fitting [4]. So the ultimate convergence condition is expressed as follow:

$$L = \sum\nolimits_{(u,i)\in V} (p_u q_i - r_{ui})^2 + \lambda(\| p_u \|^2 + \| q_i \|^2)) \tag{1}$$

$$(P^*, Q^*) = \mathrm{argmin}_{(P,Q)}(L) \tag{2}$$

Here V is the set of tuples that user u has rated on item i.

We use stochastic gradient descent to get the optimum solutions. The update formulas are as follow:

$$p'_u = \max(p_u + \alpha \cdot (\sum\nolimits_{(u,i)\in V} 2(p_u q_i - r_{ui})q_i - 2\lambda \| p_u \|), \, 0) \tag{3}$$

$$q'_i = \max(q_i + \alpha \cdot (\sum\nolimits_{(u,i)\in V} 2(p_u q_i - r_{ui})p_u - 2\lambda \| q_i \|), \, 0) \tag{4}$$

In addition, the rating matrix is very large in practice and the minimum value of convergence condition is hard to reach, we usually define threshold β of the difference of value between two updates to terminate the update.

Algorithm1: NMF

Input: $R \in R^{n \times m}$, k, t, β
Output: $P \in R^{n \times k}$, $Q \in R^{k \times m}$

1: old_err = −1
2: initialize $\hat{P}, \hat{Q}$ with random values
3: for iter $\in [t]$ do
4:     for each $R_{ui} \neq 0$ do
5:         refresh $\hat{P}_u$ and $\hat{Q}_i$ by using (3) and (4)
6:     end for
7:     calculate new_err by using (1)
8:     if (iter == 0 or old_err-new_err>β) then
9:         old_err = new_err
10:        $P = \hat{P}, Q = \hat{Q}$
11:    else
12:        $P = \hat{P}, Q = \hat{Q}$
13:        break
14:    end if
15: end for

4.2   Incremental Algorithm for Cold-Start Users

A cold-start user is a user who has never had any historical data. So when a cold-start user comes, it equals to adding a row to the original matrix. If we train the model based on all data, it will take high computational complexity and long processing time. So we consider about using auxiliary features to update the model.

Similar to [1, 6], the residual error of new data is normalized and added to the existing factorized matrices as auxiliary features. As mentioned in Sect. 2.1, we use $R_1 \in \mathbf{R}^{c \times m}$ to represent rating matrix from cold-start users, then the model can be updated as follow:

$$\begin{bmatrix} R_0 \\ R_1 \end{bmatrix} = \begin{bmatrix} P_0 & 0 \\ R_1 Q_0^{-1} & \| err \| \end{bmatrix} \begin{bmatrix} Q_0 \\ \frac{err}{\|err\|} \end{bmatrix} \tag{5}$$

We can easily get that $err = R_1 - R_1 Q_0^{-1} Q_0$. Q_0^{-1} is the pseudo-inverse of Q_0, $\| err \|$ is vector norm of err.

As [1] mentioned, this method of update is quick but also brings some problems that the feature size will increase infinitely with the new coming data and some over-fitting will occur. The precondition of matrix factorization is that feature size should be much smaller than the size of rating matrix. The feature size's increasing without restriction must broke the precondition. So we should set a point when the decomposing matrix should be shrinked. Here, this paper sets it when k' ≥ 2k (k is the initial size of feature).

When the feature size reaches 2k, we use matrix sketching to shrink the size of Q which does not change the similar relation between items. We use method in [2, 3] to guarantee that the item feature matrix after sketching $\hat{Q}_1$ satisfies $\hat{Q}_1^T \hat{Q}_1 \approx Q_1^T Q_1$, which means:

$$\hat{Q}_1^T \hat{Q}_1 < Q_1^T Q_1 \ and \ \| Q_1^T Q_1 - \hat{Q}_1^T \hat{Q}_1 \| \leq 2 \| Q_1 \|_f^2 / l \tag{6}$$

The specific algorithm is as follow:

Algorithm2: Matrix_sketching
Input: $Q_1 \in \mathbf{R}^{2k \times m}$, k
Output: $\widehat{Q_1} \in \mathbf{R}^{k \times m}$
1: $Q_{temp} \leftarrow$ first $(2k - 1)$ row $of\ Q_1$
2: $[U, S, V] \leftarrow SVD(Q_{temp})$
3: $\delta \leftarrow \sigma_k{}^2$
4: $\check{S} \leftarrow \sqrt{\max(S^2 - I_{2k-1}\delta, 0)}$
5: $\widehat{Q_1} \leftarrow \check{S}V^T$
6: $\widehat{Q_1} \leftarrow [\widehat{Q_1} ; Q_{1_{2k}}]$

The propose of Algorithm 2 is to sketch Q_1 to $\hat{Q}_1$ and ensure that $\hat{Q}_1^T \hat{Q}_1 \approx Q_1^T Q_1$. First, copy the first (2k−1) row of Q_1 to Q_{temp}. Second, use SVD on Q_{temp} to make its columns are orthogonal and in descending magnitude order. S is a nonnegative diagonal matrix and $S = \text{diag}([\sigma_1, \ldots, \sigma_{2k-1}])$, $\sigma_1 \geq \sigma_2 \ldots \ldots \geq \sigma_{2k-1}$. Third, set $\delta = \sigma_k^2$. As σ is set in decreasing order, we can make sure that $\check{S}$ has at least k all-zero rows. In the end, the feature size of $\hat{Q}_1$ can be shrinked to k.

So the incremental algorithm for cold-start users is as follow:

Algorithm3: incremental NMF for cold-start users

Input: $R_0 \in \mathbf{R}^{n \times m}$, $P_0 \in \mathbf{R}^{n \times k}$, $Q_0 \in \mathbf{R}^{k \times m}$, k, $R_1 \in \mathbf{R}^{c \times m}$, t

Output: $P_1 \in \mathbf{R}^{(n+c) \times k}$, $Q_1 \in \mathbf{R}^{k \times m}$

1: for $i \in [c]$ do
2:    $w = R_1 Q_0^{-1}$
3:    $err = R_1 - wQ_0$
4:    $Q_1 = \begin{bmatrix} Q_0 \\ err \\ \hline \|err\| \end{bmatrix}$
5:    $P_1 = \begin{bmatrix} P_0 & 0 \\ w & \|err\| \end{bmatrix}$
6:    if row of $Q_1 \geq 2k$ then
7:        $Q_{shrink} = \text{Matrix\_sketching}(Q_1)$
8:        $R = [R_0; R_{new}]$
9:        $P_1 \leftarrow$ first k column of P_1
10:        $[P_1 \, Q_1] = \text{NMF}(R, P_1, Q_{shrink}, t)$
11:    end if
12: end for

4.3   Incremental Algorithm for Existing Users

An existing user is a user who has ever visited the website and had historical data, which means it corresponds to one row in the original matrix.

The simplest way is just to change the row of the user in P and the column of the item that the user rates in Q. When a new tuple (u, i, r_{ui}) comes, this method just needs to update the row p_u in P and the column q_i in Q to get local optimum. But this way ignores influence between users and items and may cause over-fitting.

To solve this problem, we consider about the part that new ratings affect. The new ratings will affect the target user and the user set in which users share similar interests with the target user but have no influence on the users that has no similar interests with the target user.

This paper uses K-means to cluster the users and get similar user sets. The key points of K-means are as follow:

(1) The value of K: The value of K directly affects the cluster results. If K is too small, data in the same cluster is not similar. If K is too big, the distribution of data is too dispersive and does not get the goal of clustering.
(2) Measure of distance: different measures of distance also affect the cluster results. Due to the sparsity of data, we finally decided to use cosine similarity as criterion.

After user clustering, it's easy to deal with incremental problem of existing users. For each target user, we find the cluster that the target user belongs to, extract sub-matrix of the cluster from original matrix, use NMF to update the sub-matrix and fill it in the original matrix.

4.4   Theoretical Analysis of Performance

First, incremental algorithm for cold-start users. We get this algorithm by improving the method proposed in [1]. We set sketching point instead of continuous sketching. Our method does not influence the decomposing accuracy but does speed up the dealing time. Also, we can avoid making the feature size too small by setting the point. In addition, we use both P and Q in step 10 in Algorithm 3 and update them so that we can get global optimization as well as accelerate the rate of convergence.

Second, incremental algorithm for existing users. Our approach update the sub-matrix of user set in which users share similar interests with the target user. We try to get local optimization and avoid over-fitting by using this method. In addition, our approach reduce the size of matrix that need to be updated, as the part of matrix of users who are not similar to the target user does not need to be updated. However, one limitation of our approach is that it's efficient to deal with sparse matrix but inefficient to deal with dense matrix.

5   Experiment

5.1   Dataset

In this paper, we use data from a real house recommender website. By using the data, we construct the original rating matrix to train the original model and incremental rating matrix to test our incremental algorithm. The rating matrix is constructed based on user behavior and is a sparse binary matrix. If one user has ever visited the web page of one item, then the rating is 1; otherwise, the rating is 0. The original matrix contains 3000 users and 1783 items.

5.2   Parameter Setting and Experiment Environment

In this paper, we set original feature size k = 60, learning rate $\alpha = 0.0002$, $\lambda = 0.02$. All the experiments were run on machines with the same hardware (an Intel Core i5 CPU and 4 GB RAM, WIN7 64 OS). The development environment is matlab and NMF is written with C++ and transferred to mex used in matlab.

5.3   Analysis of Incremental Algorithm for Cold-Start Users

This paper set the point to do matrix sketching when $k' \geq 2k$. So we use matrix of 60×1783 to test algorithm performance.

As shown in Fig. 2(a), we set convergence condition $\beta = 0.1$. The figure is plotted from the 10th iteration. As we can see, the value of L for incremental algorithm starts from a smaller value than the original factorization and it keeps smaller than the original one consistently with the iteration going on. As shown in Table 1, for the original matrix, it takes 165.58 s to do once matrix factorization. While for incremental matrix, it only takes 68.58 s to update the whole model, that is to say it's only takes 1.14 s on average for each user to update the model.

As shown in Fig. 2(b), we set convergence condition $\beta = 0.2$, which is relatively loose compared to Fig. 2(a). It is obvious that the number of iteration needed is smaller and the running time is shorter. Also we can see in Table 1 that it takes only 0.62 s on average for each new user.

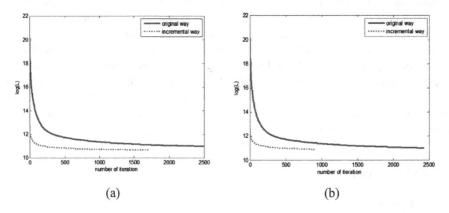

(a)   (b)

Fig. 2. The value of log(L) changing with the number of iteration. We set convergence condition $\beta = 0.1$ and $\beta = 0.2$ for (a) and (b) respectively.

Table 1. The number of iteration and running time for original NMF method and our incremental method in different convergence conditions.

	$\beta = 0.1$		$\beta = 0.2$	
	NMF	Incremental NMF	NMF	Incremental NMF
Number of iteration	4285	1705	2424	899
Running time (s)	165.57	68.58	95.30	37.49
Running time per user on average (s)	–	1.14	–	0.62

5.4   Analysis of Incremental Algorithm for Existing Users

We change some ratings in original matrix to 0 to train the original model, and use these ratings to simulate the new ratings from existing users and test the performance of our algorithm. In the experiment, we find that it just needs a few iterations and updates to get local optimization and this method can get similar results to global update.

6 Conclusion

In this paper, we propose an efficient method to solve incremental problem for both cold-start users and existing users in recommender systems. This method can solve problem in short time and give feedback in real-time.

Acknowledgements. The work was supported by NSFC (U1435214, 61432008, 61503178, 61175042, 61403208, 61321491), Jiangsu Nature Science Foundation (JSNSF) (BK20150587, DE2015213).

References

1. Song, Q., Cheng, J., Lu, H.: Incremental matrix factorization via feature space re-learning for recommendation system. In: RecSys, pp. 277–280 (2015)
2. Leng, C., Wu, J., Cheng, J., Bai, X., Lu, H.: Online sketching hashing. In: CVPR, pp. 2503–2511 (2015)
3. Liberty, E.: Simple and deterministic matrix sketching (2012). CoRR abs/1206.0594
4. Takacs, G., Pilaszy, I., Nemeth, B., Tikk, D.: Scalable collaborative filtering approaches for large recommender systems. J. Mach. Learn. Res. (JMLR) **10**, 623–656 (2009)
5. Lommatzsch, A., Albayrak, S.: Real-time recommendations for user-item streams. In: SAC, pp. 1039–1046 (2015)
6. Brand, M.: Fast online SVD revision for lightweight recommender systems. In: SDM, pp. 37–46. SIAM (2003)
7. Gogna, A., Majumdar, A.: SVD free matrix completion with online bias correction for recommender systems. In: ICAPR, pp. 1–5 (2015)
8. Achakulvisut, T., Acuna, D.E., Ruangrong, T., Kording, K.P.: Science concierge: a fast content-based recommendation system for scientific publications (2016). CoRR abs/1604.01070
9. Burke, R.: Hybrid web recommender systems. In: Brusilovsky, P., Kobsa, A., Nejdl, W. (eds.) Adaptive Web 2007. LNCS, vol. 4321, pp. 377–408. Springer, Heidelberg (2007)
10. Rendle, S., Schmidt-Thieme, L.: Online-updating regularized kernel matrix factorization models for large-scale recommender systems. In: RecSys, pp. 251–258 (2008)
11. Bilge, A., Polat, H.: A scalable privacy-preserving recommendation scheme via bisecting k-means clustering. Inf. Process. Manage. **49**(4), 912–927 (2013)
12. Subercaze, J., Gravier, C., Laforest, F.: Real-time, scalable, content-based Twitter users recommendation. Web Intell. **14**(1), 17–29 (2016)
13. Lee, D.D., Seung, H.S.: Learning the parts of objects by nonnegative matrix factorization. Nature **401**(6755), 788–791 (1999)

An Improved Outlier Detection Algorithm to Medical Insurance

Zhiping Xie[(✉)], Xiaoyu Li[(✉)], Wenyi Wu, and Xiaoling Zhang

School of Information and Software Engineering,
University of Electronic Science and Technology of China, Chengdu 610054,
Sichuan, People's Republic of China
1529464401@qq.com, xiaoyu33521@163.com

Abstract. With the development of the medical insurance industry in China, medical insurance data with complex, multidimensional and interdisciplinary feature are extremely increasing. How to mine the potential value from the vast amounts of data and improve the efficiency of data analysis are topical issues in the study of data mining. **This paper presents an improved LOF Outlier Detection Algorithm — GdiLOF, an algorithm which reduces dataset by removing the normal data and introduces information entropy to improve the accuracy of the LOF algorithm.** Platform adaptability is analyzed by running it on Hadoop platform. The experimental results show that GdiLOF algorithm has **high efficiency** and the accuracy is **6 percentage points** higher than LOF algorithm. And it also run better in the Hadoop distributed platforms, as well as having obvious advantages in processing huge amounts of data.

Keywords: LOF · Outlier detection · Information entropy · Medical insurance

1 Introduction

In recent years, with the development of the medical insurance, medical insurance data are increasing in China. According to the notice [1] in [2016] 26 of the country documents, all of the citizens will access to basic medical and health services by 2020 which means more people will have insurance making the medical insurance data more and more complex. In this scenario, data mining techniques [2] provide the technical support to discovery the potential value among the huge amounts of data. Meanwhile we believe that data mining and knowledge discovery techniques face two major challenges. First, the efficiency of data mining technique is more vulnerable. Second, more and more complex medical insurance data give rise to low accuracy of knowledge discovery. In this paper, we focus one of the most important data mining problems, outlier detection, such as medical reimbursement of patients are once a month generally. But few patients reimburse it many times a month. It is one of abnormal medical insurance data, which provides a great assistance to our further analysis of medical insurance data [3]. We presented an improved LOF outlier detection algorithm — GdiLOF. Based on the research works of Li [4], we introduced information entropy [5, 8, 9] to improve algorithm accuracy. We carried out experiments, whose results

© Springer International Publishing AG 2016
H. Yin et al. (Eds.): IDEAL 2016, LNCS 9937, pp. 436–445, 2016.
DOI: 10.1007/978-3-319-46257-8_47

prove that GdiLOF algorithm has high efficiency and better accuracy. Results indicate that GdiLOF algorithm has obvious advantages in processing huge amounts of data.

2   Related Work

Wang [6] and other researchers have carried out the basic principles, related algorithms and evaluation methods of data cleaning. They also provide the direction of analysis and algorithm evaluation for the field of outlier detection. At present, outlier detection is roughly divided into four kinds: statistical outlier detection, clustering-based outlier detection, distance-based outlier detection, density-based outlier detection [7]. With the development of the medical insurance in our country, some researchers have oriented medical field of serious illness insurance influencing factors and frequent referral problem analysis, providing some ideas for forecasting the medical expenses and medical insurance fraud. But there are still not solutions on outlier detection of medical insurance data [3].

It is Breunig [10] who presented LOF outliers detection algorithm for the first time and gave a characterization of the degree of abnormality of local outlier factor LOF for each object. Compared to the previous judgment of outliers, the algorithm has a better characterization of the effect and general applicability. However the algorithm has to search and rank all objects resulting in a large amount of calculation and inefficient operation. Moreover the algorithm calculates the distance between objects without considering the different attributes with differences, so it is also low accuracy.

Awaring the shortcomings of the original LOF algorithm, many domestic and foreign researchers have made a lot of researches based on the LOF algorithm. In reducing the LOF algorithm computation, Chen and other researchers carried on the research and improvement of LOF outlier detection algorithm based on grid [11, 12]. Their algorithm removes dense unit [4] of data set by using the grid partition and only calculates the reduced data in LOF computation outlier factor, greatly reducing the computation of the LOF algorithm. Tang [13] proposed a connection-based outlier factor algorithm (COF) through the average distance of a data object and the area ratio as the outlier factor in order to determine whether the data for outliers. Above algorithms solve the LOF computation of large amount of problems, improve the efficiency in a large extent. But operation efficiency remains to be improved with the growing data. Also algorithms applied to fields are less discussed. At the same time, those algorithms do not reflect different contributions of different data attributes to outliers.

3   LOF Algorithm

LOF is a kind of outlier detection algorithm based on local densities. The algorithm no longer considers abnormalities as a two elements (false or true), but characterizes the degree of abnormalities by the local outlier factor (LOF). The larger the local outlier factor, the greater the probability that the object is an outlier, and the smaller vice versa.

The basic steps of LOF algorithm are listed as follows [10]:

Step1: Calculate the *k-distance* of object p. Assumes that object p, o are the two data objects of data set. For any positive integer k, the *k-distance* of object p is defined as the distance $d(p, o)$ between p and an object $o \in D$, denoted as *k-distance (p)*. The object o should meet the following conditions:

(1) for at least k data objects $o' \in D - \{p\}$ it holds that $d(p, o') \leq d(p, o)$;
(2) for at most $k-1$ data objects $o' \in D - \{p\}$ it holds that $d(p, o') < d(p, o)$.

Among them, the formula of $d(p, o)$ is:

$$d(\mathrm{p,o}) = \sqrt{\sum_{j=1}^{m} \left(f\left(p_j\right) - f\left(o_j\right) \right)^2} \tag{1}$$

Where m is the dimension of set, $f\left(p_j\right)$ and $f\left(o_j\right)$ are the jth ($j = 1, 2, 3 \ldots d$) dimension attribute value of D.

Step 2: Calculate *k-distance* neighborhood of an object p, denoted as $N_{k-distance}(\mathrm{p})$. The *k-distance* neighborhood of p is the set whose distance from the p is not more than the *k-distance*. That is, the data set:

$$N_{k-distance}(\mathrm{p}) = \{\mathrm{q} | \mathrm{d}\ (\mathrm{p},\ \mathrm{q}) \leq \mathrm{k} - \mathrm{distance(p)}\} \tag{2}$$

Where $N_{k-distance}(\mathrm{p})$ is abbreviated as $N_k(\mathrm{p})$.

Step 3: Calculate *the reachability distance* of an object p with respect to object o. The reachability distance of p with respect to o is defined as:

$$reach - dist_k(p, o) = \max\{\mathrm{k} - \mathrm{distance(o)},\ \mathrm{d(p,o)}\}. \tag{3}$$

Figure 1 [10] illustrates the concept of reachability distance. Intuitively, if an object is far away from o (such as $p2$), then the distance between the two is only the actual distance. However, if an object is in *k-distance neighborhood* of o (for example $p1$), the actual distance is replaced by the *k-distance* of o.

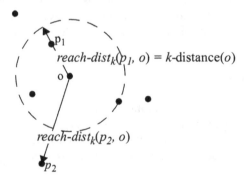

Fig. 1. $reach - dist_k(p1, o)$ and $reach - dist_k(p2, o)$, for $k = 4$

Step 4: Calculate *the local reachability density* of an object p, denoted as $lrd_k(p)$. The local reachability density of p is defined as

$$lrd_k(p) = \frac{1}{\sum reach - dist(p,o)/|N_k(p)|} \tag{4}$$

Step 5: Calculate *the local outlier factor* (LOF) of p, denoted as $LOF_k(p)$.

$$LOF_k(p) = \frac{\sum_{o|N_k(p)} lrd_k(o)/lrd_k(p)}{|N_k(p)|} \tag{5}$$

4   Improved LOF Algorithm

Based on the LOF algorithm, in view of large amount of calculation and without the consideration of the attribute values of abnormalities of LOF algorithm, this paper improves the LOF algorithm orienting medical insurance field, which can adapt to the massive data. This section will introduce the improved LOF.

4.1   Non Outliers Based on Adjacent Grid Density Factor Algorithm

GridOF algorithm proved [4] that the points of the clustering data cannot become outliers. And the GridLOF algorithm presented methods to determine interval length of girds [11]. They reduced data, which greatly reduced the amount of computation. Here is a brief description of steps of the improved GridLOF algorithm:

- Read the source data and distribute data space;
- Determine each dimension interval division length and the number. The dth dimension is divided $N^{a^{d-1}}$ parts equally. Then according to $N^{1+a+\cdots+a^{d-1}} = N^2d$, to determine the value of a. Then according to the dimension data maximum (max) and minimum (min) in each dimension and the formula $len = (max - min)/N^{a^{d-1}}$, computes the interval length;
- In every dimension, by the corresponding interval length, calculate data point corresponding mesh number and process boundary grid data. Then through the dimension reduction, map the grid number of the different dimensions to the one-dimensional space. In the map container with the label of the object count is incremented. Thus every data point has only one-dimensional label, and each dimension label correspond to a certain number of data points;
- For each grid, search the adjacent grid to determine whether it is a boundary grid;
- For the boundary grid, calculate the LOF value, and then accept the number of num input by users. Order from big to small according to the LOF value. Take larger values of the factor and label the data contained by it as outliers. Until the number of outliers in is greater than or equal to the num.

4.2   The Information Entropy Difference to Determine Weight of Data Attributes

Information entropy theory gives an effective measure of the uncertainty of a given system. Let a random variable B. Set $U(B)$ is the interval of B. The formula calculating information entropy of B is as follows [14]:

$$\mathrm{E}(\mathrm{B}) = -\sum\nolimits_{i=1}^{m} \frac{|Bi|}{|U|} log_2 \frac{|Bi|}{|U|} \tag{6}$$

Where m indicates the number of variable B in U. The greater the $E(B)$, the greater the uncertainty of the random variable B.

For data set U, the data set S which contains more than one data object is a subset of U. Object x is a data object of Set S. The set S is divided into two parts: $\{x\}$ and $S - \{x\}$, denoted $S' = \{S_1, S_2\}$. The information entropy difference of S which is partitioned for before and after is denoted as $\Delta(x, S)$. Its calculation formula can be expressed as [14]:

$$\Delta(x, S) = E(S) - E(S') \tag{7}$$

When set S is determined. $\Delta(x, S)$ will be abbreviated as $\Delta(x)$.

According to the concept above, $\Delta(x)$ denotes information entropy's change before and after the division, which is divided into before and after elimination of S in the uncertainty [14]. For the data objects, the value of $\Delta(x)$ is bigger, whose ratio is smaller in S. It is more likely to be an outlier. For an object, each of its properties contribute to the abnormal is reflected in the property's $\Delta(x)$. That is to say, the $\Delta(x)$ of properties can be used to measure the weight of attributes' contribution to the abnormity.

When calculating the distance between objects in the LOF algorithm. $\Delta(x_i)$ on each attribute is calculated as the weight when the object's distance is calculated with other objects. When the distance between two objects in data set D' recalculated, the weighted distance formula for the d dimension properties of p and q is:

$$\mathrm{d}(\mathrm{p, q}, w_j) \sqrt{\sum\nolimits_{j=1}^{m} w_j(f(p_j) - f(q_j))^2} \tag{8}$$

Where p and q are the objects in the set D'. w_j is the weight of the jth dimension attribute of p. $f(p_j)$ and $f(q_j)$ are the value of the jth dimension attribute. m is the dimension of the data set.

4.3   Improved LOF Algorithm

In order to reduce the time complexity of the LOF algorithm, we reduce dataset by removing the normal data with improved GridLOF algorithm [11]. In this way, the clustering data with no outliers are removed. We just need to calculate the reduced points of sparse data area, which reduces the amount of computation largely, can better adapt to the data under the massive data outlier detection. Algorithm 1 illustrates the procedure of data reduction.

Algorithm 1. Data Reduction

```
Input: data set;
Output: reduced set D';
1:   Read In data set D and number;
2:   if D exists then
3:       return m, where m is D's total dimensions;
4:   else return 0;
5:   end if
6:   for j=1 to m do
7:       determine the number[j] for jth dimension;
8:       end for
9:   for j=1 to m do
10:      Select max and min values of the jth dimension;
11:      Length[i]= (maxj − minj)/number[j];
12:      Calculate each interval's grid number;
13:      convert the grid number to one-dimension space;
14:      end for
15: for i=1 to lines do
16:     if grid[data[i]]≤number ;
17:        return Boundary(true);
18:     end if
19:     end for
20: return reduced D';
```

In order to improve the accuracy of LOF algorithm, information entropy is introduced when calculating the distance between data objects. Based on information entropy difference, its weight is used to measure anomalous contribution degree of data attributes. Algorithm 2 illustrates the procedure of data reduction.

Algorithm 2. Calculate weighted LOF of objects

```
Input: reduced D' ;
Output: Outliers;
1:   Read In data set D';
2:   for each object of D' do
3:       calculate each attribute's Δ(attribute);
```

4:       calculate d (p, q, w_j) $= \sqrt{\sum_{j=1}^{m} w_j (f(p_j) - f(q_j))^2}$;

6:       $\mathrm{lrd(object)} = \dfrac{1}{\sum reach\text{-}dist(object,o)/|D'_k(\text{onject})|}$;

7:       $LOF_k(object) = \dfrac{\sum_{o \in |D'_k(\text{object})} lrd_k(o)/lrd_k(object)}{|D'_k(\text{object})|}$;

```
8:      end for
9:   Select data outlier points;
10: return outliers;
```

5 Results and Analysis

In this paper, several groups of comparative experiments are designed to verify the time efficiency and accuracy of the GdiLOF algorithm to medical insurance.

This experiment is carried out on the Hadoop cluster which is built on 4 sets of identical PC units. Table 1 shows the environment configurations.

Table 1. Experimental environment configuration

OS	CentOS
CPU	Inter(R) Core(TM) i5-4460
Memory	16 g
Hadoop version	2.0

Experimental data are the data generated by Health Insurance Bureau of City in the Sichuan Province in 2014. There are many types of abnormal data in medical insurance, such as the amount of reimbursement, length of staying in hospital. In our experiments, medical insurance data are preprocessed, which makes the outliers data account for 2 % of the total data. We select three groups containing 1 million data, 2 million data, 3 million data, recorded as A, B, C respectively. We also select 20 attributes of medical insurance data to experiment. Some attributes of medical insurance are shown in Fig. 2.

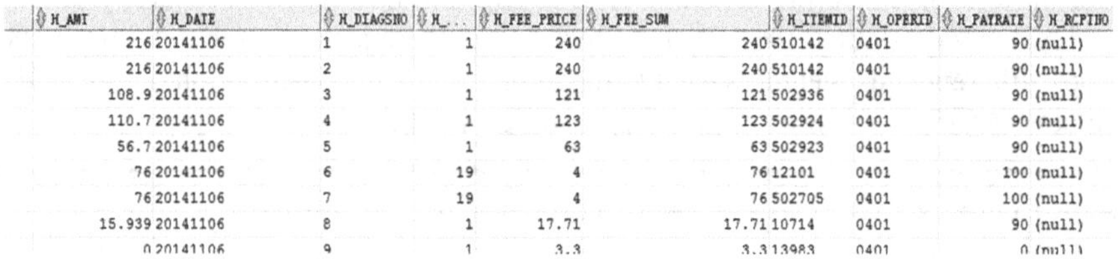

H_AMT	H_DATE	H_DIAGSNO	H_...	H_FEE_PRICE	H_FEE_SUM	H_ITEMID	H_OPERID	H_PAYRATE	H_RCPTNO
216	20141106	1	1	240	240	510142	0401	90	(null)
216	20141106	2	1	240	240	510142	0401	90	(null)
108.9	20141106	3	1	121	121	502936	0401	90	(null)
110.7	20141106	4	1	123	123	502924	0401	90	(null)
56.7	20141106	5	1	63	63	502923	0401	90	(null)
76	20141106	6	19	4	76	12101	0401	100	(null)
76	20141106	7	19	4	76	502705	0401	100	(null)
15.939	20141106	8	1	17.71	17.71	10714	0401	90	(null)
0	20141106	9	1	3.3	3.3	13983	0401	0	(null)

Fig. 2. Medical insurance data

5.1 Running Efficiency

To verify GdiLOF algorithm has higher efficiency than LOF algorithm, we make the two algorithms run the three groups data in the same experimental environment. The result is shown in Fig. 3.

As shown in Fig. 3, the horizontal coordinate represents the three sets of different sizes of data. The vertical coordinate represents the running time. From the experiment data, we can draw the following conclusions: the GdiLOF algorithm has higher efficiency than LOF, and the greater the amount of data, the more obvious this advantage will be. The GdiLOF algorithm significantly decreases the amount of calculation through reducing data, which save the time. Along with the sizes of data increasing, the improved algorithm on the computing time in the interval length determined will increase. But compared to LOF's computing time in traversal search ranking, the running time of LOF increases faster than the improved algorithm. That is, the improved can adapt to the massive data processing better. At the same time, the improved algorithm can be paralleled to the Hadoop cluster, which further improves the efficiency of outlier detection. Therefore, the GdiLOF algorithm significantly reduces the running time and adapts to the outlier detection of massive data better.

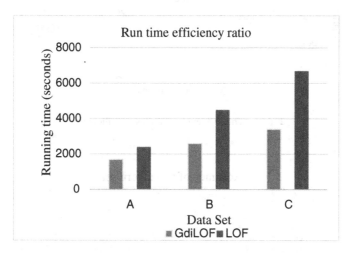

Fig. 3. The comparison of two algorithms' runtime efficiency

5.2   The Influence of Nodes on Efficiency of GdiLOF Algorithm

To prove that the GdiLOF algorithm has more obvious efficiency with more nodes, we implement experiment with different nodes. The result is shown in Fig. 4.

The running time is decreased with the increase of the node. Nodes can effectively improve the running efficiency. From the trend of polylines, the line decreased more rapidly with the amount of data. In other words, the polyline's speedup is greater if the data is in large quantity, which indicates that the GdiLOF algorithm has more advantages in dealing with massive data.

5.3   Accuracy Analysis

The accuracy is also an important criteria to evaluate algorithms. The calculation formula of accuracy is as follows:

$$\text{Accuracy} = \frac{\text{Number of outliers correctly found}}{\text{Total number of outliers}} \tag{9}$$

In this experiment, we compare the accuracy of GdiLOF algorithm with LOF's. With three groups experiment, we obtain results shown in Fig. 5.

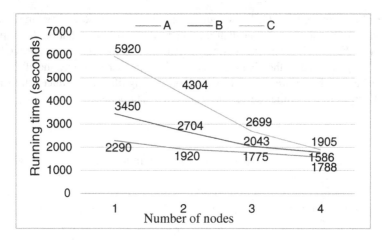

Fig. 4. The influence of nodes on efficiency

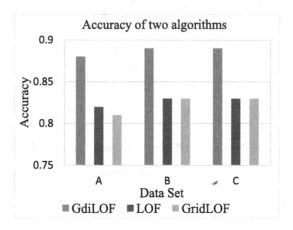

Fig. 5. The accuracy of two algorithms

As shown in Fig. 5, the accuracy of GdiLOF algorithm is higher than the accuracy of LOF algorithm. With three groups different sizes data, the accuracy of GdiLOF algorithm is about 0.89 (error rate is less than 0.01 %, within the allowable error range), but the accuracy of LOF algorithm is 0.83. It shows giving different weights to different attributes can enhance the accuracy greatly. What's more, to prove the accuracy of GridLOF algorithm is also improved, we do a contrast experiments. The experimental results show that GdiLOF also improves the accuracy of GridLOF algorithm.

6 Conclusion

Based on the LOF algorithm, we present an improved LOF outlier detection algorithm to medical insurance. The algorithm improves efficiency by using grid to reduce data. We also introduce information entropy to raise the accuracy. The experimental results indicate that the algorithm has higher efficiency and better accuracy. Meanwhile the improved algorithm can be better adapt to deal with the massive data. However, the improved algorithm uses the number of data points to measure the density of the grid. The density of the grid is used to determine whether the grid is the boundary grid, which could not adapt to the different data sets having different distributions. How to measure density of grids better will be the next step to research and improve.

Acknowledgement. This work is supported by the National Science Foundation of China (Grant Nos. 61502082) and the Fundamental Research Funds for the Central Universities, ZYGX2014J065.

References

1. General Office of the State Council of the People's Republic of China. http://www.zgylbx. com/gfwswecpnew107541_1/. Accessed 27 Apr 2016
2. Dhar, V.: Data Science and Prediction. Commun. ACM **56**(12), 64–73 (2012)
3. Tao, H.: Research and application of data mining technology in medical insurance. University of Science and Technology of China, USTC (2015)
4. Li, C.H., Sun, Z.: GridOF: efficient outlier detection algorithm for large-scale data sets. J. Comput. Res. Dev. **40**(11), 1586–1592 (2003)
5. Shannon, C.E.: A mathematical theory of communication. Bell Syst. Tech. J. **27**(3), 3–55 (1948)
6. Wang, Y.F., Zhang, C.H., Zhang, B.B., et al.: Review of data cleaning research. New Technol. Libr. Inf. Serv. **12**, 50–56 (2007)
7. Su, X., Tsai, C.L.: Outlier detection. Wiley Interdisc. Rev. Data Min. Knowl. Dis. **1**(3), 261–268 (2011)
8. Sloane, N., Wyner, A.: A Mathematical Theory of Communication, pp. 379–423. Wiley-IEEE Press, New York (2009)
9. Xie, L., Li, G., Xiao, M., et al.: Novel classification method for remote sensing images based on information entropy discretization algorithm and vector space model. Comput. Geosci. **89**, 252–259 (2015)
10. Breunig, M.M., Kriegel, H.P., Ng, R.T., et al.: LOF: identifying density-based local outliers. ACM SIGMOD Rec. **29**(2), 93–104 (2000)
11. Wang, X.X., Huang, L.W.: Research and improvement of GridLOF algorithm in data mining. Modern Computer (2007)
12. Chen, W.M.: Research and improvement of outlier mining algorithm based on GridLOF, Sun Yat-sen University (2007)
13. Tang, J., Chen, Z., Fu, A.W.-C., Cheung, D.W.: Enhancing effectiveness of outlier detections for low density patterns. In: Chen, M.-S., Yu, P.S., Liu, B. (eds.) PAKDD 2002. LNCS (LNAI), vol. 2336, pp. 535–548. Springer, Heidelberg (2002)
14. Jiang, F., Sui, Y., Cao, C.: An information entropy-based approach to outlier detection in rough sets. Expert Syst. Appl. **37**(9), 6338–6344 (2010)

Comparison of Binary Optimization Techniques for Real-Time Management of Sustainable Autonomous Microgrid

R. Hari Kumar[✉] and S. Ushakumari

College of Engineering Trivandrum, Thiruvananthapuram, Kerala, India
harikumar.cet@gmail.com

Abstract. Sustainable autonomous microgrid is an integrated power ecosystem consisting of Distributed Generators (DGs), storage devices and loads. Such microgrids are expected to become an integral part of the future power system. Existence of intermittent renewable based sources, loads with different priorities and limited generation capacity makes power balancing in an autonomous microgrid a challenging task. During real-time implementation, desired reliability and stability is achieved in such an infrastructure by utilizing a fast acting algorithm for priority based load management and network reconfiguration. Primary task of the algorithm is to identify ON/OFF status of the load breakers and the tie/sectionalizing breakers in the system. As the breaker status is represented by '1' or '0', binary version of optimization techniques need to be used to find the optimum solution. In this paper, the Binary coded Genetic Algorithm (BGA) and Binary Particle Swarm Optimization (BPSO) is used in the algorithm for real-time management of a sustainable autonomous microgrid and their performances are compared. The results show that BPSO has outperformed BGA in obtaining the solution.

Keywords: Autonomous microgrid management · Binary optimization · Genetic Algorithm · Swarm intelligence

1 Introduction

Technological developments in small capacity, environmental friendly distributed generation technologies has increased the deployment possibilities of DGs in the power system. Moreover, increased concern over reliability, efficiency and power quality has provided a boost to the installation of DGs close to the load centers in the network. With large penetration of DGs, the distribution system has to encounter many technical challenges, mainly due to its transformation from passive to active. To extract maximum benefits from such an active network, the concept of a sustainable autonomous microgrid, with sufficient generation and storage to meet the demand within it, has evolved. Major constraints in an autonomous microgrid are limited availability of generation, increased number

© Springer International Publishing AG 2016
H. Yin et al. (Eds.): IDEAL 2016, LNCS 9937, pp. 446–456, 2016.
DOI: 10.1007/978-3-319-46257-8_48

of renewable based sources with unpredictable power output and loads with definite priority. Moreover, faults may result in the formation of non-sustainable islands within the microgrid leading to a large mismatch between the generation and demand. Hence, for the successful operation of a sustainable autonomous microgrid, an efficient management strategy is inevitable.

Several studies on conventional distribution system reconfiguration, with the objective of improving voltage profile and minimizing the system losses are being carried out by researchers since past few decades [1]. However, as autonomous microgrid exists in a small area with limited length feeders, loss minimization is not the primary objective for reconfiguration. Moreover, load management to improve reliability in distribution system is usually performed without considering load priorities. Hence, algorithms developed for conventional distribution system management cannot be used for autonomous microgrid.

In the recent years, significant efforts have been put in the research on the management of microgrid for improving its reliability, considering diverse criteria [2–6]. These works do not consider load priorities and mainly focus on the management of non-autonomous microgrid. Load shedding and reconfiguration of isolated systems with prioritized loads are also considered for the investigation by researchers [7–9], where the objective function formulated is to minimize the sum of product of load priorities and load magnitudes. In microgrid, there may be loads with varying magnitudes and priorities. Hence, the optimization of objective function will fail if the load priorities are allocated in chronological order, which demands calculation of priority weights for each load separately. This will lead to different priority weights even for the loads with the same priority, preventing the identification of the load criticality. Moreover, in any optimization technique, the dimension of search space increases with the number of variables for optimization. Hence, as the number of loads in microgrid increases, the dimension of search space for optimization increases, resulting in more computation time for obtaining optimum solution. In [10] an algorithm is developed to address these challenges. However, the developed algorithm is tested in a small system which averts the identification of its potential for real-time implementations.

This paper aims to investigate the prospects of optimization techniques to obtain a solution in minimum time, for the real-time management of sustainable autonomous microgrid. The binary versions of genetic algorithm and particle swarm optimization are adopted for comparing their effectiveness in the microgrid management algorithm. The case study is carried out on modified IEEE 33-bus test system, which is assumed to be operating in autonomous mode.

2 Objectives for the Management of Microgrid

Reliable and stable operation of an autonomous microgrid with prioritized loads necessitates balancing of the generation and demand in minimum time, considering the load priorities. In order to serve maximum load in the system, load management and/or network reconfiguration has to be done, which needs to be decoupled for obtaining the solution in minimum possible time. In the first stage, the amount of load that needs to be switched ON/OFF to alleviate the power mismatch is determined considering the load priority. The objective function for the load management is defined as [10]:

$$\text{Maximize } f(x) = \sum_{i=1}^{N_L} x_{Li} L_i \tag{1}$$

$$\text{Subject to } \sum_{j=1}^{N_G} x_{Gi} P_{Gi} \geq \sum_{i=1}^{N_L} x_{Li} L_i \tag{2}$$

where, N_L is the number of loads, N_G is the number of generators, x_{Li} is the ith load breaker status, L_i is the magnitude of the ith load, x_{Gi} is the jth generator breaker status and P_{Gj} is the real power generation of the jth generator.

When a fault occurs in the system, islands may be formed within the micro-grid. In certain cases, these islands may have power deficiency, which can be mitigated by network reconfiguration. The constraint that needs to be adhered to in network reconfiguration is the preservation of radial topology of the network, with minimum number of tie/sectionalizing breaker operations. The objective function for reconfiguration is defined as [10]:

$$Minimize \ f(x) = \sum_{i=1}^{N_S} y_i \tag{3}$$

where, N_S is the number of tie/sectionalizing breakers and y_i is '1' if the tie/sectionalizing breaker status has changed and '0' otherwise.

The variables in the objective functions in Eqs. (1) and (3) are ON/OFF status of breakers, which are represented by '1' and '0'. Hence, binary versions of optimization techniques are used.

3 Optimization Techniques

The optimization techniques used in this study are BGA and BPSO. In BGA, the chromosome (solution) can be coded in binary form and hence can be used directly for implementation. However, in the basic Particle Swarm Optimization (PSO), the search space is continuous and hence, some modifications are needed to enable it to deal with the binary optimization problems.

3.1 Binary Coded Genetic Algorithm

Genetic Algorithm (GA) is considered as one of the most popular and successful technique in the family of evolutionary algorithms, proposed by Holland in 1962 [11]. GA is a multipurpose optimization technique which can be used for solving any linear or non-linear problem. GA is inspired by the principles of natural selection and genetic reproduction [11]. BGA uses the concept of chromosomes coded in binary form and the operations of crossover and mutation.

BGA starts with randomly initializing a population. A chromosome is a candidate solution in the population. At each step, called generation, fitness of all candidates are calculated. The candidate with minimum fitness is retained as a solution in the current generation and is passed to the next generation. The genetic operators, crossover and mutation, are used to form a new population of the next generation. The fitness value of each candidate is evaluated again and this process is continued until the terminating criteria are reached.

3.2 Binary Particle Swarm Optimization

PSO is a metaheuristic search technique proposed by Kennedy and Eberhart in 1995 [12]. PSO is inspired by the social behavior of bird flocking or fish schooling, where each individual particle adjusts its location (solution) according to its own experience and that of the swarm. In PSO each particle is associated with a position vector $x(k)$ and velocity vector $v(k)$. These vectors are updated using Eqs. (4) and (5) respectively.

$$v_i^d(k+1) = w(k) \cdot v_i^d(k) + c_1 \cdot r_1(p_{best_i}^d - x_i^d(k)) + c_2 \cdot r_2(g_{best_i}^d - x_i^d(k)) \quad (4)$$
$$x_i^d(k+1) = x_i^d(k) + v_i^d(k+1) \quad (5)$$

where, $w(k)$ is the inertia weight, p_{best} is the best solution of a particle, g_{best} is the global best solution, c_1, c_2 represents the tendency of particle to move towards p_{best} and g_{best} position respectively and r_1, r_2 are randomly generated numbers ranging between 0 and 1.

In BPSO the position updating means a switching between '0' and '1'. This switching should be done based on the velocity of the particles. In [13], a link between the velocity and position based on the probability of velocity is proposed. A V-shaped transfer function $S(v_i^d)$ which is bounded within the interval [0,1] and increases with increase in v_i^d is used in this work to transfer v_i^d into a probability function [14]. Further, the position vectors are updated based on the velocity, using Eq. (7).

$$S(v_i^d(k+1)) = \left| \frac{2}{\pi} arctan\left(\frac{\pi}{2} v_i^d(k+1)\right) \right| \quad (6)$$

$$x_i^d(k+1) = \begin{cases} \overline{x_i^d(k)} & \text{if } rand < S(v_i^d(k+1)) \\ x_i^d(k) & \text{if } rand \geq S(v_i^d(k+1)) \end{cases} \quad (7)$$

where, $\overline{x_i^d(k)}$ is the compliment of $x_i^d(k)$ and $rand$ is a random number ranging between 0 and 1.

4   Algorithm for the Microgrid Management

The algorithm for the management of an autonomous microgrid to maximize the load served, is described in Algorithm 1.1. The current status of the system is utilized by the algorithm for microgrid management. Inputs required for execution of the algorithm are the output of each generator, the demand of each load in the network with its priority, zone connection data with the details of interconnection between zones through tie/sectionalizing breakers and zone-breaker data with initial status of all the breakers. The breaker status may be '1' or '0', which indicates ON/close and OFF/open status of the breaker respectively. When a power mismatch or change in breaker status is detected, the algorithm will update the breaker status and identifies all the islands in the microgrid using the zone connection data. Further, generation and load data is utilized by the algorithm to identify Negative Power Islands (NPIs), where the demand is more than generation. If a NPI is detected, it identifies all the islands that can be interconnected through the tie/sectionalizing breakers and then checks whether there is any power mismatch. If power mismatch exists, it will execute the load management using Algorithm 1.2, for that island combination. After the load management, each island is again considered separately and checked for the shortage of power. If no such island is detected, reconfiguration is not done. Otherwise, the reconfiguration algorithm identifies the optimal combination of breakers to be operated for interconnecting the islands.

The procedure for the priority based load management is given in Algorithm 1.2. In this algorithm, loads with only a specific priority needs to

---

Algorithm 1.1 Management of autonomous microgrid

Input: $Generation\_Data,\ Load\_Data,\ Zone\_Connection\_Data, Breaker\_Status$
Output: Updated status of tie/sectionalizing breakers

1: $No\_Islands,\ Island\_Zones \leftarrow$ CHECK_NO_ISLANDS($Zone\_Connection\_Data,$
$\qquad Breaker\_Status)$
2: $NPI \leftarrow$ CHECK_NEGATIVE_POWER_ISLAND($No\_Islands,\ Island\_Zones$
$\qquad Generation\_Data,\ Load\_Data)$
3: **if** $(NPI \neq 0)$ **then**
4: $\qquad No\_Island\_Comb,\ Total\_Load\_To\_Be\_Shed \leftarrow$ CHECK_ISLAND_COMB(
$\qquad\qquad Zone\_Connection\_Data,\ Load\_Data,\ Breaker\_Status,\ Island\_Zones)$
5: $\quad$ **for** ($k = 1 : No\_Island\_Comb$) **do**
6: $\qquad$ **if** $(Total\_Load\_To\_Be\_Shed > 0)$ **then**
7: $\qquad\qquad$ PRIORITY_LOAD_SHEDDING($Total\_Load\_To\_Be\_Shed,\ Load\_Data$)
8: $\qquad\qquad NPI \leftarrow$ CHECK_NEGATIVE_POWER_ISLAND($Zone\_Connection\_Data,$
$\qquad\qquad\qquad Breaker\_Status,\ Island\_Zones[k])$
9: $\qquad\qquad$ **if** $(NPI \neq 0)$ **then**
10: $\qquad\qquad\qquad Tie\_Breaker\_Status \leftarrow$ RECONFIGURE_NETWORK(
$\qquad\qquad\qquad\qquad Zone\_Connection\_Data,\ Island\_Zones[k])$
11: $\qquad\qquad$ **end if**
12: $\qquad$ **else**
13: $\qquad\qquad Tie\_Breaker\_Status \leftarrow$ RECONFIGURE_NETWORK(
$\qquad\qquad\qquad Zone\_Connection\_Data,\ Island\_Zones[k])$
14: $\qquad$ **end if**
15: $\quad$ **end for**
16: **end if**

Algorithm 1.2 Priority based load management

1: **procedure** PRIORITY_LOAD_MANAGEMENT (Total_Load_To_Be_Shed, Load_Data)
2:     **if** ($Total\_Load\_To\_Be\_Shed > 0$) **then**
3:         Prepare Load Aggregate Table (LAT) based on priority
4:         $Index \leftarrow$ (Index in LAT having aggregate $\leq Total\_Load\_To\_Be\_Shed$)
5:         **if** ($Index$ exists) **then**
6:             $Aggregate \leftarrow$ (Aggregate in the LAT corresponding to $Index$)
7:             **if** ($Aggregate < Total\_Load\_To\_Be\_Shed$) **then**
8:                 $Index \leftarrow Index + 1$
9:             **end if**
10:         **else**
11:             $Index \leftarrow 1$
12:         **end if**
13:         $Priority \leftarrow$ (Priority of the load corresponding to the $Index$ from LAT)
14:         **if** ($Priority > 1$) **then**
15:             Shed the loads having priority $> Priority$
16:         **end if**
17:         $No\_Priority\_Equal \leftarrow$ (Loads having priority $= Priority$)
18:         **if** ($No\_Priority\_Equal > 1$) **then**
19:             Use optimization technique to identify the optimum load to be shed
20:             Shed the identified loads
21:         **else**
22:             Shed the load having priority $= Priority$
23:         **end if**
24:     **else**
25:         Switch ON all the loads
26:     **end if**
27: **end procedure**

---

be considered for optimization as compared to the conventional approach where the entire load is considered. This will help in reducing the dimension of search space drastically, leading to reduction in the time taken for obtaining the optimum solution.

5   Results and Discussion

In the present study, modified IEEE 33 bus radial distribution system [15] with 33 buses, 3 laterals, 3 tie lines, total real power demand of 3.715 MW and reactive power demand of 2.3 MVAr is used. The loads in this system are assigned the priorities from '1' to '12' and are given in Table 1. The system is assumed to have three DGs with capacities 789.40 kW, 1551.90 kW and 1390 kW at buses 13, 24 and 29 respectively. Test results in the event of four typical fault scenarios are tabulated in Table 2 and are explained.

In test case 1, a fault in bus 2 will result in the formation of three islands as listed in Table 2. Island 3 is having a shortage of generation and cannot be interconnected with the active part of the network using tie line, which leads

Table 1. Load Data for IEEE 33-bus test system

Load/bus number	2	3	4	5	6	7	8	9	10	11	12	13	14	15	16	17
Magnitude (kW)	100	90	120	60	60	200	200	60	60	45	60	60	120	60	60	60
Priority	1	1	2	2	3	5	4	6	7	3	5	4	4	7	6	5
Load/bus number	18	19	20	21	22	23	24	25	26	27	28	29	30	31	32	33
Magnitude (kW)	90	90	90	90	90	90	420	420	60	60	60	120	200	150	210	60
Priority	1	8	9	10	11	12	1	9	5	5	12	10	9	1	3	5

Table 2. Results of the application of microgrid management algorithm

Test Case	Faulty Bus	Islands After Fault	Negative Power Islands	Change in Load Status	Change in Breaker Status	Network Reconfiguration	kW Served
1	Bus2	Island1 : Z1, Island2 : Z3 to Z18 & Z23 to Z33, Island3 : Z19 to Z22	Island3	D : L2 I : L19, L20, L21, L22	O : BK2, BK3, BK37, BK54, BK55, BK56, BK57 C : -	No	3255
2	Bus13	Island1 : Z1 to Z12 & Z19 to Z33, Island2 : Z14 to Z18	Island1, Island2	D : L13 S : L20, L21, L22, L23, L28, L29, L30	O : BK13, BK14, BK48, BK55, BK56, BK57, BK58, BK63, BK64, BK65 C : BK36	Yes	2915
3	Bus2 & Bus13	Island1 : Z1 Island2 : Z3 to Z12 & Z23 to Z33, Island3 : Z14 to Z18, Island4 : Z19 to Z22	Island3, Island4	D : L2, L13 S : L23, L28, L29 I : L19, L20, L21, L22	O : BK2, BK3, BK13, BK14, BK37, BK48, BK54, BK55, BK57, BK56, BK58, BK63, BK64, BK65 C : BK36	Yes	2925
4	Bus5 & Bus14	Island1 : Z1 to Z4 & Z19 to Z25, Island2 : Z6 to Z13 & Z26 to Z33, Island3 : Z15 to Z18	Island1, Island3	D : L5, L14 S : -	O : BK5, BK6, BK14, BK15, BK40, BK49 C : BK34, BK36	Yes	3535

*D-Loads disconnected at faulty bus, S-Loads shed, I-Loads not supplied in islands, O-Breakers opened, C-Breakers closed, Z-Zone

to shedding of loads in this island automatically. In addition, as island 1 does not have any load or source, this island is not considered in the next step of execution of the algorithm. Now, the only island that needs to be considered is island 2, which does not have any shortage of generation. Hence, neither load shedding nor reconfiguration is performed in this test case.

In test case 2, a fault in bus 13 results in disconnection of generator G1 and the formation of two islands as given in Table 2. The algorithm identifies that the two islands formed due to the fault are negative power islands and these islands can be interconnected using tie lines. However, due to the disconnection of generator G1, there is a shortage of 690.1 kW and hence, the algorithm executes load shedding. The load shedding algorithm prepares the cumulative load priority table. The first 10 rows of the cumulative load priority table are as given in Table 3. The algorithm identifies the index corresponding to the cumulative load less than or equal to 690.1 kW as '6', having a cumulative load of 540 kW and the corresponding priority as '9'. Since the cumulative load is less than the

Table 3. Cumulative load priority table - Test case 2

Index	Load number	Load priority	Load magnitude (kW)	Cumulative load (kW)
1	23	12	60	60
2	28	12	90	150
3	22	11	90	240
4	21	10	90	330
5	29	10	120	450
6	20	9	90	540
7	25	9	420	960
8	30	9	200	1160
9	19	8	90	1250
10	10	7	60	1310

magnitude of load to be shed, the index is incremented by one, giving the new index as '7', where the cumulative load is 960. Next, all the loads with priority less than '9' are shed immediately. This results in the shedding of L21, L22, L23, L28 and L29 totaling 450 kW. Further, as there are three loads with priority '9' corresponding to the obtained index, the optimization technique identifies the minimum load to be shed as L20 and L30 with a total magnitude of 290 kW. This indicates that the algorithm sheds minimum load with the lowest priority, to bring the system back to normal. In the next step, the algorithm again checks whether the formed islands have shortage of generation even after load shedding and detect that island 2 continue to be a negative power island. Then, reconfiguration algorithm is executed and it determines the solution as closing of tie breaker BK36 to supply the loads in island 2. In this test case both load shedding and reconfiguration is performed to restore the system.

In test case 3, simultaneous fault at bus 3 and bus 4 resulted in the formation of four islands as listed in Table 2. The algorithm identifies island 3 and island 4 as negative power islands and that, even after reconfiguration there will be three island groups, viz., group-1: island 1, group-2: Island 2, Island 3 and group-3: island 4. However, island 4 cannot be interconnected with other part of the network, as there is no tie line. This results in the disconnection of all the loads in that island naturally. Island 1 does not have any loads and sources and hence, will not be considered in the next step of execution. Next, the algorithm performs load shedding in island group 2, followed by reconfiguration to restore the system.

In test case 4, simultaneous fault at bus 5 and bus 14 results in the formation of three islands as given in Table 2. Islands 1 and 3 are having shortage of generation, and can be supplied by reconfiguration of the network. Hence, reconfiguration is performed by the algorithm and tie breakers BK34 and BK36 are closed to supply the islands. In this test case, it can be noted that only reconfiguration is done to restore the system.

6   Performance Comparison of the Optimization Techniques

Performance of the optimization technique incorporated in the algorithm for management of autonomous microgrid plays a key role in determining the time taken for obtaining the optimum solution. The binary adaptation of GA and PSO were applied in the microgrid management algorithm to choose the ideal one for its real time implementation. The need for comparison arises from the fact that BGA can be directly implemented where as a transformation has to be used in PSO for implementing in binary problems.

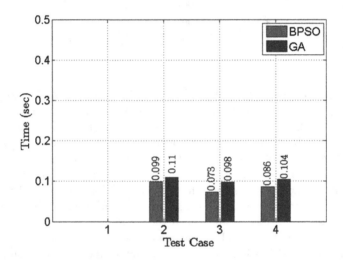

Fig. 1. Comparison of execution time taken by BGA and BPSO for load management and reconfiguration

It is observed that the optimal solution obtained using both the techniques are similar. However, as the time for obtaining the solution becomes apparent for real time implementation, especially when the number of loads, priorities and breakers increases, their computation times needs to be compared. The computation time required by BGA and BPSO for load management and reconfiguration are computed and plotted in Fig. 1. In test case 1, load management or reconfiguration is not necessary and hence, it is not done. In other test cases, based on the requirement, load management and/or reconfiguration is performed.

From the results it is evident that BPSO outperforms BGA and gives the optimum solution in the least time. Hence, BPSO is identified as the prospective tool to be used in the microgrid management algorithm for the optimization, with BGA following the suite.

7 Conclusions

Time taken for restoration is a critical element in the management of an autonomous microgrid. In this paper, the performance of two popular optimization techniques, viz., BGA and BPSO for the real-time management of a sustainable autonomous microgrid, are compared. It is found that, implementation of the microgrid management algorithm with BPSO could achieve the solution faster than that of with BGA. The results indicate that the algorithm presented in this paper, supported by BPSO can be implemented in real-time for the management of a sustainable autonomous microgrid.

References

1. Hu, Y., Hua, N., Wang, C., Gong, J., Li, X.: Research on distribution network reconfiguration. In: Proceedings of the International Conference on Computer, Mechatronics, Control and Electronic Engineering (CMCE), vol. 1, pp. 176–180, August 2010
2. Zhan, X., Xiang, T., Chen, H., Zhou, B., Yang, Z.: Vulnerability assessment and reconfiguration of microgrid through search vector artificial physics optimization algorithm. Int. J. Electr. Power Energy Syst. **62**, 679–688 (2014)
3. do Nascimento, V.C., Lambert-Torres, G., de Almeida Costa, C.I., da Silva, L.E.B.: Control model for distributed generation and network automation for microgrids operation. Electr. Power Syst. Res. **127**, 151–159 (2015)
4. Nafisi, H., Farahani, V., Askarian Abyaneh, H., Abedi, M.: Optimal daily scheduling of reconfiguration based on minimisation of the cost of energy losses and switching operations in microgrids. IET Gener. Transm. Distrib. **9**(6), 513–522 (2015)
5. Mokari-Bolhasan, A., Seyedi, H., Mohammadi-ivatloo, B., Abapour, S., Ghasemzadeh, S.: Modified centralized ROCOF based load shedding scheme in an islanded distribution network. Int. J. Electr. Power Energy Syst. **62**, 806–815 (2014)
6. Ketabi, A., Fini, M.H.: An underfrequency load shedding scheme for islanded microgrids. Int. J. Electr. Power Energy Syst. **62**, 599–607 (2014)
7. Shariatzadeh, F., Vellaithurai, C., Biswas, S., Zamora, R., Srivastava, A.: Real-time implementation of intelligent reconfiguration algorithm for microgrid. IEEE Trans. Sustain. Energy **5**(2), 598–607 (2014)
8. Mitra, P., Venayagamoorthy, G.: Implementation of an intelligent reconfiguration algorithm for an electric ship's power system. IEEE Trans. Indus. Appl. **47**(5), 2292–2300 (2011)
9. Vuppalapati, S., Srivastava, A.: Application of ant colony optimisation for reconfiguration of shipboard power system. Int. J. Eng. Sci. Technol. **2**(3), 119–131 (2010)
10. Kumar, R.H., Ushakumari, S.: Optimal management of islanded microgrid using binary particle swarm optimization. In: Proceedings of International Conference on Advances in Green Energy (ICAGE), pp. 251–257 (2014)
11. Goldberg, D.E.: Genetic Algorithms in Search, Optimization and Machine Learning, 1st edn. Addison-Wesley Longman Publishing Co., Inc., Boston (1989)
12. Kennedy, J., Eberhart, R.: Particle swarm optimization. In: Proceedings of the IEEE International Conference on Neural Networks, Perth, WA, Vol. 4, pp. 1942–1948, November 1995

13. Rashedi, E., Nezamabadi-pour, H., Saryazdi, S.: BGSA: binary gravitational search algorithm. Nat. Comput. **9**(3), 727–745 (2010)
14. Mirjalili, S., Lewis, A.: S-shaped versus V-shaped transfer functions for binary particle swarm optimization. Swarm Evol. Comput. **9**, 1–14 (2013)
15. Kirthiga, M.V., Daniel, S.A., Gurunathan, S.: A methodology for transforming an existing distribution network into a sustainable autonomous micro-grid. IEEE Trans. Sustain. Energy **4**(1), 31–41 (2013)

A Lower Bound Analysis of Population-Based Evolutionary Algorithms for Pseudo-Boolean Functions

Chao Qian[1,2]([✉]), Yang Yu[2], and Zhi-Hua Zhou[2]

[1] UBRI, School of Computer Science and Technology,
University of Science and Technology of China, Hefei 230027, China
chaoqian@ustc.edu.cn
[2] National Key Laboratory for Novel Software Technology,
Nanjing University, Nanjing 210023, China
{yuy,zhouzh}@nju.edu.cn

Abstract. Evolutionary algorithms (EAs) are population-based general-purpose optimization algorithms, and have been successfully applied in real-world optimization tasks. However, previous theoretical studies often employ EAs with only a parent or offspring population and focus on specific problems. Furthermore, they often only show upper bounds on the running time, while lower bounds are also necessary to get a complete understanding of an algorithm. In this paper, we analyze the running time of the $(\mu + \lambda)$-EA (a general population-based EA with mutation only) on the class of pseudo-Boolean functions with a unique global optimum. By applying the recently proposed switch analysis approach, we prove the lower bound $\Omega(n \ln n + \mu + \lambda n \ln \ln n / \ln n)$ for the first time. Particularly on the two widely-studied problems, OneMax and LeadingOnes, the derived lower bound reveals that the $(\mu + \lambda)$-EA will be slower than the $(1+1)$-EA when the population size μ or λ is above a moderate order. Our results imply that the increase of population size, while usually desired in practice, bears the risk of increasing the lower bound of the running time and thus should be carefully considered.

1 Introduction

Evolutionary algorithms (EAs) [2] are a kind of population-based heuristic optimization algorithm. They have been widely applied in industrial optimization problems. However, the theoretical analysis is difficult due to their complexity and randomness. In the recent decade, there has been a significant rise on the running time analysis (one essential theoretical aspect) of EAs [1,16]. For example, Droste et al. [7] proved that the expected running time of the $(1+1)$-EA on

This work was supported by the NSFC (61333014, 61375061), the Jiangsu Science Foundation (BK20160066), the Fundamental Research Funds for the Central Universities (WK2150110002), the 2015 Microsoft Research Asia Collaborative Research Program, and the Collaborative Innovation Center of Novel Software Technology and Industrialization.

© Springer International Publishing AG 2016
H. Yin et al. (Eds.): IDEAL 2016, LNCS 9937, pp. 457–467, 2016.
DOI: 10.1007/978-3-319-46257-8_49

linear pseudo-Boolean functions is $\Theta(n \ln n)$; for the $(\mu + 1)$-EA solving several artificially designed functions, a large parent population size μ was shown to be able to reduce the running time from exponential to polynomial [13,17,19,20]; for the $(1+\lambda)$-EA solving linear functions, the expected running time was proved to be $O(n \ln n + \lambda n)$ [6], and a tighter bound up to lower order terms was derived on the specific linear function OneMax [10].

Previous running time analyses often consider EAs with only a parent or offspring population, which do not fully reflect the population-based nature of real EAs. When involving both parent and offspring populations, the running time analysis gets more complex, and only a few results have been reported on the $(\lambda + \lambda)$-EA (i.e., a specific version of the $(\mu + \lambda)$-EA with $\mu = \lambda$), which maintains λ solutions and generates λ offspring solutions by only mutation in each iteration. He and Yao [12] compared the expected running time of the $(1+1)$-EA and the $(\lambda + \lambda)$-EA on two specific artificial problems, and proved that the introduction of a population can reduce the running time exponentially. On the contrary side, Chen et al. [5] found that a large population size is harmful for the $(\lambda+\lambda)$-EA solving the TrapZeros problem. Chen et al. [4] also proved that the expected running time of the $(\lambda + \lambda)$-EA on the OneMax and LeadingOnes problems is $O(\lambda n \ln \ln n + n \ln n)$ and $O(\lambda n \ln \ln n + n^2)$, respectively. Later, a low selection pressure was shown to be better for the $(\lambda + \lambda)$-EA solving a wide gap problem [3], and a proper mutation-selection balance was proved to be necessary for the effectiveness of the $(\lambda + \lambda)$-EA solving the SelPres problem [15].

The above-mentioned studies on the $(\lambda + \lambda)$-EA usually focus on specific test functions, while EAs are general purpose optimization algorithms and can be applied to all optimization problems where solutions can be represented and evaluated. Thus, it is necessary to analyze EAs over large problem classes. Meanwhile, most previous running time analyses on population-based EAs only show upper bounds. Although upper bounds are appealing for revealing the ability of an algorithm, lower bounds which reveal the limitation are also necessary for a complete understanding of the algorithm.

In this paper, we analyze the running time of the $(\mu + \lambda)$-EA solving the class of *pseudo-Boolean functions with a unique global optimum*, named UBoolean, which covers many P and NP-hard combinatorial problems. By applying the recently proposed approach switch analysis [21], we prove that the expected running time is lower bounded by $\Omega(n \ln n + \mu + \lambda n \ln \ln n / \ln n)$. Particularly, when applying this lower bound to the two specific problems, OneMax and LeadingOnes, we can have a more complete understanding of the impact of the offspring population size λ. It was known that the $(\mu + \lambda)$-EA is always not asymptotically faster than the $(1+1)$-EA on these two problems [14,18]. But it was left open that what is the range of λ where the $(\mu+\lambda)$-EA is asymptotically slower than the $(1+1)$-EA. Note that the expected running time of the $(1+1)$-EA on OneMax and LeadingOnes is $\Theta(n \ln n)$ and $\Theta(n^2)$, respectively [7]. Thus, we can easily get that the $(\mu + \lambda)$-EA is asymptotically slower than the $(1+1)$-EA when $\lambda \in \omega(\frac{(\ln n)^2}{\ln \ln n})$ on OneMax and $\lambda \in \omega(\frac{n \ln n}{\ln \ln n})$ on LeadingOnes. For the parent population size μ, we easily get obvious ranges $\omega(n \ln n)$ and $\omega(n^2)$

for the $(\mu + \lambda)$-EA being asymptotically slower on OneMax and LeadingOnes, respectively.

The rest of this paper is organized as follows. Section 2 introduces some preliminaries. Section 3 introduces the employed analysis approach. The running time analysis of the $(\mu + \lambda)$-EA on UBoolean is presented in Sect. 4. Section 5 concludes the paper.

2 Preliminaries

In this section, we first introduce the $(\mu+\lambda)$-EA and the pseudo-Boolean problem class studied in this paper, respectively, then describe how to model EAs as Markov chains.

2.1 $(\mu + \lambda)$-EA

EAs [2] are used as general heuristic randomized optimization approaches. Starting from an initial set of solutions (called a population), EAs try to improve the population by a cycle of three stages: reproducing new solutions from the current population, evaluating the newly generated solutions, and updating the population by removing bad solutions. The $(\mu + \lambda)$-EA as described in Algorithm 1 is a general population-based EA with mutation only for optimizing pseudo-Boolean problems over $\{0,1\}^n$. It maintains μ solutions. In each iteration, one solution selected from the current population is used to generate an offspring solution by bit-wise mutation (i.e., line 5); this process is repeated independently for λ times; then μ solutions out of the parent and offspring solutions are selected to be the next population. Note that the selection strategies for reproducing new solutions and updating the population can be arbitrary. Thus, the considered $(\mu + \lambda)$-EA is quite general, and covers most population-based EAs with mutation only in previous theoretical analyses, e.g., [5,12,15].

The running time of EAs is usually defined as the number of fitness evaluations until an optimal solution is found for the first time, since the fitness

---

Algorithm 1. $(\mu + \lambda)$-EA

Given solution length n and objective function f, let every population, denoted by variable ξ, contain μ solutions. The $(\mu + \lambda)$-EA consists of the following steps:

1: let $t \leftarrow 0$, and $\xi_0 \leftarrow \mu$ solutions uniformly and randomly selected from $\{0,1\}^n$.
2: **repeat until** some criterion is met
3:     **for** $i = 1$ to λ
4:         select a solution s from ξ_t according to some selection mechanism.
5:         create s'_i by flipping each bit of s with probability $1/n$.
6:     **end for**
7:     $\xi_{t+1} :=$ select μ solutions from $\xi_t \cup \{s'_1, \ldots, s'_\lambda\}$ according to some strategy.
8:     let $t \leftarrow t + 1$.

---

evaluation is the computational process with the highest cost of the algorithm [11,22]. Note that running time analysis has been a leading theoretical aspect for randomized search heuristics [1,16].

2.2 Pseudo-Boolean Function Problems

The pseudo-Boolean function class is a large function class which only requires the solution space to be $\{0,1\}^n$ and the objective space to be $\mathbb{R}$. It covers many typical P and NP-hard combinatorial problems such as minimum spanning tree and minimum set cover. We consider a subclass named UBoolean as shown in Definition 1, in which every function has a unique global optimum. Note that maximization is considered since minimizing f is equivalent to maximizing $-f$. For any function in UBoolean, we assume without loss of generality that the optimal solution is $11\ldots1$ (briefly denoted as 1^n). This is because EAs treat the bits 0 and 1 symmetrically, and thus the 0-bits in an optimal solution can be interpreted as 1-bits without affecting the behavior of EAs. The expected running time of unbiased black-box algorithms and mutation-based EAs on UBoolean has been proved to be $\Omega(n \ln n)$ [14,18].

Definition 1 (UBoolean). *A function $f : \{0,1\}^n \to \mathbb{R}$ in UBoolean satisfies that*
$$\exists s \in \{0,1\}^n, \forall s' \in \{0,1\}^n - \{s\}, f(s') < f(s).$$

Diverse pseudo-Boolean problems in UBoolean have been used for analyzing the running time of EAs, and then to disclose properties of EAs. Here, we introduce the LeadingOnes problem, which will be used in this paper. As presented in Definition 2, it is to maximize the number of consecutive 1-bits starting from the left. It has been proved that the expected running time of the $(1+1)$-EA on LeadingOnes is $\Theta(n^2)$ [7].

Definition 2 (LeadingOnes). *The LeadingOnes Problem of size n is to find an n bits binary string s^* such that, letting s_j be the j-th bit of a solution $s \in \{0,1\}^n$,*
$$s^* = \arg\max_{s \in \{0,1\}^n} \left(f(s) = \sum_{i=1}^{n} \prod_{j=1}^{i} s_j \right).$$

2.3 Markov Chain Modeling

EAs can be modeled and analyzed as Markov chains, e.g., in [11,22]. Let $\mathcal{X}$ be the population space and $\mathcal{X}^* \subseteq \mathcal{X}$ be the optimal population space. Note that an optimal population in $\mathcal{X}^*$ contains at least one optimal solution. Let $\xi_t \in \mathcal{X}$ be the population after t generations. Then, an EA can be described as a random sequence $\{\xi_0, \xi_1, \xi_2, \ldots\}$. Since ξ_{i+1} can often be decided from ξ_i and the reproduction operator of the EA (i.e., $P(\xi_{i+1} \mid \xi_i, \xi_{i-1}, \ldots, \xi_0) = P(\xi_{i+1} \mid \xi_i)$), the random sequence forms a Markov chain $\{\xi_t\}_{t=0}^{+\infty}$ with state space $\mathcal{X}$, denoted as "$\xi \in \mathcal{X}$" for simplicity. Note that all sets considered in this paper are multisets, e.g., a population can contain several copies of the same solution.

The goal of EAs is to reach the optimal space $\mathcal{X}^*$ from an initial population ξ_0. Given a Markov chain $\xi \in \mathcal{X}$ modeling an EA and $t_0 \geq 0$, we define τ as a random variable such that $\tau = \min\{t \geq 0 \mid \xi_{t_0+t} \in \mathcal{X}^*\}$. That is, τ is the number of steps needed to reach the optimal space for the first time when starting from time t_0. The mathematical expectation of τ, $\mathbb{E}[\tau \mid \xi_{t_0} = x] = \sum_{i=0}^{\infty} iP(\tau = i)$, is called the *conditional first hitting time* (CFHT) of the chain staring from $\xi_{t_0} = x$. If ξ_{t_0} is drawn from a distribution π_{t_0}, the expectation of the CFHT over π_{t_0}, $\mathbb{E}[\tau \mid \xi_{t_0} \sim \pi_{t_0}] = \sum_{x \in \mathcal{X}} \pi_{t_0}(x)\mathbb{E}[\tau \mid \xi_{t_0} = x]$, is called the *distribution-CFHT* (DCFHT) of the chain from $\xi_{t_0} \sim \pi_{t_0}$. Since the running time of EAs is counted by the number of fitness evaluations, the cost of initialization and each generation should be considered. For example, the expected running time of the $(\mu + \lambda)$-EA is $\mu + \lambda \cdot \mathbb{E}[\tau \mid \xi_0 \sim \pi_0]$.

A Markov chain $\xi \in \mathcal{X}$ is said to be absorbing, if $\forall t \geq 0 : P(\xi_{t+1} \in \mathcal{X}^* \mid \xi_t \in \mathcal{X}^*) = 1$. Note that all Markov chains modeling EAs can be transformed to be absorbing by making it unchanged once an optimal state has been found. This transformation obviously does not affect its first hitting time.

3 The Switch Analysis Approach

To derive running time bounds of the $(\mu + \lambda)$-EA on UBoolean, we first model the EA process as a Markov chain, and then apply the switch analysis approach.

Switch analysis [21,23] as presented in Theorem 1 is a recently proposed approach that compares the DCFHT of two Markov chains. Since the state spaces of the two chains may be different, an aligned mapping function $\phi : \mathcal{X} \to \mathcal{Y}$ as shown in Definition 3 is employed. Note that $\phi^{-1}(y) = \{x \in \mathcal{X} \mid \phi(x) = y\}$. Using switch analysis to derive running time bounds of a given chain $\xi \in \mathcal{X}$ (i.e., modeling a given EA running on a given problem), one needs to

1. construct a reference chain $\xi' \in \mathcal{Y}$ for comparison and design an aligned mapping function ϕ from $\mathcal{X}$ to $\mathcal{Y}$;
2. analyze their one-step transition probabilities, i.e., $P(\xi_{t+1} \mid \xi_t)$ and $P(\xi'_{t+1} \mid \xi'_t)$, the CFHT of the chain $\xi' \in \mathcal{Y}$, i.e., $\mathbb{E}[\tau' \mid \xi'_t]$, and the state distribution of the chain $\xi \in \mathcal{X}$, i.e., π_t;
3. examine Eq. (1) to get the difference ρ_t between each step of the two chains;
4. sum up ρ_t to get a running time gap ρ of the two chains, and then bounds on $\mathbb{E}[\tau \mid \xi_0]$ can be derived by combining $\mathbb{E}[\tau' \mid \xi'_0]$ with ρ.

Definition 3 (Aligned Mapping [21]). *Given two spaces $\mathcal{X}$ and $\mathcal{Y}$ with target subspaces $\mathcal{X}^*$ and $\mathcal{Y}^*$, respectively, a function $\phi : \mathcal{X} \to \mathcal{Y}$ is called*

(a) a left-aligned mapping if $\forall x \in \mathcal{X}^* : \phi(x) \in \mathcal{Y}^*$;
(b) a right-aligned mapping if $\forall x \in \mathcal{X} - \mathcal{X}^* : \phi(x) \notin \mathcal{Y}^*$;
(c) an optimal-aligned mapping if it is both left-aligned and right-aligned.

Theorem 1 (Switch Analysis [21]). *Given two absorbing Markov chains $\xi \in \mathcal{X}$ and $\xi' \in \mathcal{Y}$, let τ and τ' denote the hitting events of ξ and ξ', respectively, and let π_t denote the distribution of ξ_t. Given a series of values $\{\rho_t \in \mathbb{R}\}_{t=0}^{+\infty}$ with $\rho = \sum_{t=0}^{+\infty} \rho_t$ and a right (or left)-aligned mapping $\phi : \mathcal{X} \to \mathcal{Y}$, if $\mathbb{E}[\![\tau \mid \xi_0 \sim \pi_0]\!]$ is finite and*

$$\forall t: \sum_{x \in \mathcal{X}, y \in \mathcal{Y}} \pi_t(x) P(\xi_{t+1} \in \phi^{-1}(y) \mid \xi_t = x) \mathbb{E}[\![\tau' \mid \xi'_0 = y]\!]$$

$$\leq (or \ \geq) \sum_{u, y \in \mathcal{Y}} \pi_t^\phi(u) P(\xi'_1 = y \mid \xi'_0 = u) \mathbb{E}[\![\tau' \mid \xi'_1 = y]\!] + \rho_t, \tag{1}$$

where $\pi_t^\phi(u) = \pi_t(\phi^{-1}(u)) = \sum_{x \in \phi^{-1}(u)} \pi_t(x)$, we have

$$\mathbb{E}[\![\tau \mid \xi_0 \sim \pi_0]\!] \leq (or \ \geq) \mathbb{E}[\![\tau' \mid \xi'_0 \sim \pi_0^\phi]\!] + \rho.$$

The idea of switch analysis is to obtain the difference ρ on the DCFHT of two chains by summing up all the one-step differences ρ_t. Using Theorem 1 to compare two chains, we can waive the long-term behavior of one chain, since Eq. (1) does not involve the term $\mathbb{E}[\![\tau \mid \xi_t]\!]$. Therefore, the theorem can simplify the analysis of an EA process by comparing it with an easy-to-analyze one.

4   Running Time Analysis

In this section, we prove a lower bound on the expected running time of the $(\mu + \lambda)$-EA solving UBoolean, as shown in Theorem 2. Our proof is accomplished by using switch analysis (i.e., Theorem 1). The target EA process we are to analyze is the $(\mu + \lambda)$-EA running on any function in UBoolean. The constructed reference process for comparison is the RLS$^{\neq}$ algorithm running on the LeadingOnes problem. RLS$^{\neq}$ is a modification of the randomized local search algorithm. It maintains only one solution s. In each iteration, a new solution s' is generated by flipping a randomly chosen bit of s, and s' is accepted only if $f(s') > f(s)$. That is, RLS$^{\neq}$ searches locally and only accepts a better offspring solution.

Theorem 2. *The expected running time of the $(\mu + \lambda)$-EA on UBoolean is $\Omega(n \ln n + \mu + \lambda n \ln \ln n / \ln n)$, when μ and λ are upper bounded by a polynomial in n.*

We first give some lemmas that will be used in the proof of Theorem 2. Lemma 1 characterizes the one-step transition behavior of a Markov chain via CFHT. Lemma 2 gives the CFHT $\mathbb{E}[\![\tau' \mid \xi'_t = y]\!]$ of the reference chain ξ' (i.e., RLS$^{\neq}$ running on LeadingOnes). In the following analysis, we will use $\mathbb{E}_{rls}(j)$ to denote $\mathbb{E}[\![\tau' \mid \xi'_t = y]\!]$ with $|y|_0 = j$, i.e., $\mathbb{E}_{rls}(j) = nj$.

Lemma 1 ([9]). *Given a Markov chain $\xi \in \mathcal{X}$ and a target subspace $\mathcal{X}^* \subset \mathcal{X}$, we have, for CFHT, $\forall x \in \mathcal{X}^* : \mathbb{E}[\![\tau \mid \xi_t = x]\!] = 0$,*

$$\forall x \notin \mathcal{X}^* : \mathbb{E}[\![\tau \mid \xi_t = x]\!] = 1 + \sum_{x' \in \mathcal{X}} P(\xi_{t+1} = x' \mid \xi_t = x) \mathbb{E}[\![\tau \mid \xi_{t+1} = x']\!].$$

Lemma 2 ([21]). *For the chain $\xi' \in \mathcal{Y}$ modeling RLS$^{\neq}$ running on the LeadingOnes problem, the CFHT satisfies that $\forall y \in \mathcal{Y} = \{0,1\}^n : \mathbb{E}[\tau' \mid \xi'_t = y] = n \cdot |y|_0$, where $|y|_0$ denotes the number of 0-bits of y.*

Lemma 3. *For $m \geq i \geq 0$, $\sum_{k=0}^{i} \binom{m}{k}(\frac{1}{n})^k(1-\frac{1}{n})^{m-k}$ decreases with m.*

Proof. Let $f(m) = \sum_{k=0}^{i} \binom{m}{k}(\frac{1}{n})^k(1-\frac{1}{n})^{m-k}$. The goal is to show that $f(m+1) \leq f(m)$ for $m \geq i$. Denote $X_1, \ldots, X_{m+1}$ as independent random variables, where X_j satisfies that $P(X_j = 1) = \frac{1}{n}$ and $P(X_j = 0) = 1 - \frac{1}{n}$. Then we can express $f(m)$ and $f(m+1)$ as $f(m) = P(\sum_{j=1}^{m} X_j \leq i)$ and $f(m+1) = P(\sum_{j=1}^{m+1} X_j \leq i)$. Thus,

$$f(m+1) = P(\sum_{j=1}^{m} X_j < i) + P(\sum_{j=1}^{m} X_j = i)P(X_{m+1} = 0)$$

$$= P(\sum_{j=1}^{m} X_j < i) + P(\sum_{j=1}^{m} X_j = i)(1 - \frac{1}{n}) \leq f(m).$$

$\square$

Lemma 4. *For $\lambda \leq n^c$ where c is a positive constant, it holds that*

$$\sum_{i=0}^{n-1}\left(\sum_{k=0}^{i}\binom{n}{k}(\frac{1}{n})^k(1-\frac{1}{n})^{n-k}\right)^{\lambda} \geq n - \left\lceil\frac{e(c+1)\ln n}{\ln\ln n}\right\rceil.$$

Proof. Let $m = \lceil\frac{e(c+1)\ln n}{\ln\ln n}\rceil$. Denote $X_1, ..., X_n$ as independent random variables, where $P(X_j = 1) = \frac{1}{n}$ and $P(X_j = 0) = 1 - \frac{1}{n}$. Let $X = \sum_{j=1}^{n} X_j$, then its expectation $\mathbb{E}[X] = 1$. We thus have

$$\forall i \geq m, \sum_{k=i}^{n}\binom{n}{k}(\frac{1}{n})^k(1-\frac{1}{n})^{n-k} = P(X \geq i) \leq e^{(i-1)}/i^i, \tag{2}$$

where the inequality is by Chernoff bound. Then, we have

$$\sum_{i=0}^{n-1}(\sum_{k=0}^{i}\binom{n}{k}(\frac{1}{n})^k(1-\frac{1}{n})^{n-k})^{\lambda} \geq \sum_{i=m-1}^{n-1}(\sum_{k=0}^{i}\binom{n}{k}(\frac{1}{n})^k(1-\frac{1}{n})^{n-k})^{\lambda}$$

$$= \sum_{i=m-1}^{n-1}(1 - \sum_{k=i+1}^{n}\binom{n}{k}(\frac{1}{n})^k(1-\frac{1}{n})^{n-k})^{\lambda}$$

$$\geq \sum_{i=m-1}^{n-1}(1 - e^i/(i+1)^{(i+1)})^{\lambda} \geq \sum_{i=m-1}^{n-1}(1 - e^{m-1}/m^m)^{\lambda},$$

where the second inequality is by Eq. (2), and the last inequality can be easily derived because $e^i/(i+1)^{(i+1)}$ decreases with i when $i \geq m-1$.

Then, we evaluate e^{m-1}/m^m by taking logarithm to its reciprocal.

$$\ln(m^m/e^{m-1}) = m(\ln m - 1) + 1$$

$$\geq \frac{e(c+1)\ln n}{\ln\ln n}(1 + \ln(c+1) + \ln\ln\ln n - \ln\ln\ln\ln n - 1) + 1$$

$$\geq \frac{e(c+1)\ln n}{\ln\ln n}\frac{1}{e}\ln\ln n = (c+1)\ln n \geq \ln\lambda n. \quad \text{(by } \lambda \leq n^c)$$

This implies that $e^{m-1}/m^m \leq \frac{1}{\lambda n}$. Thus, we have

$$\sum_{i=0}^{n-1}(\sum_{k=0}^{i}\binom{n}{k}(\frac{1}{n})^k(1-\frac{1}{n})^{n-k})^\lambda \geq \sum_{i=m-1}^{n-1}(1-\frac{1}{\lambda n})^\lambda$$

$$\geq \sum_{i=m-1}^{n-1}(1-\frac{1}{n}) \geq n - m = n - \lceil\frac{e(c+1)\ln n}{\ln\ln n}\rceil,$$

where the second inequality is by $\forall 0 \leq xy \leq 1, y \geq 1 : (1-x)^y \geq 1 - xy.$ $\qquad\square$

Lemma 5 ([8]). *Let* $H(\epsilon) = -\epsilon\log\epsilon - (1-\epsilon)\log(1-\epsilon)$. *It holds that*

$$\forall n \geq 1, 0 < \epsilon < \frac{1}{2} : \sum_{k=0}^{\lfloor\epsilon n\rfloor}\binom{n}{k} \leq 2^{H(\epsilon)n}.$$

Proof of Theorem 2. We use switch analysis (i.e., Theorem 1) to prove it. Let $\xi \in \mathcal{X}$ model the analyzed EA process (i.e., the $(\mu + \lambda)$-EA running on any function in UBoolean). We use RLS$^{\neq}$ running on the LeadingOnes problem as the reference process modeled by $\xi' \in \mathcal{Y}$. Then, $\mathcal{Y} = \{0,1\}^n$, $\mathcal{X} = \{\{y_1, y_2, \ldots, y_\mu\} \mid y_i \in \{0,1\}^n\}$, $\mathcal{Y}^* = \{1^n\}$ and $\mathcal{X}^* = \{x \in \mathcal{X} \mid \max_{y\in x}|y|_1 = n\}$, where $|y|_1$ denotes the number of 1-bits of a solution $y \in \{0,1\}^n$. We construct a mapping $\phi : \mathcal{X} \to \mathcal{Y}$ as that $\forall x \in \mathcal{X} : \phi(x) = \arg\max_{y\in x}|y|_1$. It is easy to see that the mapping is an optimality-aligned mapping, because $\phi(x) \in \mathcal{Y}^*$ iff $x \in \mathcal{X}^*$.

We investigate the condition Eq. (1) of switch analysis. For any $x \notin \mathcal{X}^*$, suppose that $\min\{|y|_0 \mid y \in x\} = j > 0$. Then, $|\phi(x)|_0 = j$. By Lemmas 1 and 2, we have

$$\sum_{y\in\mathcal{Y}} P(\xi_1' = y \mid \xi_0' = \phi(x))\mathbb{E}[\tau' \mid \xi_1' = y] = \mathbb{E}_{rls}(j) - 1 = nj - 1. \qquad (3)$$

For the reproduction of the $(\mu + \lambda)$-EA (i.e., the chain $\xi \in \mathcal{X}$) on the population x, assume that the λ selected solutions from x for reproduction have the number of 0-bits $j_1, j_2, \ldots, j_\lambda$, respectively, where $j \leq j_1 \leq j_2 \leq \ldots \leq j_\lambda \leq n$. If there are at most i ($0 \leq i \leq j_1$) number of 0-bits mutating to 1-bits for each selected solution and there exists at least one selected solution which flips exactly i number of 0-bits, which happens with probability $\prod_{p=1}^{\lambda}(\sum_{k=0}^{i}\binom{j_p}{k}(\frac{1}{n})^k(1-\frac{1}{n})^{j_p-k}) - \prod_{p=1}^{\lambda}(\sum_{k=0}^{i-1}\binom{j_p}{k}(\frac{1}{n})^k(1-\frac{1}{n})^{j_p-k})$ (denoted by $p(i)$), the next population x' satisfies that $|\phi(x')|_0 \geq j_1 - i$.

Furthermore, $\mathbb{E}_{rls}(i) = ni$ increases with i. Thus, we have

$$\sum_{y \in \mathcal{Y}} P(\xi_{t+1} \in \phi^{-1}(y) \mid \xi_t = x)\mathbb{E}[\tau' \mid \xi'_0 = y] \geq \sum_{i=0}^{j_1} p(i) \cdot \mathbb{E}_{rls}(j_1 - i)$$

$$\geq \sum_{i=0}^{j} p(i) \cdot \mathbb{E}_{rls}(j - i) = n \sum_{i=0}^{j-1} (\prod_{p=1}^{\lambda} (\sum_{k=0}^{i} \binom{j_p}{k}(\frac{1}{n})^k (1 - \frac{1}{n})^{j_p - k})). \tag{4}$$

By comparing Eq. (3) with Eq. (4), we have $\forall x \notin \mathcal{X}^*$,

$$\sum_{y \in \mathcal{Y}} P(\xi_{t+1} \in \phi^{-1}(y) \mid \xi_t = x)\mathbb{E}[\tau'|\xi'_0 = y] - \sum_{y \in \mathcal{Y}} P(\xi'_1 = y \mid \xi'_0 = \phi(x))\mathbb{E}[\tau'|\xi'_1 = y]$$

$$\geq n(\sum_{i=0}^{j-1} (\prod_{p=1}^{\lambda} (\sum_{k=0}^{i} \binom{j_p}{k}(\frac{1}{n})^k (1 - \frac{1}{n})^{j_p - k})) - j) + 1$$

$$\geq n(\sum_{i=0}^{j-1} (\sum_{k=0}^{i} \binom{n}{k}(\frac{1}{n})^k (1 - \frac{1}{n})^{n-k})^\lambda - j) + 1$$

$$\geq n(\sum_{i=0}^{n-1} (\sum_{k=0}^{i} \binom{n}{k}(\frac{1}{n})^k (1 - \frac{1}{n})^{n-k})^\lambda - n) + 1,$$

where the 2nd '$\geq$' is because from Lemma 3, $\sum_{k=0}^{i} \binom{m}{k}(\frac{1}{n})^k (1 - \frac{1}{n})^{m-k}$ reaches the minimum when $m = n$, and the last '$\geq$' is by $(\sum_{k=0}^{i} \binom{n}{k}(\frac{1}{n})^k (1-\frac{1}{n})^{n-k})^\lambda \leq 1$.

When $x \in \mathcal{X}^*$, both Eqs. (3) and (4) equal 0, because both chains are absorbing and the mapping ϕ is optimality-aligned. Thus, Eq. (1) in Theorem 1 holds with $\rho_t = (n(\sum_{i=0}^{n-1}(\sum_{k=0}^{i} \binom{n}{k}(\frac{1}{n})^k (1 - \frac{1}{n})^{n-k})^\lambda - n) + 1)(1 - \pi_t(\mathcal{X}^*))$. By switch analysis,

$$\mathbb{E}[\tau|\xi_0 \sim \pi_0] \geq \mathbb{E}[\tau'|\xi'_0 \sim \pi_0^\phi]$$
$$+ (n(\sum_{i=0}^{n-1}(\sum_{k=0}^{i} \binom{n}{k}(\frac{1}{n})^k (1 - \frac{1}{n})^{n-k})^\lambda - n) + 1) \sum_{t=0}^{+\infty}(1 - \pi_t(\mathcal{X}^*)).$$

Since $\sum_{t=0}^{+\infty}(1 - \pi_t(\mathcal{X}^*)) = \mathbb{E}[\tau|\xi_0 \sim \pi_0]$, we have

$$\mathbb{E}[\tau|\xi_0 \sim \pi_0] \geq \frac{\mathbb{E}[\tau'|\xi'_0 \sim \pi_0^\phi]}{n(n - \sum_{i=0}^{n-1}(\sum_{k=0}^{i} \binom{n}{k}(\frac{1}{n})^k (1 - \frac{1}{n})^{n-k})^\lambda)} \geq \frac{\mathbb{E}[\tau'|\xi'_0 \sim \pi_0^\phi]}{n\lceil \frac{e(c+1)\ln n}{\ln \ln n} \rceil}, \tag{5}$$

where the last inequality is by Lemma 4, since $\lambda \leq n^c$ for some constant c.

We then investigate $\mathbb{E}[\tau'|\xi'_0 \sim \pi_0^\phi]$. Since each of the μ solutions in the initial population is selected uniformly and randomly from $\{0,1\}^n$, we have

$$\forall 0 \leq j \leq n : \pi_0^\phi(\{y \in \mathcal{Y} \mid |y|_0 = j\}) = \pi_0(\{x \in \mathcal{X} \mid \min_{y \in x} |y|_0 = j\})$$

$$= \frac{(\sum_{k=j}^{n} \binom{n}{k})^\mu - (\sum_{k=j+1}^{n} \binom{n}{k})^\mu}{2^{n\mu}},$$

where $\sum_{k=j}^{n} \binom{n}{k}$ is the number of solutions with not less than j number of 0-bits. Then,

$$
\mathbb{E}[\tau'|\xi_0' \sim \pi_0^\phi] = \sum_{j=0}^{n} \pi_0^\phi(\{y \in \mathcal{Y} \mid |y|_0 = j\})\mathbb{E}_{rls}(j)
$$

$$
= \frac{1}{2^{n\mu}} \sum_{j=1}^{n} ((\sum_{k=j}^{n} \binom{n}{k})^\mu - (\sum_{k=j+1}^{n} \binom{n}{k})^\mu) n j = \frac{n}{2^{n\mu}} \sum_{j=1}^{n} (\sum_{k=j}^{n} \binom{n}{k})^\mu
$$

$$
> n \sum_{j=1}^{\lfloor \frac{n}{4} \rfloor + 1} (\sum_{k=j}^{n} \binom{n}{k}/2^n)^\mu > \frac{n^2}{4} (\sum_{k=\lfloor \frac{n}{4} \rfloor + 1}^{n} \binom{n}{k}/2^n)^\mu = \frac{n^2}{4} (1 - \sum_{k=0}^{\lfloor \frac{n}{4} \rfloor} \binom{n}{k}/2^n)^\mu
$$

$$
\geq \frac{n^2}{4} (1 - 2^{H(\frac{1}{4})n - n})^\mu \geq \frac{n^2}{4} e^{-\frac{\mu}{2^{(1-H(\frac{1}{4}))n} - 1}} > \frac{n^2}{4} e^{-\frac{\mu}{1.13^n - 1}},
$$

where the third inequality is by Lemma 5, the fourth inequality is by $\forall 0 < x < 1 : (1 - x)^y \geq e^{-\frac{xy}{1-x}}$, and the last inequality is by $2^{1-H(\frac{1}{4})} > 1.13$.

Applying the above lower bound on $\mathbb{E}[\tau'|\xi_0' \sim \pi_0^\phi]$ to Eq. (5), we get, noting that μ is upper bounded by a polynomial in n,

$$
\mathbb{E}[\tau|\xi_0 \sim \pi_0] \geq \frac{n}{4\lceil \frac{e(c+1)\ln n}{\ln \ln n} \rceil} e^{-\frac{\mu}{1.13^n - 1}}, \quad \text{i.e.,} \quad \Omega(\frac{n \ln \ln n}{\ln n}).
$$

Considering the μ number of fitness evaluations for the initial population and the λ number of fitness evaluations in each generation, the expected running time of the $(\mu + \lambda)$-EA on UBoolean is lower bounded by $\Omega(\mu + \frac{\lambda n \ln \ln n}{\ln n})$. Because the $(\mu + \lambda)$-EA belongs to mutation-based EAs, we can also directly use the general lower bound $\Omega(n \ln n)$ [18]. Thus, the theorem holds.   □

5   Conclusion

This paper analyzes the expected running time of the $(\mu + \lambda)$-EA for solving a general problem class consisting of pseudo-Boolean functions with a unique global optimum. We derive the lower bound $\Omega(n \ln n + \mu + \lambda n \ln \ln n/\ln n)$ by applying the recently proposed approach switch analysis. The results partially complete the running time comparison between the $(\mu + \lambda)$-EA and the $(1+1)$-EA on the two well-studied pseudo-Boolean problems, OneMax and LeadingOnes. We can now conclude that when μ or λ is slightly large, the $(\mu + \lambda)$-EA has a worse expected running time. The investigated $(\mu + \lambda)$-EA only uses mutation, while crossover is a characterizing feature of EAs. Therefore, we will try to analyze the running time of population-based EAs with crossover operators in the future.

References

1. Auger, A., Doerr, B.: Theory of Randomized Search Heuristics: Foundations and Recent Developments. World Scientific, Singapore (2011)
2. Bäck, T.: Evolutionary Algorithms in Theory and Practice: Evolution Strategies, Evolutionary Programming, Genetic Algorithms. Oxford University Press, Oxford (1996)
3. Chen, T., He, J., Chen, G., Yao, X.: Choosing selection pressure for wide-gap problems. Theor. Comput. Sci. **411**(6), 926–934 (2010)
4. Chen, T., He, J., Sun, G., Chen, G., Yao, X.: A new approach for analyzing average time complexity of population-based evolutionary algorithms on unimodal problems. IEEE Trans. Syst. Man Cybern. Part B Cybern. **39**(5), 1092–1106 (2009)
5. Chen, T., Tang, K., Chen, G., Yao, X.: A large population size can be unhelpful in evolutionary algorithms. Theor. Comput. Sci. **436**(8), 54–70 (2012)
6. Doerr, B., Künnemann, M.: Optimizing linear functions with the $(1 + \lambda)$ evolutionary algorithm - different asymptotic runtimes for different instances. Theor. Comput. Sci. **561**, 3–23 (2015)
7. Droste, S., Jansen, T., Wegener, I.: On the analysis of the $(1 + 1)$ evolutionary algorithm. Theor. Comput. Sci. **276**(1–2), 51–81 (2002)
8. Flum, J., Grohe, M.: Parameterized Complexity Theory. Springer, New York (2006)
9. Freĭdlin, M.: Markov Processes and Differential Equations: Asymptotic Problems. Birkhäuser, Basel (1996)
10. Gießen, C., Witt, C.: Population size vs. mutation strength for the $(1+\lambda)$ EA on OneMax. In: Proceedings of GECCO 2015, Madrid, Spain, pp. 1439–1446 (2015)
11. He, J., Yao, X.: Drift analysis and average time complexity of evolutionary algorithms. Artif. Intell. **127**(1), 57–85 (2001)
12. He, J., Yao, X.: From an individual to a population: an analysis of the first hitting time of population-based evolutionary algorithms. IEEE Trans. Evol. Comput. **6**(5), 495–511 (2002)
13. Jansen, T., Wegener, I.: On the utility of populations in evolutionary algorithms. In: Proceedings of GECCO 2001, San Francisco, CA, pp. 1034–1041 (2001)
14. Lehre, P.K., Witt, C.: Black-box search by unbiased variation. Algorithmica **64**(4), 623–642 (2012)
15. Lehre, P.K., Yao, X.: On the impact of mutation-selection balance on the runtime of evolutionary algorithms. IEEE Trans. Evol. Comput. **16**(2), 225–241 (2012)
16. Neumann, F., Witt, C.: Bioinspired Computation in Combinatorial Optimization: Algorithms and Their Computational Complexity. Springer, Berlin (2010)
17. Storch, T.: On the choice of the parent population size. Evol. Comput. **16**(4), 557–578 (2008)
18. Sudholt, D.: A new method for lower bounds on the running time of evolutionary algorithms. IEEE Trans. Evol. Comput. **17**(3), 418–435 (2013)
19. Witt, C.: Runtime analysis of the $(\mu + 1)$ EA on simple pseudo-Boolean functions. Evol. Comput. **14**(1), 65–86 (2006)
20. Witt, C.: Population size versus runtime of a simple evolutionary algorithm. Theor. Comput. Sci. **403**(1), 104–120 (2008)
21. Yu, Y., Qian, C., Zhou, Z.H.: Switch analysis for running time analysis of evolutionary algorithms. IEEE Trans. Evol. Comput. **19**(6), 777–792 (2015)
22. Yu, Y., Zhou, Z.H.: A new approach to estimating the expected first hitting time of evolutionary algorithms. Artif. Intell. **172**(15), 1809–1832 (2008)
23. Yu, Y., Qian, C.: Running time analysis: convergence-based analysis reduces to switch analysis. In: Proceedings of CEC 2015, Sendai, Japan, pp. 2603–2610 (2015)

Hybrid Crossover Based Clonal Selection Algorithm and Its Applications

Hongwei Dai[✉], Yu Yang, and Cunhua Li

School of Computer Engineering, Huaihai Institute of Technology,
Lianyungang 222005, Jiangsu, China
hwdai@hhit.edu.cn

Abstract. Hybridization is confirmed as an effective way of combining the best properties of different algorithms and achieving better performances. A framework of hybrid crossover is constructed and combined with clonal selection algorithm (CSA). The new crossover solutions are generated by the mutual influence of both high affinity and low affinity solutions. Simulation results based on the traveling salesman problems demonstrate the effectiveness of the hybridization.

Keywords: Hybrid crossover · Clonal Selection Algorithm · Traveling salesman problem · Simulation

1 Introduction

During the last few decades, numerous nature-inspired methods, such as Genetic Algorithm (GA) [13], Evolutionary Algorithm (EA) [12], Artificial Bee Colony (ABC) [14], and Ant Colony Optimization (ACO) [3], have been proposed to solve optimization problems. Unlike the traditional techniques which have to face many difficulties such as multi-modality, dimensionality and differentiability associated with the optimization problems [11], the nature-inspired algorithms have better convergence speeds to the optimal or near optimal results in reasonable time.

More recently, however, one of the nature-inspired algorithms named Artificial Immune System (AIS) has received a rapid increasing interest [7]. Different algorithms such as Danger Theory (DT) models [1,2], Negative Selection Algorithms [4], Immune Network Theory-based model [8] have been proposed and successfully applied to a wide range of problems. In addition, Clonal selection algorithm (CSA), one of the most studied and used AISs, has received a rapidly increasing interest and has been verified as having a great number of useful mechanisms from the viewpoint of immune programming [10], controlling [15], optimization [9] and so on.

Although CSA has made a success triumph in solving various numerical and combinatorial optimization problems, there are still some inherent disadvantages such as no information exchanging during different antibodies. To solve this problem, a hybrid crossover is constructed and combined with CSA for solving optimization problems.

© Springer International Publishing AG 2016
H. Yin et al. (Eds.): IDEAL 2016, LNCS 9937, pp. 468–475, 2016.
DOI: 10.1007/978-3-319-46257-8_50

2   Hybrid Crossover and Novel Clonal Selection Algorithm

In our previous works, different crossovers are introduced into the CSA to improve the performance. However, these quantum crossovers are single direction. In this novel algorithm, new crossover antibodies are generated by the mutual influence of both high affinity and low affinity antibodies. The influence of high affinity antibodies on low affinity antibodies accelerate the convergence process in the first half evolution process. Then the disturbance effect of low affinity antibodies on high affinity antibodies improve the algorithm escaping from local optimum in the last half evolution process.

2.1   Hybrid Crossover Introduction

In the case of n cities traveling salesman problem, a closed tour that visits each of the n cities exactly once is needed. Let us consider the following antibody gene sequence (solution for 6 cities TSP): $\boxed{a_1\,a_2\,a_3\,a_4\,a_5\,a_6}$. This represents the following visiting tour: $a_1 \rightarrow a_2 \rightarrow a_3 \rightarrow a_4 \rightarrow a_5 \rightarrow a_6 \rightarrow a_1$.

Without loss of generality, the population size is set to equal to the number of city in TSPs. For a 6 cities TSP, 6 solutions $A_1, A_2, A_3, A_4, A_5, A_6$ are sorted by their affinity in descending order shown in Table 1.

Table 1. Example population includes six antibodies

$A_1 : a_1\ a_2\ a_3\ a_4\ a_5\ a_6$
$A_2 : a_3\ a_5\ a_1\ a_2\ a_6\ a_4$
$A_3 : a_2\ a_6\ a_4\ a_3\ a_1\ a_5$
$A_4 : a_1\ a_3\ a_5\ a_6\ a_4\ a_2$
$A_5 : a_2\ a_4\ a_6\ a_5\ a_3\ a_1$
$A_6 : a_6\ a_2\ a_1\ a_3\ a_5\ a_4$

The crossover $\theta(H2L)$ can be presented as follows:

Step 1: Select a city in solution 1 (A_1) randomly, for example the first city a_1. Because this is the first city generated by crossover, we put it in crossover antibody A_1' directly. The following city will be selected from solution A_2. The left and right cities of a_1 in solution A_2 can be obtained easily. Compare two edges $\overline{a_1 a_2}$ and $\overline{a_1 a_5}$. City a_5 will be selected if $dis(a_1,a_5) < dis(a_1,a_2)$. Then a_5 is put in the crossover antibody A_1'. Here $dis(x,y)$ is a function used to calculate the distance between two cities x and y.

Step 2: City a_5 is the current city now. The following city will be selected in solution A_3. The left and right cities of a_5 in solution A_3 can be determined easily. Compare two edges $\overline{a_5 a_1}$ and $\overline{a_5 a_2}$. City a_2 will be selected if $dis(a_5,a_2) < dis(a_5,a_1)$. Then a_2 is put in the crossover antibody A_1'.

Step 3: City a_2 is the current city now. The left and right cities of a_2 in solution A_4 can be determined similarly. Compare two edges $\overline{a_2 a_4}$ and $\overline{a_2 a_1}$. City a_4 will be selected if $dis(a_2,a_4) < dis(a_2,a_1)$. Then a_4 is pushed onto the crossover antibody A_1'.

The same way, after five steps, a new crossover solution which contains information from different antibodies in the solution population can be constructed.

The crossover $\theta(L2H)$ can be presented as follows:

Step 1: Select the 1st city a_3 in A_2. Then adjacent cities of a_3 in A_3 can be get easily. Obviously, a_4 is the left-city and a_1 is the right-city according to their positions. Then two possible inversion comparing I_1 and I_2 are performed by swapping the sub-segment $a_5 - a_1 - a_2 - a_6 - a_4$ and $a_5 - a_1$ in A_2 which start with the right adjacent city of a_3 in A_2 and end with the left or right city. Similar to the operation in above quantum crossover, there are no inversion operation occurs in A_2, it is just a comparison for finding shorter tour. a_1 will be selected if inversion I_2 generates shorter visiting tour. Then we can get the sub-segment of the new crossover tour A_2', that is, $A_2' : a_3, a_1$.

Step 2: Select the 2nd city a_1 in A_2'. Then adjacent cities of a_1 in A_4 can be obtained. Clearly, a_2 is the left-city and a_3 is the right-city. Then two possible inversion comparing I_1 and I_2 are performed by swapping the sub-segment $a_2 - a_2$ and $a_5 - a_3$ in A_2 which start with the adjacent city of a_1 in A_2 and end with the left or right city. It should be noticed that the right-city a_3 of a_1 in A_4 is located at the left of city a_1 in A_2. As a result, city a_5 is selected as the start point to generate a sub-segment $a_5 - a_3$ for inversion comparison. If I_2 can not generate a shorter tour, city a_2 will be selected to renew the crossover tour A_2'. That is, $A_2' : a_3, a_1, a_2$.

The same way, a new crossover tour $A_2' : a_3, a_1, a_2, a_4, a_5, a_6, a_3$ is constructed.

2.2   Hybrid Crossover Based Clonal Selection Algorithm (HCCSA)

The general steps of FQCCSA are illustrated in the following.

Step 1. Generate an initial population A including m antibodies randomly.

Step 2. Compute all antibodies' affinity and sort them in a descending order.

Step 3. Select n ($n \le m$) elite (fittest) antibodies based on their affinities from the m original cells.

Step 4. Place each of the n selected elites in n separate and distinct elite pools $(EP_1, EP_2, ..., EP_n)$.

Step 5. Clone the elites in each elite pool with a rate proportional to its fitness. The amount of clone generated for these antibodies is given by Eq. (1):

$$p_i = round(\frac{(n-i)}{n} \times M) \tag{1}$$

where i is the ordinal number of the elite pools, M is a multiplying factor which determines the scope of the clone and $round(.)$ is the operator that rounds its argument towards the closest integer.

Step 6. Subject the clones in each pool through either random point mutation or receptor editing processes. The mutation number (P_{hm} and P_{re} for random point mutation and receptor editing, respectively) are defined as follows:

$$P_{hm} = \lambda \cdot p_i \tag{2}$$

$$P_{re} = (1 - \lambda) \cdot p_i \tag{3}$$

where λ is a user-defined parameter which determines the complementary intensity between the random point mutation and receptor editing.

Step 7. Subject n elite solutions ($A_1, A_2, ..., A_n$) through hybrid crossover operation. QJP is a control parameter to proportion the $\theta(H2L)$ operator to $\theta(L2H)$ ones.

Step 8. Select the fittest antibody from each elite pool and new generated crossover antibodies.

Step 9. Update the parent cells in each elite pool with the fittest antibodies selected in Step 8.

The process will be terminated when the generation number matches a prespecified maximal generation number G_{max}. Otherwise, it returns to Step 3.

3   Simulation

In this section, series of tests aimed to demonstrate the performance of HCCSA are described. The algorithm is implemented with C++ and all results of each instance are replicated for 10 times. Table 2 gives the information of the TSP instances in the experiments.

The meaning of the parameters used in the proposed algorithm and their values are illustrated in Table 3.

In order to evaluate the effectiveness of HCCSA (RECSA+$\theta(H2L)$+ $\theta(L2H)$), we apply our algorithm to TSPs from eil51 to kroA200 and also compare our method with RECSA, RECSA+$\theta(H2L)$, and RECSA+$\theta(L2H)$.

Figure 1 demonstrates the convergence process of three different crossovers based CSAs for solving eil51 problem. The proposed HCCSA has a better convergence performance than other algorithms. The same conclusion on convergence ability can be observed on lin105 instance as shown in Fig. 2.

Table 2. TSP instances in simulation

Problem	City number	Optimum	G_{max}
eil51	51	426	1000
st70	70	675	1000
eil76	76	538	1000
rd100	100	7910	1000
eil101	101	629	1000
lin105	105	14379	1000
pr107	107	44303	1000
pr124	124	59030	1000
bier127	127	118282	2000
pr136	136	96772	2000
pr152	152	73682	2000
rat195	195	2323	2000
kroA200	200	29368	5000
lin318	318	42029	10000

Table 3. The meaning of the user-defined parameters and their values

Parameter	Meaning	Values
N	city number	51~200
m	number of initial antibodies	N
n	number of elite pools	N
M	proliferation rate	50
λ	complementary intensity between hypermutation and receptor editing	0.5
QJP	$\theta(H2L)$ and $\theta(L2H)$ proportion adjusting parameter	0.0~1.0
G_{max}	maximum number of generation	*

*:see Table 2

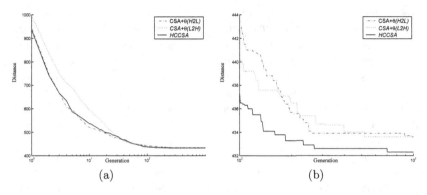

Fig. 1. (a) Convergence process of Three algorithms for eil51 and (b) the enlarged figure from 100 to 1000 generation

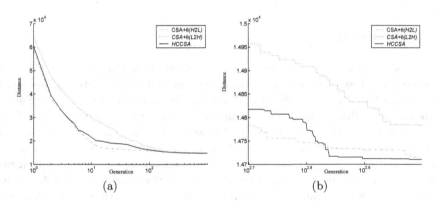

Fig. 2. (a) Convergence process of Three algorithms for lin105 and (b) the enlarged figure from 500 to 1000 generation

Table 4 shows the experimental results of the TSPs. In Table 4, parameters *PDM* and *PDB* indicate the percentage deviation from the optimal tour length D_{opt} of the mean distance D_m and the best distance D_b respectively. From this table, it can be confirmed that the hybridization is an effective way of improving performance of algorithms.

Table 4. Simulation results of different algorithms for TSP instances from eil51 to KroA200

Instance	optimum	RECSA		RECSA+$\theta(H2L)$ [5]		RECSA+$\theta(L2H)$ [6]		HCCSA	
		PDB	*PDM*	*PDB*	*PDM*	*PDB*	*PDM*	*PDB*	*PDM*
eil51	426	1.41	2.63	1.17	2.16	0.94	1.78	0.47	**1.48**
st70	675	0.30	2.39	0.44	**1.10**	0.30	1.23	0.59	1.45
eil76	538	3.35	4.54	1.86	**3.09**	3.16	3.92	2.60	3.88
rd100	7910	4.55	5.85	2.40	3.30	1.72	**3.13**	1.33	3.52
eil101	629	5.25	6.60	3.82	**4.90**	3.81	5.45	3.50	5.29
lin105	14379	2.48	5.39	1.68	2.32	1.53	2.81	1.00	**2.31**
pr107	44303	2.11	3.40	2.44	2.77	2.23	2.91	1.53	**1.99**
pr124	59030	1.65	4.42	0.64	1.64	0.64	1.51	0.65	**1.16**
bier127	118282	1.55	4.41	1.51	**2.00**	2.76	3.59	2.01	2.90
pr136	96772	4.39	6.28	5.13	6.80	2.60	4.93	3.36	**4.63**
pr152	73682	2.82	3.50	1.21	1.87	0.34	1.45	0.18	**1.39**
rat195	2323	10.55	13.12	2.54	**3.15**	7.28	9.04	7.62	8.30
kroA200	29368	4.46	6.59	1.17	**1.77**	3.48	4.41	3.06	3.77
average	–	3.45	5.32	2.00	2.84	2.37	3.55	2.15	3.24

4   Conclusion

In this paper, a hybrid crossover based clonal selection algorithm (HCCSA) is proposed. The new crossover antibodies are generated by the mutual influence of both high affinity and low affinity antibodies. The influence of high affinity antibodies on low affinity antibodies accelerate the convergence process in the first half evolution process. Then the disturbance effect of low affinity antibodies on high affinity antibodies improve the algorithm escaping from local optimum in the last half evolution process. Several experiments on combinatorial optimization problem TSP are performed to assess the performance of HCCSA. Simulation results demonstrated that hybridization is an effective way of improving performance of algorithms.

Acknowledgements. This work was supported by the Prospective Joint Research of University-Industry Cooperation of Jiangsu (No. BY2016056-02, BY2015248), the Six Talent Peaks Project of Jiangsu (No.XXRJ-013), Lianyungang Science and Technology Project (No.CG1413, CG1501), and the Natural Science Foundation of Huaihai Institute of Technology (No.z2015005, z2015012).

References

1. Aickelin, U., Bentley, P., Cayzer, S., Kim, J., Mcleod, J.: Danger theory: the link between AIS and IDS. In: Proceedings of 2nd International Conference on Artificial Immune Systems ICARIS 2003, pp. 147–155 (2003)

2. Aickelin, U., Cayzer, S.: The danger theory and its application to artificial immune systems. In: Proceedings of the 1st International Conference on Artificial Immune Systems (ICARIS-2002), pp. 141–148 (2002)
3. Angus, D., Hendtlass, T.: Dynamic ant colony optimization. Appl. Intell. **23**(1), 33–38 (2005)
4. Ayara, M., Timmis, J., de Lemos, L.N., de Castro, R., Duncan, R.: Negative selection: how to generate detectors. In: Proceedings of the 1st International Conference on Articial Immune Systems (ICARIS), pp. 89–98 (2002)
5. Dai, H.W., Yang, Y., Li, C.H., Shi, J., Gao, S.C., Tang, Z.: Quantum interference crossover-based clonal selection algorithm, its application to traveling salesman problem. IEICE Trans. Inf. Syst. **E92–D**(1), 78–85 (2009)
6. Dai, H.W., Yang, Y., Li, H., Li, C.H.: An improved clonal selection algorithm with feedback quantum interference crossover. Int. J. Adv. Comput. Technol. (IJACT) **3**(8), 181–188 (2011)
7. de Castro, L.N., Timmis, J.: Artificial Immune System: A New Computational Intelligence Approach. Springer, Heidelberg (2002)
8. Gao, S.C., Dai, H.W., Zhang, J.C., Tang, Z.: An expanded lateral interactive clonal selection algorithm, its application. IEICE Trans. Fundam. **E91–A**(8), 2223–2231 (2008)
9. Gao, S., Chai, H., Chen, B., Yang, G.: Hybrid gravitational search and clonal selection algorithm for global optimization. In: Tan, Y., Shi, Y., Mo, H. (eds.) Advances in Swarm Intelligence. LNCS, vol. 7929, pp. 1–10. Springer, Heidelberg (2013)
10. Musilek, P., Lau, A., Reformat, M., Wyard-Scott, L.: Immune programming. Inf. Sci. **176**(8), 972–1002 (2006)
11. Rao, R.V., Savsani, V.J., Vakharia, D.P.: Teaching-learning-based optimization: an optimization method for continuous non-linear large scale problems. Inf. Sci. **183**, 1–15 (2012)
12. Wang, J., Liao, J., Zhou, Y., Cai, Y.: Differential evolution enhanced with multiobjective sorting-based mutation operators. IEEE Trans. Syst. Man Cybern. **44**(12), 2792–2805 (2014)
13. Zacharia, P.T., Aspragathos, N.A.: Optimal robot task scheduling based on genetic algorithms. Robot. Comput. Integr. Manuf. **21**(1), 67–79 (2005)
14. Zhang, Y.D., Wu, L.N.: Face pose estimation by chaotic artificial bee colony. JDCTA Int. J. Digit. Content Technol. Appl. **5**(2), 55–63 (2011)
15. Zuo, X.Q., Fan, Y.S.: A chaos search immune algorithm with its application to neuro-fuzzy controller design. Chaos Solitons Fractals **30**(1), 94–109 (2006)

A Task-Oriented Self-organization Mechanism in Wireless Sensor Networks

Xiang Yin[✉], Liping Chang, Weichao Dai, Bin Li, and Chunxiao Li

Department of Information Engineering, Yangzhou University,
Yangzhou 225127, China
{yinxiang,lb,licx}@yzu.edu.cn, changyzu@163.com, daiweichao@gmail.com

Abstract. In wireless sensor networks (WSNs), one attractive and challenging issue is how to allocate tasks efficiently. Most existing studies focused on mapping and scheduling tasks to multiple sensors to ensure the task can be completed before deadline. Nevertheless, one vital aspect is neglected, that is, self-organization of WSN in task allocation. In this paper, we consider the problem of complex task allocation in which a task requires different resources for execution. A task-oriented self-organization mechanism is proposed to guarantee real time in task assignment. Toward this end, the frequently accessed sensors in previous task allocation will modify their structural links that can achieve a better allocation of tasks in the future. Simulation results illustrate significant performance improvements with our proposed mechanism.

Keywords: Wireless Sensor Networks · Task allocation · Self-organization mechanism

1 Introduction

Wireless Sensor Networks (WSNs) have emerged with significant development of the technology in tiny, low-cost sensing devices and advances of the technology in wireless communication and networks [1–3]. In general, a WSN is composed of hundreds or thousands of nodes that are deployed in an ad hoc fashion in order to sense, collect and integrate information of the target environment. Recently, the research on WSN has become a hot field due to its wide range of applications. Ranging from the military objective tracking [4], environment monitoring [5], to health related applications [6], disaster response [7], and so on.

A WSN is essentially designed for completing a specific task. As a simple individual, a sensor node has very restricted capacity and resource. Thus, nodes need to collaborate with each other for satisfying complex tasks. Following this idea, numerous studies have been conducted in task allocation and scheduling in WSN [8,9]. Most of these works focus on developing an efficient scheme with the purpose of minimizing the task execution time and reducing energy consumption. In fact, the performance, especially from the perspective of reducing completion time, to a task is affected by two main factors, one is the mechanism for allocating

© Springer International Publishing AG 2016
H. Yin et al. (Eds.): IDEAL 2016, LNCS 9937, pp. 476–483, 2016.
DOI: 10.1007/978-3-319-46257-8_51

tasks to sensor nodes, the other is the structural links among nodes, which is usually neglected.

With this motivation, in this paper we consider the task-oriented WSN in which complex tasks are to be performed. In order to guarantee the task is accomplished before deadline, nodes associated with suitable resources should be detected as quickly as possible. For this purpose, a novel self-organization algorithm is presented. By taking the previous interaction and collaboration with other nodes into account, each sensor can decide on when and with whom to adapt its structural links autonomously in a decentralized manner. In so doing, when new task arrives, sensors with required resources can be easily positioned and the task can be accomplished in time.

2   Related Work

Recently, a variety of works have been done on task allocation in WSN. Abdelhak et al. [10] proposed an energy-balancing task scheduling and allocation heuristic (EBSEL) whose aim is to extend the network's lifetime. The strategy tried to minimize the communication cost without sacrificing parallelism, and employed a thresholding technique to avoid the early death of the nodes. Jin et al. [11] proposed an adaptive intelligent task mapping and scheduling scheme, they designed a hybrid function which contains cost for processing a task and cost for communication, and Genetic Algorithm is adopted to optimize the function. Yang et al. [12] integrated various metrics, including execution time, energy consumption, and network lifetime, into a hybrid function. Based on this, a modified version of binary particle swarm optimization (MBPSO), which adopts a different transfer function and a new position updating procedure with mutation, is proposed for the task allocation problem to obtain the best solution. Building on this work, Guo et al. [13] developed a soft real-time fault-tolerant task allocation algorithm for WSNs, an improved discrete swarm optimization algorithm is utilized for minimizing task execution time, saving energy cost and balancing network load. Meanwhile, a popular fault-tolerant mechanism, namely the primary/backup (P/B) technology, is employed to improve success ratio of task execution. From aforementioned literatures, we can find that most of existing works on task allocation in WSN focus on allocating and scheduling tasks among multiple nodes while they neglect the influence of structural links on task execution.

3   Problem Formulation

Since we consider self-organization from a distinct way, some new models and definitions are presented in this section. Here, we concentrate on problem-solving WSN which means that the vital features of WSN is to satisfy specific tasks. Moreover, the task to be solved is complicated, specifically, various resources are required for accomplishing a task. Accordingly, we assume sensors in WSN are heterogeneous types that mainly refers to different resources possessed by each node, such as computation capacity, storage size and so on, while the radio characteristics are considered to be identical. It should be noted that the roles of sensors in our model is quite different from the ones in other related works, which usually suppose nodes are homogeneous, that is, the sensors can only perform similar actions. In our model, each sensor node has a certain type of resource under its disposal which represents its capability for performing a specific action. By combination of various resources, a task can be fulfilled. Another key point that is worth noting is that although our goal looks similar to previous works, i.e., distributing tasks among sensors to reduce time delay for task execution, they are distinct. Existing works normally assume that execution time of a task on a node is known in advance, which is mainly determined by the processing speed of the node. In contrast, to our model, the core factor for task execution time is how to obtain required resources, once the resource requirement is satisfied, the task can be accomplished. Thus, the key question is locating nodes with proper resources.

4   Proposed Algorithm

4.1   Resource-Oriented Task Allocation Strategy

Task allocation is the basis for self-organization. For a WSN, a large number of sensor nodes are deployed randomly and once being disposed, they can hardly change their places. In the starting time, sensors connect with each other by some routine protocol. Through the connection, they send and receive data and implement task allocation, as they have no information about resource distribution in WSN. A task enters WSN from outside, e.g., from a satellite or external Internet, then the sink node assigns a specific sensor, namely the manager node s_{mag}, for allocating the task. As mentioned before, a complex task demands multiple resources and as a sensor only has one kind of resource, various sensors need to cooperate for task execution. For the manager node, task allocation means finding nodes with proper resources such that the combination of resources of these nodes can cover the resource demands of the task. Here, we develop a task allocation algorithm in which the manager node communicates with other nodes from nearby to far-away in the network untill all requested resources are satisfied. This process is detailed in Algorithm 1.

4.2   Candidate Sensor Selection

After performing a number of tasks, the manager node should decide on which nodes to select for initiating reorganization procedures. As in our case, the main purpose of a WSN is to execute a task in real time which is significantly influenced by locating sensors with proper resources. Meanwhile, a WSN is usually designed for a kind of issues which implies that the demanded resources are similar. Therefore, to a manager node, it prefers to construct a new structural link with those sensors that can provide resources frequently used in previous tasks.

In more detail, the manager node maintains a list that records preceding successfully allocated tasks as well as the corresponding sensors. When the records

---

Algorithm 1. Resource-oriented task allocation algorithm

1: $\overline{R}_{t_j} = \overline{R}_{t_j} - R_{mag}$
2: **while** $\overline{R}_{t_j}! = \{\}$ **do**
3:    Sensor node s_i sends resource request message $q = \{mag, \overline{R}_{t_j}, TTL\}$ to its neighbor nodes sequentially
4:    **for** each neighbor node s_k **do**
5:       Match the resource it possesses R_k with $\overline{R}_{t_j}$
6:       **if** $R_k \cap \overline{R}_{t_j}! = \{\}$ **then**
7:          Send node ID k to manager node, $\overline{R}_{t_j} = \overline{R}_{t_j} - R_k$
8:       **end if**
9:       $TTL = TTL - 1$
10:   **end for**
11:   Repeat the process from s_k
12: **end while**

---

exceed a certain count, the manager node checks how many times for each node to join the previous tasks, if the number is over a predefined threshold, this node will be added to the candidate set. This can be explained by the fact that if a sensor's resource is frequently requested by previous tasks, it can be predicted that it will join other tasks in the future with high possibility. It should be noted that the manager node can determine the candidate sensors only through its local information and no global information is required.

5 Mechanism for Self-organization

Before describing the self-organization mechanism, we first differentiate two distinct concepts: routing path vs. task allocation path. For a general WSN, once it is deployed, a specific routing protocol will be operated automatically which is primarily responsible for effective data transmission. As a power-constrained network, the routing protocol in WSN is mainly concerned with how to save energy of individual node and prolong the lifetime of the whole network. In comparison, the major role for a task allocation path is assigning tasks to appropriate nodes that can cooperatively perform the task. The principal consideration is constructing an efficient path between the manager node and the target nodes, and following it, the nodes with required resources can be reached as soon as possible.

Based on the principle mentioned above, in this paper, we consider a multi-hop WSN with flat topology. In such an infrastructure, each node typically plays the same role in routes, i.e., one sensor can communicate with other sensors through intermediate nodes. For energy restriction, each sensor node has limited communication distance, other nodes within this range can send and receive information with it. On the other side, it will incur high cost to a node if it maintains too many connections with other nodes, thus, a node should constrain the maximum connectivity. In summary, in our case, one sensor node connects only with limited number of other nodes which are closest to it and within its communication scope. While for task allocation, a sensor node can view all the nodes within its communication scope as a relay node to pass through task query information.

So far, we can demonstrate the self-organization mechanism in WSN. This problem can be boiled down to the following: after completing a certain amount of tasks, the manager node will adjust the task allocation paths and establishes new paths with the candidate sensor nodes with the purpose of minimizing allocation time for future tasks. In order to achieve this goal, a shortest path from the manager node to the target node will be built up in the sense of task allocation path. As the sensor nodes communicate with each other in multi-hop mode and the message transmission delay due to distance can be ignored, this issue is equal to find a minimum hop path between the two nodes. We can map this problem to the traditional problem of searching shortest paths in graph theory by regarding manager node as the source node, and setting the weight of each edge to be identical. Obviously, there exists a few algorithms that can be adopted to solve this question, such as Dijkstra Algorithm, which is simple and efficient.

6   Simulation Results and Analysis

6.1   Experimental Setup

In order to empirically evaluate the performance of the proposed mechanism, we conduct a simulated experiment. On the one hand, the current related model of task allocation in WSN is quite different from ours, on the other hand, to the best of our knowledge, there does not exist a self-organization mechanism for task allocation in WSN. Therefore, to reveal the performance improvements for task allocation gained by the presented self-organization strategy, we compare it with that without self-organization.

Generally, in our experiment, 200 sensor nodes are deployed uniformly in a rectangle area of 200 m by 200 m. There are totally 12 kinds of resources in WSN, for an individual sensor node, its resource is generated randomly, while a task is endowed with 8 different types of resources which follows a random distribution. A task can be executed if and only if all the required resources are provided by sensor nodes. And the communication range of a sensor is set to 20 m. We use matlab7 as the simulation tool, and the experimental results are obtained by averaging 10 runs.

6.2   Results and Analysis

In the first experiment, we compare the hop for task allocation under various mechanisms. The hop is the sum of steps from the manager node to the nodes until all the resources of the task are satisfied. To the proposed mechanism, i.e., hop with self-organization, two cases with distinct number of tasks before self-organization are considered, that is, $h = 2$ and $h = 10$. The results are illustrated in Fig. 1. From the Fig, we can find that the hop for task allocation with

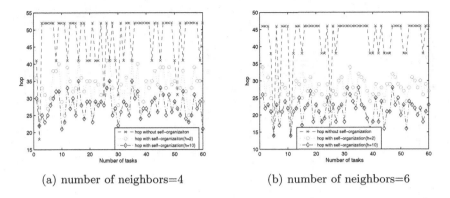

(a) number of neighbors=4   (b) number of neighbors=6

Fig. 1. Hop with the number of tasks

self-organization mechanism is much less than that without self-organization mechanism. The main reason is that by adopting self-organization mechanism, the initial routing path for data transmission is adjusted to task allocation path which focuses on constructing links between manager node and nodes that provide the needed resources for task execution. Thus, when new tasks arrive, the manager node can allocate it through appropriate path and the hop is remarkably reduced.

Furthermore, comparing two self-organization mechanisms with different parameters $h = 2$ and $h = 10$, we can see that the strategy with more allocated tasks before self-organization has superior performance than that with few allocated tasks. This can be ascribed to the fact that the self-organization strategy is based on the information of previous task assignment. If too limited tasks, e.g., h=2, are taken into account, the manager node can not construct sufficient links with proper nodes. Therefore, when new tasks arrive, some required resources may not be owned by the nodes through the established task allocation paths. In such cases, the manager node has to re-find the appropriate nodes which will obviously increase the hop for task allocation. Based on the observation, in the following experiment, we fix the number of tasks before self-organization to 10.

Figure 2 demonstrates the ratio of successful task allocation of the strategies with and without self-organization mechanism. A task is considered to be successfully allocated if and only if all its required resources are satisfied before TTL reduces to 0, otherwise, it is failed. From Fig. 2, we can see that when TTL drops from 40 to 30, the corresponding successful task allocation ratio of both methods decreases sharply. Nevertheless, task allocation strategy with self-organization mechanism performs consistently much better than that without self-organization mechanism in all situations. This trend can be explained by the fact that without self-organization mechanism, the manager node can only propagate the query via routing path which do not contain any resource information of nodes, and a great number of nodes need to be tried before obtaining the demanded resources. In comparison, to the proposed mechanism, manager

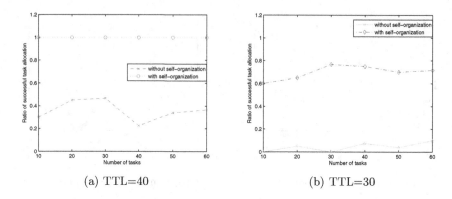

(a) TTL=40   (b) TTL=30

Fig. 2. Ratio of successful task allocation with the number of tasks

node records resources of nodes through previous task allocation, and when facing with new tasks, it can assign the subtask directly to the appropriate node and the times for spreading the query can be significantly saved.

7   Conclusion

In WSNs, the latency is a significant indicator to be investigated in allocating tasks, especially in emergent applications. This paper presents a task allocation strategy which integrates a novel self-organization mechanism. The proposed mechanism deals with complex tasks where multiple different resources are needed for execution. By utilizing resource information of sensors during previous task allocation, the structural links among sensors are self-adjusted. In so doing, shortest task allocation paths to the destination sensors can be established. When new tasks arrive, the sensors with appropriate resources can be quickly located and therefore, the latency of task allocation will be remarkably decreased. Simulation experiments show that the proposed mechanism is effective. In this paper, we are mainly concerned with allocating tasks with real-time constraint, whereas the energy consumption is ignored, thus the future work will concentrate on designing a mechanism that can strike a balance between latency and power consumption in task allocation.

Acknowledgments. This work was supported in part by the National Natural Science Foundation of China under Grant No. 61472344, No. 61401387, the Natural Science Foundation of Jiangsu Province under Grant No. BK20150460, the Natural Science Foundation of Yangzhou under Grant No. YZ2014054, and the Innovation Foundation of Yangzhou University.

References

1. Akyildiz, I.F., Su, W., Sankarasubramaniam, Y., Cayirci, E.: Wireless sensor networks: a survey. Comput. Netw. **38**, 393–422 (2002)
2. Yick, J., Mukherjee, B., Ghosal, D.: Wireless sensor network survey. Comput. Netw. **52**(12), 2292–2330 (2008)
3. Attea, B.A., Khalil, E.A.: A new evolutionary based routing protocol for clustered hetergeneous wireless sensor networks. Appl. Soft Comput. **22**(7), 1950–1957 (2011)
4. Haig, Z.: Networked unattended ground sensors for battlefield visualization. ARMS **3**(3), 387–399 (2004)
5. Corke, P., Wark, T., Jurdak, R., Hu, W.: Environmental wireless sensor networks. Proc. IEEE **98**(11), 1903–1917 (2010)
6. Koo, B., Shon, T.: Implementation of a WSN-based structural health monitoring architecture using 3D and AR mode. IEICE Trans. Commun. **93**(11), 2963–2966 (2010)
7. Chen, D., Liu, Z., Wang, L., Dou, M., Chen, J., Li, H.: Natural disaster monitoring with wireless sensor networks: a case study of data-intensive applications upon low-cost scalable systems. Mob. Netw. Appl. **18**(5), 651–663 (2013)
8. Giannecchini, S., Caccamo, M., Shih, C.S.: Collaborative resource allocation in wireless sensor networks. In: Proceedings of the 16th Euromicro Conference Real-Time System, pp. 35–44 (2004)
9. Yu, Y., Viktor, K.P.: Energy-balanced task allocation for collaborative processing in wireless sensor networks. Mob. Netw. Appl. **10**(12), 115–131 (2005)
10. Abdelhak, S., Gurram, C., Ghosh, S., Bayoumi, M.: Energy-balancing task allocation on wireless sensor networks for extending the lifetime. In: Proceedings of the 53rd IEEE International Midwest Symposium Circuits and Systems, pp. 781–784 (2010)
11. Jin, Y.C., Jin, J., Gluhak, A., Moessner, K., Palaniswami, M.: An intelligent task allocation scheme for multihop wireless networks. IEEE Trans. Parallel Distrib. Syst. **23**(3), 444–451 (2012)
12. Yang, J., Zhang, H.S., Ling, Y., Pan, C., Sun, W.: Task allocation for wireless sensor network using modified binary particle swarm optimization. IEEE Sens. J. **14**(3), 882–892 (2014)
13. Guo, W.Z., Li, J., Chen, G.L., Niu, Y.Z., Chen, C.Y.: A PSO-optimized real-time fault-tolerant task allocation algorithm in wireless sensor networks. IEEE Trans. Parallel Distrib. Syst. **26**(12), 3236–3249 (2015)

The Security Architecture and Key Technologies Research of Smart Water Resource

Ping Liu[1(✉)], Yuan-Yuan Wang[2], Wen-Ze Shi[3], and Xin-Chun Yin[3]

[1] College of Information and Engineering,
Yangzhou University, Yangzhou, China
yzuliuping@sina.com
[2] Yangzhou Polytechnic College, Yangzhou, China
[3] Guangling College, Yangzhou University, Yangzhou, China

Abstract. Smart water resource is the advanced stage of the development of water resources informatization. The following construction conceptions of it can meet the development requirements of water resources informatization under the new situation well, such as resource sharing, business collaboration, intelligent applications, etc. The emergence of Internet of Things, Big Data and Cloud Computing has profoundly changed the technical environment of water resources informatization, and brings new challenges in security assurance. On the basis of analysing the current situation and existing problems of water resource informatization and its security assurance system construction, the conception of smart water resource is defined exactly, the overall architecture design scheme of smart water resource is proposed, the main security threats to smart water resource are analyzed in detail, the security architecture of smart water resource is put forward, and also the key technologies of security assurance under the new situation are discussed.

Keywords: Smart water resource · Security architecture · Internet of Things · Cloud Computing · Big Data

1 The Content of Smart Water Resource

Water resources informatization is an important means to serve and support the trinity and mutual coordination concept of "water security, water resources and water environment" [1]. With the appearance of Internet of Things, Big Data and Cloud Computing, the informatization and modernization of water resources also get an opportunity. In the context of constructing smart city, the idea of smart water resource emerged [2].

Smart water resource uses the latest information and communication technologies to trace and monitor the whole process of water circulation and water utilization, realizing the application decision system for more comprehensive sensing, more integrated resources, more intelligent decision and more security protection in

H. Yin et al. (Eds.): IDEAL 2016, LNCS 9937, pp. 484–493, 2016.
DOI: 10.1007/978-3-319-46257-8_52

following links: raw water, production, transportation, emission, processing, utilization, protection, customer service, etc. Smart water resource covers flood control and drought relief, water resource management, soil and water conservation, irrigation, construction of water conservancy and hydropower project, urban and rural water supply, water using, water draining, wastewater treatment and recycling, farmland irrigation and water conservancy, rural hydropower, e-government of water conservancy, water policy supervising, etc. Smart water resource is aimed at improving the water use efficiency, water conservancy project benefit and the working efficiency of water sector, protecting water resources and water environment, preventing disasters and reducing damages, realizing harmony between human and water.

2 The Background and Significance of Security Architecture Research of Smart Water Resource

Water resources informatization brings many security problems as well as openness and convenience. At the initial stage of water conservancy informatization construction, the main target is to realize business function, but the construction of information security assurance system has been ignored, and the water conservancy information system is in danger of information tampering, data stealing, illegal entering all the time.

In March 2010, China's Ministry of Water Resources issued "*The basic technical requirements of the Internet of Water and Information Security system*". The architecture of the Internet of Water and information security system which is called "2 networks, 3 areas, 4 stages" was proposed as well as the security strategy of domain protection according to the classification protection requirements of national information system, and combining the reality of water resources informatization. The security factors of each security area and the related strategy to realize the security factors with high maneuverability were also put forward [3]. The water resources informatization security system is developing and improving gradually. Physical security, network security, system security, application security, data security and management security have been enhanced in a certain extent, but many problems still left, such as the unbalance of security construction, the incomprehensive of security technology measures and the unsound of security management system. Besides, with the introduction of Internet of Things, Cloud Computing, Big Data, etc., many new security risks have been brought to smart water resource.

In order to implement the security assurance work of smart water resource efficiently, advanced and controllable information security technology should be introduced actively based on the overall objectives of smart water resource development and combined with the requirement of information security assurance and the development tendency of information security technology to improve the ability of information security protection. By means of analyzing various security threats faced by the smart water resource, establishing effective running, more and more mature security assurance system is the internal requirement of smart water resource development.

3   The Overall Architecture Design of Smart Water Resource

There are many different views about the overall architecture design of smart water resource [1, 2, 4]. Based on the Internet of Things and Cloud Computing technology, referring to the three layers of Internet of Things [5], smart water resource can be divided into four logic sublayers and two security systems. Following are the four sublayers: sensing layer, transport layer, processing layer and application layer, and two security systems: information security and standard specification security. The overall architecture of smart water resource is shown in Fig. 1.

The sensing layer consists of sensing subsystem and control subsystem. By arranging a large number of sensing, measuring and controlling equipments in target areas, the status, parameters, locations and other information of monitoring targets are collected fully and timely.

The transport layer transports the water data from the sensing layer over Internet, mobile communication networks, the water private network and other networks.

The processing layer separates the network data processing technique, computing technique, middleware technique and other network supporting techniques from the three layers of Internet of Things, and then consolidates them into one layer. The processing layer is divided into three parts: virtual resources management, data center and service components library referring to the architecture of Cloud Computing [6].

The construction of the application layer is based on the requirement of the sectors of water industry, which is a set of maintainable and extensible applications.

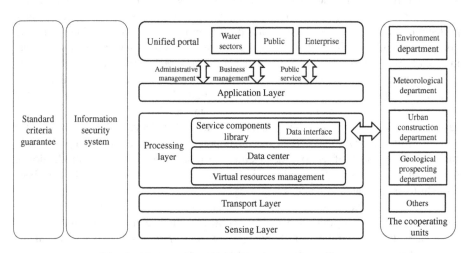

Fig. 1. The overall architecture of smart water resource

4   The Main Security Threats to Smart Water Resource

The main security threats the logic sublayers of smart water resource faces are shown in Table 1.

Table 1. The main security threats to smart water resource

Sublayers	The main security threats
Sensing Layer	① Sensing nodes (common nodes, gateway nodes) security threats: nodes are controlled or captured by enemy; nodes are under attack from the DoS of internet; threats that threaten the mark, recognition and identification of nodes (e.g., nodes disguise, nodes fake or masquerader attack, unable to be safety authentication, access authentication problems of a large number of equipments in a short time); etc.
	② Data transport security threats: interrupting, modification, faking, etc.
	③ Data consistency security threats: data in sensing networks are transferred in broadcasting or multicasting ways and the data integrity protection is limited by the ability of nodes.
	④ RFID security threats
	⑤ Physical damage or human destruction
Transport layer	① Security threats from the large size of Internet of Things: network congestion, DoS attack, DDoS attack, etc.
	② Security threats from the authentication for heterogeneous networks: the transport layer connects different network architectures, and it may be threatened by man-in-the-middle attacks, desynchronization attacks and collusion attacks.
	③ Intrinsical security threats of basic communication network
	④ Security threats of data transmission, consistency and compatibility
Processing layer	① The recognition and processing of mass data from a great deal of terminals
	② Incontrollable smart data processing
	③ Illegal attack by human (i.e., internal attack)
	④ Privacy problems of data: divulging of the data analysis and data processing in a cloud platform.
	⑤ Security problems from the virtualization management: if the data is not encrypted or separated, other users will access illegally.
	⑥ Identity impersonation
Application layer	① Loss of equipment, especially mobile equipment
	② Data privacy threats
	③ Threats from a great deal of terminals and a variety of systems
	④ Access permission threats
	⑤ Information leakage
	⑥ Software leakage
	⑦ Identity impersonation

5   Security Architecture Design of Smart Water Resource

As a very important part of smart water resource, information security assurance system is aimed at ensuring that the informatization platform and each application system can work safely, reliably and effectively in aspects such as system, management, technology, etc.

Based on the overall architecture of smart water resource, aimed at the main security threats the each logic sublayer of smart water service faces, and combined with the requirements of smart water service security and the development tendency of information security technology, the security architecture design scheme has been proposed, as shown in Fig. 2.

6 The Key Technologies of Security Assurance of Smart Water Resource

6.1 Light-Weight Security Technology

By constructing a automated and time-space sensing network, information about rainfall, flood, soil moisture, object operation, water flow, water quality, pipeline pressure, image, pump station operation, and so on is collected, and the basic sensing ability is improved comprehensively.

The smart sensing network is consists of sensor nodes, routing nodes and gate-way nodes, and the networks interconnecting them all. The networks could be short range wireless network (e.g., Zigbee, wifi) and wide area wireless network (e.g., GPS). Generally, sensor nodes include RFID labels, gate-way nodes include RFID reader-writers, and wireless networks include protocols using by RFID.

Being limited by the storage space, computing ability and communication resources, the requirement for security of Smart water sensing nodes is light-weight relatively. But there are potential security risks without any security protection, so light-weight security protections including light-weight cryptographic algorithm [7, 8] and light-weight security protocol [9] are essential, as shown in Table 2.

6.2 Network Security Technology

The transport layer of smart water resource mainly consists of internet, mobile communication network (e.g., GSM, 3G, LTE) and the private network for water sectors. There are a variety of security technologies in internet, such as network monitoring, firewall technique, network access control technique, anti-malware and anti-virus technique, etc. And mobile communication network, telecommunication network and other networks also have their own international standard, so it is not the research emphasis of smart water resource security based on Internet of Things.

6.3 Cloud Computing Security Technology

The service-oriented computing model, dynamic virtualization management style and multi-layer service pattern of Cloud Computing come up with new challenges. Cloud Computing security technology consists of reliable access control, ciphertext retrieval and processing, data existence and availability certification, data privacy protection, virtual security technique, cloud resources access control, trustworthy Cloud Computing, etc. [6, 10]. The key of Cloud Computing security assurance is to apply the

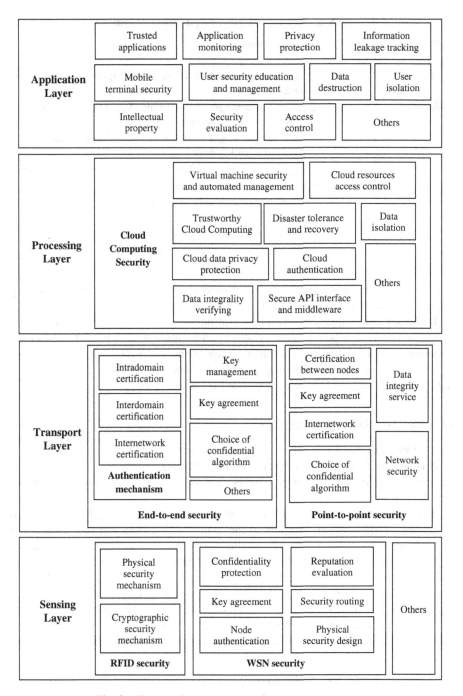

Fig. 2. The security architecture of smart water resource

Table 2. Key technologies of light-weight security

Light-weight Security protections	Key technologies
Light-weight cryptographic algorithm	Design of ultra-lightweight cryptographic algorithm that apply to RFID labels and sensor nodes with limited resource
	Design of light-weight cryptographic algorithm that can be parallelized with hardware
	Design of light-weight cryptographic algorithm that can be parallelized with software
	Design of cryptographic algorithm using unbalanced public-key
Light-weight security protocol	Light-weight security authentication protocol
	Light-weight security authentication protocol and key agreement protocol
	Light-weight authenticated encryption protocol
	Light-weight key management protocol

aforementioned technologies to Cloud Computing environment, forming a key technology system that supports the Cloud Computing security in the future.

Besides, Cloud Computing monitoring technique is needed to realize fast recognition for Cloud Computing security attack, early warning and protecting, and the monitoring of Cloud Computing content, recognizing and preventing from criminal activities based on Cloud Computing, managing technology initiative, preventing from being taken advantage of or being controlled by competitors.

6.4 Big Data Security Technology

After all kinds of water system and application have gathered a great deal of valuable information, they will become targets inevitably. Although attacks aimed at big data of water resource is not shown in papers, we can predict that this will happen unavoidably.

The challenges the big data of water resources face include user privacy protection in Big Data, verification of reliability of the content of Big Data and access control of Big Data. Aimed at the above security challenges, we have some techniques that can be used: anonymity protection technique for data release and SNS, digital watermark technique, data provenance technique, access control for risk self-adaption [11, 12].

While Big Data brings smart water resource new security risks, it also provides new opportunities for smart water resource security system. Big Data provides new possibility for security analysis, it helps to describe the abnormal behavior better in smart water resource for massive water data analysis, and to find the risk in data.

6.5 Privacy Protection Technology

In smart water resource that based on Internet of Things and Cloud Computing, privacy protection technology contains cryptology technique and other technique, including privacy protection for sensing nodes, mobile terminals, Big Data, etc.

The privacy protection for sensing nodes mainly contains data encryption and storage technique, data access control technique and bidirectional authentication technique. Privacy protection for RFID nodes is prominent, such as read access control, label encryption and label authentication [13]. With limited resource, the design of light-weight cryptographic algorithm is one of the most important methods in privacy protection of the Internet of Water [14].

ꞏ Mobile terminals including smartphone and wearable devices have abundant sensing abilities such as camera, microphone, GPS, and kinds of sensors. It can be online anytime in mobile internet environment, bringing enormous convenience for water management, but also creating many new security problems at the same time. The current mobile terminals have the continuous sensing ability for environment [15]. The equipment will turn on smart mode under default setting, while user interaction is needed when turn off. This is a big challenge to traditional access control model [16]. Besides, the sensors in mobile terminals can obtain not only function data of targets but also other environment data and sensitive data of equipments [17, 18].

When mobile terminals are missing, the value of water data that stored in the equipment is likely to eclipse the value of the equipment themselves. However, when the mobile terminals become the control terminals of smart water resource system, the damage of missing the equipment may be far greater than missing the water data. As a result, the privacy protection is an important technology challenge for mobile terminals.

Faced with such a stern fact of mobile terminal privacy protection, the monitoring and protection of privacy leakage have attracted much attention in recent years. The monitoring of privacy leakage experiences a development process of static analysis [19, 20], dynamic analysis [21, 22], combing dynamic and static analysis [23]. The mobile terminal system suppresses the unreasonable using by malware actively is an important way to prevent malware in the protection of privacy leakage [24–26].

The privacy protection technology research under the big data of smart water resource is also a problem that is worth to be in-depth discussed. A typical problem and side-effect caused by overusing big data is the privacy leakage [27]. Under the traditional collection and analysis mode, many privacies come to light by the analysis ability of big data, including the leakage of trade secrets and state secrets.

7   Conclusion

Smart water resource is a booster of improving the development of water industry, which has a vast potential for future development. At the same time, the security technology challenge it faces is unprecedented, so the researchers in water resources domain and information security domain need to work together to find solutions. In this paper, solutions for the security problems of smart water resource are explored from technical angle, and the key technologies related to the security assurance of smart water resource are discussed. In general, the research aimed at the security assurance of smart water resource is insufficient, and only by both technological means and relevant policies and rules, can the security problems of smart water resource be solved better.

Acknowledgments. This study is supported by the National Natural Science Foundation of China (grant No. 61472343).

References

1. Ma, W.Z.: Top Level Design of Water Affairs and Marine Information Technology Architecture (in Chinese). Shanghai Scientific and Technical Publishers, Shanghai (2015)
2. Liu, L.: Study on the construction of urban smart water resource (in Chinese). In: 3rd Water Conservancy Informatization and Digital Water Resources Technology Forum of China, pp. 335–341. Hohai University Press, Nanjing (2015)
3. Zhan, Q.Z.: Ministry of water resources issued "The basic technical requirements of the internet of water and information security system" (in Chinese). Netinfo Security. **6**, 85–86 (2010)
4. Liu, M., Yan, J.Z., Yu, Y.C.: Research on data resources system under Beijing's smart water framework (in Chinese). Water Resour. Informatization **4**, 5–10 (2014)
5. Uckelmann, D., Harrison, M., Michahelles, F.: The Framework of IoT-Internet of Things Technology and Its Impact on Society. Science Press, Beijing (2013)
6. Feng, D.G., Zhang, M., Zhang, Y.: Study on cloud computing security (in Chinese). J. Softw. **22**, 71–83 (2011)
7. Li, W., Gu, D.W., Zhao, C.: Security analysis of the LED lightweight cipher in the internet of things (in Chinese). Chin. J. Comput. **35**, 434–445 (2012)
8. Gong, Z.: Survey on lightweight hash functions (in Chinese). J. Cryptologic Res. **3**, 1–11 (2016)
9. Hou, C.D., Li, D., Qui, J.F., Cui, L.: EasiDEF: a horizontal lightweight data exchange protocol for internet of things (in Chinese). Chin. J. Comput. **38**, 602–613 (2015)
10. Shankarwar, M.U., Pawar, A.V.: Security and privacy in cloud computing: a survey. In: Satapathy, S.C., Biswal, B.N., Udgata, S.K., Mandal, J.K. (eds.) Proc. of the 3rd Int. Conf. on Front. of Intell. Comput. (FICTA) 2014. AISC, vol. 328, pp. 105–112. Springer International Publishing, Switzerland (2015)
11. Feng, D.G., Zhang, M., Li, H.: Big data security and privacy protection (in Chinese). Chin. J. Comput. **37**, 246–258 (2014)
12. Tankard, C.: Big Data Security. Netw Secur. **2012**, 5–8 (2012)
13. Wu, C.K.: An overview on the security techniques and challenges of the internet of things (in Chinese). J. Cryptologic Res. **2**, 40–53 (2015)
14. Guo, Y.M., Li, S.D., Chen, Z.H.: A lightweight privacy-preserving grouping proof protocol for RFID systems (in Chinese). Acta Electronica Sinica **43**, 289–292 (2015)
15. Templeman, R., Korayem, M., Crandall, D., Kapadia, A.: PlaceAvoider: steering first-person cameras away from sensitive spaces. In: 1st Annual Network and Distributed System Security Symposium, San Diego (2014)
16. Roesner, F., Molnar, D., Moshchuk, A., Kohno, T., Wang, H.J.: World-driven access control for continuous sensing. In: 2014 ACM SIGSAC Conference on Computer and Communications Security, Scottsdale, pp. 1169–1181 (2014)
17. Michalevsky, Y., Boneh, D., Nakibly, G.: Gyrophone: recognizing speech from gyroscope signals. In: 23rd USENIX Security Symposium, San Diego, pp. 1053–1067 (2014)
18. Dey, S., Roy, N., Xu, W.Y., Choudhury, R.R., Nelakuditi, S.: AccelPrint: imperfections of accelerometers make smartphones trackable. In: 21st Annual Network and Distributed System Security Symposium, San Diego (2014)

19. Gibler, C., Crussell, J., Erickson, J., Chen, H.: AndroidLeaks: automatically detecting potential privacy leaks in android applications on a large scale. In: Katzenbeisser, S., Weippl, E., Jean Camp, L., Volkamer, M., Reiter, M., Zhang, X. (eds.) Trust and Trustworthy Computing. LNCS, vol. 7344, pp. 291–307. Springer, Heidelberg (2012)
20. Fritz, C.: FlowDroid: a precise and scalable data flow analysis for Android. Master's thesis, TU Darmstadt (2013)
21. Enck, W., Gilbert, P., Han, S., et al.: Taintdroid: an information-flow tracking system for realtime privacy monitoring on smartphones. ACM Trans. Comput. Syst. **32**, 393–407 (2010)
22. Xu, R.B., Saïdi, H., Anderson, R.: Aurasium: practical policy enforcement for Android applications. In: 21st USENIX Conference on Security Symposium, pp. 27–27. USENIX Association Berkeley, California (2012)
23. Yang, Z., Yang, M., Zhang, Y., Gu, G., Ning, P., Wang. X.S.: Appintent: analyzing sensitive data transmission in Android for privacy leakage detection. In: 20th ACM Conference on Computer and Communications Security, pp. 1043–1054. ACM, New York (2013)
24. Zhang, X., Ahlawat, A., Du, W.: A frame: isolating advertisements from mobile applications in Android. In: 29th Annual Computer Security Applications Conference, pp. 9–18. ACM Press, New York (2013)
25. Pearce, P., Felt, A.P., Nunez, G., Wagner, D.: AdDroid: privilege separation for applications and advertisers in Android. In: 7th ACM Symposium on Information, Computer and Communications Security, pp. 71–72. ACM, New York (2012)
26. Shekhar, S., Dietz, M., Wallach D.S.: AdSplit: separating smartphone advertising from applications. In: 21th USENIX Conference on Security Symposium, pp. 28–28. USENIX Association Berkeley, California (2012)
27. CCF Task Force on Big Data: Predictions and interpretations about the development trend of big data in 2016 (in Chinese). Commun. CCF **12**, 40–44 (2016)

Stylized Facts of Linguistic Corpora: Exploring the Lexical Properties of Affect in News

Jason A. Cook, Zeyan Zhao[(⊠)], and Khurshid Ahmad

School of Computer Science and Statistics, Trinity College, Dublin, Ireland
{jcook,zhaoz,kahmad}@scss.tcd.ie

Abstract. Investors are often said to be driven by emotions, and studies in sentiment analysis claim that there is a causal relationship between negative affect in text and prices in financial markets. The text collections used in these studies tend to be of varying sizes and sources, with little justification of their design criteria. This is a classic data engineering problem, which requires specification of the data sources and design of the data repositories and retrieval facilities. In this paper, we explore the statistical properties of negative affect expressed in various textual corpora, differing in specification, size and provenance. The question we ask is whether there are any *stylized facts* of negative affect that are universal across all texts. We observed two main findings: (1) The frequency distribution of negative terms is generally stable across different corpus sizes and (2) The frequency of negative terms accounts for a relatively small proportion of the total terms in the corpus.

Keywords: Text analysis · Corpus linguistics · Linguistic properties · Stylized facts · Sentiment analysis

1 Introduction

Financial markets are often considered to follow a 'random walk', in the sense that the markets deviate from expectations of 'normality' or quasi-random behaviour. Yet despite the apparent random variation in financial instruments, there appears to exist a set of universal statistical properties that are consistent across instruments, markets and time periods. These *stylized facts* include properties such as the lack of autocorrelation of asset returns, non-normal distributions of returns and volatility clustering [1,2].

Markets enable buyers and sellers to act in a degree of harmony for maintaining market efficiency, by allowing both parties to discover the optimal price of goods or services. Overpriced items will deter buyers, and underpriced items will encourage sellers to seek the 'correct' price. However, scholars today (e.g. [3,4]) argue that the market is not always efficient and that at certain times behavioural and psychological factors can affect investor decision making. This can lead to economic bubbles and, in extreme situations, to boom or bust scenarios. Sentiment analysis is about detecting these instances where financial markets

© Springer International Publishing AG 2016
H. Yin et al. (Eds.): IDEAL 2016, LNCS 9937, pp. 494–502, 2016.
DOI: 10.1007/978-3-319-46257-8_53

deviate from the so-called 'norm'. By detecting the behavioural and psychological factors that affect investor decision making, it is believed that we can gain
insight into the investor thought process and discover how asset prices are likely
to be influenced.

One way of detecting investor sentiment is through the analysis of negative
affect in texts. It has been shown by scholars such as [4–6] that an increase in
negative affect in media texts has a downward pressure on market prices. In
other words, an increase in negative investor sentiment is typically followed by
a fall in asset prices and vice versa. The assumption here, of course, is that
investors read the texts that are typically used in these studies.

However, while studies of sentiment analysis have shown promising results,
researchers are often vague about the details and composition of their text collections. Consequently, in this study, we explore the statistical properties of
language typically used in these text collections in an effort to discover any
underlying *stylized facts* of negative affect. We find that while the words used
to convey negative affect typically varies between text collections, overall the
negative affect is relatively stable across different corpus sizes.

2 Literature Review

The stylized facts of financial instruments have been extensively studied in the
economics literature. Typically, researchers have investigated the properties of
returns[1] of an asset or index rather than the raw price, which allows instruments
of different sizes, markets and currencies to be easily compared. [1], for instance,
states that asset returns are non-normal and that this property is consistent
across a variety of instruments (commodities, companies, indices) and time periods. Specifically, it is claimed that returns are more peaked and exhibit heavier
tails than would be found in a normal distribution. [2] shows that returns exhibit
volatility clustering - periods in which high rises (falls) in asset prices are followed by similarly high rises (falls), giving clustered periods of variation. He also
argues that returns are not autocorrelated, in the sense that returns at a given
time period are typically independent of returns at a preceding time period.

Although the phrase 'stylized facts' has almost exclusively been used in the
financial domain, there have also been some studies conducted on the universal
properties of language (typically referred to as 'linguistic laws'). One of the most
well-known examples is Zipf's law, which states that the frequency of a word in
any given corpus is inversely proportional to its rank. The implication of this is
that the total frequency of all words can be represented by only a small number
of unique word types[2]. The distribution of word counts has also been compared
to the distribution of market capitalization [8], where the majority of companies
are of a very small size, and only a few companies are of a very large size.

---

[1] Returns denote the relative change in price of an asset between two different time
periods.

[2] Typically, the first 100 words in a ranked frequency list are responsible for 50 % of
the total words in the text collection [7].

Other examples of textual linguistic laws include the facts that the frequency of a word is inversely proportional to its number of meanings (i.e. more frequent words tend to have a greater number of meanings) and word length is inversely proportional to frequency (i.e. longer words are less frequent).

3   Methods and Data

3.1   Corpus Composition

To explore the distribution of negative affect terms we made use of six linguistic corpora (as shown in Table 1). We distinguish between *general language* corpora, meaning corpora that are representative of a language in general, and *special language* corpora, meaning those that are designed to be representative of a particular language variety (e.g. news, spoken texts, or texts about a particular subject or discipline). We make this distinction on the basis that the distribution of tokens in general language texts is different from the distribution of tokens in special language texts, as we will show empirically later in this paper.

The *general language* corpora used in this study are the Corpus of Contemporary American English (COCA corpus) [9] and the British National Corpus (BNC corpus) [10], which are composed of a variety of text types (e.g. academic, fiction, newspaper, magazines) and are designed to be representative of American and British English respectively. The *special language* corpora consist of four financial news corpora designed for use in sentiment analysis studies. The AofM corpus corresponds to the Abreast of the Market opinion column from the Wall Street Journal and represents financial news in the US market. The LEX column corresponds to the LEX opinion column from the Financial Times newspaper and represents news in the UK market. The Chinese and Danish corpora correspond to financial news in their respective countries, which was translated

Table 1. Composition of the corpora used in this study. COCA and BNC denote the Corpus of Contemporary American English and British National Corpus respectively. AofM represents the Abreast of the Market opinion column from the Wall Street Journal (American English). LEX represents the LEX opinion column from the Financial Times newspaper (British English). Danish and Chinese represent English-translated financial news from Denmark and China (possibly American English).

Corpus type	Name	Size (Words)	Dates	Reference
General language	COCA	450 million	1990–2012	[9]
	BNC	100 million	1975–1993	[10]
Special language	AofM	4.7 million	1984–1999	[4]
	LEX	7.1 million	2004–2015	[11]
	Danish	2.7 million	2002–2015	[12]
	Chinese	17.4 million	2000–2015	[13]

into English upon publication. In each case, the negative sentiment in the special language corpora has been shown to have a significant impact ($p < 0.05$) on the respective countries' market returns at the aggregate (index) level (the associated studies are shown in the reference column of Table 1).

3.2 Measuring Affect

In constructing our proxy of negative affect, we used a list of 2,291 negative terms from the General Inquirer Harvard IV dictionary [14]. This dictionary has been used in a number of sentiment analysis studies (e.g. [4, 15–19]). To measure affect, we use a content analysis program called Rocksteady, which tokenizes a given text into words and, for each word, performs a dictionary look-up against a pre-defined list of words - here, the negative terms. The output of Rocksteady is a list of the frequency counts of the number of negative terms appearing in each document. We subsequently normalise these frequencies by dividing them by the total number of words in each document and multiplying them by 10^4. The result of this normalisation gives counts of negative terms per 10,000 words of text, allowing comparison between texts of varying sizes.

4 Results

4.1 Summary Statistics

Historically, studies on sentiment analysis have used corpora of varying sizes, with little justification as to why this size was picked. In this section we explore how increasing the size of the corpus affects the distribution of negative terms using $2^7, 2^8, ..., 2^{12}$ documents (corresponding to approximately 6 months, 1 year, ... , 16 years of trading days). The results of this analysis are shown in Table 2.

While the mean and standard deviations clearly differ *between* corpora, they are relatively persistent *within* the corpora at different sizes. Increasing the size of a given corpus thus has little impact on the mean frequency or variability of negative terms. Overall, the mean number of negative words across all corpora is around 2.14 %, with the LEX corpus showing a relatively high average term frequency of near 3.6 %, and the Danish Corpus showing a relatively low frequency of about 1.3 %. By comparison, the mean percentage of returns is almost zero (around $1.45 * 10^4$). The standard deviation also shows substantial variation between corpora, varying from around 0.61 in the BNC and AofM corpora to around 1.61 in the LEX corpus.

Like financial returns, the distribution of negative terms is non-normal, generally showing higher skewness and kurtosis values than would be expected in a normal distribution. Shapiro-Wilk tests for normality rejected the null hypothesis of normality in all but two cases ($p < 0.001$)[3].

---

[3] The Shapiro-Wilk statistic for the AofM and COCA corpora at 128 documents was not significant at $p < 0.001$.

Table 2. Summary statistics for the relative frequency of negative terms across different corpora of varying sizes. Corpus size represents the number of documents in each corpus, ranging from $2^7, 2^8, ..., 2^{12}$ documents, which corresponds to approximately 6 months, 1 year, ... , 16 years of trading days. The mean frequencies are multiplied by 10^4, to give the frequency of negative terms per 10,000 words. The standard deviation units are multiplied by 10^2.

Summary statistics						
Corpus type	Corpus name	Corpus size	Mean $\times 10^4$	St.Dev $\times 10^2$	Skewness	Kurtosis
General Language	BNC	128	108	0.69	3.23	19.51
		256	107	0.58	1.18	2.89
		512	113	0.58	1.06	2.10
		1024	110	0.59	1.13	3.18
		2048	110	0.59	1.41	6.16
		4096	110	0.60	1.41	5.07
	COCA	128	247	1.05	0.97	2.06
		256	280	1.11	0.57	1.38
		512	247	1.11	0.81	1.10
		1024	253	1.07	0.65	1.26
		2048	255	1.06	0.58	1.04
		4096	276	1.09	0.69	1.48
Special Language	AofM	128	189	0.52	0.33	3.74
		256	200	0.59	0.59	4.02
		512	198	0.59	0.46	3.66
		1024	200	0.66	0.62	3.98
		2048	210	0.68	0.65	3.80
		4096	211	0.68	0.58	3.68
	LEX	128	396	1.78	0.87	3.31
		256	378	1.62	0.71	3.63
		512	358	1.51	0.61	3.84
		1024	352	1.56	0.74	4.57
		2048	352	1.61	0.78	4.63
		4096	343	1.55	0.71	4.47
	Danish	128	96	1.11	2.03	9.85
		256	91	1.07	1.91	8.76
		512	95	1.03	1.63	7.43
		1024	99	1.01	1.32	5.52
		2048	104	0.99	1.31	5.48
		4096	115	1.11	1.54	6.62
	Chinese	128	179	1.02	1.90	10.74
		256	169	0.94	1.49	8.80
		512	169	0.95	1.33	7.43
		1024	177	0.96	1.08	6.14
		2048	181	0.96	1.03	5.64
		4096	189	1.04	1.05	5.19
	Mean		214	1.12	1.04	5.43

4.2   Word Frequencies

We have already shown that the mean distribution of negative terms differs from corpus to corpus. In this section we explore these differences further, by comparing the ranks of the top 15 most frequent negative words[4]. Table 3 shows relative frequency correlations of the negative terms across the six corpora, and Table 4 shows how the negative terms are distributed in each corpus.

Table 3. Relative frequency correlation matrix of the top 15 most frequent negative terms across each corpus. Each corpus contains 2^{12} (4096) documents, corresponding to approximately 16 years of trading days. Relative frequency correlations are expressed in percentage terms.

	COCA	BNC	AofM	LEX	Danish	Chinese
COCA	100.0					
BNC	99.0	100.0				
AofM	−0.8	−3.2	100.0			
LEX	10.3	12.0	−37.2	100.0		
Danish	4.8	4.0	−11.3	19.9	100.0	
Chinese	30.5	29.1	3.8	−10.4	0.3	100.0

From Table 3 we see that the relative frequencies of the negative terms in the BNC and COCA corpora exhibit a strong correlation of 99 %. The AofM corpus exhibits a negative correlation with all but the Chinese corpus, however the anti-correlation of (−37.2 %) with the LEX corpus is particularly strong. This latter finding is surprising, given that both the AofM and LEX corpora are comprised of opinion columns. The Chinese corpus exhibits strong correlations with both the COCA and BNC corpora, but no correlation with the Danish corpus, despite the fact that both the Chinese and Danish corpora were constructed using a similar methodology and by the same author. The Danish and LEX corpus exhibit a relatively strong correlation, perhaps due to the fact that both news sources are situated in Europe and thus likely to be reporting some similar events.

Table 4 shows that, overall, the negative terms are relatively infrequent across the six corpora, accounting for less than 1 % of total words in all but the AofM corpus. The table also shows that negative words are more frequent in the special language corpora (i.e. the AofM, LEX, Danish and Chinese corpora) compared to the general language corpora (i.e. the COCA and BNC corpora), as observed by the first and fourth columns of the table. This could be due to the fact that these special language corpora comprise solely of news texts, which often focus on negative events and sensationalist stories. The high proportion of negative

---

[4] Namely the terms 'down', 'loss', 'close', 'drop', 'lose', 'against', 'crisis', 'decline', 'problem', 'concern', 'cut', 'risk', 'decline', 'inflation' and 'drop'. These words were selected based on their common occurrence across all 6 corpora.

Table 4. Rank information for the top 15 most frequent negative terms. Each corpus contains 2^{12} (4096) documents. The First and Last Neg. columns correspond to the positions of the first and last negative terms in the ranked frequency list. The Most Neg. column corresponds to the rank bin where the majority of negative terms are located. The Percentage of Total column represents the proportion of negative terms as a total of all words in the entire corpus.

	First Neg.	Last Neg.	Most Neg.	Percentage of total
COCA	186	741	300–400	0.3 %
BNC	168	834	600–700	0.3 %
AofM	60	493	100–200	1.5 %
LEX	125	430	200–300	0.6 %
Danish	85	567	300–400	0.9 %
Chinese	47	567	400–500	0.7 %

terms in the AofM corpus is also likely due to the pragmatic nature of the source: the Abreast of the Market column is written as a post-mortem of the US stock market on the previous day, describing the 'winning' and 'losing' stocks and significant market events.

5   Conclusions

In this paper, we have found that the frequency distribution of negative affect is not normal and exhibits positive skewness across corpora of varying sizes and texts. We have also discovered that negative affect is relatively stable (in terms of mean frequency and variance) across corpora of varying sizes, suggesting that more data will not necessarily alter the distribution of negative terms. This is important for computing the impact of negative terms especially when using econometric statistics such as moving averages or rolling regression which may rely upon smaller subsets of the entire data set. An additional implication of our study is that documents appear to contain some 'background' negative affect that is present in all documents, as observed by the non-zero mean for all corpora. Finally, we discovered that negative terms are relatively infrequent, with the top 15 words in each corpus accounting for less than 1 % of total words. Future work will consist of exploring alternative stylized facts of negative affect and relating these back to market fundamentals. One such property we are keen to investigate is that of autocorrelation, and we are interested to know whether there are certain periods where sentiment clusters in the same manner that financial returns do. The discoveries in this paper should also aid researchers in constructing corpora for a particular domain (e.g. economics, politics and disaster corpora).

Acknowledgments. J.A. Cook—This research is supported by Science Foundation Ireland through the CNGL Programme (Grant 07/CE/I1142) in the ADAPT Center (www.adaptcentre.ie) at Trinity College, University of Dublin. Z. Zhao—The research leading to these results has also received funding from the EU FP7 Slandail project under grant agreement no. 607691. In this study we used the text analysis system Rocksteady, developed as part of the Faireachain project for monitoring, evaluating and predicting behaviour of markets and communities (2009–2011). Support for Rocksteady's development was provided by Trinity College, University of Dublin and Enterprise Ireland (Grant IP-2009-0595).

References

1. Cont, R.: Empirical properties of asset returns: stylized facts adn statistical issues. Quant. Finance **1**, 223–236 (2001)
2. Taylor, S.J.: Asset Price Dynamics, Volatility, and Prediction. Princeton University Press, Princeton (2011)
3. Shiller, R.J., Perron, P.: Testing the random walk hypothesis: power versus frequency of observation. Econ. Lett. **18**(4), 381–386 (1985)
4. Tetlock, P.C.: Giving content to investor sentiment: the role of media in the stock-market. J. Finance **62**(3), 1139–1168 (2007)
5. Garcia, D.: Sentiment during recessions. J. Finance LXVIII **3**, 1267–1300 (2013). doi:10.1111/jofi.12027
6. Antweiler, W., Frank, M.Z.: Is all that talk just noise? the information content of internet stock message boards. J. Finance **59**(3), 1259–1294 (2004)
7. Ahmad, K.: Being in text and text in being: notes on representative texts. In: Andeman, G., Rogers, M. (eds.) Incorporating Corpora, pp. 60–91. Multilingual Matters, Clevedon (2008)
8. Loughran, T., McDonald, B.: The use of word lists in textual analysis. J. Behav. Finance **16**(1), 1–11 (2015)
9. Davies, M., The corpus of contemporary american english: 450 million words, 1990-present (2008)
10. British National Corpus. Oxford University, Humanities Computing Unit, New York (2000)
11. Kelly, S.: Signs of irrational exuberance: an investigation into the role of news and sentiment in finance. Ph.D. thesis, Trinity College, University of Dublin (2015)
12. Zhao, Z., Ahmad, K.: Qualitative and quantitative sentiment proxies: interaction between markets. In: Jackowski, K., Burduk, R., Walkowiak, K., Woźniak, M., Yin, H. (eds.) IDEAL 2015. LNCS, vol. 9375, pp. 466–474. Springer, Heidelberg (2015). doi:10.1007/978-3-319-24834-9_54
13. Zhao, Z., Ahmad, K.: A computational account of investor behaviour in chinese and US market. Int. J. Econ. Behav. Organ. **3**(6), 78–84 (2015)
14. Stone, P.J., Dunphy, D.C., Smith, M.S., Olgilvie, D.M., with associates: The General Inquirer: A Computer Approach to Content Analysis. The MIT Press, Cambridge (1966)
15. Esuli, A., Sebastiani, F.: Sentiwordnet: a publicly available lexical resource for opinion mining. In: Proceedings of LREC, vol. 6, pp. pp. 417–422. Citeseer (2006)
16. Pang, B., Lee, L.: Opinion mining and sentiment analysis. Found. Trends Inf. Retrieval **2**(1–2), 1–135 (2008)
17. Loughran, T., McDonald, B.: When is a liability not a liability. J. Finance **66**, 35–65 (2011)

18. Cook, J.A., Ahmad, K.: Behaviour and markets: the interaction between senti-ment analysis and ethical values? In: Jackowski, K., Burduk, R., Walkowiak, K., Woźniak, M., Yin, H. (eds.) IDEAL 2015. LNCS, vol. 9375, pp. 551–558. Springer, Heidelberg (2015). doi:10.1007/978-3-319-24834-9_64

19. Kelly, S., Ahmad, K.: The impact of news media and affect in financial markets. In: Jackowski, K., Burduk, R., Walkowiak, K., Woźniak, M., Yin, H. (eds.) IDEAL 2015. LNCS, vol. 9375, pp. 535–540. Springer, Heidelberg (2015). doi:10.1007/978-3-319-24834-9_62

Positive Influence Maximization Algorithm Based on Three Degrees of Influence

WeiHua Lei[1,2], Qun Yang[1,2($\boxtimes$)], and Huanhuan Wang[1,2]

[1] College of Computer Science and Technology, Nanjing University of Aeronautics and Astronautics, Nanjing, China
{leiweihuabao,qun.yang}@nuaa.edu.cn
[2] Collaborative Innovation Center for Novel Software and Industrialization, Nanjing, China

Abstract. Influence maximization aims to find a subset of nodes in social networks and make the propagation of their influence maximized. Usually, greedy algorithms for LT model have long execution time. To solve this problem, based on Three Degrees of Influence Rule (TDIR) we proposed a heuristic algorithm TDIA. We used LT-A model and change the formula of attitude weight in the model by considering the impact of three degrees of influence on attitude. We conducted extensive experiments on two real-world signed social network datasets and the experiment results showed that TDIA has much shorter execution time than LT-A Greedy algorithm and its positive influence spread is close to the greedy algorithm.

Keywords: Positive Influence Maximization · Three Degrees of Influence Rule · Heuristic algorithm · Signed social networks

1 Introduction

Influence Maximization (IM) problem is a hot topic in the field of social network. The aim of influence maximization problem is to find several influential users that will lead to the maximize propagation in social networks.

Kempe et al. first introduced the IM problem in [11]. They proposed Independent Cascade(IC) model and Linear Threshold (LT) model for it. They also proved that the influence spread function is monotonous and sub-modular and the IM problem is NP-hard. Hence, it can be solved by using greedy algorithm and can gain an approximate optimal solution with $(1-1/e)$ ratio.

Although many works have been done on IM problem, they ignore the attitude of the users. In [18], Wang considered this problem and put forward the Positive Influence Maximization (PIM) problem in signed social networks. They proposed LT-A model, they also proved the PIM problem is NP-hard and the influence spread function is monotonous and sub-modular. To get the solution to PIM, they presented the LT-A Greedy algorithm, but the time complexity of the algorithm is high and the scalability is not good. It is not suitable for large-scale social networks.

© Springer International Publishing AG 2016
H. Yin et al. (Eds.): IDEAL 2016, LNCS 9937, pp. 503–514, 2016.
DOI: 10.1007/978-3-319-46257-8_54

To solve the problem of PIM, in this paper, we proposed a heuristic algorithm TDIA. TDIA is based on Three Degrees of Influence Rule (TDIR), which is proposed by Christakis [6] in 2009. It is an influence propagation rule in social networks which states that in the social network the users will be affected by others within three degrees of influence. By applying TDIR theory in signed networks, in this paper, our contributions are showed as follows.

(1) We analyzed the TDIR in social networks and proposed a heuristic algorithm, named Three Degrees of Influence heuristic Algorithm (TDIA) to solve PIM problem.
(2) We modified the LT-A model with changing the formula of attitude weight by as considering the impact of TDIR.
(3) We conducted experiments on two real-world signed social networks and evaluate the experiment results.

The rest of this paper is organized in the following. In Sect. 2, we discussed the related work. In Sect. 3, we proposed TDIA by the analysis of TDIR in social networks. In Sect. 4, we defined the PIM problem and change the formula of attitude weight in LT-A model. In Sect. 5, we presented our experiments and analyze the experiment results. In Sect. 6, we gave the conclusion.

2 Related Work

Domimgos et al. [7] first investigated the IM problem and considered it as a probability problem. Later, Kempe et al. [11] studied the problem as a discrete optimization problem. They proposed two propagation models: IC model and LT model, they use greedy algorithm to solve it with a ratio of $(1-1/e)$ approximation.

Traditional IC model and LT model ignored the user's attitudes, if the negative attitude be ignored the influence in social network will be over-estimated, while the positive influence is also very important in many cases such as viral marketing. To better solve the influence maximization problem in social network, both positive attitudes and negative attitude should be considered.

Chen et al. [3] first took the negative influence into account, they proposed IC-N model by adding the quality factor parameter to reflect the attitude influence between users. But they didn't take trust relationship into account. Furthermore, in IC-N model the user's attitude couldn't be changed.

In [19], the individual preference was taken into account, by regarding that the users can have positive attitudes, negative attitudes and neutral attitudes due to their preferences and that attitudes of users can be changed by influence of their friends, Zhang et al. proposed a two-phase model named OC model. In [14], by considering the influence is positive or negative based on the relationships between users, Li et al. put forward a trust threshold parameter to determine the relationships between users, and proposed the LT-IO model together with a heuristic algorithm to solve the IM problem. However, they defined the equivalent trust threshold for each edge in social networks, which is not the real case in our everyday life.

Later, as the signed social networks can obviously reflect the relationships between users through the sign on edges, which can reduce the inaccuracy problem caused by parameters, in [18], Wang et al. introduced signed social networks into PIM problem and proposed the LT-A model. In LT-A model, the influence propagation function is monotonous and sub-modular and the PIM problem is NP-hard under the model. They presented LT-A Greedy algorithm. But the greedy algorithm has a long execution time and serious scalability problem, which is not suitable to apply in large-scale social networks.

To improve the efficiency of the algorithm, many studies have been done [8,12,20], among which some proposed efficient heuristic algorithms [4,5,9,10, 15,16], such as Degree Discount algorithm, PMIA and LDAG etc. Although heuristic algorithms can overcome the inefficient shortcoming of greedy algorithm, yet the accuracy is lower than that of greedy algorithms and they lack sufficient guarantee in theory. Christakis et al. [6] proposed TDI theory in social networks and a large number of experiments and analysis of data in social networks have demonstrated the TDI theory. For example, in 2011, Facebook and Milano University studied the friend relationship in 7 million users, they found that the average distance between users was 4.74 degrees, the average distance will decrease and it will be close to three degrees with the increase of social network scale. In [17], Qin et al. proposed to use TDI theory to solve the IM problem. They proposed TSCM (Three Steps Cascade Model) and TL Greedy algorithm. They proved the effectiveness of their algorithm, but they haven't mentioned the scalability problem. In this paper, we propose the Three Degrees of Influence heuristic Algorithm (TDIA) based on TDI theory in LT-A model. The algorithm can be used to solve the scalability problem and TDIA has much shorter execution time than LT-A Greedy algorithm.

3   Three Degrees of Influence Heuristic Algorithm

3.1   Formulating TDIA

An unsigned social network is modeled as a directed graph $G = (V, E, w)$, where V is the set of nodes, and E is the set of directed edges. Nodes and edges respectively represent users and relationships between users in the social networks. w is a weight function on edges and it satisfies $w \in [0,1]$, $w_{(u,v)}$ means the influence from u to v.

For each node in the graph G, $N1(v)$ denotes an out-degree neighbors set of node v, $N2(v)$ denotes an out-degree neighbors set of set $N1(v)$ and $N3(v)$ denotes an out-degree neighbors set of set $N2(v)$. Therefore, the value of $INF1(v)$, $INF2(v)$ and $INF3(v)$ can be calculated by the following formulas. $INF1(v)$ means the first degree influence of node v, $INF2(v)$ means the second degree influence of node v and $INF3(v)$ means the third degree influence of node v.

$$INF1(v) = |N1(v)| \qquad (1)$$

$$INF2(v) = |N2(v)| \tag{2}$$

$$INF3(v) = |N3(v)| \tag{3}$$

A signed social network is modeled as a directed and signed graph $G = (V, E, w, p)$, where p represents the sign of each edge, its value can be 1 or -1,1 means the positive (trust or friend) relationship, -1 means the negative (distrust or enemy) relationship. The meaning of V, E and w is the same as that in unsigned social networks.

We give three definitions in the signed social networks as follows:

Definition 1. *For each node $v \in V$ in graph $G = (V, E, w, p)$, $DF1(v)$ denotes the friend node set of v in set $N1(v)$, $DF2(v)$ denotes the friend node set of v in set $N2(v)$ and $DF3(v)$ denotes the friend node set of v in set $N3(v)$. Considering the characteristics of the signed social networks, their values can be calculated by the formula $(4 - 6)$.*

$$DF1(v) = \{u|p_{(v,u)} = 1, u \in N1(v)\} \tag{4}$$

$$DF2(v) = \{u|p_{(v,u)} = 1, u \in N2(v)\} \tag{5}$$

$$DF3(v) = \{u|p_{(v,u)} = 1, u \in N3(v)\} \tag{6}$$

As the negative relationships between nodes represent distrust or enemy relationships in signed social networks, the Definition 2 can be given as following:

Definition 2. *For each node $v \in V$ in graph $G = (V, E, w, p)$, $DE1(v)$ denotes the enemy node set of v in set $N1(v)$, $DE2(v)$ denotes the enemy node set of v in set $N2(v)$ and $DE3(v)$ denotes the enemy node set of v in set $N3(v)$. Considering the above description, their values can be calculated by the formula $(7 - 9)$.*

$$DF1(v) = \{u|p_{(v,u)} = -1, u \in N1(v)\} \tag{7}$$

$$DF2(v) = \{u|p_{(v,u)} = -1, u \in N2(v)\} \tag{8}$$

$$DF3(v) = \{u|p_{(v,u)} = -1, u \in N3(v)\} \tag{9}$$

In signed social networks, the positive relationship can promote the influence propagation, while the negative relationship can block the influence propagation. Thus, the value of $INF1(v)$ $INF2(v)$ and $INF3(v)$ in signed social networks can be calculated by the following formulas.

$$INF1(v) = |DF1(v)| - |DE1(v)| \tag{10}$$

$$INF2(v) = |DF1(v)| - |DE2(v)| \tag{11}$$

$$INF3(v) = |DF1(v)| - |DE3(v)| \tag{12}$$

According to the TDI theory, one user can be influence his friends, friends of his and friends of friends of his, we give the Definition 3 as following:

Definition 3. *For any node $v \in V$ in graph $G = (V, E, w, p)$, the TDI value of node v is defined as the sum of influence within three degrees, as shown in formula (13).*

$$TDI(v) = INF1(v) + INF2(v) + INF3(v) \tag{13}$$

Based on the Definition 3, we can calculate the TDI value in unsigned social networks by the formula (14).

$$
\begin{aligned}
TDI(v) &= INF1(v) + INF2(v) + INF3(v) \\
&= |D1(v)| + |D2(v)| + |D3(v)|
\end{aligned} \tag{14}
$$

The TDI value in signed social networks can be calculated by the formula (15).

$$
\begin{aligned}
TDI(v) = \; & INF1(v) + INF2(v) + INF3(v) \\
= \; & |DF1(v)| - |DE1(v)| + |DF2(v)| - |DE2(v)| \\
& + |DF3(v)| - |DE3(v)| \\
= \; & \sum_{u \in N1(v)} p(v, u) + \sum_{u \in N2(v)} p(v, u) + \sum_{u \in N3(v)} p(v, u)
\end{aligned} \tag{15}
$$

We give an example of to show how to calculate TDI value, it shows that different type networks have different TDI values.

In Fig. 1, $DF1(a) = 3$ and $DE1(a) = 1$, so the first degree influence $INF1(a) = 3 - 1 = 2$; as $DF2(a) = 4$ and $DE2(a) = 2$, the second degree influence $INF2(a) = 4 - 2 = 2$; as $DF3(a) = 1$ and $DE3(v) = 0$, the third degree influence $INF3(a) = 1 - 0 = 1$. According to the formula (15), $TDI(a) = 2 + 2 + 1 = 5$. Thus, the TDI value of other nodes can be calculated in accordance with the above calculating process.

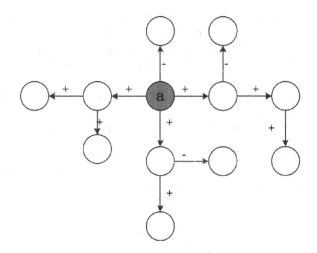

Fig. 1. An example of signed social network including 12 nodes

3.2   Three Degrees of Influence Heuristic Algorithm

In previous section, we have given the calculation method of TDI value. According to the TDI theory, the node has much more neighbors within three degrees, it will have higher influence. Based on TDI, we give the Three Degrees of Influence heuristic Algorithm (TDIA) in which we first calculate the TDI value of each node by the social network structure. Then we select the top-k nodes of TDI value as seeds. The pseudo algorithm of TDIA is shown in Algorithm 1.

Algorithm 1. Three Degrees of Influence heuristic Algorithm

Input: Graph $G = (V, E, w, p), k \in N, \theta, \eta$
Output: A seed set S of size $k, v \in S, \eta_v > 0$
 1: **Initialization** $S = \emptyset$
 2: **for** $i = 1$ to $|V|$ **do**
 3:    **for** each node $v \in V$ **do**
 4:       calculating the value of $DF1, DF2, DF3$ and the value of $DE1, DE2, DE3$
 5:       calculating the value of TDI for each node according to Definition 3
 6:    **end for**
 7:    sorting the value of TDI
 8:    adding the nodes of top-k TDI value into set S
 9: **end for**
10: return S

---

4   Positive Influence Maximization

4.1   LT-A Model

In this paper, we use the LT-A model. In the model, each node has two parameters: threshold and attitude weight, which respectively represents the ability to be influenced and attitude; each edge has two parameters: influence weight and relationship weight, which respectively represents the ability of influence and relationships between nodes.

The propagation process of the LT-A model is in the following. At time $t = 0$, the attitude weights of seeds are set to 1, and those of the others are set to 0. At time $t > 0$, if one inactivated node u satisfies the formula $\sum_{u \in A_u^t} w(v, u) \geq \theta_u$, it will be activated. Then, its attitude weight will be changed by the influence of its active neighbors. The process ends until no more activation happens.

The parameters of the model satisfy the following criteria.

Given $v \in V$, $w(u, v) \in [0, 1]$, where

$$\sum_{v \in V/v} w(v, u) \leq 1 \tag{16}$$

The value of $p_{(u,v)}$ is 1 or -1, $p_{(u,v)} = 1$ indicates the positive relationship between the two nodes and node v trust node u; $p_{(u,v)} = -1$ indicates the negative relationship between them and node v distrust node u. $\theta_v \in [0,1]$ and it denotes threshold. $\eta_v \in [-1,1]$ and it denotes the attitude weight.

In the propagation process, node can be activated if $\sum_{u \in A_u^t} w(v,u) \geq \theta_u$. In LT-A model, $\eta_v > 0$ means the attitude of node v is positive, $\eta_v < 0$ means the attitude of node v is negative, if $\eta = 0$, we assume that the node holds negative attitude through the negative dominance rule in social networks [62–65].

As three degrees of influence is applicable to the propagation of attitude, emotion and behavior, the formula for attitude weight will be changed by taking three degrees of influence into account. The original formula of calculating attitude weight in the LT-A model is as follows.

$$\eta_v^t = \sum_{v \in A_v^t} \eta_u * p_{(u,v)} * w(v,u) \tag{17}$$

The formula denotes the attitude weight of a node is the sum of the product of influence weight and relationship weight in its activated neighbors. Considering the three degrees of influence, the attitude will be influenced by their friends within three degrees. Thus, the new formula can be achieved as follows.

$$\eta_1 = \sum_{v \in A_v^t} \eta_u * p_{(u,v)} * w(u,v) \tag{18}$$

$$\eta_2 = \alpha_1 \sum_{m \in A_v^{D2}} \eta_m * w(m,u) * w(u,v) \tag{19}$$

$$\eta_3 = \alpha_2 \sum_{n \in A_{D3}^t} \eta_n * w(n,m) * w(m,u) * w(u,v) \tag{20}$$

$$\eta_v^t = \eta_1 + \eta_2 + \eta_3 \tag{21}$$

In the above formulas, A_{D2}^t denotes a set of activated nodes in the second degree neighbors of node v. A_{D3}^t denotes a set of activated nodes in the third degree neighbors of node v. In addition, η_1 denotes the attitude weight value of node v, which is influenced by its neighbors; η_2 denotes the attitude weight value of node v, which is influenced by its second degree neighbors; η_3 denotes the attitude weight value of node v, which is influenced by its third degree neighbors. Thus, the final attitude weight of node v is the sum of η_1, η_2 and η_3.

The parameters α_1 and α_2 denote decay factors of attitude influence, because the propagation of influence in social networks like a stone threw in the lake and it will vanish. The situation indicates the natural attenuation. Furthermore, the parameters α_1 and α_2 satisfy the conditions: $0 < \alpha_2 < \alpha_1 < 1$. In the model, the attitude weight $\eta_v \in [-1,1]$, we assume that if $\eta_v > 1$ we set η_v to 1, if $\eta_v < -1$ we set it to -1.

4.2 The Definition of PIM

The definition of the PIM problem in the LT-A model is as follows:

Definition 4. *(PIM problem) Given a directed graph $G = (V, E, w, p)$, in which the state of each node can be active with positive attitude, active with negative attitude and inactive, a positive integer k, $k \leq |V|$, and a propagation model M. PIM problem is to find a k-size set of seed nodes which can be used to maximize the number of the influenced nodes with positive attitude based on the model M in the given graph G.*

The PIM problem is formalized as:

$$S = max_{S \in V, |S|=k} \delta(S) \tag{22}$$

In the formula, δ is the influence spread function. Given an initial node set S, the value of $\delta(S)$ is the expected number of nodes which is activated with positive attitude by S in the LT-A model.

The process of TDIA is in the following on the basis of the above description. Firstly, we set the k-size seed with positive attitude and their attitude weights are set to 1. Then the influence of seed spread in the propagation model, and the attitude weights of other nodes can be calculated by the given formulas. Finally, we achieve the number of nodes which is with the positive attitude in the end of the propagation process.

5 Experiments

5.1 Datasets and Experiment Environment

Two real-world signed social network datasets are applied in the experiments, which are Epinions and Slashdot. The datasets can be downloaded from Standard Large Network Dataset Collection in [1,2]. The statistics of the two datasets is showed in Table 1.

Table 1. Statistics of data set

Dataset	Epinions	Slashdot
Node	131828	77350
Edge	841372	516575
Average Out-degree	6.38	6.68

Epinions: a product review social network where users trust or distrust others by their ratings and reviews of products.

Slashdot: a technology-related news network where users can submit their comments of current news. The network has Friend and foe relationships between users.

Table 2. Statistics of the sever

Name	Windows Sever 2008 R2
CPU	Intel Xeon E5420
Cores	8
Frequency	2.50 GHz
RAM	16 GB

The experiments implement the algorithm in C++ language on Microsoft Visual Studio 2012 platform, and they run on Windows Server 2008 R2. The statistics of Server is showed in Table 2.

5.2 The Design of Experiments

The parameters in the experiments are set as follows. We assume that $\alpha_1 = 0.4$ and $\alpha_2 = 0.1$ because the value of decay factors will diminish with the increase of distance and the influence will vanish in the end. For each node, the value of θ can be achieved between 0 and 1 randomly, as stated in Sect. 3, the attitude weight η of seed nodes is set to 1, and the attitude weights of other nodes can be calculated by the formula (18–22). For each edge, the value of influence weight w can be achieved between 0 and 1 randomly, which must be satisfied: $\sum_{u \in V/v} w(v, u) \leq 1$. The value of relationship weight p can be achieved from the real-world datasets. As large number of seed nodes can bring into long execution time and previous work is always set $k = 50$, we set the number of seed nodes k to be 50. Meanwhile, we make comparisons of the positive influence spread in different sizes of seeds.

We conduct experiments on the above real-world datasets. Then we evaluate the experiment results in positive influence spread and running time of algorithms. The compared algorithms are as follows.

LT-A Greedy algorithm: A greedy algorithm with CELF optimization based on LT-A model [18].

Positive Out-Degree algorithm: A heuristic algorithm of selecting top-k largest positive out-degree nodes as a seed set [13].

Random algorithm: A heuristic algorithm of selecting k nodes as a seed set randomly [13].

TDIA: A heuristic algorithm proposed in this paper.

5.3 Experiment Results

At first, we compare the positive influence spread in different algorithms on both datasets. Figure 2 respectively show the positive influence spread of different algorithms in real-world datasets: Epinions and Slashdot. As expected, we can achieve that the positive influence spread of LT-A Greedy algorithm is higher

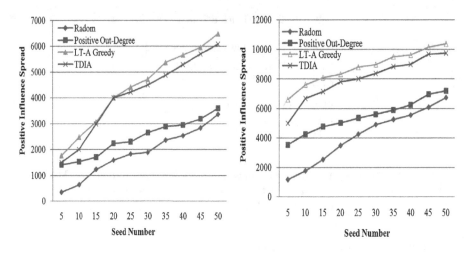

Fig. 2. Positive influence spread on Epinions and Slashdot

than the other compared algorithms. But the result of TDIA we proposed is closed to that of greedy algorithm and it outperforms other heuristic algorithms. In the first figure of Fig. 2, the positive influence spread of TDIA is about 5 % lower than that of LT-A Greedy algorithm, the positive influence spread of other algorithms is about 20 % lower than that of LT-A Greedy algorithm. As shown in second figure of Fig. 2 the positive influence spread of TDIA is also closed to that of LT-A Greedy algorithm and it significantly outperforms that of other algorithms.

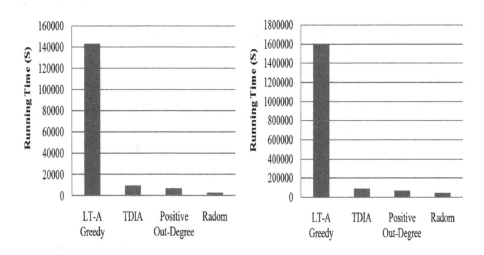

Fig. 3. Running time for selecting 50 seeds on Epinions and Slashdot

Figure 3 show the running times of different algorithms. From the two figures, we can get that the LT-A Greedy algorithm has the longest execution time, because the greedy algorithm has low efficiency. Moreover, the execution time of TDIA is closed to that of Positive Out-degree algorithm. And the two figures show the tendency obviously. Compared to the running time of the LT-A Greedy algorithm and TDIA, it is obvious that the TDIA has higher efficiency than LT-A Greedy algorithm. The other two heuristic algorithms have shorter execution time than TDIA, but their positive influence spread is much lower than that of it.

The experiment results show that the execution time of TDIA is several magnitudes faster than LT-A Greedy algorithm and its positive influence spread is close to that of LT-A Greedy algorithm. Therefore, the TDIA can be applied to solve PIM problem in large-scale social networks with high effectiveness and efficiency.

6   Conclusions

In this paper, we study the algorithm to solve PIM problem in large-scale signed social networks. We propose a new heuristic algorithm named Three Degree of Influence heuristic Algorithm by analyzing the Three Degree of Influence Rule in social networks. TDIR is proposed by Nicholas Christakis in 2009. Furthermore, we study the PIM problem based on LT-A model, and change the formula of attitude weight by considering the TDI theory. Then, we conduct experiments on two real-world signed social network datasets and evaluate the experiment results. Experiment results demonstrate that the accuracy of TDIA is closed to greedy algorithm, and TDIA has high efficiency. Thus, TDIA can be used to solve PIM problem in large-scale signed social networks.

References

1. Epinions social network. http://snap.Standford.edu/data/soc-Epinions1.html
2. Slashdot social network. http://snap.stanford.edu/data/soc-Slashdot0811.html
3. Chen, W., Collins, A., Cummings, R., Ke, T., Liu, Z., Rincon, D., Sun, X., Wang, Y., Wei, W., Yuan, Y.: Influence maximization in social networks when negative opinions may emerge and propagate. In: SDM, vol. 11, pp. 379–390. SIAM (2011)
4. Chen, W., Wang, Y., Yang, S.: Efficient influence maximization in social networks. In: Proceedings of the 15th ACM SIGKDD International Conference on Knowledge Discovery and Data Mining, pp. 199–208. ACM (2009)
5. Chen, W., Yuan, Y., Zhang, L.: Scalable influence maximization in social networks under the linear threshold model. In: 2010 IEEE International Conference on Data Mining, pp. 88–97. IEEE (2010)
6. Christakis, N.A., Fowler, J.H.: Connected: The Surprising Power of Our Social Networks and How they Shape Our Lives. Little, Brown Company, New York (2009)
7. Domingos, P., Richardson, M.: Mining the network value of customers. In: Proceedings of the Seventh ACM SIGKDD International Conference On Knowledge Discovery and Data Mining, pp. 57–66. ACM (2001)

8. Goyal, A., Lu, W., Lakshmanan, L.V.: Celf++: optimizing the greedy algorithm for influence maximization in social networks. In: Proceedings of the 20th International Conference Companion on World wide web, pp. 47–48. ACM (2011)
9. Goyal, A., Lu, W., Lakshmanan, L.V.: Simpath: an efficient algorithm for influence maximization under the linear threshold model. In: 2011 IEEE 11th International Conference on Data Mining, pp. 211–220. IEEE (2011)
10. Jung, K., Heo, W., Chen, W.: Irie: Scalable and robust influence maximization in social networks (2011). arXiv preprint arXiv:1111.4795
11. Kempe, D., Kleinberg, J., Tardos, É.: Maximizing the spread of influence through a social network. In: Proceedings of the ninth ACM SIGKDD International Conference on Knowledge Discovery and Data Mining, pp. 137–146. ACM (2003)
12. Leskovec, J., Krause, A., Guestrin, C., Faloutsos, C., VanBriesen, J., Glance, N.: Cost-effective outbreak detection in networks. In: Proceedings of the 13th ACM SIGKDD International Conference on Knowledge Discovery and Data Mining, pp. 420–429. ACM (2007)
13. Li, D., Xu, Z.M., Chakraborty, N., Gupta, A., Sycara, K., Li, S.: Polarity related influence maximization in signed social networks. PloS one 9(7), e102199 (2014)
14. Li, S., Zhu, Y., Li, D., Kim, D., Ma, H., Huang, H.: Influence maximization in social networks with user attitude modification. In: 2014 IEEE International Conference on Communications (ICC), pp. 3913–3918. IEEE (2014)
15. Liu, B., Cong, G., Zeng, Y., Xu, D., Chee, Y.M.: Influence spreading path and its application to the time constrained social influence maximization problem and beyond. IEEE Trans. Knowl. Data Eng. 26(8), 1904–1917 (2014)
16. Lv, S., Pan, L.: Influence maximization in independent cascade model with limited propagation distance. In: Han, W., Huang, Z., Hu, C., Zhang, H., Guo, L. (eds.) APWeb 2014. LNCS, vol. 8710, pp. 23–34. Springer, Heidelberg (2014). doi:10.1007/978-3-319-11119-3_3
17. Qin, Y., Ma, J., Gao, S.: Efficient influence maximization based on three degrees of influence theory. In: Dong, X.L., Yu, X., Li, J., Sun, Y. (eds.) WAIM 2015. LNCS, vol. 9098, pp. 465–468. Springer, Heidelberg (2015). doi:10.1007/978-3-319-21042-1_42
18. Wang, H., Yang, Q., Fang, L., Lei, W.: Maximizing positive influence in signed social networks. In: Huang, Z., Sun, X., Luo, J., Wang, J. (eds.) ICCCS 2015. LNCS, vol. 9483, pp. 356–367. Springer, Heidelberg (2015). doi:10.1007/978-3-319-27051-7_30
19. Zhang, H., Dinh, T.N., Thai, M.T.: Maximizing the spread of positive influence in online social networks. In: 2013 IEEE 33rd International Conference on Distributed Computing Systems (ICDCS), pp. 317–326. IEEE (2013)
20. Zhou, C., Zhang, P., Guo, J., Zhu, X., Guo, L.: Ublf: an upper bound based approach to discover influential nodes in social networks. In: 2013 IEEE 13th International Conference on Data Mining, pp. 907–916. IEEE (2013)

Faster Decision Tree Induction
with Impurity-Based Heuristic Schema

Junlong Liu[⊠], Yunfeng Liu, Jinhong Zhong, and Wangyang Shen

USTC-Birmingham Joint Research Institute in Intelligent Computation and Its
Applications, University of Science and Technology of China, Hefei, China
{junlong,jinhong,swy}@mail.ustc.edu.cn

Abstract. Decision tree is one of the most commonly-used tools in
data mining. Most popular induction algorithms construct decision trees
in top-down manner. These algorithms generally select splitting feature
only with regard to current nodes' data, while ignoring history information. This kind of approaches need to search whole feature space during
splitting each node and will be quite time-consuming in high-dimensional
cases. To tackle this problem, we propose an impurity-based heuristic
schema (IBH) to utilize history information to accelerate existing top-
down induction algorithms. In details, when child node's impurity is
smaller than parent node's, IBH takes feature performance in parent
node as the pseudo upper bound of that in child node, to cut down
unpromising computation. The feature selection of IBH biases toward
the ones that perform better in parent nodes. Both mathematical analysis and experimental results demonstrate the coherence between IBH
and original induction algorithms. Experiments show that IBH can significantly reduce induction time without accuracy degradation in both
decision tree and related ensemble methods.

Keywords: Decision tree · Impurity-based heuristic · Feature selection

1 Introduction

Decision tree (DT) is one of the most commonly-used tools in data mining. The
various heuristic methods for constructing DT can roughly be divided into three
categories: bottom-up approaches, top-down approaches, hybrid approaches.
Among these methods, top-down approaches are the most popular ones. Most
top-down DT induction algorithms are quite time consuming in high-dimensional
cases. Because it involves iterating through all features' split candidates to split
each non-leaf node.

In this paper, an impurity-based heuristic schema (IBH) is proposed to utilize
history information to cut down unpromising computation and accelerate existing top-down DT induction algorithms. IBH takes the performance of features
in parent nodes as the pseudo upper bound of that in smaller-impurity child
nodes, which avoids much computation to find the "best" split. More specifically, during splitting smaller-impurity child nodes, IBH ranks feature by their

© Springer International Publishing AG 2016
H. Yin et al. (Eds.): IDEAL 2016, LNCS 9937, pp. 515–522, 2016.
DOI: 10.1007/978-3-319-46257-8_55

Algorithm 1. Generic Top-down DT Induction

Input: node t, data partition D, splitting criterion $\mathcal{CL}$
Output: decision tree for D rooted at node t
BuildTree(Node t, partition D, splitting criterion $\mathcal{CL}$)
 1: Apply $\mathcal{CL}$ to D to calculate performance of splits
 2: Let k be the number of children of t
 3: **if** $k > 0$ **then**
 4:    Create k children $n_1, ..., n_k$ of t
 5:    Use best split to partition D into $D_1, ..., D_k$
 6:    **for** $i = 1; i \leq k; i + +$ **do**
 7:       BuildTree($n_i, D_i, \mathcal{CL}$)
 8:    **end for**
 9: **end if**

---

performance in parent nodes, then update performance in current node from 1st to the last feature, until one feature's updated performance is better than that of all updated ones and pseudo upper bound of other un-updated ones. At larger-impurity child nodes, just calculate performance of all features as conventional algorithms do. Here, impurity of node refers to variance or Shannon entropy of samples in the node. Performance of features are measured by conventional splitting criteria (like information gain (Quinlan 1986)). The feature selection of IBH biases toward the ones that perform better in parent nodes. Mathematical analysis and experimental results demonstrate the coherence between IBH and conventional algorithms. Experimental results also show that in both single DT and DT-based ensemble methods, IBH can accelerate top-down induction algorithms significantly, such as C4.5, CART etc., without any accuracy degradation.

2   Background

The generic top-down induction framework and some popular splitting criteria for classification and regression will be introduced in this section, which will be used in mathematical analysis and experiments.

The formal description of generic top-down DT induction is Algorithm 1 proposed in (Gehrke et al. 1998).

2.1   Splitting Criteria

Shannon Entropy-based information gain is used as splitting criterion in ID3 algorithm (Quinlan 1986).

Gini Index is used as the splitting criterion in CART (Breiman et al. 1984):

$$Gini(D) = \sum_I p(I) \times (1 - p(I)), \tag{1}$$

where p(I) is the proportion in D which belongs to class I.

The information gain of splitting D into D_v with splits s is:

$$Gain_g(D, s) = Gini(D) - \sum_{D_v} (\frac{|D_v|}{|D|} \times Gini(D_v)), \tag{2}$$

CART (Breiman et al. 1984) introduces a variance-based information gain:

$$Var(D) = \frac{1}{|D|} \sum_{y_i \in D} (y_i - \bar{y})^2, \tag{3}$$

$$Gain_v(D, s) = Var(D) - \sum_{D_v} (\frac{|D_v|}{|D|} \times Var(D_v)), \tag{4}$$

Friedman proposed the least-squares based criterion (Friedman 2001) to handle regression problem with binary DT.

3 Impurity-Based Heuristic Schema

Since constructing optimal DT is NP-complete (Hyafil and Rivest 1976), existing methods are heuristic in essence. Thus it seems unnecessary to search whole feature space rigorously for the "best" in terms of heuristic splitting criteria. It is natural to take advantage of history performance to cut down the number of features that need to compute performance.

3.1 Impurity-Based Heuristic DT Induction

Before introducing IBH, some mathematical analysis is needed to present. Lemma 1 demonstrates that impurity of parent node in DT is not less than weighted sum of its child nodes, which is obviously true. Proposition 1 demonstrates that performance score of splits in parent nodes are likely to be higher than that in smaller-impurity child nodes. Since problems vary a lot, it is difficult to infer concrete likelihood of this upper bound assumption.

Lemma 1. *If* $\bigcup D_v = D$, *and* $\forall i \neq j, D_i \cap D_j = \emptyset$, *then*

$$Impurity(D) \geq \sum_{D_v} \frac{|D_v|}{|D|} \times Impurity(D_v)$$

Proposition 1. *If* $Impurity(D_j) \leq Impurity(D)$ *and* $D_j \subset D$, *then* $Gain(D_j, s)$ *are likely to be smaller than* $Gain(D, s)$

Proof. We consider binary partitioning of D here. In classification case. After split with s, D is sliced into D_1 and D_2. Let α and β denote the number of samples in D_1 and D_2 respectively. p_i and q_i denote the proportions of samples belonging to class i in D_1 and D_2 respectively. The number of classes is denoted by m. And define $\lambda = \frac{\alpha}{\beta}$.

Gini Index is chosen as the splitting criterion when we prove the proposition in classification case. So, according to Eq. (2), the performance score of split s on samples D is:

$$Gain_g(D, s) = \frac{\lambda}{(1 + \lambda)^2} \sum_{i=1}^{m} (p_i - q_i)^2 \tag{5}$$

In regression case, variance-based criterion is chosen as the splitting criterion. y_1, y_2 denote the mean value of D_1 and D_2 respectively. According to Eq. (4), the performance score of split s is:

$$Gain_v(D, s) = \frac{\lambda}{(1+\lambda)^2}(y_1 - y_2)^2 \tag{6}$$

Define $\phi(\lambda) = \frac{\lambda}{(1+\lambda)^2}$, $\phi'(\lambda) = \frac{1-\lambda}{(1+\lambda)^3}$. It is obvious that $\phi(\lambda)$ monotonically increases in $(0, 1]$, and monotonically decreases in $[1, +\infty)$.

As $Impurity(D) \geq Impurity(D_j)$, samples in D_j are more homogeneous than that in D. So, the partitioning of D_j tends to be more unbalanced than that of D. In other words, λ, sample ratio of child nodes, of D tends to be closer to 1. Then, $\phi(\lambda)$ in D_j is generally smaller than that in D.

Also, as D_j is more homogeneous, samples in D_j are usually harder to discriminate than that of D. Then, D_j tends to have more alike child nodes (This is the critical problem that induction algorithms aim to solve, namely to find the split that results in least alike child nodes). As a result, $(p_i - q_i)^2$ and $(y_1 - y_2)^2$ in D_j are likely to be smaller than that in D.

$Gain_g, Gain_v$ are the results of $\phi(\lambda)$ multiplying $(p_i - q_i)^2$ or $(y_1 - y_2)^2$ respectively. So $Gain_g(D, s), Gain_v(D, s)$ are likely to be greater than $Gain_g(D_j, s)$, $Gain_v(D_j, s)$ respectively. Since feature rankings induced by various splitting criteria are very similar (Baker and Jain 1976), the proposition can be applied with other top-down splitting criteria.

Algorithm 2. Generic Top-down DT Induction with IBH

Input: node t, data partition D, splitting criterion $\mathcal{CL}$
Output: decision tree for D rooted at node t
Note: IBH: Impurity-based Heuristic Schema, Alg.3
BuildTree(Node t, partition D, splitting criterion $\mathcal{CL}$)
1: Apply IBH to D to update performance score of splits
2: Let k be the number of children of t, $k \in \mathcal{N}$
3: **if** $k > 0$ **then**
4:     Create k children $t_1, ..., t_k$ of t
5:     Use best split to partition D into $D_1, ..., D_k$
6:     **for** $i = 1; i \leq k; i + +$ **do**
7:         BuildTree($t_i, D_i, \mathcal{CL}$)
8:     **end for**
9: **end if**

Based on Algorithm 1, IBH top-down induction algorithm is described in Algorithm 2. They only differ in line 1. Instead of directly calculate performance of all splits for all features, Algorithm 2 applies IBH to update performance of partial or all features. In line 5, both algorithms select split that ranks top in terms of performance score derived in line 1. Lemma 1 ensures that there exists smaller-impurity child nodes.

As showed in Algorithm 3, IBH makes use of the pseudo upper bound assumption to accelerate the selection of promising split: for smaller-impurity child nodes, performance of features is directly inherited from their parent nodes, then the algorithm checks the "best" features' splits recursively until one feature's best split ranks top. For larger-impurity child nodes, performance of all features' splits are calculated as traditional induction algorithms do. Since there are too many split candidates to record, only performance of features' best split is recorded as features' performance.

Algorithm 3. Impurity-based Heuristic Schema (IBH)

Input: node t, data partition D, splitting criterion $\mathcal{CL}$, performance of features in parent node $gains$
Output: partially updated $gains$
1: **if** $Impurity(t) > Impurity(parent(t))$ **then**
2:     Apply $\mathcal{CL}$ to D to calculate performance of splits
3:     Update $gains$ with score of features' best splits
4:     Return $gains$
5: **end if**
6: **while** $true$ **do**
7:     Select feature f that ranks top in $gains$
8:     **if** $gains[f]$ has been updated **then**
9:         Return $gains$ //f**'s best split will split node in line 5, Algorithm 2**
10:     **end if**
11:     Apply $\mathcal{CL}$ to D to calculate performance of f's splits
12:     Update $gains[f]$ with score of f's best split
13: **end while**

Table 1. Datasets info

ID	Name	Task	Instances	Features	ID	Name	Task	Instances	Features
1	Cardiotocography	Classification	2126	10	11	Arcene	Classification	200	10000
2	Breast Cancer	Classification	569	30	12	Dexter	Classification	600	20000
3	Nomao	Classification	34465	120	13	Dorothea	Classification	1150	100000
4	Multiple Features	Classification	2000	216	14	PEMS-SF	Classification	440	138672
5	Arrhythmia	Classification	452	279	15	Compressive Strength	Regression	1030	8
6	Madelon	Classification	2600	500	16	Add10	Regression	9792	10
7	SECOM	Classification	1567	591	17	Buzz	Regression	583250	77
8	ISOLET5	Classification	1559	617	18	Communities Crime	Regression	1994	127
9	CNAE-9	Classification	1080	857	19	CT Location	Regression	53500	386
10	Gisette	Classification	7000	5000	20	UJIndoorLoc	Regression	21048	520

4   Experimental Results

As Table 1 shows, 20 datasets, with varying size, number of features, and problem domains, were selected from UCI Machine Learning Repository. All the experiments in this paper are repeated 20 times. In each run 70 % of the samples were randomly chosen as training set and the rest 30 % of the samples were regarded as validation set. In details, datasets from 1 to 14 are for classification, from 15 to 20 are for regression. And they are sorted by number of features within task domain.

In this section, the effectiveness of IBH on DT induction, bagging trees (Breiman 1996), AdaBoost trees (Freund and Schapire 1997), and gradient boosting trees (Friedman 2001) will be assessed in the aspects of time consumption and accuracy. In classification case, the accuracy is denoted as the ratio of instances in validation set that are classified correctly. As for regression tasks, coefficient of determination, namely R^2, was used to measure accuracy. In all experiments, minimal samples of leaf nodes are set as 5. The depth of ensemble trees is limited as 10.

4.1   Comparison of DT Induction Results

To check IBH algorithm's impact on DT induction, like time cost, test accuracy, we test IBH with two set of experiments on 20 datasets.

As Figs. 1(a), and 2(a) shows, IBH can significantly reduce DT induction time while maintaining accuracy with four representative splitting criteria on 20 varying datasets. The coherent results with different criteria and different datasets makes it convincible to apply IBH with other splitting criteria and on other datasets.

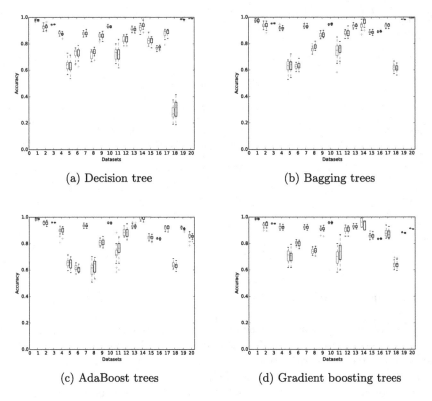

(a) Decision tree   (b) Bagging trees

(c) AdaBoost trees   (d) Gradient boosting trees

Fig. 1. Accuracy of IBH (right black box in each pair) compared to original algorithms (left blue box in each pair). (Color figure online)

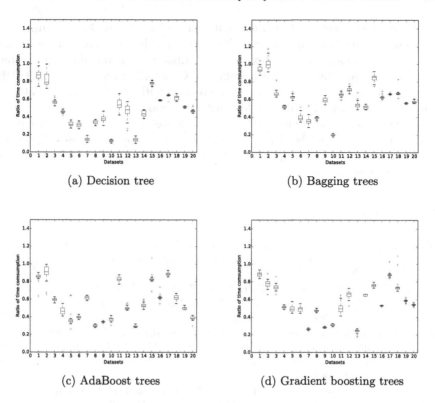

(a) Decision tree

(b) Bagging trees

(c) AdaBoost trees

(d) Gradient boosting trees

Fig. 2. $Ratio_t$, ratio of time cost of IBH compared to original algorithms.

4.2 DT Based Ensemble Methods

Since IBH might affect diversity of DTs and bring in negative impact in ensemble cases, we test IBH on bagging trees (Breiman 1996), AdaBoost trees (Freund and Schapire 1997), and gradient boosting trees (Friedman 2001). Because diversity is difficult to measure, we use accuracy to assess the impact on diversity.

As Figs. 1 and 2 IBH can significantly accelerate DT-based ensemble algorithms, without accuracy degradation.

5 Conclusion

This paper proposes an impurity-based heuristic schema (IBH) to accelerate top-down DT algorithms. Mathematical analysis and experimental results verify that IBH can significantly accelerate both decision tree induction and related ensemble methods, without accuracy degradation.

IBH utilizes history information to cut down unpromising computation during splitting non-leaf nodes. More specifically, IBH biases toward features that perform better in parent nodes, and make the pseudo upper bound assumption on features' performance. Generally, IBH can reduce nearly 30 % of time consumption with tens of features, and 60 % with a few hundreds of features. Since problems vary a lot, time complexity and likelihood of the pseudo upper bound assumption is difficult to model specifically. Generally, time cost ratio of IBH against orginal methods is negatively correlated with the number of features, namely the more feature, the smaller time cost ratio.

In the aspect of accuracy, IBH schema achieves almost the same accuracy as original methods do, both in decision tree and in ensemble cases. Coherent results on various heuristic splitting criteria and varying problems demonstrate the effectiveness of IBH.

References

Baker, E., Jain, A.: On feature ordering in practice, some finite sample effects. In: Proceedings of the Third International Joint Conference on Pattern Recognition, pp. 45–49 (1976)

Breiman, L.: Bagging predictors. Mach. Learn. **24**(2), 123–140 (1996)

Breiman, L., Friedman, J., Stone, C.J., Olshen, R.A.: Classification and Regression Trees. CRC Press, Boca Raton (1984)

Freund, Y., Schapire, R.E.: A decision-theoretic generalization of on-line learning and an application to boosting. J. Comput. Syst. Sci. **55**(1), 119–139 (1997)

Friedman, J.H.: Greedy function approximation: a gradient boosting machine. Ann. Stat., 1189–1232 (2001)

Gehrke, J., Ramakrishnan, R., Ganti, V.: Rainforest-a framework for fast decision tree construction of large datasets. In: VLDB, vol. 98, pp. 416–427 (1998)

Hyafil, L., Rivest, R.L.: Constructing optimal binary decision trees is np-complete. Inf. Process. Lett. **5**(1), 15–17 (1976)

Quinlan, J.R.: Induction of decision trees. Mach. Learn. **1**(1), 81–106 (1986)

Triaxial Accelerometer Located on the Wrist for Elderly People's Fall Detection

Armando Collado Villaverde[1], María D. R-Moreno[1], David F. Barrero[1(✉)], and Daniel Rodriguez[2]

[1] Departamento de Automática, Universidad de Alcalá, Crta. Madrid-Barcelona, Alcalá de Henares, Madrid, Spain
david@aut.uah.es
[2] Departamento de Ciencias de la Computación, Universidad de Alcalá, Crta. Madrid-Barcelona, Alcalá de Henares, Spain

Abstract. The loss of motor function in the elderly makes this population group prone to accidental falls. Actually, falls are one of the most notable concerns in elder care. Not surprisingly, there are several technical solutions to detect falls, however, none of them has achieved great acceptance. The popularization of smartwatches provides a promising tool to address this problem. In this work, we present a solution that applies *machine learning* techniques to process the output of a smartwatch accelerometer, being able to detect a fall event with high accuracy. To this end, we simulated the two most common types of falls in elders, gathering acceleration data from the wrist, then applied that data to train two classifiers. The results show high accuracy and robust classifiers able to detect falls.

Keywords: Fall detection · Accelerometer · Machine learning · Classification · Supervised learning · Care for the elderly

1 Introduction

Falls in the elderly are a public health problem [1]. They are not only a significant source of problems associated to elderly for their direct consequences (such as traumas), but also because falls are the symptom of infirmity (such as hearth attack). Therefore, it is not surprising that falls are one of the most relevant concerns for elders care professionals and their families.

The importance of this topic has motivated the rise of a notable number of solution proposals. Most of them have in common the usage of accelerometers [2]. The small size, availability in cell phones and respect to privacy explain why they are becoming so popular in falls detection. Many approaches use dedicated devices, usually placed on the trunk [3], while others exploit the capabilities and popularity of smartphones [4,5]. A similar approach was successfully used to build kinematics models of upper limps [6] and applied to home rehabilitation [7].

Image processing is another popular approach to fall detection [8–10], however, it poses some practical problems which in this context are determinant. People use to dislike having cameras in their private spaces, even if they do

© Springer International Publishing AG 2016
H. Yin et al. (Eds.): IDEAL 2016, LNCS 9937, pp. 523–532, 2016.
DOI: 10.1007/978-3-319-46257-8_56

not record or transmit images. We should also mention the need to install cameras and their limitations to the screened areas. A third group of fall detection systems uses sound or vibrations [11].

Perhaps a notable motivation for elders to reject fall detection devices is their size, which leads to inadequate ergonomics [2]. Fall detection based on cell phones do not present that problem, but raises new ones when they are used by the elderly. Perhaps the most notable one is that they do not keep the cell phone on them when being at home, where most of the falls happen. There are other additional problems, for instance, women use to keep their cell phones in a handbag, where fall detection algorithms will likely fail because they are trained to detect falls through acceleration sensors close to the body trunk.

In order to overcome the usage disadvantages of previous devices, we propose to exploit the popularity of smartwatches. Most of them are programmable devices, and include rich sensing such as accelerometers. Some advanced models even include heart monitors and communication capabilities. All those features together with the price reduction provide a good opportunity of improving elders caring. Some of the problems with cell phones do not happen when using smartwatches: they are located always in the same place, regardless of gender and age, and perhaps more important in elderly people, they are considered every day objects and thus they are not perceived as something invasive. Therefore, we believe these features will help reducing their adoption resistance. In this paper, we present a Machine Learning (ML) -based fall- detection algorithm implemented in a smartwatch. Our results show the accuracy of the classifiers used over the datasets generated under the supervision of professionals in the field.

The rest of the paper is structured as follows. Next, a discussion about the types of falls in elders, then it describes the data acquisition procedure. Section 4 describes the data preprocessing. The main contribution is located in Sect. 5, which describes the training and evaluation of the falls detection algorithm. The robustness of the proposed algorithm is evaluated in Sect. 5.4. The paper finishes with conclusions and future work.

2   Types of Falls in Elders

Our goal is to implement a falls detection algorithm in a smartwatch to monitor the elderly. In order to detect the falls, we pose the problem in terms of classification: given the acceleration values in a time window, classify them as corresponding to a fall or not. Since the final aim is to implement a classifier in a device (a smartwatch) with limited capabilities, the classifier resources consumption is an issue that needs to be taken into account.

As most ML applications, an important issue is how to gather high quality data to train and test the classifiers. In this particular application, data gathering involves people falling, which implies obvious health risks. Other approaches have used volunteers to simulate the fall, who used to be healthy young people, sometimes they were skilled in some martial art or used protections to minimize the risks. This can be a threat the validity as it can suppose a bias to the results, however, data gathering of real falls with elder people is a costly, risky and time consuming task [12], just to mention the most obvious difficulties.

The target of our work is elderly people care, whose falls follow specific dynamics. Elders suffer loose of mobility, which translates to slower motion and increased reaction time. In case of a fall, a young healthy person would react moving his/her arms to cushion the hit; on the contrary, elders do not tend to react in time, resulting in more violent hits. Another relevant issue for our work is the shift of the center of gravity in elders. With age, people tend to separate their legs, and curve the trunk forward, this implies that the center of gravity in elders tend to be lower and shifted forward. As a consequence, falls in elders rarely happen laterally or backwards.

There are usually two types of falls in elders, which we name *syncope* and *forward falls*. We refer to *syncope falls* to those falls consequence of a loss of conscience or hearth condition that prevents using the muscles to control the fall. It results in a vertical motion and a two stages fall: first the knees impact on the ground, and then the trunk moves forward until it hits the ground. The second type of fall usually found in elders is what we name *forward falls*. They are originated by the collision of a foot with an object while the person is walking, loosing the equilibrium and falling over. The trunk in this type of falls moves forward, and given the increased reaction time of elders, the trunk hits the ground without the hands cushioning.

Given the different dynamics of the falls, it seems reasonable to address them as two different, yet related, problems. To this end we will train two different classifiers, each one specialized in detecting one type of fall. In Sect. 5.4 we study the ability of the classifiers to detect falls of a different type they were trained to do so.

3 Data Acquisition

Acquisition of high quality datasets is a key process in ML. We used a smartwatch with a triaxial accelerometer, the hardware imposed a sampling period of 20 ms, yielding three measures of acceleration (X, Y and Z) each 20 ms. The variable Y stands for the longitudinal axis, X for the sideways axis and Z for the axis perpendicular to the watch display.

A key problem to consider is how to build the base class labeled as '*no fall*'. We used data captured along a basket match, removing those samples with accelerations lower than a given threshold. The idea is to keep samples containing high accelerations. This is clearly an unrealistic activity for an elderly person, but basket contains numerous vertical and horizontal motion, making it similar to a fall. In this way we avoid the naïve problem of just classing motion and lack of motion. If a classifier is correctly trained to distinguish between basket and falls, it seems fair to assume that it will be able to distinguish between a fall and normal activities in the life of an elder.

Using real falls was discarded given the risk of injuring the subject, specially when our target is a fragile population group, the elderly. Therefore we tried to capture as much data as realistically possible, taking all necessary measures to avoid risks[1]. We defined two data acquisition procedures, one for syncope and another one for forward falls.

3.1 Syncope Falls Data Capture Procedure

Syncope falls are characterized by the lack of control, with gravity as the only acting force. Other ML approaches to fall detection used volunteers to simulate this type of fall. In our opinion, this scenario is better simulated by using a nursing mannequin, which is a mannequin with the same joints mobility, size and weight than an adult human. Of course, the results will also be biased, since there is only one mannequin available for data gathering, but still the fall is more realistic than a conscious simulating it.

The center of gravity of the mannequin used did not reflect perfectly well the one found in an elderly person. For this reason, just releasing the mannequin does not generate a realistic syncope fall; the mannequin tends to stop once the knees hit the ground. In order to generate realistic falls, it was needed that the mannequin was smoothly pushed forward when it was released. The whole process was supervised by two Geriatrics experts. They helped to optimize the procedure and judged which simulated falls were realistic, and which one should be discarded.

We simulated 42 syncope falls, but the experts only validated 30 falls. There were 12 simulated falls discarded for different reasons, in some cases the mannequin did not fall on its knees, or it hit the ground with the knees, but the trunk did not move forward, or the trunk moved laterally. Falls lasted around 500 ms, but measurements begun shortly before and finished shortly after the fall. Given the fact that each sample contains three acceleration values, and samples are measured in 20 ms intervals, each fall generated between 150 and 300 measures.

3.2 Forward Fall Data Capture Procedure

Forward falls begin with the subject walking, when the subject hits an obstacle overbalancing and falling over. This scenario is poorly approximated with a mannequin. In a forward fall, the subject does have some control, and actually the reflex action is to raise the hands to cushion the hit. In elderly people this reaction can be slow, and they usually do not have enough time to raise their hands, resulting in more dangerous falls. In our opinion, this scenario is better approximated using a healthy young subjects that were trained to move slowly. This is not easy because falling over in that way seems unnatural for the subject, but the results are more accurate according to the geriatric experts consulted.

Therefore, we selected a young volunteer and placed him on a tatami for safety. The obstacle was a thick pad, which also served to safety stop the fall. The domain experts trained the volunteer not to use his hands and move slowly. Once the training was finished, the data capture begun. To simulate the fall,

---

[1] All datasets are available on http://atc1.aut.uah.es/~david/ideal2016.

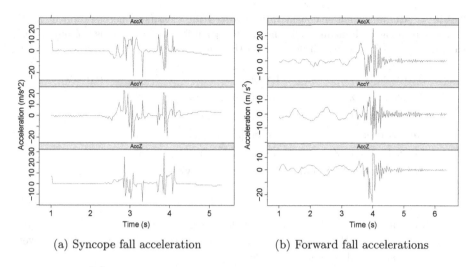

(a) Syncope fall acceleration   (b) Forward fall accelerations

Fig. 1. Example of accelerations measured on the wrist.

the volunteer begun to walk and after 4–5 steps hitting the pad with a foot and falling over. Given that data might be affected by which foot hit the pad, we repeated the process the same number of times with each foot.

The volunteer simulated 47 falls, but the experts only validated 40, 20 for each feet. In order to assess the robustness of the classifier, we gathered data from 20 valid falls of two other volunteers. The duration of each fall is around one second.

Figure 1 visualizes the acceleration in a typical (a) syncope and (b) forward fall. In the syncope fall the two hits (knees and trunk) are clearly visible, while the forward fall shows first smooth accelerations due to walking, the fall over and finally the subject laying.

4 Data Preprocessing

Data needs some preprocessing in order to feed the classifier. We grouped data in time windows, which contains samples that serve as input to the classifier. This is also an indirect way to consider history, since the time window contains historical values of acceleration. All windows containing a fall were manually labeled as 'fall', while windows coming from a basket match were labeled as 'not-fall'.

The time window width is a key parameter, we set the width to contain the fall values. We analyzed several syncope falls, observing that the average duration (see Fig. 1) is 500 ms, so we set the time window for syncope fall detection to 500 ms. Similarly, the window length for forward falls was set to 1200 ms.

In addition to raw data coming from the accelerometer, we introduced new attributes summarizing those values. In particular, for each window, we computed the mean and standard deviation of acceleration in each one of the three

Table 1. Attributes under consideration to feed the classifiers: acceleration in X, Y and Z axis along with mean and standard deviation for each axis. N is the number of samples in the window, which depends on the dataset.

Attribute	Label	No. of attributes
Acceleration X	AccelX$[X_1, ...,x_N]$	N
Acceleration Y	AccelY$[y_1, ...,y_N]$	N
Acceleration Z	AccelZ$[Z_1, ...,z_N]$	N
Mean X, Y and Z	MeanX, MeanY, MeanZ	3
Std. deviation X, Y and Z	DevX, DevY, DevZ	3

axis. Those attributes, along with raw data were used to train the classifiers. Table 1 summarizes the attributes considered, however their predictive power greatly varies, as will be analyzed later.

An important issue about data is that it is unbalanced. Since falls are hard to simulate, there were much more data coming from basket than from simulated falls. To face this issue we undersampled the 'not-fall' class, getting the same number of instances for each class. The evaluation of the classifiers was carried out using 10-fold cross-validation.

5   Detection of Fall Events

Fall detection is addressed as a binary classification problem: classify acceleration measures contained in a time window as 'fall' or 'not-fall'. The dataset was build as described in Sect. 3 and then preprocessed to generate time windows, derived attributes and labels as described in Sect. 4.

The goal is to implement the classifier in a smartwatch, which means that the classifier model generated should be as lightweight as possible (memory and computationally). We considered some classical classifiers implemented in Weka such as C4.5 (J48), 1-NN, Logistic regression, Naïve Bayes and PART. Some of these classifiers such as 1-NN are not lightweight, but its good performance motivated us to include them for comparison purposes.

5.1   Determination of Sampling Rate

An important parameter to be set is the sampling rate. Measures were taken with a hardware that imposed a maximum sampling rate of 20 ms, however, it is possible that we could use a lower rate. We should consider that there is a direct relationship between the sampling rate and the number of attributes and therefore trying to reduce the sampling rate might pay off.

We also briefly studied the influence of the sampling rate with the accuracy of the classifiers. To this end we evaluated the accuracy of several classifiers using a range of sample periods. No feature selection was used. The results are shown

Table 2. Evaluation of the influence of the sampling rate on the classifiers accuracy. The table shows the sampling rate, type of fall -syncope (S) or forward (F)-, number of attributes (as indicated in Table 1) and accuracy of the classifiers.

Sample period	Fall type	# attrib	C4.5	1-NN	Reg. Log.	Naïve Bayes	PART
20 ms	S	97	**95.38%**	**97.80%**	90.26%	**85.71%**	**95.82%**
40 ms	S	52	92.10 %	97.32 %	87.48 %	84.35 %	91.65 %
60 ms	S	37	88.38 %	92.03 %	87.24 %	83.60 %	87.70 %
20 ms	F	82	**95.66%**	**98.43%**	88.76 %	**86.74%**	**94.09%**
40 ms	F	46	91.97 %	97.21 %	**89.34%**	84.43 %	92.62 %
60 ms	F	34	88.45 %	89.50 %	84.25 %	83.46 %	85.04 %

in Table 2. We can observe a pattern regardless of the classification algorithm, high sampling rates boost performance, at least in the range of values that we have under consideration. The Nyquist Theorem states that there must be a lower bound to the sampling period from which we should not expect further performance improvements. Clearly this lower bound was not achieved as the best performance value was with 20 ms samples. Therefore, this was the chosen value.

5.2   Overview of Attributes Classification Power

The number of attributes used so far is relatively high. To illustrate this point, let us consider a syncope fall window that lasts 500 ms, 25 samples and each sample with three values of acceleration (X, Y, Z), which yields to 75 attributes. In addition, there are six derived attributes (mean and standard deviations), resulting in 81 attributes. This high number of attributes may result in higher computational costs and eventually may also degenerate the classifier performance.

To reduce the number of attributes we estimated the predictive power of the attributes. To this end we applied Correlation Feature Selection Subset Evaluation, as implemented in Weka, This method evaluates the correlation of each attribute with the class, as well as the correlation among the attributes, giving better ranks to those attributes highly correlated with the class while having low correlation among other attributes.

Table 3 ranks attributes by its predictive power suggesting that derived attributes have higher predictive power in comparison to raw acceleration values. In particular, the standard deviation on Y has the best score for syncope falls, and gets the second position for forward falls. Mean acceleration on Z has the second highest score in syncope falls, while appears as a good attribute for forward falls. Raw data seems to have less predictive power, in particular on Z. Interestingly, the mean Z acceleration get a pretty high score for syncope falls, while raw acceleration values on Z does not appear in the table. Syncope fall includes many raw acceleration values on X, with similar scores, this suggests that those attributes may be highly correlated.

Table 3. Twelve best predictive attributes ranked by its correlation to the class for syncope and forward falls.

Forward fall				Syncope fall			
% Inf.	Attrib.	% Inf.	Attrib.	% Inf.	Attrib.	% Inf.	Attrib.
0.515	DevZ	0.173	AccelX1	0.541	DevY	0.233	AccelX11
0.429	DevY	0.163	MeanZ	0.523	MeanZ	0.232	AccelX8
0.399	MeanX	0.159	AccelZ24	0.238	AccelY12	0.229	AccelY11
0.316	DevX	0.154	AccelX2	0.235	AccelX7	0.229	AccelX9
0.209	MeanY	0.144	AccelZ23	0.235	AccelX12	0.227	AccelX10
0.195	AccelX0	0.135	AccelX3	0.233	AccelX13	0.226	AccelX14

Table 4. Performance of syncope (S) and forward falls (F) detection. Attribute selection was performed using a wrapper.

		C4.5	1-NN	Log. Reg.	Naïve Bayes	PART
Accuracy	S	0.98	1	0.92	0.93	0.96
	F	0.98	1	0.89	0.91	0.95
Recall	S	0.98	1	0.94	0.90	0.97
	F	0.98	1	0.91	0.92	0.97
Attributes	S	12	7	16	9	7
	F	9	13	11	12	7

The high score that derived attributes achieve and the hint of highly correlated raw acceleration values suggest that the number of attributes may be reduced significantly. For this reason in the following section we will evaluate the classifiers integrating the feature selection.

5.3   Evaluation of Classifiers

Given the importance of feature selection, we have performed the classifier evaluation along with it using a wrapper approach. This method exploits the interaction between the classifier and the attributes, yielding, in theory, better results, specially where there are a high number of redundant attributes. We used the Weka `WrapperSubsetEval` implementation of the method with Hill Climbing for attribute search.

Table 4 summarizes the performance of the classifiers. For instance, C4.5 (J48 in Weka) with 97 attributes (Table 2) scores 95.38 % accuracy, while using wrapper feature selection it increases to 98 % with only 12 or 9 attributes, depending on the type of fall respectively. The other classifiers behave in a similar way. The 1-NN classifier has an outstanding performance, with a perfect accuracy and recall. The similarity of the instances in the training set may explain this; the

Table 5. Performance of forward falls detection classifiers shown in Table 4 evaluated with a testing set composed by unseen people simulated falls.

		C4.5	1-NN	Log. Reg.	Naïve Bayes	PART
Accuracy	F	0.91	0.66	0.97	0.98	0.98
Recall	F	0.90	0.98	0.95	0.96	0.98
Attributes	F	9	13	11	12	7

robustness analysis done in the next section supports this hypothesis. Despite the magnificent performance of 1-NN, the need to store all the training set dissuades us to implement it in a smartwatch. However, it suggests that perhaps using a Nearest Centroid Classifier may conduct to a good classifier while keeping low computational needs.

5.4 Robustness Analysis of Forward Falls

A clear weakness of the previous approach is the lack of diversity in the training set. Syncope falls used just a single mannequin, while forward falls included falls from one person. This obviously reduces the complexity of the problem, and a natural question that rises is how much the classifier degrades its performance when exposed to different people. In order to provide an insight to this question, we performed a robustness experiment.

Given the lack of alternative mannequin, we focused on forward falls. As described in Sect. 3, we captured data from three people, one of them repeated 40 times the fall simulation, while the others repeated the simulation 20. We exposed the classifiers trained with data coming from the first volunteer (whose performance is shown in Table 4), to the simulated falls of the other two volunteers. The resulting performance is shown in Table 5. As we expected, the performance drops, but in most cases remains above 0.9. The most dramatic case is with 1-NN, whose accuracy falls to 0.66. Logistic regression, Bayes and PART seems quite robust and actually they increase the performance.

6 Conclusions and Future Work

In this paper we have described an ML application to detect falls sensing the acceleration on the wrist. The aim is to implement a fall detection system in a smartwatch oriented to the elderly care. This population is prone to suffer two types of falls, *syncope* and *forward falls*. We simulated those type of falls and measured acceleration on the wrist. These data, along with measures coming from a basket match were used to train and evaluate a classifier with a wrapper-based feature selection.

We selected PART for its high accuracy (above 0.9) and the relatively low number of rules (7) it generated, which made unnecessary the use of external libraries, an interesting feature when looking for a lightweight application. We implemented the algorithm in Android Wear and tested on a Samsung Gear S, with satisfactory results. In a near future we expect to expand the detection with new sensors and an ensemble of classifiers.

Acknowledgements. The authors thank the contribution of Isabel Pascual Benito, Francisco López Martínez and Helena Hernández Martínez, from Department of Nursing and Physiotherapy of the University of Alcalá, for their help designing and supervising the simulated falls procedure. This work is supported by UAH (2015/00297/001), JCLM (PEII-2014-015-A) and MINCECO (TIN2014-56494-C4-4-P).

References

1. Sadigh, S., Reimers, A., Andersson, R., Laflamme, L.: Falls and fall-related injuries among the elderly: a survey of residential-care facilities in a swedish municipality. J. Commun. Health **29**, 129–140 (2004)
2. Noury, N., Fleury, A., Rumeau, P., Bourke, A.K., Laighin, G.O., Rialle, V., Lundy, J.E.: Fall detection - principles and Methods. In: 2007 29th Annual International Conference of the IEEE Engineering in Medicine and Biology Society, pp. 1663–1666 (2007)
3. Gibson, R.M., Amira, A., Ramzan, N., Casaseca-de-la higuera, P., Pervez, Z.: Multiple comparator classifier framework for accelerometer-based fall detection and diagnostic. Appl. Soft Comput. J. **39**, 94–103 (2016)
4. Luštrek, M., Kaluža, B.: Fall detection and activity recognition with machine learning. Informatica **33**, 205–212 (2008)
5. Albert, M.V., Kording, K., Herrmann, M., Jayaraman, A.: Fall classification by machine learning using mobile phones. PLoS ONE **7**, 3–8 (2012)
6. Zhou, H., Hu, H.: Reducing drifts in the inertial measurements of wrist and elbow positions. IEEE Trans. Instrum. Measur. **59**, 575–585 (2010)
7. Tao, Y., Hu, H., Zhou, H.: Integration of vision and inertial sensors for 3d arm motion tracking in home-based rehabilitation. Int. J. Robot. Res. **26**, 607–624 (2007)
8. Miaou, S.G., Sung, P.H., Huang, C.Y.: A customized human fall detection system using omni-camera images and personal information. In: Conference Proceedings - 1st Transdisciplinary Conference on Distributed Diagnosis and Home Healthcare, D2H2 2006, pp. 39–42 (2006)
9. Cucchiara, R., Prati, A., Vezzani, R., Emilia, R.: A multi-camera vision system for fall detection and alarm generation. Expert Syst. **24**, 334–345 (2007)
10. Auvinet, E., Multon, F., Saint-Arnaud, A., Rousseau, J., Meunier, J.: Fall detection with multiple cameras: an occlusion-resistant method based on 3-D silhouette vertical distribution. IEEE Trans. Inf. Technol. Biomed. **15**, 290–300 (2011)
11. Zigel, Y., Litvak, D., Gannot, I.: A method for automatic fall detection of elderly people using floor vibrations and sound-Proof of concept on human mimicking doll falls. IEEE Trans. Biomed. Eng. **56**, 2858–2867 (2009)
12. Bagalà, F., Becker, C., Cappello, A., Chiari, L., Aminian, K., Hausdorff, J.M., Zijlstra, W., Klenk, J.: Evaluation of accelerometer-based algorithms on real-world falls. PloS one **7**, e37062 (2012)

Mining Uplink-Downlink User Association in Wireless Heterogeneous Networks

Alfredo Cuzzocrea[1], Giorgio M. Grasso[2], Fan Jiang[3], and Carson K. Leung[3(✉)]

[1] University of Trieste and ICAR-CNR, Trieste (TS), Italy
alfredo.cuzzocrea@dia.units.it
[2] University of Messina, Messina (ME), Italy
gmgrasso@unime.it
[3] University of Manitoba, Winnipeg, MB, Canada
umjian29@myumanitoba.ca, kleung@cs.umanitoba.ca

Abstract. In the current era of big data, wide varieties of high volumes of valuable data of different veracities can be generated or collected at a high velocity. One of the popular sources of these big data is the wireless networks. Nowadays, the use of smartphones has significantly increased the traffic load in these cellular networks. Consequently, system models that are practical in real-life scenario with the significant for increasing traffic load in cellular networks have drawn attentions of researchers. Studies have been conducted to solve the related interesting research problem of user association in this complex system model. Some of these studies formulated this research problem as a many-to-one matching game, in which users and base stations evaluate each other based on well-defined utilities. In this paper, we examine how the traditional data mining techniques—in particular, the frequent pattern mining techniques—help to solve this research problem. Specifically, we examine the mining of uplink-downlink user association data in wireless heterogeneous networks.

Keywords: Association rules · Big data · Data mining · Downlink · Frequent patterns · Knowledge discovery · Uplink · Wireless heterogeneous networks

1 Introduction

In the current era of big data, wide varieties of high volumes of valuable data of different veracities can be generated or collected at a high velocity. These big data can be characterized by the well-known 5V's: (1) variety, (2) volume, (3) value, (4) veracity, and (5) velocity. Embedded in the big data are rich sets of useful information and knowledge. This leads to *data science and big data analytics*, which incorporates various techniques from a broad range of fields—including cloud computing, intelligent data engineering and automated learning, knowledge discovery and data mining, machine learning, mathematics, as well as statistics.

© Springer International Publishing AG 2016
H. Yin et al. (Eds.): IDEAL 2016, LNCS 9937, pp. 533–541, 2016.
DOI: 10.1007/978-3-319-46257-8_57

Nowadays, big data are almost everywhere, ranging from web logs to texts, documents, business transactions, banking records, financial charts, biological data, medical images, surveillance videos, to streams of advertisements, marketing, telecommunication, life science, and social media data. One popular source of big data is communication networks, including *wireless heterogeneous networks.*

The use of smartphones has significantly increased the traffic load in current cellular networks [13], and this trend is expected to continue in the next few years [5]. To meet the demand generated by this increasing traffic, significant changes to current cellular architecture is needed. The following are two common ways to address this problem:

1. One way is to apply the concept of *small cell networks (SCNs)* [12,24], which allow users to improve the capacity and coverage of wireless networks by reducing the distance between users and their serving *base stations (BSs)*. This is done by deploying *small cell base stations (SBSs)*, which overlaid on current macrocell networks. The deployment of small cells introduces numerous challenges in terms of interference management, resource allocation, and network modeling [5,12]. In particular, cell association is an important challenge in small cell networks. For instance, directly deploying classical macrocell-oriented cell association schemes in small cell networks can lead to inefficient association due to the factors such as heterogeneous capabilities and varying coverage areas [28].

2. Another way is to use full-duplex technology in small cell operation environment for improving the system spectral efficiency by using the same radio resources for simultaneous transmission and reception in different nodes of the radio system. Full-duplex technology enables the doubling of a single point-to-point link capacity in optimum case. An important challenge of using full-duplex technology is the interference management. In order to achieve the maximum gain, the base station must properly schedule a pair of the transmit user and the receive user being operated on the same resource as well. In presence of full-duplex small cell environment, associating users to their serving base station becomes a crucial problem. To improve the whole network capacity in such an environment, it is very important to associate the users to both *uplink (UL)* and *downlink (DL)* base stations simultaneously. This leads to *coupled-decoupled association.* Note the following key difference between coupled association and decoupled association:

   (a) *Coupled association* assigns a user to the *same* base station for both uplink and downlink. See Fig. 1, which shows a bi-directional full-duplex mode in which the user is assigned to the same base station for both uplink and downlink transmission.

   (b) *Decoupled association* assigns a user to *different* base stations for uplink and downlink. See Fig. 2, which shows that a three-node full-duplex mode in which the user received the downlink transmission from a macrocell base station and sent the uplink transmission to a different small cell base station.

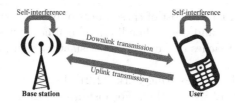

Fig. 1. Coupled user association

Fig. 2. Decoupled user association

In other words, when compared with using only coupled association or using only decoupled association, the use of coupled-decoupled association might increase the network capacity.

In this paper, we examine how the traditional data mining techniques help to analyze the uplink-downlink user association data from wireless heterogeneous networks so as to get some insights about the data and the uplink-downlink phenomena in the networks. In particular, we examine how frequent pattern mining, which aims to discover implicit, previously unknown and potentially useful knowledge in the form of frequently co-occurring events.

The remainder of this paper is organized as follows. The next section discusses related work, and Sect. 3 explains uplink-downlink user association problem. Section 4 describes how frequent pattern mining helps the system model for the joint uplink-downlink user association in wireless heterogeneous networks. Results of our evaluation via simulation are briefly presented in Sect. 5. Finally, conclusions and future work are given in Sect. 6.

2   Related Work

Although there are related work [9] on cell association, many of them are based on signal strength or *signal to interference noise ratio* (*SINR*) and are not user-centric. Most of these existing works require network-level coordination, which increases both complexity and overhead and is undesirable in small cell networks. Singh et al. [26] focused on only decoupled user association. Roth et al. [20] provided self-organizing cell association in small cell networks via the powerful tools of matching theory. While matching theory has recently attracted a lot of attention in wireless networks (e.g., associating channels in ad-hoc and cognitive networks [10,11]), most of these works only focus on the maximization of SINR-based utilities and do not handle small cell network-specific challenges. Moreover, these approaches do not offer a satisfactory solution for non-uniform user distributions.

On the one hand, some existing works [2,21] dealt with the user association problem in either uplink or downlink. For instance, Boccardi et al. [3] explained why to decouple the uplink and downlink in cellular networks from the traditional coupled downlink-uplink model, and described how to decouple the uplink and downlink in cellular networks.

On the other hand, some other existing works consider full-duplexing along with joint uplink-downlink association with a provision for coupled-decoupled association. For instance, Boostanimehr and Bhargava [4] addressed a joint downlink and uplink aware cell association problem in a multi-tier heterogeneous network, in which base stations have finite number of resource blocks to distribute among the users. They also defined an optimization problem to maximize the sum of weighted utility of long term data rate in downlink and uplink through cell association and resource block distribution while maintaining quality of service (QoS). Li et al. [18] aimed to maximize the energy efficiency, which is defined as the ratio of the achieved throughput over the energy cost by optimizing the time allocation for the downlink and multi-user uplink traffic. Sekander et al. [22,23] solved the joint uplink-downlink user association problem in presence of full-duplex small cell network environment to maximize the network rate. They used the notion of matching algorithm, and modified it to provide the users with a flexibility to choose coupled or decoupled association based on their preference. As a result, this association scheme improves the network rate over the conventional joint nearest base station association.

3   Background: Uplink-Downlink User Association Problem

Consider the joint uplink and downlink of an *orthogonal frequency-division multiple access* (*OFDMA*) small cell network having multiple macrocells overlaid with a number of small cell base stations randomly distributed in the coverage area of the *macrocell base station* (*MBS*). Consider an open access scheme, in which all users are allowed to connect to their preferred tier. Assume that all tiers use the same spectrum (i.e., co-channel deployment [17]). Here, consider a single channel, which is used by all the base stations during both uplink and downlink. Consider full-duplex users and base stations (e.g., macrocell base stations, small cell base stations) to improve system spectral efficiency through using the same radio resources for simultaneous transmission and reception.

In a conventional setting, each active user is served by a base station, which offers the highest *received signal strength index* (*RSSI*) or the nearest base station. However, in presence of small cell and full-duplexing technology, assigning users to the base station is not simple. For some users, it might be better to be associated in a joint or coupled manner (i.e., associating to same base station for uplink and downlink), whereas some users will benefit more in a decoupled manner (i.e., associating to different base stations for uplink and downlink). Recall from Sect. 1 that, when using the coupled mode (as shown in Fig. 1), a user is associated to a single base station for both uplink and downlink transmissions in the same frequency band. As such, there is self-interference at both the user side and base station sides. In contrast, when using the decoupled mode (as shown in Fig. 2), a user to is associated to *different* base stations for uplink and downlink transmissions with distinct uplink and downlink frequencies. As such self-interference occurs only at the user side, but neither the transmitter side (i.e., macrocell base station) nor the receiver side (i.e., small cell base station).

Hence, joint association to the same base station for both uplink and downlink has been a common solution to the user association problem. However, in terms of system capacity, one needs to consider the provision for decoupled uplink-downlink association, where a user can associate to different base station for uplink and downlink to maximize its data rate. Consequently, Sekander et al. [22,23] formulated the cell association problem in small cell networks as a many-to-one matching game, in which a set K of users will be assigned to a set L of base stations, where each user will be assigned to at most one base station in uplink and at most one base station in downlink. A *matching game* is defined by two sets of players that evaluate one another using well-defined preference relations [20]. Any arbitrary base station l is assumed to serve a maximum number of users q_d (i.e., quota) in the downlink, and q_u in the uplink. Depending on the channel quality or equivalently signal to interference noise ratio (SINR) values, each user builds a preference relation over the base stations. Furthermore, each base station has a preference over the subset of users based on a predefined utility function. Iteratively, the users propose to their most preferred base station according to their preferences, and base stations accept or reject the proposals based on the utilities they assign to their applicants.

4   Application of Frequent Pattern Mining to Analyzing Uplink-Downlink User Association

Data mining techniques—especially, frequent pattern mining techniques—can help in the *computation of the preference* for both base stations and users. Since the introduction of the frequent pattern mining problem—which aims to discover implicit, previously unknown and potentially useful knowledge in the form of sets of frequently purchased merchandise items in shopper market baskets or sets of collocated events or objects, there have been numerous algorithms. Examples include Apriori [1] (which finds frequent patterns in a breath-first search manner, i.e., levelwise, bottom-up approach), FP-growth [14] and TD-FP-Growth [27] (which both find frequent pattern in a depth-first search manner, i.e., building a global FP-tree for capturing the content of the original database and applying divide-and-conquer to build subsequent sub-trees each capturing the contents of subsequent projected databases in a bottom-up or top-down fashion). Moreover, H-mine [19] also finds frequent patterns in a depth-first search manner, but it does so by updating hyperlinks of hyperlinked array structures instead of updating pointers of tree structures as in FP-growth or TD-FP-Growth. Similarly, B-mine [16] first builds a bitwise table to capture contents of the database, and then recursively finds frequent patterns from this bitwise table in a depth-first search manner. In addition to these horizontal mining algorithms (i.e., transaction-centric algorithms), both Eclat [14] and VIPER [25] find frequent patterns vertically (i.e., domain item-centric). Eclat captures the IDs of transactions containing the item in a list, whereas VIPER captures the same information in a bit vector.

We adapt the aforementioned frequent pattern algorithms and their associated structures to the analysis of uplink-downlink user association. For instance, we adapt the 2-dimensional bitwise table (used in B-mine) into a *3-dimensional bitwise table* to capture the relationships or associations among M users (i.e., x-axis) and B base stations (i.e., y-axis) in both uplink and downlink transmissions (i.e., the new z-axis), for a total of $M \times B \times 2$ bits in the table.

From the resulting bitwise table, we then observe whether or not the coupled mode (i.e., using the same base station for uplink and downlink) or the decoupled mode (i.e., using different base stations) are used for any specific users and base stations. Based on the frequent patterns mined by the B-mine algorithm (adapted for mining the 3-dimensional bitwise table), the sum of uplink and downlink rates are computed. Based on the sum, each user maintains a *ranked list of preference* on which base stations to transmit. Each base station then accommodates at most $q_u + q_d$ users. If the demand is higher than the supply, then the base station keeps those top $q_u + q_d$ users—ranked based on the sum of uplink and downlink rates—so as to maximize the utilities of the wireless heterogeneous networks.

Moreover, by applying vertical mining algorithms (e.g., Eclat, VIPER) to the resulting lists of top $q_u + q_d$ users for all base stations, we find *popular and highly demanding users* who frequently send uplink transmissions to or receive downlink transmissions from the base stations. Based on the results of this frequent pattern mining application, the network designers can then get insight about users' behaviour in the wireless heterogeneous networks and take appropriate business actions (e.g., establish more base stations in the proximity of these demanding users).

Similarly, by applying horizontal mining algorithms (e.g., Apriori, FP-growth, TD-FP-Growth, H-mine, B-mine) to the resulting ranked lists of preference (of preferred base stations) for all users, we find *popular and busy base stations* in which users frequently send uplink transmissions or receive downlink transmissions. Based on the results of this frequent pattern mining application, the network designers can then get insight about the station load in the wireless heterogeneous networks and take appropriate business actions (e.g., establish more base stations in the proximity of these busy base stations so as to reduce the loads of, or the demand for, these base stations).

5   Evaluation

To evaluate the feasibility and practicality of our frequent pattern mining techniques, we conducted a simulation. The results show that (1) our adaptation of the 3-dimensional bitwise table effectively capturing the relationships or associations among users and base stations, (2) our application of vertical frequent pattern mining algorithms effectively finding sets of popular/highly demanding users, and (3) our application of horizontal frequent pattern mining algorithms effectively finding sets of popular/busy base stations. This shows how the mining of uplink-downlink user associations help to get insight about users' behaviours and station loads in wireless heterogeneous networks.

6   Conclusions and Future Work

In this paper, we examined how the data mining techniques for frequent pattern mining helps to solve the research problem of mining uplink-downlink user association. Future work is mainly oriented in exploring system models that are compliant with innovative *big data requirements* (e.g., [8,15]) as well as with *imprecise and uncertain data management* (e.g., [7,29]). Another future direction is to consider *adaptive paradigms* (e.g., [6]), as these would play a relevant role in dynamic wireless heterogeneous networks.

Acknowledgments. This project is partially supported by NSERC (Canada) and University of Manitoba. Thanks E. Hossain and S. Sekander, both from University of Manitoba, for their introduction and expertise on the uplink-downlink association problem.

References

1. Aggarwal, R., Srikant, R.: Fast algorithms for mining association rules. In: Proceedings of VLDB, pp. 487–399 (1994)
2. Bayat, S., Louie, R.H.Y., Han, Z., Vucetic, B., Li, Y.: Distributed user association and femtocell allocation in heterogeneous wireless networks. IEEE TC **62**(8), 3027–3043 (2014)
3. Boccardi, F., Andrews, J.G., Elshaer, H., Dohler, M., Parkvall, S., Popovski, P., Singh, S.: Why to decouple the uplink and downlink in cellular networks and how to do it. IEEE Commun. Mag. **54**(3), 110–117 (2016)
4. Boostanimehr, H., Bhargava, V.K.: Joint downlink and uplink aware cell association in HetNets with QoS provisioning. IEEE TWC **14**(10), 5388–5401 (2015)
5. Cisco visual networking index: global mobile data traffic forecast update, 2015–2020 white paper (2016)
6. Cannataro, M., Cuzzocrea, A., Pugliese, A.: XAHM: an adaptive hypermedia model based on XML. In: Proceedings of SEKE, pp. 627–634 (2002)
7. Cuzzocrea, A., Leung, C.K.: Computing theoretically-sound upper bounds to expected support for frequent pattern mining problems over uncertain big data. In: Carvalho, J.P., Lesot, M.-J., Kaymak, U., Vieira, S., Bouchon-Meunier, B., Yager, R.R. (eds.) IPMU 2016. CCIS, vol. 611, pp. 379–392. Springer, Heidelberg (2016). doi:10.1007/978-3-319-40581-0_31

8. Cuzzocrea, A., Leung, C.K., MacKinnon, R.K.: Mining constrained frequent itemsets from distributed uncertain data. Future Gener. Comput. Syst. **37**, 117–126 (2014)

9. Damnjanovic, A., Montojo, J., Wei, Y., Ji, T., Luo, T., Vajapeyam, M., Yoo, T., Song, O., Malladi, D.: A survey on 3GPP heterogeneous networks. IEEE Commun. Mag. **18**(3), 10–21 (2011)

10. El-Hajj, A.M., Dawy, Z., Saad, W.: A stable matching game for joint uplink/downlink resource allocation in OFDMA wireless networks. In: Proceedings of IEEE ICC, pp. 5354–5359 (2012)

11. Furquim, G., Pessin, G., Gomes, P.H., Mendiondo, E.M., Ueyama, J.: A distributed approach to flood prediction using a WSN and ML: a comparative study of ML techniques in a WSN deployed in Brazil. In: Jackowski, K., Burduk, R., Walkowiak, K., Woźniak, M., Yin, H. (eds.) IDEAL 2015. LNCS, vol. 9375, pp. 485–492. Springer, Heidelberg (2015). doi:10.1007/978-3-319-24834-9_56

12. Ghosh, A., Andrews, J.G., Mangalvedhe, N., Ratasuk, R., Mondal, B., Cudak, M., Visotsky, E., Thomas, T.A., Xia, P., Jo, H.S., Dhillon, H.S., Novlan, T.D.: Heterogeneous cellular networks: from theory to practice. IEEE Commun. Mag. **50**(6), 54–64 (2012)

13. Giri, R., Choi, H., Hoo, K.S., Rao, B.D.: User behavior modeling in a cellular network using latent dirichlet allocation. In: Corchado, E., Lozano, J.A., Quintián, H., Yin, H. (eds.) IDEAL 2014. LNCS, vol. 8669, pp. 36–44. Springer, Heidelberg (2014). doi:10.1007/978-3-319-10840-7_5

14. Han, J., Pei, J., Yin, Y.: Mining frequent patterns without candidate generation. In: Proceedings of ACM SIGMOD, pp. 1–12 (2000)

15. Jiang, F., Leung, C.K.: A data analytic algorithm for managing, querying, and processing uncertain big data in cloud environments. Algorithms **8**(4), 1175–1194 (2015)

16. Jiang, F., Leung, C.K., Zhang, H.: B-mine: frequent pattern mining and its application to knowledge discovery from social networks. In: Li, F., Shim, K., Zheng, K., Liu, G. (eds.) APWeb 2016, Part I. LNCS, vol. 9931. Springer, Heidelberg (2016). doi:10.1007/978-3-319-45814-4_27

17. Jie, X., Yan, Z., Skeie, T., Lang, X.: Downlink spectrum sharing for cognitive radio femtocell networks. IEEE Syst. J. **4**(4), 524–534 (2010)

18. Li, C., Li, Y., Song, K., Yang, L.: Energy efficient design for multiuser downlink energy and uplink information transfer in 5G. Sci. China Inf. Sci. **59**(2), 1–8 (2016)

19. Pei, J., Han, J., Lu, H., Nishio, S., Tang, S., Yang, D.: H-Mine: hyper-structure mining of frequent patterns in large databases. In: Proceedings of IEEE ICDM, pp. 441–448 (2001)

20. Roth, A.E., Oliveira Sotomayor, M.A.: Two-Sided Matching: A Study in Game-Theoretic Modeling and Analysis. Cambridge University Press, Cambridge (1992)

21. Saad, W., Han, Z., Zheng, R., Debbah, M., Vincent Poor, H.: A college admissions game for uplink user association in wireless small cell networks. In: Proceedings of IEEE INFOCOM, pp. 1096–1104 (2014)

22. Sekander, S., Tabassum, H., Hossain, E.: A matching game for decoupled uplink-downlink user association in full-duplex small cell networks. In: Proceedings of IEEE GLOBECOM Workshops (2015). doi:10.1109/GLOCOMW.2015.7414128

23. Sekander, S., Tabassum, H., Hossain, E.: Matching with externalities for decoupled uplink-downlink user association in full-duplex small cell networks. In: Proceedings of IEEE WIECON-ECE, pp. 411–414 (2015)

24. Semiari, O., Saad, W., Valentin, S., Bennis, M., Maham, B.: Matching theory for priority-based cell association in the downlink of wireless small cell networks. In: Proceedings of IEEE ICASSP, pp. 444–448 (2014)
25. Shenoy, P., Bhalotia, J.R., Bawa, M., Shah, D.: Turbo-charging vertical mining of large databases. In: Proceedings of ACM SIGMOD, pp. 22–33 (2000)
26. Singh, S., Zhang, X., Andrews, J.G.: Joint rate and SINR coverage analysis for decoupled uplink-downlink biased cell associations in HetNets. IEEE TWC **14**(10), 5360–5373 (2015)
27. Wang, K., Tang, L., Han, J., Liu, J.: Top down FP-growth for association rule mining. In: Bailey, J., Khan, L., Washio, T., Dobbie, G., Huang, J.Z., Wang, R. (eds.) PAKDD 2016. LNCS (LNAI), vol. 2336, pp. 334–340. Springer, Heidelberg (2002). doi:10.1007/3-540-47887-6_34
28. Ye, Q., Rong, B., Chen, Y., Al-Shalash, M., Caramanis, C., Andrews, J.G.: User association for load balancing in heterogeneous cellular networks. IEEE TWC **12**(6), 2706–2716 (2013)
29. Yu, B., Cuzzocrea, A., Jeong, D.H., Maydebura, S.: On managing very large sensor-network data using Bigtable. In: Proceedings of IEEE/ACM CCGrid, pp. 918–922 (2012)

Multi-label Fuzzy Similarity-Based Nearest-Neighbour Classification Using Association Rule

Yu Rong, Yanpeng Qu[✉], and Ansheng Deng

Information Science and Technology College, Dalian Maritime University,
Dalian 116026, China
yanpengqu@dlmu.edu.cn

Abstract. The demand for multi-label classification methods continues to grow in many modern applications, such as document classification, music categorisation, and semantic scene classification. This paper proposes two multi-label fuzzy similarity-based nearest-neighbour algorithms using the association rules. Specifically, in order to reduce the combination label number and avoid the label overlapping phenomenon, the association rule approach is employed to make the combination labels collapse to a set of sub-labels. Then by transforming the multi-label training data into the single-label representation data, the fuzzy similarity-based nearest-neighbour methods perform the classification label prediction. According to the extracted association rules, the resulting label set is the union of the predicted labels and their associated labels. Apparently, such result set will be more able to maintain the relevance between the labels. Empirical results suggest that the proposed approach can improve the performance and reduce the training time compared with other multi-label classification algorithms.

Keywords: Multi-label classification · Association rule · Fuzzy similarity · Nearest-neighbour

1 Introduction

Multi-label Classification (MLC) is a task to assign the instances to the class which is associated with several labels simultaneously. This topic of research originated in text categorisation, where each document may belong to several defined theme. In this case, the document, with keyword pollution, may simultaneously belong to the theme of environmental protection and policy in document classification [10]. Similarly, in other applications, each image may belong to several

Y. Qu—This work was jointly supported by the National Natural Science Foundation of China (Grant No. 61502068 and 61272171), the Fundamental Research Funds for the Central Universities (Grant No. 3132016211), and the China Postdoctoral Science Foundation (Grant No. 2013M541213 and 2015T80239).

© Springer International Publishing AG 2016
H. Yin et al. (Eds.): IDEAL 2016, LNCS 9937, pp. 542–551, 2016.
DOI: 10.1007/978-3-319-46257-8_58

semantic classes in scene classification [2] and each protein may be connected with a set of functional classes in functional protein categorisation [3]. Since the multi-label learning task is pervasive in real-world problems, much effort has been paid to develop the relevant classification algorithms. Binary relevance method (BR) [4] assumes labels are independent. Thus BR ignores correlations and inter-dependences between labels and this is not always true. Label combination (LC) can aggregate the multi-labels into a series of single-labels, however they can not predict unseen label sets. Meanwhile, LC may cause class imbalance when the number of labels is high. The pruned problem transformation (PPT) [8] method can prune label subset occurring in the training data less than certain times, nevertheless, such strategy will result in the loss of information.

In order to solve the problems of multi-label classification, this paper exploits the inter-dependence between the labels by using the association rules [5]. In this case, the labels, which can be inferred by the other ones, will be removed to reduce the size of the label sets. After combining the labels into a series of single-labels, the fuzzy similarity-based nearest-neighbour methods [7] are used to perform the classification task. Then, these single-label predictions are transformed back to the multi-label predictions. For the completeness, the labels, which are inferred by those predicted labels, will be recovered according to the forementioned association rules. Compared to the conventional methods which only use LC, since the dependencies between the classification labels are taken into account, in proposed algorithm, the unseen label sets can be predicted and the class imbalanced problems can be addressed as well. The experimental results demonstrate that the performance of proposed method is superior to other traditional multi-label learning methods.

The remainder of the paper is structured as follows. In Sect. 2, the preliminaries of multi-label classification, fuzzy similarity-based nearest-neighbour and association rules are reviewed. Section 3 introduces the proposed multi-label classification based on association rule in detail. The new multi-label classifiers are compared with others in an experimental evaluation in Sect. 4. Section 5 concludes the paper with a short discussion of future work.

2   Theoretical Background

2.1   Multi-label Classification

The conception of multi-label classification [8] is distinguished from that of single-label classification. Each training instance in the single label data can associate only one label, however, the training instance in multi-label data can have multiple labels simultaneously. The solution to multi-label classification can be divided into two different groups [12]: the problem transformation methods and the algorithm adaption methods. Commonly, the problem transformation method transforms multi-labeled training data into a single-label representation. Then one or more single-label classifiers are trained by such representation. Via reverse process, the single-label output by single-label classification is reverted into a multi-label output. The algorithm adaption methods extend specific learning algorithms in order to handle multi-label data directly. This paper focuses on the problem transformation methods for multi-label classification.

Moreover, the evaluation measures in multi-label learning need certain proper approaches, since the performance over all labels should be taken into account. For a t-instance multi-label set, $T = \{(X_i, Y_i) \mid 1 \leq i \leq t\}$, where X_i is the conditional feature set and Y_i is the label set, the performance gaugers used in this paper are as follows.

(1) *Accuracy* (Eq. (1)) is the proportion of label values correctly classified of the total number (predicted and actual) of labels for that instance averaged over all instances:

$$Accuracy = \frac{1}{t} \sum_{t}^{i=1} \frac{|Z_i \cap Y_i|}{|Z_i \cup Y_i|} \tag{1}$$

(2) *Hamming Loss* evaluates how many times, on average, an example-label pair is misclassified. This metric takes into account both prediction errors (an incorrect label is predicted) and omission errors (a correct label is not predicted). The lower the value, the better the performance of the classifier. The expression of this metric is given in Eq. (2), where Δ stands for the symmetric difference of two sets, and the 1/q factor is used to obtain a normalised value in $[0, 1]$.

$$Hamming\ Loss = \frac{1}{t} \sum_{t}^{i=1} \frac{1}{q} |Z_i \Delta Y_i| \tag{2}$$

(3) F_1-*Score* (Eq. (3)) can be interpreted as a weighted average of the precision and recall, where an F_1-*Score* reaches its best value at 1 and worst at 0.

$$F_1 = \frac{1}{t} \sum_{t}^{i=1} \frac{2 |Z_i \cap Y_i|}{|Z_i + Y_i|} \tag{3}$$

2.2   Fuzzy Similarity-Based Nearest-Neighbour Classification

The fuzzy similarity-based nearest-neighbour methods, similarity nearest-neighbour (SNN) and aggregated-similarity nearest-neighbour (ASNN), are proposed in [7]. These two approaches do not rely on the concepts or framework of fuzzy-rough sets, but are equivalent to the fuzzy-rough nearest-neighbour (FRNN) and vaguely-quantified rough sets (VQNN) methods, respectively.

The algorithm of SNN is outlined in Algorithm 1. It calculates the similarity $\mu_{R_P}(x, y)$ between the objects x and y for a feature set P. Such similarity degree is defined in Eq. (4), where T is the T-norm, $\mu_{R_a}(x, y)$ is the similarity between the objects x and y for feature a.. The value of $\mu_{R_P}(x, y)$ is then compared with the existing value (τ). If the value for the currently considered class is higher, then τ is updated with this value and the class label is assigned to this test object. If not, the algorithm continues to iterate through all of the remaining training instances.

$$\mu_{R_P}(x, y) = T_{a \in P}\{\mu_{R_a}(x, y)\}. \tag{4}$$

Algorithm 1. The similarity nearest-neighbour

Require:

    $\mathbb{U}$: the training set;

    y: the object to be classified;

1: $\tau \leftarrow 0, Class \leftarrow \emptyset$

2: $\forall x \in \mathbb{U}$

3:    **if**$(\mu_{R_P}(x, y) \geq \tau)$

4:      $Class \leftarrow l(x)$

5:      $\tau \leftarrow \mu_{R_P}(x, y)$

6: **output** $Class$

The Algorithm 2 summarise the ASNN methods. It works by iteratively examining each of the decision classes in the training data. It calculates the membership of the test data object under consideration to the greatest similarity for each class, which is then compared with the highest existing value (τ). If the greatest similarity for the currently considered class is higher, then τ is updated with this value and the class label is assigned to this test object. If not, the algorithm continues to iterate through all of the remaining decision classes.

Algorithm 2. The aggregated-similarity nearest-neighbour

Require:

    $\mathbb{U}$: the training set;

    $\mathcal{C}$: the set of decision classes;

    y: the object to be classified;

1: $N \leftarrow$ get Nearest-neighbour(y, k)

2: $\tau \leftarrow 0, Class \leftarrow \emptyset$

3: $\forall X \in \mathcal{C}$

4:    **if**$(\sum_{x \in N, l(x)=X} \mu_{R_P}(x, y) \geq \tau)$

5:      $Class \leftarrow X$

6:      $\tau \leftarrow \sum_{x \in N, l(x)=X} \mu_{R_P}(x, y)$

7: **output** $Class$

2.3 Association Rules

Association rule learning is a method that exploits the interesting relations between variables in large databases. The strong rules discovered in databases are identified by some interesting measures. The best-known constraints are minimum thresholds on the values of support and confidence. Specifically, let X be an item-set, $X \Rightarrow Y$ is an association rule and T is a set of transactions of a given database. The support value of X with respect to T is defined as the proportion of transactions in the database which contains the item-set X. The confidence value of a rule, $X \Rightarrow Y$, with respect to a set of transactions T, is the proportion of the transactions that contains X which also contains Y.

The apriori [1] algorithm is designed to operate on databases containing transactions. Apriori counts candidate item sets efficiently by the breadth-first search and a hash tree structure. Candidate item sets of length k are generated from item sets of length $k - 1$. Then the candidates are pruned which have an infrequent sub pattern. The candidate set contains all frequent the k-length item sets. After that, it scans the transaction database to determine the frequent item sets among the candidates.

In the previous work, the association rules theory has been applied in the single-label classification problem. However, there is no method that applies the association rules theory in the multi-label classification problem to exploit the inter-dependence between the labels. The mechanisms of them are distinctly different.

3   Proposed Method

The correlations and inter-dependencies between labels are ignored in the conventional methods for MLC. However, the correlation between the labels in the multi-label data can be summarised as the following three conditions:

(1) label x may only exist itself alone;
(2) label x and y may often appear together;
(3) label x and y may never occur together.

Therefore, strong correlation may exist between the labels in multi-label problems. Such relationships are revealed by association rules in this paper. In so doing, the labels, which can be inferred by the other ones, will be removed to reduce the size of the label sets. After combining the labels into a series of single-labels, the fuzzy similarity-based nearest-neighbour methods, SNN and ASNN, are used to perform the classification task. Then, these single-label predictions are transformed back to the multi-label predictions. For the completeness, the labels, which are inferred by those predicted labels, will be recovered according to the forementioned association rules. Compared to the conventional methods which only use LC, since the dependencies between the classification labels are taken into account, in proposed algorithm, the unseen label sets can be predicted and the class imbalanced problems can be addressed as well. The main procedure of the proposed algorithm consists of the following.

1. *Label initialisation*: The set of association rule (*rules*) and predicted label set (*L*) are initialised with an empty set.
2. *Generate rules*: The association rule (*rules*) on the label set of all training instances will be created by the apriori algorithm. The *rules* set must be satisfied with the minimum thresholds of support degree and confident degree.

3. *Reduce the label set*: In order to reduce the size of the label set, the labels, which can be inferred by the other ones, will be removed.
4. *Combine labels*: The multi-label set of each training instance is transformed into a single-label by LC. The labels of each instance are combined as a class.
5. *SNN/ASNN classification*: SNN and ASNN are used to train the single representatives and achieve the single-label predictions.
6. *Recovery to multi-label prediction*: Via reverse process of label combination, the single-label outputs by single-label classification are reverted into a set of multi-label outputs.
7. *Form final prediction*: In this process, the labels, which are associated with the predicted label set, will be rejoined into the label sets as the final predictions. Following the above discussion, the proposed algorithm is summarised in Algorithm 3.

Algorithm 3. Multi-label Fuzzy Similarity-based Nearest-neighbour Classification Using Association Rule

---

Require:

$U = \{\mathbb{X}, \mathbb{Y}\}$, where U: training instances, $\mathbb{X}$: feature sets, and $\mathbb{Y}$: label sets;
t: the testing instance;
$minSup$: the minimum threshold of support degree;
$minConf$: the minimum threshold of confident degree;

1: **Initialise** $rules = \varnothing$; $L = \varnothing$;
2: **Generate** $rules$ by $minSup$ and $minConf$;
3: **Order** $rules$ by $minConf$;
4: **For** $Y_i \subseteq \mathbb{Y}$ // Tune labels by $rules$
5:       **For** $y_i \in Y_i$
6:            **For** $rule_i \subseteq rules$
7:                 **If** $y_i \in rule_i$
8:                      $Y_i = Y_i - y_i$;
9:                 **End**
10:           **End**
11:      **End**
12: **End**
13: **Transform** labels into class like LC
14: $L \leftarrow SNN(t)$ or $ASNN(t)$ // Train single-label classifier
15: **Recovery** L back to multi-label predictions
16: **For** $L_i \in L$ // Join associated labels by $rules$
17:       **For** $rule_i \in rules$
18:            **If** $L_i \subseteq rule_i$
19:                 $L = L \cup rule_i$
20:           **End**
21:      **End**
22: **End**
23: **Return** L

---

4   Experimentation

4.1   Experimental Setting

Table 1 displays a collection of multi-label datasets, with certain relevant statistics, such as the number of instances, the number of attributes, the number of labels, label cardinality (LC, see Eq. (5)) and label density (LD, see Eq. (6)).

$$LC(D) = \frac{\sum_{i=1}^{|D|} |Y_i|}{|D|}. \tag{5}$$

$$LD(D) = \frac{\sum_{i=1}^{|D|} |Y_i|}{|L|}. \tag{6}$$

Table 1. Multi-label datasets used for experiments and associated properties

Dataset	Instances	Attributes	Labels	LC	LD
Yeast	2417	103	14	4.24	0.303
Scene	2407	294	6	1.074	0.179
Music	592	71	6	1.892	0.315

In this paper, the multi-label SNN (MLSNN) and multi-label ASNN (MLASNN) are compared to the alternative multi-label learning algorithms, such as, BR, LC, Classifier Chain (CC) [9], PPT, RAndom k-labELsets method (RAkEL) [11]. The settings of the competitors are implemented in meka [6]. The experiment is conducted based on the training and testing split datasets, which are 66 % and 33 % of the datasets, respectively. The proposed methods are sensitive to the parametric values. For example, the parameters of minSup and minConf are set to be 0.1 and 0.7 on Music dataset. The three representatives of the multi-label datasets are chosen from meka.

4.2   Results and Discussion

As shown in Table 2, the proposed methods of MLASNN and MLSNN have a good performance in terms of all the evaluation criteria on the music dataset. Especially, on accuracy metric, MLASNN and MLSNN outperform other methods absolutely. While in terms of Hamming Loss and F_1-score, the method of RAkEL performs better. The reason for this case may be that the mechanism of the RAkEL is creating m random sets of k label combinations and employing ensemble voting process to arrive at a decision for final classification set. However, the time complexity of the RAkEL is expensive. The performances of the

Table 2. Experiment result on music data set

Method	Accuracy	Hamming loss	F1-score
BR(IB1)	0.304	0.696	0.464
BR(J48)	0.390	0.318	0.535
LC(IB1)	0.490	0.248	0.593
LC(J48)	0.445	0.281	0.544
CC(IB1)	0.490	0.248	0.599
CC(J48)	0.408	0.295	0.521
RAkEL(IB1)	0.490	0.248	0.599
RAkEL(BayesNet)	0.489	0.243	0.608
MLASNN	0.517	0.248	0.586
MLSNN	0.501	0.252	0.543

Table 3. Experiment result on yeast data set

Method	Accuracy	Hamming loss	F1-score
BR(IB1)	0.493	0.239	0.450
BR(NavieBayes)	0.404	0.281	0.435
LC(IB1)	0.493	0.239	0.450
LC(J48)	0.403	0.278	0.383
CC(IB1)	0.493	0.239	0.608
CC(J48)	0.413	0.278	0.539
RAkEL(J48)	0.416	0.325	0.561
RAkEL(NavieBayes)	0.419	0.324	0.552
MLASNN	0.514	0.221	0.569
MLSNN	0.500	0.245	0.550

MLASNN and MLSNN on the yeast dataset are shown in Table 3. Apparently, MLASNN and MLSNN outperform most of the other algorithms. Especially, the accuracy results achieved by the proposed method are rather superior to the others. Table 4 shows that the proposed methods perform on the scene dataset in terms of all the evaluation criteria. It is also worth noting that the method of BR with IB1 performs quite poorly compared to the other algorithms.

Table 4. Experiment result on scene data set

Method	Accuracy	Hamming loss	F1-score
$BR(IB1)$	0.182	0.818	0.305
$BR(J48)$	0.506	0.143	0.618
$LC(IB1)$	0.669	0.114	0.697
$LC(J48)$	0.583	0.145	0.601
$CC(J48)$	0.581	0.147	0.593
$CC(NavieBayes)$	0.451	0.239	0.558
$RAkEL(J48)$	0.614	0.131	0.667
$RAkEL(NavieBayes)$	0.520	0.151	0.594
$MLASNN$	0.669	0.111	0.670
$MLSNN$	0.651	0.120	0.659

5   Conclusion

This paper has presented the method of multi-label fuzzy similarity-based nearest-neighbour classification using association rules. In proposed method, the dependence between labels in multi-label learning is fully considered by the association rules of labels. In this case, the labels, which can be inferred by the other ones, are removed before label combining and recovered in the final prediction sets. After combining the labels into a series of single-labels, the fuzzy similarity-based nearest-neighbour methods [7] are used to perform the classification task. Experimental results for three multi-label learning datasets show that the proposed method outperforms some state-of-the-art methods.

Topics for further investigation include the studying of using different single-label classifiers in the classification stage. An investigation into how the proposed method may be cooperated with the attribute selection methods for multi-label data to operate in a hierarchical setting remains active research.

References

1. Agrawal, R., Srikant, R.: Fast algorithms for mining association rules in large databases. In: Proceedings of the 20th International Conference on Very Large Data Bases, pp. 487–499. The Department of Computer Science, University of Wisconsin, Madison (1994)
2. Boutell, M.R., Luo, J., Shen, X., Brown, C.M.: Learning multi-label scene classification. Pattern Recogn. **37**, 1757–1771 (2004)
3. Elisseeff, A., Weston, J.: A Kernel method for multi-labelled classification. Adv. Neural Inf. Process. Syst. **14**, 681–687 (2001)
4. Godbole, S., Sarawagi, S.: Discriminative methods for multi-labeled classification. In: Dai, H., Srikant, R., Zhang, C. (eds.) PAKDD 2004. LNCS (LNAI), vol. 3056, pp. 22–30. Springer, Heidelberg (2004)

5. Hipp, J., Gntzer, U., Nakhaeizadeh, G.: Algorithms for association rule mining a general survey and comparison. ACM SIGKDD Explor. Newslett. **2**, 28–64 (2000)
6. Modi, H., Panchal, M.: Experimental comparison of different problem transformation methods for multi-label classification using meka. Int. J. Comput. Appl. **59**(15), 10–15 (2012)
7. Qu, Y., Shen, Q., Parthaláin, N.M., Shang, C., Wu, W.: Fuzzy similarity-based nearest-neighbour classification as alternatives to their fuzzy-rough parallels. Int. J. Approximate Reasoning **54**, 184–195 (2013)
8. Read, J., Pfahringer, B., Holmes, G.: Multi-label classification using ensembles of pruned sets. In: Proceedings of the Eighth IEEE International Conference on Data Mining, pp. 995–1000 (2008)
9. Read, J., Pfahringer, B., Holmes, G., Frank, E.: Classifier chains for multi-label classification. Mach. Learn. **85**, 333–359 (2011)
10. Schapire, R.E., Singer, Y.: Boostexter: a boosting-based system for text categorization. Mach. Learn. **39**, 135–168 (2000)
11. Tsoumakas, G., Vlahavas, I.: Random k-labelsets: an ensemble method for multi-label classification. In: Kok, J.N., Koronacki, J., de Mantaras, R.L., Matwin, S., Mladenič, D., Skowron, A. (eds.) Machine Learning: ECML 2007. LNCS, vol. 4701. Springer, Heidelberg (2007)
12. Zhang, M.L., Zhou, Z.H.: A review on multi-label learning algorithms. IEEE Trans. Knowl. Data Eng. **26**(8), 1819–1837 (2014)

Probabilistic Modelling for Delay Estimation in Gravitationally Lensed Photon Streams

Sultanah Al Otaibi[1,2], Peter Tiňo[1(✉)], and Somak Raychaudhury[3,4,5]

[1] School of Computer Science, University of Birmingham, Birmingham B15 2TT, UK
P.Tino@cs.bham.ac.uk
[2] College of Computer and Information Sciences, King Saud University,
Riyadh 12371, Saudi Arabia
[3] School of Physics and Astronomy, University of Birmingham,
Birmingham B15 2TT, UK
[4] Inter-University Centre for Astronomy and Astrophysics, Pune 411007, India
[5] Department of Physics, Presidency University, Kolkata 700073, India

Abstract. We test whether a more principled treatment of delay estimation in lensed photon streams, compared with the standard kernel estimation method, can have benefits of more accurate (less biased) and/or more stable (less variance) estimation. To that end, we propose a delay estimation method in which a single latent inhomogeneous Poisson process underlying the lensed photon streams is imposed. The rate function model is formulated as a linear combination of nonlinear basis functions. Such unifying rate function is then used in delay estimation based on the corresponding Innovation Process. This method is compared with a more straightforward and less principled baseline method based on kernel estimation of the rate function. Somewhat surprisingly, the overall emerging picture is that the theoretically more principled method does not bring much practical benefit in terms of the bias/variance of the delay estimation. This is in contrast to our previous findings on daily flux data.

Keywords: Gravitational lensing · Non-homogeneous Poisson process · Kernel estimation methods

1 Introduction

Time delays between images of strongly-lensed distant variable sources can serve as a valuable tool for cosmography, provided that time delays between the image fluxes can be accurately measured (e.g. [8,16]). A number of methods have been developed to accurately estimate time delays. These include the dispersion spectra method [3] and kernel-based method with variable width (K-V) [4,5]. Actively studied strong quasars with time-delay measurements include RXJ1131−1231 [21] and B1608+656 [6,8]; Q0957+561 (e.g. [9]). Available data are usually in the form of daily measurements which can be used to predict longer (days and months) delays. Current methods in astrophysics are solely

© Springer International Publishing AG 2016
H. Yin et al. (Eds.): IDEAL 2016, LNCS 9937, pp. 552–559, 2016.
DOI: 10.1007/978-3-319-46257-8_59

rooted in this scenario. However, when countering the problem of shorter delays (e.g. hours), daily measurements are insufficient and one needs to investigate the individual arrival times of photons.

Poisson process can applied as a model for photon streams [15]. To resolve the delay in gravitationally lensed photon streams one can use the standard kernel based estimation of the inhomogeneous Poisson process rate function on individual photon streams and then try to time-shift the rate function estimates so as the overlap is maximized. Another, more principled alternative is to impose that the source of the delayed photon streams is the same and we simply observe different realizations from the same inhomogeneous Poisson process, gravitationally delayed in time. We study whether, comparing with the standard kernel based baseline, such a principled approach can bring benefits in terms of more stable (less variance) estimation.

Normally, delay estimation would be done over streams of photons from a given energy band and then unified over a multitude of energy bands. The baseline and principled delay estimation methods are then compared in a controlled experimental setting using synthetic photon fluxes with known imposed delay from a variety of inhomogeneous processes assumed to come from a single energy band. To our best knowledge this is the first systematic study that addresses the problem of delay estimation on lensed photon streams. We do not perform experiments on real data, since no large real photon streams from known delayed systems with short time delay are available. Nevertheless, this study serves as a proof of concept and be readily used once appropriate lensed photon streams become available.

2   Kernel Based Delay Estimation in Lensed Photon Streams

For the sake of simplicity we will deal with the case of two lensed photon streams A and B from the same source. All techniques presented in this paper can be easily generalized to multiple streams. We assume that the observed photon streams can be accounted for by a Poisson process (e.g. [18]). In the non-homogeneous Poisson process (NHPP) the mean rate function $\lambda(s)$ varies over time s. Given a series of arrival times $s_1, s_2, ..., s_S$ over an interval $[0, T]$, the rate function is commonly estimated by imposing a (Gaussian) kernel of width r on top of each arrival time s_i,

$$K_g(s; s_i, r) = \exp\left\{-\frac{(s - s_i)^2}{2r^2}\right\}. \tag{1}$$

The rate function estimate (up to scaling) is then [12–14]

$$\hat{\lambda}(s) = \sum_{i=1}^{S} K_g(s; s_i, r). \tag{2}$$

We will refer to this method as Kernel Rate Estimation (KRE).

Suppose that we observe two lensed photon streams $\{s_i^A\}_{i=1}^{S^A}$ and $\{s_i^B\}_{i=1}^{S^B}$ from the same source. On each stream we produce a kernel based estimate of the rate function $\hat{\lambda}^A(s)$, $\hat{\lambda}^B(s)$. Given a suggested time delay Δ, the closeness of the rate estimates (under the delay Δ) can be evaluated e.g. through the mean square difference eventuated on a regular grid of time stamps $\{z_j\}_{j=1}^Z$ in the relevant time interval,

$$d_2(\hat{\lambda}^A, \hat{\lambda}^B; \Delta) = \frac{1}{Z} \sum_{j=1}^{Z} (\hat{\lambda}^A(z_j) - \hat{\lambda}^B(z_j))^2. \tag{3}$$

The delay is then estimated through minimization of $d_2(\hat{\lambda}^A, \hat{\lambda}^B; \Delta)$ with respect to Δ (e.g. via gradient descent). In the following sections we will introduce two variants of delay estimation based on innovation process corresponding to the underlying Poisson process.

3   Innovation Process Based Estimation (IPE)

Recall that if event counts can be modeled by Poisson distribution with mean rate λ, then the inter-arrival times are distributed with exponential distribution with mean λ^{-1}. We denote the differences between two consecutive arrival times by $d^A = \{d_i^A\}_{i=1}^{D^A}$ and $d^B = \{d_i^B\}_{i=1}^{D^B}$, where $d_i^A = s_{i+1}^A - s_i^A$ and $d_i^B = s_{i+1}^B - s_i^B$, respectively. Our goal is to find a probabilistic model that maximizes the probability $P(d^A, d^B | \lambda(s))$,

$$P(d^A, d^B | \lambda^A(s), \lambda^B(s)) = \prod_{i=1}^{D^A} P(d_i^A | \lambda^A(s_i; w)) \prod_{i=1}^{D^B} P(d_i^B | \lambda^B(s_i; w, \Delta)), \tag{4}$$

where $P(d | \lambda) = \lambda e^{-\lambda d}$. We impose a kernel based model on the *common rate function* underlying both streams (expressed for stream A):

$$\lambda^A(s) = \sum_{j=1}^{J} w_j K_g(s; c_j, r_o) = w^\mathsf{T} K_g(s; c, r_o), \tag{5}$$

with kernels of width r_o, centered at $c_j, j = 1, 2 \ldots J$ and the J free parameters w_j collected in vector w. $K_g(s; c, r_o)$ is a vector of kernel evaluations $K_g(s; c_j, r_o)$ at all centers of $c = (c_1, c_2, ..., c_J)$. The rate function of stream B is a time-delayed (by Δ) version of the one for stream A:

$$\lambda^B(s) = \sum_{j=1}^{J} w_j K_g(s; c_j - \Delta, r_o) = w^\mathsf{T} K_g(s; c - \Delta, r_o), \tag{6}$$

We thus obtain

$$P(d^A, d^B | \lambda^A(s), \lambda^B(s)) = \prod_{i=1}^{D^A} \lambda^A(s_i) e^{-\lambda^A(s_i) d_i^A} \prod_{i=1}^{D^B} \lambda^B(s_i) e^{-\lambda^B(s_i) d_i^B}, \tag{7}$$

leading to the error functional (negative log likelihood),

$$E = -\sum_{i=1}^{D^A}(\log \lambda^A(s_i) - \lambda^A(s_i)d_i^A) - \sum_{i=1}^{D^B}(\log \lambda^B(s_i) - \lambda^B(s_i)d_i^B). \quad (8)$$

We minimize E w.r.t two parameters $(\boldsymbol{w}, \Delta)$ via gradient descent. To that end we plug (5) and (6) into (8),

$$E = -\sum_{i=1}^{D^A}\left(\log \sum_{j=1}^{J} w_j K_g(s_i^A; c_j, r_o) - d_i^A \sum_{j=1}^{J} w_j K_g(s_i^A; c_j, r_o)\right)$$
$$-\sum_{i=1}^{D^B}\left(\log \sum_{j=1}^{J} w_j K_g(s_i^B; c_j - \Delta, r_o) - d_i^B \sum_{j=1}^{J} w_j K_g(s_i^B; c_j - \Delta, r_o)\right), \quad (9)$$

leading to,

$$\frac{\partial E}{\partial \boldsymbol{w}} = -\sum_{i=1}^{D^A}\left(\frac{K_g(s_i^A; \boldsymbol{c}, r_o)}{\boldsymbol{w}^\intercal K_g(s_i^A; \boldsymbol{c}, r_o)} - d_i^A K_g(s_i^A; \boldsymbol{c}, r_o)\right)$$
$$-\sum_{i=1}^{D^B}\left(\frac{K_g(s_i^B; \boldsymbol{c} - \Delta \cdot \mathbf{1}, r_o)}{\boldsymbol{w}^\intercal K_g(s_i^B; \boldsymbol{c} - \Delta \cdot \mathbf{1}, r_o)} - d_i^B K_g(s_i^B; \boldsymbol{c} - \Delta \cdot \mathbf{1}, r_o)\right), \quad (10)$$

where $\mathbf{1}$ is a vector of 1's and

$$\frac{\partial E}{\partial \Delta} = -\sum_{i=1}^{D^B}\left(\frac{1}{\sum_{j=1}^{J} w_j \exp\{\frac{-(s_i^B - (c_j - \Delta))^2}{2r_o^2}\}} \sum_{j=1}^{J} w_j \exp\{\frac{-(s_i^B - (c_j - \Delta))^2}{2r_o^2}\}\frac{-2(s_i^B - (c_j - \Delta))}{2r_o^2}\right.$$
$$\left. - d_i^B \sum_{j=1}^{J} w_j \exp\{\frac{-(s_i^B - (c_j - \Delta))^2}{2r_o^2}\}\frac{-2(s_i^B - (c_j - \Delta))}{2r_o^2}\right).$$
$$(11)$$

4   Parameters Initialization

Gaussian kernels have two parameters need to be determined, in particular kernel centers $\{c_j\}_{j=1}^{J}$ and the kernel width r. As explained above, in KRE, kernels are centered at each photon's arrival time, whereas in IPE, the centers c_j are uniformly distributed across the time period $[0, T]$.

The kernel width determines the degree of smoothing for the underlying rate function. For KRE, we apply a method for selecting the width based on the principle of minimizing the mean integrated square error (MISE) proposed by [19]. For IPE, the kernel width r_o is optimized using cross-validation method proposed in [4,10]. The algorithm partitions the data into 10 blocks of equal length $\mathcal{L}$. The i-th validation set $\mathcal{V}_i$, $i = 1, 2 \ldots \mathcal{L}$, is obtained by collecting the

i-th element of each block. The rest of the data is the "training set". We then fit our models on the training set and use the validation set $\mathcal{V}_i$ to calculate the cost function E over a range of suggested width values $r_o \in (L_{r_o}, U_{r_o})$. This procedure is repeated $\mathcal{L}$ times for each validation set $\mathcal{V}_i$, $i = 1, 2 \ldots \mathcal{L}$. The chosen r_o is the one yielding the smallest average cost E across the folds $i = 1, 2 \ldots \mathcal{L}$.

The IPE weight vector $\boldsymbol{w}$ is initialized using the rate function estimates readily provided by the KRE model. However, the rate functions obtained by KRE on streams A and B need to be scaled to represent the underlying rate of the non-homogeneous Poisson process[1]. Given the KRE-estimated rate functions on streams A and B, $\hat{\lambda}^A(s)$, $\hat{\lambda}^B(s)$, respectively, the overall KRE rate function is their average

$$\hat{\lambda}(s) = \frac{\hat{\lambda}^A(s) + \hat{\lambda}^B(s)}{2}. \tag{12}$$

The scaling factor ϑ is obtained by imposing the rate function $\lambda(s) = \vartheta \hat{\lambda}(s)$ and minimizing (9) with respect to ϑ. Denoting $\hat{\lambda}(s_i^A)d_i^A$ and $\hat{\lambda}(s_i^B)d_i^B$ by q_i^A and q_i^B, respectively, it can be shown that the minimum is obtained at

$$\vartheta = \frac{D^A + D^B}{\sum\limits_{i=1}^{D^A} q_i^A + \sum\limits_{i=1}^{D^B} q_i^B}. \tag{13}$$

Setting of IPE weights to match the rate function $\lambda(s)$ can then be done by imposing a regular $(s_1, s_2, ..., s_N)$ grid on $[0, T]$, evaluating the rate values on the grid, $\boldsymbol{x} = (\hat{\lambda}(s_1), \hat{\lambda}(s_2) \ldots \hat{\lambda}(s_N))^{\intercal}$, and solving

$$\boldsymbol{w} = \boldsymbol{K}^{\intercal +}\boldsymbol{x}, \tag{14}$$

where $\boldsymbol{K}$ is an $N \times N$ matrix

$$\boldsymbol{K} = [K_g(s_1; \boldsymbol{c}, r_o), K_g(s_2; \boldsymbol{c}, r_o), \ldots K_g(s_N; \boldsymbol{c}, r_o)]. \tag{15}$$

5   Experiments

To test and compare the methodologies suggested above, we performed controlled experiments on synthetic data generated from non-homogeneous Poisson processes. From each given non-homogeneous Poisson processes we generated two series A and B of arrival times, the series B was then time-shifted by a known delay Δ.

The rate functions defining non-homogeneous Poisson processes were obtained by superimposing G Gaussian functions of fixed width r_g positioned on a regular grid $\{c_g\}_{g=1}^G$ in $[0, T]$,

$$\lambda(s) = \sum\limits_{g=1}^{G} w_g \cdot K_g(s; c_g, r_g), \tag{16}$$

---

[1] Note that for the delay detection task for which the KRE method is used, no such scaling was needed - the delay is invariant to scaling the estimated rate functions by the same factor.

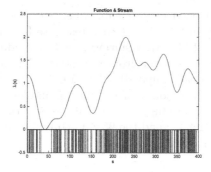

Fig. 1. An example of a test rate function and the corresponding photon stream.

where $w_g \in \mathbb{R}$ are the mixing weights generated randomly from uniform distribution on $[L_w, U_w]$. The kernel widths were set to a multiple of the kernel separation (distance between the two consecutive kernel centers) d_g, $r_g = \alpha_g \cdot d_g$. We used $T = 400$, $G = 80$, $\alpha_g = 3$, $L_w = -1$ and $U_w = 1$. The synthetic rate functions were then rescaled to the interval $[0.2]$. Given a rate function $\lambda(s)$, the arrival times were generated using the Thinning technique [2,7,11,17,20]. An example of a test rate function and the corresponding photon stream is shown in Fig. 1.

Using this process, we generate two photon streams from the same rate function: $\{s_i^A\}$ and $\{s_j^B\}$, $i = 1, 2, \cdots, S^A$ and $j = 1, 2, \cdots, S^B$. To create a pair of time shifted streams, s^B is shifted in time by a delay $\Delta > 0$

$$s_i^B \leftarrow s_i^B + \Delta, \forall i = 1, 2, \cdots, S^B \tag{17}$$

To prepare the streams for experiments, we cut the two streams to ensure they have the same start and end point in time.

Table 1. Statistical analysis of delay estimates. The results for each method and each imposed delay $\Delta \in \{20, 22, 25, 28\}$ are averaged over 100 test rate functions. The time delay trial values were taken from the interval $[10, 40]$ with increment of 10.

Delay	Method	$\mu \pm \sigma$	MAE	CI range	95 % CI
20	KRE	21.01±3.81	1.87	0.75	[20.26,21.76]
	IPE	20.88±3.93	1.73	0.77	[20.11,21.65]
22	KRE	22.68±4.50	3.49	0.88	[21.80,23.56]
	IPE	22.40±4.61	3.74	0.90	[21.50,23.30]
25	KRE	25.44±5.70	5.09	1.12	[24.32,26.56]
	IPE	25.55±5.96	5.54	1.17	[24.38,26.72]
28	KRE	28.56±4.30	3.27	0.84	[27.72,29.40]
	IPE	28.58±4.32	3.34	0.85	[27.73,29.43]

We performed controlled experiments, where 100 test rate functions were generated as described in Sect. 5. For each test rate function we imposed four delay values $\Delta \in \{20, 22, 25, 28\}$, resulting in 400 individual experiments. The time delay trial values were taken from the interval $[10, 40]$ with increment of 10. For each model and each imposed delay $\Delta \in \{20, 22, 25, 28\}$, we report the mean μ and standard deviation σ of the maximum-likelihood delay estimates across the set of 100 test rate functions. We also report the mean absolute error (MAE) of the delay estimates and the 95 % Credibility Interval (CI). Summary results are presented in Table 1.

6 Conclusion

In this paper, we tested whether a more principled treatment of delay estimation in lensed photon streams, compared with the standard kernel estimation method, can have benefits of a more accurate (less biased) and/or more stable (less variance) estimation. In particular, we formulated a baseline method (KRE) based on kernel estimation of the rate function of inhomogeneous Poisson process. The delay estimate is refined using gradient descent in the delay parameter on the error functional.

A more principled delay estimation relied on imposing a single latent inhomogeneous Poisson process underlying the lensed photon streams. The rate function model was formulated as a linear combination of nonlinear basis functions, thus making the non-linear model linear in the mixing parameters. We tested this idea in the Innovation Process Based Estimation (IPE).

Somewhat surprisingly, the overall emerging picture is that the theoretically more principled methods do not bring much practical benefit in terms of the bias/variance of the delay estimation. This is in contrast to our previous findings on daily flux data [1,4,5]. It appears that because the underlying latent rate function is represented only implicitly through the streams of arrival times weakens the stabilizing factor of the single unified intensity function that proved so useful in the case of daily flux data [1,4,5]. Indeed, in that case, knowing the amount of observational noise and observing noisy flux levels gave much better clues as to what the common source variability could be, thus stabilizing the delay estimation. Nevertheless, we propose that a study of the kind is useful and necessary for future developments of alternative methods for the delay estimation in lensed photon streams.

References

1. Al Otaibi, S., Tiňo, P., Cuevas-Tello, J.C., Mandel, I., Raychaudhury, S.: Kernel regression estimates of time delays between gravitationally lensed fluxes. MNRAS **459**(1), 573–584 (2016)
2. Bratley, P., Fox, B., Schrage, L.E.: A Guide to Simulation, 2nd edn. Springer, New York (1987)

3. Courbin, F., Chantry, V., Revaz, Y., Sluse, D., Faure, C., Tewes, M., Eulaers, E., Koleva, M., Asfandiyarov, I., Dye, S., Magain, P., van Winckel, H., Coles, J., Saha, P., Ibrahimov, M., Meylan, G.: COSMOGRAIL: the COSmological MOnitoring of GRAvItational Lenses IX. Time delays, lens dynamics and baryonic fraction in HE 0435–1223. Astron. Astrophys. **536**, A53 (2011)
4. Cuevas-Tello, J.C., Tiňo, P., Raychaudhury, S.: How accurate are the time delay estimates in gravitational lensing? Astron. Astrophys. **454**, 695–706 (2006)
5. Cuevas-Tello, J.C., Tiňo, P., Raychaudhury, S., Yao, X., Harva, M.: Uncovering delayed patterns in noisy and irregularly sampled time series: an astronomy application. Pattern Recogn. **43**(3), 1165–1179 (2009)
6. Fassnacht, C.D., Xanthopoulos, E., Koopmans, L.V.E., Rusin, D.: A determination of H_0 with the CLASS gravitational lens B1608+656 III. A significant improvement in the precision of the time delay measurements. Astrophys. J. **581**, 823–835 (2002)
7. Fathi-Vajargah, B., Khoshkar-Foshtomi, H.: Simulating nonhomogeneous poisson point process based on multi criteria intensity function and comparison with its simple form. J. Math. Comput. Sci. (JMCS) **9**(3), 133–138 (2014)
8. Greene, Z.S., Suyu, S.H., Treu, T., Hilbert, S., Auger, M.W., Collett, T.E., Marshall, P.J., Fassnacht, C.D., Blandford, R.D., Bradač, M., Koopmans, L.V.E.: Improving the precision of time-delay cosmography with observations of galaxies along the line of sight. Astrophys. J. **768**(1), 39 (2013)
9. Hainline, L.J., Morgan, C.W., Beach, J.N., Kochanek, C.S., Harris, H.C., Tilleman, T., Fadely, R., Falco, E.E., Le, T.X.: A new microlensing event in the doubly imaged Quasar Q 0957+561. Astrophys. J. **744**(2), 104 (2012)
10. Hastie, T., Tibshirani, R., Friedman, J., Franklin, J.: The elements of statistical learning: data mining, inference and prediction. Math. Intell. **27**(2), 83–85 (2005)
11. Lewis, P.A., Shedler, G.S.: Simulation of nonhomogeneous poisson processes by thinning. Nav. Res. Logistics Q. **26**(3), 403–413 (1979)
12. Nawrot, M., Aertsen, A., Rotter, S.: Single-trial estimation of neuronal firing rates: from single-neuron spike trains to population activity. J. Neurosci. Meth. **94**(1), 81–92 (1999)
13. Park, B.U., Marron, J.S.: Comparison of data-driven bandwidth selectors. J. Am. Stat. Assoc. **85**(409), 66–72 (1990)
14. Parzen, E.: On estimation of a probability density function and mode. Ann. Math. Stat. **33**(3), 1065–1076 (1962)
15. Rasch, G.: The poisson process as a model for a diversity of behavioral phenomena. In: International Congress of Psychology, vol. 2, p. 2 (1963)
16. Refsdal, S.: On the possibility of determining Hubble's parameter and the masses of galaxies from the gravitational lens effect. MNRAS **128**, 307 (1964)
17. Ross, S.M.: Introduction to Probability Models. Academic press, Boston (2014)
18. Rubinstein, R.Y., Kroese, D.P.: Simulation and the Monte Carlo Method, vol. 707. Wiley, New York (2011)
19. Shimazaki, H., Shinomoto, S.: Kernel bandwidth optimization in spike rate estimation. J. Comput. Neurosci. **29**(1–2), 171–182 (2010)
20. Sigman, K.: Poisson processes and compound (batch) poisson processes. Lecture Notes. Columbia University, USA (2007). http://www.columbia.edu/ks20/4703-Sigman/4703-07-Notes-PP-NSPP.pdf
21. Suyu, S.H., Auger, M.W., Hilbert, S., Marshall, P.J., Tewes, M., Treu, T., Fassnacht, C.D., Koopmans, L.V.E., Sluse, D., Blandford, R.D., Courbin, F., Meylan, G.: Two accurate time-delay distances from strong lensing: implications for cosmology. Astrophys. J. **766**(2), 70 (2013)

Combination of Grey Matter and White Matter Features for Early Prediction of Posttraumatic Stress Disorder

Si Wang[1], Hao Hu[2], Shanshan Su[2], Luyan Liu[1], Zhen Wang[2],
Qian Wang[1(✉)], and Dinggang Shen[3(✉)]

[1] Med-X Research Institute, School of Biomedical Engineering,
Shanghai Jiao Tong University, Shanghai, China
wang.qian@sjtu.edu.cn
[2] Department of Clinical Psychology, Shanghai Mental Health Center,
Shanghai Jiao Tong University School of Medicine, Shanghai, China
[3] Department of Radiology and BRIC, University of North Carolina at Chapel
Hill, Chapel Hill, USA
dinggang_shen@med.unc.edu

Abstract. Posttraumatic stress disorder (PTSD) is a prevalent psychiatric disorder. In previous researches, there are few studies about structural and functional alterations of the whole brain simultaneously about PTSD prediction. Early alterations could provide evidence of early diagnosis and treatment. Early diagnosis of PTSD plays an important role during the treatment. In this work, we extract discriminant features from multi-modal images and implement classification-based prediction for PTSD onset. Specifically, discriminant features are a collection of measures derived from grey matter (GM) and white matter (WM). We choose cortical thickness of GM and three descriptions of WM connection which are fiber count, fractional anisotropy (FA), and mean diffusivity (MD). After applying automated anatomical labeling (AAL) to parcellate the whole brain into 90 regions-of-interest (ROIs), the descriptions can be quantified. Then, a weighted clustering coefficient of every ROI connected with the remaining ROIs is extracted as feature. GM features and WM features are combined and selected automatically, which are later utilized by support vector machine (SVM) for early identification of the patients. The classification accuracy is around 79.86 % as the area of receiver operating characteristic (ROC) curve is 0.816 evaluated via dual leave-one-out cross-validation.

1 Introduction

Posttraumatic stress disorder (PTSD) is a psychologic anxiety disorder that can develop after a person being exposed to one or more traumatic events directly or indirectly, such as natural disaster, warfare, traffic collisions, sexual assault, terrorism or other threats. Following traumatic exposure, a proportion of trauma victims develop PTSD, and the others are trauma control (TC) without relative symptoms eventually. The prevalence of PTSD following a traffic collision ranges 10–46 % [1]. The disease has become a serious threat to survivors' mental health and causes substantial economic burden to patients and their families. Therefore, psychological debriefing and risk-targeted

© Springer International Publishing AG 2016
H. Yin et al. (Eds.): IDEAL 2016, LNCS 9937, pp. 560–567, 2016.
DOI: 10.1007/978-3-319-46257-8_60

interventions are efficient and significantly important for high-risk groups of PTSD, such as victims of vehicle accidents. It is critical to identify subjects at high risk for developing PTSD. However, the diagnosis of PTSD is dependent on clinical symptoms, such as re-experiencing, avoidance and hyperarousal. Suspected cases are screened and assessed according to clinical scales (e.g., CAPS [2]). However, when these symptoms appear, it is a relatively late start for therapy.

For this reason, researchers have spent lots of effort on early prediction of PTSD. In general, there are three categories of risk factor or predictors. First, the factors such as gender, age at trauma, and race can predict PTSD in some populations. Second, the factors such as education, trauma history, and general childhood adversity are shown to predict PTSD more consistently but to a varying extent according to the populations studied and the methods used. Last, factors such as reported childhood abuse, psychiatric history as well as family psychiatric history that had more uniform predictive effects. Individually, the effect extent of all the risk factors was modest, but factors operating during or after the trauma, such as trauma severity, lack of social support, and additional life stress, had somewhat stronger effects than pre-trauma factors. So acute reactions which used as valid references and predictions has been recognized widely. In the literature, adrenocortical activity after awakening [3] and reduced autobiographical memory [4] as well as decreased hippocampal volume [5] are also regarded as potential predictors of PTSD. Researchers later found alterations in the cortical thickness and disrupted or diminished integrity in the frontal lobe and the limbic system of patients with PTSD respectively [6, 7]. However, none above quantitatively evaluates the prediction capability of the factors to our knowledge.

In order to extract some potential predictors derived from acute reaction and study their discriminant power, we studied subjects post to traffic collisions by structural magnetic resonance imaging (MRI) and diffusional tensor imaging (DTI) in this study. Specifically, acute reactions, including alteration of subcortical grey matter and the integrity of white matter are extracted as features for PTSD prediction. We propose a multi-modality image-based classification framework to distinguish PTSD and TC. Cortical thickness of regions-of-interest (ROIs) is computed as GM feature, while fractional anisotropy (FA), mean diffusivity (MD), and fiber count are quantified for pairs of connected regions to construct networks respectively. Then a weighted clustering coefficient of every ROI connected with the remaining ROIs is extracted as WM feature. Weighted clustering coefficient vectors and cortical thickness vector are concatenated into an enriched feature vector, then selected by a feature selection method to select the most distinguished subset. Selected subset of features is utilized to train support vector machine (SVM) classifiers, and the classification performance is evaluated via leave-one-out cross validation to guarantee generalization of classifiers.

2 Methods and Materials

2.1 Data Acquisition and Pre-processing

This study involved 33 participants being exposed to traffic collision. Survivors who visited an emergency department within 48 h after the accident were recruited. 17 were eventually diagnosed of PTSD, and the others were socio-demographically matched TC. Recruitment of participants is in charge of cooperated mental health center. Subjects were excluded from the study if there were brain deformation. Diagnosis was based on CAPS.

Participants were scanned using a 3.0T MR scanner (Signa Excite; GE HealthCare, Milwaukee, WI, USA). T1-weighted images using 3D fast spoiled gradient recalled (3D-FSPGR) sequence were acquired with repetition time (TR) = 5.6 ms, echo time (TE) = 1.8 ms, flip angle = 15°, field of view = 256 × 256, matrix = 256 × 256, slice thickness = 1 mm with no gap, voxel dimensions = 0.9375 × 0.9375 × 1. A total 164 contiguous axial slices were acquired to cover the whole brain. DTI was performed using echo planar imaging (EPI) sequence with a total 20 different diffusion directions using diffusion weighting values, b = 0 and 1000 s/mm², flip angle = 90°, TR = 15000 ms, TE = 68 ms, field of view = 220 × 220, matrix = 110 × 110, slice thickness = 2 mm with no gap, voxel dimension = 0.8594 × 0.8594 × 2, 60 contiguous axial slices to cover the whole brain. All images were reviewed to screen for any clinical abnormalities. Demographic and clinical characteristics of the participants involved in this study are shown in Table 1.

Table 1. Demographic information of the participants involved in this study.

Group	PTSD	TC	P value
No. of subjects	17	16	
Male/female	6/11	6/10	>0.1
Age (years)	41.2±13.4	36.8±11.8	>0.1
Education (years)	11.5±4.5	12.9±3.6	>0.1
CAPS total score	38.8±17.9	13.4±11.6	<0.0001

2.2   Method

The proposed classification framework involves both thickness description of grey matter (GM) which is surface-based, and connection descriptions of white matter (WM) which is tractography-based. The workflow of our work is shown in Fig. 1. After extracting the WM and GM features from DTI images and T1 images respectively, we feed them into a SVM-based feature selection framework. Then, most significant WM/GM features are selected to build a set of classifiers and separate PTSD patients from the TC subjects.

2.3   Feature Extraction

Structural T1-weighted MRI images are used to extract cortical thickness features. T1 images are pre-processed by image reorientation, N4 correction, histogram matching, skull stripping, cerebellum removal and tissue segmentation. Then segmented images are used for 3D brain reconstruction, mainly including hemisphere separation, subcortical structure filling, cortical surfaces reconstruction and cortical thickness

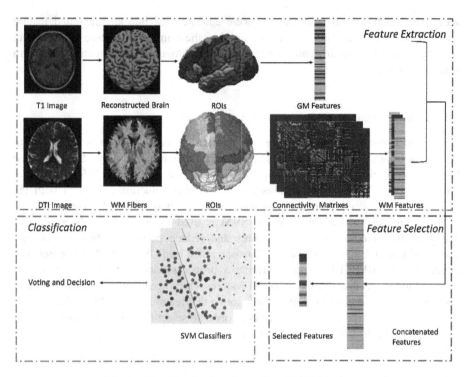

Fig. 1. Classification based on GM and WM features.

computation. Last, individual surface is aligned to a common space, then ROI's mean cortical thickness were computed based on Desikan-Killiany atlas [8].

Meanwhile, whole-brain fiber tracking is performed on each subject using diffusion spectrum MRI (DSI) Studio [9]. We utilize the automated anatomical labeling (AAL) atlas [10] to divide DTI image into 90 ROIs. For FA image, we compute the mean value for every pre-defined ROI. Then, two ROIs were considered connected if fibers pass through their respective masks simultaneously. So we can generate a 90 × 90 connectivity matrix. Based on the connectivity matrix, local weighted clustering coefficient can be calculated, which quantifies how close a node of graph and its neighbors are to being a clique, moreover, it could determine whether a graph is a small-world network of the concerned region. For a constructed network, the weighted local clustering coefficient between a ROI and its neighboring ROIs are computed as

$$c_i = \frac{\sum_{j:j\neq i\in N_i} v_{i,j}}{k_i(k_i - 1)/2} \tag{1}$$

where k_i is the number of ROIs that are connected to the i-th ROI anatomically, the neighborhood N_i for a i-th ROI is defined as its connected neighbors in the neighboring network, and $v_{i,j}$ is the measurement value between the i-th ROI and j-th ROI. For each 90 × 90 connectivity network, it could generate a 1 × 90 local coefficient feature vector. For each subject, 338 features were concatenated into an integrated feature vector. Compared with the number of samples, the number of feature is excessive. In order to avoid overfitting caused by redundant information, it's essential to perform feature selection.

2.4   Feature Selection

In order to improve the performance of classifiers, feature selection is performed to choose an optimal subset of features from a great number of features because some features are irrelevant or redundant. We applied the SVM-RFECBR for feature selection [11]. This toolbox contains correlation bias reduction compared with the well-known and effective feature selection method, namely SVM-RFE algorithm.

In the process of feature selection, SVM is used to evaluate the discriminative ability of the selected subset of features and reduce correlation bias recursively. The SVM kernel used in this study is Gaussian radial basic function (RBF) kernel. The target of SVM-RFECBR is to determine a proper subset features, whose size is expected to far less than the size of initial features. Optimized features are used to build discriminative SVM classifiers. It's worth noting that feature selection is achieved via a leave-one-out process to minimize errors in order to select the optimal combination of features, around 10 features in each phase.

2.5   Evaluation via Cross-Validation

In this study, the classification performance is evaluated using a full leave-one-out cross-validation strategy to guarantee a relatively unbiased estimate of the generalization ability of the classifiers to new subjects. Specifically, for n subjects in total, in each leave-one-out case, one subject is left out for testing, and the remaining $(n-1)$ subjects are used for feature selection and training. For these $(n-1)$ training samples, a second round of leave-one-out cross-validation is applied on the training set. That is, given $(n-1)$ different training subjects, we select $(n-2)$ to train a classifier and test the last training subject. In this way, the last training subject can be used for tuning the parameters of the classifier based on the selected $(n-2)$ subjects. Moreover, given $(n-1)$ training subjects, we can eventually build $(n-2)$ classifiers with different yet "optimal" parameters. For the testing subject, the prediction is attained by fusing all $(n-2)$ classifiers in the final.

3   Results

3.1   Results from Evaluation via Cross-Validation

By leave-one-out cross-validation, the classification accuracy with cortical thickness and connectivity features combined is 79.86 %. The score is 10 % higher than using a single type of features. The AUC of using combined features is 0.816, which is the evidence

Table 2. Classification performance and AUC values for combined GM and WM features and single cortical or connectivity matirx.

Feature	Accuracy (%)	AUC
Combined	**79.86**	**0.816**
Cortical Thickness	68.61	0.701
Fiber Count	57.15	0.584
FA	69.54	0.722
MD	54.69	0.587

that it has a satisfactory discriminant power between PTSD and TC. The classification performance of the combined features and single type of features for cross-validation is summarized in Table 2.

3.2 Comparison with Linear and Polynomial Kernel Type

Since we don't know whether the dataset is linearly separable, we want to know how well the effect of using linear or polynomial kernel of SVM classifiers would be. Linear and polynomial SVM classifiers were trained and tested in the exactly way except for kernel type. Comparison results are summarized in Table 3.

The classification performance reveals that, a higher-dimensional feature space is beneficial for a better distinguish of samples. In terms of accuracy and AUC, RBF kernel performed better than polynomial and linear kernel while the results of polynomial is better than performance of linear kernel.

Table 3. Classification performance and AUC values for linear, polynomial and nonlinear SVM classifiers.

Kernel type	Accuracy (%)	AUC
Linear	63.45	0.611
Polynomial	68.32	0.704
RBF	**79.86**	**0.816**

3.3 The Most Discriminant ROIs

We found that certain features are always selected or ranked high via SVM-RFECBR. Frequently selected regions in all leave-one-out cases can be significantly important concerning PTSD prediction regions. As shown in Fig. 2, the most discriminant regions include right parahippocampal [12], amygdala [13] and pallidum [14].

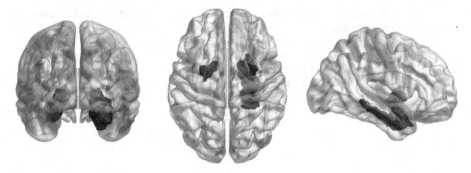

Fig. 2. The most discriminant ROIs for early prediction of PTSD (red for right parahippocampal, green for amygdala, blue for pallidum). (Color figure online)

4   Conclusion and Discussion

In this study, a set of novel predictors, which involve cortical thickness and local weighted clustering coefficient of WM connectivity map, are extracted and they contribute to early prediction of PTSD patients survived from traffic collision. With a SVM-based feature selection method, the multivariate classification framework is evaluated by leave-one-out cross-validation to guarantee generalization. Experimental results reveal that both GM and WM predictors/features provide discriminant information for identification. The performance of combined features is significantly better than the single description of WM or GM which only convey restricted information.

Besides, these discriminant regions found in our work are in consistent with previous studies. The subtle alterations in brain of acute stage are related with development of PTSD. The promising and consistent performance reveals that proposed method can be an early reliable prediction approach at the acute stage especially there are no obvious symptoms when conventional scales can't lend a hand.

Though the proposed method have some advantages, current study has some limitations. First, the number of subjects involved in is modest. Extensiveness and predictive effects in the whole populations should be evaluated in a diverse dataset. Second, the atlases utilized for parcellating cortical surface and the DTI volume follow different protocols, and it makes some inconsistency to evaluate each simple feature. We try to merge or progress a set of atlas consisting of volume atlas and surface atlas which is in line with the same protocol.

References

1. Yaşan, A., Güzel, A., Tamam, Y., Ozkan, M.: Predictive factors for acute stress disorder and posttraumatic stress disorder after motor vehicle accidents. Psychopathology **42**, 236–241 (2009)
2. Weathers, F., Blake, D., Schnurr, P., Kaloupek, D., Marx, B., Keane, T.: The Clinician-Administered PTSD Scale for DSM-5 (CAPS-5). Interview available from the National Center for PTSD (2013). www.ptsd.va.gov

3. Heinrichs, M., Wagner, D., Hellhammer, D.H., Ehlert, U.: Adrenocortical activity after awakening: a biological marker for predicting occupational stress in a high risk population for posttraumatic stress disorder? J. Psychophysiol. **15**, 131 (2001). Hogrefe & Huber Publishers Rohnsweg 25, D-37085 Gottingen, Germany

4. Kleim, B., Ehlers, A.: Reduced autobiographical memory specificity predicts depression and posttraumatic stress disorder after recent trauma. J. Consult. Clin. Psychol. **76**, 231 (2008)

5. Stein, M.B., Koverola, C., Hanna, C., Torchia, M., McClarty, B.: Hippocampal volume in women victimized by childhood sexual abuse. Psychol. Med. **27**, 951–959 (1997)

6. Wang, X., Xie, H., Cotton, A.S., Tamburrino, M.B., Brickman, K.R., Lewis, T.J., McLean, S.A., Liberzon, I.: Early cortical thickness change after mild traumatic brain injury following motor vehicle collision. J. Neurotrauma **32**, 455–463 (2015)

7. Schuff, N., Zhang, Y., Zhan, W., Lenoci, M., Ching, C., Boreta, L., Mueller, S.G., Wang, Z., Marmar, C.R., Weiner, M.W.: Patterns of altered cortical perfusion and diminished subcortical integrity in posttraumatic stress disorder: an MRI study. Neuroimage **54**, S62–S68 (2011)

8. Desikan, R.S., Ségonne, F., Fischl, B., Quinn, B.T., Dickerson, B.C., Blacker, D., Buckner, R.L., Dale, A.M., Maguire, R.P., Hyman, B.T.: An automated labeling system for subdividing the human cerebral cortex on MRI scans into gyral based regions of interest. Neuroimage **31**, 968–980 (2006)

9. Yeh, F.-C., Verstynen, T.D., Wang, Y., Fernández-Miranda, J.C., Tseng, W.-Y.I.: Deterministic diffusion fiber tracking improved by quantitative anisotropy. PLoS ONE **8**, e80713 (2013)

10. Tzourio-Mazoyer, N., Landeau, B., Papathanassiou, D., Crivello, F., Etard, O., Delcroix, N., Mazoyer, B., Joliot, M.: Automated anatomical labeling of activations in SPM using a macroscopic anatomical parcellation of the MNI MRI single-subject brain. Neuroimage **15**, 273–289 (2002)

11. Yan, K., Zhang, D.: Feature selection and analysis on correlated gas sensor data with recursive feature elimination. Sens. Actuators B Chem. **212**, 353–363 (2015)

12. Liu, Y., Li, Y.-J., Luo, E.-P., Lu, H.-B., Yin, H.: Cortical thinning in patients with recent onset post-traumatic stress disorder after a single prolonged trauma exposure. PLoS ONE **7**, e39025 (2012)

13. Stevens, J.S., Jovanovic, T., Fani, N., Ely, T.D., Glover, E.M., Bradley, B., Ressler, K.J.: Disrupted amygdala-prefrontal functional connectivity in civilian women with posttraumatic stress disorder. J. Psychiatr. Res. **47**, 1469–1478 (2013)

14. Chen, Y., Fu, K., Feng, C., Tang, L., Zhang, J., Huan, Y., Cui, J., Mu, Y., Qi, S., Xiong, L.: Different regional gray matter loss in recent onset PTSD and non PTSD after a single prolonged trauma exposure. PLoS ONE **7**, e48298 (2012)

MR-Swarm: Mining Swarms from Big Spatio-Temporal Trajectories Using MapReduce

Yanwei Yu[1(✉)], Jianpeng Qi[1], Yunhui Lu[2], Yonggang Zhang[2], and Zhaowei Liu[1]

[1] School of Computer and Control Engineering, Yantai University, Yantai, China
yuyanwei@ytu.edu.cn
[2] SCKE Key Laboratory of Ministry of Education,
Jilin University, Changchun, China

Abstract. The increasing pervasiveness of object tracking technologies has enabled collection of huge amount of spatio-temporal trajectories. Discovering the useful movement patterns from such big data is gaining in importance and challenging. In this paper we propose an distributed mining framework on Hadoop for efficiently discovering swarm patterns from big spatio-temporal trajectories in parallel. We first define the notion of maximal objectset that captures swarms by recombining clusters in timeset domain. Second, we propose a parallel model based on timeset independent property of swarm pattern to parallel the mining process. Furthermore we propose a distributed algorithm using MapReduce chain architecture based on the proposed parallel model, which features two optimization pruning strategies designed to minimize the computation costs. Our empirical study on the real Taxi dataset demonstrates its effectiveness in finding object-closed swarms. Extensive experiments on 5 network-connected workstations also validate that our proposed algorithm nearly achieves 5-fold speedups against the serial solution.

1 Introduction

The increasing availability of location-acquisition technologies including all kinds of GPS, RFID, WLAN networks, mobile phones, and the emerging location-based APPs have enabled tracking almost any kind of moving objects, which results in huge volumes of spatio-temporal trajectory data that records a variety of action features, including location, time, and velocity. Such big data provides the huge opportunity of discovering usable knowledge about movement behaviour, which fosters novel many applications and services ranging from intelligent traffic management, urban computing to location-based services [1,2].

Patrick Laube et al. [4] first propose *flock*, *leadership*, and *aggregation* patterns in geospatial lifelines. Jeung et al. [5] define the notion of *convoy* pattern, in which a set of objects that move together in a density-based cluster for at least k continuous time points, instead of the strick size and shape of group by specifying the disk radius in *flock*. A recent study by Zhenhui Li et al. [6] proposes the *swarm* pattern, which also employs the density-connected clusters

© Springer International Publishing AG 2016
H. Yin et al. (Eds.): IDEAL 2016, LNCS 9937, pp. 568–575, 2016.
DOI: 10.1007/978-3-319-46257-8_61

and relax the requirement that objects must form group for consecutive time points. In contrast to above mentioned patterns, swarm permit patterns where moving objects travel together for a number of nonconsecutive time points. In other words, moving objects could leave from the group transitorily, and then come back in swarm. Therefore, swarm is a more general and relaxed pattern that does not require k consecutive time points, also is more inline with the practical situations.

In our previous works [7,8], we propose online solutions of discovery of swarms and trajectory clusters to improve both efficiency and scalability. However the works are just suitable for trajectory data mining under streaming environment with normal rate. For the huge amounts of spatio-temporal trajectory data, the traditional serial solution is difficult to satisfy requirements of big trajectory pattern mining, which is a key issue addressed in this work. This paper focuses on efficient discovery of swarm pattern, one of useful group pattern, from high-volume moving object trajectories. We propose a distributed swarm pattern mining algorithm employing Hadoop platform, which incorporates three principles. First, we propose a notion of maximal objectset, and optimize serial method using minimal time support optimization. Second, we propose a parallel model based on timeset independent of swarm pattern, which parallelizes clustering and local swarm discovery in sub-time domain. Third, we implement an efficient distributed solution using MapReduce chain architecture on Hadoop platform. Finally, We conduct an extensive empirical study on real trajectory data to evaluate the proposed distributed framework. Our results offer insight into the effectiveness and efficiency of the proposed framework.

2 Preliminary Definition

We denote the set of trajectories TD. Let $O_{TD} = \{o_1, o_2, \ldots, o_m\}$ be the set of all moving objects and $T_{TD} = \{t_1, t_2, \ldots, t_n\}$ be the set of all time points in TD. We denote a subset of O_{TD} objectset O and a subset of T_{TD} timeset T. The size, $|O|$ and $|T|$, is the number of objects and time points in O and T respectively. A set of clusters, obtained by DBSCAN at each time point t_i, is denoted $C_{t_i} = \{C_{t_i}^1, C_{t_i}^2, \ldots, C_{t_i}^k\}$, where $C_{t_i}^j (1 \leq j \leq k)$ is a cluster at t_i.

Definition 1 (Swarm). *Given minimal thresholds min_o and min_t, a set pair $< O, T >$ is called a swarm pattern if $|O| \geq min_o$, $|T| \geq min_t$, and $\forall t \in T$, there exists a cluster $C \in C_t$ such that $O \subseteq C$.*

By Definition 1, all pairs $< O, T >$ that satisfy the minimal thresholds and all objects of O belong to a cluster at any time point in T are swarms. To avoid mining redundant swarms, Li et al. [6] further give the notion of closed swarm.

Definition 2 (Maximal Objectset or MO). *Given minimal threshold min_o and a timeset T, the objectset O is called a maximal objectset with respect to T iff $O = \bigcap_{t_i \in T} C_{t_i}^{k_i} (C_{t_i}^{k_i} \in C_{t_i})$ and $|O| \geq min_o$.*

By Definition 2, the maximal objectset can be identified by clusters of each time point in T.

Lemma 1. *Given a set pair $< O, T >$, if objectset O is a maximal objectset w.r.t. timeset T and $|T| \geq min_t$, then $< O, T >$ is an object-closed swarm (OSm).*

Lemma 1 is intuitive and easily proved by Definitions 1 and 2.

Definition 3 (Minimal Time Support). *Given minimal threshold min_t and a set pair $< O, T >$, if the objectset O is a maximal objectset w.r.t. T and $|T| = min_t$, then T is a minimal time support of $< O, T >$ to be a swarm.*

The minimal time support provides us a minimal amount of timeset for finding a swarm. Therefore, this concept of minimal time support guides us to propose the minimal time support optimization to reduce time points examination and lookup costs related to intersection operation of clusters of moving objects.

Definition 4 (Potential Swarm or PSm). *Given a set pair $< O, T >$, if the objectset O is a maximal objectset w.r.t. T and $|T| < min_t$, then we call $< O, T >$ a potential swarm.*

3   Optimization Principles

Minimal Time Support Optimization. By Definition 3, if timeset T corresponding to the maximal objectset O satisfies the condition of $|T| \geq min_t$, the we confirm that $< O, T >$ is a OSm. Next, we optimize CLUR [7] algorithm by using minimal time support instead of completed timeset (closed timeset), named *CLIP* for distinguishing from CLUR.

We also employ DBSCAN to get clusters C_t at each time point t. CLIP maintains two lists L and L_{sm}, which store potential swarms $PSms$ and object-closed swarms $OSms$ with minimal time support, respectively. PSm or OSm are stored in L and L_{sm} in form of *key-value* pairs. *key* correspond to objectset O of patterns, and *value* to timeset T.

Parallel Model. We observe the combination property of timeset corresponding to MO by Definition 2, as depicted in following Property.

Property 1. Given a maximal objectset O and its corresponding timeset T, for $\forall T_1, T_2$ $(T_1 \cap T_2 = \phi, T_1 \cup T_2 = T)$, there must exist O_1 (O_1 is a MO w.r.t. T_1) and O_2 (O_2 is a MO w.r.t. T_2) such that $O = O_1 \cap O_2$.

From Property 1, we observe that it is independent of the order of time points when examine maximal objectset by using cluster recombinant method. Thus we can adjust the order of time points or re-group time points of timeset arbitrarily. Therefore, we further conclude the parallel model based on timeset-independent for mining swarm pattern.

4 Distributed Algorithm

4.1 Framework of Distributed Mining Algorithm

In this section, we introduce our distributed algorithm called *MR-Swarm*, capable of mining swarm using MapReduce chain architecture based on proposed CLIP as a plug-in parallel method. The overall framework of our MR-Swarm is depicted in Fig. 1. The framework can be dived into four stages:

Stage 1: Preprocessing in Map1 phase. This phase reads trajectory data in parallel, and then partition data according to the attribute of time point, which provides Stage 2 load-balanced data partitions.

Stage 2: Distributed mining phase in Reducer1 phase. Stage 2 performs Reduce1 process on each reducer for mining local swarm patterns in independent timeset T_i.

Stage 3: Parallel merge phase in Map2 phase. The stage merges the local swarms from stage 2 in parallel, and generate intermediate patterns.

Stage 4: Final merge phase in Reduce2 phase. Stage 4 is executed on single machine to merge all intermediate patterns into expected swarm patterns.

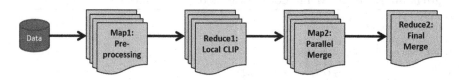

Fig. 1. Overall architecture of MR-Swarm framework

4.2 Preprocessing and Local Swarm Mining

The main task of preprocessing is load trajectory data of massive moving objects, map location points of same timestamp to same *key*, and output data partition to reducers. Therefore, the main challenges for an efficient partition are load balancing, minimized communication or shuffling cost.

To minimize traffic and shuffling cost, we employ combiner in preprocessing. In Mappers, trajectory data are mapped to *key-value* pairs according the information of time point, and then the *values* of same *key* are merged in combiner, namely, trajectory points at same time point are merged into a pair of *key-value*. This would significantly reduce the number of shuffle in Shuffle phase. Finally, we mod *key* by the number of reducers r, and output the data partition into specified reducers grouped by time point.

As shown in Fig. 1, Stage 2 performs $CLIP$ to mine local swarm candidates in subset of T_{TD} on each reducer. Each reducer receives the trajectory points of all moving objects at specified time points, So all reducers can work in parallel by the proposed parallel model. For reducer R_i, it first performs DBSCAN clustering to get density-connected clusters at each time point, and then executes $CLIP$ in any order of time points to find local swarm candidates in local timeset.

4.3 Parallel Merge Phase

Stage 3 that runs on second Map phase aims to merge the local candidates on two or more timesets. Taking two timesets as an example, pseudocode of merge process is shown in Algorithm 1. min_o, min_t and T_{TD} are global parameters.

Stage 3 first reads distributed intermediate files from reducers of Stage 2, and distributes evenly them to Mappers. So each Mapper should first merge the assigned $OSms$ stored in L_{sm} as line 1. Then, examining whether the $PSms$ in T_1 or T_2 construct OSm in $T_1 \cup T_2$. As shown on lines 2–14, the examination also employs the strategy of minimal time support optimization to identify OSm in advance. Furthermore, we also utilize predicted timeset pruning rule to prune the $PSms$ that could not become swarms in future, given as following lemma.

Lemma 2. *Given the set of all time points T_{TD} and threshold min_t, for a potential swarm $< O, T >$ in $T_1(T_1 \subset T_{TD})$, if $|T| < min_t + |T_1| - |T_{TD}|$, then $< O, T \cup (T_{TD} - T_1) >$ is certainly not a (closed) swarm.*

Algorithm 1. *Merge*: Merge Local Swarms on Two Timesets

Require: $L_1, L_{sm_1}, T_1, L_2, L_{sm_2}, T_2$
Ensure: L_1 and L_{sm_1}
1: $L_{sm_1} \leftarrow L_{sm_1} \cup L_{sm_2}$;
2: **for** each candidate $v_2 \in L_2$ **do**
3:     **for** each candidate $v_1 \in L_1$ **do**
4:         $O \leftarrow v_1.O \cap v_2.O$; $T \leftarrow v_1.T \cup v_2.T$;
5:         **if** $|O| \geq min_o$ **then**
6:             **if** $!MTS(O, T, L_{sm_1})$ **then**
7:                 **if** $|T| \geq |T_1| + |T_2| + min_t - |T_{TD}|$ **then**
8:                     $L_1.put(O, T)$;
9:                 **end if**
10:            **else**
11:                **if** $O == v_1.O$ **then**
12:                    $L_1.remove(v_1)$;
13:                **end if**
14:                **if** $O == v_2.O$ **then**
15:                    $v_2.exist \leftarrow$ TRUE;
16:                **end if**
17:            **end if**
18:        **end if**
19:     **end for**
20:     **if** $(!v_2.exist)\&\&|v_2.T| \geq |T_1| + |T_2| + min_t - |T_{TD}|$ **then**
21:         $L_1.put(v_2.O, v_2.T)$;
22:     **end if**
23: **end for**
24: **for** each candidate $v_1 \in L_1$ **do**
25:     **if** $|v_1.T| < |T_1| + |T_2| + min_t - |T_{TD}|$ **then**
26:         $L_1.remove(v_1)$;
27:     **end if**
28: **end for**
29: $T_1 \leftarrow T_1 \cup T_2$;

---

k mappers merge the local swarms on multiple subsets of time domain in parallel, however, we still need to perform State 4 on one machine to merge all intermediate swarms and output final swarms in final reduce phase.

5   Experiments

5.1   Experimental Setup

A comprehensive performance study has been conducted on 5 network connected workstations that deployed Hadoop platform of version 2.4.0 to evaluate the effectiveness and efficiency of the proposed algorithm. Each node is equipped with an Intel Core 2 Duo i5-3470 processor with 4 GB memory and runs a Centos release 6.6 operating system.

Real Dataset. We use a real spatio-temporal trajectory data *Taxi* in our experiments. The dataset is from T-drive project [3, 9] developed by Microsoft Research Asia. We divide a day into four time periods, morning and evening peak time (7 am to 10 am and 5 pm to 8 pm), work time (10 am to 4 pm) and casual time (8 pm to 2 am). We interpolate the time domain into the granularity of minute, and get 7,560 time points in T_{TD}.

Metrics. We measure the correctness of MR-Swarm by Precision and Recall as follows: $Precision = (R \cap D)/R$, $Recall = (R \cap D)/D$, where D denotes swarms obtained by the serial CLUR [7] and R is discovered swarms of MR-Swarm algorithm.

5.2   Effectiveness Evaluation

First, we evaluate the correctness of MR-Swarm algorithm by measuring the *Precision* and *Recall* on the real *Taxi* dataset. To demonstrate generality, we use the data of Monday, Friday and Sunday, three representative days. From Fig. 2(a), We can see that the *Precision* of most time is nearly 100 % except Monday work time and Friday casual time, and *Precisions* of Monday work time and Friday casual time also reach at 93.5 %, 94.5 % respectively. Figure 2(b) shows the *Recall* of MR-Swarm is also good and robust. The average *Recall* of all time intervals reaches at 97.5 %.

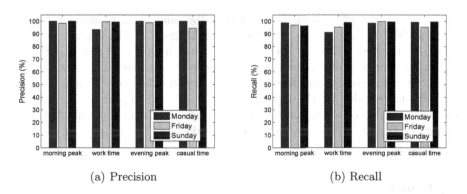

(a) Precision   (b) Recall

Fig. 2. Effectiveness evaluation on *Taxi*

5.3   Efficiency Evaluation

We also measure the CPU time of two MapReduce phase in MR-Swarm referred as MR-1 and MR-2 respectively. The default setting follows: $Eps = 0.002, MinPts = 10$, thresholds $min_o = 30, min_t = 10$ and Taxi data on Sunday.

We first evaluate the scalability of MR-Swarm algorithm in terms of the volume of trajectories. In this experiment we randomly extract four subsets from the Taxi data from 6k to 9k trajectories. Figure 3(a) shows the total running time of MR-SWARM, MR-1, MR-2 and CLUR on the five datasets. MR-Swarm exhibits much better scalability than CLUR in terms of CPU time. From the CPU time utilized by MR-1 and MR-2, we observe that the number of trajectories less affects parallel clustering and local swarm mining on subset of time domain but greatly affects the phase of merging local $PSms$. In particular, MR-Swarm nearly achieves 5-fold speedup compared against CLUR when $TD = 10\,k$.

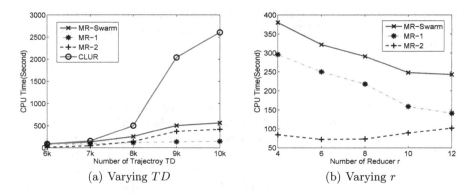

(a) Varying TD (b) Varying r

Fig. 3. Efficiency evaluation on $Taxi$

Next, we evaluate the effect of varying the number of reducers r from 4 to 12 in MR-1 phase, when the number of $Mapper$ in MR-2 phase is fixed to the number of nodes. As shown in Fig. 3(b), the CPU cost of MR-Swarm algorithm decreases as the number of reducers r increases. This is expected, MR-1 performs parallel clustering and local swarm discovery in sub timeset, thus the more reducers, the less time points assigned to each reducer, hence the average time consumption is also less. However since the number of node in our platform is limited, too many reducers does not further reducer the time consumption.

6   Conclusion

In this paper, we study the problem of discovering object-closed swarm patterns from big spatio-temporal trajectories. We propose distributed mining framework using MapReduce chain architecture scalable to big trajectories to efficiently discover swarms. At last we demonstrate the effectiveness and efficiency of proposed distributed solution by conducting comprehensive evaluations on a real large scale taxi trajectory data.

Acknowledgment. This work is partially supported by the National Natural Science Foundation of China (nos. 61403328 and 61572419), the open project program of Key Laboratory of Symbolic Computation and Knowledge Engineering of Ministry of Education, Jilin University (no. 93K172014K13), the Key Research & Development Project of Shandong Province (no. 2015GSF115009), and the Shandong Provincial Natural Science Foundation (nos. ZR2013FQ023 and ZR2013FM011).

References

1. Zheng, Y.: Trajectory data mining: an overview. ACM Trans. Intell. Syst. Technol. **6**, 1–41 (2015)
2. Yu, Y., Cao, L., Rundensteiner, E.A., et al.: Detecting moving object outliers in massive-scale trajectory streams. In: Proceedings of the 20th ACM SIGKDD International Conference on Knowledge Discovery and Data Mining, pp. 422–431. ACM (2014)
3. Yuan, J., Zheng, Y., Xie, X., et al.: T-Drive: Enhancing driving directions with taxi drivers' intelligence. IEEE Trans. Knowl. Data Eng. **25**(1), 220–232 (2013)
4. Laube, P., Imfeld, S.: Analyzing relative motion within groups oftrackable moving point objects. In: Egenhofer, M.J., Mark, D.M. (eds.) GIScience 2002. LNCS, vol. 2478, pp. 132–144. Springer, Heidelberg (2002). doi:10.1007/3-540-45799-2_10
5. Jeung, H., Yiu, M.L., Zhou, X., et al.: Discovery of convoys in trajectory databases. In: Proceedings of The 34th Very Large Databases Conference, Auckland, New Zealand, 23–28 August, pp. 1068–1080 (2008)
6. Li, Z., Ding, B., Han, J., et al.: Swarm: mining relaxed temporal moving object clusters. In: Proceedings of the 36th Very Large Databases Conference, Singapore, pp. 13–17 (2010)
7. Qi, Y., Yu, Y., Kuang, J., et al.: Efficient algorithm for real time mining swarm patterns. J. Univ. Sci. Technol. Beijing **34**(1), 32–37 (2012)
8. Yu, Y., Wang, Q., Wang, X., Wang, H.: On-line clustering for trajectory data stream of moving objects. Comput. Sci. Inf. Syst. **10**(3), 1319–1342 (2013)
9. Yuan, J., Zheng, Y., C. Zhang, et al.: T-drive: driving directions based on taxi trajectories. In: ACM SigSpatial Geographic Information Science, pp. 99–108. ACM (2010)

An Organizing System to Perform and Enable Verification and Diagnosis Activities

Vincent Leilde[1]([⊠]), Vincent Ribaud[2], and Philippe Dhaussy[1]

[1] Lab-STICC, Team MOCS, ENSTA Bretagne, rue François Verny, Brest, France
{vincent.leilde,philippe.dhaussy}@ensta-bretagne.fr
[2] Lab-STICC, Team MOCS, Université de Brest, avenue Le Gorgeu, Brest, France
ribaud@univ-brest.fr

Abstract. Model-checkers increasing performance allows engineers to apply model-checking for the verification of real-life system but little attention has been paid to the methodology of model-checking. Verification "in the large" suffers of two practical problems: the verifier has to deal with many verification objects that have to be carefully managed and often re-verified; it is often difficult to judge whether the formalized problem statement is an adequate reflection of the actual problem. An organizing system - an intentionally arranged collection of resources and the interactions they support – makes easier the management of verification objects and supports reasoning interactions that facilitates diagnosis decisions. We discuss the design of such an organizing system, we show a straightforward implementation used within our research team.

Keywords: Verification · Model-checking · Diagnosis · Organizing system

1 Introduction

System verification is used to establish that the design or product under consideration possesses certain properties. Formal verification has been advocated as a way forward to address verification tasks of complex embedded systems. Formal methods, within the field of computer science, is the formal treatment of problems related to the analysis of designs, but "it does not yet generally offer what its name seems to suggests, viz. methods for the application of formal techniques [1]."

Our research work is underlined by the observation that verification "in the large" causes a proliferation of interrelated models and verification sessions "that must be carefully managed in order to control the overall verification process [1]." The main technique discussed in this paper is verification by model-checking. "Model checking is a formal verification technique which allows for desired behavioral properties of a given system to be verified on the basis of a suitable model of the system through systematic inspection of all states of the model [2]."

Model-checking walks through different phases within an iterative process [3]: modelling, running the model-checker and analyzing the results. Moreover, the entire verification should be planned, administered, and organized.

© Springer International Publishing AG 2016
H. Yin et al. (Eds.): IDEAL 2016, LNCS 9937, pp. 576–587, 2016.
DOI: 10.1007/978-3-319-46257-8_62

The applicability of model-checking to large systems suffers of two practical problems. A verification session refines the model, and because properties are verified one by one, previously verified properties need or need not to be verified again, depending on the refinement performed. When the model-checker runs out of memory, some divide-and-conquer techniques should be employed to reduce the model. These techniques exploit regularities in the structure of the models or of the verification process itself that are difficult to understand and their performance may vary considerably.

A second practical problem arises from the difficulty to judge whether the formalized problem statement (model + properties) is an adequate reflection of the actual problem. This is known as the validation problem. If the verifier suspects the validity of a property, the property needs to be re-formalized and it starts the whole verification again. If the verifier suspects the validity of the design, the verification process restarts after an improvement of the design. The complexity of the involved system, as well as the lack of precision of the informal specification of the system's functionality, makes it hard to answer the validation problem satisfactorily [1, 3].

Both problems require an organized verification method. Organization creates or supports capabilities by intentionally imposing order and structure. In this paper, we apply the concepts of an organizing system promoted by [4]: "an Organizing System is an intentionally arranged collection of resources and the interactions they support." As an attempt to solve the problems mentioned above, we designed and built a prototype of an Organizing System for the support of verification and diagnosis activities. In Sect. 2, we precise the issues of the management of verification cycles; we introduce a general theory of diagnosis, and we presents some design decisions for our Organizing System. Section 3 deepens different aspects of an Organizing System: knowledge management and ontologies, technical aspects of the tiers of the organizing system. Section 4 relates our work with previous work, and Sect. 5 concludes.

2 Organizing System for Verification and Diagnosis Activities

2.1 Managing the Verification Trajectories

There are basically three possible outcomes of a verification run: the specified property is either valid in the given model or not, or the model is faced with the state space explosion problem (it turns out to be too large to fit within the computer memory).

If a state is encountered that violates the property under consideration, the model checker provides a counterexample that describes an execution path that leads from the initial system state to the faulty state. It is indisputable that the verification results obtained using a verification tool should always be reproducible [5]. Tool support is required and we present in Fig. 2 the objects that are significant and tool-managed during the verification phases. The specification-design-modelling-verification cycles are presented in Fig. 1. The terminology in [3] is used as a reference in the paper.

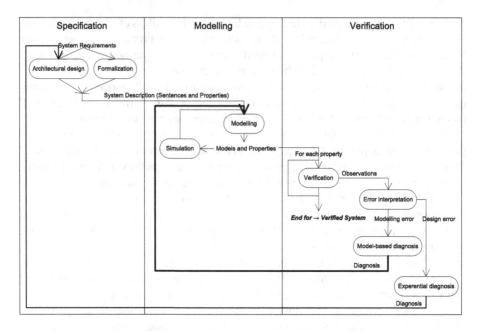

Fig. 1. The specification-design-modelling-verification process and its cycles

Modelling. Model checking inputs are a model of the system and a formal character-ization of the property to be checked. Models are mostly expressed using finite-state automata, made of a finite set of states and a set of transitions. To allow verification, properties are described using a property specification language, relying generally on temporal logic that allows people to describe properties in a precise and unambiguous manner.

Verification: running the model-checker. Current model checkers provide the user with various options and directives to optimize and tune the functionality and perform-ance of the verification run. Subsequently, the actual model checking takes place. Whenever a property is not valid, it may have different causes. A modelling error means that the model does not reflect the design of the system. After the model correction, verification has to be restarted with the improved model. This corresponds to the bolded cycle in Fig. 1. When there is no undue discrepancy between the design and its model, then either a design error has been exposed, or a property error has taken place (this case is not considered in the paper). In case of a design error, the verification is concluded with a negative result, and the design (together with its model) has to be improved. The designer will proceed through iterative refinements; a situation depicted with a plain cycle in Fig. 1.

Interpreting the error(s). The main advantage of model checking is the production of counterexamples demonstrating that a system does not satisfy a specification. Extracting the essence of an error from even a detailed source-level trace of a failing run requires a great deal of human-effort [6] and a lot of research work focus on counterexample processing to produce an error explanation. Hence, correcting the error(s) starts with an error diagnosis, and more precisely, using the fault, error, and failure nomenclature of [7], it starts with a failure diagnosis. Failure diagnosis is the process of identifying the fault that has led to an observed failure of a system or its constituent components.

Model-based diagnosis. The technique we use for failure diagnosis is a model-based diagnosis (also called reasoning from first principles) based on a theory of diagnosis established by [8]. A diagnosis is as a set of assumptions about a system component's abnormal behavior such that observations of one component's misbehavior are consistent with the assumptions that all the other components are acting correctly [8]. The computational problem is to determine all possible diagnoses for a given faulty system. The representation of the knowledge of the problem domain should achieve the desired coverage and quality of diagnoses while remaining computationally tractable [9].

Verification organization. Whether the verification trajectory is incorporated in an adaptive design strategy or focused on modelling-and-verifying cycles, the entire model-checking process should be well organized, well structured, and well planned. According to [3], different models are describing different parts of the system, various versions of the verification models are established, and plenty of verification parameters and results are available. We propose to use an organizing system to manage a practical model-checking process and to allow the reproduction of the experiments carried out by the engineers.

Robert J. Glushko and al. [4] promote The Discipline of Organizing (TDO) approach. Library and information science, informatics and other fields focus on the characteristic types of resources and collections that define those disciplines. In contrast, TDO complements the focus on specific resource and collection types with a framework that views organizing systems as existing in a multi-dimensional design space in which we can consider many types of resources at the same time and see the relationships among them. The framework assesses what is being organized, why, how much, when and by what means. It leads to an Organizing System defined as "an intentionally arranged collection of resources and the interactions they support [4]."

To sum up the problem statement of this section, managing the verification trajectories is an indispensable support for using model checkers "in the large". Moreover the proliferation of verification resources and the variety of possible interactions with them requires an Organizing System. "The concept of the Organizing System highlights the design dimensions and decisions that collectively determine the extent and nature of resource organization and the capabilities of the processes that compare, combine, transform and interact with the organized resources [4]."

2.2 A Theory of Diagnosis from First Principles

A variety of failure diagnosis techniques drawing from diverse areas of computing and mathematics such as artificial intelligence, machine learning, statistics, stochastic modelling, Bayesian inference, rule-based inference, information theory, and graph theory have been studied in the literature [10]. In the model-based diagnosis, often referred as diagnosis from first principles, one is given a description of a system, together with an observation of the system's behavior which conflicts with the way the system is meant to behave. "The diagnostic problem is to determine those components of the system which, when assumed to be functioning abnormally, will explain the discrepancy between the observed and correct system behavior [8]." At the other end of the spectrum, there are methods that do not assume any form of model information and rely only on historic process data [11]." Without denying the importance of other approaches, we use a theory of diagnosis from first principles based on Reiter's work [8] as a general theory of diagnosis.

One begins with a description of a system, including desired properties and the structure of the system's interacting components. Whatever one's choice of representation, the description will specify how that system normally behaves on the assumption that all its components are functioning correctly. We need a diagnosis if we have available an observation of the system's actual behavior and if this observation is logically inconsistent with the way the system is meant to behave. Intuitively, a diagnosis determines system components which, when assumed to be functioning abnormally, will explain the discrepancy between the observed and correct system behavior [8]. There may be several competing explanations (diagnoses) for the same faulty system, the computational problem, then, is to determine all possible diagnoses.

2.3   Design Decisions

Explicitly or by default, establishing an Organizing System (OS) requires many decisions. These decisions are deeply intertwined, but it is easier to introduce them as if they were independent. In [4], authors introduce five groups of design decisions.

What is being organized? System models, verification runs and diagnosis are our primary source of interest. There are all made of digital resources, but we can make a distinction between primary resources (such as [parts of] models, properties, verification runs, and counterexamples) and description resources about the primary resources and their relationships. Verification benchmarks (e.g. BEEM, BEnchmarks for Explicit Model checkers [12]) provide valuable inputs and need to be organized in collections. Any OS user can also organize her own verification endeavors in collections, subcollections of resources.

Why it is being organized? OS users are modelling and verification engineers, working alone or in teams, who need "to deal with the data explosion of the modelling phase and the versioned product space explosion of the verification phase [1]." The OS gathers and organizes quantitative and qualitative information to support knowledge creation and automated diagnosis reasoning. OS users share knowledge without being constrained to espouse a given formalism. OS users need to navigate efficiently through the resources space. The OS supports a reverification procedure to make sure that errors found in the model do not invalidate previous verification runs.

How much is it being organized? The simplest OS can be a software configuration management system controlling the release and change of each digital resource, leaving the burden of organization outside of the OS. At the opposite of the spectrum, the OS can be a full ontology where any relationship between any piece of information is carefully defined and controlled, allowing many reasoning possibilities. For our point of view, the OS manages essentially documents (models, results, traces). Each document organizes its knowledge structure and content, according to its document type, and the engineer writes and reads information according to this structure. The reification of the underlying knowledge structure for reasoning purposes is done automatically by the OS.

When is it being organized? The OS is intended to assist the engineer in her daily modelling and verification tasks, hence resources are organized continuously. However, the OS should offer an ingestion feature that helps to enter inputs into the OS. Ingest feature provides the services and functions to accept complex verification endeavors or benchmark collections and prepares the contents for storage and management within the OS. Conversely, an access feature provides the services and functions that support users in determining the existence, description, location and availability of information stored in the OS, and allowing users to extract information in a parametrized manner.

How or by whom, or by what computational processes, is it being organized?
Although a single verification engineer will benefit of the OS use, the OS is intended to support teamwork and to share knowledge about models and verification endeavors. Automated processes should extract as much knowledge as possible from the documents internal structure and from the collections organization. As a collaborative teamwork, organization is performed in a distributed, bottom-up manner.

3   Inside the Organizing System

3.1   Knowledge Management and Ontologies

In [13], the authors state that knowledge is an enterprise's most important assets and define the basic activities of knowledge management: identification, acquisition, development, dissemination, use, and preservation of the enterprise's knowledge. They advocate a corporate or organizational memory (OM) at the core of a learning organization, supporting sharing and reuse of individual and corporate knowledge and lessons learned [13]. The concept of an organizing system intended to knowledge management (KM) is a modern reincarnation of the organizational memory and we can benefit from the results gained in this research area. Knowledge acquisition and maintenance pose a serious challenge for organizational memories and [13] recommend adhering to the following principles: exploit easily available information sources; forgo a complete formalization of knowledge; use automatic knowledge-acquisition tools; encourage user feedback and suggestions for improvements; check the consistency of newly suggested knowledge.

An organizational memory or a KM organizing system relies substantially on existing information sources, which constitute the first tier of its architecture, called the object level in [13] and the storage tier in [4]. This level or tier is characterized by a variety of sources, heterogeneous with respect to several dimensions concerning form and content properties. An organizational memory or an information organizing system offers presentation facilities in the access tier, called the application-specific level in [13] and the presentation tier in [4]. This level or tier performs the mapping from the application-specific information needs to these heterogeneous object-level sources via a uniform access and utilization method on the basis of a logic-based, knowledge-rich level, a middle tier called the knowledge description level in [13] and the logic tier in [4]. The knowledge-rich level has the central role of a shared language to connect people to people, people to information, and information to information [14], and the level includes ontologies as a core enabler. As major knowledge-based KM applications, ontologies are used for the following three general purposes [14]: to support knowledge visualization; to support knowledge search, retrieval, and personalization; to serve as the basis for information gathering and integration.

Figure 2 represents the main concepts and relationships of our ontology. A system is referred to by several propositional objects: system requirements (sentences and formalized properties), the system model (and its components), observations generally made about verification runs that are organized within verification endeavors. A diagnosis is a conjecture that certain of the components are faulty (Abnormal) and the rest normal, stemming from an observation inconsistent with the system descriptions. All information on a particular verification run is not detailed here, and include, among others, checked properties, run outcomes, model-checker options and statistics.

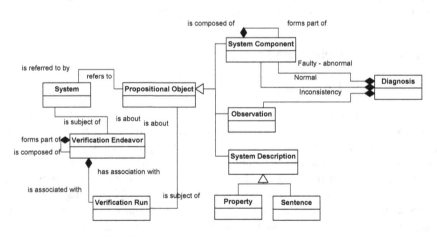

Fig. 2. Main elements of the ontological level.

3.2   Technical Aspects of the Tiers of the Organizing System

Modern applications separate the storage of data, the business logic or functions that use the data, and the user interface or presentation components through which users or other applications interact with the data. For each tier, we kept the design as simple as possible, relying on straightforward and widely-used solutions.

The storage tier. As mentioned in the introduction, a practical difficulty of using model checkers "in the large" is the management of all (generated) data during the verification endeavors. A disciplined recording of information on the different models during the verification phase becomes even more indispensable when errors are found, because, once the erroneous models have been corrected and re-verified, all models that have been verified previously and which are affected by the error should be re-verified as well [1]. We propose to use a Software Configuration Management (SCM) system to control the versioned artifacts produced during modelling and verification phases.

We do not impose any arrangement to the verification engineers. Hence, a verification endeavor is associated with a directory, with a total freedom to arrange endeavors and runs in a recursive manner. Using the combination of SCM feature and tools (e.g., the tool make) able to process automatically the building of software artifacts, each engineer organizes her endeavors in an arranged hierarchy or a rake of runs.

Complex objects such as set of properties or models decomposition are managed in the same manner, with a root directory and a freedom of organization. In order to ease the integration of each single object at the logical level, an XML description file stores information about the objects, a feature called a version description in a SCM system. The structure of the XML (its schema) is used by the software components (providing ingest and access features) to maintain an up-to-date ontological network in the logical level.

Avoiding the building of an information silo was also a concern. A silo is an insular management system that is unable to operate with any other systems. Files, directories, XML description, source version control ensure an access independent of any management system.

The access tier. One of the main goals of an Organizing System is to support the design and implementation of the actions, functions or services that make use of the resources. In classical 3-tier architecture, the presentation tier is the tier in which users interact with an application. Typical user interactions are ingestions (importing new resources into the OS), searches, browsing, tagging and annotations, retrieval, information extraction. An Organizing System is also intended to interact with other applications and should provide information exchange features with or without semantic transformations. Because any of the user interactions mentioned above might be performed in a dedicated tool, in this perspective, the presentation layer is merely an access layer and its main concern is tool interoperability. Among possible solutions, we choose a pragmatic approach, called conceptual interoperability based on the concept of the federation of models [15] and its open source tooling (http://research.openflexo.org). The Openflexo tool set provides support for building conceptual views expanding upon existing models and tools.

The ontological tier. From a conceptual perspective, the ontological layer is divided into two parts. The semantic network of types, on one hand, consists of semantic types linked by types of semantic relations, equivalent to an entity-relationship model or an UML class diagram). The semantic network of objects, on the other hand, contains a node for each fine-grained object plus nodes for the composite objects, each of which is assigned to one or more semantic types and linked to other objects by semantic relations.

The ontological tier is evolving continuously over time, because new resources types and new resources can be added at any time. People and organizations developing information and knowledge application systems are using different ontologies and there will be a need to reconcile these ontologies with a common upper ontology. We use the CIDOC CRM (http://www.cidoc-crm.org/official_release_cidoc.html), a standardized structure (ISO 21127:2014) for describing the implicit and explicit concepts and relationships used in cultural heritage documentation, as an upper ontology.

Technically, the ontological layer is stored in a TDB triple store; and we use the Apache Jena API to extract data from and write to RDF graphs (http://jena.apache.org/). There is an isomorphism between physical objects in the storage layer and semantic objects in the ontological layer; it is the access layer responsibility to maintain the isomorphism when items are updated in the storage tier.

3.3   An Organizing System to Perform and Enable the Verifier's Activities

Our team develops and maintains a model-checking toolkit. The system is described using the Fiacre language [16], which enables the specification of interacting behaviors and timing constraints through timed-automata. Our approach, called Context-aware Verification, focuses on the explicit modelling of the environment as one or more contexts. Interaction contexts are described with the Context Description Language (CDL). CDL enables also the specification of requirements through predicates and properties. The requirements are verified within the contexts that correspond to the environmental conditions in which they should be satisfied. All these developments are implemented in the OBP tool kit [17] and are freely available[1].

Recall that this research work aims to contribute to the solving of two practical problems: to deal with many verification objects (that have to be carefully managed and often re-verified); and to facilitate the judgement of the validation problem (is the formalized problem statement an adequate reflection of the actual problem?).

The first problem requires essentially a methodology of verification and a support tool. Each verification engineer work process is made of slightly different activities using their own resources through different verification tools; hence each engineer defines her methodology or uses a given one. Our aim is to provide an integration framework for the tools and methodologies; this is precisely the goal of an Organizing System to arrange resources and to support interactions with.

The second problem is part of a larger problem of computer-supported diagnosis that, as mentioned in Sect. 2.2, has been addressed with a multitude of techniques, among which we choose a model-based approach. Our hypothesis is that the logical tier of the OS, knowledge-rich and ontology-based, serves as the basis for information gathering, information integration, knowledge creation and knowledge sharing. Thanks to the

---

[1] OBP Languages and Tool kit website: http://www.obpcdl.org.

access layer, tools interoperability is made easier and tools collaboration provides the user with the required help. The OS acts as a backbone to add tools supporting techniques such as case-based reasoning.

Case-based reasoning. The main limitation of model-based diagnosis is that it requires a model. What does mean a modelling error in a model-checking approach? It means that, upon studying the error it is discovered that the model does not reflect the design of the system and that implies a correction of the model [3]. Hence it means that the design is the subject of diagnosis, and that we need a correct model of the design (that we do not have, in essence) to apply model-based reasoning. Fortunately, we can use case-based reasoning (CBR) to find this correct model. In CBR, a reasoner remembers previous situations similar to the current one and uses them to solve the new problem. So, we need to describe the old cases (called Problem Cases) in the OS using a Problem Case Template (mainly the problem statement, the formalized properties, the "correct" model and implementations for different model-checkers). Reasoning on a new case (called Sample Case) suggests "a model of reasoning that incorporates problem solving, understanding, and learning and integrates all with memory processes [18]." A work using pattern for relate Problems and Sample Cases is under submission [19].

4 Related Work

The research work of Theo C. Ruys [20] is the closest to ours, particularly for the first research problem addressed in this paper. The concept of managing the verification trajectory [1], by Ruys and Brinksma, has been a seminal paper for the understanding of the verification cycles and the need for a Software Configuration Management System for the verification "in the large" of real-life systems. We differ in scope because Ruys's work is focused on the use of the SPIN model-checker while we are looking for an agnostic view of model-checking that implies an intermediate abstract (and ontological) layer between the verification engineer and her verification objects. Because any resources ingested in the Organizing System is going through the access layer, the only extra price to pay for the verification engineer is to describe her organization documents (e.g. the version descriptor used in [1]; containing a description of the files included in a particular version) in a schema (e.g. in a XML Schema definition) and to relate schema components (e.g. element and attribute declarations and complex and simple type definitions) with semantic types and/or semantic relations of the OS ontological layer. If the verifiers' schema components does not exist in the ontological layer, the verification engineer has to indicate whether semantic constructs they refine and missing components will be added in the ontology. Thanks to this mapping, the particular view of any verification engineer will be shared with the other users of the OS.

Hence, our work and Ruys's work address the same issues and rely on the same solution scheme. However, our approach has two main advantages: it supports any model-checking tool and method and enlarges the community of OS users; the ontological layer permits shared knowledge and reasoning over different model-checking verification experiences, working across boundaries.

The research work of the Divine team [21] was a second source of inspiration. Divine verifies models in multiple input formats and has excellent execution performances using a cluster of multi-core machines and partial order reduction techniques, breaking through the limits of the state space explosion problem. However, our work has been more influenced by side products of the Divine team, mainly issues related to the BEEM benchmark management [12] and the automation of the verification process [21] by Pelanek. He agrees for the need for classifications based on a model structure and also classifications based on features of state spaces, which relate to model-based or experiential diagnosis introduced in Sect. 2.2. Pelanek's work perspectives mention a long term goal intended to develop an automated 'verification manager', which would be able to learn from experience [22]. Our approach is more humble and pragmatic: to provide the user with the bigger possible set of knowledge about verification, including an onto-logical classification of the problem and the solution spaces. Thanks to the arranged knowledge within the Organizing System, the verification engineer can plug her plug tools to address the automated verification manager issue in her way.

5   Conclusion

Verification "in the large" suffers of two practical problems: the verifier has to deal with many verification objects that have to be carefully managed and often re-veri-fied; it is often difficult to judge whether the formalized problem statement is an adequate reflection of the actual problem. We designed and built a prototype of an organizing system (OS) – an intentionally arranged collection of resources and the interactions they support – that makes easier the management of verification objects and supports reasoning interactions that facilitates diagnosis decisions.

Key points and driving issues of this research work are the ability of the OS to host a large variety of model-checking tools, techniques and methods; and the interoperability of the OS with external tools, providing the user with the freedom to use the proper approach to her problems.

However, we keep in mind that "any verification using model-based techniques is only as good as the model of the system [3]." Hence, a particular attention to the validity of the problem formalization will drive our future research efforts.

References

1. Ruys, T.C., Brinksma, E.: Managing the verification trajectory. Int. J. Softw. Tools Technol. Transf. (STTT) **4**, 246–259 (2003)
2. Larsen, K.G., Pettersson, P., Yi, W.: Model-checking for real-time systems. In: Reichel, H. (ed.) FCT 1995. LNCS, vol. 965, pp. 62–88. Springer, Heidelberg (1995)
3. Baier, C., Katoen, J.-P.: Principles of Model Checking. The MIT Press, Cambridge (2008)
4. Glushko, R.J. (ed.): The Discipline of Organizing. The MIT Press, Cambridge (2013)
5. Holzmann, G.J.: The theory and practice of a formal method: NewCoRe. In: IFIP Congress (1), pp. 35–44 (1994)
6. Groce, A., Visser, W.: What went wrong: explaining counterexamples. In: Ball, T., Rajamani, S.K. (eds.) SPIN 2003. LNCS, vol. 2648, pp. 121–135. Springer, Heidelberg (2003)

7. Avižienis, A., Laprie, J.-C., Randell, B., Landwehr, C.: Basic concepts and taxonomy of dependable and secure computing. IEEE Trans. Dependable Secure Comput. **1**, 11–33 (2004)
8. Reiter, R.: A theory of diagnosis from first principles. Artif. Intell. **32**, 57–95 (1987)
9. Peischl, B., Wotawa, F.: Model-based diagnosis or reasoning from first principles. IEEE Intell. Syst. **18**, 32–37 (2003)
10. Kavulya, S.P., Joshi, K., Giandomenico, F.D., Narasimhan, P.: Failure diagnosis of complex systems. In: Wolter, K., Avritzer, A., Vieira, M., van Moorsel, A. (eds.) Resilience Assessment and Evaluation of Computing Systems, pp. 239–261. Springer, Heidelberg (2012)
11. Venkatasubramanian, V., Rengaswamy, R., Yin, K., Kavuri, S.N.: A review of process fault detection and diagnosis: part I: quantitative model-based methods. Comput. Chem. Eng. **27**, 293–311 (2003)
12. Pelánek, R.: BEEM: Benchmarks for Explicit Model Checkers. In: Bošnački, D., Edelkamp, S. (eds.) SPIN 2007. LNCS, vol. 4595, pp. 263–267. Springer, Heidelberg (2007)
13. Abecker, A., Bernardi, A., Hinkelmann, K., Kühn, O., Sintek, M.: Toward a technology for organizational memories. IEEE Intell. Syst. **13**, 40–48 (1998)
14. Abecker, A., van Elst, L.: Ontologies for knowledge management. In: Staab, S., Studer, R. (eds.) Handbook on Ontologies, pp. 713–734. Springer, Heidelberg (2009)
15. Guychard, C., Guerin, S., Koudri, A., Beugnard, A., Dagnat, F.: Conceptual interoperability through models federation. In: Semantic Information Federation Community Workshop (2013)
16. Berthomieu, B., Bodeveix, J.-P., Farail, P., Filali, M., Garavel, H., Gaufillet, P., Lang, F., Vernadat, F.: Fiacre: an intermediate language for model verification in the Topcased environment. Presented at the ERTS 2008, January 2008
17. Dhaussy, P., Boniol, F., Roger, J.-C., Leroux, L.: Improving model checking with context modelling. Adv. Soft. Eng. **2012**, Article no. 9 (2012)
18. Kolodner, J.: Case-Based Reasoning. Kaufmann, San Mateo (1997)
19. Leilde, V., Ribaud, V., Dhaussy, P.: Organizing problem and sample cases for model-based diagnosis. In: Second International Workshop on Patterns in Model Engineering Co-located with MODELS 2016, Saint-Malo, France (2016, submitted)
20. Ruijs, T.C.: Towards effective model checking (2001). http://doc.utwente.nl/36596/
21. Barnat, J., et al.: DiVinE 3.0 – an explicit-state model checker for multithreaded C & C++ programs. In: Sharygina, N., Veith, H. (eds.) CAV 2013. LNCS, vol. 8044, pp. 863–868. Springer, Heidelberg (2013)
22. Pelánek, R.: Model classifications and automated verification. In: Leue, S., Merino, P. (eds.) FMICS 2007. LNCS, vol. 4916, pp. 149–163. Springer, Heidelberg (2008)

On Expanding Abbreviated Identifiers in the Source Code

Hui Yang[1], Xiaobing Sun[1(✉)], Yucong Duan[2], Han Zhao[1], and Bin Li[1]

[1] School of Information Engineering, Yangzhou University, Yangzhou, China
xbsun@yzu.edu.cn
[2] Hainan University, Haikou, China

Abstract. Program comprehension is an important and difficult task in software development and evolution, which is costly and time-consuming. Some software abbreviated identifiers in the source code can further increase the difficulty of the program comprehension, especially for the junior developers who have less developing expertise for the software system. Moreover, a number of studies focused on applying information retrieval (IR) techniques to analyze the source code identifiers for various software maintenance tasks. These IR techniques would have difficulty in exploring abbreviations in the program. Hence, this paper proposes a novel approach to expand the abbreviations of software identifiers. The proposed approach searches the expansions of abbreviated identifiers considering the searching resources of the program and the *Web*. An empirical study has been evaluated and demonstrates that our approach can effectively recommend the expansions, which can not only help developers comprehend the program, but also assist IR techniques in further exploiting the natural language information in the program.

Keywords: Program comprehension · Abbreviated identifier expansion · Information retrieval (IR) techniques

1 Introduction

Program comprehension is one of the most frequently performed activities during software maintenance and evolution [1]. Developers working on software maintenance tasks spend around 60 % of their time comprehending the system [2], in particular the source code. A number of recent research efforts focused on how software developers capture and express their intent in natural language embodied in the source code [3–5], such as the code identifiers used in the program [6]. These identifiers (e.g., names of classes, methods, parameters, or attributes, etc.) account for approximately more than half of linguistic information [7–9], and can often serve as a starting point in many program comprehension tasks such as traceability links recovery and feature location in source code [10–12].

Typically, informative identifiers are usually made up of full natural language words (dictionary words). It is easy for developers to read these identifiers and

© Springer International Publishing AG 2016
H. Yin et al. (Eds.): IDEAL 2016, LNCS 9937, pp. 588–595, 2016.
DOI: 10.1007/978-3-319-46257-8_63

further understand their intent. However, there are also some abbreviations for identifiers in the program. Developers who have less developing expertise cannot easily understand these abbreviated identifiers. Moreover, a number of studies focused on applying information retrieval (*IR*) techniques to analyze code identifiers for various software maintenance tasks [11–13]. But few work consider the abbreviated identifiers in the program [4,14,15], which may lose a lot valuable information, as most of the abbreviated identifiers are meaningful for program comprehension [16,17].

Hence, there is a need to expand the abbreviated identifiers as dictionary words in the source code for *IR* techniques and to better facilitate developers' comprehension of the program. In this paper, we propose a novel approach to recommend expansions for abbreviated identifiers considering the resources of the program and the *Web*. The abbreviations in the identifiers are classified as acronyms (e.g., *NMR*), abbreviations (e.g., *horiz*), and the blending words (e.g., *Strlen*). We first search the potential expansions of the abbreviations in the code. And then, we also extend our search to the web resources to find the expansion of abbreviated identifiers. An empirical study on three open-source subjects, e.g., *jEdit*, *JHotDraw* and *muCommander*, is evaluated to demonstrate that our approach can effectively expand the abbreviated identifiers.

2   Approach

The process of our approach is shown in Fig. 1. We first extract the identifiers from the source code. Then abbreviated identifiers are classified into corresponding types, and the expansions of abbreviated identifiers are searched from the program and the *Web*. Finally, all the expansions are ranked in a list for each abbreviated identifier.

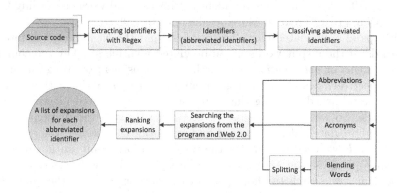

Fig. 1. The process of our approach

2.1 Extracting the Identifiers

The extracted identifier include: name of classes or interfaces (e.g., class ThrottledPushbackReader), name of functions (e.g., public void multipleSkip-CallsShouldWork), variables and constants (e.g., BLUECOLOR). Here, we use regular regrexexpressions to extract identifiers from the source code [18]. For example, to extract the names of classes, we use the regular expressions like "class ([a zA Z]|)*". In this way, we can get all the identifiers behind the key word of "class". Meanwhile, all the names of interfaces, functions, variables and constants are also extracted by regular regrexexpressions similar to the extraction of the class name.

2.2 Classifying and Expanding Abbreviated Identifiers

According to the identifier naming conventions, abbreviated identifiers are classified into three types: acronyms, abbreviations and blending words [18].

Acronym: An acronym is a word or name formed as an abbreviation from the initial components in a phrase, which is usually individual letters and sometimes syllables. For example, "cms" appears in many different identifiers and it may indicate an abbreviation of Color Management System (CMS).

Abbreviation: An abbreviation is a shortened form of a word. It consists of a group of letters taken from a word. For example, given the abbreviation "horiz", the word *horizontal* may be its full-scale form.

Blending Word: A blending word is a word formed from parts of two or more words. So we first need to split the blending word. For example, "Strlen" is split into *Str* and *len*. Then, all the split words are processed as *Acronyms* and/or *Abbreviations*. To split the *blending words*, we apply an algorithm named *processor* [19], which can split the identifier into several words according to the camel-casing (Word breaks). More details refer to [19].

To expand abbreviated identifiers, we first split the *blending words* into several words. If these split words are abbreviated identifiers[1], they are combined with *acronyms* or *abbreviations* to search the expansions from program and the *Web*, respectively. Searching from the program is easy, as all the dictionary words have well-established meaning that coincide with abbreviated identifiers in the source code. So we can use regular expression to match the dictionary words with the abbreviated identifiers, and find expansions. To search the expansions from the *Web*, we need to combine the program information (author name and class name), and use the *LocoySpider*[2] collector to collect web pages. To use the author name, we search the homepage of the author, who use the abbreviated identifiers, to match the expansions with the regular expression. To use the class

---

[1] Some split words are not abbreviated identifiers, such as *SymLink*, the split word *link* is a dictionary word, not an abbreviated identifier, so we don't need to search the expansion of *link*.

[2] http://www.locoy.com/.

name, we search the homepage of the files that include the abbreviated identifiers in the source code.

2.3   Ranking the Expansions for Each Abbreviated Identifier

The expansions extracted from the program should be ranked higher than those extracted from the *Web*, because the words in the program are more likely to be the expansions of abbreviated identifiers. If the expansions are from the same code block to the abbreviated identifier, such as the same method, the same file, and the same package, the expansions are ranked with different scores from different code blocks. That is, the expansion from the same method is ranked higher than that from the same file, and the expansion from the same package is ranked lower. If the expansion cannot be found from the program, we need to consider the expansions from the *Web*. For each abbreviated identifier, we combine the author name and class name of the source file to rank the expansions. And the expansions correlated with the class name are ranked higher than those correlated with the author name.

3   Evaluation

Our approach aims to recommend expansions of abbreviated identifiers for program comprehension. We conduct an empirical study to evaluate the effectiveness of expanding the abbreviated identifiers, and three open-source subjects are selected (in Table 1). Table 1 shows some basic information of these subjects, such as: the subject name, and the number of files, methods and abbreviated identifiers of each subject. To evaluate our approach, we have the following research questions:

RQ1: Is it Effective to Search the Expansions for Abbreviated Identifiers from the *Web*?

Our approach searches the expansions not only from the program, but also from *Web*. Since noise data exist in the web pages, it is necessary to evaluate the effectiveness of the acquired expansions from the web. To answer this research question, we classify all the abbreviated identifiers into three groups (*acronyms*, *abbreviations* and *blending words*), and some valid expansions of each group are recognized by several experienced programmers. Then, we count the ratio of

Table 1. The basic information of the evaluated subjects

Subject	File	Method	Abbreviated identifiers
jEdit	78	562	19
JHotDraw	57	342	27
muCommander	102	443	20

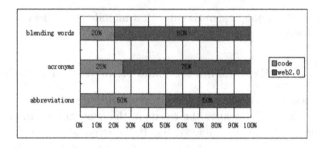

Fig. 2. Ratios of the expansions searched from the program and *Web 2.0.*

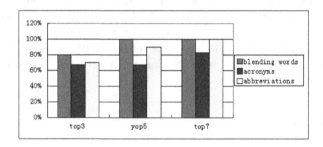

Fig. 3. The ratio of the abbreviated identifiers whose valid extensions are in top 3, 5, 7, respectively.

the valid expansions from the program and the ratio of those from the *Web*, respectively.

Figure 2 shows that 80 % of expansions for the *blending words* are identified from the *Web*, 75 % of expansions for the *acronyms* are from the *Web*. And half of expansions for the *abbreviations* are from the *Web*. In summary, searching expansions from the *Web* is effective.

RQ2: Can our Approach Effectively Recommend Expansions for Each Abbreviated Identifier?

Our approach recommends a list of expansions for each abbreviated identifier. We invited several experienced programmers to help recognize the valid expansions for each abbreviated identifier. Then we calculate the ratio of the abbreviated identifiers whose valid extensions are in top@k of the list, where k = 3, 5, 7, respectively.

Figure 3 illustrates the ratio of the abbreviated identifiers whose valid extensions are in top 3, 5, 7 for the three types of *acronyms*, *abbreviations* and *blending words*, respectively. We observe that 80 % of expansions for the *blending words* are accurately recommended in the top 3 of the list, and 66.7 % of *acronyms*, 70 % of *abbreviations* are accurately recommended with the valid expansions in the top 3, respectively. For top 5 candidate expansions, the ratio of valid expansions of *blending words* are up to 100 %, and for top 7, *abbreviations* reach 100 %,

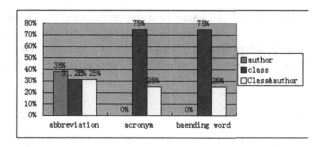

Fig. 4. The ratio of expansions' number of *author* group, *class* group and *class&author* group for *acronyms*, *abbreviations* and *blending words*, respectively.

and *acronyms* reach 82 %. In summary, our approach can effectively recommend the expansions for the abbreviated identifiers.

RQ3: Is it Effective to Combine the Messages of Class Name and Author Name to Rank the Expansions for Each Abbreviated Identifier?

To rank the expansions extracted from the *Web*, we consider the messages of class name and author name. We first select some valid expansions (which are extracted from the *Web*) from the results of *acronyms*, *abbreviations* and *blending words*, respectively. Then we classify these valid expansions into three groups: related to author name (*author* group), related to class name (*class* group), and related to author and class name (*class&author* group). Finally, we calculate the ratio of expansions for each group.

Figure 4 demonstrates the ratio of expansions' number of *author*, *class* and *class&author* group for *acronyms*, *abbreviations* and *blending words*, respectively. We observe that most expansions are extracted by combining with the *class* name, and the number of expansions extracted by combining with *author* name are relatively less. So we can rank the expansions correlated with the class name higher than those with the author name. In summary, the ranking mechanism for the expansions extracted from the *Web* is effective.

4   Related Work

Expanding identifiers has been studied widely in the state-of-art technologies. *Larkey* et al. developed a heuristic approach to expand identifiers [20]. They mined web pages to search acronyms and proposed some standard patterns to identify the correct expansion. But their approach didn't consider the expanding resource of the program and the textual patterns are not suitable for the source code. *Pakhomov* et al. worked on normalizing acronyms in medical text [21]. This approach considered numerous abbreviations and acronyms for the text, so it may be also suitable to extract expansions from source code. But the web resource are not considered. *Anquetil* et al. proposed an approach to extract information from large Pascal applications [22]. They defined the concept field of the words, and extracted the same field expansions from the web. But they did not expand the software identifiers, as they aim to process multi-field vocabularies. In this paper, we proposed a novel approach to recommend expansions for software abbreviated identifiers considering the resource of both the program and the *Web*.

5   Conclusion

This paper proposed a novel approach to expand the abbreviated identifiers from the program and the *Web*. Our approach can help developers comprehend abbreviations in the program, which can improve the effectiveness and efficiency of software development and evolution. Moreover, expanding abbreviations can also facilitate information retrieval (*IR*) techniques to efficiently explore the natural language information in the program.

Acknowledgments. This work is supported partially by Natural Science Foundation of China under Grant No. 61402396 and No. 61472344, partially by the Natural Science Foundation of the Jiangsu Higher Education Institutions of China under Grant No. 13KJB520027, and partially by the Jiangsu Qin Lan Project.

References

1. von Mayrhauser, A., Vans, A.M.: Program comprehension during software maintenance and evolution. IEEE Comput. **28**(8), 44–55 (1995)
2. Soloway, E., Ehrlich, K.: Empirical studies of programming knowledge. IEEE Trans. Softw. Eng. **10**(5), 595–609 (1984)
3. Sun, W., Sun, X., Yang, H., Li, B.: WB4SP: a tool to build the word base for specific programs. In: 24th IEEE International Conference on Program Comprehension, Austin, TX, USA, 16–17 May, pp. 1–3 (2016)
4. Binkley, D., Davis, M., Lawrie, D., Maletic, J.I., Morrell, C., Sharif, B.: The impact of identifier style on effort and comprehension. Empir. Softw. Eng. **18**(2), 219–276 (2013)
5. Wang, X., Pollock, L.L., Vijay-Shanker, K.: Automatic segmentation of method code into meaningful blocks: design and evaluation. J. Softw. Evol. Process **26**(1), 27–49 (2014)
6. Mens, T., Serebrenik, A., Cleve, A.: Evolving Software Systems. Springer, Heidelberg (2014)
7. Liu, Y., Sun, X., Duan, Y.: Analyzing program readability based on wordnet. In: Proceedings of the 19th International Conference on Evaluation, Assessment in Software Engineering, Nanjing, China, 27–29 April 2015 (2015). Article No. 27: Observation of strains. Infect Dis Ther. 3(1), 35–43.:1–27:2
8. Guerrouj, L., Bourque, D., Rigby, P.C.: Leveraging informal documentation to summarize classes and methods in context. In: 37th IEEE/ACM International Conference on Software Engineering, ICSE, Florence, Italy, 16–24 May, vol. 2, pp. 639–642 (2015)
9. Guerrouj, L.: Normalizing source code vocabulary to support program comprehension and software quality. In: 35th International Conference on Software Engineering, San Francisco, CA, USA, 18–26 May, pp. 1385–1388 (2013)

10. Sun, X., Li, B., Leung, H., Li, B., Li, Y.: MSR4SM: using topic models to effectively mining software repositories for software maintenance tasks. Inf. Softw. Technol. **66**, 1–12 (2015)

11. Fritz, T., Murphy, G.C., Murphy-Hill, E.R., Jingwen, O., Hill, E.: Degree-of-knowledge: modeling a developer's knowledge of code. ACM Trans. Softw. Eng. Methodol. **23**(2), 14 (2014)

12. Panichella, A., Dit, B., Oliveto, R., Penta, M.D., Poshyvanyk, D., De, Lucia, A.: How to effectively use topic models for software engineering tasks? an approach based on genetic algorithms. In: 35th International Conference on Software Engineering, pp. 522–531 (2013)

13. Sun, X., Li, B., Li, Y., Chen, Y.: What information in software historical repositories do we need to support software maintenance tasks? an approach based on topic model. In: Lee, R. (ed.) Computer and Information Science. SCI, vol. 566, pp. 27–37. Springer, Heidelberg (2015). doi:10.1007/978-3-319-10509-3_3

14. Hu, J., Sun, X., Lo, D., Li, B.: Modeling the evolution of development topics using dynamic topic models. In: 22nd IEEE International Conference on Software Analysis, Evolution, and Reengineering, Montreal, QC, Canada, 2–6 March, pp. 3–12 (2015)

15. Lu, M., Sun, X., Wang, S., Lo, D., Duan, Y.: Query expansion via wordnet for effective code search. In: 22nd IEEE International Conference on Software Analysis, Evolution, and Reengineering, Montreal, QC, Canada, 2–6 March, pp. 545–549 (2015)

16. Hill, E., Binkley, D., Lawrie, D.J., Pollock, L.L., Vijay-Shanker, K.: An empirical study of identifier splitting techniques. Empir. Softw. Eng. **19**(6), 1754–1780 (2014)

17. Binkley, D., Lawrie, D., Uehlinger, C.: Vocabulary normalization improves ir-based concept location. In: 28th IEEE International Conference on Software Maintenance, Trento, Italy, 23–28 September, pp. 588–591 (2012)

18. Garofolo, J.S., Lamel, L.F., Fisher, W.M., Fiscus, J.G., Pallett, D.S., Dahlgren, N.L., Zue, V.: Timit Acoustic-Phonetic Continuous Speech Corpus, vol. 33. Linguistic data consortium, Philadelphia (1993)

19. Sun, X., Liu, X., Hu, J., Zhu, J.: Empirical studies on the nlp techniques for source code data preprocessing. In: Proceedings of the International Workshop on Evidential Assessment of Software Technologies, pp. 32–39 (2014)

20. Larkey, L.S., Paul Ogilvie, M., Price, A., Tamilio, B.: Acrophile: an automated acronym extractor and server. In: Proceedings of the Fifth ACM Conference on Digital Libraries, pp. 205–214. ACM (2000)

21. Pakhomov, S.: Semi-supervised maximum entropy based approach to acronym and abbreviation normalization in medical texts. In: Proceedings of the 40th Annual Meeting on Association for Computational Linguistics, pp. 160–167. Association for Computational Linguistics (2002)

22. Anquetil, N., Kulesza, U., Mitschke, R., Moreira, A., Royer, J.-C., Rummler, A., Sousa, A.: A model-driven traceability framework for software product lines. Softw. Syst. Model. **9**(4), 427–451 (2010)

An Efficient Mining Algorithm for Maximal Weighted Frequent Patterns Based on WIdT-Trees

Qibing Qin[1] and Long Tan[1,2(✉)]

[1] College of Computer Science and Technology, Heilongjiang University,
Harbin, China
tanlong@hlju.edu.cn
[2] Key Laboratory of Database and Parallel Computing of Heilongjiang Province,
Harbin, China

Abstract. As processed data is relatively dense or the support is small in weighted frequent patterns mining process, the number of frequent patterns which meet the conditions will be exponential growth, and mining all frequent patterns will need too much computation. Hence, mining the maximal weighted frequent patterns containing all frequent patterns has less calculation, and it has more utility value. Aiming at the process of maximal weighted frequent patterns mining, an efficient algorithm, based on WIdT-Trees, is proposed to discover maximal weighted frequent patterns. In the algorithm, WIdT-Tree is optimized from WIT-Tree. The dTidset strategy is used to calculate the weighted support of frequent k-itemsets, and the nodes with equal extended weighted support are pruned off in order to reduce the computational cost and decrease the search space complexity. Algorithms are tested and compared on real and synthetic datasets and experimental results show that our algorithm is more efficient and scalable.

Keywords: Maximal weighted frequent patterns · Widt-Tree · dTidset · Extended weighted support

1 Introduction

Data mining is the process of discovering potential and useful knowledge in large amounts of data by the corresponding algorithms [1, 2]. Data mining is typically applied to find the frequent itemsets and the association rules between itemsets which hide in the database, and the frequent itemsets mining has played the vital role in the data mining. Since the maximal frequent itemsets imply all information about frequent itemsets, some frequent itemsets can be simplified into mining maximal frequent itemsets [3]. In addition, we don't need to dig up all the frequent itemsets, and just need to mine maximal frequent itemsets in some applications [4]. In some ways, the mining maxima frequent itemsets can greatly reduce the search space and time complexity, and greatly improve the mining efficiency.

Since each item represents different value and significance, frequent itemsets mining should fully consider the different characters of item in some itemsets in a real

© Springer International Publishing AG 2016
H. Yin et al. (Eds.): IDEAL 2016, LNCS 9937, pp. 596–605, 2016.
DOI: 10.1007/978-3-319-46257-8_64

world application [5]. Ramkumar, Ranka, and Tsur [6] puted forward the concept of weighted association rules(Weighted Association Rules, WAR), and proposed weighted frequent itemsets mining algorithm. And on this basis many scholars have proposed weighted association rules mining algorithm (Weighted Association Rule Mining, WARM). Wang et al. [7] proposed a more effective weighted association rules mining algorithm. Tao and Murtagh et al. [8] proved the closed down property of weighted itemsets, and proposed a more effective mining model. Khan et al. [9] proposed the WUARM framework by fully considering that the individual items have different value and significance in itemsets. According to expanding IT-tree [10] (Item-Tidset tree) into WIT-tree (Weighted Itemset-Tidset tree), Vo [11] proposed a new weighted frequent pattern mining algorithm based on WIT-tree data structure in order to reduce the candidate itemsets and optimize the mining process.

In some practical applications, we can simplify the weighted frequent itemsets mining into maximal weighted itemsets in order to simplify the weighted frequent pattern mining process [12]. As a result, the maximal weighted itemsets mining has a more meaningful and practical value compared with the existing method. Based on the above studies, Unil Yun, Hyeonil Shin [13] defined the concept of maximal weighted frequent itemsets, and proposed a maximal weighted mining frequent itemsets mining algorithm (Maximal Weighted Frequent Itemset Mining, MWFIM), which replaced weighted frequent itemsets mining. The algorithm adopted pruning mode for infrequent patterns, and employed a storage way of the prefix tree based on the inverted order weight to improve the search efficiency. However, there has been almost no consideration for the cost of computing the weight of frequent k- itemsets, and it can lead to repeated calculation of the weighted support of some certain itemsets. Therefore, the efficiency of MWFIM algorithm should be improved in the processing dense datasets or big datasets.

In the paper, we optimize the cost of computing the weight of frequent k-itemsets and reduce the space complexity. We are inspired by the way of MEI algorithm [14] for mining erasable itemsets and define the concept of dTidset(The difference of two Tidsets), WIT-Tree data structure optimize to WIT-Tree based on dTidset, and proposed a more effective maximal weighted frequent patterns mining algorithm— WIdT_MWFIM. (Maximal Weighted Frequent Itemsets Mining Algorithm based on WIdT-Tree). The proposed algorithm takes different handling methods with the weighted support calculation of nodes meeting different conditions. Some nodes adopt dTidset strategy to compute the weighted support, some nodes can directly inherit weighted support from its parent class in order to avoid computing and some nodes are pruned off according to the property of weighted extended support. Experimental results show that WIdT_MWFIM algorithm has advantages in running time.

2   System Model

In this section, we will introduce correlative conception and property of weighted frequent itemsets.

Definition 2.1 Weighted Transaction Database(D). Let a itemsets $I = \{i1, i2, i3, \ldots, im\}$, weight set $W = \{w1, w2, w3, \ldots, wm\}$, weighted Transaction Database $D = \{<TID, T|T \subseteq I>\}$, there is a one-to-one correspondence $\forall in (1 \leq n \leq m)$ bettwen $\exists wn (1 \leq n \leq m) \in W$.

Definition 2.2 The transaction weight (tw) of a transaction t_k is defined as follows $(tw(t_k))$:

$$tw(tk) = \frac{\sum_{ij \in tk} wj}{tk} \tag{2.1}$$

Definition 2.3 The weighted support of an itemset is defined as follows(ws(X)):

$$ws(X) = \frac{\sum_{tk \in t(X)} tw(tk)}{\sum_{tk \in T} tw(tk)} \tag{2.2}$$

Where T is the list of transactions in the database.

The purpose of weighted frequent itemsets (Weighted Frequent Itemsets, WFI) mining is to find all the weighted itemsets which their weighted support satisfies the minimum weighted support threshold (minws).

$$WFI = \{X \subseteq I | ws(X) \geq minws\}$$

Tao et al. [8] proved the closed down property of weighted itemsets according to the improved calculation model of weighted support.

If $X \subset Y$, then $ws(X) \geq ws(Y)$

Yun et al. [13] proposed the maximal weighted frequent itemsets mining algorithm instead of the weighted frequent itemsets, and defined the maximal weighted frequent itemsets (Maximal Weighted Frequent Itemsets, MWFI), which is defined as follows:

Let itemset X is the maximal weighted frequent itemsets, if and only if all supersets of X are infrequent.

The main purpose of the paper is to implement a more efficient algorithm for mining maximal weighted frequent itemsets.

Definition 2.4 Tidset(X). Let a item X, all TIDs which contain item X are be referred to as Tidset(X).

Definition 2.5 dTidset(The difference of two Tidsets). Let XA and XB be two itemsets with the same prefix X. The Tidset of XA is denoted by t(XA) and the Tidset of XB is denoted by t(XB). The difference Tidset of t(XA) and t(XB) is denoted as dT(XAB), is defined as follows:

$$dT(XAB) = t(XA) \backslash t(XB) \tag{2.3}$$

Property 2.1. Let XA and XB be two itemsets with the same prefix X. the difference Tidset of t(XA) and t(XB) is denoted as dT(XAB), is computed as follows:

$$dT(XAB) = dT(XB) \backslash dT(XA)$$

Proof. From [15]

$$t(XA) = t(X) \cap t(A)$$

$$t(XB) = t(X) \cap t(B)$$

According to Definition 2.3:

$$
\begin{aligned}
dT(XAB) &= t(XA) \backslash t(XB) \\
&= [t(X) \cap t(A)] \backslash [t(X) \cap t(B)] \\
&= [t(X) \cap [t(A) \backslash t(X)]] \backslash [t(X) \cap [t(B) \backslash t(X)]] \\
&= [t(X) \backslash t(B)] \backslash [t(X) \backslash t(A)] \\
&= dT(XB) \backslash dT(XA)
\end{aligned}
$$

3   WIdT-Tree Data Structure

We propose a more efficient storage structure—WIdT-Tree (Weighted Itemset-dTidset tree) by optimizing WIT-Tree data structure on the basis of dTidset. Each node in WIdT-Tree consists of three fields, a form of triplets $< X, dTidset(X), ws(X) >$. X represents itemset X; dTidset(X) represents the difference of two Tidsets; ws(X) represents weighted support of itemset X.

Property 3.1. For two given k-itemset $(k \geq 2)$ XA and XB, if $|dT(XA)| > |dT(XB)|$, then:

$$|dT(XAB)| < |dT(XBA)|$$

Proof.

$$
\begin{aligned}
dT(XAB) &= dT(XB) \backslash dT(XA) \\
&= dT(XB) \backslash [dT(XA) \cap dT(XB)]
\end{aligned}
$$

$$\because [dT(XA) \cap dT(XB)] \subseteq dT(XB) \text{ then :}$$
$$dT(XAB)| = |dT(XB)| - |dT(XA) \cap dT(XB)|$$

Similarly.

$$|dT(XBA)| = |dT(XA)| - |dT(XA) \cap dT(XB)|$$
$$\because |dT(XA)| > |dT(XB)|$$
$$|dT(XAB)| > |dT(XBA)|$$

Definition 3.1 Extended weighted support. Let a given I_i in WIdT-Tree, $Tail(Ii) = \{item \in I \wedge Ii \prec item\}, \prec$ represents the alphabetical order. $\exists Ij \in Tail(Ii)$, extended weighted support of I_i is denoted as $ws(Ii \cup Ij)$.

Definition 3.2 Extended weighted frequent itemsets. $\exists ws(Ii \cup Ij) \geq minws$, $Ii \cup Ij$ is extended weighted frequent itemsets, extended weighted frequent itemsets of I_i is denoted as:

$$EWFI(Ii) = \{Ij \in Tail(Ii) \wedge ws(Ii \cup Ij) \geq minws\}$$

Property 3.2. Let a given I_i in WIdT-Tree, $Ij \in EWFI(Ii)$, if $ws(Ij) = ws(Ii)$, In subtree rooted at node I_i, the non-I_i and the subtrees which set them as root nodes are invalid in the process of weighted frequent patterns mining.

Proof. Assuming P is the node of WIdT-Trees,

$Tail(P) = \{x1, x2, \ldots, xm\}$, $ws(P \cup xi) = ai$, $xi \in Tail(P), 1 \leq i \leq m$.
N is the child node of the P, $N = P \cup \{xj\}$.
Then $Tail(N) = \{xj + 1, xj + 2, \ldots, xm\}$
$xt \in Tail(N)$
$xt \in \{xj + 1, xj + 2, \ldots, xm\} \subset Tail(P)$

The extended weighted support of $P \cup xt$ is denoted as $ws(P \cup xt) = pt$.
The extended weighted support of $N \cup xt$ is denoted as $ws(N \cup xt) = nt$
If $pt = nt$, then the weighted itemsets including $P \cup xt$ is a subset of The extended weighted frequent itemsets of $N \cup xt$. That is, if it exists maximal weighted frequent itemsets with the parent $P \cup xt$, it contains $N \cup xt$. Therefore, we only need to calculate $ws(N \cup xt)$ in the process of maximal weighted frequent itemsets mining.

4   Proposed Algorithm

In the section, we will introduce a more effective maximal weighted frequent itemsets mining algorithm based on WIdT-Tree. Before presenting our algorithm, let us first explore an effective calculation method of weighted support.

4.1   The Calculation Method of Weighted Support

Let k-itemsets PX and PY, the Tidsets of PX and PY respectively be denoted as t(PX) and t(PY), then the calculation of ws(PXY) is shown as follows.

$$
\begin{aligned}
t(PXY) &= t(PX) \cap t(PY) \\
&= t(PX) \backslash [t(PX) \backslash t[PY] \\
&= t(PX) \backslash dT(PXY)
\end{aligned}
\tag{4.1}
$$

According to Definition 2.3:

$$
ws(PXY) = \frac{\sum_{tk \in t(PXY)} tw(tk)}{\sum_{tk \in T} tw(tk)}
\tag{4.2}
$$

Based on (4.1) and (4.2):

$$
ws(PXY) = ws(PX) - \frac{\sum_{tk \in dT(PXY)} tw(tk)}{\sum_{tk \in T} tw(tk)}
\tag{4.3}
$$

Proof.

$$
\begin{aligned}
ws(PXY) &= \frac{\sum_{tk \in t(PXY)} tw(tk)}{\sum_{tk \in T} tw(tk)} \\
&= \frac{\sum_{tk \in [t(PX) \backslash dT(PXY)]} tw(tk)}{\sum_{tk \in T} tw(tk)} \\
&= \frac{\sum_{tk \in t(PX)} tw(tk) - \sum_{tk \in dT(PXY)} tw(tk)}{\sum_{tk \in T} tw(tk)} \\
&= \frac{\sum_{tk \in t(PX)} tw(tk)}{\sum_{tk \in T} tw(tk)} - \frac{\sum_{tk \in dT(PXY)} tw(tk)}{\sum_{tk \in T} tw(tk)} \\
&= ws(PX) - \frac{\sum_{tk \in dT(PXY)} tw(tk)}{\sum_{tk \in T} tw(tk)}
\end{aligned}
$$

Property 4.1. In the WIdT-Tree, assume k-itemset PX and PXY, if dT(PXY) = Ø, then $ws(PXY) = ws(PX)$.

Proof. According to (4.3):

$$
ws(PXY) = ws(PX) - \frac{\sum_{tk \in dT(PXY)} tw(tk)}{\sum_{tk \in T} tw(tk)}
$$

4.2   WIdT_MWFIM Algorithm

In WIdT_MWFIM algorithm, WIdT-Trees are generated by **WIdT_GEN**. In the process of generating WIdT-Trees,1-itemsets which satisfy minws are sorted by |dT (XY)| descending order, and all sub_WIdT-Trees rooted as 1-itemsets which satisfy minws are generated by using the divide-and-conquer strategy. On the basis of all sub_WIdT-Trees, Candidate WIdT-Trees can be generated. Then we adopt **Procedure WIdT_MWFIM** to search the Candidate WIdT-Trees in order to all MWFI. **WIdT_MWFIM algorithm** is shown as follows.

Input: Database D and minimum weighted support threshold *minws*
Output: WIdT-Trees
Procedure WIdT_GEN(Database D, *minws*)
1) L_1=all of 1-items which their ws satisfy *minws*
2) Sorting each node $\in L_1$ in WIdT-Trees in descending order by |dTidset(X)|//**Property 3.1**
3) For each sub_WIdT rooted as $V_i \in L_1$
4)   For $Tail(V_i) \neq \emptyset$
5)   Compute dTidsets
6)   If$\exists V_j \in Tail(V_i)$ and $V_j \cup V_i$ satisfies   **Property 4.1**
7) $ws(V_j \cup V_i) = ws(V_i)$,Don't consider other nodes of the same depth in sub_WIdT // **Property 3.2**
8)   Else computing $ws(V_j \cup V_i)$
9)   // **Formula 4.3**
10)   End if
10)   End For
11)   End For
12)   Return WIdT-Trees
End Procedure

Input: Candidate WIdT-Trees
Output: MWFI(Maximal Weighted Frequent Itemsets)
Procedure WIdT_MWFIM(Candidate WIdT-Trees)
1) MWFI = $\emptyset$
2) Search Candidate WIdT-Trees
3) For each leaf node N_i in Candidate WIdT-Trees
4)   If $N_i \notin$ MWFI
5)   MWFI = $N_i \cup$ MWFI
6)   End If
7) End For
End Procedure

---

5 Performance Evaluation

In this section, we will test the running time and scalability of WIdT_MWFIM algorithm are compared with MWFIM [13] and MAFIA [16]. Note that all these algorithms discover the same weighted frequent itemsets, which confirms the result generated by any algorithms in the testing experiments is correct and complete.

5.1 Experimental Datasets

We conduct experiments on the performance of WIdT_MWFIM algorithm as compared to MAFIA and MWFIM on real datasets and synthetic datasets. Pumsb containing data related to the census is dense datasets. Retail including shopping data in a supermarket is sparse datasets. These two above datasets can be downloaded from (http://www.fimi.cs.helsinki.fi/data/). The T10I4Dx synthetic datasets are generated from the IBM dataset generator. These test datasets including dense and sparse datasets do not have item weight, and the weights of items are generated by a random generation function.

5.2 Comparison of Results

The normalized weights of items are(0.3, 0.7) and (0.3, 0.6) in Pumsb and Retail datasets. For example, the numbers of patterns in WIT_MWFIM algorithm on Pumsb is 159,417 with a minimum support of 52 %, and 185,342 with a minimum support of 50 %. The dTidset strategy can greatly improve the computational efficiency of weighted support and the property of extended weighted support is employed in order to reduce the cost of search space. WIT_MWFIM algorithm has advantages in running time than MWFIM and MAFIA algorithm. The running time comparison results are show as Figs. 1 and 2.

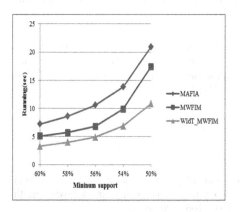

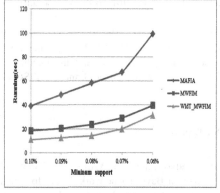

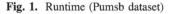

Fig. 1. Runtime (Pumsb dataset)   **Fig. 2.** Runtime (Retail dataset)

Based on the runnig time, WIT_MWFIM algorithm conduct the expriments testing scalability on synthetic datasets. We conduct a scalability test on T10I4Dx datasets, which minimum support is set as 0.1 % about 100-500 K transactions and the normalized weights of items are(0.3, 06). We set minimum support as 0.3 %,1000-4000 K transactions and the normalized weights of items are (0.3, 06). The scalability comparison results on T10I4Dx datasets are show as Fig. 3 a and b.

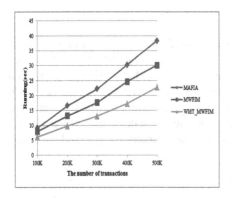 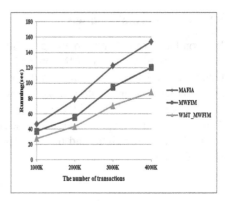

Fig. 3. a. Runtime (T10I4Dx minws = 0.1 %) b. Runtime (T10I4Dx minws = 0.3 %)

6 Conclusions

In this paper, we propose an effective maximal weighted frequent patterns mining algorithm—WIdT_MWFIM based on WIdT-Trees. We take three different ways of dealing with nodes satisfying different conditions. Some nodes adopt dTidset strategy to compute the weighted support, some nodes can directly inherit weighted support from its parent and some nodes with equal extended weighted support are pruned off. For evaluating the performance of WIdT_MWFIM, we conduct lots of experiments on real and synthetic datasets. Experimental results show that WIdT_MWFIM has advantages in running time and scalability.

Acknowledgments. This work is supported by the National Natural Science Foundation Item of China under Grant No. 81273649, Natural Science Foundation Item of Heilongjiang Province under Grant No.F201434, and the Graduate Student Innovation and Research Item of Heilongjiang University under Grant NO.YJSCX2016-018HLJU.

References

1. Bandyopadhyay, S., Wolfson, J., Vock, D.M., et al.: Data mining for censored time-to-event data: a Bayesian network model for predicting cardiovascular risk from electronic health record data. Data Min. Knowl. Disc. **29**(4), 1033–1069 (2015)

2. Solanki, S.K., Patel, J.T.: A survey on association rule mining. In: 2015 Fifth International Conference on Advanced Computing and Communication Technologies (ACCT), pp. 212–216. IEEE (2015)

3. Yun, U., Lee, G.: Incremental mining of weighted maximal frequent itemsets from dynamic databases. Expert Syst. Appl. **54**, 304–327 (2016)

4. Yadav, L., Nair, P.S.: A new data structure for finding maximalfrequent itemset in online data mining. In: 2015 International Conference on Computer, Communication and Control (IC4), pp. 1–5. IEEE (2015)

5. Zhong, Y., Liao, Y.: Research of mining effective and weighted association rules based on dual confidence. In: 2012 Fourth International Conference on Computational and Information Sciences (ICCIS), pp. 1228–1231. IEEE (2012)

6. Ramkumar, G.D., Ranka, S., Tsur, S.: Weighted association rules: model and algorithm. In: Proceedings of ACM SIGKDD, pp. 661–666 (1998)

7. Wang, W., Yang, J., Yu, P.S.: Efficient mining of weighted association rules (WAR). In: Proceedings of the Sixth ACM SIGKDD International Conference on Knowledge Discovery and Data Mining, pp. 270–274. ACM (2000)

8. Tao, F., Murtagh, F., Farid, M.: Weighted association rule mining using weighted support and significance framework. In: Proceedings of the ninth ACM SIGKDD International Conference on Knowledge Discovery and Data Mining, pp. 661–666. ACM (2003)

9. Khan, M.S., Muyeba, M., Coenen, F.: A weighted utility framework for mining association rules. In: Second UKSIM European Symposium on Computer Modeling and Simulation, EMS 2008, pp. 87–92. IEEE (2008)

10. Zaki, M.J., Hsiao, C.J.: Efficient algorithms for mining closed itemsets and their lattice structure. IEEE Trans. Knowl. Data Eng. **17**(4), 462–478 (2005)

11. Vo, B., Coenen, F., Le, B.: A new method for mining frequent weighted itemsets based on WIT-trees. Expert Syst. Appl. **40**(4), 1256–1264 (2013)

12. Tseng, V.S., Shie, B.E., Wu, C.W., et al.: Efficient algorithms for mining high utility itemsets from transactional databases. IEEE Trans. Knowl. Data Eng. **25**(8), 1772–1786 (2013)

13. Yun, U., Shin, H., Ryu, K.H., et al.: An efficient mining algorithm for maximal weighted frequent patterns in transactional databases. Knowl. Based Syst. **33**, 53–64 (2012)

14. Le, T., Vo, B.: MEI: an efficient algorithm for mining erasable itemsets. Eng. Appl. Artif. Intell. **27**, 155–166 (2014)

15. Borgelt, C.: Efficient implementations of apriori and eclat. In: FIMI 2003 Proceedings of the IEEE ICDM Workshop on Frequent Itemset Mining Implementations (2003)

16. Burdick, D., Calimlim, M., Flannick, J., et al.: MAFIA: a maximal frequent itemset algorithm. IEEE Trans. Knowl. Data Eng. **17**(11), 1490–1504 (2005)

Enhancing UML Class Diagram Abstraction with Knowledge Graph

Liang Huang[1], Yucong Duan[1(✉)], Xiaobing Sun[2], Zhaoxin Lin[3], and Chuanpu Zhu[1]

[1] College of Information Science and Technology, Hainan University, Haikou, China
{1512460987,907397737}@qq.com, duanyucong@hotmail.com
[2] School of Information Engineering, Yangzhou University, Yangzhou, China
sundomore@163.com
[3] School of Business, Iowa State University, Ames, USA
zxlin@iastate.edu

Abstract. Model-Driven Engineering (MDE) alleviates the cognitive complexity and effort spent on software development by generating codes from models. In MDE, models should be accurate, refined, reliable and efficient. Class diagram is a structural abstraction of a real system and usually used in software design. A better designed class diagram could lead to a better system. In this paper, we proposed a knowledge graph based method to improve class diagrams. We took knowledge graph as the media layer for easier information introduction, and proposed methods to map data, information and knowledge between class diagrams and knowledge graphs bidirectionally. Based on the added knowledge source, we designed hierarchical clustering algorithm to abstract the class diagram, and finally we generated abstracted class diagrams automatically.

Keywords: Class diagram · Knowledge graph · UML · OCL · Hierarchical clustering algorithm

1 Introduction

Models and modeling are essential parts of every engineering endeavors [1]. Unified Modeling Language (UML) is a de-facto standard for object-oriented modeling [2]. UML class diagram is used to describe the static structure of a system, and a well-designed class diagram can lead to a better developed program. What's more, a well-designed class diagram could be more easily for developers to read. Knowledge graph is essentially a kind of semantic network [4]. Nodes in knowledge graph stand for entities or concepts, the edges stand for semantic relationships between entities or concepts.

In this paper, we proposed a method to abstract class diagrams. We use knowledge graphs as the media layer for easier information introduction. We abstract a class diagram on the knowledge graph layer, and transform the abstracted knowledge graph to a new class diagram.

In the rest of the paper, we firstly elaborate the related works in Sect. 2 and the approach to abstract class diagram in Sect. 3. Then, we elaborate the method of mapping between class diagrams and knowledge graphs in Sect. 4, the approach of abstracting a

© Springer International Publishing AG 2016
H. Yin et al. (Eds.): IDEAL 2016, LNCS 9937, pp. 606–616, 2016.
DOI: 10.1007/978-3-319-46257-8_65

knowledge graph is elaborated in Sect. 5. After that, we make a conclusion in Sect. 6. Finally, acknowledgments are elaborated in acknowledgement section.

2 Related Works

(1) UML class diagram abstraction. UML class diagram abstraction transforms a low-level class diagram to a high-level class diagram [8]. The class diagram will be very large when a system is complex. Designers are easily overwhelmed with details when dealing with large class diagrams [13]. So the abstraction of large class diagrams is necessary. The abstraction of a large class diagram means refining the class diagram. Eyged. A find a series of abstraction rules for class diagram abstraction which contains class abstraction rules and relationship abstraction rules [9]. He concentrates on the refinement of relationships. However, by mapping class diagrams to knowledge graphs, our work can analyze and refine attributes and operations.

(2) Knowledge graph completion. Knowledge graph completion aims to extract relationships between different entities. Knowledge graph completion can find new relational facts, which is an important supplement to relationship extraction from knowledge sources [9, 11, 12]. In our work, the knowledge graph was used as a media layer to extract the attributes and operations in different classes with the same types, and we make a hierarchical clustering of classes.

3 Use Knowledge Graphs to Abstract Class Diagrams

UML class diagram is a kind of static structural model used to describe the structure of a system. It comprises classes and relationships. Classes in a class diagram include attributes and operations. Knowledge graphs contain nodes and edges. We use knowledge graphs as a media layer to abstract class diagrams. As shown in Fig. 1, our class diagram abstracting method contains four major steps as following shows.

(1) Firstly, map a class diagram to a knowledge graph and generate the knowledge graph.

(2) Then, abstract the mapped knowledge graph by hierarchical clustering algorithm.

(3) Then, map the abstracted knowledge graph to a new class diagram and generate the new class diagram.

(4) Finally, output the abstracted class diagram.

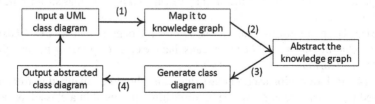

Fig. 1. The work flow of abstracting a class diagram

4   Map Between UML Class Diagrams and Knowledge Graphs

People concentrate on the transformation between models [3], however the transformation between models and other kinds of resources may also be helpful, such as the transformation between class diagrams and knowledge graphs.

4.1   Map from a UML Class Diagram to a Knowledge Graph

A UML class diagram can be mapped to a knowledge graph. A class in a class diagram can be mapped to a concept in a knowledge graph. An attribute can be mapped to an entity, and the type of an attribute can be mapped to a concept. An operation can be mapped to a concept. A concept or an entity in a knowledge graph was presented as a node.

The rules of mapping a UML class diagram to a knowledge graph are as following shows. The mapping rules conclude rules for mapping classes and rules for mapping relationships.

(1) Rules for mapping classes:

Rule (1): The node' name is *C_ClassName* if the node was mapped by a class whose name is *ClassName*. For example, the node' name is *C_Apple* if the class' name is *Apple*.

Rule (2): The node' name is *A_AttributeName* if the node was mapped by an attribute whose name is *AttributeName*. For example, the node' name is *A_Zhangsan* if the attribute' name is *Zhangsan*.

Rule (3): There is a node whose name is *C_TypeName* if the type of the attribute is *TypeName*.

Rule (4): There is a node whose name is *O_OperationName* if the name of the operation is *OperationName*.

Rule (5): The node whose name begins with *"C_" (e.g. C_XX)* can only occur once in a knowledge graph, because names of two classes can't be same in a class diagram. However, the node whose name begins with *"A_" (e.g. A_XX)* or *"O_" (e.g. O_XX)* can occur more than once, because different classes can have the same attributes and operations.

(2) Rules for mapping relationships:

Rule (1): The name of an edge is *RelationshipName* if it begins from node *C_ClassNameA* to node *C_ClassNameB*, at the same time, the relationship from class *ClassNameA* to class *ClassNameB* is *RelationshipName*

Rule (2): The name of an edge is *"Has a"* if it begins from node *C_ClassName* to node *A_AttributeName*. *"Has a"* means that the class *ClassName* has an attribute named *AttributeName*.

Rule (3): The relationship of an edge is *"Has o"* if it begins from node *C_CLassName* to node *O_OperationName*. *"Has_o"* means that the class *ClassName* has an operation named *OperationName*.

Rule (4): Relationships used in class diagrams should be mapped to the relationships used in knowledge graphs. The relationship types used in a class diagram such as "Generalization", "Aggregation", "Composition", "Dependency", "Association" need to be described as "Is a", "Aggregate", "Composite", "Depend on", "Associate with" in a knowledge graph.

Rule (5): The relationship of an edge which begins from *A_AttributeName* to *C_AttributeTypeName* is *"Is a"*. For example, Class *A* has an attribute named *Tom* and the type of *Tom* is *User*. When mapping this relationship, the relationship of the edge from *A_Tom* to *C_User* should be *"Is a"*. It means that *Tom* is a *User*.

Figure 2 shows a simple class diagram example. We map it to a knowledge graph as shown in Fig. 3. In the class diagram shown in Fig. 2, *ClassA* has two attributes named *Attribute1* and *Attribute2* and an operation named *Operation1*. When *ClassA* is mapped to the knowledge graph, there will be a node named *C_ClassA* stands for *ClassA*, a node named *A_Attribute1* stands for *Attribute1*, a node named *A_Attribute2* stands for *Attribute2*, and a node named *O_Operation* stands for *Operation1*. The relationship of the edge begins from *C_ClassA* to *A_Attribute1* is *"Has a"* which means that *ClassA* has an attribute named *Attribute1*. The relationship of the edge begins from *C_ClassA* to *O_Operation1* is *"Has o"* which means *ClassA* has an operation named *Operation1*. The relationship of the edge which begins from *A_Attribute1* to *C_Integer* is *"Is a"* which means *Attribute1* is a kind of *Integer*.

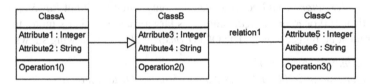

Fig. 2. A class diagram example

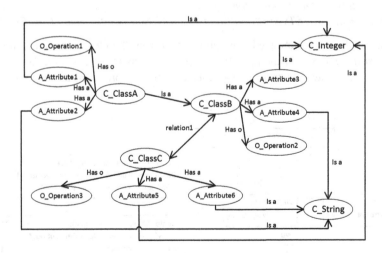

Fig. 3. The knowledge graph of the example in Fig. 2

4.2   Map from a OCL Restrained Class Diagram to a Knowledge Graph

The Object Constraint Language (OCL) was developed inside IBM in 1995 to overcome the limitation of UML and precisely specifying detailed aspects of a system [7]. In order to analysis class diagrams with OCL constraints on the knowledge graph layer, the maps between knowledge graph and OCL constraints are necessary. In this part, we will talk about how to map OCL constraints to a knowledge graph. The mapping rules are as following shows.

Rules for mapping OCL:

Rule (1): An expression statement in OCL presents as a node begin with *"E_"* in a knowledge graph. For example, the expression statement *"A>100"* is mapped to a node named *"E_A>100"* in a knowledge graph.

Rule (2): The node begin with *"E_"* should be atomic. For example, *"E_A>100"* is atomic however *"E_A>100 and E_B>50"* is not atomic.

Rule (3): An operator in OCL presents as a node in a knowledge graph.

Rule (4): The constraint of an element (e.g. Class, Attribute and Operation) starts with a node named *Constraint* in a knowledge graph.

Figure 4 is an example shows a class diagram with OCL constraints.

The *CandyFactory* could generate candies only when the input *RawSugar* amount is larger than 100 or *RefinedSugar* amount is larger than 50, what's more, the amount of *Order* must be bigger than 1. The OCL *Constraint1* is as following shows:

Contex CandyFactory.begin():
(RawSuger.amount>100 or RefinedSuger>50) and Order.amount>1

The *CandyFactory* will generate *SoftSweets* if the *Order type* is *SoftSweets*, else the *CandyFactory* will generate *HardCandies*. The OCL *Constraint2* is as following shows:

Contex CandyFactory.CandyType:
CandyType=if(Order.Type="SoftSweets") then "SoftSweets" else "HardCandies"

Figure 5 shows the knowledge graph mapped from the OCL constraints in Fig. 4. The *Constraint1* is set on the operation *begin()* in class *CandyFactory*. According to Rule(2), we separate *Constraint1* to a series of atomic expressions as shown in Fig. 5.

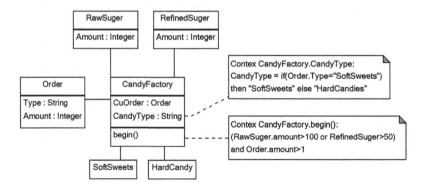

Fig. 4.   Class diagram with OCL constrains

The *Constraint2* is set on the attribute *CandyType*. We also separate *Constraint2* to a series of atomic expressions as Fig. 5 shows.

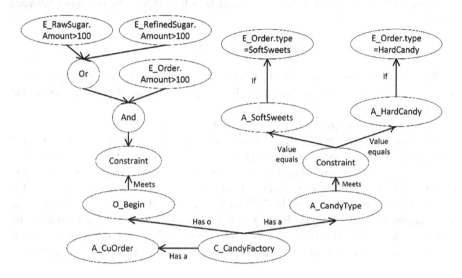

Fig. 5. The knowledge graph with the mapped OCL constraints in Fig. 4

4.3 Map from a Knowledge Graph to a UML Class Diagram

If a knowledge graph meets the rules given before, it can be mapped to a class diagram too.

5 Abstract the Mapped Knowledge Graph

A knowledge graph is used as a media layer for us to abstract our class diagrams. Firstly, we compute the similarities between classes. Then, we apply the hierarchical clustering algorithm to abstract the knowledge graph. Finally, we map the abstracted knowledge graph to a new class diagram and generate it.

5.1 Compute the Similarities Between Nodes

Define 1: Attribute *attri* in Class *i* is equal to attribute *attrj* in Class *j* if the type of *attri* is equal to the type of *attrj* except that the type of the attribute is *"String"*, *"Integer"*, *"Float"*, *"Double"* and other class types defined by system but users.

Define 2: Operation *operi* in Class *i* is equal to *operj* in *Classj* if the name of *operi* is equal to *operj*.

The similarity between two classes is associated with the similarity between their attributes and operations. The similarity between two classes is higher if their attributes and operations are more similar. Formula 1 shows how to compute the similarity between two classes.

$$S(i,j) = n_a(i,j) + n_o(i,j); \tag{1}$$

$S(i,j)$ is the similarity between class i and class j, $n_a(i,j)$ is the number of attributes that are equal to each other in class i and class j, $n_o(i,j)$ is the number of operations that are equal to each other in class i and class j.

Figure 6 is an example for explaining how to compute $n_a(i,j)$, $n_o(i,j)$ and $S(i,j)$.

As shown in Fig. 6, $n_o(C\_ClassA, C\_ClassB)$ is equal to1, because both C_ClassA and C_ClassB has the operation named $O\_Operation2$. The $n_a(C\_ClassA, C\_ClassB)$ is equal to 1, because the type of $A\_Attribute3$ in C_ClassA is C_Type1 which is equal to the type of $A\_Attribute4$ in C_ClassB. Though the type of $A\_Attribute2$ in C_ClassA is equal to the type of $A\_Attribute4$ in C_ClassB, $n_a(C\_ClassA, C\_ClassB)$ can't increase once because the type of $A\_Attribute2$ and $A\_Attribute4$ is "String" which is defined by system. Finally, the similarity between C_ClassA and C_ClassB is $S(C\_ClassA, C\_ClassB)$ which is equal to $n_a(C\_ClassA, C\_ClassB)+n_o(C\_ClassA, C\_ClassB) = 2$.

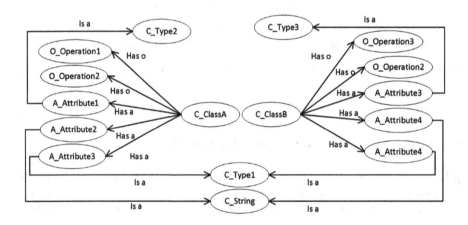

Fig. 6. An example of computing similarity between two classes

5.2   Generate the Abstracted Knowledge Graph

The Hierarchical clustering algorithm was used in our method to abstract our knowledge graphs [5, 6]. A class diagram with a hierarchical structure is easier to read and more refined [13].

For the first step, the similarity between any two classes in a class diagram was computed. Then, we choose two classes. The similarity between the two classes is the highest. We merge the two classes into a new class New_Class. The new class contains the same attributes and operations among the two classes. Then, we compute the similarity between the *New_Class* and other classes which are not merged. Then, we repeat the second and third step until all the classes in the class diagram are merged.

For example, Fig. 7 shows an original class diagram, and Fig. 8 shows the mapped knowledge graph of the class diagram shown in Fig. 7.

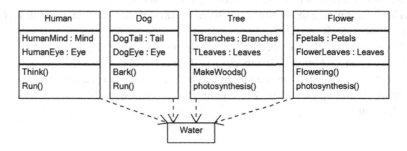

Fig. 7. The original class diagram

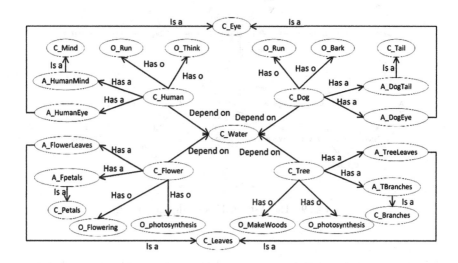

Fig. 8. The knowledge graph of the class diagram in Fig. 7

The hierarchical clustering algorithm is used to cluster the knowledge graph shown in Fig. 8.

The process of using hierarchical clustering method to abstract the knowledge graph in Fig. 8 is as following shows.

(1) For the first step, we compute the similarity between *C_Human, C_Dog, C_Flower, C_Tree* and *C_Water*.

(2) Then, we can find that *S(C_Human, C_Dog)* is equal to *S(C_Tree, C_Flower)* and
      is equal to 2. *S(C_Human, C_Dog)* and *S(C_Tree, C_Flower)* are the highest
      similarities.

(3) Then, we choose *C_Human* and *C_Dog* (or choose *C_Tree* and *C_Flower*. It is
      random), and merge them to a new class *C_Animal*. *C_Animal* contains the attrib-
      utes and methods which are same in *C_Human* and *C_Dog*.

(4) Then, we compute the similarity between *C_Animal* and other classes (e.g. *C_Tree*,
      C_Flower, *C_Water*).

(5) Then, we still choose the highest similarity *S(C_Tree, C_Flower)* and merge
      C_Tree and *C_Flower* to a new class *C_Plant*.

(6) Finally, we find there are no more classes that can be merged (there are no equal
      attributes and operations in two different classes) and we get the abstracted knowl-
      edge graph as shown in Fig. 9.

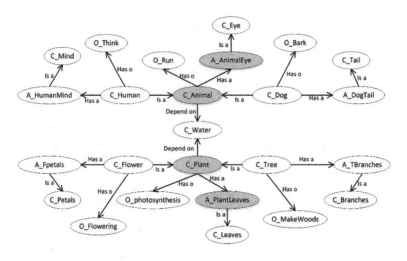

Fig. 9. The abstracted knowledge graph

It is obviously that the number of nodes and edges in Fig. 9 is less than in Fig. 8. The
result shows that the knowledge graph is abstracted and refined.

5.3 Generate the Abstracted Class Diagram

The abstracted class diagram can be generated by mapping the abstracted knowledge
graph. Figure 10 shows the abstracted class diagram of the original one shown in Fig. 7.
The abstracted class diagram is more hierarchical and refined.

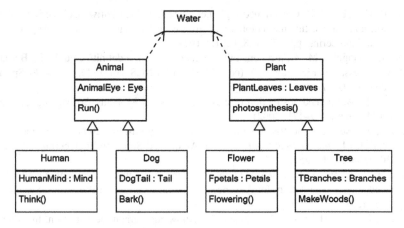

Fig. 10. The abstracted class diagram

6   Conclusion

In this paper, we proposed a method to abstract class diagrams. We used the knowledge graph as a media layer to analysis and deal with class diagrams. We proposed a method to generate the map between class diagrams and knowledge graphs. We used the hierarchical clustering method to abstract original class diagrams. Our method can generate more hierarchical and refined class diagrams.

In the future, we are going to study the methods of mapping the navigation in OCL to knowledge graph. So that, we can get a more precise knowledge graph mapped by the class diagram. Besides, we are going to analyze class diagrams on the knowledge graph layer, for example, we are going to check class diagrams based on knowledge graph.

Acknowledgments. The authors acknowledge the support of the NSFC of China (No. 61363007, 61662021 and No. 61462022) and Hainan NSF (No. 20156234).

References

1. Tom, B.D.M.: Techniques of event history modeling: new approaches to causal analysis. Mahwah N. J. Lawrence Erlbaum Assoc. **52**(2), 236–238 (1995)
2. France, R.B., Kim, D.-K., Ghosh, S., Song, E.: A UML-based pattern specification technique. IEEE Trans. Softw. Eng. **30**(3), 193–206 (2004)
3. Duan, Y., Cheung, S.-C., Fu, X., Gu, Y.: A metamodel based model transformation approach. In: 2005 Third ACIS International Conference on Software Engineering Research, Management and Applications, pp. 184–191. IEEE (2005)
4. Chein, M., Mugnier, M.L.: Graph-Based Knowledge Representation. Springer, London (2009)
5. Navarro, J.F., Frenk, C.S., White, S.D.M.: A universal density profile from hierarchical clustering. Astrophys. J. **490**(2), 493 (1997)

6. Chong, C.Y., Lee, S.P.: Constrained agglomerative hierarchical software clustering with hard and soft constraints. In: International Conference on Evaluation of Novel Approaches to Software Engineering, pp. 177–188. IEEE (2015)

7. Cabot, J., Gogolla, M.: Object constraint language (OCL): a definitive guide. In: Bernardo, M., Cortellessa, V., Pierantonio, A. (eds.) SFM 2012. LNCS, vol. 7320, pp. 58–90. Springer, Heidelberg (2012)

8. Egyed, A.: Automated abstraction of class diagrams. ACM Trans. Softw. Eng. Methodol. (TOSEM) **11**(4), 449–491 (2002)

9. Egyed, A.: Semantic abstraction rules for class diagrams. In: 2000 Proceedings of the Fifteenth IEEE International Conference on Automated Software Engineering, ASE 2000, pp. 301–304. IEEE (2000)

10. Socher, R., Chen, D., Manning, C.D., Ng, A.: Reasoning with neural tensor networks for knowledge base completion. In: Advances in Neural Information Processing Systems, pp. 926–934 (2013)

11. Pujara, J., Miao, H., Getoor, L., Cohen, W.: Knowledge graph identification. In: Alani, H. (ed.) ISWC 2013, Part I. LNCS, vol. 8218, pp. 542–557. Springer, Heidelberg (2013)

12. Sugiyama, K., Tagawa, S., Toda, M.: Methods for visual understanding of hierarchical system structures. IEEE Trans. Syst. Man Cybern. **11**(2), 109–125 (1981)

13. Storrle, H.: On the impact of layout quality to understanding UML diagrams. In: 2011 IEEE Symposium on Visual Languages and Human-Centric Computing (VL/HCC), pp. 135–142. IEEE (2011)

Improved Fuzzy C-Means Clustering Algorithm Based on Particle Swarm Optimization Algorithm

Quan Hu[✉], Kai Zheng, and Zheng Wang

School of Computer Science and Software Engineering,
East China Normal University, Shanghai 200062, China
51141211005@ecnu.cn

Abstract. Traditional FCM clustering algorithm has some problems, including sensitivity to initial values, local optimum and wrong division. In this paper, we proposed an improved fuzzy C-means clustering algorithm based on particle swarm algorithm. Firstly, we use PSO to determine the initial clustering center. Then define a new distance to reassign the fuzzy points. The experimental results show that this new algorithm not only reduces the number of iterations and makes the objective function value smaller, it also corrects some points with wrong division and improves the accuracy of classification results.

Keywords: Fuzzy C-means clustering · Particle swarm · Fuzzy point · Second clustering

1 Introduction

Cluster analysis has been widely used in data analysis, pattern recognition, image processing and so on. Traditional clustering analysis requires that each data point is accurately divided into a certain class, with the nature of "either A or B". However, there are many cluster points with vague boundaries [1–3].

The fuzzy C-means (FCM) algorithm uses membership to determine the degree that each data point belongs to a certain class. FCM algorithm uses gradient descent method to find the optimal solution in essence, so there is a local optimization problem, and the convergence speed greatly depends on the initial value [4].

Particle swarm optimization (PSO) algorithm is based on swarm intelligence with global optimization tools. It can get the optimal solution by the swarm intelligence optimization search guidance produced by the cooperation and competition between particles. PSO algorithm has the advantages of fast convergence, global optimization and simple program implementation [5,6].

In order to achieve global optimization and fast convergence, we combined FCM and PSO algorithm. At first, we code for initial particle swarm, and do particle swarm optimization search to get clustering result and the cluster centre; then we reprocess fuzzy point in result set to improve the search capability and make the result more accurate.

© Springer International Publishing AG 2016
H. Yin et al. (Eds.): IDEAL 2016, LNCS 9937, pp. 617–623, 2016.
DOI: 10.1007/978-3-319-46257-8_66

2   Fuzzy C-Means Clustering Algorithm

Assuming $X = \{x_1, x_2, \cdots, x_n\}$ as the sample space, FCM divides n vectors $x_i (i = 1, 2, ..., n)$ into c fuzzy classes, and $V = \{v_1, v_2, ..., v_n\}$ represents the class center. u_{ij} represents the fitness value of j belonging to the class i. The greater u_{ij}, the higher degree of the sample belonging to the class. FCM uses error square sum function as the objective function of clustering:

$$J(U, V) = \sum_{j=1}^{n} \sum_{i=1}^{c} u_{ij}^m d_{ij}^2 \tag{1}$$

$u_{ij} \in [0, 1], \sum_{i=1}^{c} u_{ij} = 1, j = 1, 2, ..., n, d_{ij} = \|\bullet\|$ is the similarity distance between the sample and the center, and m is fuzzy parameter. Using the Lagrange multiplier method, the necessary condition for objective function J to achieve the minimum value [7] is

$$u_{ij} = \frac{(\|x_i - v_i\|)^{-\frac{2}{m-1}}}{\sum_{k=1}^{c} (\|x_i - v_k\|)^{-\frac{2}{m-1}}}, j = 1, 2, ..., n \tag{2}$$

$$v_i = \sum_{j=1}^{n} u_{ij}^m x_j \bullet (\sum_{j=1}^{n} u_{ij}^m)^{-1}, i = 1, 2, ..., c \tag{3}$$

FCM iterates formula (2) and (3) repeatedly, which is a kind of local search optimization technology in essence and greatly depends on the initial cluster center. Improper initialization may cause converge to a local minimum value, limit the clustering and reduce the effectiveness of the FCM clustering. When the number of clusters is larger, it becomes more obvious [8,9].

3   PSO Algorithm

PSO is based on the group and moves the individuals to a good region according to the fitness. It takes each individual as a particle with no volume in D dimensional search space and all the particles fly at a certain speed. The speed is adjusted dynamically according to the experience of its own and its partners. All particles have a fitness value that is determined by optimization function. Each particle updates their own location by tracking two "best location". One is found by itself (pbest), the other is currently found by all the particles (gbest), and gbest is the maximum value of all pbest [10,11]. For the Kth iteration, each particle updates their position and speed in accordance with the following formula:

$$v_i^{k+1} = w * v_i^k + c1 * r1^k * (pbest_i^k - x_i^k) + c2 * r2^k * (gbest_i^k - x_i^k) \tag{4}$$

$\omega \geq 0$, the inertia weight, the larger ω is more conducive to jumping out of the local larger point, and the smaller ω is more conducive to the convergence of the algorithm. $c1$ and $c2$ are the acceleration constant, usually taken as 2. $r1^k$ and $r1^k$ are random number with uniform distribution in (0,1). In this way, PSO algorithm can keep track of pbest and global gbest until it reaches the prescribed number of iterations or satisfies the prescribed error criteria.

4   Improved PSO-FCM Clustering

The key of FCM algorithm is to assign its initial cluster center. We apply PSO in FCM to guarantee the global optimal solution, which includes two important steps: the encoding of the original particle and the determination of the fitness function.

(a) Individual Determination and Particle Coding. The cluster center is chosen as the individual particle in PSO algorithm. Presume that the initial sample dimension, the number of cluster centers C, then the dimension of each particle is d*C, and the specific structure is as follows:

$$\underbrace{\underbrace{x_{11}x_{12}\cdots x_{1d}}_{d} x_{21}x_{22}\cdots x_{2d} \cdots x_{C1}x_{C2}\cdots x_{Cd}}_{C} \tag{5}$$

(B) The Fitness Function. PSO usually takes the maximum fitness as the optimal solution. So we set $f(x_i) = \frac{1}{J(U,V)}$. The smaller J, the higher individual fitness $f(x_i)$, and the clustering effect is better.

FCM improved by PSO can get the specific class of each data point, but there will be vague point in the boundary and we define it as fuzzy point. We propose a new distance formula, which can extend the original cluster center point to a region and recalculate the fitness value of the fuzzy points. Specific steps are as follows:

(1) Assign a value to the parameters in the algorithm, including the number of cluster C, fuzzy index m, learning factor $c1$ and $c2$, inertia weight ω, and calculate the population size M according to the input data set X.
(2) Encode the initial input data set X to generate M first generation particles,calculate the membership degree of each particle, calculate the fitness value of each particle p_i and the global optimal value of the best position it has experienced p_g.
(3) Update the position of the particles according to the PSO algorithm. If it reaches the maximum number of iterations or the change of the optimal adaptive value is less than the preset threshold, then stop the iteration.
(4) Get C initial clustering centers according to the coding string of the best particle.

(5) Use the initial cluster centers in step (4) and use FCM algorithm to get the global optimal cluster centers and the optimal objective function values.

(6) Classify the processing results of original data set into three types according to the initial clustering results obtained in step (5), the point with membership degree less than 0.65 belonging to fuzzy set φ, greater than 0.85 belonging to clear point set ξ and between them belonging to second-clear point set τ.

(7) Do a reprocess to the point in fuzzy set φ.

(8) Calculate the average distance between each clear point in ξ and the corresponding cluster center

$$\overline{L}_k = \frac{1}{N_k} \sum_{i=1}^{N_k} d_{ik} \tag{6}$$

$d_{ik} = \|x_i - v_k\|, k = 1, 2, ..., c, x_i \in \tau, N_k$ is the number of clear points in class k.

(9) Calculate the distance between the fuzzy point in φ and the hypersphere taking $\overline{L}_k$ as radius

$$\widehat{d}_{ik} = d_{ik} - L_k, k = 1, 2, ..., c \tag{7}$$

(10) Calculate the new membership degree of the fuzzy point by the following formula, and normalize the new membership degree.

$$u_{ik} = \frac{\left(\widehat{d}_{ik}\right)^{2^{-1}}}{\sum_{i=1}^{c} \left(\widehat{d}_{ik}\right)^{2^{-1}}} \tag{8}$$

(11) Determine the final membership class, that is, the class of the clear point is the same and the class of fuzzy point is re-determined by step (10).

(12) Combine clustering results of the clear points with the fuzzy points to obtain the final classification results.

5   Experiment and Analysis

This paper uses two-dimensional data generated by Matlab to test the performance of the improved algorithm. It contains 90 data points, $C = 3$, 30 data points in each class. But it is hard to tell which class the data points on the boundary belong to. The original data is shown in Fig. 1, where the horizontal axis is X, the vertical axis is Y, and different shapes represent different classes.

The clustering results obtained by FCM are shown in Fig. 2. FCM divided the data set into three classes and put some boundary points into the wrong category. We use clear points in each class to calculate its clear radius and distance between data points and the clear radius. As shown in Fig. 3, v_1, v_2, v_3

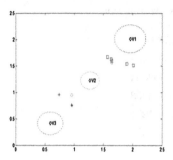

Fig. 1. Original data point distribution

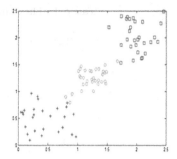

Fig. 2. Classification result of FCM

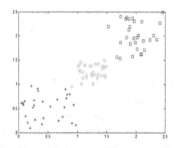

Fig. 3. Clear radius

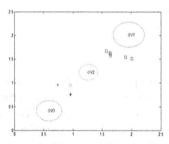

Fig. 4. Classification result of the improved algorithm

represent three corresponding categories and the circle is the corresponding clear radius. The final classification results are shown in Fig. 4.

Through the comparison in Figs. 2 and 4, it can be seen that some points with fuzzy classification have achieved a correct classification. We use the two-dimensional data set to run the FCM and the improved algorithm 5 times. The experimental results are shown in Table 1. It shows that the improved PSO-FCM algorithm is better than the traditional FCM algorithm because it can gain smaller objective function value.

We take the Iris and Wine data set in the UCI machine learning database to verify the performance of the proposed algorithm. Iris contains 3 classes of 50 instances each, where each instance have 4 dimensional version. Wine contains 3 classes of 50, 71 and 48 instances, where each instance have 13 dimensional version. Run the proposed algorithm 20 times and take the average value of the clustering results. Compare the performance of the improved PSO-FCM algorithm with FCM and K-means algorithm. It can be seen from Table 2 that the improved algorithm has certain advantages. In general, the larger density difference between data sets, the better processing effect of the clustering algorithm based on density.

Table 1. The value of the objective function

Running times	FCM	PSO-FCM
1	0.001434	0.000398
2	0.079561	0.002654
3	0.044055	0.013203
4	0.200856	0.021271
5	0.181401	0.008675

Table 2. Clustering results (Accuracy)

Data set	K-means	FCM	improved PSO-FCM
Iris	0.780	0.880	**0.904**
Wine	0.680	0.740	**0.926**

The improved PSO-FCM algorithm use PSO to update particle positions in order to gain the initial cluster center, then it introduces the new calculation formula of distance, which is not only an effective solution to solve the problem of local minimum problem, but also improves the accuracy. It increases the PSO pretreatment, data set resolution distance and the re-calculation of membership, so it needs more time to run the algorithm than the traditional FCM algorithm.

6   Conclusion

To sum up, this paper proposes an improved fuzzy C-means clustering algorithm based on PSO algorithm. Compared with the traditional FCM algorithm, it ensures the optimal initial clustering centers approaching the global optimization with greater probability, and can rapidly converge to the global optimal solution. At last, we do a re-classification for the fuzzy points in the result set, which improves the accuracy of the algorithm. The experimental results show that the improved PSO-FCM algorithm not only makes up for the defects of the FCM algorithm for the initial value selection, but also effectively prevents the algorithm from falling into local optimum, and improves the accuracy. The algorithm is suitable for different data sets with different data points, and it has certain generalization in fuzzy clustering.

References

1. Wang, W., Wang, Y., Huang, S.: Fuzzy clustering analysis. Comput. Eng. Sci. **30**(5), 75–77 (2008)
2. Zadeh, L.A.: Fuzzy sets and their application to pattern classification and clusteringanalysis. In: Classification & Clustering, pp. 251–299 (2015)

3. Zhang, Y., Wang, W., Zhang, X., Li, Y.: A cluster validity index for fuzzy clustering. Inf. Sci. **178**(4), 1205–1218 (2008)
4. Kantardzic, M., Han, J.: Data mining: concepts and techniques. Data Min. Concepts Models Methods Algorithms **5**(4), 1–18 (2000). 2nd edition
5. Zhou, F.H., Liao, Z.Z.: A particle swarm optimization algorithm. Appl. Mech. Mater. **303–306**, 1369–1372 (2013)
6. Wang, S., Xu, Y., Pang, Y.: A fast underwater optical image segmentation algorithm based on a histogram weighted fuzzy c-means improved by PSO. J. Marine Sci. Appl. **10**(1), 70–75 (2011)
7. Bezdek, J.C., Ehrlich, R., Full, W.: FCM: the fuzzy c-means clustering algorithm. Comput. Geosci. **10**(2/3), 191–203 (1984)
8. Klawonn, F., Kruse, R., Winkler, R.: Fuzzy clustering. Fuzzy Sets Syst. **281**(C), 272–279 (2015)
9. Li, Y., Mao, L., Xu, W.: Research of improved fuzzy c-means algorithm based on quantum-behavior particle swarm optimization. Comput. Eng. Appl. **48**(35), 151–155 (2012)
10. Chen, D.-B., Zou, F., Wang, J.-T.: A multi-objective endocrine pso algorithm and application. Appl. Soft Comput. **11**(8), 4508–4520 (2011)
11. Chen, D., Chen, J., Jiang, H., Zou, F., Liu, T.: An improved pso algorithm based on particle exploration for function optimization and the modeling of chaotic systems. Soft Comput. **19**(11), 3071–3081 (2015)

Patent Trend Mining for Internet
of Things in Logistics

Jae Un Jung[1(✉)], Hyun Soo Kim[2], and Hyung Rim Choi[2]

[1] BK21Plus Groups, Dong-A University, Busan, Korea
imhere@dau.ac.kr
[2] Department of MIS, Dong-A University, Busan, Korea
{hskim,hrchoi}@dau.ac.kr

Abstract. The philosophy of the *internet of things* (IoT) provides an ideal environment for the logistics industry, which seeks the seamless integration and management of physically scattered logistics assets (freight, facilities, equipment, etc.). The great attention to IoT in the field of logistics stimulates the development of relevant technologies, leading to a high rate of patent applications. Our research aims to trace patent trends in IoT for the field of logistics by utilizing patent information. We analyzed International Patent Classification (IPC) codes and abstracts of patent documents using text mining techniques. Our results indicate that technologies for the IoT are in the early stages of being applied to the logistics industry and are in the process of being integrated with cloud computing and big data technologies.

Keywords: Patent trend · Text mining · Internet of things · Logistics · Trend analysis

1 Introduction

The term *internet of things* (IoT) was first coined in 1999 to describe internet-connected devices with RFID (radio frequency identification) in a supply chain, and now it is regarded as a mainstream concept in information communications technology (ICT) [1, 2]. Gartner expects that the approximately 4 billion interconnected devices used in the consumer and business sectors in 2014 will increase to about 20 billion by 2020 [3].

The IoT philosophy underlies what the logistics industry seeks: to be smarter and more agile by integrating and tracking dispersed logistics assets seamlessly. Accordingly, stakeholders in the field of logistics show great interest in deploying IoT enablers (technologies) to save time and cost, to ensure security for their services, and ultimately to increase profitability [4, 5]. Their keen attention to the IoT leads to competitive development of relevant technology and high-tech patent applications in the logistics industry [6].

With regard to this issue, we aim to identify the notable trends and the technologies being developed and applied to enable IoT in the field of logistics by using patent information, which is actually a good source of information for understanding

© Springer International Publishing AG 2016
H. Yin et al. (Eds.): IDEAL 2016, LNCS 9937, pp. 624–634, 2016.
DOI: 10.1007/978-3-319-46257-8_67

technology and business trends, even with the recognition that patents may be invalidated by rejection, expiration, or renunciation [7–10].

To analyze technology trends from the patent documents, we employed text mining techniques on International Patent Classification (IPC) codes and patent abstracts collected from the Korea Intellectual Property Rights Information Service (KIPRIS) linked with global patent databases [11].

Our research, as an instance of the application of diverse scientific approaches to the analysis of patent trends, shows IoT trends specific to the field of logistics. It helps in understanding the direction of industrial technologies.

In Sect. 2, we describe the collection of patent information for IoT in logistics from the patent search engine, and in Sects. 3 and 4, we describe the analysis of the patent information and identify technology trends. Lastly, in Sect. 5 we discuss our contribution and the limitations of the work.

2   Patent Information for Internet of Things in Logistics

Patent information includes all documents and administrative archives produced during the patent application process for achieving patent rights (see Table 1), and it is easy to collect through online patent search engines [7].

Table 1. Kinds of patent information

Category	Composition
Bibliographical Information	Dates (Application, Publication, Registration, etc.)
	Persons (Assignee, Inventor, Attorney, etc.)
	Classifications (IPC, UPC, etc.)
	References (Cited Patents, Foreign References, etc.)
Technology Content	Abstract, Background of Invention, Detailed Description, Claims, etc.
Other	Patent Family Information, Legal Status, etc.

Utilizing patent information, Trappey et al. analyzed Chinese technology trends in RFID [12], and LexInnova investigated the top assignees of IoT patents along with IoT technologies [6]. Tseng et al. suggested text mining techniques such as clustering and topic mapping for analyzing atypical texts in patent information [13].

KIPRIS is a patent search engine in Korea; it is connected with the patent databases of 12 other countries or organizations (Europe, USA, Japan, China, World Intellectual Property Organization (WIPO), etc.) [11]. WIPO's system enables access to patent information in 148 countries around the world [14]; this means that KIPRIS is linked with practically the entire range of patent systems.

In selecting keywords (queries) for collecting patent information for IoT in logistics, we assumed that

- The term "internet of things" represents synonyms such as "internet-of-things" and "IoT."
- The term "logistics" also represents synonyms such as "distribution" and "transportation."
- The two terms are comprehensive and thus cover subordinate concepts or words.

Then, we retrieved the patent information containing both of the two keywords from KIPRIS.

The result was 127 patent information collected as of April 2016. Among the 127 patents, 126 were applied for between 2010 and 2015, and 1 in 2008. The number of patent applications increased sharply by a factor of about four, from 10 cases in 2010 to 43 cases in 2015, as shown in Fig. 1.

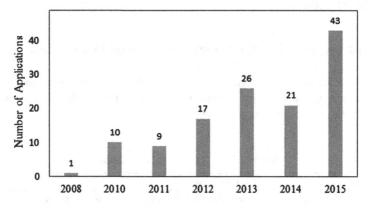

Fig. 1. Trend in patent applications in IoT for logistics

On the other hand, although KIPRIS conveniently offers bibliographical information including IPC codes and abstracts (see Table 1) in Excel spreadsheet file format, the text field for each abstract is limited to 600 characters. In most abstracts, all text after the 600th character was eliminated, which may cause inaccuracies in the data.

Thus, the IPC codes analyzed in Sect. 3 were collected through this service, whereas the 127 complete abstracts (with no truncation) were gathered through an additional search process using the patent lists retrieved from KIPRIS.

3   Patent Mining for Trends Using IPC Codes

The International Patent Classification (IPC) system divides technology into eight sections (A to H) with approximately 70,000 subdivisions [14]. One or more IPC codes can be assigned for each patent application.

In our collected datasets, the IPC codes were spread over the four sections B, E, G, and H (see Table 2). To facilitate understanding of technology trends, we categorized each IPC code to the subclass level (e.g., B60R), as shown in Table 3.

The number of patent applications for IoT in logistics shows an upward trend in sections B and G (See Fig. 2). The highest number of applications was recorded in

Table 2. IPC codes retrieved from datasets

Section	Content
B	Performing Operations, Transporting
E	Fixed Constructions
G	Physics
H	Electricity

section G, specifically in subclasses G06K (recognition of data, presentation of data, record carriers, handling record carriers) and G06Q (data processing systems or methods, specially adapted for administrative, commercial, financial, managerial, supervisory or forecasting purposes, etc.). Subclasses G06K and G06Q are essential technology areas for the implementation of IoT services, along with subclass H04L (telephonic communication). These show the highest numbers of applications.

Table 3. Patent application trends by technology (IPC code subclasses)

Section & Subclass		2010–2011	2012–2013	2014–2015	Total
B	60R	0	1	2	3
	65D	2	0	1	3
	65G	0	0	2	2
	66F	0	2	0	2
E	04F	0	0	1	1
	05B	2	0	0	2
G	01N	0	1	0	1
	01S	0	1	0	1
	05B	0	2	7	9
	06F	2	2	2	6
	06K	**11**	**9**	**20**	**40**
	06Q	**11**	**63**	**83**	**157**
	07C	0	0	1	1
	07F	2	0	2	4
	08B	0	0	1	1
	08C	1	0	1	2
	09B	1	2	0	3
H	02J	4	0	0	4
	04L	3	4	8	**15**
	04M	1	0	0	1
	04N	0	2	0	2
	04W	2	1	1	4
Total		42	90	132	264

In section B, which is closely related to the logistics industry, subclasses B60R (safety belts or body harnesses used in land vehicles), B65D (containers for storage or transport of articles or materials), B65G (transport or storage devices), and B66F (hoisting, lifting, hauling, or pushing devices) were targeted for applications, along with subclass E05B (locks, accessories therefor, handcuffs). (See Table 3 and Appendix for further details.)

From this result, we can infer that diverse technologies for the IoT are being applied in the field of logistics.

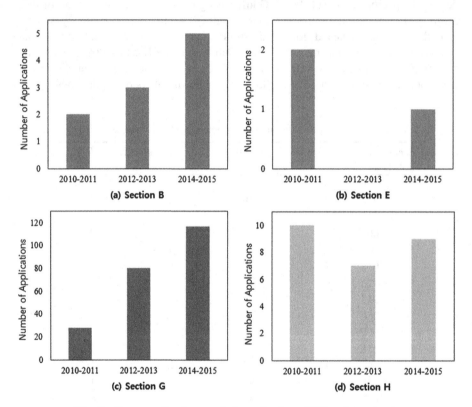

Fig. 2. Patent application trends by technology (IPC code sections)

4   Patent Mining for Trends Using Patent Abstracts

4.1   Preprocessing of Patent Abstracts

To more easily identify changes in patent trends in IoT for the logistics industry, we excluded the single case in 2008 and utilized only the 126 abstracts from 2010 to 2015 among the initial 127 patent cases. Then, we split these 126 patent abstracts into three periods as shown in Table 4.

Table 4. Datasets for patent trend analysis

Period	Number of abstracts
2010–2011	19
2012–2013	43
2014–2015	64
Total	126

Next, we preprocessed the patent datasets in R [15] to identify the characteristics of our datasets and to analyze the patent trends, using code as shown in Fig. 3.

- We first made corpora of the datasets for text mining.
- We then filtered out stopwords (special characters, articles, and punctuation) from the datasets and transformed words to their stem or root words for effective text mining.
- We converted the refined text data to matrix format.

```
43  # preprocessing
44  abs.cor <- tm_map(abs.cor, tolower) #uppercase -> lowercase
45  abs.cor <- tm_map(abs.cor, removeWords, stopwords("en"))#remove special characters & articles
46  abs.cor <- tm_map(abs.cor, removePunctuation) #remove punctuation
47  abs.cor <- tm_map(abs.cor, stripWhitespace) #remove space
48  abs.cor <- tm_map(abs.cor, stemDocument)#remove affixes of words
49  abs.cor <- tm_map(abs.cor, PlainTextDocument)#text type
50  abs.cor <- DocumentTermMatrix(abs.cor)# word matrix
51  inspect(abs.cor)
52  list(abs.cor)
```

Fig. 3. Sample of code for preprocessing datasets

To identify the extent to which we should refine the datasets thereafter, we analyzed the sparsity of the word matrix. The size of the word matrix for the total period from 2010 to 2015 was 126 (number of abstracts) × 2,015 (number of distinct words). To calculate the sparsity of the matrix, the number of entries having a value greater than 1 was divided by the number of entries having a value of 0 (7,395/246,495). The result, word density of 3 %, can be interpreted to mean 97 % sparsity. There were 915 words that were found in more than two different abstracts.

In the same way, each of the three data subsets was analyzed, with the results as shown in Table 5.

Table 5. Characteristics of word matrices by period

Period	Matrix size	Sparsity	Number of words occurring in multiple abstracts
2010–2011	19 × 2,015	97% (1,127/37,158)	195
2012–2013	43 × 2,015	97% (2,645/84,000)	392
2014–2015	64 × 2,015	97% (3,623/125,337)	516

4.2   Tracking Trends in Patent Abstracts

Term Frequency Analysis. Using the preprocessed datasets, we analyzed term (word) frequency using code as shown in Fig. 4 and aimed to infer or make conjectures about technology and business issues from the trends in term frequencies.

Table 6 shows the result of the term frequency analysis. For the whole period, "information" and "system" were found to be the most frequent terms. Thus, we exclude these two words and list the next ten most frequent words.

```
85  # term frequency
86  ## with frequency
87  inspect(abs.cor[1:19,])
88  inspect(abs.cor[20:62,])
89  inspect(abs.cor[63:126,])
90  ## without frequency
91  findFreqTerms(abs.cor[1:19,],11)
92  findFreqTerms(abs.cor[20:62,],33)
93  findFreqTerms(abs.cor[63:126,],36)
```

Fig. 4. Sample of code for analyzing term frequency

In 2010–2011, the terms "safety," "monitoring," and "tracing" for logistics are notable exceptions to the words related to technologies for the IoT. In 2012–2013, the technology terms "cloud" and "platform" emerge in the top ten. For 2014–2015, "storage" appears for the first time.

Table 6. Analysis of term frequency (excluding terms "information" and "system")

Frequency rank	2010–2011	2012–2013	2014–2015
1	RFID	management	data
2	data	data	management
3	wireless	monitoring	monitoring
4	management	transportation	cargo
5	label	wireless	goods
6	**safety**	communication	RFID
7	communication	**cloud**	tag
8	**monitoring**	**platform**	control
9	terminal	positioning	**storage**
10	**tracing**	server	terminal

These changes in the three segmented periods can be understood as showing that IoT technologies, which are integrated with cloud computing and big data over time, are reflected in the realm of logistics [16–18].

Analysis of Associated Terms. In this phase, our intent was to analyze patent trends in terms of associated words. Figure 5 shows a sample of the code that we programmed in R.

```
 96  # Associated terms
 97  findAssocs(abs.cor[1:19,], "logistics",0.3)
 98  findAssocs(abs.cor[1:19,], "internetthings",0.3)
 99  findAssocs(abs.cor[20:62,], "logistics",0.3)
100  findAssocs(abs.cor[20:62,], "internetthings",0.3)
101  findAssocs(abs.cor[63:126,], "logistics",0.3)
102  findAssocs(abs.cor[63:126,], "internetthings",0.3)
```

Fig. 5. Sample of code for analyzing associated terms

Though all abstracts contained the two terms "logistics" and "internet of things," words associated with the two terms were different, as shown in Table 7, which lists the top ten terms associated with "logistics" and with "internet of things."

Table 7. Analysis of associated terms

Frequency rank	2010–2011		2012–2013		2014–2015	
	Logistics	IoT	Logistics	IoT	Logistics	IoT
1	assembly	documents	cost	task	**time**	handover
2	freight	goods	**realtime**	distribution	connection	inquiry
3	generator	palletization	destination	vehicle	delivery	maintenance
4	store	recognizing	receipt	integration	GRPS	planning
5	transportation	transport	shipment	low-carbon	PLC	machine
6	generator	pallet	location	resource	recipient	dispatch
7	self-supply	management	GPS	transmit	transit	oversea
8	sensor	computer	**tracking**	load	bracket	statistic
9	terminal	**tracking**	cargo	**realtime**	buffer	decision
10	power	transmission	task	automatic	cabinet	database

In 2010–2011, terms related to assembly, transportation, and tracking are present, and in 2012–2013, the term "realtime" stands out. In 2014–2015, "time" (realtime) is emphasized relatively more than before, and detailed technologies for freight tracking, data storage, and analytics ("planning", "decision") are listed.

So far, we have performed analyses of patent trends in IoT for logistics from various angles. Though the queries we used to search the patent information were limited in order to cover the broad range of IoT and logistics, we have identified the fact that IoT for the logistics industry is in the process of combining with cloud computing and big data technologies that are megatrends of ICT.

5 Conclusion

The IoT, cloud, and big data are all notable issues that are expected to bring a new paradigm for the logistics industry. With this in view, our research was intended to deepen the understanding of patent (technology) trends in IoT for logistics.

Through the analysis of IPC codes and patent abstracts, we demonstrated that technologies in the IoT for the logistics industry are evolving by combining with the cloud and big data technologies.

To embrace and trace the new issues that we identified in this study, in follow-up research we plan to expand our scope to include cloud and big data technologies specific to logistics.

On the other hand, we found that concrete technological terms (those having a low level of abstraction) appeared with low frequency; this is regarded as characteristic of patent information. In response to this, in future research we will employ an additional process to classify low-frequency words for more elaborate mining of patent data.

Appendix: Contents of IPC Codes in Table 3

Section		Contents
B	60R	safety belts or body harnesses used in all types of land vehicles
	65D	containers for storage or transport of articles or materials, e.g. bags, barrels, bottles, boxes, cans, cartons, crates, drums, jars, tanks, hoppers, forwarding containers; accessories, closures, or fittings therefor; packaging elements; packages
	65G	transport or storage devices, e.g. conveyors for loading or tipping, shop conveyor systems or pneumatic tube conveyors
	66F	hoisting, lifting, hauling, or pushing, not otherwise provided for, e.g. devices which apply a lifting or pushing force directly to the surface of a load
E	04F	finishing work on buildings, e.g. stairs, floors
	05B	locks; accessories therefor; handcuffs
G	01N	investigating or analyzing materials by determining their chemical or physical properties
	01S	radio direction-finding; radio navigation; determining distance or velocity by use of radio waves; locating or presence-detecting by use of the reflection or reradiation of radio waves; analogous arrangements using other waves
	05B	control or regulating systems in general; functional elements of such systems; monitoring or testing arrangements for such systems or elements
	06F	electric digital data processing
	06K	recognition of data; presentation of data; record carriers; handling record carriers
	06Q	data processing systems or methods, specially adapted for administrative, commercial, financial, managerial, supervisory or forecasting purposes; systems or methods specially adapted for administrative, commercial, financial, managerial, supervisory or forecasting purposes, not otherwise provided for
	07C	time or attendance registers; registering or indicating the working of machines; generating random numbers; voting or lottery apparatus; arrangements, systems or apparatus for checking not provided for elsewhere
	07F	coin-freed or like apparatus
	08B	signaling or calling systems; order telegraphs; alarm systems
	08C	transmission systems for measured values, control or similar signals

(Continued)

(Continued)

Section		Contents
	09B	educational or demonstration appliances; appliances for teaching, or communicating with, the blind, deaf or mute; models; planetaria; globes; maps; diagrams
H	02J	circuit arrangements or systems for supplying or distributing electric power; systems for storing electric energy
	04L	transmission of digital information, e.g. telegraphic communication
	04M	telephonic communication
	04N	pictorial communication, e.g. television
	04W	wireless communication networks

Source: [14]

References

1. The guardian. www.theguardian.com/media-network/2015/mar/31/the-internet-of-things-is-revolutionising-our-lives-but-standards-are-a-must
2. RFID Journal. www.rfidjournal.com/articles/view?4986
3. Gartner. http://www.gartner.com/newsroom/id/3165317
4. Macaulay, J., Buckalew, L., Chung, G.: Internet of Things in Logistics. DHL Customer Solutions & Innovation, Troisdorf (2015)
5. PRNewswire. http://www.prnewswire.com/news-releases/connected-logistics-market-by-internet-of-things-technologies-in-fleet-management-warehouse-and-inventory-management-by-connectivity-technologies-by-devices-transportation-mode-services-global-forecast-to-2020-reportlinke-300152078.html
6. LexInnova: Internet of Things – Patent Landscape Analysis (2015)
7. Jung, S.: Importance of using patent information. In: WIPO-MOST Intermediate Training Course on Practical Intellectual Property Issues in Business, World Intellectual Property Organization (WIPO), Geneva (2003)
8. Daim, T.U., Rueda, G., Martin, H., Gerdsri, P.: Forecasting emerging technologies: use of bibliometrics and patent analysis. Technol. Forecast. Soc. Change **73**(8), 981–1012 (2006)
9. Lee, S., Yoon, B., Park, Y.: An approach to discovering new technology opportunities: keyword-based patent map approach. Technovation **29**, 481–497 (2009)
10. Albino, V., Ardito, L., Dangelico, R.M., Petruzzelli, A.M.: Understanding the development trends of low-carbon energy technologies: a patent analysis. Appl. Energy **135**, 836–854 (2014)
11. Korea Intellectual Property Rights Information Service. http://eng.kipris.or.kr/enghome/main.jsp
12. Trappey, C.V., Wu, H.Y., Taghaboni-Dutta, F., Trappey, A.J.C.: Using patent data for technology forecasting: China RFID patent analysis. Adv. Eng. Inf. **25**, 53–64 (2011)
13. Tseng, Y.H., Lin, C.J., Lin, Y.I.: Text Mining Techniques for Patent Analysis. Info. Proc. Mang. **43**, 1216–1247 (2007)
14. Word Intellectual Property Organization. http://www.wipo.int
15. The R Project for Statistical Computing. https://www.r-project.org
16. IBM Center for Applied Insights. https://ibmcai.com/2014/11/20/iot-internet-of-things-will-go-nowhere-without-cloud-computing-and-big-data-analytics/

17. Bughin, J., Chui, M., Manyika, J.: Clouds, big data, and smart assets: ten tech-enabled business trends to watch. McKinsey Q. **56**, 75–86 (2010)
18. InfoWorld.   http://www.infoworld.com/article/2867978/cloud-computing/thank-the-cloud-for-making-big-data-and-internet-of-things-possible.html

Expression Classification and Intensity Estimation by Expression Manifold Synthesis

Yao Peng[✉] and Hujun Yin

School of Electrical and Electronic Engineering, The University of Manchester,
Manchester M13 9PL, UK
yao.peng@postgrad.manchester.ac.uk, hujun.yin@manchester.ac.uk

Abstract. Facial expression and its dynamic property play an important role in interpreting and conveying emotions. Recently facial expression analysis has been an active topic in both psychology and computer vision. Most previous investigations have focused on the recognition of static images with intense expressions. Different from the previous work, we present an expression synthesis method for both expression classification and intensity estimation. By means of synthesising expression manifolds from neutral faces, the dynamic variations in facial expression can be modelled and analysed. Eigentransformation is utilised on both shape and expression details in generating novel expressions. Expression classification is performed on the expanded training sets with synthesised expression landmarks, and the intensity can be estimated with synthesised expression manifolds. Comprehensive experimental results conducted on the extended Cohn-Kanade database are reported.

Keywords: Expression synthesis · Manifold · Eigentansformation · Expression intensity estimation

1 Introduction

Facial expression refers to one or more movements of muscles beneath the facial skin that convey emotions of an individual. As a form of nonverbal communication, facial expression provides rich information in depicting its meaning and takes a significant position in applications such as human-computer interaction and affective computing systems.

Facial expression analysis aims at automatically analysing and measuring facial behaviours from visual information [1]. Since as early as 1970s, many studies have been carried out attempting to interpret facial expressions. Psychologists have developed the Facial Action Coding System (FACS) [2] that provides an objective measurement for facial expression by muscle movements. Over the last two decades, facial expression analysis has attracted much attention in computer vision and much effort has been made [3]. Most work concentrated on identifying expressions into six prototypic expressions based on static displays of intense expressions. Automatic expression recognition remains to be difficult due to the subtlety, complexity and variability of facial expressions [4,5].

© Springer International Publishing AG 2016
H. Yin et al. (Eds.): IDEAL 2016, LNCS 9937, pp. 635–644, 2016.
DOI: 10.1007/978-3-319-46257-8_68

Besides the categorisation of facial expressions, recent research has indicated the importance of dynamic variations in expression intensity in temporal domain [6]. On the one hand, emotions are often communicated by the dynamics of expressions lying in the small subtle changes. On the other hand, expression intensity information would be critical for applications such as patient monitoring, security surveillance, and human-computer interaction.

In this paper, we introduce an expression synthesis method for expression classification and intensity estimation. Synthetic expressions are generated from neutral faces for both geometry attributes and expression details to expand the gallery set. Since dynamic variations are important in interpreting facial expressions, we then further extend the synthesis to expression manifolds, which provide continuous expression sequences. For a probe image with arbitrary expression, its expression is first classified and intensity estimated using synthesised expression manifolds. The effectiveness of the proposed method has been verified on the extended AU-Coded Cohn-Kanade (CK+) facial expression database [7,8].

The remainder of this paper will be structured as follows: Related work is reviewed in Sect. 2. Section 3 describes the eigentransformation algorithm, the proposed process of expression manifold synthesis, and how to perform expression classification and intensity estimation. Section 4 presents experimental results, followed by conclusions in Sect. 5.

2   Related Work

Facial expression analysis dates back to Darwin's demonstration of general principals of expression and how human expressions link with human actions. In 1978, Ekman and Friesen [9] proposed FACS that comprehensively describes all observable facial movements in terms of 44 anatomically separated and visually distinguishable Action Units (AUs). According to FACS, the facial expression intensity can be categorised into five levels as *trace*, *slight*, *marked* or *pronounced*, *extreme* or *severe*, and *maximum*. In 1992, Ekman also defined six basic expressions as *anger*, *disgust*, *fear*, *happiness*, *sadness* and *surprise* [10].

In order to automatically analyse and recognise facial expressions, many approaches have been developed to derive and represent facial changes caused by expressions. Tian *et al.* [11] developed a multi-state facial component models to analyse expressions and recognise AUs based on both permanent and transient facial features in image sequences. Pantic and Rothkrantz [12,13] reported multiple effort on automatically recognising AUs occurring alone or in combination with extracted fiducial points of facial components in static images and video sequences. Uddin *et al.* [14] applied enhanced independent component analysis to extract independent features and performed expression classification by hidden Markov models. Song *et al.* [15] derived image ratio features to observe skin deformation caused by expression in eight facial patches. Facial animation parameters were used to describe facial feature movements and further improve expression recognition accuracy. Jabid *et al.* [16] proposed local directional pattern to represent facial expression by measuring the edge response in eight directions for each pixel. Gu *et al.* [17] adopted radial encoding strategy to downsample the Gabor features derived from local patches of an input image. The local features were fed to local classifiers to be integrated into global features for representation of facial expressions. Happy and Routray [18] presented an expression recognition framework by deriving features on selected facial patches and recognising expressions on salient patches which stayed active during emotion elicitation.

In addition to expression classification, expression intensity estimation has been another active topic in computer vision and artificial intelligence. Yang *et al.* [19] proposed a ranking-based framework to perform expression recognition and intensity estimation by building ranking models using RankBoost classifiers. Delannoy and McDonald [20] used local linear embedding to extract the underlying manifolds of expressions and categorise into three levels using support vector machines. Chang *et al.* [21] developed a rank-based method that utilised relative order information and scattering transformation to infer the discrete expression ranks. Recently Mohammadi *et al.* [22] adopted robust principal component analysis to decompose expression from face identity for an input image. Expression intensity of the image was estimated using a regression model based on dictionary learning and sparse representation.

3   Expression Synthesis and Expression Manifold

3.1   Shape Alignment and Texture Normalisation

To analyse and minimise the distribution variation of landmarks that describe the shape of a face, it is necessary to align all the shapes as closely as possible. In each image, shape is defined by a set of landmarks located around key areas such as eyes, eyebrows, nose, mouse and contour. In order to align faces, translation, rotation and scaling are applied on all face images so that eyes are located at the same coordinates.

Furthermore, each image is warped so that its landmarks match the mean shape. Delaunay triangulation strategy is used to divide the face texture into small patches, and a piece-wise affine transformation is applied to warp the texture from a triangle to its corresponding triangle.

3.2   Eigentransformation Approach

Many studies have shown that an unseen face image can be reconstructed by eigenfaces using principal component analysis as

$$x = \psi + E\omega, \tag{1}$$

where ψ is the mean of the training set, E denotes a set of eigenvectors of the covariance matrix $C = \phi\phi^T$, and ω is a set of projection coefficients. Based on the eigenface approach, eigentransformation algorithm indicates that an image can also be represented by a weighted combination of training images instead of eigenfaces [23]. According to singular value decomposition [24],

$$E = \phi V \Lambda^{-\frac{1}{2}}, \tag{2}$$

where V and Λ are the eigenvectors and eigenvalues of matrix $\phi^T \phi$. Thus, the reconstructed image can also be represented as a linear combination of training samples instead of eigenfaces by

$$x = \psi + \phi(V \Lambda^{-\frac{1}{2}} \omega) = \psi + \phi A. \tag{3}$$

To extend this principle for expressions, for the same group of subjects with two different expressions, we assume the reconstruction coefficients A stay the same.

For the sake of simplicity, we assume that only the neutral expression is used as the training images for synthesising other expression images. But this process can be generalised to other expressions being used as the training images.

3.3   Shape Deformation and Expression Details Generation

With the AAM, to describe expression variations containing both shape deformation and subtle appearance changes, shape and texture information from an unseen face can be transformed respectively by using statistical shape model [25] and shape-free texture [26].

Given a set of aligned shapes, we divide the landmarks of each shape into four groups corresponding to facial features such as eyebrows and eyes, nose, mouth and face contour. For an unseen shape, eigentransformation is applied to each group of landmarks to derive the weight matrices. Under the assumption that the weight matrices remain unchanged with different expressions, we can synthesis its corresponding expressive shapes.

To synthesise the appearance variations, we warp a neutral expression into a non-neutral expressions and calculate the differences for each pair of neutral and non-neutral expressions. To remove the effect of noise and small image misalignment, a 2-D Wiener filter is applied to the difference image by estimating local mean and variance around each pixel. With the eigentransformation, we can also obtain the weight matrix for texture attributes. Based on the fact that similar persons perform similar expressions or expression in a large extent is universal, the weight matrix is used to synthesise expression appearance variations. Synthesised texture is warped to the reconstructed shape of the target face to obtain the final expression image.

3.4   Expression Manifold Synthesis

In the above static expression synthesis process, we obtain the deformed shape and synthesised expression texture for a target subject. To extend this work for expression manifolds with varying expression degrees, we need a series of shapes and expression details describing the variations. We first calculate the shape differences between neutral expression and its synthesised expressions, then we generate a continuous expression manifold by interpolating both shape and texture attributes between neutral and a synthesised expression by multiplying a ratio r to the shape differences and expression details. It is noted that the value r controls the expression intensity and can be greater than 1. However, large value may cause unnatural artefacts.

3.5 Expression Classification Using Synthesised Landmarks

For simplicity and generalisation, we consider only synthesised facial landmarks in expression classification. Synthesised landmarks are added to expand the training set. We applied generalised discriminant analysis (GDA) [27] with Gaussian kernels as the classification method, and shapes of gallery images and synthesised landmarks are then used to train GDA classifier to find nonlinear projection directions from which classes are separated.

3.6 Expression Intensity Estimation Using Synthesised Expression Manifolds

To describe the dynamics of facial expression, we explore supervised locality preserving projections (SLPP) [28] to learn the expression manifolds. Unlike LPP that only preserves the local structures, SLPP combines the locality preserving property with discriminant class information when constructing the neighbourhood graph. Synthesised expression images with varying degrees are assigned different labels according to the intensity controlling factor r, and the local neighbourhood of an image should be the images with the same expression intensity only. For an probe image with known expression determined by the expression classification process, the expression intensity is estimated by finding the nearest neighbour of its projection in the embedded subspace of that expression.

4 Experiments and Discussions

4.1 Dataset

To evaluate the performance of the proposed method, experiments were conducted on the extended AU-Coded Cohn-Kanade (CK+) facial expression database [7,8]. The CK+ database contains 593 video sequences collected from 123 subjects displaying various expressions. Each subject was instructed to perform a series of one to seven expressions and each video sequence starts with a neutral expression (first frame) and ends with the most expressive image (last frame). For this paper, a subset of 212 image sequences were selected, where 83 facial landmarks were detected automatically [29]. We selected videos labelled with six prototypic expressions: Anger (33), Disgust (46), Fear (20), Happiness (50), Sadness (18) and Surprise (45).

4.2 Results of Expression Synthesis

We adopted leave-one-subject-out strategy for expression synthesis due to limited number of subjects. For example, the synthesised smile face of subject ξ was synthesised from a combination of smile faces from all subjects except ξ.

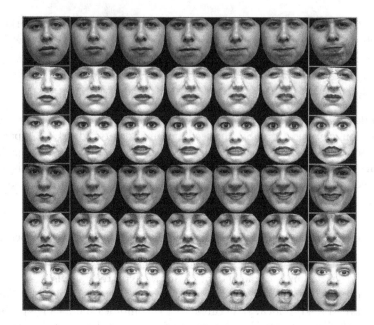

Fig. 1. Input neutral expressions (in red boxes), synthesised anger, disgust, fear, happiness, sad and surprise faces, and corresponding ground truth images (in green boxes) from first to last rows on the CK+ database. (Color figure online)

In each video sequence, the first and last frames were selected as the input neutral images and the expression at the maximal level respectively. Figure 1 illustrates the input neutral images (in red boxes), synthesised expressions at different intensity levels, and the expression images at its peak intensity (in green boxes). The synthesised expressions are anger, disgust, fear, happiness, sadness and surprise from first to last rows.

4.3   Results of Expression Classification

Expression classification was performed by expanding training set with synthesised shapes on seven expressions as neutral, anger, disgust, fear, happiness, sadness and surprise. The expanded training set contains both neutral (first frame) and expressive (last frame) shapes from the video sequences and synthesised shapes. We applied 10-fold cross validation and repeated the process 50 times to achieve satisfactory precision. Table 1 shows the expression classification results of seven emotions. It is noted that the proposed method achieves the best results among these methods.

4.4   Results of Expression Intensity Estimation

To estimate expression intensity, for all subjects, 60 synthesised expression images with continuously varying degrees from level 1 (low) to level 5 (maximal)

Table 1. Expression classification accuracy on the CK+ database.

Expressions	Classification accuracy (%)					
	[14]	[15]	[16]	[17]	[18]	Proposed
Neutral	-	90.74	89.30	94.12	-	**98.17**
Anger	82.50	**90.57**	86.90	75.33	87.80	89.58
Disgust	**97.50**	86.04	94.20	95.93	93.33	94.27
Fear	95.00	84.61	94.40	85.47	94.33	**100**
Happiness	**100**	93.62	98.90	93.33	94.20	99.68
Sadness	92.50	90.24	92.60	90.03	**96.42**	83.34
Surprise	92.50	92.31	99.00	97.67	98.46	**100**
Average	93.33	89.73	93.61	90.27	94.09	**95.01**

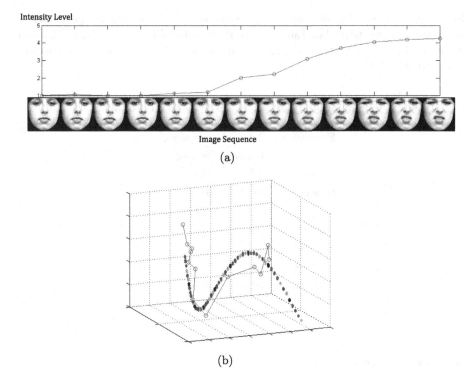

Fig. 2. An exemplar image sequence with (a) its estimated expression intensity, and (b) its embeddings in the SLPP manifold.

were used to construct the expression manifolds, and image sequences from the CK+ database were considered as the testing sets. By applying SLPP to extract expression manifolds, image sequences of facial expressions of an probe individual were mapped into the embedded space. The intensity of each image was estimated by the intensity label of its nearest neighbour. Figure 2(a) shows an exemplar image sequence with its estimated changing intensity from the CK+ database. As shown in Fig. 2(b), the image sequence representing disgust expression with increasing intensity was embedded as a curve on the manifold in the embedding space spanned by the first three dimensions of SLPP.

In order to provide a quantitative analysis, relevant accuracy (RA) [30] was used to evaluate the performance of proposed expression intensity method. Given an image sequence whose intensity increases monotonically, we built relevant pairs by randomly selecting two images and compared their estimated intensity according to the relative order of two expressions in the sequence. To avoid undistinguished subtle changes, each relevant pair was constrained with an interval over than three frames. RA was calculated as

$$RA = \frac{number\ of\ correctly\ ranked\ relevant\ pairs}{number\ of\ all\ relevant\ pairs}. \tag{4}$$

In this experiment, 50 relevant pairs were constructed for each image sequence and results were reported in Table 2. To test the reliability of intensity estimation method, experiment was also conducted on the training subjects using leave-one-out strategy.

Table 2. Relevant accuracy of different expressions on the CK+ database.

Expressions	Relevant accuracy (%)	
	Training	Testing
Anger	100	93.58
Disgust	99.91	96.87
Fear	98.90	97.80
Happiness	100	95.52
Sadness	100	95.78
Surprise	99.69	93.16
Average	99.75	95.45

5   Conclusions

This paper presents a framework for automatic facial expression classification and expression intensity estimation by synthesising expression manifolds. We apply eigentransformation to synthesis shapes and expression details for the input subject, and extend this method to expression manifolds. The proposed method uses synthesised landmarks along with gallery set to train GDA classifier for expression classification. To estimate expression intensity, SLPP is utilised to learn the expression manifolds, which quantify the intensity variations of images sequences.

The proposed method yields realistic synthesised expressions and expression manifolds, and the experimental results validate marked improvements in classification accuracy over those without synthesised landmarks. In addition, the synthesised expression manifolds describes the dynamic variations of facial expressions and can be used to estimate expression intensity of image sequences. Future work will entail applying this method to more comprehensive datasets in real-time applications.

References

1. Jain, A.K., Li, S.Z.: Handbook of Face Recognition, vol. 1. Springer, New York (2005)
2. Ekman, P., Friesen, W.V.: Facial Action Coding System: A Technique for the Measurement of Facial Movement. Consulting Psychologists Press, Palo Alto (1978)
3. Pantic, M., Rothkrantz, L.: Automatic analysis of facial expressions: the state of the art. IEEE Trans. Pattern Anal. Mach. Intell. **22**(12), 1424–1445 (2000)
4. Fasel, B., Luettin, J.: Automatic facial expression analysis: a survey. Pattern Recogn. **36**(1), 259–275 (2003)
5. Shan, C., Gong, S., McOwan, P.W.: Facial expression recognition based on local binary patterns: a comprehensive study. Image Vis. Comput. **27**(6), 803–816 (2009)
6. Ambadar, Z., Schooler, J.W., Cohn, J.F.: Deciphering the enigmatic face: the importance of facial dynamics in interpreting subtle facial expressions. Psychol. Sci. **16**(5), 403–410 (2005)
7. Kanade, T., Cohn, J.F., Tian, Y.: Comprehensive database for facial expression analysis. In: Proceedings of the IEEE International Conference on Automatic Face and Gesture Recognition, pp. 46–53 (2000)
8. Lucey, P., Cohn, J.F., Kanade, T., Saragih, J., Ambadar, Z., Matthews, I.: The extended Cohn-Kanade dataset (CK+): a complete dataset for action unit and emotion-specified expression. In: Proceedings of the IEEE International Conference on Computer Vision and Pattern Recognition Workshop, pp. 94–101 (2010)
9. Ekman, P., Friesen, W.V.: The Facial Action Coding System. Consulting Psychologists Press, Palo Alto (1982)
10. Ekman, P., Rolls, E., Perrett, D., Ellis, H.: Facial expressions of emotion: an old controversy and new findings. Philos. Trans. R. Soc. B **335**(1273), 63–69 (1992)
11. Tian, Y., Kanade, T., Cohn, J.F.: Recognizing action units for facial expression analysis. IEEE Trans. Pattern Anal. Mach. Intell. **23**(2), 97–115 (2001)
12. Pantic, M., Rothkrantz, L.J.: Facial action recognition for facial expression analysis from static face images. IEEE Trans. Syst. Man Cybern. B Cybern. **34**(3), 1449–1461 (2004)
13. Pantic, M., Patras, I.: Dynamics of facial expression: recognition of facial actions and their temporal segments from face profile image sequences. IEEE Trans. Syst. Man Cybern. B Cybern. **36**(2), 433–449 (2006)
14. Uddin, M.Z., Lee, J., Kim, T.: An enhanced independent component-based human facial expression recognition from video. IEEE Trans. Syst. Man Cybern. B Cybern. **55**(4), 2216–2224 (2009)
15. Song, M., Tao, D., Liu, Z., Li, X., Zhou, M.: Image ratio features for facial expression recognition application. IEEE Trans. Syst. Man Cybern. B Cybern. **40**(3), 779–788 (2010)

16. Jabid, T., Kabir, M.H., Chae, O.: Facial expression recognition using local directional pattern. In: Proceedings of the International Conference on Image Processing, pp. 1605–1608 (2010)

17. Gu, W., Xiang, C., Venkatesh, Y.V., Huang, D., Lin, H.: Facial expression recognition using radial encoding of local Gabor features and classifier synthesis. Pattern Recogn. **45**(1), 80–91 (2012)

18. Happy, S.L., Routray, A.: Automatic facial expression recognition using features of salient facial patches. IEEE Trans. Affect. Comput. **6**(1), 1–12 (2015)

19. Yang, P., Liu, Q., Metaxas, D.N.: Rankboost with l1 regularization for facial expression recognition and intensity estimation. In: Proceedings of the International Conference on Computer Vision, pp. 1018–1025 (2009)

20. Delannoy, J.R., McDonald, J.: Automatic estimation of the dynamics of facial expression using a three-level model of intensity. In: Proceedings of the IEEE International Conference on Automatic Face and Gesture Recognition, pp. 1–6 (2008)

21. Chang, K.Y., Chen, C.S., Hung, Y.P.: Intensity rank estimation of facial expressions based on a single image. In: Proceedings of the IEEE International Conference on Systems, Man, and Cybernetics, pp. 3157–3162 (2013)

22. Mohammadi, M.R., Fatemizadeh, E., Mahoor, M.H.: Intensity estimation of spontaneous facial action units based on their sparsity properties. IEEE Trans. Cybern. **46**(3), 817–826 (2016)

23. Tang, X., Wang, X.: Face sketch synthesis and recognition. In: Proceedings of the IEEE International Conference on Computer Vision, pp. 687–694 (2003)

24. Golub, G.H., Reinsch, C.: Singular value decomposition and least squares solutions. Numer. Math. **14**(5), 403–420 (1970)

25. Dryden, I.L., Mardia, K.V.: Statistical Shape Analysis, vol. 4. Wiley, Chichester (1998)

26. Zhao, W., Chellappa, R., Phillips, P.J., Rosenfeld, A.: Face recognition: a literature survey. ACM Comput. Surv. **35**(4), 399–458 (2003)

27. Baudat, G., Anouar, F.: Generalized discriminant analysis using a kernel approach. Neural Comput. **12**(10), 2385–2404 (2000)

28. Zheng, Z., Yang, F., Tan, W., Jia, J., Yang, J.: Gabor feature-based face recognition using supervised locality preserving projection. Signal Process. **87**(10), 2473–2483 (2007)

29. Megvii, Inc.: Face++ research toolkit (2013). www.faceplusplus.com

30. Yao, B., Ai, H., Lao, S.: Logit-rankboost with pruning for face recognition. In: Proceedings of the IEEE International Conference on Automatic Face and Gesture Recognition, pp. 1–8 (2008)

Author Index

Printed in the United States
By Bookmasters